Fourth Edition

Complete Solutions Guide

Chemistry

Steven S. Zumdahl

Thomas J. Hummel
Susan Arena Zumdahl
Steven S. Zumdahl

University of Illinois at Urbana-Champaign

HOUGHTON MIFFLIN COMPANY   **Boston   New York**

Senior Sponsoring Editor: Richard Stratton
Assistant Editor: Marianne Stepanian
Director of Manufacturing: Michael O'Dea
Executive Marketing Manager: Karen Natale

Cover design: Harold Burch, Harold Burch Design, New York City.
Cover image: Pete Turner, The Image Bank.

Printed in the U.S.A.

ISBN: 0-669-41799-8

3456789-VG-00 99 98

TABLE OF CONTENTS

TO THE STUDENT: HOW TO USE THIS GUIDE

Solutions to all of the end of chapter questions and exercises are in this manual.   This "Solutions Guide" can be very valuable if you use it properly. The way <u>NOT</u> to use it is to look at an exercise in the book and then immediately check the solution, often saying to yourself, "That's easy, I can do it." Chemistry is easy once you get the hang of it, but it takes practice. Don't look up a solution to a problem until you have tried to work it on your own. If you are completely stuck, see if you can find a similar problem in the Sample Exercises in the chapter. Only look up the solution as a last resort. If you do this for a problem, look for a similar problem in the end of chapter exercises and try working it. The more problems you do, the easier chemistry becomes. It is also in your self interest to try to work as many problems as possible. Most exams that you will take in chemistry will involve a lot of problem solving. If you have worked several problems similar to the ones on an exam, you will do much better than if the exam is the first time you try to solve a particular type of problem. No matter how much you read and study the text, or how well you think you understand the material, you don't really understand it until you have taken the information in the text and applied the principles to problem solving. You will make mistakes, but the good students learn from their mistakes.

In this manual we have worked problems as in the textbook. We have shown intermediate answers to the correct number of significant figures and used the rounded answer in later calculations. Thus, some of your answers may differ slightly from ours. When we have not followed this convention, we have usually noted this in the solution.

We are grateful to Delores Wyatt for her outstanding effort in preparing the manuscript of this manual. We also thank Robert Pfaff for his careful and thorough accuracy review of the Solutions Guide.

TJH
SAZ
SSZ

CHAPTER ONE

CHEMICAL FOUNDATIONS

Questions

12.   a.   Law versus theory: A law is a concise statement or equation that summarizes observed behavior. A theory is a set of hypotheses that gives an overall explanation of some phenomenon. A law summarizes what happens; a theory (or model) attempts to explain why it happens.

       b.   Theory versus experiment: A theory is an explanation of why things behave the way they do, while an experiment is the process of observing that behavior. Theories attempt to explain the results of experiments and are, in turn, tested by further experiments.

       c.   Qualitative versus quantitative: A qualitative observation only describes a quality while a quantitative observation attaches a number to the observation. Examples: qualitative observations: The water was hot to the touch. Mercury was found in the drinking water. quantitative observations: The temperature of the water was $62\,^{\circ}C$. The concentration of mercury in the drinking water was 1.5 ppm.

       d.   Hypothesis versus theory: Both are explanations of experimental observation. A theory is a set of hypotheses that has been tested over time and found to still be valid, with (perhaps) some modifications.

13.   No, it is useful whenever a systematic approach of observation and hypothesis testing can be used.

14.   a.   No   b.   Yes   c.   Yes

       Only statements b and c can be determined from experiment.

15.   Precision is related to how many significant figures one can associate with a measurement. Consider weighing an object on three different balances with the following results: 11 g; 11.25 g; 11.2456 g. Since the assumed uncertainty in all measurements is ± 1 in the last digit, then the uncertainty of the three balances are: ± 1 g; ± 0.01 g; ± 0.0001 g, respectively. The balance with the smallest uncertainty is the balance with the most significant figures associated with the measurement, which is also the balance that is assumed most precise.

16.   Accuracy: How close a measurement or series of measurements are to an accepted or true value. Precision: How close a series of measurements of the same thing are to each other. The results, average = 14.91% ± 0.03% are precise (close to each other) but are not accurate (not close to the true value).

17.   Chemical changes involve the making and breaking of chemical forces (bonds). Physical changes do not. The identity of a substance changes after a chemical change, but not after a physical change.

18.   Many techniques of chemical analysis need to be performed on relatively pure materials. Thus, many times a separation step is necessary to remove materials that will interfere with the analytical measurement.

Exercises

Significant Figures and Unit Conversions

19.   a.   inexact   b.   exact   c.   exact

For c, $\dfrac{36 \text{ in}}{\text{yd}} \times \dfrac{2.54 \text{ cm}}{\text{in}} \times \dfrac{1 \text{ m}}{100 \text{ cm}} = \dfrac{0.9144 \text{ m}}{\text{yd}}$ (All conversion factors used are exact.)

   d.   inexact; Although this number appears to be exact, it probably isn't. The announced attendance may be tickets sold but not the number who were actually in the stadium. Some people who paid may not have gone, some may leave early, or arrive late, some may sneak in without paying, etc.

   e.   exact   f.   inexact

20.   a.   exact   b.   inexact; 0.9144 m/yd is exact (see Exercise 1.19c)
                                       Thus, there are 1/0.9144 = 1.093613 ... yd/m.

   c.   exact   d.   inexact (π has an infinite number of decimal places.)

21.   a.   0.00$\underline{12}$; 2 S.F., 1.2×10^{-3} b. $\underline{437},000$; 3 S.F., 4.37×10^5

   c. $\underline{900.0}$; 4 S.F., 9.000×10^2 d. $\underline{106}$; 3 S.F.; 1.06×10^2

   e. $\underline{125,904},000$; 6 S.F., 1.25904×10^8 f. $\underline{1.0012}$; 5 S.F., 1.0012×10^0

   g. $\underline{2006}$; 4 S.F., 2.006×10^3 h. $\underline{3050}$; 3 S.F., 3.05×10^3

   i.   0.001$\underline{060}$; 4 S.F., 1.060×10^{-3}

22.   a. $\underline{1}00$; 1 S.F.   b. $\underline{1.0} \times 10^2$; 2 S.F.

   c. $\underline{1.00} \times 10^3$; 3 S.F.   d. $\underline{100.}$; 3 S.F.

   e.   0.00$\underline{48}$; 2 S.F.   f.   0.00$\underline{480}$; 3 S.F.

   g. $\underline{4.80} \times 10^{-3}$; 3 S.F.   h. $\underline{4.800} \times 10^{-3}$; 4 S.F.

23.   a. 6×10^8 b. 5.8×10^8 c. 5.82×10^8

        d. 5.8200×10^8 e. 5.820000×10^8

24.   a. 5×10^2 b. 4.8×10^2 c. 4.80×10^2 d. 4.800×10^2

25.   a.   467; The difference of 25.27 - 24.16 = 1.11 has only three significant figures, so the answer will
             only have 3 significant figures. For this problem and for subsequent problems, the addition/
             subtraction rule is applied separately from the multiplication/division rule.

        b.   0.24; The difference of 8.925 - 8.904 = 0.021 has only 2 significant figures.

        c. $(9.04 - 8.23 + 21.954 + 81.0) \div 3.1416 = 103.8 \div 3.1416 = 33.04$

             We will generally round off at intermediate steps in order to show the correct number of
             significant figures. However, you should round off at the end of all the mathematical operations
             in order to avoid round off error. Make sure you keep track of the correct number of significant
             figures during intermediate steps, but round off at the end.

        d. $\dfrac{9.2 \times 100.65}{8.321 + 4.026} = \dfrac{9.2 \times 100.65}{12.347} = 75$

        e. $0.1654 + 2.07 - 2.114 = 0.12$

             Uncertainty begins to appear in the second decimal place. Numbers were added as written and
             the answer was rounded off to 2 decimal places at the end. If you round to 2 decimal places and
             add you get 0.13. Always round off at the end of the operation to avoid round off error.

        f. $8.27(4.987 - 4.962) = 8.27(0.025) = 0.21$

        g. $\dfrac{9.5 + 4.1 + 2.8 + 3.175}{4} = \dfrac{19.6}{4} = 4.90 = 4.9$

             Uncertainty appears in the first decimal place. The average of several numbers can only be as
             precise as the least precise number. Averages can be exceptions to the significant figure rules.

        h. $\dfrac{9.025 - 9.024}{9.025} \times 100 = \dfrac{0.001}{9.025} \times 100 = 0.01$

26.   a. $6.022 \times 10^{23} \times 1.05 \times 10^2 = 6.32 \times 10^{25}$

        b. $\dfrac{6.6262 \times 10^{-34} \times 2.998 \times 10^8}{2.54 \times 10^{-9}} = 7.82 \times 10^{-17}$

        c. $1.285 \times 10^{-2} + 1.24 \times 10^{-3} + 1.879 \times 10^{-1}$

$$= 0.1285 \times 10^{-1} + 0.0124 \times 10^{-1} + 1.879 \times 10^{-1} = 2.020 \times 10^{-1}$$

             When the exponents are different, it is easiest to apply the addition/subtraction rule when all
             numbers are based on the same power of 10.

d. $1.285 \times 10^{-2} - 1.24 \times 10^{-3} = 1.285 \times 10^{-2} - 0.124 \times 10^{-2} = 1.161 \times 10^{-2}$

e. $\dfrac{(1.00866 - 1.00728)}{6.02205 \times 10^{23}} = \dfrac{0.00138}{6.02205 \times 10^{23}} = 2.29 \times 10^{-27}$

f. $\dfrac{9.875 \times 10^{2} - 9.795 \times 10^{2}}{9.875 \times 10^{2}} \times 100 = \dfrac{0.080 \times 10^{2}}{9.875 \times 10^{2}} \times 100 = 8.1 \times 10^{-1}$

g. $\dfrac{9.42 \times 10^{2} + 8.234 \times 10^{2} + 1.625 \times 10^{3}}{3}$

$$= \dfrac{0.942 \times 10^{3} + 0.8234 \times 10^{3} + 1.625 \times 10^{3}}{3} = 1.130 \times 10^{3}$$

27.   a. $1 \text{ km} = 10^{3} \text{ m} = 10^{6} \text{ mm} = 10^{15} \text{ pm}$

      b. $1 \text{ g} = 10^{-3} \text{ kg} = 10^{3} \text{ mg} = 10^{9} \text{ ng};$ c. $1 \text{ mL} = 10^{-3} \text{ L} = 10^{-3} \text{ dm}^{3} = 1 \text{ cm}^{3}$

      d. $1 \text{ mg} = 10^{-6} \text{ kg} = 10^{-3} \text{ g} = 10^{3} \text{ } \mu\text{g} = 10^{6} \text{ ng} = 10^{9} \text{ pg} = 10^{12} \text{ fg}$

      e. $1 \text{ s} = 10^{3} \text{ ms} = 10^{9} \text{ ns}$

28.   a. $1 \text{ Tg} \times \dfrac{10^{12} \text{ g}}{\text{Tg}} \times \dfrac{1 \text{ kg}}{1000 \text{ g}} = 10^{9} \text{ kg}$

      b. $6.50 \times 10^{2} \text{ Tm} \times \dfrac{10^{12} \text{ m}}{\text{Tm}} \times \dfrac{10^{9} \text{ nm}}{\text{m}} = 6.50 \times 10^{23} \text{ nm}$

      c. $25 \text{ fg} \times \dfrac{1 \text{ g}}{10^{15} \text{ fg}} \times \dfrac{1 \text{ kg}}{1000 \text{ g}} = 25 \times 10^{-18} \text{ kg} = 2.5 \times 10^{-17} \text{ kg}$

      d. $8.0 \text{ dm}^{3} \times \dfrac{1 \text{ L}}{\text{dm}^{3}} = 8.0 \text{ L} \quad (1 \text{ L} = 1 \text{ dm}^{3} = 1000 \text{ cm}^{3} = 1000 \text{ mL})$

      e. $1 \text{ mL} \times \dfrac{1 \text{ L}}{1000 \text{ mL}} \times \dfrac{10^{6} \text{ } \mu\text{L}}{\text{L}} = 10^{3} \text{ } \mu\text{L};$ f. $1 \text{ } \mu\text{g} \times \dfrac{1 \text{ g}}{10^{6} \text{ } \mu\text{g}} \times \dfrac{10^{12} \text{ pg}}{\text{g}} = 10^{6} \text{ pg}$

29. $1 \text{ Å} \times \dfrac{10^{-8} \text{ cm}}{\text{Å}} \times \dfrac{1 \text{ m}}{100 \text{ cm}} \times \dfrac{10^{9} \text{ nm}}{\text{m}} = 1 \times 10^{-1} \text{ nm}$

      $1 \text{ Å} \times \dfrac{1 \times 10^{-1} \text{ nm}}{\text{Å}} \times \dfrac{1 \text{ m}}{10^{9} \text{ nm}} \times \dfrac{10^{12} \text{ pm}}{\text{m}} = 1 \times 10^{2} \text{ pm}$

30. $134 \text{ pm} \times \dfrac{1 \text{ m}}{10^{12} \text{ pm}} \times \dfrac{10^{9} \text{ nm}}{\text{m}} = 0.134 \text{ nm}$

      $1 \text{ Å} = 1 \times 10^{-8} \text{ cm}; \quad 134 \text{ pm} \times \dfrac{1 \text{ m}}{10^{12} \text{ pm}} \times \dfrac{100 \text{ cm}}{\text{m}} \times \dfrac{1 \text{ Å}}{10^{-8} \text{ cm}} = 1.34 \text{ Å}$

31.   a.   Appropriate conversion factors are found on the back cover of the text. In general, the number of significant figures we use in the conversion factors will be one more than the number of significant figures from the numbers given in the problem. This is usually sufficient to avoid round off error.

$$3.91 \text{ kg} \times \frac{1 \text{ lb}}{0.4536 \text{ kg}} = 8.62 \text{ lb}; \quad 0.62 \text{ lb} \times \frac{16 \text{ oz}}{\text{lb}} = 9.9 \text{ oz}$$

Baby's weight = 8 lb and 9.9 oz or to the nearest ounce, 8 lb and 10 oz.

$$51.4 \text{ cm} \times \frac{1 \text{ in}}{2.54 \text{ cm}} = 20.2 \text{ in} \approx 20 \text{ 1/4 in} = \text{baby's height}$$

   b. $$25,000 \text{ mi} \times \frac{1.61 \text{ km}}{\text{mi}} = 4.0 \times 10^4 \text{ km}; \quad 4.0 \times 10^4 \text{ km} \times \frac{1000 \text{ m}}{\text{km}} = 4.0 \times 10^7 \text{ m}$$

   c. $$V = 1 \times w \times h = 1.0 \text{ m} \times \left(5.6 \text{ cm} \times \frac{1 \text{ m}}{100 \text{ cm}} \right) \times \left(2.1 \text{ dm} \times \frac{1 \text{ m}}{10 \text{ dm}} \right) = 1.2 \times 10^{-2} \text{ m}^3$$

$$1.2 \times 10^{-2} \text{ m}^3 \times \left(\frac{10 \text{ dm}}{\text{m}} \right)^3 \times \frac{1 \text{ L}}{\text{dm}^3} = 12 \text{ L}$$

$$12 \text{ L} \times \frac{1000 \text{ cm}^3}{\text{L}} \times \left(\frac{1 \text{ in}}{2.54 \text{ cm}} \right)^3 = 730 \text{ in}^3; \quad 730 \text{ in}^3 \times \left(\frac{1 \text{ ft}}{12 \text{ in}} \right)^3 = 0.42 \text{ ft}^3$$

32.   a. $$18 \text{ gal} \times \frac{4 \text{ qt}}{\text{gal}} \times \frac{0.946 \text{ L}}{\text{qt}} = 68 \text{ L} = 68 \text{ dm}^3$$

   b. $$4.00 \times 10^2 \text{ in}^3 \times \left(\frac{2.54 \text{ cm}}{\text{in}} \right)^3 = 6.55 \times 10^3 \text{ cm}^3 = 6.55 \text{ L} \quad (1000 \text{ cm}^3 = 1000 \text{ mL} = 1 \text{ L})$$

   c. $$0.25 \text{ in} \times \frac{2.54 \text{ cm}}{\text{in}} = 0.64 \text{ cm}$$

   d. $$14,110 \text{ ft} \times \frac{1 \text{ yd}}{3 \text{ ft}} \times \frac{1 \text{ m}}{1.0936 \text{ yd}} = 4301 \text{ m}$$ e. $$198 \text{ lb} \times \frac{1 \text{ kg}}{2.205 \text{ lb}} = 89.8 \text{ kg}$$

33.   a. $$928 \text{ mi} \times \frac{5280 \text{ ft}}{\text{mi}} \times \frac{1 \text{ fathom}}{6 \text{ ft}} \times \frac{1 \text{ cable length}}{100 \text{ fathoms}} \times \frac{1 \text{ nautical mi}}{10 \text{ cable lengths}}$$

$$\times \frac{1 \text{ league}}{3 \text{ nautical miles}} = 272 \text{ leagues}$$

$$928 \text{ mi} \times \frac{1.609 \text{ km}}{1 \text{ mi}} = 1.49 \times 10^3 \text{ km}$$

   b. $$1.0 \text{ cable length} \times \frac{100 \text{ fathom}}{\text{cable length}} \times \frac{6 \text{ ft}}{\text{fathom}} \times \frac{1 \text{ yd}}{3 \text{ ft}} \times \frac{1 \text{ m}}{1.09 \text{ yd}} \times \frac{1 \text{ km}}{1000 \text{ m}} = 0.18 \text{ km}$$

$$1.0 \text{ cable length} = 0.18 \text{ km} \times \frac{1000 \text{ m}}{\text{km}} \times \frac{100 \text{ cm}}{\text{m}} = 1.8 \times 10^4 \text{ cm}$$

c. $$315 \text{ ft} \times \frac{12 \text{ in}}{\text{ft}} \times \frac{2.54 \text{ cm}}{\text{in}} \times \frac{1 \text{ m}}{100 \text{ cm}} = 96.0 \text{ m}$$

$$37 \text{ ft} \times \frac{12 \text{ in}}{\text{ft}} \times \frac{2.54 \text{ cm}}{\text{in}} \times \frac{1 \text{ m}}{100 \text{ cm}} = 11 \text{ m}$$

$$315 \text{ ft} \times \frac{1 \text{ fathom}}{6 \text{ ft}} \times \frac{1 \text{ cable length}}{100 \text{ fathoms}} = 5.25 \times 10^{-1} \text{ cable lengths}$$

$$37 \text{ ft} \times \frac{1 \text{ fathom}}{6 \text{ ft}} = 6.2 \text{ fathoms}$$

34.   a. $$1.25 \text{ mi} \times \frac{8 \text{ furlongs}}{\text{mi}} = 10.0 \text{ furlongs}; \quad 10.0 \text{ furlongs} \times \frac{40 \text{ rods}}{\text{furlong}} = 4.00 \times 10^2 \text{ rods}$$

$$4.00 \times 10^2 \text{ rods} \times \frac{5.5 \text{ yd}}{\text{rod}} \times \frac{36 \text{ in}}{\text{yd}} \times \frac{2.54 \text{ cm}}{\text{in}} \times \frac{1 \text{ m}}{100 \text{ cm}} = 2.01 \times 10^3 \text{ m}$$

$$2.01 \times 10^3 \text{ m} \times \frac{1 \text{ km}}{1000 \text{ m}} = 2.01 \text{ km}$$

b.   Let's assume we know this distance to ± 1 yard. First convert 26 miles to yards.

$$26 \text{ mi} \times \frac{5280 \text{ ft}}{\text{mi}} \times \frac{1 \text{ yd}}{3 \text{ ft}} = 45,760. \text{ yd}$$

$$26 \text{ mi} + 385 \text{ yd} = 45,760. \text{ yd} + 385 \text{ yd} = 46,145 \text{ yards}$$

$$46,145 \text{ yard} \times \frac{1 \text{ rod}}{5.5 \text{ yd}} = 8390.0 \text{ rods}; \quad 8390.0 \text{ rods} \times \frac{1 \text{ furlong}}{40 \text{ rods}} = 209.75 \text{ furlongs}$$

$$46,145 \text{ yard} \times \frac{36 \text{ in}}{\text{yd}} \times \frac{2.54 \text{ cm}}{\text{in}} \times \frac{1 \text{ m}}{100 \text{ cm}} = 42,195 \text{ m}; \quad 42,195 \text{ m} \times \frac{1 \text{ km}}{1000 \text{ m}} = 42.195 \text{ km}$$

35.   a. $$1 \text{ troy lb} \times \frac{12 \text{ troy oz}}{\text{troy lb}} \times \frac{20 \text{ pw}}{\text{troy oz}} \times \frac{24 \text{ grains}}{\text{pw}} \times \frac{0.0648 \text{ g}}{\text{grain}} \times \frac{1 \text{ kg}}{1000 \text{ g}} = 0.373 \text{ kg}$$

$$1 \text{ troy lb} = 0.373 \text{ kg} \times \frac{2.205 \text{ lb}}{\text{kg}} = 0.822 \text{ lb}$$

b. $$1 \text{ troy oz} \times \frac{20 \text{ pw}}{\text{troy oz}} \times \frac{24 \text{ grains}}{\text{pw}} \times \frac{0.0648 \text{ g}}{\text{grain}} = 31.1 \text{ g}$$

$$1 \text{ troy oz} = 31.1 \text{ g} \times \frac{1 \text{ carat}}{0.200 \text{ g}} = 156 \text{ carats}$$

c. 1 troy lb = 0.373 kg; $0.373 \text{ kg} \times \dfrac{1000 \text{ g}}{\text{kg}} \times \dfrac{1 \text{ cm}^3}{19.3 \text{ g}} = 19.3 \text{ cm}^3$

36. a. $1 \text{ grain ap} \times \dfrac{1 \text{ scruple}}{20 \text{ grain ap}} \times \dfrac{1 \text{ dram ap}}{3 \text{ scruples}} \times \dfrac{3.888 \text{ g}}{\text{dram ap}} = 0.06480 \text{ g}$

From the previous question we are given that 1 grain troy = 0.0648 g = 1 grain ap. So, the two are the same.

b. $1 \text{ oz ap} \times \dfrac{8 \text{ dram ap}}{\text{oz ap}} \times \dfrac{3.888 \text{ g}}{\text{dram ap}} \times \dfrac{1 \text{ oz troy*}}{31.1 \text{ g}} = 1.00 \text{ oz troy}$ *See Exercise 35b.

c. $5.00 \times 10^2 \text{ mg} \times \dfrac{1 \text{ g}}{1000 \text{ mg}} \times \dfrac{1 \text{ dram ap}}{3.888 \text{ g}} \times \dfrac{3 \text{ scruples}}{\text{dram ap}} = 0.386 \text{ scruple}$

$0.386 \text{ scruple} \times \dfrac{20 \text{ grains ap}}{\text{scruple}} = 7.72 \text{ grains ap}$

d. $1 \text{ scruple} \times \dfrac{1 \text{ dram ap}}{3 \text{ scruples}} \times \dfrac{3.888 \text{ g}}{\text{dram ap}} = 1.296 \text{ g}$

37. $\dfrac{4.4 \times 10^9 \text{ mi}}{2.0 \text{ yr}} \times \dfrac{1.61 \text{ km}}{\text{mi}} \times \dfrac{1 \text{ yr}}{365 \text{ d}} \times \dfrac{1 \text{ d}}{24 \text{ hr}} \times \dfrac{1 \text{ hr}}{60 \text{ min}} \times \dfrac{1 \text{ min}}{60 \text{ s}} = 110 \text{ km/s}$

38. $\dfrac{100. \text{ m}}{9.84 \text{ s}} = 10.2 \text{ m/s}; \quad \dfrac{100. \text{ m}}{9.84 \text{ s}} \times \dfrac{1 \text{ km}}{1000 \text{ m}} \times \dfrac{60 \text{ s}}{\text{min}} \times \dfrac{60 \text{ min}}{\text{hr}} = 36.6 \text{ km/hr}$

$\dfrac{100. \text{ m}}{9.84 \text{ s}} \times \dfrac{1.0936 \text{ yd}}{\text{m}} \times \dfrac{3 \text{ ft}}{\text{yd}} = 33.3 \text{ ft/s}; \quad \dfrac{33.3 \text{ ft}}{\text{s}} \times \dfrac{1 \text{ mi}}{5280 \text{ ft}} \times \dfrac{60 \text{ s}}{\text{min}} \times \dfrac{60 \text{ min}}{\text{hr}} = 22.7 \text{ mi/hr}$

$1.00 \times 10^2 \text{ yds} \times \dfrac{1 \text{ m}}{1.0936 \text{ yd}} \times \dfrac{9.84 \text{ s}}{100. \text{ m}} = 9.00 \text{ s}$

39. $\dfrac{14 \text{ km}}{\text{L}} \times \dfrac{1 \text{ mi}}{1.61 \text{ km}} \times \dfrac{3.79 \text{ L}}{\text{gal}} = 33 \text{ mi/gal}; \quad$ The spouse's car has the better gas mileage.

40. $\dfrac{36.1 \text{ mi}}{\text{gal}} \times \dfrac{1.609 \text{ km}}{\text{mi}} \times \dfrac{1 \text{ gal}}{3.785 \text{ L}} = 15.3 \text{ km/L}; \quad 1.0 \text{ L} \times \dfrac{15.3 \text{ km}}{\text{L}} = 15 \text{ km}$

Temperature

41. $T_C = \dfrac{5}{9}(T_F - 32) = \dfrac{5}{9}(102.5 - 32) = 39.2 °C; \quad T_K = T_C + 273.2 = 312.4 \text{ K}$ (Note: 32 is exact)

42. $T_C = \dfrac{5}{9}(74 - 32) = 23 °C; \quad T_K = 23 + 273 = 296 \text{ K}$

43. $T_F = \dfrac{9}{5} \times T_C + 32 = \dfrac{9}{5} \times 25 + 32 = 77 °F; \quad T_K = 25 + 273 = 298 \text{ K}$

44. $T_K = T_C + 273$; $T_C = 4 - 273 = -269°C$; $T_F = \dfrac{9}{5} \times T_C + 32 = \dfrac{9}{5} \times (-269) + 32 = -452°F$

45.   We can do this two ways. First, we calculate the high and low temperature and get the uncertainty from the range. $20.6°C \pm 0.1°C$ means the temperature can range from $20.5°C$ to $20.7°C$.

$T_F = \dfrac{9}{5} \times T_c + 32 \leftarrow$ (exact); $T_F = \dfrac{9}{5} \times 20.6 + 32 = 69.1°F$

$T_F(\text{min}) = \dfrac{9}{5} \times 20.5 + 32 = 68.9°F$; $T_F(\text{max}) = \dfrac{9}{5} \times 20.7 + 32 = 69.3°F$

So the temperature ranges from $68.9°F$ to $69.3°F$ which we can express as $69.1 \pm 0.2°F$.

An alternative way is to treat the uncertainty and the temperature in °C separately.

$T_F = \dfrac{9}{5} \times T_C + 32 = \dfrac{9}{5} \times 20.6 + 32 = 69.1°F$; $\pm 0.1°C \times \dfrac{9°F}{5°C} = \pm 0.18°F \approx \pm 0.2°F$

Combining the two calculations: $T_F = 69.1 \pm 0.2°F$

46. $96.1°F \pm 0.2°F$; First convert $96.1°F$ to °C. $T_C = \dfrac{5}{9}(T_F - 32) = \dfrac{5}{9}(96.1 - 32) = 35.6°C$

A change in temperature of $9°F$ is equal to a change in temperature of $5°C$. So the uncertainty is:

$\pm 0.2°F \times \dfrac{5°C}{9°F} = \pm 0.1°C$. Thus, $96.1 \pm 0.2°F = 35.6 \pm 0.1°C$

Density

47. $\dfrac{2.70\text{ g}}{\text{cm}^3} \times \dfrac{1\text{ kg}}{1000\text{ g}} \times \left(\dfrac{100\text{ cm}}{\text{m}}\right)^3 = \dfrac{2.70 \times 10^3\text{ kg}}{\text{m}^3}$

$\dfrac{2.70\text{ g}}{\text{cm}^3} \times \dfrac{1\text{ lb}}{453.6\text{ g}} \times \left(\dfrac{2.54\text{ cm}}{\text{in}}\right)^3 \times \left(\dfrac{12\text{ in}}{\text{ft}}\right)^3 = \dfrac{169\text{ lb}}{\text{ft}^3}$

48. $\dfrac{1.0\text{ g}}{\text{cm}^3} \times \dfrac{1\text{ kg}}{1000\text{ g}} \times \left(\dfrac{100\text{ cm}}{\text{m}}\right)^3 = 1.0 \times 10^3\text{ kg/m}^3$

$\dfrac{1.0\text{ g}}{\text{cm}^3} \times \dfrac{1\text{ lb}}{454\text{ g}} \times \left(\dfrac{2.54\text{ cm}}{\text{in}} \times \dfrac{12\text{ in}}{\text{ft}}\right)^3 = 62\text{ lb/ft}^3$

49. $d = \text{density} = \dfrac{\text{mass}}{\text{volume}}$; mass $= 1.67 \times 10^{-24}$ g; radius $=$ diameter/2 $= 5.0 \times 10^{-4}$ pm

$V = \dfrac{4}{3}\pi r^3 = \dfrac{4}{3} \times 3.14 \times \left(5.0 \times 10^{-4}\text{ pm} \times \dfrac{1\text{ m}}{10^{12}\text{ pm}} \times \dfrac{100\text{ cm}}{\text{m}}\right)^3 = 5.2 \times 10^{-40}\text{ cm}^3$

$d = \dfrac{1.67 \times 10^{-24}\text{ g}}{5.2 \times 10^{-40}\text{ cm}^3} = \dfrac{3.2 \times 10^{15}\text{ g}}{\text{cm}^3}$

50. $V = l \times w \times h = 2.9\text{ cm} \times 3.5\text{ cm} \times 10.0\text{ cm} = 1.0 \times 10^2\text{ cm}^3$

$$d = \text{density} = \frac{615.0 \text{ g}}{1.0 \times 10^2 \text{ cm}^3} = \frac{6.2 \text{ g}}{\text{cm}^3}$$

51. $5.0 \text{ carat} \times \dfrac{0.200 \text{ g}}{\text{carat}} \times \dfrac{1 \text{ cm}^3}{3.51 \text{ g}} = 0.28 \text{ cm}^3$

52. $2.8 \text{ mL} \times \dfrac{1 \text{ cm}^3}{\text{mL}} \times \dfrac{3.51 \text{ g}}{\text{cm}^3} \times \dfrac{1 \text{ carat}}{0.200 \text{ g}} = 49 \text{ carats}$

53.   a.   Both have the same mass.

        b.   1.0 mL of mercury; Mercury has a larger density than water.

        $1.0 \text{ mL} \times \dfrac{13.6 \text{ g}}{\text{mL}} = 13.6 \text{ g of mercury}; \quad 1.0 \text{ mL} \times \dfrac{0.998 \text{ g}}{\text{mL}} = 1.0 \text{ g of water}$

        c.   Same; Both represent 19.3 g of substance.   d.   1.0 L of benzene (880 g vs 670 g)

54.   a.   1.0 kg feathers; Feathers are less dense than lead.

        b.   100 g water since water is less dense than gold.   c.   Same; Both volumes are 1.0 L.

Classification and Separation of Matter

55.   Solid: own volume, own shape, does not flow; Liquid: own volume, takes shape of container, flows; Gas: takes volume and shape of container, flows

56.   Homogeneous: Having visibly indistinguishable parts (the same throughout). Heterogeneous: Having visibly distinguishable parts (not uniform throughout).

        a.   heterogeneous (Due to mulch, water, roots, etc which can all be present.)

        b.   heterogeneous: There is usually a fair amount of particulate matter present in the atmosphere (dirt, smog) in addition to condensed water (rain, clouds). However, a clean atmosphere consisting of only clean air can be considered homogeneous.

        c.   heterogeneous (due to bubbles)   d.   homogeneous

        e.   homogeneous   f.   homogeneous

57.   a.   pure   b.   mixture   c.   mixture   d.   pure   e.   mixture (copper and zinc)

        f.   pure   g.   mixture   h.   mixture   i.   pure

        Iron and uranium are elements. Water and table salt are compounds. Water is H_2O and table salt is NaCl. Compounds are composed of two or more elements.

58.   a.   Distillation separates components of a mixture, so the orange liquid is a mixture (has an average color of the yellow liquid and the red solid). Distillation utilizes boiling point differences to separate out the components of a mixture. Distillation is a physical change since the components of the mixture do not become different compounds or elements.

      b.   Decomposition is a type of chemical reaction. The crystalline solid is a compound and decomposition is a chemical change where new substances are formed.

      c.   Tea is a mixture of tea compounds dissolved in water. The process of mixing sugar into tea is a physical change. Sugar doesn't react with the tea compounds, it just makes the solution sweeter.

Additional Exercises

59.   a.   8.41 (2.16 has only three significant figures.)

      b.   16.1 (Uncertainty appears in the first decimal place for 8.1.)

      c.   52.5 (All numbers have 3 significant figures.)

      d.   5 (2 contains one significant figure.)   e.   0.009

      f.   429.59 (Uncertainty appears in 2nd decimal place for 2.17 and 4.32.)

60.   a. $\dfrac{0.30\text{ g}}{\text{mL}} \times \dfrac{1000\text{ mg}}{\text{g}} = \dfrac{3.0 \times 10^2\text{ mg}}{\text{mL}}$

      b. $\dfrac{0.30\text{ g}}{\text{mL}} \times \dfrac{1\text{ kg}}{10^3\text{ g}} \times \dfrac{1\text{ mL}}{\text{cm}^3} \times \left(\dfrac{100\text{ cm}}{\text{m}}\right)^3 = \dfrac{3.0 \times 10^2\text{ kg}}{\text{m}^3}$

      c. $\dfrac{0.30\text{ g}}{\text{mL}} \times \dfrac{10^6\text{ μg}}{\text{g}} = \dfrac{3.0 \times 10^5\text{ μg}}{\text{mL}}$;   d. $\dfrac{0.30\text{ g}}{\text{mL}} \times \dfrac{10^9\text{ ng}}{\text{g}} = \dfrac{3.0 \times 10^8\text{ ng}}{\text{mL}}$

      e. $\dfrac{0.30\text{ g}}{\text{mL}} \times \dfrac{10^6\text{ μg}}{\text{g}} \times \dfrac{1000\text{ mL}}{\text{L}} \times \dfrac{1\text{ L}}{10^6\text{ μL}} = \dfrac{3.0 \times 10^2\text{ μg}}{\text{μL}}$

61. $1.5\text{ teaspoons} \times \dfrac{80.\text{ mg acet}}{0.50\text{ teaspoons}} = 240\text{ mg acetaminophen}$

      $\dfrac{240\text{ mg acet}}{24\text{ lb}} \times \dfrac{1\text{ lb}}{0.454\text{ kg}} = 22\text{ mg acetaminophen/kg}$

      $\dfrac{240\text{ mg acet}}{35\text{ lb}} \times \dfrac{1\text{ lb}}{0.454\text{ kg}} = 15\text{ mg acetaminophen/kg}$

      The range is from 15 mg to 22 mg acetaminophen per kg of body weight.

62. $1\text{ mmole He} \times \dfrac{1\text{ mole He}}{1000\text{ mmole He}} \times \dfrac{6.02 \times 10^{23}\text{ He atoms}}{\text{mole He}} = 6.02 \times 10^{20}\text{ He atoms}$

$$1 \text{ kmol He} \times \frac{1000 \text{ mole He}}{\text{kmole He}} \times \frac{6.02 \times 10^{23} \text{ He atoms}}{\text{mole He}} = 6.02 \times 10^{26} \text{ He atoms}$$

63. $126 \text{ gal} \times \dfrac{4 \text{ qt}}{\text{gal}} \times \dfrac{1 \text{ L}}{1.057 \text{ qt}} = 477 \text{ L}$

64.   a. $1 \text{ ha} \times \dfrac{10{,}000 \text{ m}^2}{\text{ha}} \times \left(\dfrac{1 \text{ km}}{1000 \text{ m}}\right)^2 = 1 \times 10^{-2} \text{ km}^2$

b. $5.5 \text{ acre} \times \dfrac{160 \text{ rod}^2}{\text{acre}} \times \left(\dfrac{5.5 \text{ yd}}{\text{rod}} \times \dfrac{36 \text{ in}}{\text{yd}} \times \dfrac{2.54 \text{ cm}}{\text{in}} \times \dfrac{1 \text{ m}}{100 \text{ cm}}\right)^2 = 2.2 \times 10^4 \text{ m}^2$

$2.2 \times 10^4 \text{ m}^2 \times \dfrac{1 \text{ ha}}{10^4 \text{ m}^2} = 2.2 \text{ ha};\ \ 2.2 \times 10^4 \text{ m}^2 \times \left(\dfrac{1 \text{ km}}{1000 \text{ m}}\right)^2 = 0.022 \text{ km}^2$

c.   Area of lot $= 120 \text{ ft} \times 75 \text{ ft} = 9.0 \times 10^3 \text{ ft}^2$

$9.0 \times 10^3 \text{ ft}^2 \times \left(\dfrac{1 \text{ yd}}{3 \text{ ft}} \times \dfrac{1 \text{ rod}}{5.5 \text{ yd}}\right)^2 \times \dfrac{1 \text{ acre}}{160 \text{ rod}^2} = 0.21 \text{ acre},\ \dfrac{\$6{,}500}{0.21 \text{ acre}} = \dfrac{\$31{,}000}{\text{acre}}$

We can use our result from (b) to get the conversion factor between acres and ha (5.5 acre = 2.2 ha.). Thus, 1 ha = 2.5 acre.

$0.21 \text{ acre} \times \dfrac{1 \text{ ha}}{2.5 \text{ acre}} = 0.084 \text{ ha};\quad$ The price is: $\dfrac{\$6{,}500}{0.084 \text{ ha}} = \dfrac{\$77{,}000}{\text{ha}}$

65.   Total volume $= \left(200.\text{ m} \times \dfrac{100 \text{ cm}}{\text{m}}\right) \times \left(300.\text{ m} \times \dfrac{100 \text{ cm}}{\text{m}}\right) \times 4.0 \text{ cm} = 2.4 \times 10^9 \text{ cm}^3$

Vol. covered by 1 bag of topsoil $= \left[10 \text{ ft}^2 \times \left(\dfrac{12 \text{ in}}{\text{ft}}\right)^2 \times \left(\dfrac{2.54 \text{ cm}}{\text{in}}\right)^2\right] \times \left(1.0 \text{ in} \times \dfrac{2.54 \text{ cm}}{\text{in}}\right) = 2.4 \times 10^4 \text{ cm}^3$

$2.4 \times 10^9 \text{ cm}^3 \times \dfrac{1 \text{ bag}}{2.4 \times 10^4 \text{ cm}^3} = 1.0 \times 10^5 \text{ bags topsoil}$

66. $1.71 \text{ warp factor} = \left(5.00 \times \dfrac{3.00 \times 10^8 \text{ m}}{\text{s}}\right) \times \dfrac{1.094 \text{ yd}}{\text{m}} \times \dfrac{60 \text{ s}}{\text{min}} \times \dfrac{60 \text{ min}}{\text{hr}}$

$\times \dfrac{1 \text{ knot}}{2000 \text{ yd/hr}} = 2.95 \times 10^9 \text{ knots}$

67. $T_c = (68°F - 32°F) \times \dfrac{5°C}{9°F} = 20.°C$

Gallium is a solid at 20.°C. From the melting point, gallium doesn't convert to the liquid state until 29.8°C.

68.   a.   No; If the volumes were the same, then the gold idol would have a much greater mass since gold is much more dense than sand.

       b.   Mass $= 1.0 \text{ L} \times \dfrac{1000 \text{ cm}^3}{\text{L}} \times \dfrac{19.32 \text{ g}}{\text{cm}^3} \times \dfrac{1 \text{ kg}}{1000 \text{ g}} = 19.32 \text{ kg} \ (= 42.59 \text{ lb})$

       It wouldn't be easy to play catch with the idol since it would have a mass of over 40 pounds.

69.   Volume of lake $= 100 \text{ mi}^2 \times \left(\dfrac{5280 \text{ ft}}{\text{mi}} \right)^2 \times 20 \text{ ft} = 6 \times 10^{10} \text{ ft}^3$

       $6 \times 10^{10} \text{ ft}^3 \times \left(\dfrac{12 \text{ in}}{\text{ft}} \times \dfrac{2.54 \text{ cm}}{\text{in}} \right)^3 \times \dfrac{1 \text{ mL}}{\text{cm}^3} \times \dfrac{0.4 \text{ μg}}{\text{mL}} = 7 \times 10^{14} \text{ μg}$

       $7 \times 10^{14} \text{ μg} \times \dfrac{1 \text{ g}}{10^6 \text{ μg}} \times \dfrac{1 \text{ kg}}{10^3 \text{ g}} = 7 \times 10^5 \text{ kg of mercury}$

70.   The object that sinks has a greater density than water. The other object has a density less than water since it floats. Both objects have the same mass. Thus, the sphere that sinks has the smaller volume since it is more dense. The sphere that floats has the larger diameter.

71. $V = 5.00 \text{ cm} \times 4.00 \text{ cm} \times 2.50 \text{ cm} = 50.0 \text{ cm}^3$; $50.0 \text{ cm}^3 \times \dfrac{22.57 \text{ g}}{\text{cm}^3} = 1130 \text{ g}$

       $1.00 \text{ kg} \times \dfrac{1000 \text{ g}}{\text{kg}} \times \dfrac{1 \text{ cm}^3}{22.57 \text{ g}} = 44.3 \text{ cm}^3$

72. d = density; $d_{cube} = \dfrac{140.4 \text{ g}}{(3.00 \text{ cm})^3} = \dfrac{5.20 \text{ g}}{\text{cm}^3}$

       If this is correct to 1.00%, then the density is: $5.20 \pm 0.05 \text{ g/cm}^3$

       $V_{sphere} = \dfrac{4}{3} \pi r^3 = \dfrac{4}{3} \pi (1.42 \text{ cm})^3 = 12.0 \text{ cm}^3$

       $d_{sphere} = \dfrac{61.6 \text{ g}}{12.0 \text{ cm}^3} = \dfrac{5.13 \text{ g}}{\text{cm}^3}$; $d_{sphere} = 5.13 \pm 0.05 \text{ g/cm}^3$

       Since d_{cube} is between 5.15 g/cm^3 and 5.25 g/cm^3 and d_{sphere} is between 5.08 g/cm^3 and 5.18 g/cm^3, then we can't decisively say if they are the same or different. The data are not precise enough to say.

73.   Circumference $= c = 2\pi r$; $V = \dfrac{4\pi r^3}{3} = \dfrac{4\pi}{3} \left(\dfrac{c}{2\pi} \right)^3 = \dfrac{c^3}{6\pi^2}$

       Largest density $= \dfrac{5.25 \text{ oz}}{\dfrac{(9.00 \text{ in})^3}{6\pi^2}} = \dfrac{5.25 \text{ oz}}{12.3 \text{ in}^3} = \dfrac{0.427 \text{ oz}}{\text{in}^3}$

       Smallest density $= \dfrac{5.00 \text{ oz}}{\dfrac{(9.25 \text{ in})^3}{6\pi^2}} = \dfrac{5.00 \text{ oz}}{13.4 \text{ in}^3} = \dfrac{0.373 \text{ oz}}{\text{in}^3}$

Maximum range is: $\dfrac{(0.373 - 0.427)\ \text{oz}}{\text{in}^3}$ or 0.40 ± 0.03 oz/in^3 (Uncertainty in 2nd decimal place.)

74.   a.   For $\dfrac{103 \pm 1}{101 \pm 1}$: Maximum $= \dfrac{104}{100} = 1.04$; Minimum $= \dfrac{102}{102} = 1.00$

So: $\dfrac{103 \pm 1}{101 \pm 1} = 1.02 \pm 0.02$

b.   For $\dfrac{101 \pm 1}{99 \pm 1}$: Maximum $= \dfrac{102}{98} = 1.04$; Minimum $= \dfrac{100}{100} = 1.00$

So: $\dfrac{101 \pm 1}{99 \pm 1} = 1.02 \pm 0.02$

c.   For $\dfrac{99 \pm 1}{101 \pm 1}$: Maximum $= \dfrac{100}{100} = 1.00$; Minimum $= \dfrac{98}{102} = 0.96$

So: $\dfrac{99 \pm 1}{101 \pm 1} = 0.98 \pm 0.02$

Considering the error limits, (a) and (b) should be expressed to three significant figures and (c) to two significant figures. The division rule differs in (b). The rule says (b) should be expressed to two significant figures. If we do this for (b), we imply that the answer is 1.0 or between 0.95 and 1.05. The actual range is less than this, so we should use the more precise way of expressing uncertainty. The significant figure rules only give us guidelines for estimating uncertainty. When we have a better handle on the uncertainty, we should use it in precedence of the significant figure guidelines.

75.   We need to calculate the maximum and minimum values of the density, given the uncertainty in each measurement. The maximum value is:

$$d_{max} = \frac{19.625\ \text{g} + 0.002\ \text{g}}{25.00\ \text{cm}^3 - 0.03\ \text{cm}^3} = \frac{19.627\ \text{g}}{24.97\ \text{cm}^3} = 0.7860\ \text{g/cm}^3$$

The minimum value of the density is:

$$d_{min} = \frac{19.625\ \text{g} - 0.002\ \text{g}}{25.00\ \text{cm}^3 + 0.03\ \text{cm}^3} = \frac{19.623\ \text{g}}{25.03\ \text{cm}^3} = 0.7840\ \text{g/cm}^3$$

The density of the liquid is between 0.7840 g/cm^3 and 0.7860 g/cm^3. These measurements are sufficiently precise to distinguish between ethanol (d = 0.789 g/cm^3) and isopropyl alcohol (d = 0.785 g/cm^3).

76.   We will calculate the largest and smallest values that the density can be.

$$d = \frac{\text{mass}}{V} = \frac{16.50\ \text{g}}{15.5\ \text{cm}^3} = \frac{1.06\ \text{g}}{\text{cm}^3}; \quad d_{max} = \frac{\text{mass}_{max}}{V_{min}} = \frac{16.52\ \text{g}}{15.3\ \text{cm}^3} = \frac{1.08\ \text{g}}{\text{cm}^3}$$

$$d_{min} = \frac{\text{mass}_{min}}{V_{max}} = \frac{16.48\ \text{g}}{15.7\ \text{cm}^3} = \frac{1.05\ \text{g}}{\text{cm}^3}$$

Combining these results, we can express the density as: 1.06 ± 0.02 g/cm^3

77. $V = V(\text{final}) - V(\text{initial}); \quad d = \dfrac{28.90 \text{ g}}{9.8 \text{ cm}^3 - 6.4 \text{ cm}^3} = \dfrac{28.90 \text{ g}}{3.4 \text{ cm}^3} = 8.5 \text{ g/cm}^3$

$d_{max} = \dfrac{mass_{max}}{V_{min}}$, We get V_{min} from 9.7 cm^3 - 6.5 cm^3 = 3.2 cm^3

$d_{max} = \dfrac{28.93 \text{ g}}{3.2 \text{ cm}^3} = \dfrac{9.0 \text{ g}}{\text{cm}^3}; \quad d_{min} = \dfrac{mass_{min}}{V_{max}} = \dfrac{28.87 \text{ g}}{9.9 \text{ cm}^3 - 6.3 \text{ cm}^3} = \dfrac{8.0 \text{ g}}{\text{cm}^3}$

The density is: 8.5 ± 0.5 g/cm^3.

Challenge Problems

78.   a. $\dfrac{2.70 - 2.64}{2.70} \times 100 = 2\%$ b. $\dfrac{|16.12 - 16.48|}{16.12} \times 100 = 2.2\%$

       c. $\dfrac{1.000 - 0.9981}{1.000} \times 100 = \dfrac{0.002}{1.000} \times 100 = 0.2\%$

79.   In a subtraction, the result gets smaller but the uncertainties add. If the two numbers are very close together, the uncertainty may be larger than the result. For example, let us assume we want to take the difference of the following two measured quantities, $999,999 \pm 2$ and $999,996 \pm 2$. The difference is 3 ± 4. Because of the large uncertainty, subtracting two similar numbers is bad practice.

80.   a.   At some point in 1982, the composition of the metal used in minting pennies was changed since the mass changed during this year (assuming the volume of the pennies were constant).

       b.   It should be expressed as 3.08 ± 0.05 g. The uncertainty in the second decimal place will swamp any effect of the next decimal places.

81.   Heavy pennies (old): mean mass = 3.08 ± 0.05 g

       Light pennies (new): mean mass = $\dfrac{(2.467 + 2.545 + 2.518)}{3} = 2.51 \pm 0.04$ g

       Average density of old pennies: $d_{old} = \dfrac{\dfrac{95 \times 8.96 \text{ g}}{\text{cm}^3} + \dfrac{5 \times 7.14 \text{ g}}{\text{cm}^3}}{100} = \dfrac{8.9 \text{ g}}{\text{cm}^3}$

       Average density of new pennies: $d_{new} = \dfrac{\dfrac{2.4 \times 8.96 \text{ g}}{\text{cm}^3} + \dfrac{97.6 \times 7.14 \text{ g}}{\text{cm}^3}}{100} = \dfrac{7.18 \text{ g}}{\text{cm}^3}$

       Since $d = \dfrac{mass}{volume}$ and the volume of old and new pennies are the same, then:

$$\frac{d_{new}}{d_{old}} = \frac{mass_{new}}{mass_{old}}; \quad \frac{d_{new}}{d_{old}} = \frac{7.18}{8.9} = 0.81; \quad \frac{mass_{new}}{mass_{old}} = \frac{2.51}{3.08} = 0.815$$

To the first two decimal places, the ratios are the same. We can reasonably conclude that yes, the difference in mass is accounted for by the difference in the alloy used.

82.   a.

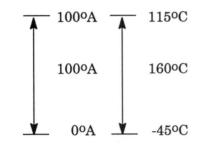

A change in temperature of 160°C equals a change in temperature of 100°A.

So, $\dfrac{160°C}{100°A}$ is our unit conversion for a degree change in temperature.

At the freezing point: 0°A = -45°C

Combining the two pieces of information:

$$T_A = (T_C + 45°C) \times \frac{100°A}{160°C} = (T_C + 45°C) \times \frac{5°A}{8°C} \text{ or } T_C = T_A \times \frac{8°C}{5°A} - 45°C$$

b. $T_C = (T_F - 32) \times \dfrac{5}{9}; \quad T_C = T_A \times \dfrac{8}{5} - 45 = (T_F - 32) \times \dfrac{5}{9}$

$$T_F - 32 = \frac{9}{5} \times \left[T_A \times \frac{8}{5} - 45 \right] = T_A \times \frac{72}{25} - 81, \quad T_F = T_A \times \frac{72°F}{25°A} - 49°F$$

c. $T_C = T_A \times \dfrac{8}{5} - 45$ and $T_C = T_A;$ So, $T_C = T_C \times \dfrac{8}{5} - 45,$ $\dfrac{3 \times T_C}{5} = 45, \; T_C = 75°C = 75°A$

d. $T_C = 86°A \times \dfrac{8°C}{5°A} - 45°C = 93°C; \quad T_F = 86°A \times \dfrac{72°F}{25°A} - 49°F = 199°F = 2.0 \times 10^2 °F$

e. $T_A = (45°C + 45°C) \times \dfrac{5°A}{8°C} = 56°A$

83.   a.   One possibility is that rope B is not attached to anything and rope A and rope C are connected via a pair of pulleys and/or gears.

       b.   Try to pull rope B out of the box. Measure the distance moved by C for a given movement of A. Hold either A or C firmly while pulling on the other.

84.   The bubbles of gas is air in the sand that is escaping. We will assume that the mass of trapped air is insignificant.

mass of dry sand = 37.3488 g - 22.8317 g = 14.5171 g

mass of methanol = 45.2613 g - 37.3488 g = 7.9125 g

Volume of sand particles (air absent) = volume of sand and methanol - volume of methanol

Volume of sand particles (air absent) = 17.6 mL - 10.00 mL = 7.6 mL

density of dry sand (air present) = $\dfrac{14.5171 \text{ g}}{10.0 \text{ mL}}$ = 1.45 g/mL

density of methanol = $\dfrac{7.9125 \text{ g}}{10.00 \text{ mL}}$ = 0.7913 g/mL

density of sand particles (air absent) = $\dfrac{14.5171 \text{ g}}{7.6 \text{ mL}}$ = 1.9 g/mL

CHAPTER TWO

ATOMS, MOLECULES AND IONS

Questions

11.   a.   Atoms have mass and are neither destroyed nor created by chemical reactions. Therefore, mass is neither created nor destroyed by chemical reactions. Mass is conserved.

       b.   The composition of a substance depends on the number and kinds of atoms that form it.

       c.   Compounds of the same elements differ only in the numbers of atoms of the elements forming them, i.e., NO, N_2O, NO_2.

12.   Some elements exist as molecular substances. That is, hydrogen normally exists as H_2 molecules, not single hydrogen atoms. The same is true for N_2, O_2, F_2, Cl_2, etc.

13.   Deflection of cathode rays by magnetic and electric fields led to the conclusion that they were negatively charged. The cathode ray was produced at the negative electrode and repelled by the negative pole of the applied electric field.

14.   J. J. Thomson discovered the electrons. Henri Becqueral discovered radioactivity. Lord Rutherford proposed the nuclear model of the atom. Dalton's original model proposed that atoms were indivisible particles (that is, atoms had no internal structure). Thomson and Becqueral discovered subatomic particles and Rutherford's model attempted to describe the internal structure of the atom composed of these subatomic particles. In addition, the existence of isotopes, atoms of the same element but with different mass, had to be included in the model.

15.   The atomic number of an element is equal to the number of protons in the nucleus of an atom of that element. The mass number is the sum of the number of protons plus neutrons in the nucleus. The atomic mass is the actual mass of a particular isotope (including electrons). As we will see in chapter three, the average mass of an atom is taken from a measurement made on a large number of atoms. The average atomic mass value is listed in the periodic table.

16.   A family is a set of elements in the same vertical column. A family is also called a group. A period is a set of elements in the same horizontal row.

17.   A compound will always contain the same numbers (and types) of atoms. A given amount of hydrogen will react only with a specific amount of oxygen. Any excess oxygen will remain unreacted.

18.   The halogens have a high affinity for electrons and one important way they react is to form anions of the type X^-. The alkali metals tend to give up electrons easily and in most of their compounds exist as M^+ cations. Note: These two very reactive groups are only one electron away (in the periodic table) from the least reactive family of elements, the noble gases.

Exercises

Development of the Atomic Theory

19.   a.   The composition of a substance depends on the numbers of atoms of each element making up
       the compound (i.e., the formula of the compound) and not on the composition of the mixture
       from which it was formed.

       b.   Avogadro's hypothesis implies that volume ratios are equal to molecule ratios at constant
       temperature and pressure. $H_2 + Cl_2 \rightarrow 2\,HCl$. From the balanced equation, the volume of HCl
       produced will be twice the volume of H_2 (or Cl_2) reacted.

20.   From Avogadro's hypothesis, volume ratios are equal to molecule ratios at constant temperature and
       pressure. Therefore, we can write a balanced equation using the volume data, $Cl_2 + 3\,F_2 \rightarrow 2\,X$.
       Two molecules of X contain 6 atoms of F and two atoms of Cl. The formula of X is ClF_3 for a
       balanced reaction.

21. $\dfrac{1.188}{1.188} = 1.000; \quad \dfrac{2.375}{1.188} = 1.999; \quad \dfrac{3.563}{1.188} = 2.999$

       The masses of fluorine are simple ratios of whole numbers to each other, 1:2:3.

22.   Hydrazine: 1.44×10^{-1} g H/g N; Ammonia: 2.16×10^{-1} g H/g N

       Hydrogen azide: 2.40×10^{-2} g H/g N

       Let's try all of the ratios:

       $\dfrac{0.216}{0.144} = 1.50 = \dfrac{3}{2}; \quad \dfrac{0.144}{0.0240} = 6.00; \quad \dfrac{0.216}{0.0240} = 9.00$

       All the masses of hydrogen in these three compounds can be expressed as simple whole number
       ratios. The g H/g N in hydrazine, ammonia, and hydrogen azide are in the ratios 6:9:1.

23.   To get the atomic mass of H to be 1.00, we divide the mass of hydrogen that reacts with 1.00 g of
       oxygen by 0.126, i.e., $\dfrac{0.126}{0.126} = 1.00$. To get Na, Mg and O on the same scale, we do the same
       division.

       Na: $\dfrac{2.875}{0.126} = 22.8$; Mg: $\dfrac{1.500}{0.126} = 11.9$; O: $\dfrac{1.00}{0.126} = 7.94$

	H	O	Na	Mg
Relative Value	1.00	7.94	22.8	11.9
Accepted Value	1.008	16.00	22.99	24.31

The atomic masses of O and Mg are incorrect. The atomic masses of H and Na are close. Something must be wrong about the assumed formulas of the compounds. It turns out the correct formulas are H_2O, Na_2O, and MgO. The smaller discrepancies result from the error in the atomic mass of H.

24.   If the formula was Be_2O_3, then 2 times the atomic mass of Be would combine with three times the atomic mass of oxygen, or:

$$\frac{2\,A}{3(16.00)} = \frac{0.5633}{1.000}$$

Atomic mass of Be = A = 13.52. The accepted value is 9.01 and the discrepancy is due to the assumed formula. The actual formula is BeO.

The Nature of the Atom

25.   Density of hydrogen nucleus (contains one proton only):

$$V_{nucleus} = \frac{4}{3}\pi r^3 = \frac{4}{3}(3.14)(5 \times 10^{-14}\ cm)^3 = 5 \times 10^{-40}\ cm^3$$

$$d = \frac{1.67 \times 10^{-24}\ g}{5 \times 10^{-40}\ cm^3} = 3 \times 10^{15}\ g/cm^3$$

Density of H-atom (contains one proton and one electron):

$$V_{atom} = \frac{4}{3}(3.14)(1 \times 10^{-8}\ cm)^3 = 4 \times 10^{-24}\ cm^3$$

$$d = \frac{1.67 \times 10^{-24} + 9 \times 10^{-28}\ g}{4 \times 10^{-24}\ cm^3} = 0.4\ g/cm^3$$

26.   Since electrons move about the nucleus at an average distance of about 1×10^{-8} cm, then the diameter of an atom is about 2×10^{-8} cm. Let's set up a ratio:

$$\frac{\text{diameter of nucleus}}{\text{diameter of atom}} = \frac{1\ mm}{\text{diameter of model}} = \frac{1 \times 10^{-13}\ cm}{2 \times 10^{-8}\ cm},\ \text{Solving:}$$

diameter of model = 2×10^5 mm = 200 m

27. $5.93 \times 10^{-18}\ C \times \dfrac{1\ \text{electron charge}}{1.602 \times 10^{-19}\ C} = 37$ negative (electron) charges on the oil drop

28.   First, divide all charges by the smallest quantity, 6.40×10^{-13}.

$$\frac{2.56 \times 10^{-12}}{6.40 \times 10^{-13}} = 4.00; \quad \frac{7.68}{0.640} = 12.00; \quad \frac{3.84}{0.640} = 6.00$$

Since all charges are whole number multiples of 6.40×10^{-13} zirkombs then the charge on one electron could be 6.40×10^{-13} zirkombs. However, 6.40×10^{-13} zirkombs could be the charge of two electrons (or three electrons, etc.). All one can conclude is that the charge of an electron is 6.40×10^{-13} zirkombs or an integer fraction of 6.40×10^{-13}.

29.   gold - Au; silver - Ag; mercury - Hg; potassium - K; iron - Fe; antimony - Sb; tungsten - W

30.   sodium - Na; beryllium - Be; manganese - Mn; chromium - Cr; uranium - U

31.   fluorine - F; chlorine - Cl; bromine - Br; sulfur - S; oxygen - O; phosphorus - P

32.   titanium - Ti; selenium - Se; plutonium - Pu; nitrogen - N; silicon - Si

33.   Sn - tin; Pt - platinum; Co - cobalt; Ni - nickel; Mg - magnesium; Ba - barium; K - potassium

34.   As - arsenic; I - iodine; Xe - xenon; He - helium; C - carbon; Si - silicon

35.   The noble gases are He, Ne, Ar, Kr, Xe, and Rn (helium, neon, argon, krypton, xenon, and radon). Radon has only radioactive isotopes. In the periodic table the whole number enclosed in parenthesis is the mass number of the longest lived isotope of the element.

36.   promethium (Pm) and technetium (Tc)

37.   a. Eight; Li to Ne   b. Eight; Na to Ar

       c. Eighteen; K to Kr   d. Five; N, P, As, Sb, Bi

38.   a. Six; Be, Mg, Ca, Sr, Ba, Ra   b. Five; O, S, Se, Te, Po

       c. Three; Ni, Pd, Pt   d. Six; He, Ne, Ar, Kr, Xe, Rn

39.   a. $^{238}_{94}$Pu; 94 protons, 238 - 94 = 144 neutrons   b. $^{65}_{29}$Cu; 29 protons, 65 - 29 = 36 neutrons

       c. $^{52}_{24}$Cr; 24 protons, 28 neutrons   d. $^{4}_{2}$He; 2 protons, 2 neutrons

       e. $^{60}_{27}$Co; 27 protons, 33 neutrons   f. $^{54}_{24}$Cr; 24 protons, 30 neutrons

40.   a. ^{15}N; 7 protons, 8 neutrons   b. ^{3}H; 1 proton, 2 neutrons

       c. ^{207}Pb; 82 protons, 125 neutrons   d. ^{151}Eu; 63 protons, 88 neutrons

       e. ^{107}Ag; 47 protons, 60 neutrons   f. ^{109}Ag; 47 protons, 62 neutrons

41.   9 protons means the atomic number is 9. The mass number is 9 + 10 = 19; Symbol: $^{19}_{9}$F

42.   a. P   b. I   c. K   d. Yb

43.   Atomic number = 63 (Eu); Charge = +63 - 60 = +3; Mass number = 63 + 88 = 151;
       Symbol: $^{151}_{63}$Eu^{3+}

44.   Atomic number = 50 (Sn); Mass number = 50 + 68 = 118; Net charge = +50 - 48 = +2; The symbol is $_{50}^{118}Sn^{2+}$.

45.   Atomic number = 16 (S); Charge = +16 - 18 = -2; Mass number = 16 + 18 = 34; Symbol: $_{16}^{34}S^{2-}$

46.   Atomic number = 16 (S); Charge = +16 - 18 = -2; Mass number = 16 + 16 = 32; Symbol: $_{16}^{32}S^{2-}$

47.

Symbol	Number of protons in nucleus	Number of neutrons in nucleus	Number of electrons	Net charge
$_{33}^{75}As^{3+}$	33	42	30	3+
$_{52}^{128}Te^{2-}$	52	76	54	2-
$_{16}^{32}S$	16	16	16	0
$_{81}^{204}Tl^{+}$	81	123	80	1+
$_{78}^{195}Pt$	78	117	78	0

48.

Symbol	Number of protons in nucleus	Number of neutrons in nucleus	Number of electrons	Net charge
$_{92}^{238}U$	92	146	92	0
$_{20}^{40}Ca^{2+}$	20	20	18	2+
$_{23}^{51}V^{3+}$	23	28	20	3+
$_{39}^{89}Y$	39	50	39	0
$_{35}^{79}Br^{-}$	35	44	36	1-
$_{15}^{31}P^{3-}$	15	16	18	3-

49.   Metals: Mg, Ti, Au, Bi, Ge, Eu, Am;   Nonmetals: Si, B, At, Rn, Br

50.   Si, Ge, B, At;   The elements at the boundary between the metals and the non-metals: B, Si, Ge, As, Sb, Te, Po, At. Aluminum has mostly properties of metals.

51.   a and d; A group is a vertical column of elements in the periodic table. Elements in the same family (group) have similar chemical properties.

52.   b (Group 4A) and d (Group 2A)

53.   Carbon is a nonmetal. Silicon and germanium are metalloids. Tin and lead are metals. Thus, metallic character increases as one goes down a family in the periodic table.

54.   The metallic character decreases from left to right.

55.   Metals lose electrons to form cations and nonmetals gain electrons to form anions. Group IA, IIA and IIIA metals form stable +1,+2 and +3 charged cations, repectively. Group VA, VIA and VIIA nonmetals form -3, -2 and -1 charged anions, respectively.

        a.   Lose 1 e^- to form Na^+.   b.   Lose 2 e^- to form Sr^{2+}.   c.   Lose two e^- to form Ba^{2+}.

        d.   Gain 1 e^- to form I^-.   e.   Lose 3 e^- to form Al^{3+}.   f.   Gain 2 e^- to form S^{2-}.

56.   a.   Gain 1 e^- to form Cl^-.   b.   Lose 1 e^- to form Cs^+.   c.   Gain 2 e^- to form Se^{2-}.

        d.   Gain 3 e^- to form N^{3-}.   e.   Gain 2 e^- to form O^{2-}.   f.   Lose 2 e^- to form Mg^{2+}.

Nomenclature

57.   a.   sodium chloride   b.   rubidium oxide
        c.   calcium sulfide   d.   aluminum iodide

58.   a.   cobalt(III) oxide   b.   copper(I) oxide
        c.   iron(II) bromide   d.   lead(IV) sulfide

59.   a.   chromium(VI) oxide   b.   chromium(III) oxide   c.   aluminum oxide
        d.   sodium hydride   e.   calcium bromide
        f.   zinc chloride (Zinc only forms +2 ions so no roman numerals are needed for zinc compounds.)

60.   a.   cesium fluoride   b.   lithium nitride
        c.   silver sulfide (Silver only forms +1 ions so no roman numerals are needed.)
        d.   manganese(IV) oxide   e.   titanium(IV) oxide   f.   strontium phosphide

61.   a.   potassium perchlorate   b.   calcium phosphate
        c.   aluminum sulfate   d.   lead(II) nitrate

62.   a.   barium sulfite   b.   sodium nitrite
        c.   potassium permanganate   d.   potassium dichromate

63.   a.   nitrogen triiodide   b.   phosphorus trichloride
      c.   sulfur difluoride   d.   dinitrogen tetrafluoride

64.   a.   silicon tetrafluoride   b.   tetraphosphorus hexoxide
      c.   nitrogen monoxide   d.   selenium trioxide

65.   a.   copper(I) iodide   b.   copper(II) iodide   c.   cobalt(II) iodide
      d.   sodium carbonate   e.   sodium hydrogen carbonate or sodium bicarbonate
      f.   tetrasulfur tetranitride g.   sulfur hexafluoride   h.   sodium hypochlorite
      i.   barium chromate   j.   ammonium nitrate

66.   a.   nitric acid   b.   nitrous acid   c.   phosphoric acid
      d.   nitrogen trifluoride   e.   sodium hydrogen sulfate or sodium bisulfate (common name)
      f.   dichlorine heptoxide   g.   sodium bromate   h.   iron(III) periodate
      i.   ruthenium(III) nitrate   j.   vanadium(V) oxide   k.   platinum(IV) chloride
      l.   magnesium phosphate

67.   a. $CsBr$ b. $BaSO_4$ c. NH_4Cl d. ClO
      e. $SiCl_4$ f. ClF_3 g. BeO h. MgF_2

68.   a. SF_2 b. SF_6 c. NaH_2PO_4
      d. Li_3N e. $Cr_2(CO_3)_3$ f. SnF_2
      g. $NH_4C_2H_3O_2$ h. NH_4HSO_4 i. $Co(NO_3)_3$
      j. Hg_2Cl_2; Mercury(I) exists as Hg_2^{2+} ions.   k. $KClO_3$ l. NaH

69.   a. $NaOH$ b. $Al(OH)_3$ c. HCN
      d. Na_2O_2 e. $Cu(C_2H_3O_2)_2$ f. CF_4
      g. PbO h. PbO_2 i. $HC_2H_3O_2$
      j. $CuBr$ k. H_2SO_3 l. $GaAs$ (Ga^{3+} and As^{3-} ions)

70.   a. $(NH_4)_2HPO_4$ b. Hg_2S c. SiO_2
      d. Na_2SO_3 e. $Al(HSO_4)_3$ f. NCl_3
      g. HBr h. $HBrO_2$ i. $HBrO_4$
      j. KHS k. CaI_2 l. $CsClO_4$

Additional Exercises

71.   There should be no difference. The composition of insulin from both sources will be the same and therefore, it will have the same activity regardless of the source. As a practical note, trace contaminants in the two types of insulin may be different. These trace contaminates may be important.

72.   a. $^{24}_{12}Mg$: 12 protons, 12 neutrons, 12 electrons

      b. $^{24}_{12}Mg^{2+}$: 12 p, 12 n, 10 e   c. $^{59}_{27}Co^{2+}$: 27 p, 32 n, 25 e

      d. $^{59}_{27}Co^{3+}$: 27 p, 32 n, 24 e   e. $^{59}_{27}Co$: 27 p, 32 n, 27 e

f. $^{79}_{34}$Se: 34 p, 45 n, 34 e   g. $^{79}_{34}$Se^{2-}: 34 p, 45 n, 36 e

h. $^{63}_{28}$Ni: 28 p, 35 n, 28 e   i. $^{59}_{28}$Ni^{2+}: 28 p, 31 n, 26 e

73.   a. $Pb(C_2H_3O_2)_2$: lead(II) acetate   b. $CuSO_4$: copper(II) sulfate

       c. CaO: calcium oxide   d. $MgSO_4$: magnesium sulfate

       e. $Mg(OH)_2$: magnesium hydroxide   f. $CaSO_4$: calcium sulfate

       g. N_2O: dinitrogen monoxide or nitrous oxide

74.   a. Atomic number is 54. Xe   b. Atomic number is 34. Se
       c. Atomic number is 20. Ca   d. Atomic number is 42. Mo
       e. Atomic number is 94. Pu

75.   A chemical formula gives the actual number and kind of atoms in a compound. In all cases, 12 hydrogen atoms are present. For example:

$$4 \text{ molecules } H_3PO_4 \times \frac{3 \text{ atoms H}}{\text{molecule } H_3PO_4} = 12 \text{ atoms H}$$

76.   In the case of sulfur, SO_4^{2-} is sulfate and SO_3^{2-} is sulfite. By analogy:

SeO_4^{2-}: selenate; SeO_3^{2-}: selenite; TeO_4^{2-}: tellurate; TeO_3^{2-}: tellurite

Challenge Problems

77.   Copper(Cu), silver(Ag) and gold(Au) make up the coinage metals.

78.   a. Ba^{2+} and O^{2-}: BaO, barium oxide   b. Li^+ and N^{3-}: Li_3N, lithium nitride

       c. Al^{3+} and F^-: AlF_3, aluminum fluoride   d. K^+ and S^{2-}: K_2S, potassium sulfide

       e. In^{3+} and O^{2-}: In_2O_3, indium(III) oxide; In also forms In^+ ions, but you would predict In^{3+} ions from the periodic table.

       f. Ca^{2+} and Br^-: $CaBr_2$, calcium bromide   g. Al^{3+} and P^{3-}: AlP, aluminum phosphide

       h. Mg^{2+} and N^{3-}: Mg_3N_2, magnesium nitride

79.   Avogadro proposed that equal volumes of gases (at constant temperature and pressure) contains equal numbers of molecules. In terms of balanced equations, Avogadro's hypothesis implies that volume ratios will be identical to molecule ratios. Assuming one molecule of octane reacting, then 1 molecule of C_xH_y produces 8 molecules of CO_2 and 9 molecules of H_2O. $C_xH_y + O_2 \rightarrow$ $8 CO_2 + 9 H_2O$. Since all the carbon in octane ends up as carbon in CO_2, then octane contains 8 atoms of C. Similarly, all hydrogen in octane ends up as hydrogen in H_2O, so one molecule of octane contains $9 \times 2 = 18$ atoms of H. Octane formula = C_8H_{18} and the ratio of C:H = 8:18 or 4:9.

80.   These compounds are similar to phosphate (PO_4^{3-}) compounds. Na_3AsO_4 contain Na^+ ions and AsO_4^{3-} ions. The name would be sodium arsenate. H_3AsO_4 is analogous to phosphoric acid, H_3PO_4. H_3AsO_4 would be arsenic acid. $Mg_3(SbO_4)_2$ contains Mg^{2+} ions and SbO_4^{3-} ions and the name would be magnesium antimonate.

81.   Compound I: $\dfrac{14.0\,g\,R}{3.00\,g\,Q} = \dfrac{4.67\,g\,R}{1.00\,g\,Q}$;   Compound II: $\dfrac{7.00\,g\,R}{4.50\,g\,Q} = \dfrac{1.56\,g\,R}{1.00\,g\,Q}$

The ratio of the masses of R that combines with 1.00 g Q is: $\dfrac{4.67}{1.56} = 2.99 \approx 3$

As expected from the law of multiple proportions, this ratio is a small whole number.

Since Compound I contains three times the mass of R per gram of Q as compared to Compound II (RQ), then the formula of Compound I should be R_3Q.

82.   The solid residue must have come from the flask.

83.   a. The compounds have the same number and types of atoms (some formula), but the atoms in the molecules are bonded together differently. Therefore, the two compounds are different compounds with different properties. The compounds are called isomers of each other.

        b. When wood burns, most of the solid material in wood is converted to gases, which escape. The gases produced are most likely CO_2 and H_2O.

        c. The atom is not an indivisible particle, but is instead composed of other smaller particles, e.g., electrons, neutrons, protons.

        d. The two hydride samples contain different isotopes of either hydrogen and/or lithium. Although the compounds are composed of different isotopes, their properties are similar because different isotopes of the same element have similar properties (except, of course, their mass).

CHAPTER THREE

STOICHIOMETRY

Questions

12. The two major isotopes of boron are ^{10}B and ^{11}B. The listed mass of 10.81 is the average mass of a very large number of boron atoms.

13. The molecular formula tells us the actual number of atoms of each element in a molecule (or formula unit) of a compound. The empirical formula tells only the simplest whole number ratio of atoms of each element in a molecule. The molecular formula is a whole number multiple of the empirical formula. If that multiplier is one, the molecular and empirical formulas are the same. For example, both the molecular and empirical formulas of water are H_2O. They are the same. For hydrogen peroxide, the empirical formula is OH; the molecular formula is H_2O_2.

14. Side reactions may occur. For example, in the combustion of CH_4 (methane) to CO_2 and H_2O, some CO is also formed. Also, some reactions only go part way to completion and reach a state of equilibrium where both reactants and products are present (see Ch. 13).

Exercises

Atomic Masses and the Mass Spectrometer

15. $A = $ atomic mass $= 0.7899(23.9850 \text{ amu}) + 0.1000(24.9858 \text{ amu}) + 0.1101(25.9826 \text{ amu})$

    $A = 18.95 \text{ amu} + 2.499 \text{ amu} + 2.861 \text{ amu} = 24.31 \text{ amu}$

16. $0.3460(283.4 \text{ amu}) + 0.2120(284.7 \text{ amu}) + 0.4420(287.8 \text{ amu}) = 285.6 \text{ amu}$

17. $A = $ atomic mass $= 0.7553 (34.96885 \text{ amu}) + 0.2447 (36.96590 \text{ amu})$

    $A = 26.41 \text{ amu} + 9.046 = 35.46 \text{ amu}$; From atomic masses in the period table, this is chlorine.

18. Atomic mass $= \dfrac{90.51(19.992) + 0.27(20.994) + 9.22(21.990)}{100} = 20.18 \text{ amu}$. The element is neon.

19. Let x = % of ^{151}Eu and y = % of ^{153}Eu, then $x + y = 100$ and $y = 100 - x$.

    $$151.96 = \frac{x(150.9196) + (100 - x)(152.9209)}{100}$$

$15196 = 150.9196\ x + 15292.09 - 152.9209\ x,\ \ -96 = -2.0013\ x$

$x = 48\%;\ \ 48\%\ ^{151}\text{Eu}$ and $100 - 48 = 52\%\ ^{153}\text{Eu}$

20.   If Ag is 51.82% ^{107}Ag, then the remainder is ^{109}Ag or 48.18%. The average atomic mass is then:

$$107.868 = \frac{51.82\ (106.905) + 48.18\ (A)}{100}$$

$10786.8 = 5540. + 48.18\ A;\ \ A = 108.9\ \text{amu} = $ atomic mass of ^{109}Ag

21.   There are three peaks in the mass spectrum, each 2 mass units apart. This is consistent with two isotopes, differing in mass by two mass units. The peak at 157.84 corresponds to a Br_2 molecule composed of two atoms of the lighter isotope. This isotope has mass equal to 157.84/2 or 78.92. This corresponds to ^{79}Br. The second isotope is ^{81}Br with mass equal to 161.84/2 = 80.92. The peaks in the mass spectrum correspond to $^{79}\text{Br}_2$, $^{79}\text{Br}^{81}\text{Br}$ and $^{81}\text{Br}_2$ in order of increasing mass. The intensities of the highest and lowest mass tell us the two isotopes are present at about equal abundance. The actual abundance is 50.69% ^{79}Br and 49.31% ^{81}Br. The calculation of the abundance from the mass spectrum is beyond the scope of this text.

22.

Compound	Mass	Intensity	Scaled Intensity Largest Peak = 100
$\text{H}_2{}^{120}\text{Te}$	121.92	0.09	0.3
$\text{H}_2{}^{122}\text{Te}$	123.92	2.46	7.1
$\text{H}_2{}^{123}\text{Te}$	124.92	0.87	2.5
$\text{H}_2{}^{124}\text{Te}$	125.92	4.61	13.4
$\text{H}_2{}^{125}\text{Te}$	126.92	6.99	20.3
$\text{H}_2{}^{126}\text{Te}$	127.92	18.71	54.3
$\text{H}_2{}^{128}\text{Te}$	129.92	31.79	92.2
$\text{H}_2{}^{130}\text{Te}$	131.93	34.48	100.0

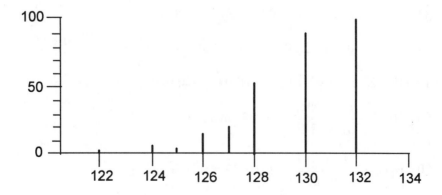

Moles and Molar Masses

23.   When more than one conversion factor is necessary to determine the answer, we will apply the conversion factors into one calculation instead of determining intermediate answers. This method reduces round-off error and is a time saver.

$$500. \text{ atoms Fe} \times \frac{1 \text{ mol Fe}}{6.022 \times 10^{23} \text{ atoms Fe}} \times \frac{55.85 \text{ g Fe}}{\text{mol Fe}} = 4.64 \times 10^{-20} \text{ g Fe}$$

24. $$500.0 \text{ g Fe} \times \frac{1 \text{ mol Fe}}{55.85 \text{ g Fe}} = 8.953 \text{ mol Fe}$$

$$8.953 \text{ mol Fe} \times \frac{6.022 \times 10^{23} \text{ atoms Fe}}{\text{mol Fe}} = 5.391 \times 10^{24} \text{ atoms Fe}$$

25. $$1.00 \text{ carat} \times \frac{0.200 \text{ g C}}{\text{carat}} \times \frac{1 \text{ mol C}}{12.01 \text{ g C}} \times \frac{6.022 \times 10^{23} \text{ atoms C}}{\text{mol C}} = 1.00 \times 10^{22} \text{ atoms C}$$

26. $$5.0 \times 10^{21} \text{ atoms C} \times \frac{1 \text{ mol C}}{6.022 \times 10^{23} \text{ atoms C}} = 8.3 \times 10^{-3} \text{ mol C}$$

$$8.3 \times 10^{-3} \text{ mol C} \times \frac{12.01 \text{ g C}}{\text{mol C}} = 0.10 \text{ g C}$$

27. Al_2O_3: $2(26.98) + 3(16.00) = 101.96$ g/mol

Na_3AlF_6: $3(22.99) + 1(26.98) + 6(19.00) = 209.95$ g/mol

28.   HFC - 134a, CH_2FCF_3: $2(12.01) + 2(1.008) + 4(19.00) = 102.04$ g/mol

HCFC-124, $CHClFCF_3$: $2(12.01) + 1(1.008) + 1(35.45) + 4(19.00) = 136.48$ g/mol

29.   a. NH_3: 14.01 g/mol $+ 3(1.008$ g/mol$) = 17.03$ g/mol

b. N_2H_4: $2(14.01) + 4(1.008) = 32.05$ g/mol

c. $(NH_4)_2Cr_2O_7$: $2(14.01) + 8(1.008) + 2(52.00) + 7(16.00) = 252.08$ g/mol

30.   a. P_4O_6: $4(30.97$ g/mol$) + 6(16.00$ g/mol$) = 219.88$ g/mol

b. $Ca_3(PO_4)_2$: $3(40.08) + 2(30.97) + 8(16.00) = 310.18$ g/mol

c. Na_2HPO_4: $2(22.99) + 1(1.008) + 1(30.97) + 4(16.00) = 141.96$ g/mol

31.   a. $$1.00 \text{ g NH}_3 \times \frac{1 \text{ mol NH}_3}{17.03 \text{ g NH}_3} = 0.0587 \text{ mol NH}_3$$

b. $$1.00 \text{ g N}_2\text{H}_4 \times \frac{1 \text{ mol N}_2\text{H}_4}{32.05 \text{ g N}_2\text{H}_4} = 0.0312 \text{ mol N}_2\text{H}_4$$

c. $1.00 \text{ g (NH}_4)_2\text{Cr}_2\text{O}_7 \times \dfrac{1 \text{ mol (NH}_4)_2\text{Cr}_2\text{O}_7}{252.08 \text{ g (NH}_4)_2\text{Cr}_2\text{O}_7} = 3.97 \times 10^{-3} \text{ mol (NH}_4)_2\text{Cr}_2\text{O}_7$

32.   a. $1.00 \text{ g P}_4\text{O}_6 \times \dfrac{1 \text{ mol P}_4\text{O}_6}{219.88 \text{ g}} = 4.55 \times 10^{-3} \text{ mol P}_4\text{O}_6$

b. $1.00 \text{ g Ca}_3(\text{PO}_4)_2 \times \dfrac{1 \text{ mol Ca}_3(\text{PO}_4)_2}{310.18 \text{ g}} = 3.22 \times 10^{-3} \text{ mol Ca}_3(\text{PO}_4)_2$

c. $1.00 \text{ g Na}_2\text{HPO}_4 \times \dfrac{1 \text{ mol Na}_2\text{HPO}_4}{141.96 \text{ g}} = 7.04 \times 10^{-3} \text{ mol Na}_2\text{HPO}_4$

33.   a. $5.00 \text{ mol NH}_3 \times \dfrac{17.03 \text{ g NH}_4}{\text{mol NH}_3} = 85.2 \text{ g NH}_3$

b. $5.00 \text{ mol N}_2\text{H}_4 \times \dfrac{32.05 \text{ g N}_2\text{H}_4}{\text{mol N}_2\text{H}_4} = 160. \text{ g N}_2\text{H}_4$

c. $5.00 \text{ mol (NH}_4)_2\text{Cr}_2\text{O}_7 \times \dfrac{252.08 \text{ g (NH}_4)_2\text{Cr}_2\text{O}_7}{\text{mol (NH}_4)_2\text{Cr}_2\text{O}_7} = 1260 \text{ g (NH}_4)_2\text{Cr}_2\text{O}_7$

34.   a. $5.00 \text{ mol P}_4\text{O}_6 \times \dfrac{219.88 \text{ g}}{\text{mol P}_4\text{O}_6} = 1.10 \times 10^3 \text{ g P}_4\text{O}_6$

b. $5.00 \text{ mol Ca}_3(\text{PO}_4)_2 \times \dfrac{310.18 \text{ g}}{\text{mol Ca}_3(\text{PO}_4)_2} = 1.55 \times 10^3 \text{ g Ca}_3(\text{PO}_4)_2$

c. $5.00 \text{ mol Na}_2\text{HPO}_4 \times \dfrac{141.96 \text{ g}}{\text{mol Na}_2\text{HPO}_4} = 7.10 \times 10^2 \text{ g Na}_2\text{HPO}_4$

35.   Chemical formulas give atom ratios as well as mol ratios.

a. $5.00 \text{ mol NH}_3 \times \dfrac{1 \text{ mol N}}{\text{mol NH}_3} \times \dfrac{14.01 \text{ g N}}{\text{mol N}} = 70.1 \text{ g N}$

b. $5.00 \text{ mol N}_2\text{H}_4 \times \dfrac{2 \text{ mol N}}{\text{mol N}_2\text{H}_4} \times \dfrac{14.01 \text{ g N}}{\text{mol N}} = 140. \text{ g N}$

c. $5.00 \text{ mol (NH}_4)_2\text{Cr}_2\text{O}_7 \times \dfrac{2 \text{ mol N}}{\text{mol (NH}_4)_2\text{Cr}_2\text{O}_7} \times \dfrac{14.01 \text{ g N}}{\text{mol N}} = 140. \text{ g N}$

36.   a. $5.00 \text{ mol P}_4\text{O}_6 \times \dfrac{4 \text{ mol P}}{\text{mol P}_4\text{O}_6} \times \dfrac{30.97 \text{ g P}}{\text{mol P}} = 619 \text{ g P}$

b. $5.00 \text{ mol Ca}_3(\text{PO}_4)_2 \times \dfrac{2 \text{ mol P}}{\text{mol Ca}_3(\text{PO}_4)_2} \times \dfrac{30.97 \text{ g P}}{\text{mol P}} = 310. \text{ g P}$

c. $5.00 \text{ mol Na}_2\text{HPO}_4 \times \dfrac{1 \text{ mol P}}{\text{mol Na}_2\text{HPO}_4} \times \dfrac{30.97 \text{ g P}}{\text{mol P}} = 155 \text{ g P}$

37.   a. $1.00 \text{ g NH}_3 \times \dfrac{1 \text{ mol NH}_3}{17.03 \text{ g NH}_3} \times \dfrac{6.022 \times 10^{23} \text{ molecules NH}_3}{\text{mol NH}_3} = 3.54 \times 10^{22} \text{ molecules NH}_3$

       b. $1.00 \text{ g N}_2\text{H}_4 \times \dfrac{1 \text{ mol N}_2\text{H}_4}{32.05 \text{ g N}_2\text{H}_4} \times \dfrac{6.022 \times 10^{23} \text{ molecules N}_2\text{H}_4}{\text{mol N}_2\text{H}_4} = 1.88 \times 10^{22} \text{ molecules N}_2\text{H}_4$

       c. $1.00 \text{ g (NH}_4)\text{Cr}_2\text{O}_7 \times \dfrac{1 \text{ mol (NH}_4)_2\text{Cr}_2\text{O}_7}{252.08 \text{ g (NH}_4)_2\text{Cr}_2\text{O}_7} \times \dfrac{6.022 \times 10^{23} \text{ molecules (NH}_4)_2\text{Cr}_2\text{O}_7}{\text{mol (NH}_4)_2\text{Cr}_2\text{O}_7}$

            $= 2.39 \times 10^{21} \text{ molecules (NH}_4)_2\text{Cr}_2\text{O}_7$

38.   a. $1.00 \text{ g P}_4\text{O}_6 \times \dfrac{1 \text{ mol P}_4\text{O}_6}{219.88 \text{ g}} \times \dfrac{6.022 \times 10^{23} \text{ molecules}}{\text{mol P}_4\text{O}_6} = 2.74 \times 10^{21} \text{ molecules P}_4\text{O}_6$

       b. $1.00 \text{ g Ca}_3(\text{PO}_4)_2 \times \dfrac{1 \text{ mol Ca}_3(\text{PO}_4)_2}{310.18 \text{ g}} \times \dfrac{6.022 \times 10^{23} \text{ molecules}}{\text{mol Ca}_3(\text{PO}_4)_2}$

            $= 1.94 \times 10^{21} \text{ molecules Ca}_3(\text{PO}_4)_2$

       c. $1.00 \text{ g Na}_2\text{HPO}_4 \times \dfrac{\text{mol Na}_2\text{HPO}_4}{141.96 \text{ g}} \times \dfrac{6.022 \times 10^{23} \text{ molecules}}{\text{mol Na}_2\text{HPO}_4} = 4.24 \times 10^{21} \text{ molecules Na}_2\text{HPO}_4$

39.   Using answers from Exercise 37:

       a. $3.54 \times 10^{22} \text{ molecules NH}_3 \times \dfrac{1 \text{ atom N}}{\text{molecule NH}_3} = 3.54 \times 10^{22} \text{ atoms N}$

       b. $1.88 \times 10^{22} \text{ molecules N}_2\text{H}_4 \times \dfrac{2 \text{ atoms N}}{\text{molecule N}_2\text{H}_4} = 3.76 \times 10^{22} \text{ atoms N}$

       c. $2.39 \times 10^{21} \text{ molecules (NH}_4)_2\text{Cr}_2\text{O}_7 \times \dfrac{2 \text{ atoms N}}{\text{molecule (NH}_4)_2\text{Cr}_2\text{O}_7} = 4.78 \times 10^{21} \text{ atoms N}$

40.   Using answers from Exercise 38:

       a. $2.74 \times 10^{21} \text{ molecules P}_4\text{O}_6 \times \dfrac{4 \text{ atoms P}}{\text{molecule P}_4\text{O}_6} = 1.10 \times 10^{22} \text{ atoms P}$

       b. $1.94 \times 10^{21} \text{ molecules Ca}_3(\text{PO}_4)_2 \times \dfrac{2 \text{ atoms P}}{\text{molecule Ca}_3(\text{PO}_4)_2} = 3.88 \times 10^{21} \text{ atoms P}$

       c. $4.24 \times 10^{21} \text{ molecules Na}_2\text{HPO}_4 \times \dfrac{1 \text{ atom P}}{\text{molecule Na}_2\text{HPO}_4} = 4.24 \times 10^{21} \text{ atoms P}$

41.   Molar mass of $\text{C}_6\text{H}_8\text{O}_6 = 6(12.01) + 8(1.008) + 6(16.00) = 176.12 \text{ g/mol}$

       $500.0 \text{ mg} \times \dfrac{1 \text{ g}}{1000 \text{ mg}} \times \dfrac{1 \text{ mol}}{176.12 \text{ g}} = 2.839 \times 10^{-3} \text{ mol}$

$$2.839 \times 10^{-3} \text{ mol} \times \frac{6.022 \times 10^{23} \text{ molecules}}{\text{mol}} = 1.710 \times 10^{21} \text{ molecules}$$

42.   a. $9(12.01) + 8(1.008) + 4(16.00) = 180.15$ g/mol

      b. $500. \text{ mg} \times \dfrac{1 \text{ g}}{1000 \text{ mg}} \times \dfrac{1 \text{ mol}}{180.15 \text{ g}} = 2.78 \times 10^{-3}$ mol

      $$2.78 \times 10^{-3} \text{ mol} \times \frac{6.022 \times 10^{23} \text{ molecules}}{\text{mol}} = 1.67 \times 10^{21} \text{ molecules}$$

43.   a. $100 \text{ molecules } H_2O \times \dfrac{1 \text{ mol } H_2O}{6.022 \times 10^{23} \text{ molecules } H_2O} = 1.661 \times 10^{-22} \text{ mol } H_2O$

      b. $100.0 \text{ g } H_2O \times \dfrac{1 \text{ mol } H_2O}{18.02 \text{ g } H_2O} = 5.549 \text{ mol } H_2O$

      c. $150 \text{ molecules } O_2 \times \dfrac{1 \text{ mol } O_2}{6.022 \times 10^{23} \text{ molecules } O_2} = 2.491 \times 10^{-22} \text{ mol } O_2$

44.   a. $150.0 \text{ g } Fe_2O_3 \times \dfrac{1 \text{ mol}}{159.70 \text{ g}} = 0.9393 \text{ mol } Fe_2O_3$

      b. $10.0 \text{ mg } NO_2 \times \dfrac{1 \text{ g}}{10^3 \text{ mg}} \times \dfrac{1 \text{ mol}}{46.01 \text{ g}} = 2.17 \times 10^{-4} \text{ mol } NO_2$

      c. $1.5 \times 10^{16} \text{ molecules } BF_3 \times \dfrac{1 \text{ mol}}{6.02 \times 10^{23} \text{ molecules}} = 2.5 \times 10^{-8} \text{ mol } BF_3$

45.   a. $3.00 \times 10^{20} \text{ molecules } N_2 \times \dfrac{1 \text{ mol } N_2}{6.02 \times 10^{23} \text{ molecules}} \times \dfrac{28.02 \text{ g } N_2}{\text{mol } N_2} = 1.40 \times 10^{-2} \text{ g } N_2$

      b. $3.00 \times 10^{-3} \text{ mol } N_2 \times \dfrac{28.02 \text{ g } N_2}{\text{mol } N_2} = 8.41 \times 10^{-2} \text{ g } N_2$

      c. $1.5 \times 10^{2} \text{ mol } N_2 \times \dfrac{28.02 \text{ g } N_2}{\text{mol } N_2} = 4.2 \times 10^{3} \text{ g } N_2$

      d. $1 \text{ molecule } N_2 \times \dfrac{1 \text{ mol } N_2}{6.022 \times 10^{23} \text{ molecules } N_2} \times \dfrac{28.02 \text{ g } N_2}{\text{mol } N_2} = 4.653 \times 10^{-23} \text{ g } N_2$

      e. $2.00 \times 10^{-15} \text{ mol } N_2 \times \dfrac{28.02 \text{ g } N_2}{\text{mol } N_2} = 5.60 \times 10^{-14} \text{ g } N_2 = 56.0 \text{ fg } N_2$

      f. $18.0 \text{ pmol } N_2 \times \dfrac{1 \text{ mol } N_2}{10^{12} \text{ pmol}} \times \dfrac{28.02 \text{ g } N_2}{\text{mol } N_2} = 5.04 \times 10^{-10} \text{ g } N_2 = 504 \text{ pg } N_2$

      g. $5.0 \text{ nmol } N_2 \times \dfrac{1 \text{ mol } N_2}{10^{9} \text{ nmol}} \times \dfrac{28.02 \text{ g } N_2}{\text{mol } N_2} = 1.4 \times 10^{-7} \text{ g } N_2 = 140 \text{ ng}$

46.   a.   A chemical formula gives atom ratios as well as mole ratios. We will use both ideas to show how these conversion factors can be used.

Molar mass of $CH_4O = 1(12.01) + 4(1.008) + 1(16.00) = 32.04$ g/mol

$$1.0 \text{ g } CH_4O \times \frac{1 \text{ mol } CH_4O}{32.04 \text{ g } CH_4O} \times \frac{6.02 \times 10^{23} \text{ molecules } CH_4O}{\text{mol } CH_4O} \times \frac{1 \text{ atom C}}{\text{molecule } CH_4O}$$

$$= 1.9 \times 10^{22} \text{ atoms C}$$

b.   Molar mass of $CH_3CO_2H = 2(12.01) + 4(1.008) + 2(16.00) = 60.05$ g/mol

$$1.0 \text{ g } CH_3CO_2H \times \frac{1 \text{ mol}}{60.05 \text{ g}} \times \frac{6.02 \times 10^{23} \text{ molecules}}{\text{mol}} \times \frac{2 \text{ atoms C}}{\text{molecule } CH_3CO_2H} = 2.0 \times 10^{22} \text{ atoms C}$$

c.   Molar mass of $Na_2CO_3 = 2(22.99) + 1(12.01) + 3(16.00) = 105.99$ g/mol

$$1.0 \text{ g } Na_2CO_3 \times \frac{1 \text{ mol } Na_2CO_3}{105.99 \text{ g } Na_2CO_3} \times \frac{1 \text{ mol C}}{\text{mol } Na_2CO_3} \times \frac{6.02 \times 10^{23} \text{ atoms C}}{\text{mol C}} = 5.7 \times 10^{21} \text{ atoms C}$$

d. $1.0 \text{ g } C_6H_{12}O_6 \times \dfrac{1 \text{ mol}}{180.16 \text{ g}} \times \dfrac{6 \text{ mol C}}{\text{mol } C_6H_{12}O_6} \times \dfrac{6.02 \times 10^{23} \text{ atoms C}}{\text{mol C}} = 2.0 \times 10^{22} \text{ atoms C}$

e. $1.0 \text{ g } C_{12}H_{22}O_{11} \times \dfrac{1 \text{ mol}}{342.30 \text{ g}} \times \dfrac{12 \text{ mol C}}{\text{mol } C_{12}H_{22}O_{11}} \times \dfrac{6.02 \times 10^{23} \text{ atoms C}}{\text{mol C}} = 2.1 \times 10^{22} \text{ atoms C}$

f. $1.0 \text{ g } CHCl_3 \times \dfrac{1 \text{ mol } CHCl_3}{119.37 \text{ g } CHCl_3} \times \dfrac{1 \text{ mol C}}{\text{mol } CHCl_3} \times \dfrac{6.02 \times 10^{23} \text{ atoms C}}{\text{mol C}} = 5.0 \times 10^{21} \text{ atoms C}$

47.   a. $14 \text{ mol C} \left(\dfrac{12.01 \text{ g}}{\text{mol C}} \right) + 18 \text{ mol H} \left(\dfrac{1.008 \text{ g}}{\text{mol H}} \right) + 2 \text{ mol N} \left(\dfrac{14.01 \text{ g}}{\text{mol N}} \right) + 5 \text{ mol O} \left(\dfrac{16.00 \text{ g}}{\text{mol O}} \right)$

$$= 294.30 \text{ g/mol}$$

b. $10.0 \text{ g aspartame} \times \dfrac{1 \text{ mol}}{294.30 \text{ g}} = 3.40 \times 10^{-2} \text{ mol}$

c. $1.56 \text{ mol} \times \dfrac{294.30 \text{ g}}{\text{mol}} = 459 \text{ g}$

d. $5.0 \text{ mg} \times \dfrac{1 \text{ g}}{1000 \text{ mg}} \times \dfrac{1 \text{ mol}}{294.30 \text{ g}} \times \dfrac{6.02 \times 10^{23} \text{ molecules}}{\text{mol}} = 1.0 \times 10^{19} \text{ molecules}$

e.   The chemical formula tells us that 1 molecule of aspartame contains two atoms of N. The chemical formula also says that 1 mol of aspartame contains two mol of N.

$$1.2 \text{ g aspartame} \times \frac{1 \text{ mol aspartame}}{294.30 \text{ g aspartame}} \times \frac{2 \text{ mol N}}{\text{mol aspartame}} \times \frac{6.02 \times 10^{23} \text{ atoms N}}{\text{mol N}}$$

$$= 4.9 \times 10^{21} \text{ atoms of nitrogen}$$

f. $1.0 \times 10^9 \text{ molecules} \times \dfrac{1 \text{ mol}}{6.02 \times 10^{23} \text{ molecules}} \times \dfrac{294.30 \text{ g}}{\text{mol}} = 4.9 \times 10^{-13} \text{ g or } 490 \text{ fg}$

g. $1 \text{ molecule aspartame} \times \dfrac{1 \text{ mol}}{6.022 \times 10^{23} \text{ molecules}} \times \dfrac{294.30 \text{ g}}{\text{mol}} = 4.887 \times 10^{-22} \text{ g}$

48.   As we shall see later, the formula written as $(CH_3)_2N_2O$ tries to tell us something about how the atoms are attached to each other. For our purposes in this problem, we can write the formula as $C_2H_6N_2O$.

a. $2(12.01) + 6(1.008) + 2(14.01) + 1(16.00) = 74.09 \text{ g/mol}$

b. $250 \text{ mg} \times \dfrac{1 \text{ g}}{1000 \text{ mg}} \times \dfrac{1 \text{ mol}}{74.09 \text{ g}} = 3.4 \times 10^{-3} \text{ mol}$ c. $0.050 \text{ mol} \times \dfrac{74.09 \text{ g}}{\text{mol}} = 3.7 \text{ g}$

d. $1.0 \text{ mol } C_2H_6N_2O \times \dfrac{6.02 \times 10^{23} \text{ molecules } C_2H_6N_2O}{\text{mol } C_2H_6N_2O} \times \dfrac{6 \text{ atoms of H}}{\text{molecule } C_2H_6N_2O}$

$$= 3.6 \times 10^{24} \text{ atoms of hydrogen}$$

e. $1.0 \times 10^6 \text{ molecules} \times \dfrac{1 \text{ mol}}{6.02 \times 10^{23} \text{ molecules}} \times \dfrac{74.09 \text{ g}}{\text{mol}} = 1.2 \times 10^{-16} \text{ g} = 0.12 \text{ fg}$

f. $1 \text{ molecule} \times \dfrac{1 \text{ mol}}{6.022 \times 10^{23} \text{ molecules}} \times \dfrac{74.09 \text{ g}}{\text{mol}} = 1.230 \times 10^{-22} \text{ g}$

Percent Composition

49. $\text{mass \%Cd} = \dfrac{\text{mass of Cd in 1 mol compound}}{\text{molar mass of compound}} \times 100$

CdS: $\%Cd = \dfrac{112.4 \text{ g Cd}}{144.5 \text{ g CdS}} \times 100 = 77.79\% \text{ Cd}$

$CdSe$: $\%Cd = \dfrac{112.4 \text{ g}}{191.4 \text{ g}} \times 100 = 58.73\% \text{ Cd}$; $CdTe$: $\%Cd = \dfrac{112.4 \text{ g}}{240.0 \text{ g}} \times 100 = 46.83\% \text{ Cd}$

50.   a. $C_3H_4O_2$: Molar mass $= 3(12.01) + 4(1.008) + 2(16.00) = 36.03 + 4.032 + 32.00 = 72.06 \text{ g/mol}$

$\%C = \dfrac{36.03 \text{ g C}}{72.06 \text{ g compound}} \times 100 = 50.00\% \text{ C}$; $\%H = \dfrac{4.032 \text{ g H}}{72.06 \text{ g compound}} \times 100 = 5.595\% \text{ H}$

$\%O = 100.00 - (50.00 + 5.595) = 44.41\% \text{ O or } \%O = \dfrac{32.00 \text{ g}}{72.06 \text{ g}} \times 100 = 44.41\% \text{ O}$

b. $C_4H_6O_2$: Molar mass $= 4(12.01) + 6(1.008) + 2(16.00) = 48.04 + 6.048 + 32.00 = 86.09$ g/mol

$$\%C = \frac{48.04\,g}{86.09\,g} \times 100 = 55.80\%\,C; \quad \%H = \frac{6.048\,g}{86.09\,g} \times 100 = 7.025\%\,H$$

$$\%O = 100.00 - (55.80 + 7.025) = 37.18\%\,O$$

c. C_3H_3N: Molar mass $= 3(12.01) + 3(1.008) + 1(14.01) = 36.03 + 3.024 + 14.01 = 53.06$ g/mol

$$\%C = \frac{36.03\,g}{53.06\,g} \times 100 = 67.90\%\,C; \qquad \%H = \frac{3.024\,g}{53.06\,g} \times 100 = 5.699\%\,H$$

$$\%N = \frac{14.01\,g}{53.06\,g} \times 100 = 26.40\%\,N \;\; or \;\; \%N = 100.00 - (67.90 + 5.699) = 26.40\%\,N$$

51. In 1 mole of $YBa_2Cu_3O_7$, there are 1 mole of Y, 2 moles of Ba, 3 moles of Cu, and 7 moles of O.

$$Molar\ mass = 1\ mol\ Y \left(\frac{88.91\,g\,Y}{mol\,Y} \right) + 2\ mol\ Ba \left(\frac{137.3\,g\,Ba}{mol\,Ba} \right)$$

$$+ 3\ mol\ Cu \left(\frac{63.55\,g\,Cu}{mol\,Cu} \right) + 7\ mol\ O \left(\frac{16.00\,g\,O}{mol\,O} \right)$$

Molar mass $= 88.91 + 274.6 + 190.65 + 112.00 = 666.2$ g/mol

$$\%Y = \frac{88.91\,g}{666.2\,g} \times 100 = 13.35\%\,Y; \;\; \%Ba = \frac{274.6\,g}{666.2\,g} \times 100 = 41.22\%\,Ba$$

$$\%Cu = \frac{190.65\,g}{666.2\,g} \times 100 = 28.62\%\,Cu; \;\; \%O = \frac{112.0\,g}{666.2\,g} \times 100 = 16.81\%\,O$$

52. a. NO: $\%N = \dfrac{14.01\,g\,N}{30.01\,g\,NO} \times 100 = 46.68\%\,N$

b. NO_2: $\%N = \dfrac{14.01\,g\,N}{46.01\,g\,NO_2} \times 100 = 30.45\%\,N$

c. N_2O_4: $\%N = \dfrac{28.02\,g\,N}{92.02\,g\,N_2O_4} \times 100 = 30.45\%\,N$

d. N_2O: $\%N = \dfrac{28.02\,g\,N}{44.02\,g\,N_2O} \times 100 = 63.65\%\,N$

53. Na_3PO_4: molar mass $= 3(22.99) + 30.97 + 4(16.00) = 163.94$ g/mol

$$\%P = \frac{30.97\,g}{163.94\,g} \times 100 = 18.89\%\,P$$

PH_3: molar mass $= 30.97 + 3(1.008) = 33.99$ g/mol

$$\%P = \frac{30.97 \text{ g}}{33.99 \text{ g}} \times 100 = 91.12\% \text{ P}$$

P_4O_{10}: molar mass = 4(30.97) + 10(16.00) = 283.88 g/mol

$$\%P = \frac{123.88 \text{ g}}{283.88 \text{ g}} \times 100 = 43.638\% \text{ P}$$

$(NPCl_2)_3$: molar mass = 3(14.01) + 3(30.97) + 6(35.45) = 347.64 g/mol

$$\%P = \frac{92.91 \text{ g}}{347.64 \text{ g}} \times 100 = 26.73\% \text{ P}$$

The order from lowest to highest percentage of phosphorus is:

$$Na_3PO_4 < (NPCl_2)_3 < P_4O_{10} < PH_3$$

54.   From results in Exercise 52: $NO_2 = N_2O_4 < NO < N_2O$

55.   There are many valid methods to solve this problem. We will assume 100.00 g of compound, then determine from the information in the problem how many mol of compound equals 100.00 g of compound. From this information, we can determine the mass of one mol of compound (the molar mass) by setting up a ratio. Assuming 100.00 g cyanocobalamin:

$$\text{mol cyanocobalamin} = 4.34 \text{ g Co} \times \frac{1 \text{ mol Co}}{58.93 \text{ g Co}} \times \frac{1 \text{ mol cyanocobalamin}}{\text{mol Co}}$$

$$= 7.36 \times 10^{-2} \text{ mol cyanocobalamin}$$

$$\frac{\text{x g cyanocobalamin}}{1 \text{ mol cyanocobalamin}} = \frac{100.00 \text{ g}}{7.36 \times 10^{-2} \text{ mol}}, \text{ x = molar mass = 1360 g/mol}$$

56.   There are 0.342 g Fe for every 100.000 g hemoglobin (Hb). Assuming 100.000 g hemoglobin:

$$\text{mol Hb} = 0.342 \text{ g Fe} \times \frac{1 \text{ mol Fe}}{55.85 \text{ g Fe}} \times \frac{1 \text{ mol Hb}}{4 \text{ mol Fe}} = 1.53 \times 10^{-3} \text{ mol Hb}$$

$$\frac{\text{x g Hb}}{\text{mol Hb}} = \frac{100.000 \text{ g Hb}}{1.53 \times 10^{-3} \text{ mol Hb}}, \text{ x = molar mass = 65,400 g/mol}$$

Empirical and Molecular Formulas

57.   a.   Molar mass of CH_2O = 1 mol C $\left(\dfrac{12.01 \text{ g}}{\text{mol C}} \right)$ + 2 mol H $\left(\dfrac{1.008 \text{ g H}}{\text{mol H}} \right)$

$$+ 1 \text{ mol O} \left(\frac{16.00 \text{ g}}{\text{mol O}} \right) = 30.03 \text{ g/mol}$$

$$\%C = \frac{12.01 \text{ g C}}{30.03 \text{ g CH}_2O} \times 100 = 39.99\% \text{ C}; \ \%H = \frac{2.016 \text{ g H}}{30.03 \text{ g CH}_2O} \times 100 = 6.713\% \text{ H}$$

$$\%O = \frac{16.00 \text{ g O}}{30.03 \text{ g CH}_2\text{O}} \times 100 = 53.28\% \text{ O} \qquad \text{or } \%O = 100.00 - (39.99 + 6.713) = 53.30\%$$

b. Molar Mass of $C_6H_{12}O_6 = 6(12.01) + 12(1.008) + 6(16.00) = 180.16$ g/mol

$$\%C = \frac{72.06 \text{ g C}}{180.16 \text{ g C}_6\text{H}_{12}\text{O}_6} \times 100 = 40.00\%; \quad \%H = \frac{12.096 \text{ g}}{180.16 \text{ g}} \times 100 = 6.7140\%$$

$\%O = 100.00 - (40.00 + 6.714) = 53.29\%$

c. Molar mass of $HC_2H_3O_2 = 2(12.01) + 4(1.008) + 2(16.00) = 60.05$ g/mol

$$\%C = \frac{24.02 \text{ g}}{60.05 \text{ g}} \times 100 = 40.00\%; \quad \%H = \frac{4.032 \text{ g}}{60.05 \text{ g}} \times 100 = 6.714\%$$

$\%O = 100.00 - (40.00 + 6.714) = 53.29\%$

58.   All three compounds have the same empirical formula, CH_2O, and different molecular formulas. The composition of all three in mass percent is also the same. Therefore, elemental analysis will give us only the empirical formula.

59.   a. $C_3H_4O_3$ b. CH   c. CH   d. P_2O_5 e. CH_2O f. CH_2O

60.   a. SNH: Empirical formula mass $= 32.07 + 14.01 + 1.008 = 47.09$ g

$$\frac{188.35}{47.09} = 4.000; \text{ So the molecular formula is } (SNH)_4 \text{ or } S_4N_4H_4.$$

b. $NPCl_2$: Empirical formula mass $= 14.01 + 30.97 + 2(35.45) = 115.88$ g/mol

$$\frac{347.64}{115.88} = 3.0000; \text{ Molecular formula is } (NPCl_2)_3 \text{ or } N_3P_3Cl_6.$$

c. CoC_4O_4: $58.93 + 4(12.01) + 4(16.00) = 170.97$ g/mol

$$\frac{341.94}{170.97} = 2.0000; \text{ Molecular formula: } Co_2C_8O_8$$

d. SN: $32.07 + 14.01 = 46.08$ g/mol; $\dfrac{184.32}{46.08} = 4.000$; Molecular formula: S_4N_4

61.   Out of 100.00 g of compound, there are:

$$48.64 \text{ g C} \times \frac{1 \text{ mol C}}{12.01 \text{ g C}} = 4.050 \text{ mol C}; \quad 8.16 \text{ g H} \times \frac{1 \text{ mol H}}{1.008 \text{ g H}} = 8.10 \text{ mol H}$$

$$\%O = 100.00 - 48.64 - 8.16 = 43.20\%; \quad 43.20 \text{ g O} \times \frac{1 \text{ mol O}}{16.00 \text{ g O}} = 2.700 \text{ mol O}$$

Dividing each mol value by the smallest number:

$$\frac{4.050}{2.700} = 1.500, \quad \frac{8.10}{2.700} = 3.00, \quad \frac{2.700}{2.700} = 1.000$$

Since a whole number ratio is required, the C:H:O ratio is 1.5:3:1 or 3:6:2. So the empirical formula is $C_3H_6O_2$.

62.   Out of 100.0 g of the pigment, there are:

$$59.9 \text{ g Ti} \times \frac{1 \text{ mol Ti}}{47.88 \text{ g Ti}} = 1.25 \text{ mol Ti}; \quad 40.1 \text{ g O} \times \frac{1 \text{ mol O}}{16.00 \text{ g O}} = 2.51 \text{ mol O}$$

Empirical formula = TiO_2 since mol O to mol Ti is in a 2:1 mol ratio.

63. $0.979 \text{ g Na} \times \dfrac{1 \text{ mol Na}}{22.99 \text{ g Na}} = 4.26 \times 10^{-2} \text{ mol Na}$

$1.365 \text{ g S} \times \dfrac{1 \text{ mol S}}{32.07 \text{ g S}} = 4.256 \times 10^{-2} \text{ mol S}$

$1.021 \text{ g O} \times \dfrac{1 \text{ mol O}}{16.00 \text{ g O}} = 6.381 \times 10^{-2} \text{ mol O}$

Determine the mol ratios by dividing by mol S (the smallest number):

$$\frac{6.381 \times 10^{-2} \text{ mol O}}{4.256 \times 10^{-2} \text{ mol S}} = 1.499 \approx \frac{1.5 \text{ mol O}}{\text{mol S}} = \frac{3 \text{ mol O}}{2 \text{ mol S}}; \quad \frac{4.26 \times 10^{-2} \text{ mol Na}}{4.256 \times 10^{-2} \text{ mol S}} = 1.00$$

From mol ratios, the empirical formula = $Na_2S_2O_3$.

64. $1.121 \text{ g N} \times \dfrac{1 \text{ mol N}}{14.01 \text{ g N}} = 8.001 \times 10^{-2} \text{ mol N}; \quad 0.161 \text{ g H} \times \dfrac{1 \text{ mol H}}{1.008 \text{ g H}} = 1.60 \times 10^{-1} \text{ mol H}$

$0.480 \text{ g C} \times \dfrac{1 \text{ mol C}}{12.01 \text{ g C}} = 4.00 \times 10^{-2} \text{ mol C}; \quad 0.640 \text{ g O} \times \dfrac{1 \text{ mol O}}{16.00 \text{ g O}} = 4.00 \times 10^{-2} \text{ mol O}$

Dividing all mol values by the smallest number:

$$\frac{8.001 \times 10^{-2}}{4.00 \times 10^{-2}} = 2.00; \quad \frac{1.60 \times 10^{-1}}{4.00 \times 10^{-2}} = 4.00; \quad \frac{4.00 \times 10^{-2}}{4.00 \times 10^{-2}} = 1.00$$

Empirical formula = N_2H_4CO

65.   Out of 100.0 g compound: $30.4 \text{ g N} \times \dfrac{1 \text{ mol N}}{14.01 \text{ g N}} = 2.17 \text{ mol N}$

$\%O = 100.0 - 30.4 = 69.6\% \text{ O}; \quad 69.6 \text{ g O} \times \dfrac{1 \text{ mol O}}{16.00 \text{ g O}} = 4.35 \text{ mol O}$

$\dfrac{2.17}{2.17} = 1.00;$ $\dfrac{4.35}{2.17} = 2.00;$ Empirical formula is NO_2.

The empirical formula mass of $NO_2 \approx 14 + 2(16) = 46$ g/mol.

$\dfrac{92\,g}{46\,g} = 2.0;$ Therefore, the molecular formula is N_2O_4.

66.   Out of 100.0 g, there are:

$69.6\,g\,S \times \dfrac{1\,mol\,S}{32.07\,g\,S} = 2.17\,mol\,S;$ $30.4\,g\,N \times \dfrac{1\,mol\,N}{14.01\,g\,N} = 2.17\,mol\,N$

Empirical formula is SN since mol values are in a 1:1 mol ratio.

The empirical formula mass of SN is ~ 46 g.   Since $\dfrac{184}{46} = 4.0$, then the molecular formula is S_4N_4.

67.   Assuming 100.00 g of compound:

$7.74\,g\,H \times \dfrac{1\,mol\,H}{1.008\,g\,H} = 7.68\,mol\,H;$ $92.26\,g\,C \times \dfrac{1\,mol\,C}{12.01\,g\,C} = 7.682\,mol\,C$

The mole ratio between C and H is 1:1 so the empirical formula is CH.

empirical formula mass = $12.01 + 1.008 = 13.02$ g/mol

$\dfrac{molar\ mass}{empirical\ formula\ mass} = \dfrac{78.1}{13.02} = 6.00$

molecular formula = $(CH)_6 = C_6H_6$

68.   Assuming 100.00 g of compound:

$47.08\,g\,C \times \dfrac{1\,mol\,C}{12.01\,g\,C} = 3.920\,mol\,C;$ $6.59\,g\,H \times \dfrac{1\,mol\,H}{1.008\,g\,H} = 6.54\,mol\,H$

$46.33\,g\,Cl \times \dfrac{1\,mol\,Cl}{35.45\,g\,Cl} = 1.307\,mol\,Cl$

Dividing all mole values by 1.307 gives:

$\dfrac{3.920}{1.307} = 2.999;$ $\dfrac{6.54}{1.307} = 5.00;$ $\dfrac{1.307}{1.307} = 1.000$

The empirical formula is C_3H_5Cl.

The empirical formula mass is: $3(12.01) + 5(1.008) + 1(35.45) = 76.52$ g/mol

$\dfrac{molar\ mass}{empirical\ formula\ mass} = \dfrac{153}{76.52} = 2.00;$ The molecular formula = $(C_3H_5Cl)_2 = C_6H_{10}Cl_2$

69.   When combustion data is given, it is assumed that all the carbon in the compound ends up as carbon in CO_2 and all the hydrogen in the compound ends up a hydrogen in H_2O. In the sample of propane combusted, the moles of C and H are:

$$\text{mol C} = 2.641 \text{ g CO}_2 \times \frac{1 \text{ mol CO}_2}{44.01 \text{ g CO}_2} \times \frac{1 \text{ mol C}}{\text{mol CO}_2} = 0.06001 \text{ mol C}$$

$$\text{mol H} = 1.442 \text{ g H}_2\text{O} \times \frac{1 \text{ mol H}_2\text{O}}{18.02 \text{ g H}_2\text{O}} \times \frac{2 \text{ mol H}}{\text{mol H}_2\text{O}} = 0.1600 \text{ mol H}$$

$$\frac{\text{mol H}}{\text{mol C}} = \frac{0.1600}{0.06001} = 2.666$$

Multiplying this ratio by three gives the empirical formula of C_3H_8.

70.   Glucose contains oxygen and one way to determine the amount of oxygen in this sample of glucose combusted is to calculate composition by mass percent. We assume that all of the carbon in 0.528 mg CO_2 come from 0.360 mg glucose and all of the hydrogen in 0.216 mg H_2O came from 0.360 mg glucose.

$$5.28 \times 10^{-4} \text{ g CO}_2 \times \frac{1 \text{ mol CO}_2}{44.01 \text{ g CO}_2} \times \frac{1 \text{ mol C}}{\text{mol CO}_2} \times \frac{12.01 \text{ g C}}{\text{mol C}} = 1.44 \times 10^{-4} \text{ g C}$$

$$\%\text{C} = \frac{1.44 \times 10^{-4} \text{ g C}}{3.60 \times 10^{-4} \text{ g glucose}} \times 100 = 40.0\% \text{ C}$$

$$2.16 \times 10^{-4} \text{ g H}_2\text{O} \times \frac{1 \text{ mol H}_2\text{O}}{18.02 \text{ g H}_2\text{O}} \times \frac{2 \text{ mol H}}{\text{mol H}_2\text{O}} \times \frac{1.008 \text{ g H}}{\text{mol H}} = 2.42 \times 10^{-5} \text{ g H}$$

$$\%\text{H} = \frac{2.42 \times 10^{-5} \text{ g H}}{3.60 \times 10^{-4} \text{ g glucose}} \times 100 = 6.72\% \text{ H}$$

The mass percent of oxygen is obtained by difference:

$$\% \text{ O} = 100.00 - (40.0 + 6.72) = 53.3\% \text{ O}$$

Now perform the empirical formula determination by first assuming 100.00 g of glucose. Out of 100.00 g of glucose, there are:

$$40.0 \text{ g C} \times \frac{1 \text{ mol C}}{12.01 \text{ g C}} = 3.33 \text{ mol C}; \ 6.72 \text{ g H} \times \frac{1 \text{ mol H}}{1.008 \text{ g H}} = 6.67 \text{ mol H}$$

$$53.3 \text{ g O} \times \frac{1 \text{ mol O}}{16.00 \text{ g O}} = 3.33 \text{ mol O}$$

The mol ratio of C:H:O is 1:2:1 so the empirical formula of glucose is CH_2O.

71.   Since the compound contains only carbon and hydrogen, we could solve this problem using the same procedure as in Exercise 3.69. Instead, we will solve this problem using the procedure outlined in the text, i.e., we will determine the composition by mass percent and then solve for the empirical formula.

$$156.8 \text{ mg } CO_2 \times \frac{1 \text{ g}}{1000 \text{ mg}} \times \frac{12.01 \text{ g C}}{44.01 \text{ g } CO_2} \times \frac{1000 \text{ mg}}{\text{g}} = 42.79 \text{ mg C}$$

$$42.8 \text{ mg } H_2O \times \frac{1 \text{ g}}{1000 \text{ mg}} \times \frac{2.016 \text{ g H}}{18.02 \text{ g } H_2O} \times \frac{1000 \text{ mg}}{\text{g}} = 4.79 \text{ mg H}$$

$$\%C = \frac{42.79 \text{ mg}}{47.6 \text{ mg}} \times 100 = 89.9\% \text{ C}; \quad \%H = 100.0 - 89.9 = 10.1\% \text{ H}$$

Out of 100.0 g Cumene, we have:

$$89.9 \text{ g C} \times \frac{1 \text{ mol C}}{12.01 \text{ g C}} = 7.49 \text{ mol C}; \quad 10.1 \text{ g H} \times \frac{1 \text{ mol H}}{1.008 \text{ g H}} = 10.0 \text{ mol H}$$

$$\frac{10.0}{7.49} = 1.34 \approx \frac{4}{3}, \text{ i.e., mol H to mol C are in a 4:3 ratio. Empirical formula} = C_3H_4$$

Empirical formula mass $\approx$ 3(12) + 4(1) = 40 g/mol

The molecular formula is $(C_3H_4)_3$ or C_9H_{12} since the molar mass will be between 115 and 125 g/mol.

72.   First, we will determine composition by mass percent:

$$16.01 \text{ mg } CO_2 \times \frac{1 \text{ g}}{1000 \text{ mg}} \times \frac{12.01 \text{ g C}}{44.01 \text{ g } CO_2} \times \frac{1000 \text{ mg}}{\text{g}} = 4.369 \text{ mg C}$$

$$\%C = \frac{4.369 \text{ mg C}}{10.68 \text{ mg compound}} \times 100 = 40.91\% \text{ C}$$

$$4.37 \text{ mg } H_2O \times \frac{1 \text{ g}}{1000 \text{ mg}} \times \frac{2.016 \text{ g H}}{18.02 \text{ g } H_2O} \times \frac{1000 \text{ mg}}{\text{g}} = 0.489 \text{ mg H}$$

$$\%H = \frac{0.489 \text{ mg}}{10.68 \text{ mg}} \times 100 = 4.58\% \text{ H}; \quad \%O = 100.00 - (40.91 + 4.58) = 54.51\% \text{ O}$$

So, if we have 100.00 g of the compound, we have:

$$40.91 \text{ g C} \times \frac{1 \text{ mol C}}{12.01 \text{ g C}} = 3.406 \text{ mol C}; \quad 4.58 \text{ g H} \times \frac{1 \text{ mol H}}{1.008 \text{ g H}} = 4.54 \text{ mol H}$$

$$54.51 \text{ g O} \times \frac{1 \text{ mol O}}{16.00 \text{ g O}} = 3.407 \text{ mol O}$$

Dividing by smallest number: $\dfrac{4.54}{3.406} = 1.33 = \dfrac{4}{3}$; Therefore, empirical formula is $C_3H_4O_3$.

The empirical formula mass of $C_3H_4O_3$ is $\approx 3(12) + 4(1) + 3(16) = 88$ g.

Since $\dfrac{176.1}{88} = 2.0$, then the molecuar formula is $C_6H_8O_6$.

Balancing Chemical Equations

73.   a. $In + O_2 \rightarrow In_2O_3$, $4\ In(s) + 3\ O_2(g) \rightarrow 2\ In_2O_3(s)$

    b. $C_6H_{12}O_6 \rightarrow C_2H_6O + CO_2$

        Balance C-atoms first.   6-C on left, 3-C on right.

        First, try multiplying both products by 2:

            $C_6H_{12}O_6 \rightarrow 2\ C_2H_6O + 2\ CO_2$

        O and H are also balanced, and the balanced equation is:

            $C_6H_{12}O_6(aq) \rightarrow 2\ C_2H_6O(aq) + 2\ CO_2(g)$

    c. $K + H_2O \rightarrow KOH + H_2$, $2\ K(s) + 2\ H_2O(l) \rightarrow 2\ KOH(aq) + H_2(g)$

74.   a. $C_6H_{12}O_6(s) + O_2(g) \rightarrow CO_2(g) + H_2O(g)$

        Balance C atoms: $C_6H_{12}O_6 + O_2 \rightarrow 6\ CO_2 + H_2O$

        Balance H atoms: $C_6H_{12}O_6 + O_2 \rightarrow 6\ CO_2 + 6\ H_2O$

        Lastly, balance O atoms: $C_6H_{12}O_6(s) + 6\ O_2(g) \rightarrow 6\ CO_2(g) + 6\ H_2O(g)$

        Equation is balanced.

    b. $Fe_2S_3(s) + HCl(g) \rightarrow FeCl_3(s) + H_2S(g)$

        Balance Fe atoms: $Fe_2S_3 + HCl \rightarrow 2\ FeCl_3 + H_2S$

        There are 6 H and 6 Cl on right, so balance with 6 HCl on left:

            $Fe_2S_3(s) + 6\ HCl(g) \rightarrow 2\ FeCl_3(s) + 3\ H_2S(g)$. Equation is balanced.

    c. $CS_2(l) + NH_3(g) \rightarrow H_2S(g) + NH_4SCN(s)$

        C and S balanced; balance N:

            $CS_2 + 2\ NH_3 \rightarrow H_2S + NH_4SCN$

H is also balanced. So: $CS_2(l) + 2\ NH_3(g) \rightarrow H_2S(g) + NH_4SCN(s)$

75.   a. $Cu(s) + 2\ AgNO_3(aq) \rightarrow 2\ Ag(s) + Cu(NO_3)_2(aq)$

b. $Zn(s) + 2\ HCl(aq) \rightarrow ZnCl_2(aq) + H_2(g)$

c. $Au_2S_3(s) + 3\ H_2(g) \rightarrow 2\ Au(s) + 3\ H_2S(g);$ d. $Ca(s) + 2\ H_2O(l) \rightarrow Ca(OH)_2(aq) + H_2(g)$

76.   a. $3\ Ca(OH)_2(aq) + 2\ H_3PO_4(aq) \rightarrow 6\ H_2O(l) + Ca_3(PO_4)_2(s)$

b. $Al(OH)_3(s) + 3\ HCl(aq) \rightarrow AlCl_3(aq) + 3\ H_2O(l)$

c. $2\ AgNO_3(aq) + H_2SO_4(aq) \rightarrow Ag_2SO_4(s) + 2\ HNO_3(aq)$

77.   a. $C_{12}H_{22}O_{11}(s) + 12\ O_2(g) \rightarrow 12\ CO_2(g) + 11\ H_2O(g)$

b. $C_6H_6(l) + \dfrac{15}{2}O_2(g) \rightarrow 6\ CO_2(g) + 3\ H_2O(g);$ Multiply by two to give whole numbers.

$2\ C_6H_6(l) + 15\ O_2(g) \rightarrow 12\ CO_2(g) + 6\ H_2O(g)$

c. $2\ Fe + \dfrac{3}{2}O_2 \rightarrow Fe_2O_3;$ For whole numbers: $4\ Fe(s) + 3\ O_2(g) \rightarrow 2\ Fe_2O_3(s)$

d. $C_4H_{10} + \dfrac{13}{2}O_2 \rightarrow 4\ CO_2 + 5\ H_2O;$ Multiply by two to give whole numbers.

$2\ C_4H_{10}(g) + 13\ O_2(g) \rightarrow 8\ CO_2(g) + 10\ H_2O(g)$

e. $2\ FeO(s) + \dfrac{1}{2}O_2(g) \rightarrow Fe_2O_3(s);$ For whole numbers, multiply by two.

$4\ FeO(s) + O_2(g) \rightarrow 2\ Fe_2O_3(s)$

78.   a. $16\ Cr(s) + 3\ S_8(s) \rightarrow 8\ Cr_2S_3(s)$

b. $2\ NaHCO_3(s) \rightarrow Na_2CO_3(s) + CO_2(g) + H_2O(g)$

c. $2\ KClO_3(s) \rightarrow 2\ KCl(s) + 3\ O_2(g)$

d. $2\ Eu(s) + 6\ HF(g) \rightarrow 2\ EuF_3(s) + 3\ H_2(g)$

79.   a. $SiO_2(s) + C(s) \rightarrow Si(s) + CO(g)$

Balance oxygen atoms: $SiO_2 + C \rightarrow Si + 2\ CO$

Balance carbon atoms: $SiO_2(s) + 2\ C(s) \rightarrow Si(s) + 2\ CO(g)$

b. $SiCl_4(l) + Mg(s) \rightarrow Si(s) + MgCl_2(s)$

Balance Cl atoms: $SiCl_4 + Mg \rightarrow Si + 2\ MgCl_2$

Balance Mg atoms: $SiCl_4(l) + 2\ Mg(s) \rightarrow Si(s) + 2\ MgCl_2(s)$

c. $Na_2SiF_6(s) + Na(s) \rightarrow Si(s) + NaF(s)$

Balance F atoms: $Na_2SiF_6 + Na \rightarrow Si + 6\ NaF$

Balance Na atoms: $Na_2SiF_6(s) + 4\ Na(s) \rightarrow Si(s) + 6\ NaF(s)$

80.   Unbalanced equation:

$$CaF_2 \cdot 3Ca_3(PO_4)_2(s) + H_2SO_4(aq) \rightarrow H_3PO_4(aq) + HF(aq) + CaSO_4 \cdot 2H_2O(s)$$

Balancing Ca^{2+}, F^-, and $PO_4{}^{3-}$:

$$CaF_2 \cdot 3Ca_3(PO_4)_2(s) + H_2SO_4(aq) \rightarrow 6\ H_3PO_4(aq) + 2\ HF(aq) + 10\ CaSO_4 \cdot 2H_2O(s)$$

On the right hand side there are 20 extra hydrogen atoms, 10 extra sulfates, and 20 extra water molecules. We can balance the hydrogen and sulfate with 10 sulfuric acid molecules. The extra waters came from the water in the sulfuric acid solution. The balanced equation is:

$$CaF_2 \cdot 3Ca_3(PO_4)_2(s) + 10\ H_2SO_4(aq) + 20\ H_2O(l) \rightarrow 6\ H_3PO_4(aq) + 2\ HF(aq) + 10\ CaSO_4 \cdot 2H_2O(s)$$

81. $Pb(NO_3)_2(aq) + H_3AsO_4(aq) \rightarrow PbHAsO_4(s) + 2\ HNO_3(aq)$

Note: The insecticide used is $PbHAsO_4$ and is commonly called lead arsenate. This is not the correct name, however. Correctly, lead arsenate would be $Pb_3(AsO_4)_2$ and $PbHAsO_4$ should be named lead hydrogen arsenate.

82. $2\ NaCl(s) + 2\ H_2O(l) \rightarrow Cl_2(g) + H_2(g) + 2\ NaOH(aq)$

Reaction Stoichiometry

83.   The stepwise method to solve stoichiometry problems is outlined in the text. Instead of calculating intermediate answers for each step, we will combine conversion factors into one calculation. This practice reduces round-off error and saves time.

The balanced reaction is: $(NH_4)_2Cr_2O_7(s) \rightarrow Cr_2O_3(s) + N_2(g) + 4\ H_2O(g)$

$$10.8\ g\ (NH_4)_2Cr_2O_7 \times \frac{1\ mol\ (NH_4)_2Cr_2O_7}{252.08\ g} = 4.28 \times 10^{-2}\ mol\ (NH_4)_2Cr_2O_7$$

$$4.28 \times 10^{-2}\ mol\ (NH_4)_2Cr_2O_7 \times \frac{1\ mol\ Cr_2O_3}{mol\ (NH_4)_2Cr_2O_7} \times \frac{152.00\ g\ Cr_2O_3}{mol\ Cr_2O_3} = 6.51\ g\ Cr_2O_3$$

$$4.28 \times 10^{-2}\ mol\ (NH_4)_2Cr_2O_7 \times \frac{1\ mol\ N_2}{mol\ (NH_4)_2Cr_2O_7} \times \frac{28.02\ g\ N_2}{mol\ N_2} = 1.20\ g\ N_2$$

$$4.28 \times 10^{-2}\ mol\ (NH_4)_2Cr_2O_7 \times \frac{4\ mol\ H_2O}{mol\ (NH_4)_2Cr_2O_7} \times \frac{18.02\ g\ H_2O}{mol\ H_2O} = 3.09\ g\ H_2O$$

84. $Fe_2O_3(s) + 2\ Al(s) \rightarrow 2\ Fe(l) + Al_2O_3(s)$

$15.0\ g\ Fe \times \dfrac{1\ mol\ Fe}{55.85\ g\ Fe} = 0.269\ mol\ Fe;\qquad 0.269\ mol\ Fe \times \dfrac{2\ mol\ Al}{2\ mol\ Fe} \times \dfrac{26.98\ g\ Al}{mol\ Al} = 7.26\ g\ Al$

$0.269\ mol\ Fe \times \dfrac{1\ mol\ Fe_2O_3}{2\ mol\ Fe} \times \dfrac{159.70\ g\ Fe_2O_3}{mol\ Fe_2O_3} = 21.5\ g\ Fe_2O_3$

$0.269\ mol\ Fe \times \dfrac{1\ mol\ Al_2O_3}{2\ mol\ Fe} \times \dfrac{101.96\ g\ Al_2O_3}{mol\ Al_2O_3} = 13.7\ g\ Al_2O_3$

85. $1.000\ kg\ Al \times \dfrac{1000\ g\ Al}{kg\ Al} \times \dfrac{1\ mol\ Al}{26.98\ g\ Al} \times \dfrac{3\ mol\ NH_4ClO_4}{3\ mol\ Al} \times \dfrac{117.49\ g\ NH_4ClO_4}{mol\ NH_4ClO_4} = 4355\ g$

86.   a. $Ba(OH)_2 \bullet 8H_2O(s) + 2\ NH_4SCN(s) \rightarrow Ba(SCN)_2(s) + 10\ H_2O(l) + 2\ NH_3(g)$

b. $6.5\ g\ Ba(OH)_2 \bullet 8H_2O \times \dfrac{1\ mol\ Ba(OH)_2 \bullet 8H_2O}{315.4\ g} = 0.0206\ mol = 0.021\ mol$

$0.021\ mol\ Ba(OH)_2 \bullet 8H_2O \times \dfrac{2\ mol\ NH_4SCN}{1\ mol\ Ba(OH)_2 \bullet 8H_2O} \times \dfrac{76.13\ g\ NH_4SCN}{mol\ NH_4SCN} = 3.2\ g\ NH_4SCN$

87. $1.0 \times 10^2\ g\ Ca_3(PO_4)_2 \times \dfrac{1\ mol\ Ca_3(PO_4)_2}{310.2\ g\ Ca_3(PO_4)_2} \times \dfrac{3\ mol\ H_2SO_4}{mol\ Ca_3(PO_4)_2} \times \dfrac{98.09\ g\ H_2SO_4}{mol\ H_2SO_4}$

$= 95\ g\ H_2SO_4\ are\ needed$

$95\ g\ H_2SO_4 \times \dfrac{100\ g\ concentrated\ reagent}{98\ g\ H_2SO_4} = 97\ g\ of\ concentrated\ sulfuric\ acid$

88. $1.0\ ton\ CuO \times \dfrac{907\ kg}{ton} \times \dfrac{1000\ g}{kg} \times \dfrac{1\ mol\ CuO}{79.55\ g\ CuO} \times \dfrac{1\ mol\ C}{2\ mol\ CuO} \times \dfrac{12.01\ g\ C}{mol\ C} \times \dfrac{100.\ g\ coke}{95\ g\ C}$

$= 7.2 \times 10^4\ g\ coke$

89.   a. $1.0 \times 10^2\ mg\ NaHCO_3 \times \dfrac{1\ g}{1000\ mg} \times \dfrac{1\ mol\ NaHCO_3}{84.01\ g\ NaHCO_3} \times \dfrac{1\ mol\ C_6H_8O_7}{3\ mol\ NaHCO_3} \times \dfrac{192.12\ g\ C_6H_8O_7}{mol\ C_6H_8O_7}$

$= 0.076\ g\ or\ 76\ mg\ C_6H_8O_7$

b. $0.10\ g\ NaHCO_3 \times \dfrac{1\ mol\ NaHCO_3}{84.01\ g\ NaHCO_3} \times \dfrac{3\ mol\ CO_2}{3\ mol\ NaHCO_2} \times \dfrac{44.01\ g\ CO_2}{mol\ CO_2} = 0.052\ g\ or\ 52\ mg\ CO_2$

90.   a. $1.00 \times 10^2\ g\ C_7H_6O_3 \times \dfrac{1\ mol\ C_7H_6O_3}{138.12\ g\ C_7H_6O_3} \times \dfrac{1\ mol\ C_4H_6O_3}{1\ mol\ C_7H_6O_3} \times \dfrac{102.09\ g\ C_4H_6O_3}{mol\ C_4H_6O_3} = 73.9\ g\ C_4H_6O_3$

b. $1.00 \times 10^2 \text{ g C}_7\text{H}_6\text{O}_3 \times \dfrac{1 \text{ mol C}_7\text{H}_6\text{O}_3}{138.12 \text{ g C}_7\text{H}_6\text{O}_3} \times \dfrac{1 \text{ mol C}_9\text{H}_8\text{O}_4}{1 \text{ mol C}_7\text{H}_6\text{O}_3} \times \dfrac{180.15 \text{ g C}_9\text{H}_8\text{O}_4}{\text{mol C}_9\text{H}_8\text{O}_4}$

$$= 1.30 \times 10^2 \text{ g aspirin}$$

Limiting Reactants and Percent Yield

91.   a. $\text{Mg(s)} + \text{I}_2\text{(s)} \rightarrow \text{MgI}_2\text{(s)}$

From the balanced equation, 100 molecules of I_2 reacts completely with 100 atoms of Mg. We have a stoichiometric mixture. Neither is limiting.

b. $150 \text{ atoms Mg} \times \dfrac{1 \text{ molecule I}_2}{1 \text{ atom Mg}} = 150 \text{ molecules I}_2 \text{ needed}$

We need 150 molecules I_2 to react completely with 150 atoms Mg; we only have 100 molecules I_2. Therefore, I_2 is limiting.

c. $200 \text{ atoms Mg} \times \dfrac{1 \text{ molecule I}_2}{1 \text{ atom Mg}} = 200 \text{ molecules I}_2;$ Mg is limiting since 300 molecules I_2 are present.

d. $0.16 \text{ mol Mg} \times \dfrac{1 \text{ mol I}_2}{1 \text{ mol Mg}} = 0.16 \text{ mol I}_2;$ Mg is limiting since 0.25 mol I_2 are present.

e. $0.14 \text{ mol Mg} \times \dfrac{1 \text{ mol I}_2}{1 \text{ mol Mg}} = 0.14 \text{ mol I}_2 \text{ needed};$ Stoichiometric mixture. Neither is limiting.

f. $0.12 \text{ mol Mg} \times \dfrac{1 \text{ mol I}_2}{1 \text{ mol Mg}} = 0.12 \text{ mol I}_2 \text{ needed};$ I_2 is limiting since only 0.08 mol I_2 are present.

g. $6.078 \text{ g Mg} \times \dfrac{1 \text{ mol Mg}}{24.31 \text{ g Mg}} \times \dfrac{1 \text{ mol I}_2}{1 \text{ mol Mg}} \times \dfrac{253.8 \text{ g I}_2}{\text{mol I}_2} = 63.46 \text{ g I}_2$

Stoichiometric mixture. Neither is limiting.

h. $1.00 \text{ g Mg} \times \dfrac{1 \text{ mol Mg}}{24.31 \text{ g Mg}} \times \dfrac{1 \text{ mol I}_2}{1 \text{ mol Mg}} \times \dfrac{253.8 \text{ g I}_2}{\text{mol I}_2} = 10.4 \text{ g I}_2$

10.4 g I_2 needed, but we only have 2.00 g. I_2 is limiting.

i.   From h above, we calculated that 10.4 g I_2 will react completely with 1.00 g Mg. We have 20.00 g I_2. I_2 is in excess. Mg is limiting.

92. $2 \text{ H}_2\text{(g)} + \text{O}_2\text{(g)} \rightarrow 2 \text{ H}_2\text{O(g)}$

a. $50 \text{ molecules H}_2 \times \dfrac{1 \text{ molecule O}_2}{2 \text{ molecules H}_2} = 25 \text{ molecules O}_2$

Stoichiometric mixture. Neither is limiting.

b.   100 molecules $H_2 \times \dfrac{1\text{ molecule } O_2}{2\text{ molecules } H_2}$ = 50 molecules O_2; O_2 is limiting since only 40 molecules O_2 are present.

c.   From b, 50 molecules of O_2 will react completely with 100 molecules of H_2. We have 100 molecules (an excess) of O_2. So, H_2 is limiting.

d.   0.50 mol $H_2 \times \dfrac{1\text{ mol } O_2}{2\text{ mol } H_2}$ = 0.25 mol O_2; H_2 is limiting since 0.75 mol O_2 are present.

e.   0.80 mol $H_2 \times \dfrac{1\text{ mol } O_2}{2\text{ mol } H_2}$ = 0.40 mol O_2; H_2 is limiting since 0.75 mol O_2 are present.

f.   1.0 g $H_2 \times \dfrac{1\text{ mol } H_2}{2.016\text{ g } H_2} \times \dfrac{1\text{ mol } O_2}{2\text{ mol } H_2}$ = 0.25 mol O_2

Stoichiometric mixture, neither is limiting.

g.   5.00 g $H_2 \times \dfrac{1\text{ mol } H_2}{2.016\text{ g } H_2} \times \dfrac{1\text{ mol } O_2}{2\text{ mol } H_2} \times \dfrac{32.00\text{ g } O_2}{\text{mol } O_2}$ = 39.7 g O_2; H_2 is limiting since 56.00 g O_2 are present.

93.   a.   mol Ag = 2.0 g Ag $\times \dfrac{1\text{ mol Ag}}{107.9\text{ g Ag}}$ = 1.9×10^{-2} mol Ag

mol S_8 = 2.0 g $S_8 \times \dfrac{1\text{ mol } S_8}{256.56\text{ g } S_8}$ = 7.8×10^{-3} mol S_8

From the balanced equation the required mol Ag to mol S_8 ratio is 16:1. The actual mol ratio is:

$$\dfrac{1.9 \times 10^{-2}\text{ mol Ag}}{7.8 \times 10^{-3}\text{ mol } S_8} = 2.4$$

This is well below the required ratio so Ag is the limiting reagent.

1.9×10^{-2} mol Ag $\times \dfrac{8\text{ mol } Ag_2S}{16\text{ mol Ag}} \times \dfrac{247.9\text{ g } Ag_2S}{\text{mol } Ag_2S}$ = 2.4 g Ag_2S

b. 1.9×10^{-2} mol Ag $\times \dfrac{1\text{ mol } S_8}{16\text{ mol Ag}} \times \dfrac{256.56\text{ g } S_8}{\text{mol } S_8}$ = 0.30 g S_8

0.30 g S_8 are required to react with all of the Ag present.

S_8 in excess = 2.0 g S_8 - 0.30 g S_8 = 1.7 g S_8 in excess

94.   a. 1.00×10^3 g $N_2 \times \dfrac{1\text{ mol } N_2}{28.02\text{ g } N_2}$ = 35.7 mol N_2

5.00×10^2 g $H_2 \times \dfrac{1\text{ mol } H_2}{2.016\text{ g } H_2}$ = 248 mol H_2

The required mol ratio from the balanced reaction is 3 mol H_2 to 1 mol N_2. The actual ratio is:

$$\frac{248 \text{ mol } H_2}{35.7 \text{ mol } N_2} = 6.95$$

This is well above the required mole ratio so N_2 is the limiting reagent.

$$35.7 \text{ mol } N_2 \times \frac{2 \text{ mol } NH_3}{\text{mol } N_2} \times \frac{17.03 \text{ g } NH_3}{\text{mol } NH_3} = 1.22 \times 10^3 \text{ g } NH_3 \text{ produced}$$

b. $$35.7 \text{ mol } N_2 \times \frac{3 \text{ mol } H_2}{\text{mol } N_2} \times \frac{2.016 \text{ g } H_2}{\text{mol } H_2} = 216 \text{ g } H_2 \text{ reacted}$$

excess H_2 = 500. g H_2 - 216 g H_2 = 284 g H_2 unreacted

95. $Ca_3(PO_4)_2 + 3 H_2SO_4 \rightarrow 3 CaSO_4 + 2 H_3PO_4$

$$1.0 \times 10^3 \text{ g } Ca_3(PO_4)_2 \times \frac{1 \text{ mol } Ca_3(PO_4)_2}{310.18 \text{ g } Ca_3(PO_4)_2} = 3.2 \text{ mol } Ca_3(PO_4)_2$$

$$1.0 \times 10^3 \text{ g conc. } H_2SO_4 \times \frac{98 \text{ g } H_2SO_4}{100 \text{ g conc. } H_2SO_4} \times \frac{1 \text{ mol } H_2SO_4}{98.09 \text{ g } H_2SO_4} = 10. \text{ mol } H_2SO_4$$

The required mol ratio from the balanced equation is 3 mol H_2SO_4 to 1 mol $Ca_3(PO_4)_2$. The actual
ratio is: $$\frac{10. \text{ mol } H_2SO_4}{3.2 \text{ mol } Ca_3(PO_4)_2} = 3.1$$

This is higher than the required mol ratio so $Ca_3(PO_4)_2$ is the limiting reagent.

$$3.2 \text{ mol } Ca_3(PO_4)_2 \times \frac{3 \text{ mol } CaSO_4}{\text{mol } Ca_3(PO_4)_2} \times \frac{136.15 \text{ g } CaSO_4}{\text{mol } CaSO_4} = 1300 \text{ g } CaSO_4 \text{ produced}$$

$$3.2 \text{ mol } Ca_3(PO_4)_2 \times \frac{2 \text{ mol } H_3PO_4}{\text{mol } Ca_3(PO_4)_2} \times \frac{97.99 \text{ g } H_3PO_4}{\text{mol } H_3PO_4} = 630 \text{ g } H_3PO_4 \text{ produced}$$

96.   a. $$10.0 \text{ g } Hg \times \frac{1 \text{ mol } Hg}{200.6 \text{ g } Hg} = 4.99 \times 10^{-2} \text{ mol } Hg$$

$$9.00 \text{ g } Br_2 \times \frac{1 \text{ mol } Br_2}{159.80 \text{ g } Br_2} = 5.63 \times 10^{-2} \text{ mol } Br_2$$

The required mol ratio from the balanced equation is 1 mol Br_2 to 1 mol Hg. The actual mol
ratio is:

$$\frac{5.63 \times 10^{-2} \text{ mol } Br}{4.99 \times 10^{-2} \text{ mol } Hg} = 1.13$$

This is higher than the required ratio so Hg is the limiting reagent.

$$4.99 \times 10^{-2} \text{ mol Hg} \times \frac{1 \text{ mol HgBr}_2}{\text{mol Hg}} \times \frac{360.4 \text{ g HgBr}_2}{\text{mol HgBr}_2} = 18.0 \text{ g HgBr}_2 \text{ produced}$$

$$4.99 \times 10^{-2} \text{ mol Hg} \times \frac{1 \text{ mol Br}_2}{1 \text{ mol Hg}} \times \frac{159.80 \text{ g Br}_2}{\text{mol Br}_2} = 7.97 \text{ g Br}_2 \text{ reacted}$$

excess Br_2 = 9.00 g Br_2 - 7.97 g Br_2 = 1.03 g Br_2

b. $$5.00 \text{ mL Hg} \times \frac{13.6 \text{ g Hg}}{\text{mL Hg}} \times \frac{1 \text{ mol Hg}}{200.6 \text{ g Hg}} = 0.339 \text{ mol Hg}$$

$$5.00 \text{ mL Br}_2 \times \frac{3.10 \text{ g Br}_2}{\text{mL Br}_2} \times \frac{1 \text{ mol Br}_2}{159.80 \text{ g Br}_2} = 0.0970 \text{ mol Br}_2$$

Br_2 is limiting since the actual moles of Br_2 present is well below the required 1:1 mol ratio.

$$0.0970 \text{ mol Br}_2 \times \frac{1 \text{ mol HgBr}_2}{\text{mol Br}_2} \times \frac{360.4 \text{ g HgBr}_2}{\text{mol HgBr}_2} = 35.0 \text{ g HgBr}_2 \text{ produced}$$

97. $2 \text{ Cu}(s) + \text{S}(s) \rightarrow \text{Cu}_2\text{S}(s)$ or $16 \text{ Cu}(s) + \text{S}_8(s) \rightarrow 8 \text{ Cu}_2\text{S}(s)$

$$1.50 \text{ g Cu} \times \frac{1 \text{ mol Cu}}{63.55 \text{ g Cu}} \times \frac{1 \text{ mol Cu}_2\text{S}}{2 \text{ mol Cu}} \times \frac{159.17 \text{ g Cu}_2\text{S}}{\text{mol Cu}_2\text{S}} = 1.88 \text{ g Cu}_2\text{S is theoretical yield.}$$

$$\% \text{ yield} = \frac{\text{actual yield}}{\text{theoretical yield}} \times 100 = \frac{1.76 \text{ g}}{1.88 \text{ g}} \times 100 = 93.6\%$$

98. $$6.0 \text{ g Al} \times \frac{1 \text{ mol Al}}{26.98 \text{ mol Al}} \times \frac{2 \text{ mol AlBr}_3}{2 \text{ mol Al}} \times \frac{266.68 \text{ g AlBr}_3}{\text{mol AlBr}_3} = 59 \text{ g AlBr}_3$$

$$\% \text{ yield} = \frac{50.3 \text{ g}}{59 \text{ g}} \times 100 = 85\%$$

99. $\text{C}_2\text{H}_6 + \text{Cl}_2 \rightarrow \text{C}_2\text{H}_5\text{Cl} + \text{HCl}$

$$300. \text{ g C}_2\text{H}_6 \times \frac{1 \text{ mol C}_2\text{H}_6}{30.07 \text{ g C}_2\text{H}_6} = 9.98 \text{ mol C}_2\text{H}_6; \ 650. \text{ g Cl}_2 \times \frac{1 \text{ mol Cl}_2}{70.90 \text{ g Cl}_2} = 9.17 \text{ mol Cl}_2$$

The balanced equation requires a 1:1 mol ratio between reactants. 9.17 mol of C_2H_6 will react with all of the Cl_2 present (9.17 mol). Since 9.98 mol C_2H_6 is present, then Cl_2 is the limiting reagent.

The theoretical yield of C_2H_5Cl is:

$$9.17 \text{ mol Cl}_2 \times \frac{1 \text{ mol C}_2\text{H}_5\text{Cl}}{\text{mol Cl}_2} \times \frac{64.51 \text{ g C}_2\text{H}_5\text{Cl}}{\text{mol C}_2\text{H}_5\text{Cl}} = 592 \text{ g C}_2\text{H}_5\text{Cl}$$

$$\text{Percent yield} = \frac{\text{actual}}{\text{theoretical}} \times 100 = \frac{490. \text{ g}}{592 \text{ g}} \times 100 = 82.8\%$$

100. $C_7H_6O_3 + C_4H_6O_3 \rightarrow C_9H_8O_4 + HC_2H_3O_2$

$$1.50 \text{ g } C_7H_6O_3 \times \frac{1 \text{ mol } C_7H_6O_3}{138.12 \text{ g } C_7H_6O_3} = 1.09 \times 10^{-2} \text{ mol } C_7H_6O_3$$

$$2.00 \text{ g } C_4H_6O_3 \times \frac{1 \text{ mol } C_4H_6O_3}{102.09 \text{ g } C_4H_6O_3} = 1.96 \times 10^{-2} \text{ mol } C_4H_6O_3$$

$C_7H_6O_3$ is the limiting reagent since the actual moles of $C_7H_6O_2$ is below the required 1:1 mol ratio. The theoretical yield of aspirin is:

$$1.09 \times 10^{-2} \text{ mol } C_7H_6O_3 \times \frac{1 \text{ mol } C_9H_8O_4}{\text{mol } C_7H_6O_3} \times \frac{180.15 \text{ g } C_9H_8O_4}{\text{mol } C_9H_8O_4} = 1.96 \text{ g } C_9H_8O_4$$

$$\% \text{ yield} = \frac{1.50 \text{ g}}{1.96 \text{ g}} \times 100 = 76.5\%$$

Additional Exercises

101. $$\frac{9.123 \times 10^{-23} \text{ g}}{\text{atom}} \times \frac{6.022 \times 10^{23} \text{ atom}}{\text{mol}} = \frac{54.94 \text{ g}}{\text{mol}}$$

The atomic mass is 54.94. From the periodic table, the element is maganese (Mn).

102.   a. $2(12.01) + 3(1.008) + 3(35.45) + 2(16.00) = 165.39 \text{ g/mol}$

b. $500.0 \text{ g} \times \dfrac{1 \text{ mol}}{165.39 \text{ g}} = 3.023 \text{ mol};$ c. $2.0 \times 10^{-2} \text{ mol} \times \dfrac{165.39 \text{ g}}{\text{mol}} = 3.3 \text{ g}$

d. $5.0 \text{ g } C_2H_3Cl_3O_2 \times \dfrac{1 \text{ mol}}{165.39 \text{ g}} \times \dfrac{6.02 \times 10^{23} \text{ molecules}}{\text{mol}} \times \dfrac{3 \text{ atoms Cl}}{\text{molecule}}$

$$= 5.5 \times 10^{22} \text{ atoms of chlorine}$$

e. $1.0 \text{ g Cl} \times \dfrac{1 \text{ mol Cl}}{35.45 \text{ g}} \times \dfrac{1 \text{ mol } C_2H_3Cl_3O_2}{3 \text{ mol Cl}} \times \dfrac{165.39 \text{ g } C_2H_3Cl_3O_2}{\text{mol } C_2H_3Cl_3O_2} = 1.6 \text{ g chloral hydrate}$

f. $500 \text{ molecules} \times \dfrac{1 \text{ mol}}{6.022 \times 10^{23} \text{ molecules}} \times \dfrac{165.39 \text{ g}}{\text{mol}} = 1.373 \times 10^{-19} \text{ g}$

103.   Out of 100.00 g of compound there are:

$$83.53 \text{ g Sb} \times \frac{1 \text{ mol Sb}}{121.8 \text{ g Sb}} = 0.6858 \text{ mol Sb}; \quad 16.47 \text{ g O} \times \frac{1 \text{ mol O}}{16.00 \text{ g O}} = 1.029 \text{ mol O}$$

$$\frac{1.029}{0.6858} = 1.500 = \frac{3}{2}; \quad \text{Empirical formula: } Sb_2O_3; \quad \text{Empirical formula mass} = 291.6 \text{ g/mol}$$

Mass of Sb_4O_6 is 583.2, which is in the correct range. Molecular formula: Sb_4O_6.

104. $41.98 \text{ mg CO}_2 \times \dfrac{12.01 \text{ mg C}}{44.01 \text{ mg CO}_2} = 11.46 \text{ mg C}$; $\%C = \dfrac{11.46 \text{ mg}}{19.81 \text{ mg}} \times 100 = 57.85\% \text{ C}$

$6.45 \text{ mg H}_2O \times \dfrac{2.016 \text{ mg H}}{18.02 \text{ mg H}_2O} = 0.722 \text{ mg H}$; $\%H = \dfrac{0.722 \text{ mg}}{19.81 \text{ mg}} \times 100 = 3.64\% \text{ H}$

$\%O = 100.00 - (57.85 + 3.64) = 38.51\% \text{ O}$

Out of 100.00 g terephthalic acid, there are:

$57.85 \text{ g C} \times \dfrac{1 \text{ mol C}}{12.01 \text{ g C}} = 4.817 \text{ mol C}$; $3.64 \text{ g H} \times \dfrac{1 \text{ mol H}}{1.008 \text{ g H}} = 3.61 \text{ mol H}$

$38.51 \text{ g O} \times \dfrac{1 \text{ mol O}}{16.00 \text{ g O}} = 2.407 \text{ mol O}$

$\dfrac{4.817}{2.407} = 2.001$; $\dfrac{3.61}{2.407} = 1.50$; $\dfrac{2.407}{2.407} = 1.000$

C:H:O ratio is 2:1.5:1 or 4:3:2. Empirical formula: $C_4H_3O_2$

Mass of $C_4H_3O_2 \approx 4(12) + 3(1) + 2(16) = 83$; $\dfrac{166}{83} = 2$; Molecular formula: $C_8H_6O_4$

105.   Compound I: mass O = 0.6498 g Hg_xO_y - 0.6018 g Hg = 0.0480 g O

$0.6018 \text{ g Hg} \times \dfrac{1 \text{ mol Hg}}{200.6 \text{ g Hg}} = 3.000 \times 10^{-3} \text{ mol Hg}$

$0.0480 \text{ g O} \times \dfrac{1 \text{ mol O}}{16.00 \text{ g O}} = 3.00 \times 10^{-3} \text{ mol O}$

The mol ratio between Hg and O is 1:1, so the empirical formula of compound I is HgO.

Compound II: mass Hg = 0.4172 g Hg_xO_y - 0.016 g O = 0.401 g Hg

$0.401 \text{ g Hg} \times \dfrac{1 \text{ mol Hg}}{200.6 \text{ g Hg}} = 2.00 \times 10^{-3} \text{ mol Hg}$; $0.016 \text{ g O} \times \dfrac{1 \text{ mol O}}{16.00 \text{ g O}} = 1.0 \times 10^{-3} \text{ mol O}$

The mol ratio between Hg and O is 2:1, so the empirical formula is Hg_2O.

106.   Mass of H_2O = 0.755 g $CuSO_4 \cdot xH_2O$ - 0.483 g $CuSO_4$ = 0.272 g H_2O

$0.483 \text{ g CuSO}_4 \times \dfrac{1 \text{ mol CuSO}_4}{159.62 \text{ g CuSO}_4} = 0.00303 \text{ mol CuSO}_4$

$$0.272 \text{ g H}_2\text{O} \times \frac{1 \text{ mol H}_2\text{O}}{18.02 \text{ g H}_2\text{O}} = 0.0151 \text{ mol H}_2\text{O}$$

$$\frac{0.0151 \text{ mol H}_2\text{O}}{0.00303 \text{ mol CuSO}_4} = \frac{4.98 \text{ mol H}_2\text{O}}{1 \text{ mol CuSO}_4}; \text{ Compound formula} = \text{CuSO}_4\cdot5\text{H}_2\text{O}, x = 5$$

107. $2.00 \times 10^6 \text{ g CaCO}_3 \times \dfrac{1 \text{ mol CaCO}_3}{100.09 \text{ g CaCO}_3} \times \dfrac{1 \text{ mol CaO}}{\text{mol CaCO}_3} \times \dfrac{56.08 \text{ g CaO}}{\text{mol CaO}} = 1.12 \times 10^6 \text{ g CaO}$

108. a. $\text{C}_8\text{H}_{18}(l) + \dfrac{25}{2}\text{O}_2(g) \rightarrow 8 \text{ CO}_2(g) + 9 \text{ H}_2\text{O}(g)$

 or $2 \text{ C}_8\text{H}_{18}(l) + 25 \text{ O}_2(g) \rightarrow 16 \text{ CO}_2(g) + 18 \text{ H}_2\text{O}(g)$

 b. $1.2 \times 10^{10} \text{ gallon} \times \dfrac{4 \text{ qt}}{\text{gal}} \times \dfrac{946 \text{ mL}}{\text{qt}} \times \dfrac{0.692 \text{ g}}{\text{mL}} = 3.1 \times 10^{13} \text{ g of gasoline}$

 $3.1 \times 10^{13} \text{ g C}_8\text{H}_{18} \times \dfrac{1 \text{ mol C}_8\text{H}_{18}}{114.22 \text{ g C}_8\text{H}_{18}} \times \dfrac{16 \text{ mol CO}_2}{2 \text{ mol C}_8\text{H}_{18}} \times \dfrac{44.01 \text{ g CO}_2}{\text{mol CO}_2} = 9.6 \times 10^{13} \text{ g of CO}_2$

109. $4 \text{ Al} + 3 \text{ O}_2 \rightarrow 2 \text{ Al}_2\text{O}_3$

 a. $1.0 \text{ mol Al} \times \dfrac{3 \text{ mol O}_2}{4 \text{ mol Al}} = 0.75 \text{ mol O}_2$; Al is limiting since 1.0 mol O$_2$ is present.

 b. $2.0 \text{ mol Al} \times \dfrac{3 \text{ mol O}_2}{4 \text{ mol Al}} = 1.5 \text{ mol O}_2$; Al is limiting since 4.0 mol O$_2$ is present.

 c. $0.50 \text{ mol Al} \times \dfrac{3 \text{ mol O}_2}{4 \text{ mol Al}} = 0.38 \text{ mol O}_2$; Al is limiting since 0.75 mol O$_2$ is present.

 d. $64.75 \text{ g Al} \times \dfrac{1 \text{ mol Al}}{26.98 \text{ g Al}} \times \dfrac{3 \text{ mol O}_2}{4 \text{ mol Al}} \times \dfrac{32.00 \text{ g O}_2}{\text{mol O}_2} = 57.60 \text{ g O}_2$; Al is limiting.

 e. $75.89 \text{ g Al} \times \dfrac{1 \text{ mol Al}}{26.98 \text{ g Al}} \times \dfrac{3 \text{ mol O}_2}{4 \text{ mol Al}} \times \dfrac{32.00 \text{ g O}_2}{\text{mol O}_2} = 67.51 \text{ g O}_2$; Al is limiting.

 f. $51.28 \text{ g Al} \times \dfrac{1 \text{ mol Al}}{26.98 \text{ g Al}} \times \dfrac{3 \text{ mol O}_2}{4 \text{ mol Al}} \times \dfrac{32.00 \text{ g O}_2}{\text{mol O}_2} = 45.62 \text{ g O}_2$; Al is limiting.

110. a. $\text{CH}_4(g) + 4 \text{ S}(s) \rightarrow \text{CS}_2(l) + 2 \text{ H}_2\text{S}(g)$ or $2 \text{ CH}_4(g) + \text{S}_8(s) \rightarrow 2 \text{ CS}_2(l) + 4 \text{ H}_2\text{S}(g)$

 b. $120. \text{ g CH}_4 \times \dfrac{1 \text{ mol CH}_4}{16.04 \text{ g CH}_4} = 7.48 \text{ mol CH}_4$; $120. \text{ g S} \times \dfrac{1 \text{ mol S}}{32.07 \text{ g S}} = 3.74 \text{ mol S}$

 The required S to CH$_4$ mol ratio is 4:1. The actual S to CH$_4$ mol ratio is:

 $$\frac{3.74 \text{ mol S}}{7.48 \text{ mol CH}_4} = 0.500$$

This is well below the required ratio so sulfur is the limiting reagent.

The theoretical yield of CS_2 is: $3.74 \text{ mol S} \times \dfrac{1 \text{ mol } CS_2}{4 \text{ mol S}} \times \dfrac{76.15 \text{ g } CS_2}{\text{mol } CS_2} = 71.2 \text{ g } CS_2$

The same amount of CS_2 would be produced using the balanced equation with S_8.

111. $C_6H_{10}O_4 + 2 NH_3 + 4 H_2 \rightarrow C_6H_{16}N_2 + 4 H_2O$

Adip. (Adipic acid)   HMD

a. $1.00 \times 10^3 \text{ g Adip.} \times \dfrac{1 \text{ mol Adip.}}{146.14 \text{ g Adip.}} \times \dfrac{1 \text{ mol HMD}}{\text{mol Adip.}} \times \dfrac{116.21 \text{ g HMD}}{\text{mol HMD}} = 795 \text{ g HMD}$

b.   % yield $= \dfrac{765 \text{ g}}{795 \text{ g}} \times 100 = 96.2\%$

112.   a.   Mass of Zn in alloy $= 0.0985 \text{ g } ZnCl_2 \times \dfrac{65.38 \text{ g Zn}}{136.28 \text{ g } ZnCl_2} = 0.0473 \text{ g Zn}$

%Zn $= \dfrac{0.0473 \text{ g Zn}}{0.5065 \text{ g brass}} \times 100 = 9.34\% \text{ Zn}$; %Cu $= 100.00 - 9.34 = 90.66\% \text{ Cu}$

b.   The Cu remains unreacted. After filtering, washing, and drying, the mass of the unreacted copper could be measured.

113. $2 NaNO_3(s) \rightarrow 2 NaNO_2(s) + O_2(g)$

The amount of $NaNO_3$ in the impure sample is:

$0.2864 \text{ g } NaNO_2 \times \dfrac{1 \text{ mol } NaNO_2}{69.00 \text{ g } NaNO_2} \times \dfrac{2 \text{ mol } NaNO_3}{2 \text{ mol } NaNO_2} \times \dfrac{85.00 \text{ g } NaNO_3}{\text{mol } NaNO_3} = 0.3528 \text{ g } NaNO_3$

% $NaNO_3 = \dfrac{0.3528 \text{ g } NaNO_3}{0.4230 \text{ g sample}} \times 100 = 83.40\%$

Challenge Problems

114. $\dfrac{^{85}Rb}{^{87}Rb} = 2.591$; Assuming 100 atoms, let x = number of ^{85}Rb atoms and 100 - x = number of ^{87}Rb atoms.

$\dfrac{x}{100 - x} = 2.591$, $x = 259.1 - 2.591 \, x$, $x = \dfrac{259.1}{3.591} = 72.15\% \; ^{85}Rb$

$0.7215 \, (84.9117) + 0.2785 \, (A) = 85.4678$, $A = \dfrac{85.4678 - 61.26}{0.2785} = 86.92 \text{ amu}$

115.   First, we will determine composition in mass percent. We assume all of the carbon in 0.213 g CO_2 came from 0.157 g of the compound and that all of the hydrogen in the 0.0310 g H_2O came from the 0.157 g of the compound.

$$0.213 \text{ g } CO_2 \times \frac{12.01 \text{ g C}}{44.01 \text{ g } CO_2} = 0.0581 \text{ g C}; \quad \%C = \frac{0.0581 \text{ g C}}{0.157 \text{ g compound}} \times 100 = 37.0\% \text{ C}$$

$$0.0310 \text{ g } H_2O \times \frac{2.016 \text{ g H}}{18.02 \text{ g } H_2O} = 3.47 \times 10^{-3} \text{ g H}; \quad \%H = \frac{3.47 \times 10^{-3} \text{ g}}{0.157 \text{ g}} = 2.21\% \text{ H}$$

We get %N from the second experiment:

$$0.0230 \text{ g } NH_3 \times \frac{14.01 \text{ g N}}{17.03 \text{ g } NH_3} = 1.89 \times 10^{-2} \text{ g N}$$

$$\%N = \frac{1.89 \times 10^{-2} \text{ g}}{0.103 \text{ g}} \times 100 = 18.3\% \text{ N}$$

The mass percent of oxygen is obtained by difference:

$$\%O = 100.00 - (37.0 + 2.21 + 18.3) = 42.5\%$$

So out of 100.00 g of compound, there are:

$$37.0 \text{ g C} \times \frac{1 \text{ mol C}}{12.01 \text{ g C}} = 3.08 \text{ mol C}; \quad 2.21 \text{ g H} \times \frac{1 \text{ mol H}}{1.008 \text{ g H}} = 2.19 \text{ mol H}$$

$$18.3 \text{ g N} \times \frac{1 \text{ mol N}}{14.01 \text{ g N}} = 1.31 \text{ mol N}; \quad 42.5 \text{ g O} \times \frac{1 \text{ mol O}}{16.00 \text{ g O}} = 2.66 \text{ mol O}$$

Lastly, and often the hardest part, we need to find simple whole number ratios. Divide all mole values by the smallest number:

$$\frac{3.08}{1.31} = 2.35; \quad \frac{2.19}{1.31} = 1.67; \quad \frac{1.31}{1.31} = 1.00; \quad \frac{2.66}{1.31} = 2.03$$

Multiplying all these ratios by 3 gives an empirical formula of $C_7H_5N_3O_6$.

116. $$1.0 \times 10^6 \text{ kg } HNO_3 \times \frac{1000 \text{ g } HNO_3}{\text{kg } HNO_3} \times \frac{1 \text{ mol } HNO_3}{63.02 \text{ g } HNO_3} = 1.6 \times 10^7 \text{ mol } HNO_3$$

We need to get the relationship between moles of HNO_3 and moles of NH_3. We have to use all 3 equations.

$$\frac{2 \text{ mol } HNO_3}{3 \text{ mol } NO_2} \times \frac{2 \text{ mol } NO_2}{2 \text{ mol } NO} \times \frac{4 \text{ mol } NO}{4 \text{ mol } NH_3} = \frac{16 \text{ mol } HNO_3}{24 \text{ mol } NH_3}$$

Thus, we can produce 16 mol HNO_3 for every 24 mol NH_3 that we begin with:

$$1.6 \times 10^7 \text{ mol HNO}_3 \times \frac{24 \text{ mol NH}_3}{16 \text{ mol HNO}_3} \times \frac{17.03 \text{ g NH}_3}{\text{mol NH}_3} = 4.1 \times 10^8 \text{ g or } 4.1 \times 10^5 \text{ kg}$$

This is an oversimplified answer. In practice the NO produced in the 3rd step is recycled back into the process in the second step.

117. $2 \text{ C}_3\text{H}_6 + 2 \text{ NH}_3 + 3 \text{ O}_2 \rightarrow 2 \text{ C}_3\text{H}_3\text{N} + 6 \text{ H}_2\text{O}$

a.   An alternative method to determine the amount of product produced in a limiting reagent problem is to assume each reactant is limiting and calculate the amount of product produced from each reactant. The reactant that produces the smallest amount of product is the limiting reagent. Applying this method to the problem:

$$5.00 \times 10^2 \text{ g C}_3\text{H}_6 \times \frac{1 \text{ mol C}_3\text{H}_6}{42.08 \text{ g C}_3\text{H}_6} \times \frac{2 \text{ mol C}_3\text{H}_3\text{N}}{2 \text{ mol C}_3\text{H}_6} = 11.9 \text{ mol C}_3\text{H}_3\text{N}$$

$$5.00 \times 10^2 \text{ g NH}_3 \times \frac{1 \text{ mol NH}_3}{17.03 \text{ g NH}_3} \times \frac{2 \text{ mol C}_3\text{H}_3\text{N}}{2 \text{ mol NH}_3} = 29.4 \text{ mol C}_3\text{H}_3\text{N}$$

$$1.00 \times 10^3 \text{ g O}_2 \times \frac{1 \text{ mol O}_2}{32.00 \text{ g O}_2} \times \frac{2 \text{ mol C}_3\text{H}_3\text{N}}{3 \text{ mol O}_2} = 20.8 \text{ mol C}_3\text{H}_3\text{N}$$

Since C_3H_6 produces the smallest amount of product, then C_3H_6 is limiting and the mass of acrylonitrile produced is:

$$11.9 \text{ mol} \times \frac{53.06 \text{ g C}_3\text{H}_3\text{N}}{\text{mol}} = 631 \text{ g C}_3\text{H}_3\text{N}$$

b. $11.9 \text{ mol C}_3\text{H}_3\text{N} \times \dfrac{6 \text{ mol H}_2\text{O}}{2 \text{ mol C}_3\text{H}_3\text{N}} \times \dfrac{18.02 \text{ g H}_2\text{O}}{\text{mol H}_2\text{O}} = 643 \text{ g H}_2\text{O}$

Amount of NH_3 needed:

$$11.9 \text{ mol C}_3\text{H}_3\text{N} \times \frac{2 \text{ mol NH}_3}{2 \text{ mol C}_3\text{H}_3\text{N}} \times \frac{17.03 \text{ g NH}_3}{\text{mol H}_2\text{O}} = 203 \text{ g NH}_3$$

Amount NH_3 left = 500. g - 203 g = 297 g

Amount O_2 needed: $11.9 \text{ mol C}_3\text{H}_3\text{N} \times \dfrac{3 \text{ mol O}_2}{2 \text{ mol C}_3\text{H}_3\text{N}} \times \dfrac{32.00 \text{ g O}_2}{\text{mol O}_2} = 571 \text{ g O}_2$

Amount O_2 left = 1.00×10^3 g - 571 g = 430 g; 297 g NH_3 and 430 g O_2 left unreacted.

118.   a.   From the reaction stoichiometry we would expect to produce 4 mol of acetaminophen for every 4 mol of $\text{C}_6\text{H}_5\text{O}_3\text{N}$ reacted. The actual yield is 3 moles of acetaminophen compared to a theoretical yield of 4 moles of acetaminophen. Solving for percent yield by mass (where M = molar mass acetaminophen):

$$\% \text{ yield} = \frac{3 \text{ mol} \times \text{M}}{4 \text{ mol} \times \text{M}} \times 100 = 75\%$$

b.   The product of the percent yields of the individual steps must equal the overall yield, 75%.

$(0.87)(0.98)(x) = 0.75$, $x = 0.88$; Step III has a % yield = 88%.

119.   10.00 g XCl_2 + excess $Cl_2 \rightarrow$ 12.55 g XCl_4; 2.55 g Cl reacted with XCl_2 to form XCl_4. XCl_4 contains 2.55 g Cl and 10.00 g XCl_2. From mol ratios, 10.00 g XCl_2 must also contain 2.55 g Cl with 10.00 - 2.45 = 7.45 g X.

$$2.55 \text{ g Cl} \times \frac{1 \text{ mol Cl}}{35.45 \text{ g Cl}} \times \frac{1 \text{ mol } XCl_2}{2 \text{ mol Cl}} \times \frac{1 \text{ mol X}}{\text{mol } XCl_2} = 3.60 \times 10^{-2} \text{ mol X}$$

So, 3.60×10^{-2} mol X must equal 7.45 g X. The molar mass of X is:

$$\frac{7.45 \text{ g X}}{3.60 \times 10^{-2} \text{ mol X}} = \frac{207 \text{ g}}{\text{mol X}}; \text{ X is Pb.}$$

120.   4.000 g $M_2S_3 \rightarrow$ 3.723 g MO_2

There must be twice as many mol of MO_2 as mol of M_2S_3 in order to balance M in the reaction. Setting up an equation for 2 mol MO_2 = mol M_2S_3 where A = atomic mass M:

$$2\left(\frac{4.000 \text{ g}}{2 \text{ A} + 3(32.07)}\right) = \frac{3.723 \text{ g}}{\text{A} + 2(16.00)}, \frac{8.000}{2 \text{ A} + 96.21} = \frac{3.723}{\text{A} + 32.00}$$

8.000 A + 256.0 = 7.446 A + 358.2, 0.554 A = 102.2, A = 184 g/mol; The metal is W.

121.   Consider the case of aluminum plus oxygen. Aluminum forms Al^{3+} ions; oxygen forms O^{2-} anions. The simplest compound of the two elements is Al_2O_3. Similarly we would expect the formula of any group 6A element with Al to be Al_2X_3. Assuming this, out of 100.00 g of compound there are 18.56 g Al and 81.44 g of the unknown element, X. Let's use this information to determine the molar mass of X which will allow us to identify X from the periodic table.

$$18.56 \text{ g Al} \times \frac{1 \text{ mol Al}}{26.98 \text{ g Al}} \times \frac{3 \text{ mol X}}{2 \text{ mol Al}} = 1.032 \text{ mol X}$$

81.44 g of X must equal 1.032 mol of X.

The molar mass of X = $\dfrac{81.44 \text{ g X}}{1.032 \text{ mol X}} = 78.91 \text{ g/mol X}$

From the periodic table, the unknown element is selenium and the formula is Al_2Se_3.

122.   The reaction is: $BaX_2(aq) + H_2SO_4(aq) \rightarrow BaSO_4(s) + 2 HX(aq)$

$$0.124 \text{ g BaSO}_4 \times \frac{137.3 \text{ g Ba}}{233.4 \text{ g BaSO}_4} = 0.0729 \text{ g Ba}; \%\text{Ba} = \frac{0.0729 \text{ g Ba}}{0.158 \text{ g BaX}_2} \times 100 = 46.1\%$$

The formula is BaX_2 (from positions of the elements in the periodic table) and 100.0 g of compound contains 46.1 g Ba and 53.9 g of the unknown halogen. There must also be:

$$46.1 \text{ g Ba} \times \frac{1 \text{ mol Ba}}{137.3 \text{ g Ba}} \times \frac{2 \text{ mol X}}{\text{mol Ba}} = 0.672 \text{ mol of the halogen in 53.9 g of halogen}$$

Therefore, the atomic mass of the halogen is: $\dfrac{53.9 \text{ g}}{0.672 \text{ mol}} = 80.2 \text{ g/mol}$

This atomic mass is close to that of bromine. Thus, the formula of the compound is $BaBr_2$.

123. $1.252 \text{ g Cu} \times \dfrac{1 \text{ mol Cu}}{63.55 \text{ g Cu}} = 1.970 \times 10^{-2} \text{ mol Cu}$

The molar mass of Cu_2O is 143.10 g/mol and the molar mass of CuO is 79.55 g/mol and note that Cu_2O contains twice the mol Cu as compared to CuO. Let x = g Cu_2O and y = g CuO, then x + y = 1.500 and:

$$2\left(\frac{x}{143.10}\right) + \frac{y}{79.55} = 1.970 \times 10^{-2} \text{ total mol Cu or } 1.112\, x + y = 1.567$$

Solving by the method of simultaneous equations:

$$\begin{array}{r} 1.112\, x + y = 1.567 \\ \underline{-x\ -\ y = -1.500} \\ 0.112\, x = 0.067 \end{array}$$

x = 0.067/0.112 = 0.60 g = mass Cu_2O

$\%Cu_2O = \dfrac{0.60 \text{ g}}{1.500 \text{ g}} = 40.\%; \ \ \%CuO = 100. - 40. = 60.\%$

CHAPTER FOUR

TYPES OF CHEMICAL REACTIONS AND SOLUTION STOICHIOMETRY

Questions

5. "Slightly soluble" refers to substances that dissolve only to a small extent. A slightly soluble salt may still dissociate completely to ions and, hence, be a strong electrolyte. An example of such a substance is $Mg(OH)_2$. It is a strong electrolyte, but not very soluble. A weak electrolyte is a substance that doesn't dissociate completely to produce ions. A weak electrolyte may be very soluble in water, or it may not be very soluble. Acetic acid is an example of a weak electrolyte that is very soluble in water.

6. Measure the electrical conductivity of a solution and compare it to the conductivity of a solution of equal concentration of a strong electrolyte.

Exercises

Aqueous solutions: Strong and Weak Electrolytes

7. a. $NaBr(s) \rightarrow Na^+(aq) + Br^-(aq)$ b. $MgCl_2(s) \rightarrow Mg^{2+}(aq) + 2\ Cl^-(aq)$

     c. $Al(NO_3)_3(s) \rightarrow Al^{3+}(aq) + 3\ NO_3^-(aq)$ d. $(NH_4)_2SO_4(s) \rightarrow 2NH_4^+(aq) + SO_4^{2-}(aq)$

     e. $HI(g) \rightarrow H^+(aq) + I^-(aq)$ f. $FeSO_4(s) \rightarrow Fe^{2+}(aq) + SO_4^{2-}(aq)$

     g. $KMnO_4(s) \rightarrow K^+(aq) + MnO_4^-(aq)$ h. $HClO_4(s) \rightarrow H^+(aq) + ClO_4^-(aq)$

     i. $NH_4C_2H_3O_2(s) \rightarrow NH_4^+(aq) + C_2H_3O_2^-(aq)$

8. a. $HCl(g) \rightarrow H^+(aq) + Cl^-(aq)$ b. $HNO_3(l) \rightarrow H^+(aq) + NO_3^-(aq)$

     c. $Ca(OH)_2(s) \rightarrow Ca^{2+}(aq) + 2\ OH^-(aq)$ d. $KOH(s) \rightarrow K^+(aq) + OH^-(aq)$

9. $CaCl_2(s) \rightarrow Ca^{2+}(aq) + 2\ Cl^-(aq)$

10. $MgSO_4(s) \rightarrow Mg^{2+}(aq) + SO_4^{2-}(aq);\ \ NH_4NO_3(s) \rightarrow NH_4^+(aq) + NO_3^-(aq)$

Solution Concentration: Molarity

11.   a. $5.623 \text{ g NaHCO}_3 \times \dfrac{1 \text{ mol NaHCO}_3}{84.01 \text{ g NaHCO}_3} = 6.693 \times 10^{-2} \text{ mol NaHCO}_3$

$$M = \dfrac{6.693 \times 10^{-2} \text{ mol}}{250.0 \text{ mL}} \times \dfrac{1000 \text{ mL}}{\text{L}} = 0.2677 \, M \text{ NaHCO}_3$$

b. $0.1846 \text{ g K}_2\text{Cr}_2\text{O}_7 \times \dfrac{1 \text{ mol K}_2\text{Cr}_2\text{O}_7}{294.20 \text{ g K}_2\text{Cr}_2\text{O}_7} = 6.275 \times 10^{-4} \text{ mol K}_2\text{Cr}_2\text{O}_7$

$$M = \dfrac{6.275 \times 10^{-4} \text{ mol}}{500.0 \times 10^{-3} \text{ L}} = 1.255 \times 10^{-3} \, M \text{ K}_2\text{Cr}_2\text{O}_7$$

c. $0.1025 \text{ g Cu} \times \dfrac{1 \text{ mol Cu}}{63.55 \text{ g Cu}} = 1.613 \times 10^{-3} \text{ mol Cu} = 1.613 \times 10^{-3} \text{ mol Cu}^{2+}$

$$M = \dfrac{1.613 \times 10^{-3} \text{ mol Cu}^{2+}}{200.0 \text{ mL}} \times \dfrac{1000 \text{ mL}}{\text{L}} = 8.065 \times 10^{-3} \, M \text{ Cu}^{2+}$$

12.   a. $\dfrac{16.45 \text{ g NaCl}}{1.000 \text{ L}} \times \dfrac{1 \text{ mol NaCl}}{58.44 \text{ g NaCl}} = 0.2815 \, M \text{ NaCl}$

b. $853.5 \text{ mg KIO}_3 \times \dfrac{1 \text{ g}}{1000 \text{ mg}} \times \dfrac{1 \text{ mol KIO}_3}{214.0 \text{ g KIO}_3} = 3.988 \times 10^{-3} \text{ mol KIO}_3$

$$\dfrac{3.988 \times 10^{-3} \text{ mol}}{250.0 \text{ mL}} \times \dfrac{1000 \text{ mL}}{\text{L}} = 1.595 \times 10^{-2} \, M \text{ KIO}_3$$

c. $0.4508 \text{ g Fe} \times \dfrac{1 \text{ mol Fe}}{55.85 \text{ g Fe}} = 8.072 \times 10^{-3} \text{ mol Fe} = 8.072 \times 10^{-3} \text{ mol Fe}^{3+}$

$$\dfrac{8.072 \times 10^{-3} \text{ mol Fe}^{2+}}{500.0 \text{ mL}} \times \dfrac{1000 \text{ mL}}{\text{L}} = 1.614 \times 10^{-2} \, M \text{ Fe}^{3+}$$

13.   a. $\text{CaCl}_2(s) \rightarrow \text{Ca}^{2+}(aq) + 2 \text{ Cl}^-(aq); \; M_{\text{Ca}^{2+}} = 0.15 \, M; \; M_{\text{Cl}^-} = 2(0.15) = 0.30 \, M$

b. $\text{Al(NO}_3)_3(s) \rightarrow \text{Al}^{3+}(aq) + 3 \text{ NO}_3^-(aq); \; M_{\text{Al}^{3+}} = 0.26 \, M; \; M_{\text{NO}_3^-} = 3(0.26) = 0.78 \, M$

c. $\text{K}_2\text{Cr}_2\text{O}_7(s) \rightarrow 2 \text{ K}^+(aq) + \text{Cr}_2\text{O}_7^{2-}(aq); \; M_{\text{K}^+} = 2(0.25) = 0.50 \, M; \; M_{\text{Cr}_2\text{O}_7^{2-}} = 0.25 \, M$

d. $\text{Al}_2(\text{SO}_4)_3(s) \rightarrow 2 \text{ Al}^{3+}(aq) + 3 \text{ SO}_4^{2-}(aq)$

$$M_{\text{Al}^{3+}} = \dfrac{2.0 \times 10^{-3} \text{ mol Al}_2(\text{SO}_4)_3}{\text{L}} \times \dfrac{2 \text{ mol Al}^{3+}}{\text{mol Al}_2(\text{SO}_4)_3} = 4.0 \times 10^{-3} \, M$$

$$M_{\text{SO}_4^{2-}} = \dfrac{2.0 \times 10^{-3} \text{ mol Al}_2(\text{SO}_4)_3}{\text{L}} \times \dfrac{3 \text{ mol SO}_4^{2-}}{\text{mol Al}_2(\text{SO}_4)_3} = 6.0 \times 10^{-3} \, M$$

14.   a. $0.100 \text{ g MgCl}_2 \times \dfrac{1 \text{ mol MgCl}_2}{95.21 \text{ g MgCl}_2} = 1.05 \times 10^{-3} \text{ mol MgCl}_2$

$$M = \dfrac{1.05 \times 10^{-3} \text{ mol MgCl}_2}{0.1000 \text{ L}} = 1.05 \times 10^{-2} \, M \text{ MgCl}_2$$

$\text{MgCl}_2(s) \rightarrow \text{Mg}^{2+}(aq) + 2 \text{ Cl}^-(aq); \ M_{\text{Mg}^{2+}} = 1.05 \times 10^{-2} \, M; \ M_{\text{Cl}^-} = 2.10 \times 10^{-2} \, M$

   b. $55.1 \times 10^{-3} \text{ g NH}_4\text{Br} \times \dfrac{1 \text{ mol}}{97.94 \text{ g}} = 5.63 \times 10^{-4} \text{ mol NH}_4\text{Br}$

$$M = \dfrac{5.63 \times 10^{-4} \text{ mol}}{0.5000 \text{ L}} = 1.13 \times 10^{-3} \, M \ \text{NH}_4\text{Br}$$

$\text{NH}_4\text{Br}(s) \rightarrow \text{NH}_4^+(aq) + \text{Br}^-(aq); \ M_{\text{NH}_4^+} = M_{\text{Br}^-} = 1.13 \times 10^{-3} \, M$

   c. $M_{\text{Na}_2\text{S}} = \dfrac{5.47 \text{ g Na}_2\text{S} \times \dfrac{1 \text{ mol Na}_2\text{S}}{78.05 \text{ g Na}_2\text{S}}}{1.00 \text{ L}} = 7.01 \times 10^{-2} \, M \text{ Na}_2\text{S}$

$\text{Na}_2\text{S}(s) \rightarrow 2 \text{ Na}^+(aq) + \text{S}^{2-}(aq); \ M_{\text{Na}^+} = 1.40 \times 10^{-1} \, M; \ M_{\text{S}^{2-}} = 7.01 \times 10^{-2} \, M$

   d. $0.208 \text{ g AlCl}_3 \times \dfrac{1 \text{ mol}}{133.33 \text{ g}} = 1.56 \times 10^{-3} \text{ mol AlCl}_3$

$$M_{\text{AlCl}_3} = \dfrac{1.56 \times 10^{-3} \text{ mol}}{250.0 \text{ mL}} \times \dfrac{1000 \text{ mL}}{\text{L}} = 6.24 \times 10^{-3} \, M \text{ AlCl}_3$$

$\text{AlCl}_3(s) \rightarrow \text{Al}^{3+}(aq) + 3 \text{ Cl}^-(aq); \ M_{\text{Al}^{3+}} = 6.24 \times 10^{-3} \, M; \ M_{\text{Cl}^-} = 1.87 \times 10^{-2} \, M$

15. $\text{mol solute} = \text{volume (L)} \times \text{molarity} \left(\dfrac{\text{mol}}{\text{L}} \right); \ \text{AlCl}_3(s) \rightarrow \text{Al}^{3+}(aq) + 3 \text{ Cl}^-(aq)$

$$\text{mol Cl}^- = 0.1000 \text{ L} \times \dfrac{0.30 \text{ mol AlCl}_3}{\text{L}} \times \dfrac{3 \text{ mol Cl}^-}{\text{mol AlCl}_3} = 9.0 \times 10^{-2} \text{ mol Cl}^-$$

$\text{MgCl}_2(s) \rightarrow \text{Mg}^{2+}(aq) + 2 \text{ Cl}^-(aq)$

$$\text{mol Cl}^- = 0.0500 \text{ L} \times \dfrac{0.60 \text{ mol MgCl}_2}{\text{L}} \times \dfrac{2 \text{ mol Cl}^-}{\text{mol MgCl}_2} = 6.0 \times 10^{-2} \text{ mol Cl}^-$$

$\text{NaCl}(s) \rightarrow \text{Na}^+(aq) + \text{Cl}^-(aq)$

$$\text{mol Cl}^- = 0.2000 \text{ L} \times \dfrac{0.40 \text{ mol NaCl}}{\text{L}} \times \dfrac{1 \text{ mol Cl}^-}{\text{mol NaCl}} = 8.0 \times 10^{-2} \text{ mol Cl}^-$$

100.0 mL of 0.30 M AlCl$_3$ contains the largest moles of Cl$^-$ ions.

16. $NaOH(s) \rightarrow Na^+(aq) + OH^-(aq)$, 2 total mol of ions (1 mol Na^+ and 1 mol Cl^-) per mol NaOH.

$$0.1000 \text{ L} \times \frac{0.100 \text{ mol NaOH}}{\text{L}} \times \frac{2 \text{ mol ions}}{\text{mol NaOH}} = 2.0 \times 10^{-2} \text{ mol ions}$$

$BaCl_2(s) \rightarrow Ba^{2+}(aq) + 2 Cl^-(aq)$, 3 total mol of ions per mol $BaCl_2$.

$$0.0500 \text{ L} \times \frac{0.200 \text{ mol}}{\text{L}} \times \frac{3 \text{ mol ions}}{\text{mol BaCl}_2} = 3.0 \times 10^{-2} \text{ mol ions}$$

$Na_3PO_4(s) \rightarrow 3 Na^+(aq) + PO_4^{3-}(aq)$, 4 total mol of ions per mol Na_3PO_4.

$$0.0750 \text{ L} \times \frac{0.150 \text{ mol Na}_3\text{PO}_4}{\text{L}} \times \frac{4 \text{ mol ions}}{\text{mol Na}_3\text{PO}_4} = 4.50 \times 10^{-2} \text{ mol ions}$$

75.0 mL of 0.150 M Na_3PO_4 contains the largest number of ions.

17.   Molar mass of $NaHCO_3$ = 22.99 + 1.008 + 12.01 + 3(16.00) = 84.01 g/mol

$$\text{Volume} = 0.350 \text{ g NaHCO}_3 \times \frac{1 \text{ mol NaHCO}_3}{84.01 \text{ g NaHCO}_3} \times \frac{1 \text{ L}}{0.100 \text{ mol NaHCO}_3} = 0.0417 \text{ L} = 41.7 \text{ mL}$$

41.7 mL of 0.100 M $NaHCO_3$ contains 0.350 g $NaHCO_3$.

18.   Molar mass of NaOH = 22.99 + 16.00 + 1.008 = 40.00 g/mol

$$\text{Mass NaOH} = 0.2500 \text{ L} \times \frac{0.400 \text{ mol NaOH}}{\text{L}} \times \frac{40.00 \text{ g NaOH}}{\text{mol NaOH}} = 4.00 \text{ g NaOH}$$

19.   a. $1.0 \text{ L} \times \dfrac{0.10 \text{ mol NaCl}}{\text{L}} \times \dfrac{58.44 \text{ g NaCl}}{\text{mol}} = 5.8 \text{ g NaCl}$

Place 5.8 g NaCl in a 1 L volumetric flask; add water to dissolve the NaCl and fill to the mark.

b. $1.0 \text{ L} \times \dfrac{0.10 \text{ mol NaCl}}{\text{L}} \times \dfrac{1 \text{ L stock}}{2.5 \text{ mol NaCl}} = 4.0 \times 10^{-2} \text{ L}$

Add 40. mL of 2.5 M solutiuon to a 1 L volumetric flask; fill to the mark with water.

c. $1.0 \text{ L} \times \dfrac{0.20 \text{ mol NaIO}_3}{\text{L}} \times \dfrac{197.9 \text{ g NaIO}_3}{\text{mol NaIO}_3} = 4.0 \times 10^{1} \text{ g NaIO}_3$

As in a, instead using 40. g $NaIO_3$.

d. $1.0 \text{ L} \times \dfrac{0.010 \text{ mol NaIO}_3}{\text{L}} \times \dfrac{1 \text{ L stock}}{0.20 \text{ mol NaIO}_3} = 0.050 \text{ L}$

As in b, instead using 50. mL of the 0.20 M sodium iodate stock solution.

e. $1.0 \text{ L} \times \dfrac{0.050 \text{ mol KHP}}{\text{L}} \times \dfrac{204.22 \text{ g KHP}}{\text{mol KHP}} = 10.2 \text{ g KHP} = 10. \text{ g KHP}$

As in a, instead using 10. g KHP.

f. $1.0 \text{ L} \times \dfrac{0.040 \text{ mol KHP}}{\text{L}} \times \dfrac{1 \text{ L stock}}{0.50 \text{ mol KHP}} = 0.080 \text{ L}$

As in b, instead using 80. mL of the 0.50 M KHP stock solution.

20.   a. $1.00 \text{ L solution} \times \dfrac{0.50 \text{ mol H}_2\text{SO}_4}{\text{L}} = 0.50 \text{ mol H}_2\text{SO}_4$

$0.50 \text{ mol H}_2\text{SO}_4 \times \dfrac{1 \text{ L}}{18 \text{ mol H}_2\text{SO}_4} = 2.8 \times 10^{-2} \text{ L conc. H}_2\text{SO}_4 \text{ or } 28 \text{ mL}$

Dilute 28 mL of concentrated H_2SO_4 to a total volume of 1.00 L with water.

b.   We will need 0.50 mol HCl.

$0.50 \text{ mol HCl} \times \dfrac{1 \text{ L}}{12 \text{ mol HCl}} = 4.2 \times 10^{-2} \text{ L} = 42 \text{ mL}$

Dilute 42 mL of concentrated HCl to a final volume of 1.00 L .

c.   We need 0.50 mol $NiCl_2$.

$0.50 \text{ mol NiCl}_2 \times \dfrac{1 \text{ mol NiCl}_2\bullet6\text{H}_2\text{O}}{\text{mol NiCl}_2} \times \dfrac{237.69 \text{ g NiCl}_2\bullet6\text{H}_2\text{O}}{\text{mol NiCl}_2\bullet6\text{H}_2\text{O}} = 118.8 \text{ g NiCl}_2\bullet6\text{H}_2\text{O} \approx 120 \text{ g}$

Dissolve 120 g $NiCl_2\bullet6H_2O$ in water, and add water until the total volume of the solution is 1.00 L.

d. $1.00 \text{ L} \times \dfrac{0.50 \text{ mol HNO}_3}{\text{L}} = 0.50 \text{ mol HNO}_3$

$0.50 \text{ mol HNO}_3 \times \dfrac{1 \text{ L}}{16 \text{ mol HNO}_3} = 0.031 \text{ L} = 31 \text{ mL}$

Dissolve 31 mL of concentrated reagent in water. Dilute to a total volume of 1.00 L.

e.   We need 0.50 mol Na_2CO_3.

$0.50 \text{ mol Na}_2\text{CO}_3 \times \dfrac{105.99 \text{ g Na}_2\text{CO}_3}{\text{mol}} = 53 \text{ g Na}_2\text{CO}_3$

Dissolve 53 g Na_2CO_3 in water, dilute to 1.00 L.

21. $10.8 \text{ g (NH}_4)_2\text{SO}_4 \times \dfrac{1 \text{ mol}}{132.15 \text{ g}} = 8.17 \times 10^{-2} \text{ mol (NH}_4)_2\text{SO}_4$

$$\text{Molarity} = \frac{8.17 \times 10^{-2} \, \text{mol}}{100.0 \, \text{mL}} \times \frac{1000 \, \text{mL}}{\text{L}} = 0.817 \, M \, (NH_4)_2SO_4$$

Moles of $(NH_4)_2SO_4$ in final solution:

$$10.00 \times 10^{-3} \, \text{L} \times \frac{0.817 \, \text{mol}}{\text{L}} = 8.17 \times 10^{-3} \, \text{mol}$$

$$\text{Molarity of final solution} = \frac{8.17 \times 10^{-3} \, \text{mol}}{(10.00 + 50.00) \, \text{mL}} \times \frac{1000 \, \text{mL}}{\text{L}} = 0.136 \, M \, (NH_4)_2SO_4$$

$$(NH_4)_2SO_4(s) \rightarrow 2 \, NH_4^+(aq) + SO_4^{2-}(aq); \quad M_{NH_4^+} = 2(0.136) = 0.272 \, M; \quad M_{SO_4^{2-}} = 0.136 \, M$$

22. $\text{Molarity} = \dfrac{\text{total mol } HNO_3}{\text{total volume}};$ Total volume = 0.05000 L + 0.1000 L = 0.1500 L

$$\text{Total mol } HNO_3 = 0.0500 \, \text{L} \times \frac{0.100 \, \text{mol } HNO_3}{\text{L}} + 0.1000 \, \text{L} \times \frac{0.200 \, \text{mol } HNO_3}{\text{L}}$$

$$\text{Total mol } HNO_3 = 5.00 \times 10^{-3} \, \text{mol } HNO_3 + 2.00 \times 10^{-2} \, \text{mol } HNO_3 = 2.50 \times 10^{-2} \, \text{mol } HNO_3$$

$$M = \frac{2.50 \times 10^{-2} \, \text{mol } HNO_3}{0.1500 \, \text{L}} = 0.167 \, M \, HNO_3$$

As expected, the molarity of HNO_3 is between 0.100 M and 0.200 M.

23. $0.5842 \, \text{g} \times \dfrac{1 \, \text{mol}}{90.04 \, \text{g}} = 6.488 \times 10^{-3} \, \text{mol } H_2C_2O_4$

$\dfrac{6.488 \times 10^{-3} \, \text{mol}}{100.0 \, \text{mL}} \times \dfrac{1000 \, \text{mL}}{\text{L}} = 6.488 \times 10^{-2} \, M$ This is the concentration of the initial oxalic acid solution.

Consider, next, the dilution step:

$$10.00 \times 10^{-3} \, \text{L} \times \frac{6.488 \times 10^{-2} \, \text{mol}}{\text{L}} = 6.488 \times 10^{-4} \, \text{mol } H_2C_2O_4$$

The final solution contains 6.488×10^{-4} mol of oxalic acid in 250.0 mL of solution:

$$M = \frac{6.488 \times 10^{-4} \, \text{mol}}{0.2500 \, \text{L}} = 2.595 \times 10^{-3} \, M \, H_2C_2O_4$$

24. Stock solution:

$$1.584 \, \text{g } Mn^{2+} \times \frac{1 \, \text{mol } Mn^{2+}}{54.94 \, \text{g } Mn^{2+}} = 2.883 \times 10^{-2} \, \text{mol } Mn^{2+}; \quad \frac{2.883 \times 10^{-2} \, \text{mol } Mn^{2+}}{1.000 \, \text{L}} = 2.883 \times 10^{-2} \, M$$

Solution A contains:

$$50.00 \, \text{mL} \times \frac{1 \, \text{L}}{1000 \, \text{mL}} \times \frac{2.883 \times 10^{-2} \, \text{mol}}{\text{L}} = 1.442 \times 10^{-3} \, \text{mol } Mn^{2+}$$

$$\text{Molarity} = \frac{1.442 \times 10^{-3}\ \text{mol}}{1000.0\ \text{mL}} \times \frac{1000\ \text{mL}}{\text{L}} = 1.442 \times 10^{-3}\ M$$

Solution B contains:

$$10.0\ \text{mL} \times \frac{1\ \text{L}}{1000\ \text{mL}} \times \frac{1.442 \times 10^{-3}\ \text{mol}}{\text{L}} = 1.442 \times 10^{-5}\ \text{mol}\ Mn^{2+}$$

$$\text{Molarity} = \frac{1.442 \times 10^{-5}\ \text{mol}}{0.2500\ \text{L}} = 5.768 \times 10^{-5}\ M$$

Solution C contains:

$$10.00 \times 10^{-3}\ \text{L} \times \frac{5.768 \times 10^{-5}\ \text{mol}}{\text{L}} = 5.768 \times 10^{-7}\ \text{mol}\ Mn^{2+}$$

$$\text{Molarity} = \frac{5.768 \times 10^{-7}\ \text{mol}}{0.5000\ \text{L}} = 1.154 \times 10^{-6}\ M$$

Precipitation Reactions

25.   In all these reactions, soluble ionic compounds are mixed together. To predict the precipitate, switch the anions and cations in the two reactant compounds to predict possible products, then use the solubility rules in Table 4.1 to predict if any of these possible products are insoluble (are the precipitate).

a.   Possible products = $BaSO_4$ and NaCl; precipitate = $BaSO_4(s)$
b.   Possible products = $PbCl_2$ and KNO_3; precipitate = $PbCl_2(s)$
c.   Possible products = Ag_3PO_4 and $NaNO_3$; precipitate = $Ag_3PO_4(s)$
d.   Possible products = $NaNO_3$ and $Fe(OH)_3$; precipitate = $Fe(OH)_3(s)$

26.   a.   Possible products = CuS and NaCl; precipitate = CuS(s)
b.   Possible products = $Ni(OH)_2$ and K_2SO_4; precipitate = $Ni(OH)_2(s)$
c   Possible products = KNO_3 and NaOH; Both salts are soluble so no precipitate forms in this reaction.
d.   Possible products = Na_2SO_4 and $Mn(OH)_2$; precipitate = $Mn(OH)_2(s)$

27.   For the following answers, the balanced molecular equation is first, followed by the complete ionic equation, then the net ionic equation.

a. $BaCl_2(aq) + Na_2SO_4(aq) \rightarrow BaSO_4(s) + 2\ NaCl(aq)$

$Ba^{2+}(aq) + 2\ Cl^-(aq) + 2\ Na^+(aq) + SO_4^{2-}(aq) \rightarrow BaSO_4(s) + 2\ Na^+(aq) + 2\ Cl^-(aq)$

$Ba^{2+}(aq) + SO_4^{2-}(aq) \rightarrow BaSO_4(s)$

b. $Pb(NO_3)_2(aq) + 2\ KCl(aq) \rightarrow PbCl_2(s) + 2\ KNO_3(aq)$

$Pb^{2+}(aq) + 2\ NO_3^-(aq) + 2\ K^+(aq) + 2\ Cl^-(aq) \rightarrow PbCl_2(s) + 2\ K^+(aq) + 2\ NO_3^-(aq)$

$Pb^{2+}(aq) + 2\ Cl^-(aq) \rightarrow PbCl_2(s)$

c. $3\ AgNO_3(aq) + Na_3PO_4(aq) \rightarrow Ag_3PO_4(s) + 3\ NaNO_3(aq)$

$3\ Ag^+(aq) + 3\ NO_3^-(aq) + 3\ Na^+(aq) + PO_4^{3-}(aq) \rightarrow Ag_3PO_4(s) + 3\ Na^+(aq) + 3\ NO_3^-(aq)$

$3\ Ag^+(aq) + PO_4^{3-}(aq) \rightarrow Ag_3PO_4(s)$

d. $3\ NaOH(aq) + Fe(NO_3)_3(aq) \rightarrow Fe(OH)_3(s) + 3\ NaNO_3(aq)$

$3\ Na^+(aq) + 3\ OH^-(aq) + Fe^{3+}(aq) + 3\ NO_3^-(aq) \rightarrow Fe(OH)_3(s) + 3\ Na^+(aq) + 3\ NO_3^-(aq)$

$Fe^{3+}(aq) + 3\ OH^-(aq) \rightarrow Fe(OH)_3(s)$

28.   a. $CuCl_2(aq) + Na_2S(aq) \rightarrow CuS(s) + 2\ NaCl(aq)$

$Cu^{2+}(aq) + 2\ Cl^-(aq) + 2\ Na^+(aq) + S^{2-}(aq) \rightarrow CuS(s) + 2\ Na^+(aq) + 2\ Cl^-(aq)$

$Cu^{2+}(aq) + S^{2-}(aq) \rightarrow CuS(s)$

b. $NiSO_4(aq) + 2\ KOH(aq) \rightarrow Ni(OH)_2(s) + K_2SO_4(aq)$

$Ni^{2+}(aq) + SO_4^{2-}(aq) + 2\ K^+(aq) + 2\ OH^-(aq) \rightarrow Ni(OH)_2(s) + 2\ K^+(aq) + SO_4^{2-}(aq)$

$Ni^{2+}(aq) + 2\ OH^-(aq) \rightarrow Ni(OH)_2(s)$

c. $KOH(aq) + NaNO_3(aq) \rightarrow$ No reaction, all possible products are soluble.

d. $2\ NaOH(aq) + MnSO_4(aq) \rightarrow Mn(OH)_2(s) + Na_2SO_4(aq)$

$2\ Na^+(aq) + 2\ OH^-(aq) + Mn^{2+}(aq) + SO_4^{2-}(aq) \rightarrow Mn(OH)_2(s) + 2\ Na^+(aq) + SO_4^{2-}(aq)$

$Mn^{2+}(aq) + 2\ OH^-(aq) \rightarrow Mn(OH)_2(s)$

29.   a.   Silver iodide is insoluble. $AgNO_3(aq) + KI(aq) \rightarrow AgI(s) + KNO_3(aq)$

$Ag^+(aq) + I^-(aq) \rightarrow AgI(s)$

b.   Copper(II) sulfide is insoluble. $CuSO_4(aq) + Na_2S(aq) \rightarrow CuS(s) + Na_2SO_4(aq)$

$Cu^{2+}(aq) + S^{2-}(aq) \rightarrow CuS(s)$

c. $CoCl_2(aq) + 2\ NaOH(aq) \rightarrow Co(OH)_2(s) + 2\ NaCl(aq)$

$Co^{2+}(aq) + 2\ OH^-(aq) \rightarrow Co(OH)_2(s)$

d.   The potential products are $Ni(NO_3)_2$ and KCl. Both are soluble in water. Thus, no reaction occurs.

30. a. AgCl is insoluble. $Ag^+(aq) + Cl^-(aq) \rightarrow AgCl(s)$

   b. FeS is insoluble. $Fe^{2+}(aq) + S^{2-}(aq) \rightarrow FeS(s)$

   c. No reaction

   d. $Hg_2(NO_3)_2$ is made up of Hg_2^{2+} and NO_3^- ions. Hg_2Cl_2, mercury(I) chloride or mercurous chloride, is insoluble.

       $Hg_2^{2+}(aq) + 2\ Cl^-(aq) \rightarrow Hg_2Cl_2(s)$

31. a. $(NH_4)_2SO_4(aq) + Ba(NO_3)_2(aq) \rightarrow 2\ NH_4NO_3(aq) + BaSO_4(s)$

       $Ba^{2+}(aq) + SO_4^{2-}(aq) \rightarrow BaSO_4(s)$

   b. $Pb(NO_3)_2(aq) + 2\ NaCl(aq) \rightarrow PbCl_2(s) + 2\ NaNO_3(aq)$

       $Pb^{2+}(aq) + 2\ Cl^-(aq) \rightarrow PbCl_2(s)$

   c. Potassium phosphate and sodium nitrate are both soluble in water. No reaction occurs.

   d. No reaction occurs since all possible products are soluble.

   e. $CuCl_2(aq) + 2\ NaOH(aq) \rightarrow Cu(OH)_2(s) + 2\ NaCl(aq)$

       $Cu^{2+}(aq) + 2\ OH^-(aq) \rightarrow Cu(OH)_2(s)$

32. a. $Fe(NO_3)_3(aq) + 3\ NaOH(aq) \rightarrow Fe(OH)_3(s) + 3\ NaNO_3(aq)$

       $Fe^{3+}(aq) + 3\ OH^-(aq) \rightarrow Fe(OH)_3(s)$

   b. $CdCl_2(aq) + Na_2S(aq) \rightarrow CdS(s) + 2\ NaCl(aq);\ \ Cd^{2+}(aq) + S^{2-}(aq) \rightarrow CdS(s)$

   c. $AgNO_3(aq) + RbBr(aq) \rightarrow AgBr(s) + RbNO_3(aq);\ \ Ag^+(aq) + Br^-(aq) \rightarrow AgBr(s)$

   d. $CuCl_2(aq) + Ca(OH)_2(aq) \rightarrow Cu(OH)_2(s) + CaCl_2(aq)$

       $Cu^{2+}(aq) + 2\ OH^-(aq) \rightarrow Cu(OH)_2(s)$

33. Three possibilities are:

     Addition of K_2SO_4 solution to give a white ppt. of $PbSO_4$. Addition of NaCl solution to give a white ppt. of $PbCl_2$. Addition of K_2CrO_4 solution to give a bright yellow ppt. of $PbCrO_4$.

34. Since no precipitates formed upon addition of NaCl or Na_2SO_4, we can conclude that Hg_2^{2+} and Ba^{2+} are not present since Hg_2Cl_2 and $BaSO_4$ are insoluble salts. Since a precipitate formed with NaOH, the solution must contain Mn^{2+} which forms $Mn(OH)_2(s)$.

35. The reaction is: $AgNO_3(aq) + NaCl(aq) \rightarrow AgCl(s) + NaNO_3(aq)$

$$50.0 \times 10^{-3} \text{ L AgNO}_3 \times \frac{0.0500 \text{ mol AgNO}_3}{\text{L AgNO}_3} \times \frac{1 \text{ mol NaCl}}{1 \text{ mol AgNO}_3} \times \frac{58.44 \text{ g NaCl}}{\text{mol NaCl}} = 0.146 \text{ g NaCl}$$

36. The reaction is: $Ni(NO_3)_2(aq) + 2 \text{ NaOH}(aq) \rightarrow Ni(OH)_2(s) + 2 \text{ NaNO}_3(aq)$

$$150.0 \text{ mL Ni(NO}_3)_2(aq) \times \frac{1 \text{ L}}{1000 \text{ mL}} \times \frac{0.250 \text{ mol Ni(NO}_3)_2}{\text{L Ni(NO}_3)_2} \times \frac{2 \text{ mol NaOH}}{1 \text{ mol Ni(NO}_3)_2} \times \frac{1 \text{ L NaOH}}{0.100 \text{ mol NaOH}}$$

$$= 0.750 \text{ L or } 750. \text{ mL}$$

37. The reaction is: $AgNO_3(aq) + NaOH(aq) \rightarrow AgBr(s) + NaNO_3(aq)$

$$100.0 \text{ mL AgNO}_3 \times \frac{1 \text{ L}}{1000 \text{ mL}} \times \frac{0.150 \text{ mol AgNO}_3}{\text{L AgNO}_3} = 1.50 \times 10^{-2} \text{ mol AgNO}_3$$

$$20.0 \text{ mL NaBr} \times \frac{1 \text{ L}}{1000 \text{ mL}} \times \frac{1.00 \text{ mol NaBr}}{\text{L NaBr}} = 2.00 \times 10^{-2} \text{ mol NaBr}$$

From the balanced reaction, 1 mol of $AgNO_3$ is required to react with 1 mol of NaBr (1:1 mol ratio). The actual $AgNO_3$ to NaBr mol ratio is $1.50 \times 10^{-2}/2.00 \times 10^{-2} = 0.750$. Since the actual mol ratio is less than the required mol ratio, then $AgNO_3$ is the limiting reagent ($AgNO_3$ runs out first with NaBr in excess).

$$1.50 \times 10^{-2} \text{ mol AgNO}_3 \times \frac{1 \text{ mol AgBr}}{1 \text{ mol AgNO}_3} \times \frac{187.8 \text{ g AgBr}}{\text{mol AgBr}} = 2.82 \text{ g AgBr}$$

38. The balanced reaction is: $3 \text{ BaCl}_2(aq) + Fe_2(SO_4)_3(aq) \rightarrow 3 \text{ BaSO}_4(s) + 2 \text{ FeCl}_3(aq)$

$$100.0 \text{ mL BaCl}_2 \times \frac{1 \text{ L}}{1000 \text{ mL}} \times \frac{0.100 \text{ mol BaCl}_2}{\text{L}} = 1.00 \times 10^{-2} \text{ mol BaCl}_2$$

$$100.0 \text{ mL Fe}_2(SO_4)_3 \times \frac{1 \text{ L}}{1000 \text{ mL}} \times \frac{0.100 \text{ mol Fe}_2(SO_4)_3}{\text{L Fe}_2(SO_4)_3} = 1.00 \times 10^{-2} \text{ mol Fe}_2(SO_4)_3$$

The required mol $BaCl_2$ to mol $Fe_2(SO_4)_3$ ratio from the balanced reaction is 3:1. The actual mol ratio is $0.0100/0.0100 = 1$ (1:1). This is well below the required mol ratio so $BaCl_2$ is the limiting reagent.

$$0.0100 \text{ mol BaCl}_2 \times \frac{3 \text{ mol BaSO}_4}{3 \text{ mol BaCl}_2} \times \frac{233.4 \text{ g BaSO}_4}{\text{mol BaSO}_4} = 2.33 \text{ g BaSO}_4$$

39. a. The balanced reaction is: $2 \text{ KOH}(aq) + Mg(NO_3)_2(aq) \rightarrow Mg(OH)_2(s) + 2 \text{ KNO}_3(aq)$

b. The precipitate is magnesium hydroxide.

c. $0.1000 \text{ L KOH} \times \dfrac{0.200 \text{ mol KOH}}{\text{L KOH}} = 2.00 \times 10^{-2} \text{ mol KOH}$

$$0.1000 \text{ L Mg(NO}_3)_2 \times \frac{0.200 \text{ mol Mg(NO}_3)_2}{\text{L Mg(NO}_3)_2} = 2.00 \times 10^{-2} \text{ mol Mg(NO}_3)_2$$

From the balanced equation, the required mol KOH to mol $Mg(NO_3)_2$ ratio is 2:1. The actual mol ratio present is 1:1. Not enough KOH is present to react with all of the $Mg(NO_3)_2$ present, so KOH is the limiting reagent.

$$0.0200 \text{ mol KOH} \times \frac{1 \text{ mol Mg(OH)}_2}{2 \text{ mol KOH}} \times \frac{58.33 \text{ g Mg(OH)}_2}{\text{mol Mg(OH)}_2} = 0.583 \text{ g Mg(OH)}_2$$

d.   The net ionic equation for this reaction is: $Mg^{2+}(aq) + 2 \text{ OH}^-(aq) \rightarrow Mg(OH)_2(s)$

Since KOH was the limiting reagent, then all of the OH^- was used up in the reaction. So, $M_{OH^-} = 0 \, M$. Note that K^+ is a spectator ion, so it is still present in solution after precipitation was complete. Also present will be the excess Mg^{2+} and NO_3^- (the other spectator ion).

$$\text{total Mg}^{2+} = 0.0200 \text{ mol Mg(NO}_3)_2 \times \frac{1 \text{ mol Mg}^{2+}}{\text{mol Mg(NO}_3)_2} = 0.0200 \text{ mol Mg}^{2+}$$

$$\text{mol Mg}^{2+} \text{ reacted} = 0.0200 \text{ mol KOH} \times \frac{1 \text{ mol Mg(NO}_3)_2}{2 \text{ mol KOH}} \times \frac{1 \text{ mol Mg}^{2+}}{\text{mol Mg(NO}_3)_2} = 0.0100 \text{ mol Mg}^{2+}$$

$$M_{Mg^{2+}} = \frac{\text{mol excess Mg}^{2+}}{\text{total volume}} = \frac{0.0200 - 0.0100 \text{ mol Mg}^{2+}}{0.1000 \text{ L} + 0.1000 \text{ L}} = 5.00 \times 10^{-2} \, M \, Mg^{2+}$$

The spectator ions are K^+ and NO_3^-. The moles of each present are:

$$\text{mol K}^+ = 0.0200 \text{ mol KOH} \times \frac{1 \text{ mol K}^+}{\text{mol KOH}} = 0.0200 \text{ mol K}^+$$

$$\text{mol NO}_3^- = 0.0200 \text{ mol Mg(NO}_3)_2 \times \frac{2 \text{ mol NO}_3^-}{\text{mol Mg(NO}_3)_2} = 0.0400 \text{ mol NO}_3^-$$

The concentrations are:

$$M_{K^+} = \frac{0.0200 \text{ mol K}^+}{0.2000 \text{ L}} = 0.100 \, M \, K^+; \quad M_{NO_3^-} = \frac{0.0400 \text{ mol NO}_3^-}{0.2000 \text{ L}} = 0.200 \, M \, NO_3^-$$

40. $AgNO_3(aq) + NaCl(aq) \rightarrow AgCl(s) + NaNO_3(aq)$

$$\text{mol AgNO}_3 = 0.1000 \text{ L} \times \frac{0.10 \text{ mol AgNO}_3}{\text{L}} = 0.010 \text{ mol AgNO}_3$$

$$\text{mol NaCl} = 0.1000 \text{ L} \times \frac{0.20 \text{ mol NaCl}}{\text{L}} = 0.020 \text{ mol NaCl}$$

The required mol $AgNO_3$ to mol NaCl is 1:1 (from the balanced equation). The actual mol ratio is 1:2. $AgNO_3$ is the limiting reagent.

$$\text{mass AgCl} = 0.010 \text{ mol AgNO}_3 \times \frac{1 \text{ mol AgCl}}{1 \text{ mol AgNO}_3} \times \frac{143.4 \text{ g AgCl}}{\text{mol AgCl}} = 1.4 \text{ g AgCl}$$

The net ionic equation is: $Ag^+(aq) + Cl^-(aq) \rightarrow AgCl(s)$. The ions remaining in solution are the unreacted Cl^- ions and the spectator ions, NO_3^- and Na^+. Since all the salts have 1 cation per 1 anion, then the mol of each ion present initially (before reaction) is equal to the mol of each reactant compound calculated previously.

mol unreacted $Cl^- = 0.020$ mol Cl^- initially - 0.010 mol Cl^- reacted = 0.010 mol Cl^- unreacted

$$M_{Cl^-} = \frac{0.010 \text{ mol Cl}^-}{\text{total volume}} = \frac{0.010 \text{ mol Cl}^-}{0.1000 \text{ L} + 0.1000 \text{ L}} = 0.050 \, M \, Cl^-$$

The molarity of the spectator ions are:

$$M_{NO_3^-} = \frac{0.010 \text{ mol NO}_3^-}{0.2000 \text{ L}} = 0.050 \, M \, NO_3^-; \quad M_{Na^+} = \frac{0.020 \text{ mol Na}^+}{0.2000 \text{ L}} = 0.10 \, M \, Na^+$$

Acid-Base Reactions

41.   All the bases in this problem are soluble ionic compounds containing OH^-. The acids are either strong or weak electrolytes. The best way to determine if an acid is a strong or weak electrolyte is to memorize all the strong electrolytes (strong acids). Any other acid you encounter that is not a strong acid will be a weak electrolyte and should be kept together in a balanced equation. The strong acids to recognize are HCl, HBr, HI, HNO_3, $HClO_4$ and H_2SO_4. For the answers below, the order of the reactions are molecular, complete ionic and net ionic.

   a. $2 \, HClO_4(aq) + Mg(OH_2)(s) \rightarrow 2 \, H_2O(l) + Mg(ClO_4)_2(aq)$

        $2 \, H^+(aq) + 2 \, ClO_4^-(aq) + Mg(OH)_2(s) \rightarrow 2 \, H_2O(l) + Mg^{2+}(aq) + 2 \, ClO_4^-(aq)$

        $2 \, H^+(aq) + Mg(OH)_2(s) \rightarrow 2 \, H_2O(l) + Mg^{2+}(aq)$

   b. $HCN(aq) + NaOH(aq) \rightarrow H_2O(l) + NaCN(aq)$

        $HCN(aq) + Na^+(aq) + OH^-(aq) \rightarrow H_2O(l) + Na^+(aq) + CN^-(aq)$

        $HCN(aq) + OH^-(aq) \rightarrow H_2O(l) + CN^-(aq)$

   c. $HCl(aq) + NaOH(aq) \rightarrow H_2O(l) + NaCl(aq)$

        $H^+(aq) + Cl^-(aq) + Na^+(aq) + OH^-(aq) \rightarrow H_2O(l) + Na^+(aq) + Cl^-(aq)$

        $H^+(aq) + OH^-(aq) \rightarrow H_2O(l)$

42.   a. $3 \, HNO_3(aq) + Al(OH)_3(s) \rightarrow 3 \, H_2O(l) + Al(NO_3)_3(aq)$

        $3 \, H^+(aq) + 3 \, NO_3^-(aq) + Al(OH)_3(s) \rightarrow 3 \, H_2O(l) + Al^{3+}(aq) + 3 \, NO_3^-(aq)$

$3 H^+(aq) + Al(OH)_3(s) \rightarrow 3 H_2O(l) + Al^{3+}(aq)$

b. $HC_2H_3O_2(aq) + KOH(aq) \rightarrow H_2O(l) + KC_2H_3O_2(aq)$

$HC_2H_3O_2(aq) + K^+(aq) + OH^-(aq) \rightarrow H_2O(l) + K^+(aq) + C_2H_3O_2^-(aq)$

$HC_2H_3O_2(aq) + OH^-(aq) \rightarrow H_2O(l) + C_2H_3O_2^-(aq)$

c. $Ca(OH)_2(aq) + 2 HCl(aq) \rightarrow 2 H_2O(l) + CaCl_2(aq)$

$Ca^{2+}(aq) + 2 OH^-(aq) + 2 H^+(aq) + 2 Cl^-(aq) \rightarrow 2 H_2O(l) + Ca^{2+}(aq) + 2 Cl^-(aq)$

$2 H^+(aq) + 2 OH^-(aq) \rightarrow 2 H_2O(l)$ or $H^+(aq) + OH^-(aq) \rightarrow H_2O(l)$

43.   All the acids in this problem are strong electrolytes. The acids to recognize as strong electrolytes are $HCl, HBr, HI, HNO_3, HClO_4$ and H_2SO_4.

a. $KOH(aq) + HNO_3(aq) \rightarrow H_2O(l) + KNO_3(aq)$

$K^+(aq) + OH^-(aq) + H^+(aq) + NO_3^-(aq) \rightarrow H_2O(l) + K^+(aq) + NO_3^-(aq)$

$OH^-(aq) + H^+(aq) \rightarrow H_2O(l)$

b. $Ba(OH)_2(aq) + 2 HCl(aq) \rightarrow 2 H_2O(l) + BaCl_2(aq)$

$Ba^{2+}(aq) + 2 OH^-(aq) + 2 H^+(aq) + 2 Cl^-(aq) \rightarrow 2 H_2O(l) + Ba^{2+}(aq) + 2 Cl^-(aq)$

$2 OH^-(aq) + 2 H^+(aq) \rightarrow 2 H_2O(l)$ or $OH^-(aq) + H^+(aq) \rightarrow H_2O(l)$

c. $3 HClO_4(aq) + Fe(OH)_3(s) \rightarrow 3 H_2O(l) + Fe(ClO_4)_3(aq)$

$3 H^+(aq) + 3 ClO_4^-(aq) + Fe(OH)_3(s) \rightarrow 3 H_2O(l) + Fe^{3+}(aq) + 3 ClO_4^-(aq)$

$3 H^+(aq) + Fe(OH)_3(s) \rightarrow 3 H_2O(l) + Fe^{3+}(aq)$

44.   a. $AgOH(s) + HBr(aq) \rightarrow AgBr(s) + H_2O(l)$

$AgOH(s) + H^+(aq) + Br^-(aq) \rightarrow AgBr(s) + H_2O(l)$

$AgOH(s) + H^+(aq) + Br^-(aq) \rightarrow AgBr(s) + H_2O(l)$

b. $Sr(OH)_2(aq) + 2 HI(aq) \rightarrow 2 H_2O(l) + SrI_2(aq)$

$Sr^{2+}(aq) + 2 OH^-(aq) + 2 H^+(aq) + 2 I^-(aq) \rightarrow 2 H_2O(l) + Sr^{2+}(aq) + 2 I^-(aq)$

$2 OH^-(aq) + 2 H^+(aq) \rightarrow 2 H_2O(l)$ or $OH^-(aq) + H^+(aq) \rightarrow H_2O(l)$

c. $Fe(OH)_3(s) + 3 HNO_3(aq) \rightarrow 3 H_2O(l) + Fe(NO_3)_3(aq)$

$$Fe(OH)_3(s) + 3 H^+(aq) + 3 NO_3^-(aq) \rightarrow 3 H_2O(l) + Fe^{3+}(aq) + 3 NO_3^-(aq)$$

$$Fe(OH)_3(s) + 3 H^+(aq) \rightarrow 3 H_2O(l) + Fe^{3+}(aq)$$

45.   If we begin with 50.00 mL of 0.200 M NaOH, then:

$$50.00 \times 10^{-3} \text{ L} \times \frac{0.200 \text{ mol}}{\text{L}} = 1.00 \times 10^{-2} \text{ mol NaOH is to be neutralized.}$$

a. $NaOH(aq) + HCl(aq) \rightarrow NaCl(aq) + H_2O(l)$

$$1.00 \times 10^{-2} \text{ mol NaOH} \times \frac{1 \text{ mol HCl}}{\text{mol NaOH}} \times \frac{1 \text{ L soln}}{0.100 \text{ mol}} = 0.100 \text{ L or } 100. \text{ mL}$$

b. $HNO_3(aq) + NaOH(aq) \rightarrow H_2O(l) + NaNO_3(aq)$

$$1.00 \times 10^{-2} \text{ mol NaOH} \times \frac{1 \text{ mol HNO}_3}{\text{mol NaOH}} \times \frac{1 \text{ L}}{0.150 \text{ mol HNO}_3} = 6.67 \times 10^{-2} \text{ L or } 66.7 \text{ mL}$$

c. $HC_2H_3O_2(aq) + NaOH(aq) \rightarrow H_2O(l) + NaC_2H_3O_2(aq)$

$$1.00 \times 10^{-2} \text{ mol NaOH} \times \frac{1 \text{ mol HC}_2\text{H}_3\text{O}_2}{\text{mol NaOH}} \times \frac{1 \text{ L}}{0.200 \text{ mol HC}_2\text{H}_3\text{O}_2} = 5.00 \times 10^{-2} \text{ L or } 50.0 \text{ mL}$$

46.   We begin with 25.00 mL of 0.200 M HCl or $25.00 \times 10^{-3} \text{ L} \times 0.200 \text{ mol/L} = 5.00 \times 10^{-3} \text{ mol HCl}$.

a. $HCl(aq) + NaOH(aq) \rightarrow H_2O(l) + NaCl(aq)$

$$5.00 \times 10^{-3} \text{ mol HCl} \times \frac{1 \text{ mol NaOH}}{\text{mol HCl}} \times \frac{1 \text{ L}}{0.100 \text{ mol NaOH}} = 5.00 \times 10^{-2} \text{ L or } 50.0 \text{ mL}$$

b. $2 HCl(aq) + Ba(OH)_2(aq) \rightarrow 2 H_2O(l) + BaCl_2(aq)$

$$5.00 \times 10^{-3} \text{ mol HCl} \times \frac{1 \text{ mol Ba(OH)}_2}{2 \text{ mol HCl}} \times \frac{1 \text{ L}}{0.0500 \text{ mol Ba(OH)}_2} = 5.00 \times 10^{-2} \text{ L} = 50.0 \text{ mL}$$

c. $HCl(aq) + KOH(aq) \rightarrow H_2O(l) + KCl(aq)$

$$5.00 \times 10^{-3} \text{ mol HCl} \times \frac{1 \text{ mol KOH}}{\text{mol HCl}} \times \frac{1 \text{ L}}{0.250 \text{ mol KOH}} = 2.00 \times 10^{-2} \text{ L or } 20.0 \text{ mL}$$

47. $HNO_3(aq) + NaOH(aq) \rightarrow NaNO_3(aq) + H_2O(l)$

$$15.0 \text{ g NaOH} \times \frac{1 \text{ mol NaOH}}{40.00 \text{ g}} = 0.375 \text{ mol NaOH}$$

$$0.1500 \text{ L} \times \frac{0.250 \text{ mol HNO}_3}{\text{L}} = 0.0375 \text{ mol HNO}_3$$

We have added more moles of NaOH than mol of HNO_3 present. Since NaOH and HNO_3 react in a 1:1 mol ratio then NaOH is in excess and the solution will be basic. The ions present after reaction will be the excess OH^- ions and the spectator ions, Na^+ and NO_3^+. The moles of ions present initially are:

mol NaOH = mol Na^+ = mol OH^- = 0.375 mol

mol HNO_3 = mol H^+ = mol NO_3^- = 0.0375 mol

The net ionic reaction occurring is: $H^+(aq) + OH^-(aq) \rightarrow H_2O(l)$

The mol of excess OH^- remaining after reaction will be the initial mol of OH^- minus the amount of OH^- neutralized by reaction with H^+:

mol excess OH^- = 0.375 mol - 0.0375 mol = 0.338 mol OH^- excess

The concentration of ions present is:

$$M_{OH^-} = \frac{\text{mol OH}^- \text{ excess}}{\text{volume}} = \frac{0.338 \text{ mol OH}^-}{0.1500 \text{ L}} = 2.25 \, M \, OH^-$$

$$M_{NO_3^-} = \frac{0.0375 \text{ mol NO}_3^-}{0.1500 \text{ L}} = 0.250 \, M \, NO_3^-; \quad M_{Na^+} = \frac{0.375 \text{ mol}}{0.1500 \text{ L}} = 2.50 \, M \, Na^+$$

48. $Ba(OH)_2(aq) + 2 \text{ HCl}(aq) \rightarrow BaCl_2(aq) + 2 \text{ H}_2O(l); \; H^+(aq) + OH^-(aq) \rightarrow H_2O(l)$

$$75.0 \times 10^{-3} \text{ L} \times \frac{0.250 \text{ mol HCl}}{\text{L}} = 1.88 \times 10^{-2} \text{ mol HCl} = 1.88 \times 10^{-2} \text{ mol H}^+ + 1.88 \times 10^{-2} \text{ mol Cl}^-$$

$$225.0 \times 10^{-3} \text{ L} \times \frac{0.0550 \text{ mol Ba(OH)}_2}{\text{L}} = 1.24 \times 10^{-2} \text{ mol Ba(OH)}_2 = 1.24 \times 10^{-2} \text{ mol Ba}^{2+}$$
$$+ 2.48 \times 10^{-2} \text{ mol OH}^-$$

The net ionic equation requires a 1:1 mol ratio between OH^- and H^+. The actual mol OH^- to mol H^+ ratio is greater than 1:1 so OH^- is in excess.

Since 1.88×10^{-2} mol OH^- will be neutralized by the H^+, then we have $(2.48 - 1.88) \times 10^{-2} = 0.60 \times 10^{-2}$ mol OH^- remaining in excess.

$$M_{OH^-} = \frac{\text{mol OH}^- \text{ excess}}{\text{total volume}} = \frac{6.0 \times 10^{-3} \text{ mol OH}^-}{0.0750 \text{ L} + 0.2250 \text{ L}} = 2.0 \times 10^{-2} \, M \, OH^-$$

49. $HCl(aq) + NaOH(aq) \rightarrow H_2O(l) + NaCl(aq)$

$$24.16 \times 10^{-3} \text{ L NaOH} \times \frac{0.106 \text{ mol NaOH}}{\text{L NaOH}} \times \frac{1 \text{ mol HCl}}{\text{mol NaOH}} = 2.56 \times 10^{-3} \text{ mol HCl}$$

$$\text{Molarity of HCl} = \frac{2.56 \times 10^{-3} \text{ mol}}{25.00 \times 10^{-3} \text{ L}} = 0.102 \, M \text{ HCl}$$

50. $2 \text{ HNO}_3(aq) + \text{Ca(OH)}_2(aq) \rightarrow 2 \text{ H}_2\text{O}(l) + \text{Ca(NO}_3)_2(aq)$

$$34.66 \times 10^{-3} \text{ L HNO}_3 \times \frac{0.0980 \text{ mol HNO}_3}{\text{L HNO}_3} \times \frac{1 \text{ mol Ca(OH)}_2}{2 \text{ mol HNO}_3} = 1.70 \times 10^{-3} \text{ mol Ca(OH)}_2$$

$$\text{Molarity of Ca(OH)}_2 = \frac{1.70 \times 10^{-3} \text{ mol}}{50.00 \times 10^{-3} \text{ L}} = 0.0340 \, M \text{ Ca(OH)}_2$$

51.   Since KHP is a monoprotic acid, the reaction is: $\text{NaOH}(aq) + \text{KHP}(aq) \rightarrow \text{H}_2\text{O}(l) + \text{NaKP}(aq)$

$$\text{mass KHP} = 0.02046 \text{ L NaOH} \times \frac{0.1000 \text{ mol NaOH}}{\text{L NaOH}} \times \frac{1 \text{ mol KHP}}{\text{mol NaOH}} \times \frac{204.22 \text{ g KHP}}{\text{mol KHP}}$$

$$= 0.4178 \text{ g KHP}$$

52. $\text{HNO}_3(aq) + \text{KOH}(aq) \rightarrow \text{H}_2\text{O}(l) + \text{KNO}_3(aq)$

$$\text{Volume HNO}_3 = 0.200 \text{ g KOH} \times \frac{1 \text{ mol KOH}}{56.11 \text{ g KOH}} \times \frac{1 \text{ mol HNO}_3}{\text{mol KOH}} \times \frac{1 \text{ L HNO}_3}{0.250 \text{ mol HNO}_3}$$

$$= 1.43 \times 10^{-2} \text{ L} = 14.3 \text{ mL HNO}_3$$

Oxidation-Reduction Reactions

53.   Apply rules in Table 4.2.

  a. KMnO_4 is composed of K^+ and MnO_4^- ions. Assign oxygen a value of -2 which gives manganese at +7 oxidation state since the sum of oxidation states for all atoms in MnO_4^- must equal the -1 charge on MnO_4^-. K, +1; O, -2; Mn, +7.

  b. Assign O a -2 oxidation state, which gives nickle a +4 oxidation state. Ni, +4; O, -2.

  c. $\text{K}_4\text{Fe(CN)}_6$ is composed of K^+ cations and Fe(CN)_6^{4-} anions. Fe(CN)_6^{4-} is composed of iron and CN^- anions. For an overall anion charge of -4, iron must have a +2 oxidation state.

  d. $(\text{NH}_4)_2\text{HPO}_4$ is made of NH_4^+ cations and HPO_4^{2-} anions. Assign +1 as oxidation state of H and -2 as oxidation state of O. In NH_4^+, $x + 4(+1) = +1$, $x = -3$ = oxidation state of N. In HPO_4^{2-}, $+1 + y + 4(-2) = -2$, $y = +5$ = oxidation state of P.

  e. O, -2; P, +3   f. O, -2; Fe, + 8/3

  g. O, -2; F, -1; Xe, +6   h. F, -1; S, +4

  i. O, -2; C, +2   j. Na, +1; O, -2; C, +3

54.   a. UO_2^{2+}: O, -2; For U, $x + 2(-2) = +2$, $x = \underline{+6}$

       b. As_2O_3: O, -2; For As, $2(x) + 3(-2) = 0$, $x = \underline{+3}$

       c. $NaBiO_3$: Na, +1; O, -2; For Bi, $+1 + x + 3(-2) = 0$, $x = \underline{+5}$

       d. As_4: As, 0

       e. $HAsO_2$: assign H = +1 and O = -2; For As, $+1 + x + 2(-2) = 0$; $x = \underline{+3}$

       f. $Mg_2P_2O_7$: Composed of Mg^{2+} ions and $P_2O_7^{4-}$ ions. Oxidation states are:

              Mg, +2; O, -2; P, +5

       g. $Na_2S_2O_3$: Composed of Na^+ ions and $S_2O_3^{2-}$ ions.   Na, +1; O, -2; S, +2

       h. Hg_2Cl_2: Hg, +1; Cl, -1

       i. $Ca(NO_3)_2$: Composed of Ca^{2+} ions and NO_3^- ions.   Ca, +2; O, -2; N, +5

55. OCl^-:   Oxidation state of oxygen is (-2).

              $-2 + x = -1$, $x = +1$;   The oxidation state of Cl in OCl^- is +1.

       ClO_2^-: $2(-2) + x = -1$, $x = +3$ ClO_3^-: $3(-2) + x = -1$, $x = +5$ ClO_4^-: $4(-2) + x = -1$, $x = +7$

56.   a. -3   b. -3   c. $2(x) + 4(+1) = 0$, $x = -2$
       d. +2   e. +1   f. +4
       g. +3   h. +5   i. 0

57.   To determine if the reaction is an oxidation-reduction reaction, assign oxidation numbers. If the
       oxidation numbers change for some elements, then the reaction is a redox reaction. If the oxidation
       numbers do not change, then the reaction is not a redox reaction. In redox reactions the species
       oxidized (called the reducing agent) shows an increase in the oxidation numbers and the species
       reduced (called the oxidizing agent) shows a decrease in oxidation numbers.

	Redox?	Oxidizing Agent	Reducing Agent	Substance Oxidized	Substance Reduced
a.	Yes	O_2	CH_4	CH_4 (C)	O_2 (O)
b.	Yes	HCl	Zn	Zn	HCl (H)
c.	No	-	-	-	-
d.	Yes	O_3	NO	NO (N)	O_3 (O)
e.	Yes	H_2O_2	H_2O_2	H_2O_2 (O)	H_2O_2 (O)
f.	Yes	CuCl	CuCl	CuCl (Cu)	CuCl (Cu)

       In c, no oxidation numbers change from reactants to products.

58.   Redox?   Oxidizing   Reducing   Substance   Substance
                          Agent   Agent   Oxidized   Reduced

	Redox?	Oxidizing Agent	Reducing Agent	Substance Oxidized	Substance Reduced
a.	Yes	Ag^+	Cu	Cu	Ag^+
b.	No	-	-	-	-
c.	No	-	-	-	-
d.	Yes	$SiCl_4$	Mg	Mg	$SiCl_4$ (Si)
e.	No	-	-	-	-

In b, c, and e, no oxidation numbers change.

59.   Use the method of half-reactions described in Section 4.10 of the text to balance these redox
       reactions. The first step always is to separate the reaction into the two half-reactions, then balance
       each half-reaction separately.

a. $Zn \rightarrow Zn^{2+} + 2\ e^-$ $2e^- + 2\ HCl \rightarrow H_2 + 2Cl^-$

      Adding the two balanced half-reactions, $Zn(s) + 2\ HCl(aq) \rightarrow H_2(g) + Zn^{2+}(aq) + 2Cl^-(aq)$

b. $3\ I^- \rightarrow I_3^- + 2e^-$ $ClO^- \rightarrow Cl^-$
                                                         $2e^- + 2H^+ + ClO^- \rightarrow Cl^- + H_2O$

      Adding the two balanced half-reactions so electrons cancel:

            $3\ I^-(aq) + 2\ H^+(aq) + ClO^-(aq) \rightarrow I_3^-(aq) + Cl^-(aq) + H_2O(l)$

c. $As_2O_3 \rightarrow H_3AsO_4$ $NO_3^- \rightarrow NO + 2\ H_2O$
      $As_2O_3 \rightarrow 2\ H_3AsO_4$ $4\ H^+ + NO_3^- \rightarrow NO + 2\ H_2O$
      Left 3 - O; Right 8 - O $(3\ e^- + 4\ H^+ + NO_3^- \rightarrow NO + 2\ H_2O) \times 4$
      Right hand side has 5 extra O.
      Balance the oxygen atoms first using H_2O, then balance H using H^+, and finally balance charge
      using electrons.
                  $(5\ H_2O + As_2O_3 \rightarrow 2\ H_3AsO_4 + 4\ H^+ + 4\ e^-) \times 3$

      Common factor is a transfer of 12 e^-. Add half-reactions so electrons cancel.

                       $12\ e^- + 16\ H^+ + 4\ NO_3^- \rightarrow 4\ NO + 8\ H_2O$
                       $15\ H_2O + 3\ As_2O_3 \rightarrow 6\ H_3AsO_4 + 12\ H^+ + 12\ e^-$
      _____

      $7\ H_2O(l) + 4\ H^+(aq) + 3\ As_2O_3(s) + 4\ NO_3^-(aq) \rightarrow 4\ NO(g) + 6\ H_3AsO_4(aq)$

d. $(2\ Br^- \rightarrow Br_2 + 2\ e^-) \times 5$ $MnO_4^- \rightarrow Mn^{2+} + 4\ H_2O$
                                                     $(5\ e^- + 8\ H^+ + MnO_4^- \rightarrow Mn^{2+} + 4\ H_2O) \times 2$

      Common factor is a transfer of 10 e^-.

                          $10\ Br^- \rightarrow 5\ Br_2 + 10\ e^-$
                 $10\ e^- + 16\ H^+ + 2\ MnO_4^- \rightarrow 2\ Mn^{2+} + 8\ H_2O$
      _____

      $16\ H^+(aq) + 2\ MnO_4^-(aq) + 10\ Br^-(aq) \rightarrow 5\ Br_2(l) + 2\ Mn^{2+}(aq) + 8\ H_2O(l)$

e. $CH_3OH \rightarrow CH_2O$ $Cr_2O_7^{2-} \rightarrow Cr^{3+}$
     $(CH_3OH \rightarrow CH_2O + 2 H^+ + 2 e^-) \times 3$ $14 H^+ + Cr_2O_7^{2-} \rightarrow 2 Cr^{3+} + 7 H_2O$
                                                                 $6 e^- + 14 H^+ + Cr_2O_7^{2-} \rightarrow 2 Cr^{3+} + 7 H_2O$

Common factor is a transfer of 6 e⁻.

$$3 CH_3OH \rightarrow 3 CH_2O + 6 H^+ + 6 e^-$$
$$6 e^- + 14 H^+ + Cr_2O_7^{2-} \rightarrow 2 Cr^{3+} + 7 H_2O$$

_____

$$8 H^+(aq) + 3 CH_3OH(aq) + Cr_2O_7^{2-}(aq) \rightarrow 2 Cr^{3+}(aq) + 3 CH_2O(aq) + 7 H_2O(l)$$

60.   a. $(Cu \rightarrow Cu^{2+} + 2 e^-) \times 3$ $NO_3^- \rightarrow NO + 2 H_2O$
                                                                 $(3 e^- + 4 H^+ + NO_3^- \rightarrow NO + 2 H_2O) \times 2$

Adding the two balanced half-reactions so electrons cancel:

$$3 Cu \rightarrow 3 Cu^{2+} + 6 e^-$$
$$6 e^- + 8 H^+ + 2 NO_3^- \rightarrow 2 NO + 4 H_2O$$

_____

$$3 Cu(s) + 8 H^+(aq) + 2 NO_3^- (aq) \rightarrow 3 Cu^{2+}(aq) + 2 NO(g) + 4 H_2O(l)$$

b. $(2 Cl^- \rightarrow Cl_2 + 2 e^-) \times 3$ $Cr_2O_7^{2-} \rightarrow 2 Cr^{3+} + 7 H_2O$
                                                                 $6 e^- + 14 H^+ + Cr_2O_7^{2-} \rightarrow 2 Cr^{3+} + 7 H_2O$

Add the two half-reactions with six electrons transfered:

$$6 Cl^- \rightarrow 3 Cl_2 + 6 e^-$$
$$6 e^- + 14 H^+ + Cr_2O_7^{2-} \rightarrow 2 Cr^{3+} + 7 H_2O$$

_____

$$14 H^+(aq) + Cr_2O_7^{2-}(aq) + 6 Cl^-(aq) \rightarrow 3 Cl_2(g) + 2 Cr^{3+}(aq) + 7 H_2O(l)$$

c. $Pb \rightarrow PbSO_4$ $PbO_2 \rightarrow PbSO_4$
     $Pb + H_2SO_4 \rightarrow PbSO_4 + 2 H^+$ $PbO_2 + H_2SO_4 \rightarrow PbSO_4 + 2 H^+$
     $Pb + H_2SO_4 \rightarrow PbSO_4 + 2 H^+ + 2 e^-$ $2 e^- + 2 H^+ + PbO_2 + H_2SO_4 \rightarrow PbSO_4 + 2 H_2O$

Add the two half-reactions with two electrons transfered:

$$2 e^- + 2 H^+ + PbO_2 + H_2SO_4 \rightarrow PbSO_4 + 2 H_2O$$
$$Pb + H_2SO_4 \rightarrow PbSO_4 + 2 H^+ + 2 e^-$$

_____

$$Pb(s) + H_2SO_4(aq) + PbO_2(s) \rightarrow 2 PbSO_4(s) + 2 H_2O(l)$$

This is the reaction that occurs in an automobile lead storage battery.

d. $Mn^{2+} \rightarrow MnO_4^-$
     $(4 H_2O + Mn^{2+} \rightarrow MnO_4^- + 8 H^+ + 5 e^-) \times 2$

$$NaBiO_3 \rightarrow Bi^{3+} + Na^+$$
$$6\,H^+ + NaBiO_3 \rightarrow Bi^{3+} + Na^+ + 3\,H_2O$$
$$(2\,e^- + 6\,H^+ + NaBiO_3 \rightarrow Bi^{3+} + Na^+ + 3\,H_2O) \times 5$$

$$8\,H_2O + 2\,Mn^{2+} \rightarrow 2\,MnO_4^- + 16\,H^+ + 10\,e^-$$
$$10\,e^- + 30\,H^+ + 5\,NaBiO_3 \rightarrow 5\,Bi^{3+} + 5\,Na^+ + 15\,H_2O$$
$$\overline{}$$
$$8\,H_2O + 30\,H^+ + 2\,Mn^{2+} + 5\,NaBiO_3 \rightarrow 2\,MnO_4^- + 5\,Bi^{3+} + 5\,Na^+ + 15\,H_2O + 16\,H^+$$

Simplifying :

$$14\,H^+(aq) + 2\,Mn^{2+}(aq) + 5\,NaBiO_3(s) \rightarrow 2\,MnO_4^-(aq) + 5\,Bi^{3+}(aq) + 5\,Na^+(aq) + 7\,H_2O(l)$$

e.
$$H_3AsO_4 \rightarrow AsH_3 \qquad\qquad\qquad (Zn \rightarrow Zn^{2+} + 2\,e^-) \times 4$$
$$H_3AsO_4 \rightarrow AsH_3 + 4\,H_2O$$
$$8\,e^- + 8\,H^+ + H_3AsO_4 \rightarrow AsH_3 + 4\,H_2O$$

$$8\,e^- + 8\,H^+ + H_3AsO_4 \rightarrow AsH_3 + 4\,H_2O$$
$$4\,Zn \rightarrow 4\,Zn^{2+} + 8\,e^-$$
$$\overline{}$$
$$8\,H^+(aq) + H_3AsO_4(aq) + 4\,Zn(s) \rightarrow 4\,Zn^{2+}(aq) + AsH_3(g) + 4\,H_2O(l)$$

61.   Use the same method as with acidic solutions. After the final balanced equation, then convert H^+ to OH^- as described in section 14.10 of the text. The extra step involves converting H^+ into H_2O by adding equal moles of OH^- to each side of the reaction. This converts the reaction to a basic solution while keeping it balanced.

a.
$$Al \rightarrow Al(OH)_4^- \qquad\qquad\qquad MnO_4^- \rightarrow MnO_2$$
$$4\,H_2O + Al \rightarrow Al(OH)_4^- + 4\,H^+ \qquad\qquad 3\,e^- + 4\,H^+ + MnO_4^- \rightarrow MnO_2 + 2\,H_2O$$
$$4\,H_2O + Al \rightarrow Al(OH)_4^- + 4\,H^+ + 3\,e^-$$

$$4\,H_2O + Al \rightarrow Al(OH)_4^- + 4\,H^+ + 3\,e^-$$
$$3\,e^- + 4\,H^+ + MnO_4^- \rightarrow MnO_2 + 2\,H_2O$$
$$\overline{}$$
$$2\,H_2O(l) + Al(s) + MnO_4^-(aq) \rightarrow Al(OH)_4^-(aq) + MnO_2(s)$$

Since H^+ doesn't appear in the final balanced reaction, we are done.

b.
$$Cl_2 \rightarrow Cl^- \qquad\qquad\qquad\qquad Cl_2 \rightarrow ClO^-$$
$$2\,e^- + Cl_2 \rightarrow 2\,Cl^- \qquad\qquad\qquad 2\,H_2O + Cl_2 \rightarrow 2\,ClO^- + 4\,H^+ + 2\,e^-$$

$$2\,e^- + Cl_2 \rightarrow 2\,Cl^-$$
$$2\,H_2O + Cl_2 \rightarrow 2\,ClO^- + 4\,H^+ + 2\,e^-$$
$$\overline{}$$
$$2\,H_2O + 2\,Cl_2 \rightarrow 2\,Cl^- + 2\,ClO^- + 4\,H^+$$

Now convert to a basic solution. Add $4\,OH^-$ to both sides of the equation. The $4\,OH^-$ will react with the $4\,H^+$ on the product side to give $4\,H_2O$. After this step, cancel identical species on both sides ($2\,H_2O$). Applying these steps gives: $4\,OH^- + 2\,Cl_2 \rightarrow 2\,Cl^- + 2\,ClO^- + 2\,H_2O$, which can be further simplified to:

$$2\ OH^-(aq) + Cl_2(g) \rightarrow Cl^-(aq) + ClO^-(aq) + H_2O(l)$$

c.
$$NO_2^- \rightarrow NH_3$$
$$6\ e^- + 7\ H^+ + NO_2^- \rightarrow NH_3 + 2\ H_2O$$

$$Al \rightarrow AlO_2^-$$
$$(2\ H_2O + Al \rightarrow AlO_2^- + 4\ H^+ + 3\ e^-) \times 2$$

Common factor is a transfer of 6 e^-.

$$6e^- + 7\ H^+ + NO_2^- \rightarrow NH_3 + 2\ H_2O$$
$$4\ H_2O + 2\ Al \rightarrow 2\ AlO_2^- + 8\ H^+ + 6\ e^-$$
_____
$$OH^- + 2\ H_2O + NO_2^- + 2\ Al \rightarrow NH_3 + 2\ AlO_2^- + H^+ + OH^-$$

Reducing gives: $OH^-(aq) + H_2O(l) + NO_2^-(aq) + 2\ Al(s) \rightarrow NH_3(g) + 2\ AlO_2^-(aq)$

62. a.
$$Cr \rightarrow Cr(OH)_3$$
$$3\ H_2O + Cr \rightarrow Cr(OH)_3 + 3\ H^+ + 3\ e^-$$

$$CrO_4^{2-} \rightarrow Cr(OH)_3$$
$$3\ e^- + 5\ H^+ + CrO_4^{2-} \rightarrow Cr(OH)_3 + H_2O$$

$$3\ H_2O + Cr \rightarrow Cr(OH)_3 + 3\ H^+ + 3\ e^-$$
$$3\ e^- + 5\ H^+ + CrO_4^{2-} \rightarrow Cr(OH)_3 + H_2O$$
_____
$$2\ OH^- + 2\ H^+ + 2\ H_2O + Cr + CrO_4^{2-} \rightarrow 2\ Cr(OH)_3 + 2\ OH^-$$

Two OH^- were added above to each side to convert to basic solution. The two OH^- react with the 2 H^+ on the reactant side to produce 2 H_2O. The overall balanced equation is:

$$4\ H_2O(l) + Cr(s) + CrO_4^{2-}(aq) \rightarrow 2\ Cr(OH)_3(s) + 2\ OH^-(aq)$$

b. $S^{2-} \rightarrow S$
$$(S^{2-} \rightarrow S + 2\ e^-) \times 5$$

$$MnO_4^- \rightarrow MnS$$
$$MnO_4^- + S^{2-} \rightarrow MnS$$
$$(5\ e^- + 8\ H^+ + MnO_4^- + S^{2-} \rightarrow MnS + 4\ H_2O) \times 2$$

Common factor is a transfer of 10 e^-.

$$5\ S^{2-} \rightarrow 5\ S + 10\ e^-$$
$$10\ e^- + 16\ H^+ + 2\ MnO_4^- + 2\ S^{2-} \rightarrow 2\ MnS + 8\ H_2O$$
_____
$$16\ OH^- + 16\ H^+ + 7\ S^{2-} + 2\ MnO_4^- \rightarrow 5\ S + 2\ MnS + 8\ H_2O + 16\ OH^-$$

$$16\ H_2O + 7\ S^{2-} + 2\ MnO_4^- \rightarrow 5\ S + 2\ MnS + 8\ H_2O + 16\ OH^-$$

Reducing gives: $8\ H_2O(l) + 7\ S^{2-}(aq) + 2\ MnO_4^-(aq) \rightarrow 5\ S(s) + 2\ MnS(s) + 16\ OH^-(aq)$

c.
$$CN^- \rightarrow CNO^-$$
$$(H_2O + CN^- \rightarrow CNO^- + 2\ H^+ + 2\ e^-) \times 3$$

$$MnO_4^- \rightarrow MnO_2$$
$$(3\ e^- + 4\ H^+ + MnO_4^- \rightarrow MnO_2 + 2\ H_2O) \times 2$$

Common factor is a transfer of 6 electrons.

$$3\,H_2O + 3\,CN^- \rightarrow 3\,CNO^- + 6\,H^+ + 6\,e^-$$
$$6\,e^- + 8\,H^+ + 2\,MnO_4^- \rightarrow 2\,MnO_2 + 4\,H_2O$$

$$2\,OH^- + 2\,H^+ + 3\,CN^- + 2\,MnO_4^- \rightarrow 3\,CNO^- + 2\,MnO_2 + H_2O + 2\,OH^-$$

Reducing gives:

$$H_2O(l) + 3\,CN^-(aq) + 2\,MnO_4^-(aq) + H_2O(l) \rightarrow 3\,CNO^-(aq) + 2\,MnO_2(s) + 2\,OH^-(aq)$$

63. $NaCl + H_2SO_4 + MnO_2 \rightarrow Na_2SO_4 + MnCl_2 + Cl_2 + H_2O$

We could balance this reaction by the half-reaction method or by inspection. Lets try inspection. To balance Cl^-, we need 4 NaCl:

$$4\,NaCl + H_2SO_4 + MnO_2 \rightarrow Na_2SO_4 + MnCl_2 + Cl_2 + H_2O$$

Balance the Na^+ and SO_4^{2-} ions next:

$$4\,NaCl + 2\,H_2SO_4 + MnO_2 \rightarrow 2\,Na_2SO_4 + MnCl_2 + Cl_2 + H_2O$$

On the left side: 4-H and 10-O; On the right side: 8-O not counting H_2O

We need 2 H_2O on the right side to balance H and O:

$$4\,NaCl(aq) + 2\,H_2SO_4(aq) + MnO_2(s) \rightarrow 2\,Na_2SO_4(aq) + MnCl_2(aq) + Cl_2(g) + 2\,H_2O(l)$$

64. $Au + HNO_3 + HCl \rightarrow AuCl_4^- + NO$

Only deal with ions that are reacting (omit H^+): $Au + NO_3^- + Cl^- \rightarrow AuCl_4^- + NO$

The balanced half-reactions are:

$$Au + 4\,Cl^- \rightarrow AuCl_4^- + 3\,e^- \qquad\qquad\qquad 3\,e^- + 4\,H^+ + NO_3^- \rightarrow NO + 2\,H_2O$$

Adding the two balanced half-reactions:

$$Au(s) + 4\,Cl^-(aq) + 4\,H^+(aq) + NO_3^-(aq) \rightarrow AuCl_4^-(aq) + NO(g) + 2\,H_2O(l)$$

Additional Exercises

65. $4.25\text{ g Ca} \times \dfrac{1\text{ mol Ca}}{40.08\text{ g Ca}} \times \dfrac{1\text{ mol Ca(OH)}_2}{\text{mol Ca}} \times \dfrac{2\text{ mol OH}^-}{\text{mol Ca(OH)}_2} = 0.212\text{ mol OH}^-$

Molarity $= \dfrac{0.212\text{ mol}}{225 \times 10^{-3}\text{ L}} = 0.942\ M\text{ OH}^-$

66.   mol $CaCl_2$ present $= 0.230\text{ L CaCl}_2 \times \dfrac{0.275\text{ mol CaCl}_2}{\text{L CaCl}_2} = 6.33 \times 10^{-2}\text{ mol CaCl}_2$

The volume of $CaCl_2$ solution after evaporation is:

$$6.33 \times 10^{-2} \text{ mol } CaCl_2 \times \frac{1 \text{ L } CaCl_2}{1.10 \text{ mol } CaCl_2} = 5.75 \times 10^{-2} \text{ L} = 57.5 \text{ mL } CaCl_2$$

Volume H_2O evaporated = 230. mL - 57.5 mL = 173 mL H_2O evaporated

67.   For the following answers, the balanced molecular equation is first, followed by the complete ionic equation with the net ionic equation last.

a. $2 AgNO_3(aq) + BaCl_2(aq) \rightarrow 2 AgCl(s) + Ba(NO_3)_2(aq)$

$2 Ag^+(aq) + 2 NO_3^-(aq) + Ba^{2+}(aq) + 2 Cl^-(aq) \rightarrow 2 AgCl(s) + Ba^{2+}(aq) + 2 NO_3^-(aq)$

$Ag^+(aq) + Cl^-(aq) \rightarrow AgCl(s)$

b.   No reaction occurs since all of the possible products (NH_4Cl and KNO_3) are soluble.

c. $(NH_4)_2S(aq) + FeCl_2(aq) \rightarrow FeS(s) + 2 NH_4Cl(aq)$

$2 NH_4^+(aq) + S^{2-}(aq) + Fe^{2+}(aq) + 2 Cl^-(aq) \rightarrow FeS(s) + 2 NH_4^+(aq) + 2 Cl^-(aq)$

$Fe^{2+}(aq) + S^{2-}(aq) \rightarrow FeS(s)$

d. $K_2CO_3(aq) + CuSO_4(aq) \rightarrow CuCO_3(s) + K_2SO_4(aq)$

$2 K^+(aq) + CO_3^{2-}(aq) + Cu^{2+}(aq) + SO_4^{2-}(aq) \rightarrow CuCO_3(s) + 2 K^+(aq) + SO_4^{2-}(aq)$

$Cu^{2+}(aq) + CO_3^{2-}(aq) \rightarrow CuCO_3(s)$

68. $Fe(NO_3)_3(aq) + 3 NaOH(aq) \rightarrow Fe(OH)_3(s) + 3 NaNO_3(aq)$

$$75.0 \times 10^{-3} \text{ L} \times \frac{0.105 \text{ mol } Fe(NO_3)_3}{L} = 7.88 \times 10^{-3} \text{ mol } Fe(NO_3)_3$$

$$125 \times 10^{-3} \text{ L} \times \frac{0.150 \text{ mol } NaOH}{L} = 1.88 \times 10^{-2} \text{ mol } NaOH$$

The balanced reaction requires 3 mol of NaOH for every one mol of $Fe(NO_3)_3$. The actual mol ratio present is:

$$\frac{1.88 \times 10^{-2} \text{ mol } NaOH}{7.88 \times 10^{-3} \text{ mol } Fe(NO_3)_3} = 2.39$$

Not enough NaOH is present to react with all of the $Fe(NO_3)_3$ since the actual mol ratio is less than the required mol ratio. NaOH is the limiting reagent.

$$\text{Mass } Fe(OH)_3 = 1.88 \times 10^{-2} \text{ mol } NaOH \times \frac{1 \text{ mol } Fe(OH)_3}{3 \text{ mol } NaOH} \times \frac{106.87 \text{ g } Fe(OH)_3}{\text{mol } Fe(OH)_3} = 0.670 \text{ g } Fe(OH)_3$$

69. a. No; No element shows a change in oxidation number.

b. $1.0 \text{ L oxalic acid} \times \dfrac{0.14 \text{ mol oxalic acid}}{\text{L}} \times \dfrac{1 \text{ mol Fe}_2\text{O}_3}{6 \text{ mol oxalic acid}} \times \dfrac{159.70 \text{ g Fe}_2\text{O}_3}{\text{mol Fe}_2\text{O}_3} = 3.7 \text{ g Fe}_2\text{O}_3$

70. $M_2\text{SO}_4(aq) + \text{BaCl}_2(aq) \rightarrow \text{BaSO}_4(s) + 2 \text{ MCl}(aq)$

$2.33 \text{ g BaSO}_4 \times \dfrac{1 \text{ mol BaSO}_4}{233.4 \text{ g BaSO}_4} \times \dfrac{1 \text{ mol M}_2\text{SO}_4}{\text{mol BaSO}_4} = 9.98 \times 10^{-3} \text{ mol M}_2\text{SO}_4$

From the problem, 1.42 g $M_2\text{SO}_4$ was reacted so:

$1.42 \text{ g M}_2\text{SO}_4 = 9.98 \times 10^{-3} \text{ mol M}_2\text{SO}_4$, molar mass $= \dfrac{1.42 \text{ g M}_2\text{SO}_4}{9.98 \times 10^{-3} \text{ mol M}_2\text{SO}_4} = 142 \text{ g/mol}$

142 g/mol = 2(atomic mass M) + 32.07 + 4(16.00), atomic mass M = 23. g/mol

From periodic table, M = Na(sodium).

71. Use the silver nitrate data to calculate the mol Cl⁻ present, then use the formula of douglasite to convert from Cl⁻ to douglasite. The net ionic reaction is: $\text{Ag}^+ + \text{Cl}^- \rightarrow \text{AgCl}(s)$.

$0.03720 \text{ L} \times \dfrac{0.1000 \text{ mol Ag}^+}{\text{L}} \times \dfrac{1 \text{ mol Cl}^-}{\text{mol Ag}^+} \times \dfrac{1 \text{ mol douglasite}}{4 \text{ mol Cl}^-}$

$\times \dfrac{311.88 \text{ g douglasite}}{\text{mol}} = 0.2900 \text{ g douglasite}$

Mass % douglasite $= \dfrac{0.2900 \text{ g}}{0.4550 \text{ g}} \times 100 = 63.74\%$

72. All the sulfur in BaSO_4 came from the saccharin. The conversion from BaSO_4 to saccharin utilizes the molar masses of each.

$0.5032 \text{ g BaSO}_4 \times \dfrac{32.07 \text{ g S}}{233.4 \text{ g BaSO}_4} \times \dfrac{183.19 \text{ g saccharin}}{32.07 \text{ g S}} = 0.3949 \text{ g saccharin}$

$\dfrac{\text{Avg. mass}}{\text{Tablet}} = \dfrac{0.3949 \text{ g}}{10 \text{ tablets}} = \dfrac{3.949 \times 10^{-2} \text{ g}}{\text{tablet}} = \dfrac{39.49 \text{ mg}}{\text{tablet}}$

Avg. mass % $= \dfrac{0.3949 \text{ g saccharin}}{0.5894 \text{ g}} \times 100 = 67.00\%$ saccharin by mass

73. $0.104 \text{ g AgCl} \times \dfrac{35.45 \text{ g Cl}^-}{143.4 \text{ g AgCl}} = 2.57 \times 10^{-2} \text{ g Cl}^- = \text{Cl}^-$ in chlorisondiamine

Molar mass of chlorisondiamine = 14(12.01) + 18(1.008) + 6(35.45) + 2(14.01) = 427.00 g/mol

There are 6(35.45) = 212.70 g chlorine for every mole (427.00 g) of chlorisondiamine.

$$2.57 \times 10^{-2} \text{ g Cl}^- \times \frac{427.00 \text{ g drug}}{212.70 \text{ g Cl}^-} = 5.16 \times 10^{-2} \text{ g drug}; \quad \% \text{ drug} = \frac{5.16 \times 10^{-2} \text{ g}}{1.28 \text{ g}} \times 100 = 4.03\%$$

74. $Zn_2P_2O_7$: $2(65.38) + 2(30.97) + 7(16.00) = 304.69/\text{mol}$

All the zinc in $Zn_2P_2O_7$ came from zinc in the foot powder.

$$0.4089 \text{ g Zn}_2\text{P}_2\text{O}_7 \times \frac{130.76 \text{ g Zn}}{304.69 \text{ g Zn}_2\text{P}_2\text{O}_7} = 0.1755 \text{ g Zn}$$

$$\% \text{ Zn} = \frac{0.1755 \text{ g}}{1.200 \text{ g}} \times 100 = 14.63\% \text{ Zn}$$

75. $HC_2H_3O_2(aq) + NaOH(aq) \rightarrow H_2O(l) + NaC_2H_3O_2(aq)$

a. $16.58 \times 10^{-3} \text{ L soln} \times \dfrac{0.5062 \text{ mol NaOH}}{\text{L soln}} \times \dfrac{1 \text{ mol acetic acid}}{\text{mol NaOH}} = 8.393 \times 10^{-3} \text{ mol acetic acid}$

Concentration of acetic acid $= \dfrac{8.393 \times 10^{-3} \text{ mol}}{0.01000 \text{ L}} = 0.8393 \, M$

b.   If we have 1.000 L of solution: total mass $= 1000. \text{ mL} \times \dfrac{1.006 \text{ g}}{\text{mL}} = 1006 \text{ g}$

Mass of $HC_2H_3O_2 = 0.8393 \text{ mol} \times \dfrac{60.05 \text{ g}}{\text{mol}} = 50.40 \text{ g}$

Mass % acetic acid $= \dfrac{50.40 \text{ g}}{1006 \text{ g}} \times 100 = 5.010\%$

76.   Since KHP is a monprotic acid, the reaction is: $NaOH + KHP \rightarrow NaKP + H_2O$ where KHP is shorthand for potassium hydrogen phthalate.

$$0.1082 \text{ g KHP} \times \frac{1 \text{ mol KHP}}{204.22 \text{ g KHP}} \times \frac{1 \text{ mol NaOH}}{\text{mol KHP}} = 5.298 \times 10^{-4} \text{ mol NaOH}$$

There is 5.298×10^{-4} mol of sodium hydroxide in 20.46 mL of solution. Therefore, the concentration of sodium hydroxide is:

$$\frac{5.298 \times 10^{-4} \text{ mol}}{20.46 \times 10^{-3} \text{ L}} = 2.589 \times 10^{-2} \, M \text{ NaOH}$$

77.   Using HA as an abbreviation for acetylsalicylic acid:

$$HA(aq) + NaOH(aq) \rightarrow H_2O(l) + NaA(aq)$$

mol HA present $= 0.03517 \text{ L NaOH} \times \dfrac{0.5065 \text{ mol NaOH}}{\text{L NaOH}} \times \dfrac{1 \text{ mol HA}}{\text{mol NaOH}} = 1.781 \times 10^{-2} \text{ mol HA}$

From the problem, 3.210 g HA was reacted so:

$$3.210 \text{ g HA} = 1.781 \times 10^{-2} \text{ mol HA, molar mass} = \frac{3.210 \text{ g HA}}{1.781 \times 10^{-2} \text{ mol HA}} = 180.2 \text{ g/mol}$$

78.   a. $Al(s) + 3 \text{ HCl}(aq) \rightarrow AlCl_3(aq) + 3/2 \text{ H}_2(g)$ or $2 Al(s) + 6 \text{ HCl}(aq) \rightarrow 2 AlCl_3(aq) + 3 \text{ H}_2(g)$

Hydrogen is reduced (goes from +1 oxidation state to 0 oxidation state) and aluminum Al is oxidized ($0 \rightarrow +3$).

b.   Balancing S is most complicated since sulfur is in both products. Balance C and H first then worry about S.

$$CH_4(g) + 4 \text{ S}(s) \rightarrow CS_2(l) + 2 \text{ H}_2S(g)$$

Sulfur is reduced ($0 \rightarrow -2$) and carbon is oxidized ($-4 \rightarrow +4$).

c.   Balance C and H first then balance O.

$$C_3H_8(g) + 5 \text{ O}_2(g) \rightarrow 3 \text{ CO}_2(g) + 4 \text{ H}_2O(l)$$

Oxygen is reduced ($0 \rightarrow -2$) and carbon is oxidized ($-8/3 \rightarrow +4$).

d.   Although this reaction is mass balanced, it is not charge balanced. We need 2 mol of silver on each side to balance the charge.

$$Cu(s) + 2 \text{ Ag}^+(aq) \rightarrow 2 \text{ Ag}(s) + Cu^{2+}(aq)$$

Silver is reduced ($+1 \rightarrow 0$) and copper is oxidized ($0 \rightarrow +2$).

79.   a. $4 \text{ NH}_3(g) + 5 \text{ O}_2(g) \rightarrow 4 \text{ NO}(g) + 6 \text{ H}_2O(g)$
          -3 +1   0   +2 -2   +1 -2 oxidation numbers

$2 \text{ NO}(g) + O_2(g) \rightarrow 2 \text{ NO}_2(g)$
   +2 -2   0   +4 -2

$3 \text{ NO}_2(g) + H_2O(l) \rightarrow 2 \text{ HNO}_3(aq) + NO(g)$
  +4 -2   +1 -2   +1 +5 -2   +2 -2

All three reactions are oxidation-reduction reactions since there is a change in oxidation numbers of some of the elements in each reaction.

b. $4 \text{ NH}_3 + 5 \text{ O}_2 \rightarrow 4 \text{ NO} + 6 \text{ H}_2O$; O_2 is the oxidizing agent and NH_3 is the reducing agent.

$2 \text{ NO} + O_2 \rightarrow 2 \text{ NO}_2$; O_2 is the oxidizing agent and NO is the reducing agent.

$3 \text{ NO}_2 + H_2O \rightarrow 2 \text{ HNO}_3 + NO$; NO_2 is both the oxidizing and reducing agent.

80. $Mn + HNO_3 \rightarrow Mn^{2+} + NO_2$

$Mn \rightarrow Mn^{2+} + 2\ e^-$ $\qquad\qquad\qquad\qquad HNO_3 \rightarrow NO_2$
$\qquad\qquad\qquad\qquad\qquad\qquad\qquad\qquad\qquad HNO_3 \rightarrow NO_2 + H_2O$
$\qquad\qquad\qquad\qquad\qquad\qquad (e^- + H^+ + HNO_3 \rightarrow NO_2 + H_2O) \times 2$

$$Mn \rightarrow Mn^{2+} + 2\ e^-$$
$$2\ e^- + 2\ H^+ + 2\ HNO_3 \rightarrow 2\ NO_2 + 2\ H_2O$$

$$\overline{2\ H^+(aq) + Mn(s) + 2\ HNO_3(aq) \rightarrow Mn^{2+}(aq) + 2\ NO_2(g) + 2\ H_2O(l)}$$

$Mn^{2+} + IO_4^- \rightarrow MnO_4^- + IO_3^-$

$(4\ H_2O + Mn^{2+} \rightarrow MnO_4^- + 8\ H^+ + 5\ e^-) \times 2$ $\qquad\qquad (2\ e^- + 2\ H^+ + IO_4^- \rightarrow IO_3^- + H_2O) \times 5$

$$8\ H_2O + 2\ Mn^{2+} \rightarrow 2\ MnO_4^- + 16\ H^+ + 10\ e^-$$
$$10\ e^- + 10\ H^+ + 5\ IO_4^- \rightarrow 5\ IO_3^- + 5\ H_2O$$

$$\overline{3\ H_2O(l) + 2\ Mn^{2+}(aq) + 5\ IO_4^-(aq) \rightarrow 2\ MnO_4^-(aq) + 5\ IO_3^-(aq) + 6\ H^+(aq)}$$

81. Fe^{2+} will react with MnO_4^- (purple) producing Fe^{3+} and Mn^{2+}(almost colorless). There is no reaction between MnO_4^- and Fe^{3+}. Therefore, add a few drops of the potassium permanganate solution. If the purple color persists, the solution contains iron(III) sulfate. If the color disappears, iron(II) sulfate is present.

Challenge Problems

82.   a. 5.0 ppb Hg in water $= \dfrac{5.0\ \text{ng Hg}}{\text{g soln}} = \dfrac{5.0 \times 10^{-9}\ \text{g Hg}}{\text{mL soln}}$

$\dfrac{5.0 \times 10^{-9}\ \text{g Hg}}{\text{mL soln}} \times \dfrac{1\ \text{mol Hg}}{200.6\ \text{g Hg}} \times \dfrac{1000\ \text{mL}}{\text{L}} = 2.5 \times 10^{-8}\ M\ \text{Hg}$

b. $\dfrac{1.0 \times 10^{-9}\ \text{g CHCl}_3}{\text{mL}} \times \dfrac{1\ \text{mol CHCl}_3}{119.37\ \text{g CHCl}_3} \times \dfrac{1000\ \text{mL}}{\text{L}} = 8.4 \times 10^{-9}\ M\ \text{CHCl}_3$

c. 10.0 ppm As $= \dfrac{10.0\ \mu\text{g As}}{\text{g soln}} = \dfrac{10.0 \times 10^{-6}\ \text{g As}}{\text{mL soln}}$

$\dfrac{10.0 \times 10^{-6}\ \text{g As}}{\text{mL soln}} \times \dfrac{1\ \text{mol As}}{74.92\ \text{g As}} \times \dfrac{1000\ \text{mL}}{\text{L}} = 1.33 \times 10^{-4}\ M\ \text{As}$

d. $\dfrac{0.10 \times 10^{-6}\ \text{g DDT}}{\text{mL}} \times \dfrac{1\ \text{mol DDT}}{354.46\ \text{g DDT}} \times \dfrac{1000\ \text{mL}}{\text{L}} = 2.8 \times 10^{-7}\ M\ \text{DDT}$

83.   a. 0.308 g AgCl $\times \dfrac{35.45\ \text{g Cl}}{143.4\ \text{g AgCl}} = 0.0761$ g Cl; $\%\text{Cl} = \dfrac{0.0761\ \text{g}}{0.256\ \text{g}} \times 100 = 29.7\%\ \text{Cl}$

Cobalt(III) oxide, Co_2O_3: $2(58.93) + 3(16.00) = 165.86$ g/mol

$$0.145 \text{ g } Co_2O_3 \times \frac{117.86 \text{ g Co}}{165.86 \text{ g } Co_2O_3} = 0.103 \text{ g Co}; \quad \%Co = \frac{0.103 \text{ g}}{0.416 \text{ g}} \times 100 = 24.8\% \text{ Co}$$

The remainder, $100.0 - (29.7 + 24.8) = 45.5\%$, is water. Assuming 100.0 g of compound:

$$45.5 \text{ g } H_2O \times \frac{2.016 \text{ g H}}{18.02 \text{ g } H_2O} = 5.09 \text{ g H}; \quad \%H = \frac{5.09 \text{ g H}}{100.0 \text{ g compound}} \times 100 = 5.09\% \text{ H}$$

$$45.5 \text{ g } H_2O \times \frac{16.00 \text{ g O}}{18.02 \text{ g } H_2O} = 40.4 \text{ g O}; \quad \%O = \frac{40.4 \text{ g O}}{100.0 \text{ g compound}} \times 100 = 40.4\% \text{ O}$$

The mass percent composition is 24.8% Co, 29.7% Cl, 5.09% H and 40.4% O.

b.   Out of 100.0 g of compound, there are:

$$24.8 \text{ g Co} \times \frac{1 \text{ mol}}{58.93 \text{ g Co}} = 0.421 \text{ mol Co}; \quad 29.7 \text{ g Cl} \times \frac{1 \text{ mol}}{35.45 \text{ g Cl}} = 0.838 \text{ mol Cl}$$

$$5.09 \text{ g H} \times \frac{1 \text{ mol}}{1.008 \text{ g H}} = 5.05 \text{ mol H}; \quad 40.4 \text{ g O} \times \frac{1 \text{ mol}}{16.00 \text{ g O}} = 2.53 \text{ mol O}$$

Dividing all results by 0.421, we get $CoCl_2 \cdot 6H_2O$.

c. $CoCl_2 \cdot 6H_2O(aq) + 2 \text{ } AgNO_3(aq) \rightarrow 2 \text{ } AgCl(s) + Co(NO_3)_2(aq) + 6 \text{ } H_2O(l)$

$CoCl_2 \cdot 6H_2O(aq) + 2 \text{ } NaOH(aq) \rightarrow Co(OH)_2(s) + 2 \text{ } NaCl(aq) + 6 \text{ } H_2O(l)$

$Co(OH)_2 \rightarrow Co_2O_3$ This is an oxidation-reduction reaction. Thus, we also need to include an oxidizing agent. The obvious choice is O_2.

$4 \text{ } Co(OH)_2(s) + O_2(g) \rightarrow 2 \text{ } Co_2O_3(s) + 4 \text{ } H_2O(l)$

84.   Molar masses: KCl, $39.10 + 35.45 = 74.55$ g/mol; KBr, $39.10 + 79.90 = 119.00$ g/mol

AgCl, $107.9 + 35.45 = 143.4$ g/mol; AgBr, $107.9 + 79.90 = 187.8$ g/mol

Let x = number of moles of KCl in mixture and y = number of moles of KBr in mixture.
Since $Ag^+ + Cl^- \rightarrow AgCl$ and $Ag^+ + Br^- \rightarrow AgBr$, then x = moles AgCl and y = moles AgBr.

Setting up two equations:

$0.1024 \text{ g} = 74.55 \text{ x} + 119.0 \text{ y}$ and $0.1889 \text{ g} = 143.4 \text{ x} + 187.8 \text{ y}$

Multiply the first equation by $\frac{187.8}{119.0}$, and subtract from the second.

$$
\begin{array}{rl}
0.1889 = & 143.4 \text{ x} + 187.8 \text{ y} \\
-0.1616 = & -117.7 \text{ x} - 187.8 \text{ y} \\
\hline
0.0273 = & 25.7 \text{ x}, \qquad x = 1.06 \times 10^{-3} \text{ mol KCl}
\end{array}
$$

$$1.06 \times 10^{-3} \text{ mol KCl} \times \frac{74.55 \text{ g KCl}}{\text{mol KCl}} = 0.0790 \text{ g KCl}$$

$$\% \text{ KCl} = \frac{0.0790 \text{ g}}{0.1024 \text{ g}} \times 100 = 77.1\%, \quad \% \text{ KBr} = 100.0 - 77.1 = 22.9\%$$

85. $0.298 \text{ g BaSO}_4 \times \dfrac{96.07 \text{ g SO}_4^{2-}}{233.4 \text{ g BaSO}_4} = 0.123 \text{ g SO}_4^{2-}; \quad \% \text{ sulfate} = \dfrac{0.123 \text{ g SO}_4^{2-}}{0.205 \text{ g}} = 60.0\%$

Assume we have 100.0 g of the mixture of Na_2SO_4 and K_2SO_4. There is:

$$60.0 \text{ g SO}_4^{2-} \times \frac{1 \text{ mol}}{96.07 \text{ g}} = 0.625 \text{ mol SO}_4^{2-}$$

There must be $2 \times 0.625 = 1.25$ mol of +1 cations to balance the 2- charge of SO_4^{2-}.

Let x = number of moles of K^+ and y = number of moles of Na^+, then $x + y = 1.25$.

The total mass of Na^+ and K^+ must be 40.0 g in the assumed 100.0 g of mixture. Setting up an equation:

$$x \text{ mol K}^+ \times \frac{39.10 \text{ g}}{\text{mol}} + y \text{ mol Na}^+ \times \frac{22.99 \text{ g}}{\text{mol}} = 40.0 \text{ g}$$

So, we have two equations with two unknowns: $x + y = 1.25$ and $39.10 x + 22.99 y = 40.0$

Since $x = 1.25 - y$, then $39.10(1.25 - y) + 22.99 y = 40.0$

$48.9 - 39.10 y + 22.99 y = 40.0, \quad -16.11 y = -8.9$

$y = 0.55 \text{ mol Na}^+$ and $x = 1.25 - 0.55 = 0.70 \text{ mol K}^+$

Therefore:

$$0.70 \text{ mol K}^+ \times \frac{1 \text{ mol K}_2\text{SO}_4}{2 \text{ mol K}^+} = 0.35 \text{ mol K}_2\text{SO}_4; \quad 0.35 \text{ mol K}_2\text{SO}_4 \times \frac{174.27 \text{ g}}{\text{mol}} = 61 \text{ g K}_2\text{SO}_4$$

Since we assumed 100.0 g, then the mixture is 61% K_2SO_4 and 39% Na_2SO_4.

86.   a. $H_3PO_4(aq) + 3 \text{ NaOH}(aq) \rightarrow 3 H_2O(l) + Na_3PO_4(aq)$

      b. $2 \text{ Al(OH)}_3(s) + 3 H_2SO_4(aq) \rightarrow 6 H_2O(l) + Al_2(SO_4)_3(aq)$

      c. $H_2Se(aq) + Ba(OH)_2(s) \rightarrow 2 H_2O(l) + BaSe(s)$

      d. $H_2C_2O_4(aq) + 2 \text{ NaOH}(aq) \rightarrow 2 H_2O(l) + Na_2C_2O_4(aq)$

87. $CaCO_3(s) + H_2SO_4(aq) \rightarrow CaSO_4(aq) + H_2O(l) + CO_2(g)$

88. $35.08 \text{ mL NaOH} \times \dfrac{1 \text{ L}}{1000 \text{ mL}} \times \dfrac{2.12 \text{ mol NaOH}}{\text{L NaOH}} \times \dfrac{1 \text{ mol H}_2\text{SO}_4}{2 \text{ mol NaOH}} = 3.72 \times 10^{-2} \text{ mol H}_2\text{SO}_4$

       $\text{Molarity} = \dfrac{3.72 \times 10^{-2} \text{ mol}}{10.00 \text{ mL}} \times \dfrac{1000 \text{ mol}}{\text{L}} = 3.72 \, M \, \text{H}_2\text{SO}_4$

89.   a. $\text{MgO(s)} + 2 \text{ HCl(aq)} \rightarrow \text{MgCl}_2\text{(aq)} + \text{H}_2\text{O(l)}$

             $\text{Mg(OH)}_2\text{(s)} + 2 \text{ HCl(aq)} \rightarrow \text{MgCl}_2\text{(aq)} + 2 \text{ H}_2\text{O(l)}$

             $\text{Al(OH)}_3\text{(s)} + 3 \text{ HCl(aq)} \rightarrow \text{AlCl}_3\text{(aq)} + 3 \text{ H}_2\text{O(l)}$

       b.   Let's calculate the number of moles of HCl neutralized per gram of substance. We can get these directly from the balanced equations and the molar masses of the substances.

             $\dfrac{2 \text{ mol HCl}}{\text{mol MgO}} \times \dfrac{1 \text{ mol MgO}}{40.31 \text{ g MgO}} = \dfrac{4.962 \times 10^{-2} \text{ mol HCl}}{\text{g MgO}}$

             $\dfrac{2 \text{ mol HCl}}{\text{mol Mg(OH)}_2} \times \dfrac{1 \text{ mol Mg(OH)}_2}{58.33 \text{ g Mg(OH)}_2} = \dfrac{3.429 \times 10^{-2} \text{ mol HCl}}{\text{g Mg(OH)}_2}$

             $\dfrac{3 \text{ mol HCl}}{\text{mol Al(OH)}_3} \times \dfrac{1 \text{ mol Al(OH)}_3}{78.00 \text{ g Al(OH)}_3} = \dfrac{3.846 \times 10^{-2} \text{ mol HCl}}{\text{g Al(OH)}_3}$

             Therefore, one gram of magnesium oxide would neutralize the most $0.10 \, M$ HCl.

90.   We get the empirical formula from the elemental analysis. Out of 100.00 g carminic acid there are:

       $53.66 \text{ g C} \times \dfrac{1 \text{ mol C}}{12.01 \text{ g C}} = 4.468 \text{ mol C}; \quad 4.09 \text{ g H} \times \dfrac{1 \text{ mol H}}{1.008 \text{ g H}} = 4.06 \text{ mol H}$

       $42.25 \text{ g O} \times \dfrac{1 \text{ mol O}}{16.00 \text{ g O}} = 2.641 \text{ mol O}$

       Dividing the moles by the smallest number gives:

       $\dfrac{4.468}{2.641} = 1.692; \quad \dfrac{4.06}{2.641} = 1.54$

       These numbers don't give obvious mol ratios. Lets determine the mol C to mol H ratio.

       $\dfrac{4.468}{4.06} = 1.10 = \dfrac{11}{10}$

       So let's try $\dfrac{4.06}{10} = 0.406$ as a common factor: $\quad \dfrac{4.468}{4.06} = 11.0; \quad \dfrac{4.06}{0.406} = 10.0; \quad \dfrac{2.641}{0.406} = 6.50$

       Therefore, $\text{C}_{22}\text{H}_{20}\text{O}_{13}$ is the empirical formula.

We can get molar mass from the titration data.

$$18.02 \times 10^{-3} \text{ L soln} \times \frac{0.0406 \text{ mol NaOH}}{\text{L soln}} \times \frac{1 \text{ mol carminic acid}}{\text{mol NaOH}} = 7.32 \times 10^{-4} \text{ mol carminic acid}$$

$$\text{Molar mass} = \frac{0.3602 \text{ g}}{7.32 \times 10^{-4} \text{ mol}} = \frac{492 \text{ g}}{\text{mol}}$$

The empirical formula mass of $C_{22}H_{20}O_{13} \approx 22(12) + 20(1) + 13(16) = 492$ g.

Therefore, the molecular formula of carminic acid is also $C_{22}H_{20}O_{13}$.

91. $H_2SO_4(aq) + 2 \text{ NaOH}(aq) \rightarrow Na_2SO_4(aq) + 2 H_2O(l)$

$$0.02844 \text{ L} \times \frac{0.1000 \text{ mol NaOH}}{\text{L}} \times \frac{1 \text{ mol } H_2SO_4}{2 \text{ mol NaOH}} \times \frac{1 \text{ mol } SO_2}{\text{mol } H_2SO_4} \times \frac{32.07 \text{ g S}}{\text{mol } SO_2} = 4.560 \times 10^{-2} \text{ g S}$$

$$\%S = \frac{0.04560 \text{ g}}{1.325 \text{ g}} \times 100 = 3.442\% \text{ by mass}$$

92.   a. $HCl(aq)$ dissociates to $H^+(aq) + Cl^-(aq)$. For simplicity let's use H^+ and Cl^- separately.

$$H^+ \rightarrow H_2 \qquad\qquad\qquad\qquad Fe \rightarrow HFeCl_4$$
$$(2 H^+ + 2 e^- \rightarrow H_2) \times 3 \qquad\qquad (H^+ + 4 Cl^- + Fe \rightarrow HFeCl_4 + 3 e^-) \times 2$$

$$6 H^+ + 6 e^- \rightarrow 3 H_2$$
$$2 H^+ + 8 Cl^- + 2 Fe \rightarrow 2 HFeCl_4 + 6 e^-$$
$$\overline{}$$
$$8 H^+ + 8 Cl^- + 2 Fe \rightarrow 2 HFeCl_4 + 3 H_2$$

or $8 HCl(aq) + 2 Fe(s) \rightarrow 2 HFeCl_4(aq) + 3 H_2(g)$

b.
$$IO_3^- \rightarrow I_3^- \qquad\qquad\qquad\qquad I^- \rightarrow I_3^-$$
$$3 IO_3^- \rightarrow I_3^- \qquad\qquad\qquad\qquad (3 I^- \rightarrow I_3^- + 2 e^-) \times 8$$
$$3 IO_3^- \rightarrow I_3^- + 9 H_2O$$
$$16 e^- + 18 H^+ + 3 IO_3^- \rightarrow I_3^- + 9 H_2O$$

$$16 e^- + 18 H^+ + 3 IO_3^- \rightarrow I_3^- + 9 H_2O$$
$$24 I^- \rightarrow 8 I_3^- + 16 e^-$$
$$\overline{}$$
$$18 H^+ + 24 I^- + 3 IO_3^- \rightarrow 9 I_3^- + 9 H_2O$$

Reducing: $6 H^+(aq) + 8 I^-(aq) + IO_3^-(aq) \rightarrow 3 I_3^-(aq) + 3 H_2O(l)$

c. $(Ce^{4+} + e^- \rightarrow Ce^{3+}) \times 97$

$$Cr(NCS)_6^{4-} \rightarrow Cr^{3+} + NO_3^- + CO_2 + SO_4^{2-}$$
$$54 H_2O + Cr(NCS)_6^{4-} \rightarrow Cr^{3+} + 6 NO_3^- + 6 CO_2 + 6 SO_4^{2-} + 108 H^+$$

Charge on left -4. Charge on right = +3 + 6(-1) + 6(-2) + 108(+1) = +93. Add 97 e⁻ to the right, then add the two balanced half-reactions with a common factor of 97 e⁻ transfered.

$$54 \, H_2O + Cr(NCS)_6{}^{4-} \rightarrow Cr^{3+} + 6 \, NO_3{}^- + 6 \, CO_2 + 6 \, SO_4{}^{2-} + 108 \, H^+ + 97 \, e^-$$
$$97 \, e^- + 97 \, Ce^{4+} \rightarrow 97 \, Ce^{3+}$$

$$97 \, Ce^{4+}(aq) + 54 \, H_2O(l) + Cr(NCS)_6{}^{4-}(aq) \rightarrow 97 \, Ce^{3+}(aq) + Cr^{3+}(aq) + 6 \, NO_3{}^-(aq) + 6 \, CO_2(g)$$
$$+ 6 \, SO_4{}^{2-}(aq) + 108 \, H^+(aq)$$

This is very complicated. A check of the net charge is a good check to see if the equation is balanced. Left: charge = 97(+4) - 4 = +384. Right: charge = 97(+3) + 3 + 6(-1) + 6(-2) + 108(+1) = +384.

d.
$$CrI_3 \rightarrow CrO_4{}^{2-} + IO_4{}^-$$
$$(16 \, H_2O + CrI_3 \rightarrow CrO_4{}^{2-} + 3 \, IO_4{}^- + 32 \, H^+ + 27 \, e^-) \times 2$$

$$Cl_2 \rightarrow Cl^-$$
$$(2 \, e^- + Cl_2 \rightarrow 2 \, Cl^-) \times 27$$

Common factor is a transfer of 54 e⁻.

$$54 \, e^- + 27 \, Cl_2 \rightarrow 54 \, Cl^-$$
$$32 \, H_2O + 2 \, CrI_3 \rightarrow 2 \, CrO_4{}^{2-} + 6 \, IO_4{}^- + 64 \, H^+ + 54 \, e^-$$

$$32 \, H_2O + 2 \, CrI_3 + 27 \, Cl_2 \rightarrow 54 \, Cl^- + 2 \, CrO_4{}^{2-} + 6 \, IO_4{}^- + 64 \, H^+$$

Add 64 OH⁻ to both sides and convert 64 H⁺ into 64 H₂O.

$$64 \, OH^- + 32 \, H_2O + 2 \, CrI_3 + 27 \, Cl_2 \rightarrow 54 \, Cl^- + 2 \, CrO_4{}^{2-} + 6 \, IO_4{}^- + 64 \, H_2O$$

Reducing gives:

$$64 \, OH^-(aq) + 2 \, CrI_3(s) + 27 \, Cl_2(g) \rightarrow 54 \, Cl^-(aq) + 2 \, CrO_4{}^{2-}(aq) + 6 \, IO_4{}^-(aq) + 32 \, H_2O(l)$$

e.
$$Ce^{4+} \rightarrow Ce(OH)_3$$
$$(e^- + 3 \, H_2O + Ce^{4+} \rightarrow Ce(OH)_3 + 3 \, H^+) \times 61$$

$$Fe(CN)_6{}^{4-} \rightarrow Fe(OH)_3 + CO_3{}^{2-} + NO_3{}^-$$
$$Fe(CN)_6{}^{4-} \rightarrow Fe(OH)_3 + 6 \, CO_3{}^{2-} + 6 \, NO_3{}^-$$

There are 39 extra O atoms on right. Add 39 H₂O to left, then add 75 H⁺ to right to balance H⁺.

$$39 \, H_2O + Fe(CN)_6{}^{4-} \rightarrow Fe(OH)_3 + 6 \, CO_3{}^{2-} + 6 \, NO_3{}^- + 75 \, H^+$$
$$\text{net charge} = -4 \qquad\qquad \text{net charge} = +57$$

Add 61 e⁻ to the right then add the two balanced half-reactions with a common factor of 61 e⁻ transferred.

$$39\ H_2O + Fe(CN)_6{}^{4-} \rightarrow Fe(OH)_3 + 6\ CO_3{}^{2-} + 6\ NO_3{}^- + 75\ H^+ + 61\ e^-$$
$$61\ e^- + 183\ H_2O + 61\ Ce^{4+} \rightarrow 61\ Ce(OH)_3 + 183\ H^+$$

$$\overline{222\ H_2O + Fe(CN)_6{}^{4-} + 61\ Ce^{4+} \rightarrow 61\ Ce(OH)_3 + Fe(OH)_3 + 6\ CO_3{}^{2-} + 6\ NO_3{}^- + 258\ H^+}$$

Adding 258 OH$^-$ to each side then reducing gives:

$$258\ OH^-(aq) + Fe(CN)_6{}^{4-}(aq) + 61\ Ce^{4+}(aq) \rightarrow 61\ Ce(OH)_3(s) + Fe(OH)_3(s)$$

$$+\ 6\ CO_3{}^{2-}(aq) + 6\ NO_3{}^-(aq) + 36\ H_2O(l)$$

f. $\qquad Fe(OH)_2 \rightarrow Fe(OH)_3 \qquad\qquad\qquad\qquad\qquad H_2O_2 \rightarrow H_2O$
$\quad (H_2O + Fe(OH)_2 \rightarrow Fe(OH)_3 + H^+ + e^-) \times 2 \qquad\qquad 2\ e^- + 2\ H^+ + H_2O_2 \rightarrow 2\ H_2O$

$$2\ H_2O + 2\ Fe(OH)_2 \rightarrow 2\ Fe(OH)_3 + 2\ H^+ + 2\ e^-$$
$$2\ e^- + 2\ H^+ + H_2O_2 \rightarrow 2\ H_2O$$

$$\overline{2\ H_2O + 2\ H^+ + 2\ Fe(OH)_2 + H_2O_2 \rightarrow 2\ Fe(OH)_3 + 2\ H_2O + 2\ H^+}$$

Reducing gives: $2\ Fe(OH)_2(s) + H_2O_2(aq) \rightarrow 2\ Fe(OH)_3(s)$

93. First we will calculate the molarity of NaCl while ignoring the uncertainty.

$$0.150\ g \times \frac{1\ mol}{58.44\ g} = 2.57 \times 10^{-3}\ moles;\ \ Molarity = \frac{2.57 \times 10^{-3}\ mol}{0.1000\ L} = \frac{2.57 \times 10^{-2}\ mol}{L}$$

The maximum value for the molarity is $= \dfrac{0.153\ g \times \dfrac{1\ mol}{58.44\ g}}{0.0995\ L} = \dfrac{2.63 \times 10^{-2}\ mol}{L}$

The minimum value for the molarity is $= \dfrac{0.147\ g \times \dfrac{1\ mol}{58.44\ g}}{0.1005\ L} = \dfrac{2.50 \times 10^{-2}\ mol}{L}$

The range of the NaCl molarity is 0.0250 M to 0.0263 M or we can express this range as 0.0257 $\pm$ 0.0007 M.

94. The amount of KHP used $= 0.4016\ g \times \dfrac{1\ mol}{204.22\ g} = 1.967 \times 10^{-3}$ mol KHP

Since one mole of NaOH reacts completely with one mole of KHP, then the NaOH solution contains 1.967×10^{-3} mol NaOH.

Molarity of NaOH $= \dfrac{1.967 \times 10^{-3}\ mol}{25.06 \times 10^{-3}\ L} = \dfrac{7.849 \times 10^{-2}\ mol}{L}$

Maximum molarity $= \dfrac{1.967 \times 10^{-3}\ mol}{25.01 \times 10^{-3}\ L} = \dfrac{7.865 \times 10^{-2}\ mol}{L}$

Minimum molarity $= \dfrac{1.967 \times 10^{-3}\ mol}{25.11 \times 10^{-3}\ L} = \dfrac{7.834 \times 10^{-2}\ mol}{L}$

We can express this as $0.07849 \pm 0.00016 \, M$. An alternate is to express the molarity as $0.0785 \pm 0.0002 \, M$. This second way shows the actual number of significant figures in the molarity. The advantage of the first method is that it shows that we made all of our individual measurements to four significant figures.

95.   Desired uncertainty is 1% of 0.02 or ± 0.0002. So we want the solution to be $0.0200 \pm 0.0002 \, M$ or the concentration should be between 0.0198 and 0.0202 M. We should use a 1-L volumetric flask to make the solution. They are good to $\pm 0.1\%$. We want to weigh out between 0.0198 mol and 0.0202 mol of KIO_3.

Molar mass of $KIO_3 = 39.10 + 126.9 + 3(16.00) = 214.0$ g/mol

$$0.0198 \text{ mol} \times \frac{214.0 \text{ g}}{\text{mol}} = 4.24 \text{ g}; \ \ 0.0202 \text{ mol} \times \frac{214.0 \text{ g}}{\text{mol}} = 4.32 \text{ g}$$

We should weigh out between 4.24 and 4.32 g of KIO_3. We should weigh it to the nearest mg or 0.1 mg. Dissolve the KIO_3 in water and dilute to the mark in a one liter volumetric flask. This will produce a solution whose concentration is within the limits and is known to at least the fourth decimal place.

CHAPTER FIVE

GASES

Questions

14.   a.   Heating the can will increase the pressure of the gas inside the can, $P \propto T$, V and n constant. As the pressure increases it may be enough to rupture the can.

     b.   As you draw a vacuum in your mouth, atmopsheric pressure pushes the liquid up the straw.

     c.   The external atmospheric pressure pushes on the can. Since there is no opposing pressure from the air in the inside, the can collapses.

     d.   How "hard" the tennis ball is depends on the difference between the pressure of the air inside the tennis ball and atmospheric pressure. A "sea level" ball will be much "harder" at high altitude since the external pressure is lower at high altitude. A high altitude ball will be "soft" at sea level.

15. $PV = nRT$ = constant at constant n and T. At two sets of conditions, P_1V_1 = constant = P_2V_2.

     $P_1V_1 = P_2V_2$ (Boyle's law).

     $\dfrac{V}{T} = \dfrac{nR}{P}$ = constant at constant n and P. At two sets of conditions, $\dfrac{V_1}{T_1}$ = constant = $\dfrac{V_2}{T_2}$.

     $\dfrac{V_1}{T_1} = \dfrac{V_2}{T_2}$ (Charles's law)

16.   Boyle's law: $P \propto 1/V$ at constant n and T

     In the kinetic molecular theory (kmt), P is proportional to the collision frequency which is proportional to 1/V. As the volume increases there will be fewer collisions with the walls of the container and pressure decreases (Boyle's Law).

     Charles's law: $V \propto T$ at constant n and P

     Pressure is proportional to collision frequency. If pressure is constant, then the collision frequency of the gas molecules with the walls of the container is constant. Volume is inversely proportional to collision frequency and temperature is directly proportional. If the temperature increases, to keep the pressure (and collision frequency) constant, the volume of the container must increase. Therefore, volume and temperature are directly related at constant n and P (Charles's Law).

17.   For an ideal gas at constant n and T, the PV product should equal a constant value no matter what pressure or volume combination is used. The dashed line in Figure 5.6 is the PV vs P plot for an ideal gas. The real gas closest to the ideal plot is Ne, although O_2 is also fairly close to the ideal plot.

18.   Molecules in the condensed phases (liquids and solids) are very close together. Molecules in the gaseous phase are very far apart. A sample of gas is mostly empty space. Therefore, one would expect 1 mol of $H_2O(g)$ to occupy a huge volume as compared to 1 mol of $H_2O(l)$.

19.   Method 1: molar mass = $\dfrac{dRT}{P}$

Determine the density of a gas at a measurable temperature and pressure then use the above equation to determine the molar mass.

Method 2: $\dfrac{\text{effusion rate for gas 1}}{\text{effusion rate for gas 2}} = \sqrt{\dfrac{\text{(molar mass)}_1}{\text{(molar mass)}_2}}$

Determine the relative effusion rate of the unknown gas to some known gas, then use Graham's law of effusion (the above equation) to determine the molar mass.

20.   a. $(KE)_{avg} = (3/2)RT$; As temperature increases, the average kinetic energy will increase.

        b. $KE = 1/2 \times \text{mass} \times (\text{velocity})^2$; As temperature increases, the average kinetic energy increases which occurs because the average velocity of the gas molecules increase. As T increases, the average velocity of the gas molecules increases.

        c.   For two different gases at the same temperature, the average kinetic energy of each gas will be equal since temperature is constant. From the equation for kinetic energy in b, the gas sample with the lightest molecules must have the larger average velocity to keep the kinetic energy constant. Therefore, at constant temperature, the smaller the gas molecules the faster the average velocity.

Exercises

Pressure

21.   a. $4.8 \text{ atm} \times \dfrac{760 \text{ mm Hg}}{\text{atm}} = 3.6 \times 10^3 \text{ mm Hg}$; b. $3.6 \times 10^3 \text{ mm Hg} \times \dfrac{1 \text{ torr}}{\text{mm Hg}} = 3.6 \times 10^3 \text{ torr}$

        c. $4.8 \text{ atm} \times \dfrac{1.013 \times 10^5 \text{ Pa}}{\text{atm}} = 4.9 \times 10^5 \text{ Pa}$;   d. $4.8 \text{ atm} \times \dfrac{14.7 \text{ psi}}{\text{atm}} = 71 \text{ psi}$

22.   a. $2200 \text{ psi} \times \dfrac{1 \text{ atm}}{14.7 \text{ psi}} = 150 \text{ atm}$;   b. $150 \text{ atm} \times \dfrac{1.013 \times 10^5 \text{ Pa}}{\text{atm}} \times \dfrac{1 \text{ MPa}}{10^6 \text{ Pa}} = 15 \text{ MPa}$

        c. $150 \text{ atm} \times \dfrac{760 \text{ torr}}{\text{atm}} = 1.1 \times 10^5 \text{ torr}$

23. $6.5 \text{ cm} \times \dfrac{10 \text{ mm}}{\text{cm}} = 65 \text{ mm Hg or } 65 \text{ torr}; \quad 65 \text{ torr} \times \dfrac{1 \text{ atm}}{760 \text{ torr}} = 8.6 \times 10^{-2} \text{ atm}$

$8.6 \times 10^{-2} \text{ atm} = \dfrac{1.013 \times 10^5 \text{ Pa}}{\text{atm}} = 8.7 \times 10^3 \text{ Pa}$

24. $20.0 \text{ in Hg} \times \dfrac{2.54 \text{ cm}}{\text{in}} \times \dfrac{10 \text{ mm}}{\text{cm}} = 508 \text{ mm Hg} = 508 \text{ torr}; \quad 508 \text{ torr} \times \dfrac{1 \text{ atm}}{760 \text{ torr}} = 0.668 \text{ atm}$

25. If the levels of Hg in each arm of the manometer are equal, then the pressure in the flask is equal to atmospheric pressure. When they are unequal, the difference in height in mm will be equal to the difference in pressure in mm Hg between the flask and the atmosphere. Which level is higher will tell us whether the pressure in the flask is less than or greater than atmospheric.

a. $P_{flask} < P_{atm}; \ P_{flask} = 760. - 140. = 620. \text{ mm Hg} = 620. \text{ torr}; \ 620. \text{ torr} \times \dfrac{1 \text{ atm}}{760 \text{ torr}} = 0.816 \text{ atm}$

$0.816 \text{ atm} \times \dfrac{1.013 \times 10^5 \text{ Pa}}{\text{atm}} = 8.27 \times 10^4 \text{ Pa}$

b. $P_{flask} > P_{atm}; \ P_{flask} = 760. \text{ torr} + 175 \text{ torr} = 935 \text{ torr}; \ 935 \text{ torr} \times \dfrac{1 \text{ atm}}{760 \text{ torr}} = 1.23 \text{ atm}$

$1.23 \text{ atm} \times \dfrac{1.013 \times 10^5 \text{ Pa}}{\text{atm}} = 1.25 \times 10^5 \text{ Pa}$

c. $P_{flask} = 635 - 140. = 495 \text{ torr}; \ P_{flask} = 635 + 175 = 810. \text{ torr}$

26. a. The pressure is proportional to the mass of the fluid. The mass is proportional to the volume of the column of fluid (or to the height of the column assuming the area of the column of fluid is constant).

$d = \dfrac{\text{mass}}{\text{volume}};$ In this case the volume of silicon oil will be the same as the volume of Hg in Exercise 5.25.

$V = \dfrac{m}{d}; \ V_{Hg} = V_{oil}, \ \dfrac{m_{Hg}}{d_{Hg}} = \dfrac{m_{oil}}{d_{oil}}, \ m_{oil} = \dfrac{m_{Hg} d_{oil}}{d_{Hg}}$

Since P is proportional to the mass of liquid:

$P_{oil} = P_{Hg} \left(\dfrac{d_{oil}}{d_{Hg}} \right) = P_{Hg} \left(\dfrac{1.30}{13.6} \right) = 0.0956 \ P_{Hg}$

This conversion applies only to the column of liquid.

$P_{flask} = 760. \text{ torr} - (140. \times 0.0956) \text{ torr} = 760. - 13.4 = 747 \text{ torr}$

$747 \text{ torr} \times \dfrac{1 \text{ atm}}{760 \text{ torr}} = 0.983 \text{ atm}; \ 0.983 \text{ atm} \times \dfrac{1.013 \times 10^5 \text{ Pa}}{\text{atm}} = 9.96 \times 10^4 \text{ Pa}$

$$P_{flask} = 760. \text{ torr} + (175 \times 0.0956) \text{ torr} = 760. + 16.7 = 777 \text{ torr}$$

$$777 \text{ torr} \times \frac{1 \text{ atm}}{760 \text{ torr}} = 1.02 \text{ atm}; \quad 1.02 \text{ atm} \times \frac{1.013 \times 10^5 \text{ Pa}}{\text{atm}} = 1.03 \times 10^5 \text{ Pa}$$

b.   If we are measuring the same pressure, the height of the silicon oil column would be $13.6 \div 1.30$ = 10.5 times the height of a mercury column. The advantage of using a less dense fluid than mercury is in measuring small pressures. The quantity measured (length) will be larger for the less dense fluid. Thus, the measurement will be more precise.

Gas Laws

27.   From Boyle's law, $P_1V_1 = P_2V_2$ at constant n and T.

$$P_2 = \frac{P_1V_1}{V_2} = \frac{5.20 \text{ atm} \times 0.400 \text{ L}}{2.14 \text{ L}} = 0.972 \text{ atm}$$

As expected, as the volume increased, the pressure decreased.

28.   The pressure exerted on the balloon is constant and the moles of gas present is constant. From Charles's law, $V_1/T_1 = V_2/T_2$ at constant P and n.

$$V_2 = \frac{V_1T_2}{T_1} = \frac{700. \text{ mL} \times 100. \text{ K}}{(273.2 + 20.0) \text{ K}} = 239 \text{ mL}$$

As expected, as the temperature decreased, the volume decreased.

29.   From Avogadro's law, $V_1/n_1 = V_2/n_2$ at constant T and P.

$$V_2 = \frac{V_1 n_2}{n_1} = \frac{11.2 \text{ L} \times 2.00 \text{ mol}}{0.500 \text{ mol}} = 44.8 \text{ L}$$

As expected, as the mol of gas present increases, volume increases.

30.   As NO_2 is converted completely into N_2O_4, the mol of gas present will decrease by one-half (from the 2:1 mol ratio in the balanced equation). Using Avogadro's law,

$$\frac{V_1}{n_1} = \frac{V_2}{n_2}, \quad V_2 = V_1 \times \frac{n_2}{n_1} = 25.0 \text{ mL} \times \frac{1}{2} = 12.5 \text{ mL}$$

$N_2O_4(g)$ will occupy one-half the original volume of $NO_2(g)$. This is expected since the mol of gas present decreases by one-half when NO_2 is converted into N_2O_4.

31.   a. $PV = nRT, \quad V = \dfrac{nRT}{P} = \dfrac{2.00 \text{ mol} \times \dfrac{0.08206 \text{ L atm}}{\text{mol K}} \times (155 + 273) \text{ K}}{5.00 \text{ atm}} = 14.0 \text{ L}$

b. $PV = nRT, \quad n = \dfrac{PV}{RT} = \dfrac{0.300 \text{ atm} \times 2.00 \text{ L}}{\dfrac{0.08206 \text{ L atm}}{\text{mol K}} \times 155 \text{ K}} = 4.72 \times 10^{-2} \text{ mol}$

c. $PV = nRT$, $T = \dfrac{PV}{nR} = \dfrac{4.47 \text{ atm} \times 25.0 \text{ L}}{2.01 \text{ mol} \times \dfrac{0.08206 \text{ L atm}}{\text{mol K}}} = 678 \text{ K} = 405\,°C$

d. $PV = nRT$, $P = \dfrac{nRT}{V} = \dfrac{10.5 \text{ mol} \times \dfrac{0.08206 \text{ L atm}}{\text{mol K}} \times (273 + 75) \text{ K}}{2.25 \text{ L}} = 133 \text{ atm}$

32. a. $PV = nRT$, $T = \dfrac{PV}{nR} = \dfrac{(875 \text{ torr} \times \dfrac{1 \text{ atm}}{760 \text{ torr}}) \times 275 \times 10^{-3} \text{ L}}{0.0105 \text{ mol} \times \dfrac{0.08206 \text{ L atm}}{\text{mol K}}} = 367 \text{ K} = 94\,°C$

b. $P = \dfrac{nRT}{V} = \dfrac{0.200 \text{ mol} \times \dfrac{0.08206 \text{ L atm}}{\text{mol K}} \times 311 \text{ K}}{0.100 \text{ L}} = 51.0 \text{ atm}$

c. $V = \dfrac{nRT}{P} = \dfrac{3.00 \text{ mol} \times \dfrac{0.08206 \text{ L atm}}{\text{mol K}} \times 838 \text{ K}}{2.50 \text{ atm}} = 82.5 \text{ L}$

d. $n = \dfrac{PV}{RT} = \dfrac{(688 \text{ torr} \times \dfrac{1 \text{ atm}}{760 \text{ torr}}) \times 986 \times 10^{-3} \text{ L}}{\dfrac{0.08206 \text{ L atm}}{\text{mol K}} \times 565 \text{ K}} = 1.93 \times 10^{-2} \text{ mol}$

33. $PV = nRT$; $n = \dfrac{PV}{RT} = \dfrac{145 \text{ atm} \times 75.0 \times 10^{-3} \text{ L}}{\dfrac{0.08206 \text{ L atm}}{\text{mol K}} \times 295 \text{ K}} = 0.449 \text{ mol } O_2$

34. $P = \dfrac{nRT}{V} = \dfrac{\left(0.60 \text{ g} \times \dfrac{1 \text{ mol}}{32.00 \text{ g}}\right) \times \dfrac{0.08206 \text{ L atm}}{\text{mol K}} \times (273 + 22) \text{ K}}{5.0 \text{ L}} = 0.091 \text{ atm}$

35. a. $PV = nRT$; $175 \text{ g Ar} \times \dfrac{1 \text{ mol Ar}}{39.95 \text{ g Ar}} = 4.38 \text{ mol Ar}$

$T = \dfrac{PV}{nR} = \dfrac{10.0 \text{ atm} \times 2.50 \text{ L}}{4.38 \text{ mol} \times \dfrac{0.08206 \text{ L atm}}{\text{mol K}}} = 69.6 \text{ K}$

b. $PV = nRT$; $P = \dfrac{nRT}{V} = \dfrac{4.38 \text{ mol} \times \dfrac{0.08206 \text{ L atm}}{\text{mol K}} \times 225 \text{ K}}{2.50 \text{ L}} = 32.3 \text{ atm}$

36. $PV = nRT$, $V = \dfrac{nRT}{P}$; $2.0 \text{ g He} \times \dfrac{1 \text{ mol He}}{4.003} = 0.50 \text{ mol He}$

$$V = \frac{0.50 \text{ mol} \times \dfrac{0.08206 \text{ L atm}}{\text{mol K}} \times (273 + 25) \text{ K}}{775 \text{ mm Hg} \times \dfrac{1 \text{ atm}}{760 \text{ mm Hg}}} = 12 \text{ L}$$

37.   At constant n and T, PV = nRT = constant, $P_1V_1 = P_2V_2$; At sea level, P = 1.00 atm = 760. mm Hg.

$$V_2 = \frac{P_1V_1}{P_2} = \frac{760. \text{ mm Hg} \times 2.0 \text{ L}}{500. \text{ mm Hg}} = 3.0 \text{ L}$$

The balloon will burst at this pressure since the volume must expand beyond the 2.5 L limit of the balloon.

Note: To solve this problem, we did not have to convert the pressure units into atm; the units of mm Hg cancelled each other. In general, only convert units if you have to. Whenever the gas constant R is not used to solve a problem, pressure and volume units must only be consistent, and not necessarily in units of atm and L. The exception is temperature as T must <u>always</u> be converted to the Kelvin scale.

38.   At constant n and P, $\dfrac{V_1}{T_1} = \dfrac{V_2}{T_2}$. From the problem, $V_2 = 125.0\%$ of $V_1 = 1.250 \ V_1$.

$$T_2 = \frac{T_1V_2}{V_1} = \frac{(273 + 19) \text{ K} \times 1.250 \ V_1}{V_1} = 365 \text{ K or } 92\,^\circ\text{C}$$

39.   PV = nRT, V and n constant, so $\dfrac{P}{T} = \dfrac{nR}{V}$ = constant and $\dfrac{P_1}{T_1} = \dfrac{P_2}{T_2}$.

$$P_2 = \frac{P_1T_2}{T_1} = 13.7 \text{ MPa} \times \frac{(273 + 450.) \text{ K}}{(273 + 23) \text{ K}} = 33.5 \text{ MPa}$$

40.   a.   At constant n and V, $\dfrac{P_1}{T_1} = \dfrac{P_2}{T_2}$, $P_2 = \dfrac{P_1T_2}{T_1} = 40.0 \text{ atm} \times \dfrac{318 \text{ K}}{273 \text{ K}} = 46.6 \text{ atm}$

       b. $\dfrac{P_1}{T_1} = \dfrac{P_2}{T_2}$, $T_2 = \dfrac{T_1P_2}{P_1} = 273 \text{ K} \times \dfrac{150. \text{ atm}}{40.0 \text{ atm}} = 1.02 \times 10^3 \text{ K}$

       c. $T_2 = \dfrac{T_1P_2}{P_1} = 273 \text{ K} \times \dfrac{25.0 \text{ atm}}{40.0 \text{ atm}} = 171 \text{ K}$

41.   PV = nRT, n constant; $\dfrac{PV}{T} = nR$ = constant, $\dfrac{P_1V_1}{T_1} = \dfrac{P_2V_2}{T_2}$

$$P_2 = \frac{P_1V_1T_2}{V_2T_1} = 710. \text{ torr} \times \frac{5.0 \times 10^2 \text{ mL}}{25 \text{ mL}} \times \frac{(273 + 820.) \text{ K}}{(273 + 30.) \text{ K}} = 5.1 \times 10^4 \text{ torr}$$

42.   Begin with the ideal gas law, $PV = nRT$. From the information given, we know that the volume is constant, i.e., the gas is in a steel cylinder, and P, n and T are changing. So, we can say that:

$$\frac{nT}{P} = \frac{V}{R} = \text{a constant or } \frac{n_1 T_1}{P_1} = \frac{n_2 T_2}{P_2}$$

$$n_2 = \frac{n_1 T_1 P_2}{T_2 P_1} = 150 \text{ mol} \times \frac{298 \text{ K}}{292 \text{ K}} \times \frac{2.0 \text{ MPa}}{7.5 \text{ MPa}} = 41 \text{ mol Ar}; \; 41 \text{ mol Ar} \times \frac{39.95 \text{ g}}{\text{mol}} = 1600 \text{ g Ar}$$

43. $PV = nRT$, Assume n is constant. $\dfrac{PV}{T} = nR = \text{constant}$, $\dfrac{P_1 V_1}{T_1} = \dfrac{P_2 V_2}{T_2}$

$$\frac{V_2}{V_1} = \frac{T_2 P_1}{T_1 P_2} = \frac{(273 + 15) \text{ K} \times 720. \text{ torr}}{(273 + 25) \text{ K} \times 605 \text{ torr}} = 1.15; \; V_2 = 1.15 \times 855 \text{ L} = 983 \text{ L}$$

$V_2 = 1.15 \, V_1$ or the volume has increased by 15% or $\Delta V = 983 \text{ L} - 855 \text{ L} = 128 \text{ L}$.

44. $PV = nRT$, n is constant. $\dfrac{PV}{T} = nR = \text{constant}$, $\dfrac{P_1 V_1}{T_1} = \dfrac{P_2 V_2}{T_2}$

$$\frac{V_2}{V_1} = \frac{P_1 T_2}{P_2 T_1} = \frac{760. \text{ torr} \times (273 + 127) \text{ K}}{7600. \text{ torr} \times (273 - 73) \text{ K}} = 0.200$$

$V_2 = 0.200 \, V_1$ or the volume of the gas sample decreased to one-fifth of the initial volume.

Gas Density, Molar Mass, and Reaction Stoichiometry

45.   STP: T = 273 K and P = 1.00 atm; $n = \dfrac{PV}{RT} = \dfrac{1.00 \text{ atm} \times 1.5 \text{ L}}{\dfrac{0.08206 \text{ L atm}}{\text{mol K}} \times 273 \text{ K}} = 6.7 \times 10^{-2} \text{ mol He}$

Or we can use the fact that at STP, 1 mol of an ideal gas occupies 22.42 L.

$$1.5 \text{ L} \times \frac{1 \text{ mol He}}{22.42 \text{ L}} = 6.7 \times 10^{-2} \text{ mol He}; \; 6.7 \times 10^{-2} \text{ mol He} \times \frac{4.003 \text{ g He}}{\text{mol He}} = 0.27 \text{ g He}$$

46. $CO_2(s) \rightarrow CO_2(g)$; $4.00 \text{ g } CO_2 \times \dfrac{1 \text{ mol } CO_2}{44.01 \text{ g } CO_2} = 9.09 \times 10^{-2} \text{ mol } CO_2$

At STP, the molar volume of a gas is 22.42 L. $9.09 \times 10^{-2} \text{ mol } CO_2 \times \dfrac{22.42 \text{ L}}{\text{mol } CO_2} = 2.04 \text{ L}$

47.   The balanced equation is: $2 \, C_8 H_{18}(l) + 25 \, O_2(g) \rightarrow 16 \, CO_2(g) + 18 \, H_2O(g)$

$$125 \text{ g } C_8 H_{18} \times \frac{1 \text{ mol } C_8 H_{18}}{114.22 \text{ g } C_8 H_{18}} \times \frac{25 \text{ mol } O_2}{2 \text{ mol } C_8 H_{18}} = 13.7 \text{ mol } O_2$$

$$V = \frac{nRT}{P} = \frac{13.7 \text{ mol} \times \dfrac{0.08206 \text{ L atm}}{\text{mol K}} \times 273 \text{ K}}{1.00 \text{ atm}} = 307 \text{ L O}_2$$

Or we can make use of the fact that at STP one mole of an ideal gas occupies a volume of 22.42 L. This can be calculated using the ideal gas law.

So: $13.7 \text{ mol O}_2 \times \dfrac{22.42 \text{ L}}{\text{mol}} = 307 \text{ L O}_2$

48. $4.10 \text{ g HgO} \times \dfrac{1 \text{ mol HgO}}{216.6 \text{ g HgO}} \times \dfrac{1 \text{ mol O}_2}{2 \text{ mol HgO}} = 9.46 \times 10^{-3} \text{ mol O}_2$

$$V = \frac{nRT}{P} = \frac{9.46 \times 10^{-3} \text{ mol} \times \dfrac{0.08206 \text{ L atm}}{\text{mol K}} \times 303 \text{ K}}{725 \text{ torr} \times \dfrac{1 \text{ atm}}{760 \text{ torr}}} = 0.247 \text{ L}$$

49. $n_{H_2} = \dfrac{PV}{RT} = \dfrac{1.0 \text{ atm} \times \left[4800 \text{ m}^3 \times \left(\dfrac{100 \text{ cm}}{\text{m}} \right)^3 \times \dfrac{1 \text{ L}}{1000 \text{ cm}^3} \right]}{\dfrac{0.08206 \text{ L atm}}{\text{mol K}} \times 273 \text{ K}} = 2.1 \times 10^5 \text{ mol}$

2.1×10^5 mol H_2 are in the balloon. This is 80.% of the total amount of H_2 that had to be generated:

0.80 (total mol H_2) = 2.1×10^5, total mol H_2 = 2.6×10^5 mol H_2

$2.6 \times 10^5 \text{ mol H}_2 \times \dfrac{1 \text{ mol Fe}}{\text{mol H}_2} \times \dfrac{55.85 \text{ g Fe}}{\text{mol Fe}} = 1.5 \times 10^7 \text{ g Fe}$

$2.6 \times 10^5 \text{ mol H}_2 \times \dfrac{1 \text{ mol H}_2SO_4}{\text{mol H}_2} \times \dfrac{98.09 \text{ g H}_2SO_4}{\text{mol H}_2SO_4} \times \dfrac{100 \text{ g reagent}}{98 \text{ g H}_2SO_4} = 2.6 \times 10^7 \text{ g of 98\%}$ sulfuric acid

50. $CH_3OH + 3/2 \ O_2 \rightarrow CO_2 + 2 \ H_2O$ or $2 \ CH_3OH(l) + 3 \ O_2(g) \rightarrow 2 \ CO_2(g) + 4 \ H_2O(g)$

$50.0 \text{ mL} \times \dfrac{0.850 \text{ g}}{\text{mL}} \times \dfrac{1 \text{ mol}}{32.04 \text{ g}} = 1.33 \text{ mol CH}_3OH(l)$ available

$n_{O_2} = \dfrac{PV}{RT} = \dfrac{2.00 \text{ atm} \times 22.8 \text{ L}}{\dfrac{0.08206 \text{ L atm}}{\text{mol K}} \times 300. \text{ K}} = 1.85 \text{ mol O}_2$ available

$1.33 \text{ mol CH}_3OH \times \dfrac{3 \text{ mol O}_2}{2 \text{ mol CH}_3OH} = 2.00 \text{ mol O}_2$ required for complete reaction. We only have 1.85 mol O_2, so O_2 is limiting.

$1.85 \text{ mol O}_2 \times \dfrac{4 \text{ mol H}_2O}{3 \text{ mol O}_2} = 2.47 \text{ mol H}_2O$

51. $Xe(g) + 2 F_2(g) \rightarrow XeF_4(s)$; $n_{Xe} = \dfrac{PV}{RT} = \dfrac{0.500 \text{ atm} \times 20.0 \text{ L}}{\dfrac{0.08206 \text{ L atm}}{\text{mol K}} \times 673 \text{ K}} = 0.181 \text{ mol Xe}$

We could do the same calculation for F_2. However, the only variable that changed is the pressure. Since the partial pressure of F_2 is triple that of Xe, then mol $F_2 = 3(0.181) = 0.543$ mol F_2. The balanced equation requires 2 mol of F_2 for every mol of Xe. The actual mol ratio is 3 mol F_2:1 mol Xe. Xe is the limiting reagent.

$0.181 \text{ mol Xe} \times \dfrac{1 \text{ mol XeF}_4}{\text{mol Xe}} \times \dfrac{207.3 \text{ g XeF}_4}{\text{mol Xe}} = 37.5 \text{ g XeF}_4$

52.   For ammonia (in one minute):

$n_{NH_3} = \dfrac{PV}{RT} = \dfrac{90. \text{ atm} \times 500. \text{ L}}{\dfrac{0.08206 \text{ L atm}}{\text{mol K}} \times 496 \text{ K}} = 1.1 \times 10^3 \text{ mol NH}_3$

NH_3 flows into the reactor at a rate of 1.1×10^3 mol/min.

For CO_2 (in one minute):

$n_{CO_2} = \dfrac{PV}{RT} = \dfrac{45 \text{ atm} \times 600. \text{ L}}{\dfrac{0.08206 \text{ L atm}}{\text{mol K}} \times 496 \text{ K}} = 6.6 \times 10^2 \text{ mol CO}_2$

CO_2 flows into the reactor at 6.6×10^2 mol/min.

To react completely with 1.1×10^3 mol NH_3/min, we need:

$\dfrac{1.1 \times 10^3 \text{ mol NH}_3}{\text{min}} \times \dfrac{1 \text{ mol CO}_2}{2 \text{ mol NH}_3} = 5.5 \times 10^2 \text{ mol CO}_2/\text{min}$

Since 660 mol CO_2/min are present, ammonia is the limiting reagent.

$\dfrac{1.1 \times 10^3 \text{ mol NH}_3}{\text{min}} \times \dfrac{1 \text{ mol urea}}{2 \text{ mol NH}_3} \times \dfrac{60.06 \text{ g urea}}{\text{mol urea}} = 3.3 \times 10^4 \text{ g urea/min}$

53.   a. $CH_4(g) + NH_3(g) + O_2(g) \rightarrow HCN(g) + H_2O(g)$; Balancing H first then O gives:

$CH_4 + NH_3 + \dfrac{3}{2}O_2 \rightarrow HCN + 3 H_2O$ or $2 CH_4(g) + 2 NH_3(g) + 3 O_2(g) \rightarrow 2 HCN(g) + 6 H_2O(g)$

        b. $PV = nRT$, T and P constant; $\dfrac{V_1}{n_1} = \dfrac{V_2}{n_2}$, $\dfrac{V_1}{V_2} = \dfrac{n_1}{n_2}$

Since the volumes are all measured at constant T and P, then the volumes of gas present are directly proportional to the mol of gas present (Avogadro's law). Because Avogadro's law applies, the balanced reaction gives mol relationships as well as volume relationships. Therefore, 2 L of CH_4, 2 L of NH_3 and 3 L of O_2 are required by the balanced equation for the production of 2 L of HCN. The actual volume ratio is 20 L CH_4:20 L NH_3:20 L O_2 (or 1:1:1). The volume of O_2 required to react with all of the CH_4 and NH_3 present is 20 L ×(3/2) = 30 L. Since only 20.0 L of O_2 are present, then O_2 is the limiting reagent. The volume of HCN produced is:

$$20.0 \text{ L } O_2 \times \frac{2 \text{ L HCN}}{3 \text{ L } O_2} = 13.3 \text{ L HCN}$$

54.   The reaction is 1:1 between ethene and hydrogen. Since T and P are constant, a greater volume of H_2 and thus, more moles of H_2 are flowing into the reactor. Ethene is the limiting reagent.

In one minute:

$$n_{C_2H_4} = \frac{PV}{RT} = \frac{25.0 \text{ atm} \times 1000. \text{ L}}{\dfrac{0.08206 \text{ L atm}}{\text{mol K}} \times 573 \text{ K}} = 532 \text{ mol } C_2H_4 \text{ reacted}$$

$$\text{Theoretical yield} = \frac{532 \text{ mol } C_2H_4}{\text{min}} \times \frac{1 \text{ mol } C_2H_6}{\text{mol } C_2H_4} \times \frac{30.07 \text{ g } C_2H_6}{\text{mol } C_2H_6} \times \frac{1 \text{ kg}}{1000 \text{ g}} = 16.0 \text{ kg } C_2H_6/\text{min}$$

$$\text{\% yield} = \frac{15.0 \text{ kg/min}}{16.0 \text{ kg/min}} \times 100 = 93.8\%$$

55.   One of these equations developed in the text to determine molar mass is:

$$\text{molar mass} = \frac{dRT}{P} \text{ where } d = \text{density in units of g/L}$$

$$\text{molar mass} = \frac{1.65 \text{ g/L} \times \dfrac{0.08206 \text{ L atm}}{\text{mol K}} \times (273 + 27) \text{ K}}{734 \text{ torr} \times \dfrac{1 \text{ atm}}{760 \text{ torr}}} = 42.1 \text{ g/mol}$$

The empirical formula mass of CH_2 = 12.01 + 2(1.008) = 14.03 g/mol.

$$\frac{42.1}{14.03} = 3.00; \text{ Molecular formula} = C_3H_6$$

56. $P \times (\text{molar mass}) = dRT, d = \dfrac{\text{mass}}{\text{volume}}, P \times (\text{molar mass}) = \dfrac{\text{mass}}{V} \times RT$

$$\text{Molar mass} = M = \frac{\text{mass} \times RT}{PV} = \frac{0.800 \text{ g} \times \dfrac{0.08206 \text{ L atm}}{\text{mol K}} \times 373 \text{ K}}{(750. \text{ torr} \times \dfrac{1 \text{ atm}}{760 \text{ torr}}) \times 0.256 \text{ L}} = 96.9 \text{ g/mol}$$

Empirical mass of CHCl ≈ 12.0 + 1.0 + 35.5 = 48.5; Molecular formula is $C_2H_2Cl_2$.

57. $P \times (\text{molar mass}) = dRT$, $d = \text{density} = \dfrac{P \times (\text{molar mass})}{RT}$

For $SiCl_4$, molar mass = M = 28.09 + 4(35.45) = 169.89 g/mol

$$d = \frac{\left(758 \text{ torr} \times \dfrac{1 \text{ atm}}{760 \text{ torr}}\right) \times \dfrac{169.89 \text{ g}}{\text{mol}}}{\dfrac{0.08206 \text{ L atm}}{\text{mol K}} \times 358 \text{ K}} = 5.77 \text{ g/L for } SiCl_4$$

For $SiHCl_3$, molar mass = M = 28.09 + 1.008 + 3(35.45) = 135.45 g/mol

$$d = \frac{PM}{RT} = \frac{\left(758 \text{ torr} \times \dfrac{1 \text{ atm}}{760 \text{ torr}}\right) \times \dfrac{135.45 \text{ g}}{\text{mol}}}{\dfrac{0.08206 \text{ L atm}}{\text{mol K}} \times 358 \text{ K}} = 4.60 \text{ g/L for } SiHCl_3$$

58. $$d = \frac{P \times (\text{molar mass})}{RT} = \frac{\left(635 \text{ torr} \times \dfrac{1 \text{ atm}}{760 \text{ torr}}\right) \times \dfrac{17.03 \text{ g}}{\text{mol}}}{\dfrac{0.08206 \text{ L atm}}{\text{mol K}} \times 300. \text{ K}} = 0.578 \text{ g/L for } NH_3$$

Partial Pressure

59. $$P_{CO_2} = \frac{nRT}{V} = \frac{\left(7.8 \text{ g} \times \dfrac{1 \text{ mol}}{44.01 \text{ g}}\right) \times \dfrac{0.08206 \text{ L atm}}{\text{mol K}} \times 300. \text{ K}}{4.0 \text{ L}} = 1.1 \text{ atm}$$

With air present, the partial pressure of CO_2 will still be 1.1 atm. The total pressure will be the sum of the partial pressures, $P_{total} = P_{CO_2} + P_{air}$.

$$P_{total} = 1.1 \text{ atm} + \left(740 \text{ torr} \times \frac{1 \text{ atm}}{760 \text{ torr}}\right) = 1.1 + 0.97 = 2.1 \text{ atm}$$

60. $n_{H_2} = 1.00 \text{ g } H_2 \times \dfrac{1 \text{ mol } H_2}{2.016 \text{ g } H_2} = 0.496 \text{ mol } H_2$; $n_{He} = 1.00 \text{ g He} \times \dfrac{1 \text{ mol He}}{4.003 \text{ g He}} = 0.250 \text{ mol He}$

$$P_{H_2} = \frac{n_{H_2} \times RT}{V} = \frac{0.496 \text{ mol} \times \dfrac{0.08206 \text{ L atm}}{\text{mol K}} \times (273 + 27) \text{ K}}{1.00 \text{ L}} = 12.2 \text{ atm}$$

$$P_{He} = \frac{n_{He} \times RT}{V} = 6.15 \text{ atm}; P_{total} = P_{H_2} + P_{He} = 12.2 \text{ atm} + 6.15 \text{ atm} = 18.4 \text{ atm}$$

61.   Use the relationship $P_1V_1 = P_2V_2$ for each gas, since T and n for each gas is constant.

For H_2: $P_2 = \dfrac{P_1V_1}{V_2} = 475$ torr $\times \dfrac{2.00\,L}{3.00\,L} = 317$ torr

For N_2: $P_2 = 0.200$ atm $\times \dfrac{1.00\,L}{3.00\,L} = 0.0667$ atm; 0.0667 atm $\times \dfrac{760\,torr}{atm} = 50.7$ torr

$P_{total} = P_{H_2} + P_{N_2} = 317 + 50.7 = 368$ torr

62. $P_1V_1 = P_2V_2$, $P_2 = \dfrac{P_1V_1}{V_2}$;   For H_2: $P_2 = 360.$ torr $\times \left(\dfrac{2.00\,L}{3.00\,L}\right) = 240.$ torr

For N_2: $P_2 = 240.$ torr $\times \left(\dfrac{1.00\,L}{3.00\,L}\right) = 80.0$ torr; $P_{total} = P_{H_2} + P_{N_2} = 240. + 80.0 = 320.$ torr

63.   a.   mol fraction $CH_4 = \chi_{CH_4} = \dfrac{P_{CH_4}}{P_{total}} = \dfrac{0.175\,atm}{0.175\,atm + 0.250\,atm} = 0.412$; $\chi_{O_2} = 1.000 - 0.412 = 0.588$

b. $PV = nRT$, $n_{total} = \dfrac{P_{total} \times V}{RT} = \dfrac{0.425\,atm \times 10.5\,L}{\dfrac{0.08206\,L\,atm}{mol\,K} \times 338\,K} = 0.161$ mol

c. $\chi_{CH_4} = \dfrac{n_{CH_4}}{n_{total}}$, $n_{CH_4} = \chi_{CH_4} \times n_{total} = 0.412 \times 0.161$ mol $= 6.63 \times 10^{-2}$ mol CH_4

6.63×10^{-2} mol $CH_4 \times \dfrac{16.04\,g\,CH_4}{mol\,CH_4} = 1.06$ g CH_4

$n_{O_2} = 0.588 \times 0.161$ mol $= 9.47 \times 10^{-2}$ mol O_2; 9.47×10^{-2} mol $O_2 \times \dfrac{32.00\,g\,O_2}{mol\,O_2} = 3.03$ g O_2

64.   We can use the ideal gas law to calculate the partial pressure of each gas or to calculate the total pressure. There will be less math if we calculate the total pressure from the ideal gas law.

$n_{O_2} = 1.5 \times 10^2$ mg $O_2 \times \dfrac{1\,g}{1000\,mg} \times \dfrac{1\,mol\,O_2}{32.00\,g\,O_2} = 4.7 \times 10^{-3}$ mol O_2

$n_{NH_3} = 5.0 \times 10^{21}$ molecules $NH_3 \times \dfrac{1\,mol\,NH_3}{6.022 \times 10^{23}\,molecules\,NH_3} = 8.3 \times 10^{-3}$ mol NH_3

$n_{total} = n_{N_2} + n_{O_2} + n_{NH_3} = 5.0 \times 10^{-2} + 4.7 \times 10^{-3} + 8.3 \times 10^{-3} = 6.3 \times 10^{-2}$ mol

$P_{total} = \dfrac{n_{total} \times RT}{V} = \dfrac{6.3 \times 10^{-2}\,mol \times \dfrac{0.08206\,L\,atm}{mol\,K} \times 273\,K}{1.0\,L} = 1.4$ atm

$$P_{N_2} = \chi_{N_2} \times P_{total}, \quad \chi_{N_2} = \frac{n_{N_2}}{n_{total}}; \quad P_{N_2} = \frac{5.0 \times 10^{-2} \text{ mol}}{6.3 \times 10^{-2} \text{ mol}} \times 1.4 \text{ atm} = 1.1 \text{ atm}$$

$$P_{O_2} = \frac{4.7 \times 10^{-3}}{6.3 \times 10^{-2}} \times 1.4 \text{ atm} = 0.10 \text{ atm}; \qquad P_{NH_3} = \frac{8.3 \times 10^{-3}}{6.3 \times 10^{-2}} \times 1.4 \text{ atm} = 0.18 \text{ atm}$$

65. $P_{total} = 1.00 \text{ atm} = 760. \text{ torr} = P_{N_2} + P_{H_2O} = P_{N_2} + 17.5 \text{ torr}, \quad P_{N_2} = 743 \text{ torr}$

$$PV = nRT; \quad n_{N_2} = \frac{P_{N_2} \times V}{RT} = \frac{(743 \text{ torr} \times \frac{1 \text{ atm}}{760 \text{ torr}}) \times (2.50 \times 10^2 \text{ mL} \times \frac{1 \text{ L}}{1000 \text{ mL}})}{\frac{0.08206 \text{ L atm}}{\text{mol K}} \times 293 \text{ K}} = 1.02 \times 10^{-2} \text{ mol } N_2$$

$$1.02 \times 10^{-2} \text{ mol } N_2 \times \frac{28.02 \text{ g } N_2}{\text{mol } N_2} = 0.286 \text{ g } N_2$$

66. $P_{total} = P_{O_2} + P_{H_2O}, \quad P_{O_2} + P_{total} - P_{H_2O} = 641 \text{ torr} - 23.8 \text{ torr} = 617 \text{ torr}$

$$PV = nRT; \quad n_{O_2} = \frac{P_{O_2} \times V}{RT} = \frac{\frac{617}{760} \text{ atm} \times 0.5000 \text{ L}}{\frac{0.08206 \text{ L atm}}{\text{mol K}} \times 298 \text{ K}} = 1.66 \times 10^{-2} \text{ mol } O_2$$

$$1.66 \times 10^{-2} \text{ mol } O_2 \times \frac{32.00 \text{ g}}{\text{mol}} = 0.531 \text{ g } O_2$$

67. $3.70 \text{ g } KClO_3 \times \dfrac{1 \text{ mol } KClO_3}{122.55 \text{ g } KClO_3} \times \dfrac{3 \text{ mol } O_2}{2 \text{ mol } KClO_3} = 4.53 \times 10^{-2} \text{ mol } O_2$

$$P_{total} = P_{O_2} + P_{H_2O}, \quad P_{O_2} = P_{total} - P_{H_2O} = 735 - 26.7 = 708 \text{ torr} \times \frac{1 \text{ atm}}{760 \text{ torr}} = 0.932 \text{ atm}$$

$$V = \frac{n_{O_2} \times RT}{P_{O_2}} = \frac{4.53 \times 10^{-2} \text{ mol} \times \frac{0.08206 \text{ L atm}}{\text{mol K}} \times 300. \text{ K}}{0.932 \text{ atm}} = 1.20 \text{ L}$$

68. $2 \text{ NaClO}_3(s) \rightarrow 2 \text{ NaCl}(s) + 3 \text{ O}_2(g)$

$$P_{total} = P_{O_2} + P_{H_2O}, \quad P_{O_2} = P_{total} - P_{H_2O} = 734 \text{ torr} - 19.8 \text{ torr} = 714 \text{ torr}$$

$$n_{O_2} = \frac{P_{O_2} \times V}{RT} = \frac{\left(714 \text{ torr} \times \frac{1 \text{ atm}}{760 \text{ torr}}\right) \times 0.0572 \text{ L}}{\frac{0.08206 \text{ L atm}}{\text{mol K}} \times (273 + 22) \text{ K}} = 2.22 \times 10^{-3} \text{ mol } O_2$$

$$\text{Mass NaClO}_3 \text{ decomposed} = 2.22 \times 10^{-3} \text{ mol } O_2 \times \frac{2 \text{ mol NaClO}_3}{3 \text{ mol } O_2} \times \frac{106.44 \text{ g NaClO}_3}{\text{mol NaClO}_3} = 0.158 \text{ g NaClO}_3$$

$$\text{Mass \% NaClO}_3 = \frac{0.158 \text{ g}}{0.8765 \text{ g}} \times 100 = 18.0\%$$

Kinetic Molecular Theory and Real Gases

69. $(KE)_{avg} = (3/2) \, RT$; At 273 K: $(KE)_{avg} = \dfrac{3}{2} \times \dfrac{8.3145 \text{ J}}{\text{mol K}} \times 273 \text{ K} = 3.40 \times 10^3 \text{ J/mol}$

At 546 K: $(KE)_{avg} = \dfrac{3}{2} \times \dfrac{8.3145 \text{ J}}{\text{mol K}} \times 546 \text{ K} = 6.81 \times 10^3 \text{ J/mol}$

70. $(KE)_{avg} = (3/2) \, RT$. Since the kinetic energy depends only on temperature, CH_4 (Exercise 5.69) and N_2 at the same temperature will have the same average kinetic energy. So for N_2 the average kinetic energy is 3.40×10^3 J/mol (at 273 K) and 6.81×10^3 J/mol (at 546 K).

71. $u_{rms} = \left(\dfrac{3RT}{M} \right)^{1/2}$, where $R = \dfrac{8.3145 \text{ J}}{\text{mol K}}$ and M = molar mass in kg = 1.604×10^{-2} kg/mol for CH_4

For CH_4 at 273 K: $u_{rms} = \left(\dfrac{\dfrac{3 \times 8.3145 \text{ J}}{\text{mol K}} \times 273 \text{ K}}{1.604 \times 10^{-2} \text{ kg/mol}} \right)^{1/2} = 652 \text{ m/s}$

Similarly u_{rms} for CH_4 at 546 K is 921 m/s.

72. $u_{rms} = \left(\dfrac{3RT}{M} \right)^{1/2}$, where $R = \dfrac{8.3145 \text{ J}}{\text{mol K}}$ and M = 2.802×10^{-2} kg/mol for N_2

For N_2 at 273 K: $u_{rms} = \left(\dfrac{\dfrac{3 \times 8.3145 \text{ J}}{\text{mol K}} \times 273 \text{ K}}{2.802 \times 10^{-2} \text{ kg/mol}} \right)^{1/2} = 493 \text{ m/s}$

Similarly for N_2 at 546 K, u_{rms} = 697 m/s.

73.   No, the number calculated in 5.69 is the average kinetic energy. There is a distribution of energies.

74.   No, the number calculated in 5.71 is an average velocity. There is a distribution of velocities.

75. $KE_{ave} = (3/2) \, RT$ and $KE = (1/2) \, mv^2$; As the temperature increases, the average kinetic energy of the gas sample will increase. The average kinetic energy increases because the increased temperature results in an increase in the average velocity of the gas molecules.

76.

	a	b	c	d
avg. KE	inc	dec	same (KE $\propto$ T)	same
avg. velocity	inc	dec	same ($\frac{1}{2}\, mv^2$ = KE $\propto$ T)	same
coll. freq wall	inc	dec	inc	inc

Average kinetic energy and average velocity depend on T. As T increases, both average kinetic energy and average velocity increase. At constant T, both average kinetic energy and average velocity are constant. The collision frequency is proportional to the average velocity (as velocity increases it takes less time to move to the next collision) and to the quantity n/V (as molecules per volume increases, collision frequency increases).

77.   a.   They will all have the same average kinetic energy since they are all at the same temperature.

b.   Flask C; H_2 has the smallest molar mass. At constant T, the lightest molecules are the fastest (on the average). This must be true in order for the average kinetic energies to be constant.

78.   a.   All the gases have the same average kinetic energy since they are all at the same temperature.

b.   At constant T, the lighter the gas molecule, the faster the average velocity.

Ar (39.95 g/mol) < Ne (20.18 g/mol) < CH_4 (16.04 g/mol) < He (4.003 g/mol).
slowest   fastest

c.   At constant T, the lighter He atoms have a faster average velocity than the heavier Ne atoms. As temperature increases, the average velocity of the gas molecules increase. Separate samples of He and Ne can only have the same average velocities if the temperature of the Ne sample is greater than the temperature of the He sample.

79.   Graham's law of effusion:

$$\frac{Rate_1}{Rate_2} = \left(\frac{M_2}{M_1}\right)^{1/2} \text{ where M = molar mass; } \frac{31.50}{30.50} = \left(\frac{32.00}{M}\right)^{1/2} = 1.033$$

$\frac{32.00}{M}$ = 1.067, so M = 29.99 g/mol; Of the choices, the gas would be NO, nitrogen monoxide.

80. $$\frac{Rate_1}{Rate_2} = \left(\frac{M_2}{M_1}\right)^{1/2}; \text{ } Rate_1 = \frac{24.0 \text{ mL}}{min}, \text{ } Rate_2 = \frac{47.8 \text{ mL}}{min}, \text{ } M_2 = \frac{16.04 \text{ g}}{mol} \text{ and } M_1 = ?$$

$$\frac{24.0}{47.8} = \left(\frac{16.04}{M_1}\right)^{1/2} = 0.502; \text{ } 16.04 = (0.502)^2 \times M_1; \text{ } M_1 = \frac{16.04}{0.252} = \frac{63.7 \text{ g}}{mol}$$

81. $$\frac{Rate_1}{Rate_2} = \left(\frac{M_2}{M_1}\right)^{1/2}, \text{ } \frac{^{12}C\,^{17}O}{^{12}C\,^{18}O} = \left(\frac{30.0}{29.0}\right)^{1/2} = 1.02; \text{ } \frac{^{12}C\,^{16}O}{^{12}C\,^{18}O} = \left(\frac{30.0}{28.0}\right)^{1/2} = 1.04$$

The relative rates of effusion of $^{12}C^{16}O$: $^{12}C^{17}O$: $^{12}C^{18}O$ are 1.04: 1.02: 1.00.

Advantage: CO_2 isn't as toxic as CO.

Major disadvantages of using CO_2 instead of CO:

1.   Can get a mixture of oxygen isotopes in CO_2.

2. Some species, e.g., $^{12}C^{16}O^{18}O$ and $^{12}C^{17}O_2$, would effuse at about the same rate since the masses are about equal. Thus, some species cannot be separated from each other.

82. $\dfrac{\text{Rate}_1}{\text{Rate}_2} = \left(\dfrac{M_2}{M_1}\right)^{1/2}$ where M = molar mass; Let Gas (1) = He, Gas (2) = Cl_2

$\dfrac{\dfrac{1.0\ L}{4.5\ \text{min}}}{\dfrac{1.0\ L}{t}} = \left(\dfrac{70.90}{4.003}\right)^{1/2}, \quad \dfrac{t}{4.5\ \text{min}} = 4.209, \ t = 19\ \text{min}$

83.   a. $P = \dfrac{nRT}{V} = \dfrac{0.5000\ \text{mol} \times \dfrac{0.08206\ L\ \text{atm}}{\text{mol K}} \times (25.0 + 273.2)\ K}{1.0000\ L} = 12.24\ \text{atm}$

b. $\left[P + a\left(\dfrac{n}{V}\right)^2\right] \times (V - nb) = nRT$; For N_2: a = 1.39 atm L^2/mol^2 and b = 0.0391 L/mol

$\left[P + 1.39\left(\dfrac{0.5000}{1.0000}\right)^2 \text{atm}\right] \times (1.0000\ L - 0.5000 \times 0.0391\ L) = 12.24\ L\ \text{atm}$

(P + 0.348 atm) × (0.9805 L) = 12.24 L atm

$P = \dfrac{12.24\ L\ \text{atm}}{0.9805\ L}$ - 0.348 atm = 12.48 - 0.348 = 12.13 atm

c.   The ideal gas law is high by 0.11 atm or $\dfrac{0.11}{12.13} \times 100 = 0.91\%$.

84.   a. $P = \dfrac{nRT}{V} = \dfrac{0.5000\ \text{mol} \times \dfrac{0.08206\ L\ \text{atm}}{\text{mol K}} \times 298.2\ K}{10.000\ L} = 1.224\ \text{atm}$

b. $\left[P + a\left(\dfrac{n}{V}\right)^2\right] \times (V - nb) = nRT$; For N_2: a = 1.39 atm L^2/mol^2 and b = 0.0391 L/mol

$\left[P + 1.39\left(\dfrac{0.5000}{10.000}\right)^2 \text{atm}\right] \times (10.000\ L - 0.5000 \times 0.0391\ L) = 12.24\ L\ \text{atm}$

(P + 0.00348 atm) × (10.000 L - 0.0196 L) = 12.24 L atm

$P + 0.00348\ \text{atm} = \dfrac{12.24\ L\ \text{atm}}{9.980\ L}$ = 1.226 atm, P = 1.226 - 0.00348 = 1.223 atm

c.   The results agree to ± 0.001 atm (0.08%).

d. In 5.83 the pressure is relatively high and there is a significant disagreement. In 5.84 the pressure is around 1 atm and both gas laws show better agreement. The ideal gas law is valid at relatively low pressures.

Atmospheric Chemistry

85. $\chi_{NO} = 5 \times 10^{-7}$ from Table 5.4. $P_{NO} = \chi_{NO} \times P_{total} = 5 \times 10^{-7} \times 1.0$ atm $= 5 \times 10^{-7}$ atm

$$PV = nRT, \frac{n}{V} = \frac{P}{RT} = \frac{5 \times 10^{-7} \text{ atm}}{\dfrac{0.08206 \text{ L atm}}{\text{mol K}} \times 273 \text{ K}} = 2 \times 10^{-8} \text{ mol NO/L}$$

$$\frac{2 \times 10^{-8} \text{ mol}}{L} \times \frac{1 \text{ L}}{1000 \text{ cm}^3} \times \frac{6.022 \times 10^{23} \text{ molecules}}{\text{mol}} = 1 \times 10^{13} \text{ molecules NO/cm}^3$$

86. $\chi_{He} = 5.24 \times 10^{-6}$ from Table 5.4. $P_{He} = \chi_{He} \times P_{total} = 5.24 \times 10^{-6} \times 1.0$ atm $= 5.2 \times 10^{-6}$ atm

$$\frac{n}{V} = \frac{P}{RT} = \frac{5.2 \times 10^{-6} \text{ atm}}{\dfrac{0.08206 \text{ L atm}}{\text{mol K}} \times 298 \text{ K}} = 2.1 \times 10^{-7} \text{ mol He/L}$$

$$\frac{2 \times 10^{-7} \text{ mol}}{L} \times \frac{1 \text{ L}}{1000 \text{ cm}^3} \times \frac{6.022 \times 10^{23} \text{ molecules}}{\text{mol}} = 1.2 \times 10^{14} \text{ molecules He/cm}^3$$

87.   At 100. km, T $\approx$ - 75°C and P $\approx 10^{-4.5} \approx 3 \times 10^{-5}$ atm.

$$PV = nRT, \quad \frac{PV}{T} = nR = \text{Constant}, \quad \frac{P_1 V_1}{T_1} = \frac{P_2 V_2}{T_2}$$

$$V_2 = \frac{V_1 P_1 T_2}{P_2 T_1} = \frac{10.0 \text{ L} \times (3 \times 10^{-5} \text{ atm}) \times 273 \text{ K}}{1.0 \text{ atm} \times 198 \text{ K}} = 4 \times 10^{-4} \text{ L} = 0.4 \text{ mL}$$

88.   At 15 km, T $\approx$ -50°C and P = 0.1 atm. Use $\dfrac{P_1 V_1}{T_1} = \dfrac{P_2 V_2}{T_2}$ since n is constant.

$$V_2 = \frac{V_1 P_1 T_2}{P_2 T_1} = \frac{1.0 \text{ L} \times 1.00 \text{ atm} \times 223 \text{ K}}{0.1 \text{ atm} \times 298 \text{ K}} = 7 \text{ L}$$

89. $N_2(g) + O_2(g) \rightarrow 2 \text{ NO}(g)$, automobile combustion or formed by lightning

$2 \text{ NO}(g) + O_2(g) \rightarrow 2 \text{ NO}_2(g)$, reaction with atmospheric O_2

$2 \text{ NO}_2(g) + H_2O(l) \rightarrow HNO_3(aq) + HNO_2(aq)$, reaction with atmospheric H_2O

$S(s) + O_2(g) \rightarrow SO_2(g)$, combustion of coal

$2 SO_2(g) + O_2(g) \rightarrow 2SO_3(g)$, reaction with atmospheric O_2

$H_2O(l) + SO_3(g) \rightarrow H_2SO_4(aq)$, reaction with atmospheric H_2O

90. $2 HNO_3(aq) + CaCO_3(s) \rightarrow Ca(NO_3)_2(aq) + H_2O(l) + CO_2(g)$

$H_2SO_4(aq) + CaCO_3(s) \rightarrow CaSO_4(aq) + H_2O(l) + CO_2(g)$

Additional Exercises

91.   a. $PV = nRT$ b. $PV = nRT$ c. $PV = nRT$

$PV = $ Constant $P = \left(\dfrac{nR}{V} \right) \times T = \text{Const} \times T$ $T = \left(\dfrac{P}{nR} \right) \times V = \text{Const} \times V$

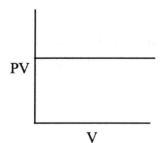   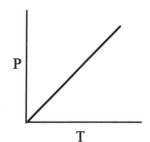   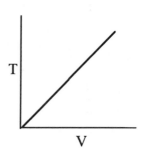

d. $PV = nRT$ e. $P = \dfrac{nRT}{V} = \dfrac{\text{Constant}}{V}$ f. $PV = nRT$

$PV = $ Constant $P = \text{Constant} \times \dfrac{1}{V}$ $\dfrac{PV}{T} = nR = \text{Constant}$

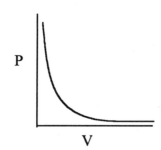   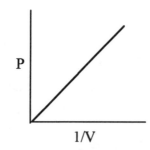   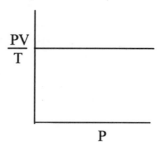

92.   At constant T and P, Avogadro's law applies, that is, equal volumes contain equal moles of molecules. In terms of balanced equations, we can say that mol ratios and volume ratios between the various reactants and products will be equal to each other. $Br_2 + 3 F_2 \rightarrow 2 X$; Two moles of X must contain two moles of Br and 6 moles of F; X must have the formula BrF_3.

93.   Processes a, c and e will all result in a doubling of the pressure. Process a has the effect of halving the volume, which would double the pressure (Boyle's law). Process c doubles the pressure because the absolute temperature is doubled (from 200. K to 400. K). Process e doubles the pressure because

the moles of gas are doubled (28 g N_2 is 1 mol of N_2 and 32 g O_2 is 1 mol of O_2). Process b won't double the pressure since the absolute temperature is not doubled (303 K to 333 K). Process d won't double the pressure because 28 g O_2 is less than one mol of gas.

94. $$n = \frac{PV}{RT} = \frac{\left(1.00 \times 10^{-6} \text{ torr} \times \frac{1 \text{ atm}}{760 \text{ torr}}\right) \times 1.00 \text{ L}}{\frac{0.08206 \text{ L atm}}{\text{mol K}} \times 295 \text{ K}} = 5.44 \times 10^{-11} \text{ mol}$$

$$5.44 \times 10^{-11} \text{ mol} \times \frac{6.022 \times 10^{23} \text{ molecules}}{\text{mol}} = 3.28 \times 10^{13} \text{ molecules}$$

95.   We will apply Boyle's law to solve. $PV = nRT = $ constant, $P_1V_1 = P_2V_2$

Let condition (1) correspond to He from the tank that can be used to fill balloons. We must leave 1.0 atm of He in the tank, so $P_1 = 200.$ atm - 1.00 = 199 atm and $V_1 = 15.0$ L. Condition (2) will correspond to the filled balloons with $P_2 = 1.00$ atm and $V_2 = N(2.00 \text{ L})$ where N is the number of filled balloons, each at a volume of 2.00 L.

199 atm $\times$ 15.0 L = 1.00 atm $\times$ N(2.00 L), N = 1492.5; We can't fill 0.5 of a balloon. So N = 1492 balloons or to 3 significant figures, 1490 balloons.

96. $PV = nRT$, n is constant. $\dfrac{PV}{T} = nR = $ constant, $\dfrac{P_1V_1}{T_1} = \dfrac{P_2V_2}{T_2}$

$V_2 = 1.040 \; V_1$, so $\dfrac{V_1}{V_2} = \dfrac{1.000}{1.040}$

$$P_2 = \frac{P_1V_1T_2}{V_2T_1} = 100. \text{ psi} \times \frac{1.000}{1.040} \times \frac{(273 + 58) \text{ K}}{(273 + 19) \text{ K}} = 109 \text{ psi}$$

97. $P_1V_1 = P_2V_2$; The total volume is 1.00 L + 1.00 L + 2.00 L = 4.00 L.

For He: $P_2 = \dfrac{P_1V_1}{V_2} = 180. \text{ torr} \times \dfrac{1.00 \text{ L}}{4.00 \text{ L}} = 45.0 \text{ torr He}$

For Ne: $P_2 = 0.450 \text{ atm} \times \dfrac{1.00 \text{ L}}{4.00 \text{ L}} = 0.113 \text{ atm};$ $0.113 \text{ atm} \times \dfrac{760 \text{ torr}}{\text{atm}} = 85.9 \text{ torr Ne}$

For Ar: $P_2 = 25.0 \text{ kPa} \times \dfrac{2.0 \text{ L}}{4.0 \text{ L}} = 12.5 \text{ kPa};$ $12.5 \text{ kPa} \times \dfrac{1 \text{ atm}}{101.3 \text{ kPa}} \times \dfrac{760 \text{ torr}}{\text{atm}} = 93.8 \text{ torr Ar}$

$P_{\text{total}} = P_{\text{He}} + P_{\text{Ne}} + P_{\text{Ar}} = 45.0 + 85.9 + 93.8 = 224.7 \text{ torr}$

98.   mol of He removed $= \dfrac{PV}{RT} = \dfrac{1.00 \text{ atm} \times 1.75 \times 10^{-3} \text{ L}}{\dfrac{0.08206 \text{ L atm}}{\text{mol K}} \times 298 \text{ K}} = 7.16 \times 10^{-5} \text{ mol}$

In the original flask, 7.16×10^{-5} mol of He exerted a partial pressure of $1.960 - 1.710 = 0.250$ atm.

$$V = \frac{nRT}{P} = \frac{7.16 \times 10^{-5} \text{ mol} \times 0.08206 \times 298 \text{ K}}{0.250 \text{ atm}} = 7.00 \times 10^{-3} \text{ L} = 7.00 \text{ mL}$$

99.   For O_2, n and T are constant, so $P_1V_1 = P_2V_2$.

$$P_1 = \frac{P_2V_2}{V_1} = 785 \text{ torr} \times \frac{1.94 \text{ L}}{2.00 \text{ L}} = 761 \text{ torr} = P_{O_2}$$

$$P_{total} = P_{O_2} + P_{H_2O}, \quad P_{H_2O} = 785 - 761 = 24 \text{ torr}$$

100.   The balanced equation is: $CaC_2(s) + 2 H_2O(l) \rightarrow C_2H_2(g) + Ca(OH)_2(aq)$

$$5.20 \text{ g CaC}_2 \times \frac{1 \text{ mol CaC}_2}{64.10 \text{ g CaC}_2} \times \frac{1 \text{ mol C}_2\text{H}_2}{\text{mol CaC}_2} = 8.11 \times 10^{-2} \text{ mol C}_2\text{H}_2$$

$$P_{C_2H_2} = 715 - 23.8 = 691 \text{ torr}; \quad \chi_{C_2H_2} = \frac{P_{C_2H_2}}{P_{total}} = \frac{691}{715} = 0.966$$

$$\chi_{C_2H_2} = 0.966 = \frac{n_{C_2H_2}}{n_{total}} = \frac{8.11 \times 10^{-2} \text{ mol}}{n_{total}}, \quad n_{total} = 8.40 \times 10^{-2} \text{ mol}$$

$$V_{wet} = V_{total} = \frac{n_{total} \times RT}{P_{total}} = \frac{8.40 \times 10^{-2} \text{ mol} \times 0.08206 \times 298 \text{ K}}{715/760 \text{ atm}} = 2.18 \text{ L}$$

101. $750. \text{ mL juice} \times \dfrac{12 \text{ mL C}_2\text{H}_5\text{OH}}{100 \text{ mL juice}} = 90. \text{ mL C}_2\text{H}_5\text{OH present}$

$$90. \text{ mL C}_2\text{H}_5\text{OH} \times \frac{0.79 \text{ g C}_2\text{H}_5\text{OH}}{\text{mL C}_2\text{H}_5\text{OH}} \times \frac{1 \text{ mol C}_2\text{H}_5\text{OH}}{46.07 \text{ g C}_2\text{H}_5\text{OH}} \times \frac{2 \text{ mol CO}_2}{2 \text{ mol C}_2\text{H}_5\text{OH}} = 1.5 \text{ mol CO}_2$$

The CO_2 will occupy $(825 - 750. =)$ 75 mL not occupied by the liquid (headspace).

$$P_{CO_2} = \frac{n_{CO_2} \times RT}{V} = \frac{1.5 \text{ mol} \times \dfrac{0.08206 \text{ L atm}}{\text{mol K}} \times 298 \text{ K}}{75 \times 10^{-3} \text{ L}} = 490 \text{ atm}$$

Actually, enough CO_2 will dissolve in the wine to lower the pressure of CO_2 to a much more reasonable value.

102. $PV = nRT$, V and T are constant. $\dfrac{P_1}{n_1} = \dfrac{P_2}{n_2}$ or $\dfrac{P_1}{P_2} = \dfrac{n_1}{n_2}$

When V and T are constant, then pressure is directly proportional to moles of gas present and pressure ratios are indentical to mol ratios.

At 25°C: $2 H_2(g) + O_2(g) \rightarrow 2 H_2O(l)$, $H_2O(l)$ is produced.

The balanced equation requires 2 mol H_2 for every mol O_2 reacted. The same ratio (2:1) holds true for pressure units. The actual pressure ratio present is 2 atm H_2 to 3 atm O_2, well below the required 2:1 ratio. Therefore H_2 is the limiting reagent. The only gas present at 25 °C after the reaction goes to completion will be the excess O_2.

$$P_{O_2} \text{(reacted)} = 2.00 \text{ atm } H_2 \times \frac{1 \text{ atm } O_2}{2 \text{ atm } H_2} = 1.00 \text{ atm } O_2$$

$$P_{O_2} \text{(excess)} = P_{O_2} \text{(initially)} - P_{O_2} \text{(reacted)} = 3.00 \text{ atm} - 1.00 \text{ atm} = 2.00 \text{ atm } O_2$$

At 125 °C: $2 H_2(g) + O_2(g) \rightarrow 2 H_2O(g)$, $H_2O(g)$ is produced.

The major difference in the problem is that gaseous water is now a product, which will increase the total pressure.

$$P_{H_2O} \text{(produced)} = 2.00 \text{ atm } H_2 \times \frac{2 \text{ atm } H_2O}{2 \text{ atm } H_2} = 2.00 \text{ atm } H_2O$$

$$P_{total} = P_{O_2} \text{(excess)} + P_{H_2O} \text{(produced)} = 2.00 \text{ atm } O_2 + 2.00 \text{ atm } H_2O = 4.00 \text{ atm}$$

103. $$d_{UF_6} = \frac{P \times \text{(molar mass)}}{RT} = \frac{\left(745 \text{ torr} \times \dfrac{1 \text{ atm}}{760 \text{ torr}} \right) \times 352.0 \text{ g/mol}}{\dfrac{0.08206 \text{ L atm}}{\text{mol K}} \times 333 \text{ K}} = 12.6 \text{ g/L}$$

104.   If Be^{3+}, the formula is $Be(C_5H_7O_2)_3$ and $M \approx 13.5 + 15(12) + 21(1) + 6(16) = 311$ g/mol.

If Be^{2+}, the formula is $Be(C_5H_7O_2)_2$ and $M \approx 9.0 + 10(12) + 14(1) + 4(16) = 207$ g/mol.

Data Set I   ($M = dRT/P$ and $d = \text{mass}/V$):

$$M = \frac{\text{mass} \times RT}{PV} = \frac{0.2022 \text{ g} \times \dfrac{0.08206 \text{ L atm}}{\text{mol K}} \times 286 \text{ K}}{(765.2 \text{ torr} \times \dfrac{1 \text{ atm}}{760 \text{ torr}}) \times 22.6 \times 10^{-3} \text{ L}} = 209 \text{ g/mol}$$

Data Set II:

$$M = \frac{\text{mass} \times RT}{PV} = \frac{0.2224 \text{ g} \times \dfrac{0.08206 \text{ L atm}}{\text{mol K}} \times 290. \text{ K}}{(764.6 \text{ torr} \times \dfrac{1 \text{ atm}}{760 \text{ torr}}) \times 26.0 \times 10^{-3} \text{ L}} = 202 \text{ g/mol}$$

These results are closed to the expected value of 207 g/mol for $Be(C_5H_7O_2)_2$. Thus, we conclude from these data that beryllium is a divalent element with an atomic weight (mass) of 9.0 g/mol.

105.   Out of 100.0 g of compound, there are:

$$87.4 \text{ g N} \times \frac{1 \text{ mol N}}{14.01 \text{ g N}} = 6.24 \text{ mol N}; \quad \frac{6.24}{6.24} = 1.00$$

$$12.6 \text{ g H} \times \frac{1 \text{ mol H}}{1.008 \text{ g H}} = 12.5 \text{ mol H}; \quad \frac{12.5}{6.24} = 2.00$$

Empirical formula is NH_2.

$$\text{molar mass} = M = \frac{dRT}{P} = \frac{\dfrac{0.977 \text{ g}}{L} \times \dfrac{0.08206 \text{ L atm}}{\text{mol K}} \times 373 \text{ K}}{710. \text{ torr} \times \dfrac{1 \text{ atm}}{760 \text{ torr}}} = 32.0 \text{ g/mol}$$

Empirical formula mass of $NH_2 = 16.0$ g. Therefore, molecular formula is N_2H_4.

106.   Out of 100.00 g compounds, there are:

$$58.51 \text{ g C} \times \frac{1 \text{ mol C}}{12.01 \text{ g C}} = 4.872 \text{ mol C}; \quad \frac{4.872}{2.435} = 2.001$$

$$7.37 \text{ g H} \times \frac{1 \text{ mol H}}{1.008 \text{ g H}} = 7.31 \text{ mol H}; \quad \frac{7.31}{2.435} = 3.00$$

$$34.12 \text{ g N} \times \frac{1 \text{ mol N}}{14.01 \text{ g N}} = 2.435 \text{ mol N}; \quad \frac{2.435}{2.435} = 1.000$$

Empirical formula: C_2H_3N

$$\frac{\text{Rate}_1}{\text{Rate}_2} = \left(\frac{M_2}{M_1} \right)^{1/2}; \quad \text{Let Gas (1) = He}; \quad 3.20 = \left(\frac{M_2}{4.003} \right)^{1/2}, \quad M_2 = 41.0 \text{ g/mol}$$

Empirical formula mass of $C_2H_3N \approx 2(12.0) + 3(1.0) + 1(14.0) = 41.0$. So molecular formula is also C_2H_3N.

107. $P_{total} = P_{N_2} + P_{H_2O}$, $P_{N_2} = 726 \text{ torr} - 23.8 \text{ torr} = 702 \text{ torr} \times \dfrac{1 \text{ atm}}{760 \text{ torr}} = 0.924 \text{ atm}$

$$PV = nRT, \quad n_{N_2} = \frac{P_{N_2} \times V}{RT} = \frac{0.924 \text{ atm} \times 31.8 \times 10^{-3} \text{ L}}{\dfrac{0.08206 \text{ L atm}}{\text{mol K}} \times 298 \text{ K}} = 1.20 \times 10^{-3} \text{ mol N}_2$$

Mass of N in compound $= 1.20 \times 10^{-3} \text{ mol} \times \dfrac{28.02 \text{ g N}_2}{\text{mol}} = 3.36 \times 10^{-2} \text{ g}$

$$\% \text{ N} = \frac{3.36 \times 10^{-2} \text{ g}}{0.253 \text{ g}} \times 100 = 13.3\% \text{ N}$$

108. $33.5 \text{ mg } CO_2 \times \dfrac{12.01 \text{ mg C}}{44.01 \text{ mg } CO_2} = 9.14 \text{ mg C};\ \% \text{ C} = \dfrac{9.14 \text{ mg}}{35.0 \text{ mg}} \times 100 = 26.1\% \text{ C}$

$41.1 \text{ mg } H_2O \times \dfrac{2.016 \text{ mg H}}{18.02 \text{ mg } H_2O} = 4.60 \text{ mg H};\ \% \text{ H} = \dfrac{4.60 \text{ mg}}{35.0 \text{ mg}} \times 100 = 13.1\% \text{ H}$

$n_{N_2} = \dfrac{P_{N_2} \times V}{RT} = \dfrac{\dfrac{740.}{760} \text{ atm} \times 35.6 \times 10^{-3} \text{ L}}{\dfrac{0.08206 \text{ L atm}}{\text{mol K}} \times 298 \text{ K}} = 1.42 \times 10^{-3} \text{ mol } N_2$

$1.42 \times 10^{-3} \text{ mol } N_2 \times \dfrac{28.02 \text{ g } N_2}{\text{mol } N_2} = 3.98 \times 10^{-2} \text{ g N} = 39.8 \text{ mg N}$

$\% \text{ N} = \dfrac{39.8 \text{ mg}}{65.2 \text{ mg}} \times 100 = 61.0\% \text{ N}$

Or we can get % N by difference: % N = 100.0 - (26.1 + 13.1) = 60.8%

Out of 100.0 g:

$26.1 \text{ g C} \times \dfrac{1 \text{ mol}}{12.01 \text{ g}} = 2.17 \text{ mol C};\ \dfrac{2.17}{2.17} = 1.00$

$13.1 \text{ g H} \times \dfrac{1 \text{ mol}}{1.008 \text{ g}} = 13.0 \text{ mol H};\ \dfrac{13.0}{2.17} = 5.99$

$60.8 \text{ g N} \times \dfrac{1 \text{ mol}}{14.01 \text{ g}} = 4.34 \text{ mol N};\ \dfrac{4.34}{2.17} = 2.00$

Empirical formula is CH_6N_2.

$\dfrac{\text{Rate}_1}{\text{Rate}_2} = \left(\dfrac{M}{39.95} \right)^{1/2} = \dfrac{26.4}{24.6} = 1.07,\ M = (1.07)^2 \times 39.95 = 45.7 \text{ g/mol}$

Empirical formula mass of $CH_6N_2 \approx 12 + 6 + 28 = 46$. Thus, molecular formula is also CH_6N_2.

109.   The pressure will increase because the lighter H_2 molecules will effuse into container A faster than air will escape.

110.   The van der Waals' constant b is a measure of the size of the molecule. Thus, C_3H_8 should have the largest value of b since it has the largest molar mass (size).

111.   The values of a are: H_2, $\dfrac{0.244 \text{ L}^2 \text{ atm}}{\text{mol}^2}$; CO_2, 3.59; N_2, 1.39; CH_4, 2.25

Since a is a measure interparticle attractions, the attractions are greatest for CO_2.

Challenge Problems

112. $PV = nRT$, V and T are constant. $\dfrac{P_1}{n_1} = \dfrac{P_2}{n_2}, \ \dfrac{P_2}{P_1} = \dfrac{n_2}{n_1}$

We will do this limiting reagent problem using an alternative method. Let's calculate the partial pressure of C_3H_3N that can be produced from each of the starting materials assuming each reactant is limiting. The reactant that produces the smallest amount of product will run out first and is the limiting reagent.

$$P_{C_3H_3N} = 0.500 \text{ MPa} \times \frac{2 \text{ mol } C_3H_3N}{2 \text{ mol } C_3H_6} = 0.500 \text{ MPa if } C_3H_6 \text{ is limiting.}$$

$$P_{C_3H_3N} = 0.800 \text{ MPa} \times \frac{2 \text{ mol } C_3H_3N}{2 \text{ mol } NH_3} = 0.800 \text{ MPa if } NH_3 \text{ is limiting.}$$

$$P_{C_3H_3N} = 1.500 \text{ MPa} \times \frac{2 \text{ mol } C_3H_3N}{3 \text{ mol } O_2} = 1.000 \text{ MPa if } O_2 \text{ is limiting.}$$

Thus, C_3H_6 is limiting. Although more product could be produced from NH_3 and O_2, there is only enough C_3H_6 to produce 0.500 MPa of C_3H_3N. The partial pressure of C_3H_3N after the reaction is:

$$0.500 \times 10^6 \text{ Pa} \times \frac{1 \text{ atm}}{1.013 \times 10^5 \text{ Pa}} = 4.94 \text{ atm}$$

$$n = \frac{PV}{RT} = \frac{4.94 \text{ atm} \times 150. \text{ L}}{\dfrac{0.08206 \text{ L atm}}{\text{mol K}} \times 298 \text{ K}} = 30.3 \text{ mol } C_3H_3N$$

$$30.3 \text{ mol} \times \frac{53.06 \text{ g}}{\text{mol}} = 1.61 \times 10^3 \text{ g } C_3H_3N \text{ can be produced.}$$

113. $BaO(s) + CO_2(g) \rightarrow BaCO_3(s); \ CaO(s) + CO_2(g) \rightarrow CaCO_3(s)$

$$n_i = \frac{P_iV}{RT} = \text{initial moles of } CO_2 = \frac{\dfrac{750.}{760} \text{ atm} \times 1.50 \text{ L}}{\dfrac{0.08206 \text{ L atm}}{\text{mol K}} \times 303.2 \text{ K}} = 0.0595 \text{ mol } CO_2$$

$$n_f = \frac{P_fV}{RT} = \text{final moles of } CO_2 = \frac{\dfrac{230.}{760} \text{ atm} \times 1.50 \text{ L}}{\dfrac{0.08206 \text{ L atm}}{\text{mol K}} \times 303.2 \text{ K}} = 0.0182 \text{ mol } CO_2$$

$0.0595 - 0.0182 = 0.0413 \text{ mol } CO_2$ reacted.

Since each metal reacts 1:1 with CO_2, then the mixture contains 0.0413 mol of BaO and CaO. The molar masses of BaO and CaO are 153.3 g/mol and 56.08 g/mol, respectively.

Let x = g BaO and y = g CaO, so:

$$x + y = 5.14 \text{ g and } \frac{x}{153.3} + \frac{y}{56.08} = 0.0413 \text{ mol}$$

Solving by simultaneous equations:

$$
\begin{array}{r}
x + 2.734\,y = 6.33 \\
\underline{-x -y = -5.14} \\
1.734\,y = 1.19
\end{array}
$$

y = 0.686 g CaO and 5.14 - y = x = 4.45 g BaO

$$\% \text{ BaO} = \frac{4.45 \text{ g BaO}}{5.14 \text{ g}} \times 100 = 86.6\% \text{ BaO}; \quad \% \text{ CaO} = 100.0 - 86.6 = 13.4\% \text{ CaO}$$

114. $Cr(s) + 3HCl(aq) \rightarrow CrCl_3(s) + 3/2\ H_2(g)$; $Zn(s) + 2HCl(aq) \rightarrow ZnCl_2(s) + H_2(g)$

$$\text{mol } H_2 \text{ produced} = n = \frac{PV}{RT} = \frac{\left(750.\text{ torr} \times \dfrac{1 \text{ atm}}{760 \text{ torr}}\right) \times 0.225 \text{ L}}{\dfrac{0.08206 \text{ L atm}}{\text{mol K}} \times (273 + 27) \text{ K}} = 9.02 \times 10^{-3} \text{ mol } H_2$$

9.02×10^{-3} mol H_2 = mol H_2 from Cr reaction + mol H_2 from Zn reaction

From the balanced equation: 9.02×10^{-3} mol H_2 = mol Cr × (3/2) + mol Zn × 1

Let x = mass of Cr and y = mass of Zn, then:

$$x + y = 0.362 \text{ g and } 9.02 \times 10^{-3} = \frac{1.5\,x}{52.00} + \frac{y}{65.38}$$

We have two equations and two unknowns. Solving by simultaneous equations:

$$
\begin{array}{l}
9.02 \times 10^{-3} = 0.02885\,x + 0.01530\,y \\
\underline{-0.01530 \times 0.362 = -0.01530\,x - 0.01530\,y} \\
3.48 \times 10^{-3} = 0.01355\,x
\end{array}
$$

$$x = \text{mass Cr} = \frac{3.48 \times 10^{-3}}{0.01355} = 0.257 \text{ g}$$

y = mass Zn = 0.362 g - 0.257 g = 0.105 g Zn; $\text{mass \% Zn} = \dfrac{0.105 \text{ g}}{0.362 \text{ g}} \times 100 = 29.0\% \text{ Zn}$

115.   a.   The reaction is: $CH_4(g) + 2\ O_2(g) \rightarrow CO_2(g) + 2\ H_2O(g)$

$$PV = nRT; \quad \frac{PV}{n} = RT = \text{constant}; \quad \frac{P_{CH_4} V_{CH_4}}{n_{CH_4}} = \frac{P_{O_2} V_{O_2}}{n_{O_2}}$$

For three fold excess of O_2: $n_{O_2} = 6\,n_{CH_4}$, $\dfrac{n_{O_2}}{n_{CH_4}} = 6$; $P_{O_2} = 0.21\ P_{air} = 0.21 \text{ atm}$

In one minute:

$$V_{O_2} = V_{CH_4} \times \frac{n_{O_2}}{n_{CH_4}} \times \frac{P_{CH_4}}{P_{O_2}} = 200. \text{ L} \times 6 \times \frac{1.50 \text{ atm}}{0.21 \text{ atm}} = 8.6 \times 10^3 \text{ L O}_2$$

We need: $\dfrac{8.6 \times 10^3 \text{ L O}_2}{\text{min}} \times \dfrac{100 \text{ L air}}{21 \text{ L O}_2} = 4.1 \times 10^4 \text{ L air/min}$

b. If n moles of CH_4 were reacted, then 6 n mol O_2 were added, producing 0.950 n mol CO_2 and 0.050 n mol of CO. In addition, 2 n mol H_2O must be produced to balance the hydrogens.

$$CH_4 + 2 \, O_2 \rightarrow CO_2 + 2 \, H_2O; \ \ CH_4 + 3/2 \, O_2 \rightarrow CO + 2 \, H_2O$$

Amount O_2 reacted:

$$0.950 \text{ n mol CO}_2 \times \frac{2 \text{ mol O}_2}{\text{mol CO}_2} = 1.90 \text{ n mol O}_2$$

$$0.050 \text{ n mol CO} \times \frac{1.5 \text{ mol O}_2}{\text{mol CO}} = 0.075 \text{ n mol O}_2$$

Amount of O_2 left in reaction mixture = 6.00 n - 1.90 n - 0.075 n = 4.03 n mol O_2

Amount of N_2 remaining = $6.00 \text{ n mol O}_2 \times \dfrac{79 \text{ mol N}_2}{21 \text{ mol O}_2} = 22.6 \text{ n} \approx 23 \text{ n mol N}_2$

The reaction mixture contains:

$$0.950 \text{ n mol CO}_2 + 0.050 \text{ n mol CO} + 4.03 \text{ n mol O}_2 + 2.00 \text{ n mol H}_2\text{O}$$

$$+ \ 23 \text{ n mol N}_2 = 30. \text{ n total mol of gas}$$

$$\chi_{CO} = \frac{0.050 \text{ n}}{30. \text{ n}} = 0.0017; \ \ \chi_{CO_2} = \frac{0.950 \text{ n}}{30. \text{ n}} = 0.032; \ \ \chi_{O_2} = \frac{4.03 \text{ n}}{30. \text{ n}} = 0.13;$$

$$\chi_{H_2O} = \frac{2.00 \text{ n}}{30. \text{ n}} = 0.067; \ \ \ \chi_{N_2} = \frac{23 \text{ n}}{30. \text{ n}} = 0.77$$

116. $M = \dfrac{dRT}{P} = \dfrac{\dfrac{0.70902 \text{ g}}{\text{L}} \times \dfrac{0.08206 \text{ L atm}}{\text{mol K}} \times 273.2 \text{ K}}{1.000 \text{ atm}} = 15.90 \text{ g/mol}$

15.90 g/mol is the average molar mass of the mixture of methane and helium. Let x = mol % of CH_4. This is also the volume % of CH_4 since T and P are constant.

$15.90 = \dfrac{x \, (16.04) + (100 - x) \, (4.003)}{100}$, 1590. = 16.04 x + 400.3 - 4.003 x, 1190. = 12.04 x

x = 98.84 % CH_4 by volume; % He = 100.00 - x = 1.16 % He by volume

117.   a.   Volume of hot air: $V = \frac{4}{3}\pi^3 = \frac{4}{3}(2.50 \text{ m})\pi^3 = 65.4 \text{ m}^3$

(Note: radius = diameter/2 = 5.00/2 = 2.50 m)

$$65.4 \text{ m}^3 \times \left(\frac{10 \text{ dm}}{\text{m}}\right)^3 \times \frac{1 \text{ L}}{\text{dm}^3} = 6.54 \times 10^4 \text{ L}$$

$$n = \frac{PV}{RT} = \frac{\left(745 \text{ torr} \times \dfrac{1 \text{ atm}}{760 \text{ torr}}\right) \times 6.54 \times 10^4 \text{ L}}{\dfrac{0.08206 \text{ L atm}}{\text{mol K}} \times (273 + 65) \text{ K}} = 2.31 \times 10^3 \text{ mol air}$$

Mass of hot air $= 2.31 \times 10^3 \text{ mol} \times \dfrac{29.0 \text{ g}}{\text{mol}} = 6.70 \times 10^4 \text{ g}$

Mass of air displaced:

$$n = \frac{PV}{RT} = \frac{\dfrac{745}{760} \text{ atm} \times 6.54 \times 10^4 \text{ L}}{\dfrac{0.08206 \text{ L atm}}{\text{mol K}} \times (273 + 21) \text{ K}} = 2.66 \times 10^3 \text{ mol air}$$

Mass $= 2.66 \times 10^3 \text{ mol} \times \dfrac{29.0 \text{ g}}{\text{mol}} = 7.71 \times 10^4 \text{ g of air displaced}$

Lift $= 7.71 \times 10^4 \text{ g} - 6.70 \times 10^4 \text{ g} = 1.01 \times 10^4 \text{ g}$

b.   Mass of air displaced is the same, 7.71×10^4 g. Moles of He in balloon will be the same as moles of air displaced, 2.66×10^3 mol, since P, V and T are the same.

Mass of He $= 2.66 \times 10^3 \text{ mol} \times \dfrac{4.003 \text{ g}}{\text{mol}} = 1.06 \times 10^4 \text{ g}$

Lift $= 7.71 \times 10^4 \text{ g} - 1.06 \times 10^4 \text{ g} = 6.65 \times 10^4 \text{ g}$

c.   Mass of hot air:

$$n = \frac{PV}{RT} = \frac{\dfrac{630.}{760} \text{ atm} \times 6.54 \times 10^4 \text{ L}}{\dfrac{0.08206 \text{ L atm}}{\text{mol K}} \times 338 \text{ K}} = 1.95 \times 10^3 \text{ mol air}$$

$$1.95 \times 10^3 \text{ mol} \times \dfrac{29.0 \text{ g}}{\text{mol}} = 5.66 \times 10^4 \text{ g of hot air}$$

Mass of air displaced:

$$n = \frac{PV}{RT} = \frac{\dfrac{630.}{760} \text{ atm} \times 6.54 \times 10^4 \text{ L}}{\dfrac{0.08206 \text{ L atm}}{\text{mol K}} \times 294 \text{ K}} = 2.25 \times 10^3 \text{ mol air}$$

$$2.25 \times 10^3 \text{ mol} \times \frac{29.0 \text{ g}}{\text{mol}} = 6.53 \times 10^4 \text{ g of air displaced}$$

$$\text{Lift} = 6.53 \times 10^4 \text{ g} - 5.66 \times 10^4 \text{ g} = 8.7 \times 10^3 \text{ g}$$

118. $\left(P + \dfrac{an^2}{V^2}\right) \times (V - nb) = nRT, \quad PV + \dfrac{an^2V}{V^2} - nbP - \dfrac{an^3b}{V^2} = nRT, \quad PV + \dfrac{an^2}{V} - nbP - \dfrac{an^3b}{V^2} = nRT$

At low P and high T, the molar volume of a gas will be relatively large. The an^2/V and an^3b/V^2 terms become negligible because V is large. Since nb is the actual volume of the gas molecules themselves, then $nb \ll V$ and the -nbP term is negligible compared to PV. Thus PV = nRT.

119.   a.   If we have 1.0×10^6 L of air, then there are 3.0×10^2 L of CO.

$$P_{CO} = \chi_{CO} \times P_{total}; \quad \chi_{CO} = \frac{V_{CO}}{V_{total}} \text{ since } V \propto n; \quad P_{CO} = \frac{3.0 \times 10^2 \text{ L}}{1.0 \times 10^6 \text{ L}} \times 628 \text{ torr} = 0.19 \text{ torr}$$

b. $n_{CO} = \dfrac{P_{CO} \times V}{RT};$ Assuming 1.0 cm^3 of air = $1.0 \text{ mL} = 1.0 \times 10^{-3}$ L:

$$n_{CO} = \frac{\dfrac{0.19}{760} \text{ atm} \times 1.0 \times 10^{-3} \text{ L}}{\dfrac{0.08206 \text{ L atm}}{\text{mol K}} \times 273 \text{ K}} = 1.1 \times 10^{-8} \text{ mol CO}$$

$$1.1 \times 10^{-8} \text{ mol} \times \frac{6.022 \times 10^{23} \text{ molecules}}{\text{mol}} = 6.6 \times 10^{15} \text{ molecules CO in the } 1.0 \text{ cm}^3 \text{ of air}$$

120.   A concentration of 1.0 ppbv means there is 1.0 L of CH_2O for every 1.0×10^9 L of air measured at the same temperature and pressure. The molar volume of an ideal gas is 22.42 L at STP. Assuming 1.0×10^9 L air:

$$\frac{1.0 \text{ L CH}_2\text{O}}{1.0 \times 10^9 \text{ L}} \times \frac{1 \text{ mol CH}_2\text{O}}{22.42 \text{ L}} \times \frac{6.022 \times 10^{23} \text{ molecules}}{\text{mol}} \times \frac{1 \text{ L}}{1000 \text{ cm}^3} = \frac{2.7 \times 10^{10} \text{ molecules}}{\text{cm}^3}$$

$$V = \left(18.0 \text{ ft} \times \frac{12 \text{ in}}{\text{ft}} \times \frac{2.54 \text{ cm}}{\text{in}}\right) \times \left(24.0 \text{ ft} \times \frac{12 \text{ in}}{\text{ft}} \times \frac{2.54 \text{ cm}}{\text{in}}\right)$$

$$\times \left(8.0 \text{ ft} \times \frac{12 \text{ in}}{\text{ft}} \times \frac{2.54 \text{ cm}}{\text{in}}\right) = 9.8 \times 10^7 \text{ cm}^3$$

$$9.8 \times 10^7 \text{ cm}^3 \times \frac{2.7 \times 10^{10} \text{ molecules}}{\text{cm}^3} \times \frac{1 \text{ mol}}{6.022 \times 10^{23} \text{ molecules}} \times \frac{30.03 \text{ g}}{\text{mol}} = 1.3 \times 10^{-4} \text{ g CH}_2\text{O}$$

CHAPTER SIX

THERMOCHEMISTRY

Questions

9.   A coffee-cup calorimeter is at constant (atmospheric) pressure. The heat released or gained at constant pressure is ΔH. A bomb calorimeter is at constant volume. The heat released or gained at constant volume is ΔE.

10.   Water has a relatively large heat capacity. Because of this, the temperature fluctuations of large bodies of water (oceans) are small as compared to the temperature fluctuations of air. Hence, oceans act as a heat reservoir for areas close to them which results in smaller temperature fluctuations as compared to areas farther away from the oceans.

11.   The specific heat capacities are: 0.89 J/g•°C (Al) and 0.45 J/g•°C (Fe)
        Al would be the better choice. It has a higher heat capacity and a lower density than Fe. Using Al, the same amount of heat could be dissipated by a smaller mass, keeping the mass of the amplifier down.

12.   No matter how insulated your thermos bottle is, some heat will always escape into the surroundings. If the temperature of the thermos bottle (the surroundings) is high, less heat initially will escape from the coffee (the system) which results in your coffee staying hotter for a longer period of time.

13.   A state function is a function whose change depends only on the initial and final states and not on how one got from the initial to the final state. If H and E were not state functions, the law of conservation of energy (first law) would not be true.

14.   If Hess's law were not true it would be possible to create energy by reversing a reaction using a different series of steps. This violates the law of conservation of energy (first law). Thus, Hess's law is another statement of the law of conservation of energy.

15.   In order to compare values of ΔH to each other, a common reference (or zero) point must be chosen. The definition of ΔH_f° establishes the pure elements in their standard states as that common reference point.

16.   Advantages: H_2 burns cleanly (less pollution) and gives a lot of energy per gram of fuel.

        Disadvantages: Expense and storage.

Exercises

Potential and Kinetic Energy

17. $KE = \frac{1}{2}mv^2$; Convert mass and velocity to SI units. $1 \text{ J} = \frac{1 \text{ kg m}^2}{s^2}$

Mass $= 5.25 \text{ oz} \times \frac{1 \text{ lb}}{16 \text{ oz}} \times \frac{1 \text{ kg}}{2.205 \text{ lb}} = 0.149 \text{ kg}$

Velocity $= \frac{1.0 \times 10^2 \text{ mi}}{\text{hr}} \times \frac{1 \text{ hr}}{60 \text{ min}} \times \frac{1 \text{ min}}{60 \text{ s}} \times \frac{1760 \text{ yd}}{\text{mi}} \times \frac{1 \text{ m}}{1.094 \text{ yd}} = \frac{45 \text{ m}}{\text{s}}$

$KE = \frac{1}{2}mv^2 = \frac{1}{2} \times 0.149 \text{ kg} \times \left(\frac{45 \text{ m}}{\text{s}}\right)^2 = 150 \text{ J}$

18. $KE = \frac{1}{2}mv^2 = \frac{1}{2} \times \left(1.0 \times 10^{-5} \text{ g} \times \frac{1 \text{ kg}}{1000 \text{ g}}\right) \times \left(\frac{2.0 \times 10^5 \text{ cm}}{\text{sec}} \times \frac{1 \text{ m}}{100 \text{ cm}}\right)^2 = 2.0 \times 10^{-2} \text{ J}$

19. $KE = \frac{1}{2}mv^2 = \frac{1}{2} \times 2.0 \text{ kg} \times \left(\frac{1.0 \text{ m}}{\text{s}}\right)^2 = 1.0 \text{ J}; KE = \frac{1}{2}mv^2 = \frac{1}{2} \times 1.0 \text{ kg} \times \left(\frac{2.0 \text{ m}}{\text{s}}\right)^2 = 2.0 \text{ J}$

The 1.0 kg object with a velocity of 2.0 m/s has the greater kinetic energy.

20.   Ball A: $PE = mgz = 5.0 \text{ kg} \times \frac{9.8 \text{ m}}{s^2} \times 5.0 \text{ m} = \frac{250 \text{ kg m}^2}{s^2} = 250 \text{ J}$

At Point I: All of this energy is transferred to Ball B. All of B's energy is kinetic energy at this point. $E_{total} = KE = 250 \text{ J}$. At point II, the sum of the total energy will equal 250 J.

At Point II: $PE = mgz = 1.0 \text{ kg} \times \frac{9.8 \text{ m}}{s^2} \times 2.0 \text{ m} = 20. \text{ J}$

$KE = E_{total} - PE = 250 \text{ J} - 20. \text{ J} = 230 \text{ J}$

Heat and Work

21.   a. $\Delta E = q + w = 51 \text{ kJ} + (-15 \text{ kJ}) = 36 \text{ kJ}$

b. $\Delta E = 100. \text{ kJ} + (-65 \text{ kJ}) = 35 \text{ kJ}$ c. $\Delta E = -65 + (-20.) = -85 \text{ kJ}$

d.   When the system delivers work to the surroundings, $w < 0$. This is the case in all these examples, a, b and c.

22.   a. $\Delta E = q + w = -47 \text{ kJ} + 88 \text{ kJ} = 41 \text{ kJ}$

b. $\Delta E = 82 + 47 = 129 \text{ kJ}$ c. $\Delta E = 47 + 0 = 47 \text{ kJ}$

d.   When the surroundings deliver work to the system, $w > 0$. This is the case for a and b.

23. $\Delta E = q + w = 45 \text{ kJ} + (-29 \text{ kJ}) = 16 \text{ kJ}$

24. $\Delta E = q + w = -125 + 104 = -21 \text{ kJ}$

25. $w = -P\Delta V = -P \times (V_f - V_i) = -2.0 \text{ atm} \times (5.0 \times 10^{-3} \text{ L} - 5.0 \text{ L}) = -2.0 \text{ atm} \times (-5.0 \text{ L}) = 10. \text{ L atm}$

We can also calculate the work in Joules.

$$1 \text{ atm} = 1.013 \times 10^5 \text{ Pa} = 1.013 \times 10^5 \frac{\text{kg}}{\text{m s}^2}; \quad 1 \text{ L} = 1000 \text{ cm}^3 = 1 \times 10^{-3} \text{ m}^3$$

$$1 \text{ L atm} = 1 \times 10^{-3} \text{ m}^3 \times 1.013 \times 10^5 \frac{\text{kg}}{\text{m s}^2} = 101.3 \frac{\text{kg m}^2}{\text{s}^2} = 101.3 \text{ J}$$

$$w = 10. \text{ L atm} \times \frac{101.3 \text{ J}}{\text{L atm}} = 1013 \text{ J} = 1.0 \times 10^3 \text{ J}$$

26. In this problem $q = w = -950. \text{ J}$

$$-950. \text{ J} \times \frac{1 \text{ L atm}}{101.3 \text{ J}} = -9.38 \text{ L atm of work done by the gases.}$$

$$w = -P\Delta V, \quad -9.38 \text{ L atm} = \frac{-650.}{760} \text{ atm} \times (V_f - 0.040 \text{ L}), \quad V_f - 0.040 = 11.0 \text{ L}, \quad V_f = 11.0 \text{ L}$$

27. $q = \text{molar heat capacity} \times \text{mol} \times \Delta T = \dfrac{20.8 \text{ J}}{\text{°C mol}} \times 39.1 \text{ mol} \times (38.0 - 0.0) \text{ °C} = 30,900 \text{ J} = 30.9 \text{ kJ}$

$$w = -P\Delta V = -1.00 \text{ atm} \times (998 \text{ L} - 876 \text{ L}) = -122 \text{ L atm} \times \frac{101.3 \text{ J}}{\text{L atm}} = -12,400 \text{ J} = -12.4 \text{ kJ}$$

$$\Delta E = q + w = 30.9 \text{ kJ} + (-12.4 \text{ kJ}) = 18.5 \text{ kJ}$$

28. $313 \text{ g He} \times \dfrac{1 \text{ mol He}}{4.003 \text{ g He}} = 78.2 \text{ mol He}$

$$q = \frac{20.8 \text{ J}}{\text{mol °C}} \times 78.2 \text{ mol He} \times (-15 \text{°C}) = -2.4 \times 10^4 \text{ J or } -24 \text{ kJ}$$

$$w = -P\Delta V = -1.00 \text{ atm} (1814 - 1910.) \text{ L} = 96 \text{ L atm} \times \frac{101.3 \text{ J}}{\text{L atm}} = 9.7 \times 10^3 \text{ J} = 9.7 \text{ kJ}$$

$$\Delta E = q + w = -24 \text{ kJ} + 9.7 \text{ kJ} = -14 \text{ kJ}$$

Properties of Enthalpy

29. Since the sign of ΔH is negative, the reaction is exothermic. Heat is evolved by the system to the surroundings.

30.   One should try to cool the reaction mixture or provide some means of removing heat since the reaction is very exothermic (heat is released). The $H_2SO_4(aq)$ will get very hot and possibly boil, unless cooling is provided.

31.   Heat is absorbed in endothermic processes and heat is released in exothermic processes.

        a. endothermic   b. exothermic   c. exothermic   d. endothermic

32.   The only processes in this question where heat is a product is the conversion of a gas to a solid (process b) and the freezing of water (process d). In both cases, the reactant molecules must slow down in order to form the products, which occurs by a temperature decrease. The other processes (a and c) require heat as a reactant and are all endothermic. Process a is bond breaking which always requires outside energy to occur. Process c refers to the boiling of liquid which also always requires energy from the surroundings to occur.

33. $S(s) + O_2(g) \rightarrow SO_2(g)$ $\Delta H = \dfrac{-296\ kJ}{mol}$; Molar mass of SO_2 = 64.07 g/mol

        a. $275\ g\ S \times \dfrac{1\ mol\ S}{32.07\ g\ S} \times \dfrac{-296\ kJ}{mol\ S} = -2.54 \times 10^3\ kJ$ heat released

        b. $25\ mol\ S \times \dfrac{-296\ kJ}{mol\ S} = -7.4 \times 10^3\ kJ$;   c. $150.\ g\ SO_2 \times \dfrac{1\ mol\ SO_2}{64.07\ g\ SO_2} \times \dfrac{-296\ kJ}{mol\ SO_2} = -693\ kJ$

34. $B_2H_6(g) + 3\ O_2(g) \rightarrow B_2O_3(s) + 3\ H_2O(g)$ $\Delta H = -2035\ kJ$

        Molar mass of B_2H_6 = 2(10.81) + 6(1.008) = 27.67 g/mol

        a. $1.0\ g\ B_2H_6 \times \dfrac{1\ mol\ B_2H_6}{27.67\ g\ B_2H_6} \times \dfrac{-2035\ kJ}{mol\ B_2H_6} = -74\ kJ$ heat released

        b. $1.0\ mol\ B_2H_6 \times \dfrac{-2035\ kJ}{mol\ B_2H_6} = -2.0 \times 10^3\ kJ$

        c. $1.0 \times 10^2\ mol\ B_2H_6 \times \dfrac{-2035\ kJ}{mol} = -2.0 \times 10^5\ kJ$

        d. $10.0\ g\ O_2 \times \dfrac{1\ mol\ O_2}{32.00\ g\ O_2} = 0.313\ mol\ O_2$; $10.0\ g\ B_2H_6 \times \dfrac{1\ mol\ B_2H_6}{27.67\ g\ B_2H_6} = 0.361\ mol\ B_2H_6$

        $0.313\ mol\ O_2 \times \dfrac{1\ mol\ B_2H_6}{3\ mol\ O_2} = 0.104\ mol\ B_2H_6$

        We only have enough O_2 to completely react with 0.104 mol B_2H_6. We have 0.361 mol B_2H_6 present, so O_2 is the limiting reactant.

        $0.313\ mol\ O_2 \times \dfrac{-2035\ kJ}{3\ mol\ O_2} = -212\ kJ$

35.   From Sample Exercise 6.3, q = 1.3×10^8 J. Molar mass of C_3H_8 = 44.09 g/mol

mass C_3H_8 = 1.3×10^8 J $\times \dfrac{1 \text{ mol } C_3H_8}{2221 \times 10^3 \text{ J}} \times \dfrac{44.09 \text{ g } C_3H_8}{\text{mol } C_3H_8}$ = 2.6×10^3 g C_3H_8

36.   a.   1.00 g $CH_4 \times \dfrac{1 \text{ mol } CH_4}{16.04 \text{ g } CH_4} \times \dfrac{-891 \text{ kJ}}{\text{mol } CH_4}$ = -55.5 kJ

b. $PV = nRT$, $n = \dfrac{PV}{RT} = \dfrac{\dfrac{740.}{760} \text{ atm} \times 1.00 \times 10^3 \text{ L}}{\dfrac{0.08206 \text{ L atm}}{\text{mol K}} \times 298 \text{ K}}$ = 39.8 mol

39.8 mol $\times \dfrac{-891 \text{ kJ}}{\text{mol}}$ = -3.55×10^4 kJ

Calorimetry and Heat Capacity

37.   a.   energy = s $\times$ m $\times \Delta T = \dfrac{0.900 \text{ J}}{\text{g }^\circ C} \times 850.$ g $\times (94.6 - 22.8)^\circ C = 5.49 \times 10^4$ J or 54.9 kJ

b. $\dfrac{0.900 \text{ J}}{\text{g }^\circ C} \times \dfrac{26.98 \text{ g}}{\text{mol Al}} = \dfrac{24.3 \text{ J}}{\text{mol }^\circ C}$

38.   a.   energy = s $\times$ m $\times \Delta T = \dfrac{0.71 \text{ J}}{\text{g }^\circ C} \times \left(1.0 \text{ mol C} \times \dfrac{12.01 \text{ g C}}{\text{mol C}} \right) \times 1.0^\circ C = 8.5$ J

b. $\dfrac{0.71 \text{ J}}{\text{g }^\circ C} \times 850$ g C $\times 150^\circ C = 9.1 \times 10^4$ J = 91 kJ

c. $\dfrac{0.71 \text{ J}}{\text{g }^\circ K} \times \left(75 \text{ kg C} \times \dfrac{1000 \text{ g}}{\text{kg}} \right) \times (348 - 294) \text{ K} = 2.9 \times 10^6$ J = 2.9×10^3 kJ

Note: $\Delta T(K) = \Delta T(^\circ C)$

39.   The units for specific heat capacity (s) are J/g•°C. s = $\dfrac{78.2 \text{ J}}{45.6 \text{ g} \times 13.3^\circ C} = \dfrac{0.129 \text{ J}}{\text{g }^\circ C}$

Molar heat capacity = $\dfrac{0.129 \text{ J}}{\text{g }^\circ C} \times \dfrac{207.2 \text{ g}}{\text{mol Pb}} = \dfrac{26.7 \text{ J}}{\text{mol }^\circ C}$

40.   s = $\dfrac{585 \text{ J}}{125.6 \text{ g} \times (53.5 - 20.0)^\circ C}$ = 0.139 J/g•°C

Molar heat capacity = $\dfrac{0.139 \text{ J}}{\text{g }^\circ C} \times \dfrac{200.6 \text{ g}}{\text{mol Hg}} = \dfrac{27.9 \text{ J}}{\text{mol }^\circ C}$

41. | Heat loss by hot water | = | Heat gain by cooler water |

The magnitude of heat loss and heat gain are equal in calorimetry problem. The only difference is the sign (positive and negative). To avoid sign errors, keep all quantities positive and, if necessary, deduce the correct signs at the end of the problem. Water has a specific heat capacity = s = 4.18 J/°C•g = 4.18 J/K•g (ΔT in °C = ΔT in K).

Heat loss by hot water = s × m × ΔT = $\dfrac{4.18\ J}{g\ K}$ × 50.0 g × (330. K - T_f)

Heat gain by cooler water = $\dfrac{4.18\ J}{g\ K}$ × 30.0 g × (T_f - 280. K); Heat loss = Heat gain, so:

$\dfrac{209\ J}{K}$ × (330. K - T_f) = $\dfrac{125\ J}{K}$ × (T_f - 280. K), 6.90 × 10^4 - 209 T_f = 125 T_f - 3.50 × 10^4

334 T_f = 1.040 × 10^5, T_f = 311 K

Note that the final temperature is closer to the temperature of the more massive hot water, which is as it should be.

42.   Heat loss by Ni = Heat gain by water; Keeping ΔT values positive to avoid sign errors:

$\dfrac{0.444\ J}{g\ °C}$ × 15.0 g × (100.0°C - T_f) = $\dfrac{4.18\ J}{g\ °C}$ × 55.0 g × (T_f - 23.0°C)

666 - 6.66 T_f = 230. T_f - 5290, T_f = $\dfrac{5960}{237}$ = 25.1°C

43.   Heat gained by water = Heat lost by nickel = s × m × ΔT where s = specific heat capacity

Heat gain = $\dfrac{4.18\ J}{g\ °C}$ × 150.0 g × (25.0°C - 23.5°C) = 940 J

Note: A temperature <u>change</u> of one Kelvin is the same as a temperature change of one degree Celsius.

A common error in calorimetery problems are sign errors. Keeping all quantities positive helps eliminate sign errors. Therefore:

Heat loss = 940 J = s × 28.2 g × (99.8 - 25.0)°C, s = $\dfrac{940\ J}{28.2\ g × 74.8°C}$ = $\dfrac{0.45\ J}{g\ °C}$

44.   Heat gained by water = Heat lost by copper

$\dfrac{4.18\ J}{g\ °C}$ × 75.0 g × 2.2°C = s × 46.2 g × 73.6°C, s = 0.20 J/g•°C

45.   Heat lost by solution = Heat gained by KBr; Mass of solution = 125 g + 10.5 g = 136 g

Note: Sign errors are common with calorimetry problems. However, the correct sign for ΔH can easily be obtained from the ΔT data. When working calorimetry problems, keep all quantities positive (ignore signs). When finished, deduce the correct sign for ΔH. For this problem, T decreases as KBr dissolves so ΔH is positive; the dissolution of KBr is endothermic (absorbs heat).

Heat lost by solution $= \dfrac{4.18 \text{ J}}{\text{g} \,^\circ\text{C}} \times 136 \text{ g} \times (24.2\,^\circ\text{C} - 21.1\,^\circ\text{C}) = 1800 \text{ J} = $ Heat gained by KBr

ΔH in units of J/g $= \dfrac{1800 \text{ J}}{10.5 \text{ g KBr}} = 170 \text{ J/g}$

ΔH in units of kJ/mol $= \dfrac{170 \text{ J}}{\text{g KBr}} \times \dfrac{119.0 \text{ g KBr}}{\text{mol KBr}} \times \dfrac{1 \text{ kJ}}{1000 \text{ J}} = 20. \text{ kJ/mol}$

46. $NaOH(aq) + HCl(aq) \rightarrow NaCl(aq) + H_2O(l)$

We have a stoichiometric mixture. All of the NaOH and HCl will react.

$0.10 \text{ L} \times \dfrac{1.0 \text{ mol}}{\text{L}} = 0.10 \text{ mol of HCl is neutralized by } 0.10 \text{ mol NaOH.}$

Heat lost by chemicals = Heat gained by solution;   Volume of solution = 100.0 + 100.0 = 200.0 mL

Heat gain $= \dfrac{4.18 \text{ J}}{\text{g} \,^\circ\text{C}} \times \left(200.0 \text{ mL} \times \dfrac{1.0 \text{ g}}{\text{mL}} \right) \times (31.3 - 24.6)\,^\circ\text{C} = 5.6 \times 10^3 \text{ J} = 5.6 \text{ kJ}$

Heat loss = 5.6 kJ;   This is the heat released by the neutralization of 0.10 mol HCl. Since the temperature increased, the sign for ΔH must be negative, i.e., the reaction is exothermic. For calorimetry problems, keep all quantities positive until the end of the calculation, then decide the sign for ΔH.

$\Delta H = \dfrac{-5.6 \text{ kJ}}{0.10 \text{ mol}} = -56 \text{ kJ/mol}$

47. $50.0 \times 10^{-3} \text{ L} \times 0.100 \text{ mol/L} = 5.00 \times 10^{-3}$ mol of both $AgNO_3$ and HCl are reacted. Thus, 5.00×10^{-3} mol of AgCl will be produced since there is a 1:1 mol ratio between reactants.

Heat lost by chemicals = Heat gained by solution

Heat gain $= \dfrac{4.18 \text{ J}}{\text{g} \,^\circ\text{C}} \times 100.0 \text{ g} \times (23.40 - 22.60)\,^\circ\text{C} = 330 \text{ J}$

Heat loss = 330 J;   This is the heat evolved (exothermic reaction) when 5.00×10^{-3} mol of AgCl is produced. So q = -330 J and ΔH (heat per mol AgCl formed) is negative with a value of:

$\Delta H = \dfrac{-330 \text{ J}}{5.00 \times 10^{-3} \text{ mol}} \times \dfrac{1 \text{ kJ}}{1000 \text{ J}} = -66 \text{ kJ/mol}$

48. $0.100 \text{ L} \times \dfrac{0.500 \text{ mol HCl}}{\text{L}} = 5.00 \times 10^{-2} \text{ mol HCl}$

$0.300 \text{ L} \times \dfrac{0.500 \text{ mol Ba(OH)}_2}{\text{L}} = 0.150 \text{ mol Ba(OH)}_2$

To react with all the HCl present, $5.00 \times 10^{-2}/2 = 2.50 \times 10^{-2}$ mol $Ba(OH)_2$ are required. Since 0.150 mol $Ba(OH)_2$ are present, then HCl is the limiting reactant.

$5.00 \times 10^{-2} \text{ mol HCl} \times \dfrac{118 \text{ kJ}}{2 \text{ mol HCl}} = 2.95 \text{ kJ of heat is evolved by reaction.}$

Heat gained by solution $= 2.95 \times 10^3 \text{ J} = \dfrac{4.18 \text{ J}}{\text{g °C}} \times 400.0 \text{ g} \times \Delta T$

$\Delta T = 1.76°C = T_f - T_i = T_f - 25.0°C, \ \ T_f = 26.8°C$

49.   Heat lost by camphor = Heat gained by calorimeter

Heat lost by combustion of camphor $= 0.1204 \text{ g} \times \dfrac{1 \text{ mol}}{152.23 \text{ g}} \times \dfrac{5903.6 \text{ kJ}}{\text{mol}} = 4.669 \text{ kJ}$

Let C_{cal} = heat capacity of the calorimeter in units of kJ/°C, then:

Heat gained by calorimter $= C_{cal} \times \Delta T, \ \ 4.669 \text{ kJ} = C_{cal} \times 2.28°C, \ \ C_{cal} = 2.05 \text{ kJ/°C}$

50.   Heat gain by calorimeter $= \dfrac{1.56 \text{ kJ}}{°C} \times 3.2°C = 5.0 \text{ kJ} = $ heat loss by quinone

Heat loss = 5.0 kJ which is the heat evolved (exothermic reaction) by the combustion of 0.1964 g of quinone.

$\Delta E_{comb} = \dfrac{-5.0 \text{ kJ}}{0.1964 \text{ g}} = -25 \text{ kJ/g;} \qquad \Delta E_{comb} = \dfrac{-25 \text{ kJ}}{\text{g}} \times \dfrac{108.09 \text{ g}}{\text{mol}} = -2700 \text{ kJ/mol}$

Hess's Law

51.   Information given:

$$C(s) + O_2(g) \rightarrow CO_2(g) \qquad\qquad \Delta H = -393.7 \text{ kJ}$$
$$CO(g) + 1/2 \ O_2(g) \rightarrow CO_2(g) \qquad\qquad \Delta H = -283.3 \text{ kJ}$$

Using Hess's Law:

$$2 \ C(s) + 2 \ O_2(g) \rightarrow 2 \ CO_2(g) \qquad\qquad \Delta H_1 = 2(-393.7 \text{ kJ})$$
$$2 \ CO_2(g) \rightarrow 2 \ CO(g) + O_2(g) \qquad\qquad \Delta H_2 = -2(-283.3 \text{ kJ})$$

$$\overline{\quad 2 \ C(s) + O_2(g) \rightarrow 2 \ CO(g) \qquad\qquad \Delta H = \Delta H_1 + \Delta H_2 = -220.8 \text{ kJ} \quad}$$

Note: The enthalpy change for a reaction that is reversed is the negative quantity of the enthalpy change for the original reaction. If the coefficients in a balanced reaction are multiplied by an integer, then the value of ΔH is multiplied by the same integer.

52. $C_4H_4(g) + 5 O_2(g) \rightarrow 4 CO_2(g) + 2 H_2O(l)$ $\Delta H_{comb} = -2341$ kJ
      $C_4H_8(g) + 6 O_2(g) \rightarrow 4 CO_2(g) + 4 H_2O(l)$ $\Delta H_{comb} = -2755$ kJ
      $H_2(g) + 1/2 O_2(g) \rightarrow H_2O(l)$ $\Delta H_{comb} = -286$ kJ

By convention, $H_2O(l)$ is produced when enthalpies of combustion are given and since per mol quantities are given, the combustion reaction refers to 1 mol of that quantity reacting with $O_2(g)$.

   Using Hess's Law to solve:

         $C_4H_4(g) + 5 O_2(g) \rightarrow 4 CO_2(g) + 2 H_2O(l)$ $\Delta H_1 = -2341$ kJ
         $4 CO_2(g) + 4 H_2O(l) \rightarrow C_4H_8(g) + 6 O_2(g)$ $\Delta H_2 = -(2755$ kJ$)$
         $2 H_2(g) + O_2(g) \rightarrow 2 H_2O(l)$ $\Delta H_3 = 2(-286$ kJ$)$
         _____
         $C_4H_4(g) + 2 H_2(g) \rightarrow C_4H_8(g)$ $\Delta H = \Delta H_1 + \Delta H_2 + \Delta H_3$
                                                                        $\Delta H = -158$ kJ

53. $S + 3/2 O_2 \rightarrow SO_3$ $\Delta H = -395.2$ kJ
      $SO_3 \rightarrow SO_2 + 1/2 O_2$ $\Delta H = -1/2(-198.2$ kJ$) = 99.1$ kJ
      _____
      $S(s) + O_2(g) \rightarrow SO_2(g)$ $\Delta H = -296.1$ kJ

54. $2 C + 2 O_2 \rightarrow 2 CO_2$ $\Delta H = 2(-394$ kJ$)$
        $H_2 + 1/2 O_2 \rightarrow H_2O$ $\Delta H = -286$ kJ
        $2 CO_2 + H_2O \rightarrow C_2H_2 + 5/2 O_2$ $\Delta H = -(-1300.$ kJ$)$
      _____
      $2 C(s) + H_2(g) \rightarrow C_2H_2(g)$ $\Delta H = 226$ kJ

55. $NO + O_3 \rightarrow NO_2 + O_2$ $\Delta H = -199$ kJ
        $3/2 O_2 \rightarrow O_3$ $\Delta H = -1/2(-427$ kJ$)$
        $O \rightarrow 1/2 O_2$ $\Delta H = -1/2(495$ kJ$)$
      _____
      $NO(g) + O(g) \rightarrow NO_2(g)$ $\Delta H = -233$ kJ

56. $C_6H_4(OH)_2 \rightarrow C_6H_4O_2 + H_2$ $\Delta H = 177.4$ kJ
        $H_2O_2 \rightarrow H_2 + O_2$ $\Delta H = -(-191.2$ kJ$)$
        $2 H_2 + O_2 \rightarrow 2 H_2O(g)$ $\Delta H = 2(-241.8$ kJ$)$
        $2 H_2O(g) \rightarrow 2 H_2O(l)$ $\Delta H = 2(-43.8$ kJ$)$
      _____
      $C_6H_4(OH)_2(aq) + H_2O_2(aq) \rightarrow C_6H_4O_2(aq) + 2 H_2O(l)$ $\Delta H = -202.6$ kJ

57. $4 HNO_3 \rightarrow 2 N_2O_5 + 2 H_2O$ $\Delta H = -2(-76.6$ kJ$)$
        $2 N_2 + 6 O_2 + 2 H_2 \rightarrow 4 HNO_3$ $\Delta H = 4(-174.1$ kJ$)$
                $2 H_2O \rightarrow 2 H_2 + O_2$ $\Delta H = -2(-285.8$ kJ$)$
      _____
      $2 N_2(g) + 5 O_2(g) \rightarrow 2 N_2O_5(g)$ $\Delta H = 28.4$ kJ

58.   We want ΔH for $N_2H_4(l) + O_2(g) \rightarrow N_2(g) + 2\ H_2O(l)$. It will be easier to calculate ΔH for the combustion of four moles of N_2H_4 since we will avoid fractions.

$$
\begin{array}{ll}
9\ H_2 + 9/2\ O_2 \rightarrow 9\ H_2O & \Delta H = 9(-286\ kJ) \\
3\ N_2H_4 + 3\ H_2O \rightarrow 3\ N_2O + 9\ H_2 & \Delta H = -3(-317\ kJ) \\
2\ NH_3 + 3\ N_2O \rightarrow 4\ N_2 + 3\ H_2O & \Delta H = -1010.\ kJ \\
N_2H_4 + H_2O \rightarrow 2\ NH_3 + 1/2\ O_2 & \Delta H = -(-143\ kJ)
\end{array}
$$

$$
4\ N_2H_4(l) + 4\ O_2(g) \rightarrow 4\ N_2(g) + 8\ H_2O(l) \qquad \Delta H = -2490.\ kJ
$$

For: $N_2H_4(l) + O_2(g) \rightarrow N_2(g) + 2\ H_2O(l)$ $\Delta H = \dfrac{-2490.\ kJ}{4} = -623\ kJ$

Note: By the significant figure rules, we could report this answer to four significant figures. However, since the ΔH values given in the problem are only known to ± 1 kJ, then our final answer will at best be ± 1 kJ.

Standard Enthalpies of Formation

59.   The change in enthalpy that accompanies the formation of one mole of a compound from its elements, with all substances in their standard states is the standard enthalpy of formation for a compound. The reactions that refer to ΔH_f° are:

$$
Na(s) + 1/2\ Cl_2(g) \rightarrow NaCl(s);\ \ H_2(g) + 1/2\ O_2(g) \rightarrow H_2O(l)
$$

$$
6\ C(graphite, s) + 6\ H_2(g) + 3\ O_2(g) \rightarrow C_6H_{12}O_6(s);\ \ Pb(s) + S(s) + 2\ O_2(g) \rightarrow PbSO_4(s)
$$

60.   a.   aluminum oxide = Al_2O_3; $2\ Al(s) + 3/2\ O_2(g) \rightarrow Al_2O_3(s)$

       b. $C_2H_5OH(l) + 3\ O_2(g) \rightarrow 2\ CO_2(g) + 3\ H_2O(l)$

       c. $NaOH(aq) + HCl(aq) \rightarrow H_2O(l) + NaCl(aq)$

       d. $2\ C\ (graphite, s) + 3/2\ H_2(g) + 1/2\ Cl_2(g) \rightarrow C_2H_3Cl(g)$

       e. $C_6H_6(l) + 15/2\ O_2(g) \rightarrow 6\ CO_2(g) + 3\ H_2O(l)$

             Note: ΔH_{comb} values assume one mol of compound combusted.

       f. $NH_4Br(s) \rightarrow NH_4^+(aq) + Br^-(aq)$

61.   In general: $\Delta H^\circ = \Sigma n_p \Delta H_{f,\ products}^\circ - \Sigma n_r \Delta H_{f,\ reactants}^\circ$ and all elements in their standard state have $\Delta H_f^\circ = 0$ by definition.

       a. $2\ NH_3(g) + 3\ O_2(g) + 2\ CH_4(g) \rightarrow 2\ HCN(g) + 6\ H_2O(g)$

             $\Delta H^\circ = [2\ mol\ HCN \times \Delta H_{f,\ HCN}^\circ + 6\ mol\ H_2O(g) \times \Delta H_{f,\ H_2O}^\circ]$

$$
- [2\ mol\ NH_3 \times \Delta H_{f,\ NH_3}^\circ + 2\ mol\ CH_4 \times \Delta H_{f,\ CH_4}^\circ]
$$

$$\Delta H° = [2(135.1) + 6(-242)] - [2(-46) + 2(-75)] = -940. \text{ kJ}$$

b. $Ca_3(PO_4)_2(s) + 3\ H_2SO_4(l) \rightarrow 3\ CaSO_4(s) + 2\ H_3PO_4(l)$

$$\Delta H° = \left[3 \text{ mol CaSO}_4\left(\frac{-1433 \text{ kJ}}{\text{mol}}\right) + 2 \text{ mol H}_3PO_4(l)\left(\frac{-1267 \text{ kJ}}{\text{mol}}\right)\right]$$

$$- \left[1 \text{ mol Ca}_3(PO_4)_2\left(\frac{-4126 \text{ kJ}}{\text{mol}}\right) + 3 \text{ mol H}_2SO_4(l)\left(\frac{-814 \text{ kJ}}{\text{mol}}\right)\right]$$

$$\Delta H° = -6833 \text{ kJ} - (-6568 \text{ kJ}) = -265 \text{ kJ}$$

c. $NH_3(g) + HCl(g) \rightarrow NH_4Cl(s)$

$$\Delta H° = [1 \text{ mol NH}_4Cl \times \Delta H°_{f,\ NH_4Cl}] - [1 \text{ mol NH}_3 \times \Delta H°_{f,\ NH_3} + 1 \text{ mol HCl} \times \Delta H°_{f,\ HCl}]$$

$$\Delta H° = \left[1 \text{ mol}\left(\frac{-314 \text{ kJ}}{\text{mol}}\right)\right] - \left[1 \text{ mol}\left(\frac{-46 \text{ kJ}}{\text{mol}}\right) + 1 \text{ mol}\left(\frac{-92 \text{ kJ}}{\text{mol}}\right)\right]$$

$$\Delta H° = -314 \text{ kJ} + 138 \text{ kJ} = -176 \text{ kJ}$$

62.   a. $C_2H_5OH(l) + 3\ O_2(g) \rightarrow 2\ CO_2(g) + 3\ H_2O(g)$

$$\Delta H° = \left[2 \text{ mol}\left(\frac{-393.5 \text{ kJ}}{\text{mol}}\right) + 3 \text{ mol}\left(\frac{-242 \text{ kJ}}{\text{mol}}\right)\right] - \left[1 \text{ mol}\left(\frac{-278 \text{ kJ}}{\text{mol}}\right)\right]$$

$$= -1513 \text{ kJ} - (-278 \text{ kJ}) = -1235 \text{ kJ}$$

b. $SiCl_4(l) + 2\ H_2O(l) \rightarrow SiO_2(s) + 4\ HCl(aq)$

Since $HCl(aq)$ is $H^+(aq) + Cl^-(aq)$, then $\Delta H°_f = 0 - 167 = -167 \text{ kJ/mol}$.

$$\Delta H° = \left[4 \text{ mol}\left(\frac{-167 \text{ kJ}}{\text{mol}}\right) + 1 \text{ mol}\left(\frac{-911 \text{ kJ}}{\text{mol}}\right)\right] - \left[1 \text{ mol}\left(\frac{-687 \text{ kJ}}{\text{mol}}\right) + 2 \text{ mol}\left(\frac{-286 \text{ kJ}}{\text{mol}}\right)\right]$$

$$\Delta H° = -1579 \text{ kJ} - (-1259 \text{ kJ}) = -320. \text{ kJ}$$

c. $MgO(s) + H_2O(l) \rightarrow Mg(OH)_2(s)$

$$\Delta H° = \left[1 \text{ mol}\left(\frac{-925 \text{ kJ}}{\text{mol}}\right)\right] - \left[1 \text{ mol}\left(\frac{-602 \text{ kJ}}{\text{mol}}\right) + 1 \text{ mol}\left(\frac{-286 \text{ kJ}}{\text{mol}}\right)\right]$$

$$\Delta H° = -925 \text{ kJ} - (-888 \text{ kJ}) = -37 \text{ kJ}$$

63.   a. $4 NH_3(g) + 5 O_2(g) \rightarrow 4 NO(g) + 6 H_2O(g)$; $\Delta H^\circ = \Sigma n_p \Delta H^\circ_{f, \text{products}} - \Sigma n_r \Delta H^\circ_{f, \text{reactants}}$

$$\Delta H^\circ = \left[4 \text{ mol} \left(\frac{90.\, kJ}{mol} \right) + 6 \text{ mol} \left(\frac{-242 \, kJ}{mol} \right) \right] - \left[4 \text{ mol} \left(\frac{-46 \, kJ}{mol} \right) \right] = -908 \text{ kJ}$$

$2 NO(g) + O_2(g) \rightarrow 2 NO_2(g)$

$$\Delta H^\circ = \left[2 \text{ mol} \left(\frac{34 \, kJ}{mol} \right) \right] - \left[2 \text{ mol} \left(\frac{90.\, kJ}{mol} \right) \right] = -112 \text{ kJ}$$

$3 NO_2(g) + H_2O(l) \rightarrow 2 HNO_3(aq) + NO(g)$

$$\Delta H^\circ = \left[2 \text{ mol} \left(\frac{-207 \, kJ}{mol} \right) + 1 \text{ mol} \left(\frac{90.\, kJ}{mol} \right) \right]$$

$$- \left[3 \text{ mol} \left(\frac{34 \, kJ}{mol} \right) + 1 \text{ mol} \left(\frac{-286 \, kJ}{mol} \right) \right] = -140.\text{ kJ}$$

Note: All ΔH°_f values are assumed ± 1 kJ.

b. $12 NH_3(g) + 15 O_2(g) \rightarrow 12 NO(g) + 18 H_2O(g)$
        $12 NO(g) + 6 O_2(g) \rightarrow 12 NO_2(g)$
        $12 NO_2(g) + 4 H_2O(l) \rightarrow 8 HNO_3(aq) + 4 NO(g)$
             $4 H_2O(g) \rightarrow 4 H_2O(l)$

$12 NH_3(g) + 21 O_2(g) \rightarrow 8 HNO_3(aq) + 4 NO(g) + 14 H_2O(g)$

The overall reaction is exothermic since each step is exothermic.

64. $4 Na(s) + O_2(g) \rightarrow 2 Na_2O(s)$, $\Delta H^\circ = 2 \text{ mol} \left(\frac{-416 \, kJ}{mol} \right) = -832 \text{ kJ}$

$2 Na(s) + 2 H_2O(l) \rightarrow 2 NaOH(aq) + H_2(g)$

$$\Delta H^\circ = \left[2 \text{ mol} \left(\frac{-470.\, kJ}{mol} \right) \right] - \left[2 \text{ mol} \left(\frac{-286 \, kJ}{mol} \right) \right] = -368 \text{ kJ}$$

$2 Na(s) + CO_2(g) \rightarrow Na_2O(s) + CO(g)$

$$\Delta H^\circ = \left[1 \text{ mol} \left(\frac{-416 \, kJ}{mol} \right) + 1 \text{ mol} \left(\frac{-110.5 \, kJ}{mol} \right) \right] - \left[1 \text{ mol} \left(\frac{-393.5 \, kJ}{mol} \right) \right] = -133 \text{ kJ}$$

In both cases sodium metal reacts with the "extinguishing agent." Both reactions are exothermic and each reaction produces a flammable gas, H_2 and CO, respectively.

65. $3 \text{ Al(s)} + 3 \text{ NH}_4\text{ClO}_4\text{(s)} \rightarrow \text{Al}_2\text{O}_3\text{(s)} + \text{AlCl}_3\text{(s)} + 3 \text{ NO(g)} + 6 \text{ H}_2\text{O(g)}$

$$\Delta H° = \left[6 \text{ mol} \left(\frac{-242 \text{ kJ}}{\text{mol}} \right) + 3 \text{ mol} \left(\frac{90. \text{ kJ}}{\text{mol}} \right) + 1 \text{ mol} \left(\frac{-704 \text{ kJ}}{\text{mol}} \right) + 1 \text{ mol} \left(\frac{-1676 \text{ kJ}}{\text{mol}} \right) \right]$$

$$- \left[3 \text{ mol} \left(\frac{-295 \text{ kJ}}{\text{mol}} \right) \right] = -2677 \text{ kJ}$$

66. $5 \text{ N}_2\text{O}_4\text{(l)} + 4 \text{ N}_2\text{H}_3\text{CH}_3\text{(l)} \rightarrow 12 \text{ H}_2\text{O(g)} + 9 \text{ N}_2\text{(g)} + 4 \text{ CO}_2\text{(g)}$

$$\Delta H° = \left[12 \text{ mol} \left(\frac{-242 \text{ kJ}}{\text{mol}} \right) + 4 \text{ mol} \left(\frac{-393.5 \text{ kJ}}{\text{mol}} \right) \right]$$

$$- \left[5 \text{ mol} \left(\frac{-20. \text{ kJ}}{\text{mol}} \right) + 4 \text{ mol} \left(\frac{54 \text{ kJ}}{\text{mol}} \right) \right] = -4594 \text{ kJ}$$

67. $2 \text{ ClF}_3\text{(g)} + 2 \text{ NH}_3\text{(g)} \rightarrow \text{N}_2\text{(g)} + 6 \text{ HF(g)} + \text{Cl}_2\text{(g)} \qquad \Delta H° = -1196 \text{ kJ}$

$\Delta H° = [6 \, \Delta H°_{f, \text{HF}}] - [2 \, \Delta H°_{f, \text{ClF}_3} + 2 \, \Delta H°_{f, \text{NH}_3}]$

$-1196 \text{ kJ} = 6 \text{ mol} \left(\frac{-271 \text{ kJ}}{\text{mol}} \right) - 2 \, \Delta H°_{f, \text{ClF}_3} - 2 \text{ mol} \left(\frac{-46 \text{ kJ}}{\text{mol}} \right)$

$-1196 \text{ kJ} = -1626 \text{ kJ} - 2 \, \Delta H°_{f, \text{ClF}_3} + 92 \text{ kJ}, \ \Delta H°_{f, \text{ClF}_3} = \dfrac{(-1626 + 92 + 1196) \text{ kJ}}{2 \text{ mol}} = \dfrac{-169 \text{ kJ}}{\text{mol}}$

68. $\text{C}_2\text{H}_4\text{(g)} + 3 \text{ O}_2\text{(g)} \rightarrow 2 \text{ CO}_2\text{(g)} + 2 \text{ H}_2\text{O(l)} \quad \Delta H° = -1411 \text{ kJ}$

$\Delta H° = -1411.1 \text{ kJ} = 2(-393.5) \text{ kJ} + 2(-285.9) \text{ kJ} - \Delta H°_{f, \text{C}_2\text{H}_4}$

$-1411.1 \text{ kJ} = -1358.8 \text{ kJ} - \Delta H°_{f, \text{C}_2\text{H}_4}, \ \Delta H°_{f, \text{C}_2\text{H}_4} = 52.3 \text{ kJ/mol}$

Energy Consumption and Sources

69. $\text{C(s)} + \text{H}_2\text{O(g)} \rightarrow \text{H}_2\text{(g)} + \text{CO(g)}, \ \Delta H° = -110.5 \text{ kJ} - (-242 \text{ kJ}) = 132 \text{ kJ}$

70. $\text{CO(g)} + 2 \text{ H}_2\text{(g)} \rightarrow \text{CH}_3\text{OH(l)}, \ \Delta H° = -239 \text{ kJ} - (-110.5 \text{ kJ}) = -129 \text{ kJ}$

71. $\text{C}_3\text{H}_8\text{(g)} + 5 \text{ O}_2\text{(g)} \rightarrow 3 \text{ CO}_2\text{(g)} + 4 \text{ H}_2\text{O(l)}$

$\Delta H° = [3(-393.5 \text{ kJ}) + 4(-286 \text{ kJ})] - [-104 \text{ kJ}] = -2221 \text{ kJ/mol C}_3\text{H}_8$

$\dfrac{-2221 \text{ kJ}}{\text{mol}} \times \dfrac{1 \text{ mol}}{44.09 \text{ g}} = \dfrac{-50.37 \text{ kJ}}{\text{g}}$ vs. -47.7 kJ/g for octane (Sample Exercise 6.11)

The fuel values are very close. An advantage of propane is that it burns more cleanly. The boiling point of propane is -42°C. Thus, it is more difficult to store propane and there are extra safety hazards associated with using high pressure compressed gas tanks.

72.   Since 1 mol of $C_2H_2(g)$ and 1 mol of $C_4H_{10}(g)$ have equivalent volumes at the same T and P, then:

$$\frac{\text{enthalpy of combustion per volume of } C_2H_2}{\text{enthalpy of combustion per volume of } C_4H_{10}} = \frac{\text{enthalpy of combustion per mol } C_2H_2}{\text{enthalpy of combustion per mol } C_4H_{10}}$$

$$\frac{\text{enthalpy of combustion per volume of } C_2H_2}{\text{enthalpy of combustion per volume of } C_4H_{10}} = \frac{\dfrac{-49.9\text{ kJ}}{\text{g } C_2H_2} \times \dfrac{26.04\text{ g } C_2H_2}{\text{mol } C_2H_2}}{\dfrac{-49.5\text{ kJ}}{\text{g } C_4H_{10}} \times \dfrac{58.12\text{ g } C_4H_{10}}{\text{mol } C_4H_{10}}} = 0.452$$

Almost twice the volume of acetylene is needed to furnish the same energy as a given volume of butane.

73.   The molar volume of a gas at STP is 22.42 L.

$$4.19 \times 10^6 \text{ kJ} \times \frac{1 \text{ mol CH}_4}{891 \text{ kJ}} \times \frac{22.42 \text{ L CH}_4}{\text{mol CH}_4} = 1.05 \times 10^5 \text{ L CH}_4$$

74.   Mass of $H_2O = 1.00 \text{ gal} \times \dfrac{3.785 \text{ L}}{\text{gal}} \times \dfrac{1000 \text{ mL}}{\text{L}} \times \dfrac{1.00 \text{ g}}{\text{mL}} = 3790 \text{ g } H_2O$

Energy required (theoretical) $= s \times m \times \Delta T = \dfrac{4.18 \text{ J}}{\text{g °C}} \times 3790 \text{ g} \times 10.0 \text{ °C} = 1.58 \times 10^5 \text{ J}$

For an actual (80.0% efficient) process, more than this quantity of energy is needed since heat is always lost in any transfer of energy. The energy required is:

$$1.58 \times 10^5 \text{ J} \times \frac{100. \text{ J}}{80.0 \text{ J}} = 1.98 \times 10^5 \text{ J}$$

Mass of $C_2H_2 = 1.98 \times 10^5 \text{ J} \times \dfrac{1 \text{ mol } C_2H_2}{1300. \times 10^3 \text{ J}} \times \dfrac{26.04 \text{ g } C_2H_2}{\text{mol } C_2H_2} = 3.97 \text{ g } C_2H_2$

Additional Exercises

75. $w = -P\Delta V$; $\Delta n = $ mol gaseous products - mol gaseous reactants. Only gases can do PV work. When a balanced reaction has more mol of product gases than mol of reactant gases (Δn positive), then the reaction will expand in volume (ΔV positive) and the system does work on the surroundings. For example, in reaction e, $\Delta n = 6 - 0 = 6$ mol and this reaction would do expansion work against the surroundings. When a balanced reaction has a decrease in mol gas from reactants to products (Δn negative), then the reaction will contract in volume (ΔV negative) and the surroundings does compression work on the system. When there is no change in mol of gas from reactants to products, then $\Delta V = 0$ and $w = 0$.

When $\Delta V > 0$ ($\Delta n > 0$), then $w < 0$ and system does work on the surroundings (e and f).

When $\Delta V < 0$ ($\Delta n < 0$), then $w > 0$ and the surroundings does work on the system (a, c and d).

When $\Delta V = 0$ ($\Delta n = 0$), then $w = 0$ (b).

76. $V = 10.0 \text{ m} \times 4.0 \text{ m} \times 3.0 \text{ m} = 1.2 \times 10^2 \text{ m}^3 \times (100 \text{ cm/m})^3 = 1.2 \times 10^8 \text{ cm}^3$

Mass of water $= 1.2 \times 10^8$ g since the density of water is 1.0 g/cm^3. $\Delta T = 30.0 - 20.2 = 9.8°C$

Energy $= \dfrac{4.18 \text{ J}}{\text{g °C}} \times (1.2 \times 10^8 \text{ g}) \times 9.8°C = 4.9 \times 10^9$ J or 4.9×10^6 kJ

77. $2 \text{ K(s)} + 2 \text{ H}_2\text{O(l)} \rightarrow 2 \text{ KOH(aq)} + \text{H}_2\text{(g)}$, $\Delta H° = 2(-481 \text{ kJ}) - 2(-286 \text{ kJ}) = -390. \text{ kJ}$

$5.00 \text{ g K} \times \dfrac{1 \text{ mol K}}{39.10 \text{ g K}} \times \dfrac{-390. \text{ kJ}}{2 \text{ mol K}} = -24.9$ kJ of heat released upon reaction of 5.00 g of potassium.

$24,900 \text{ J} = \dfrac{4.18 \text{ J}}{\text{g °C}} \times (1.00 \times 10^3 \text{ g}) \times \Delta T$, $\Delta T = \dfrac{24,900}{4.18 \times 1.00 \times 10^3} = 5.96°C$

Final temperature $= 24.0 + 5.96 = 30.0°C$

78. $\text{HCl(aq)} + \text{NaOH(aq)} \rightarrow \text{H}_2\text{O(l)} + \text{NaCl(aq)}$ $\Delta H = -56 \text{ kJ}$

$0.2000 \text{ L} \times \dfrac{0.400 \text{ mol HCl}}{\text{L}} = 8.00 \times 10^{-2}$ mol HCl

$0.1500 \text{ L} \times \dfrac{0.500 \text{ mol NaOH}}{\text{L}} = 7.50 \times 10^{-2}$ mol NaOH

Since the balanced reaction requires a 1:1 mol ratio between HCl and NaOH, and since fewer mol of NaOH are actually present, then NaOH is the limiting reagent.

Heat released $= 7.50 \times 10^{-2}$ mol NaOH $\times \dfrac{-56 \text{ kJ}}{\text{mol NaOH}} = -4.2$ kJ heat released

79.   The specific heat of water is 4.18 J/g•°C, which is equal to 4.18 kJ/kg•°C.

We have 1.00 kg of H_2O, so: $1.00 \text{ kg} \times \dfrac{4.18 \text{ kJ}}{\text{kg °C}} = 4.18$ kJ/°C

This is the portion of the heat capacity that can be attributed to H_2O.

Total heat capacity $= C_{cal} + C_{\text{H}_2\text{O}}$, $C_{cal} = 10.84 - 4.18 = 6.66$ kJ/°C

80.   Heat released $= 1.056 \text{ g} \times 26.42 \text{ kJ/g} = 27.90 \text{ kJ} =$ Heat gain by water and calorimeter

Heat gain $= 27.90 \text{ kJ} = \dfrac{4.18 \text{ kJ}}{\text{kg °C}} \times 0.987 \text{ kg} \times \Delta T + \dfrac{6.66 \text{ kJ}}{°C} \times \Delta T$

$27.90 = (4.13 + 6.66) \Delta T = 10.79 \Delta T, \; \Delta T = 2.586°C$

$2.586°C = T_f - 23.32°C, \; T_f = 25.91°C$

81. First, we need to get the heat capacity of the calorimeter from the combustion of benzoic acid.

Heat lost by combustion = Heat gained by calorimeter

Heat loss $= 0.1584 \text{ g} \times \dfrac{26.42 \text{ kJ}}{\text{g}} = 4.185 \text{ kJ}$

Heat gain $= 4.185 \text{ kJ} = C_{cal} \times \Delta T, \; C_{cal} = \dfrac{4.185 \text{ kJ}}{2.54°C} = 1.65 \text{ kJ}/°C$

Now we can calculate the heat of combustion of vanillin. Heat loss = Heat gain

Heat gain by calorimeter $= \dfrac{1.65 \text{ kJ}}{°C} \times 3.25°C = 5.36 \text{ kJ}$

Heat loss = 5.36 kJ which is the heat evolved by the combustion of the vanillin.

$\Delta E_{comb} = \dfrac{-5.36 \text{ kJ}}{0.2130 \text{ g}} = -25.2 \text{ kJ/g}; \; \Delta E_{comb} = \dfrac{-25.2 \text{ kJ}}{\text{g}} \times \dfrac{152.14 \text{ g}}{\text{mol}} = -3830 \text{ kJ/mol}$

82. To avoid fractions, let's first calculate ΔH for the reaction:

$6 \text{ FeO}(s) + 6 \text{ CO}(g) \rightarrow 6 \text{ Fe}(s) + 6 \text{ CO}_2(g)$

$6 \text{ FeO} + 2 \text{ CO}_2 \rightarrow 2 \text{ Fe}_3\text{O}_4 + 2 \text{ CO}$	$\Delta H° = -2(18 \text{ kJ})$
$2 \text{ Fe}_3\text{O}_4 + \text{CO}_2 \rightarrow 3 \text{ Fe}_2\text{O}_3 + \text{CO}$	$\Delta H° = -(-39 \text{ kJ})$
$3 \text{ Fe}_2\text{O}_3 + 9 \text{ CO} \rightarrow 6 \text{ Fe} + 9 \text{ CO}_2$	$\Delta H° = 3(-23 \text{ kJ})$

$6 \text{ FeO}(s) + 6 \text{ CO}(g) \rightarrow 6 \text{ Fe}(s) + 6 \text{ CO}_2(g) \quad\quad \Delta H° = -66 \text{ kJ}$

So for: $\text{FeO}(s) + \text{CO}(g) \rightarrow \text{Fe}(s) + \text{CO}_2(g) \quad\quad \Delta H° = \dfrac{-66 \text{ kJ}}{6} = -11 \text{ kJ}$

83. The combustion of phosphorus is exothermic. Thus the product, P_4O_{10}, is lower in energy than either red or white phosphorus. Since the conversion of white phosphorus to red phosphorus is exothermic, red phosphorus is lower in energy than white phosphorus. Thus, white phosphorus will release more heat when burned in air since there is a larger energy difference between white phosphorus and products.

84. The reaction of $\Delta H_f°$ for a substance always refers to that substance being produced from its elements in their standard state. Therefore, $\Delta H_f°$ is equal to zero for any element in their standard state and is equal to a nonzero value for any substance not in its standard state. Only c has an element in its standard state, $H_2(g)$, so only process c has $\Delta H_f° = 0$. The standard state of oxygen is $O_2(g)$, water is not an element and the standard state of chlorine is $Cl_2(g)$.

85.   a. $C_2H_4(g) + O_3(g) \rightarrow CH_3CHO(g) + O_2(g)$, $\Delta H° = -166$ kJ - [143 kJ + 52 kJ] = -361 kJ

       b. $O_3(g) + NO(g) \rightarrow NO_2(g) + O_2(g)$, $\Delta H° = 34$ kJ - [90. kJ + 143 kJ] = -199 kJ

       c. $SO_3(g) + H_2O(l) \rightarrow H_2SO_4(aq)$, $\Delta H° = -909$ kJ - [-396 kJ + (-286 kJ)] = -227 kJ

       d. $2 NO(g) + O_2(g) \rightarrow 2 NO_2(g)$, $\Delta H° = 2(34)$ kJ - 2(90.) kJ = -112 kJ

Challenge Problems

86.   Only when there is a volume change can PV work be done. In pathway 1, only the first step does PV work. In pathway 2, only the second step does PV work.

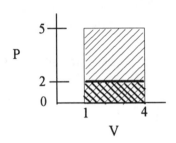

Pathway 1: w = complete shaded area

w = $-P\Delta V$ = -5.0(4.0 -1.0) = -15 L atm

Pathway 2: w = area of cross hatched square

w = $-P\Delta V$ = -2.0(4.0 - 1.0) = -6.0 L atm

Sign is (-) because the system is doing work on the surroundings (an expansion).

Since w depends on the pathway, then work cannot be a state function.

87.   a. $C_{12}H_{22}O_{11}(s) + 12 O_2(g) \rightarrow 12 CO_2(g) + 11 H_2O(l)$

       b.   A bomb calorimeter is at constant volume, so heat released = q_v = ΔE:

$$\Delta E = \frac{-24.00 \text{ kJ}}{1.46 \text{ g}} \times \frac{342.30 \text{ g}}{\text{mol}} = -5630 \text{ kJ/mol } C_{12}H_{22}O_{11}$$

       c.   Since PV = nRT, then $P\Delta V = RT\Delta n$ where Δn = mol gaseous products - mol gaseous reactants.

$\Delta H = \Delta E + P\Delta V = \Delta E + RT\Delta n$

For this reaction Δn = 12 -12 = 0, so $\Delta H = \Delta E$ = -5630 kJ/mol.

88.   Energy needed = $\dfrac{20. \times 10^3 \text{ g } C_{12}H_{22}O_{11}}{\text{hr}} \times \dfrac{1 \text{ mol } C_{12}H_{22}O_{11}}{342.30 \text{ g } C_{12}H_{22}O_{11}} \times \dfrac{5640 \text{ kJ}}{\text{mol}} = 3.3 \times 10^5$ kJ/hr

Energy from sun = 1.0 kW/m^2 = 1000 W/m^2 = $\dfrac{1000 \text{ J}}{\text{s m}^2} = \dfrac{1.0 \text{ kJ}}{\text{s m}^2}$

$10{,}000 \text{ m}^2 \times \dfrac{1.0 \text{ kJ}}{\text{s m}^2} \times \dfrac{60 \text{ s}}{\text{min}} \times \dfrac{60 \text{ min}}{\text{hr}} = 3.6 \times 10^7$ kJ/hr

$$\% \text{ efficiency} = \frac{\text{Energy used per hour}}{\text{Total energy per hour}} \times 100 = \frac{3.3 \times 10^5 \text{ kJ}}{3.6 \times 10^7 \text{ kJ}} \times 100 = 0.92\%$$

89. Energy used in 8.0 hours = 40. kWh = $\dfrac{40. \text{ kJ h}}{\text{s}} \times \dfrac{3600 \text{ s}}{\text{h}} = 1.4 \times 10^5 \text{ kJ}$

Energy from the sun in 8.0 hours = $\dfrac{1.0 \text{ kJ}}{\text{s m}^2} \times \dfrac{60 \text{ s}}{\text{min}} \times \dfrac{60 \text{ min}}{\text{hr}} \times 8.0 \text{ h} = 2.9 \times 10^4 \text{ kJ/m}^2$

Only 13% of the sunlight is converted into electricity:

$$0.13 \times (2.9 \times 10^4 \text{ kJ/m}^2) \times \text{Area} = 1.4 \times 10^5 \text{ kJ}, \quad \text{Area} = 37 \text{ m}^2$$

90. a. $2 \text{ HNO}_3(aq) + \text{Na}_2\text{CO}_3(s) \rightarrow 2 \text{ NaNO}_3(aq) + \text{H}_2\text{O}(l) + \text{CO}_2(g)$

$\Delta H° = [2(-467 \text{ kJ}) + (-286 \text{ kJ}) + (-393.5 \text{ kJ})] - [2(-207 \text{ kJ}) + (-1131 \text{ kJ})] = -69 \text{ kJ}$

$2.0 \times 10^4 \text{ gallons} \times \dfrac{4 \text{ qt}}{\text{gal}} \times \dfrac{946 \text{ mL}}{\text{qt}} \times \dfrac{1.42 \text{ g}}{\text{mL}} = 1.1 \times 10^8 \text{ g of concentrated nitric acid solution}$

$1.1 \times 10^8 \text{ g solution} \times \dfrac{70.0 \text{ g HNO}_3}{100 \text{ g solution}} = 7.7 \times 10^7 \text{ g HNO}_3$

$7.7 \times 10^7 \text{ g HNO}_3 \times \dfrac{1 \text{ mol}}{63.02 \text{ g}} \times \dfrac{1 \text{ mol Na}_2\text{CO}_3}{2 \text{ mol HNO}_3} \times \dfrac{105.99 \text{ g Na}_2\text{CO}_3}{\text{mol Na}_2\text{CO}_3} = 6.5 \times 10^7 \text{ g Na}_2\text{CO}_3$

There are $(7.7 \times 10^7/63.02)$ mol of HNO_3 from the previous calculation. There are 69 kJ of heat evolved for every two moles of nitric acid neutralized. Combining these two results:

$$7.7 \times 10^7 \text{ g HNO}_3 \times \frac{1 \text{ mol HNO}_3}{63.02 \text{ g HNO}_3} \times \frac{-69 \text{ kJ}}{2 \text{ mol HNO}_3} = -4.2 \times 10^7 \text{ kJ}$$

b. They feared the heat generated by the neutralization reaction would vaporize the unreacted nitric acid, causing widespread airborne contamination.

91. $400 \text{ kcal} \times \dfrac{4.18 \text{ kJ}}{\text{kcal}} = 1.67 \times 10^3 \text{ kJ} \approx 2 \times 10^3 \text{ kJ}$

$$\text{PE} = mgz = \left(180 \text{ lb} \times \frac{1 \text{ kg}}{2.205 \text{ lb}}\right) \times \frac{9.8 \text{ m}}{\text{s}^2} \times \left(8 \text{ in} \times \frac{2.54 \text{ cm}}{\text{in}} \times \frac{1 \text{ m}}{100 \text{ cm}}\right) = 160 \text{ J} \approx 200 \text{ J}$$

200 J of energy are needed to climb one step. The total numer of steps to climb are:

$$2 \times 10^6 \text{ J} \times \frac{1 \text{ step}}{200 \text{ J}} = 1 \times 10^4 \text{ steps}$$

CHAPTER SEVEN

ATOMIC STRUCTURE AND PERIODICITY

Questions

13. Planck found that heated bodies only give off certain frequencies of light and Einstein's study of the photoelectric effect.

14. When something is quantized, it can only have certain discrete values. In the Bohr model of the H-atom, the energy of the electron is quantized.

15. Only very small particles with a minuscule mass exhibit quantum effects, e.g., an electron. These tiny particles generally have large velocities.

16. 1) Electrons can be diffracted like light.

    2) The electron microscope uses electrons in a fashion similar to the way in which light is used in a light microscope.

17. a. A discrete bundle of light energy.

    b. A number describing a discrete energy state of an electron.

    c. The lowest energy state of the electron(s) in an atom or ion.

    d. An allowed energy state that is higher in energy than the ground state.

18. n: Gives the energy (it completely specifies the energy only for the H-atom or ions with one electron) and the relative size of the orbitals.

    ℓ: Gives the type (shape) of orbital.

    m_ℓ: Gives information about the direction in which the orbital is pointing.

19. The 2p orbitals differ from each other in the direction in which they point in space.

20. The 2p and 3p orbitals differ from each other in their size and number of nodes.

21.   A nodal surface in an atomic orbital is a surface in which the probability of finding an electron is zero.

22.   The electrostatic energy of repulsion, from Coulomb's Law, will be of the form Q^2/r where Q is the charge of the electron and r is the distance between the two electrons. From the Heisenberg uncertainty principle, we cannot know precisely the position of each electron. Thus, we cannot precisely know the distance between the electrons nor the value of the electrostatic repulsions.

23.   No, the spin is a convenient model. Since we cannot locate or "see" the electron, we cannot see if it is spinning.

24.   There is a higher probability of finding the 4s electron very close to the nucleus than for the 3d electron.

25.   The outermost electrons are the valence electrons. When atoms interact with each other, it will be the outermost electrons that are involved in these interactions. In addition, how tightly the nucleus holds these outermost electrons determines atomic size, ionization energy and other properties of atoms.

26.   Elements in the same group have similar valence electron configurations and, as a result, have similar chemical properties.

27.   If one more electron is added to a half-filled subshell, electron-electron repulsions will increase since two electrons must now occupy the same atomic orbital. This may slightly decrease the stability of the atom. Hence, half-filled subshells minimize electron-electron repulsions.

28.   As one goes across a period, the positive charge from the nucleus increases as protons are added. The number of electrons also increase, but these outer electrons do not completely shield the increasing nuclear charge from each other. The general result is that the outer electrons are more strongly bound as one goes across a period which results in larger ionization energies (and smaller size).

        Aluminum is out of order because the electrons in the filled 3s orbital do shield some of the nuclear charge from the 3p electron. Hence, the 3p electron is less tightly bound than a 3s electron resulting in a lower ionization energy for aluminum as compared to magnesium. The ionization energy of sulfur is lower than phosphorus because of the extra electron-electron repulsions in the doubly occupied sulfur 3p orbital. These added repulsions, which are not present in phosphorus, makes it slightly easier to remove an electron from sulfur as compared to phosphorus.

29.   As electrons are removed, the nuclear charge exerted on the remaining electrons increases. Since the remaining electrons are 'held' more strongly by the nucleus, the energy required to remove these electrons increases.

30.   Yes; The electron configuration for Si is $1s^2 2s^2 2p^6 3s^2 3p^2$. There should be another big jump when the thirteenth electron is removed, i.e., when the 1s electrons are removed.

31.   For hydrogen all atomic orbitals with the same n value have the same energy. For polyatomic atoms/ions, the energy of the atomic orbitals also depends on ℓ. Since there are more nondegenerate energy levels for polyatomic atoms/ions as compared to hydrogen, then there are many more possible electronic transitions resulting in more complicated line spectra.

32.   Each element has a characteristic spectrum. Thus, the presence of the characteristic spectral lines of an element confirms its present in any particular sample.

33.   Yes, the maximum number of unpaired electrons in any configuration corresponds to a minimum in electron-electron repulsions.

34.   The electron is no longer part of that atom. The proton and electron are completely separated.

Exercises

Light and Matter

35. $\nu = \dfrac{c}{\lambda} = \dfrac{2.998 \times 10^8 \text{ m/s}}{780. \times 10^{-9} \text{ m}} = 3.84 \times 10^{14} \text{ s}^{-1}$

36. $99.5 \text{ MHz} = 99.5 \times 10^6 \text{ Hz} = 99.5 \times 10^6 \text{ s}^{-1}$; $\lambda = \dfrac{c}{\nu} = \dfrac{2.998 \times 10^8 \text{ m/s}}{99.5 \times 10^6 \text{ s}^{-1}} = 3.01 \text{ m}$

37. $\nu = \dfrac{c}{\lambda} = \dfrac{3.00 \times 10^8 \text{ m/s}}{1.0 \times 10^{-2} \text{ m}} = 3.0 \times 10^{10} \text{ s}^{-1}$

        $E = h\nu = 6.63 \times 10^{-34} \text{ J s} \times 3.0 \times 10^{10} \text{ s}^{-1} = 2.0 \times 10^{-23} \text{ J/photon}$

        $\dfrac{2.0 \times 10^{-23} \text{ J}}{\text{photon}} \times \dfrac{6.02 \times 10^{23} \text{ photons}}{\text{mol}} = 12 \text{ J/mol}$

38.   For 404.7 nm light: $\nu = \dfrac{c}{\lambda} = \dfrac{2.9979 \times 10^8 \text{ m/s}}{404.7 \times 10^{-9} \text{ m}} = 7.408 \times 10^{14} \text{ s}^{-1}$

        $E = h\nu = 6.6261 \times 10^{-34} \text{ J s} \times 7.408 \times 10^{14} \text{ s}^{-1} = 4.909 \times 10^{-19} \text{ J}$

        $\dfrac{4.909 \times 10^{-19} \text{ J}}{\text{photon}} \times \dfrac{6.0221 \times 10^{23} \text{ photons}}{\text{mol}} = 2.956 \times 10^5 \text{ J/mol} = 295.6 \text{ kJ/mol}$

        For 435.8 nm light: $\nu = \dfrac{c}{\lambda} = \dfrac{2.9979 \times 10^8 \text{ m/s}}{435.8 \times 10^{-9} \text{ m}} = 6.879 \times 10^{14} \text{ s}^{-1}$

        $E = h\nu = 6.6261 \times 10^{-34} \text{ J s} \times 6.879 \times 10^{14} \text{ s}^{-1} = 4.558 \times 10^{-19} \text{ J}$

        $\dfrac{4.558 \times 10^{-19} \text{ J}}{\text{photon}} \times \dfrac{6.0221 \times 10^{23} \text{ photons}}{\text{mol}} = 2.745 \times 10^5 \text{ J/mol} = 274.5 \text{ kJ/mol}$

39.   The energy needed to remove a single electron is:

        $\dfrac{279.7 \text{ kJ}}{\text{mol}} \times \dfrac{1 \text{ mol}}{6.0221 \times 10^{23}} = 4.645 \times 10^{-22} \text{ kJ} = 4.645 \times 10^{-19} \text{ J}$

$$E = \frac{hc}{\lambda}, \quad \lambda = \frac{hc}{E} = \frac{6.6261 \times 10^{-34} \text{ J s} \times 2.9979 \times 10^8 \text{ m/s}}{4.645 \times 10^{-19} \text{ J}} = 4.277 \times 10^{-7} \text{ m} = 427.7 \text{ nm}$$

40. $\frac{492 \text{ kJ}}{\text{mol}} \times \frac{1 \text{ mol}}{6.022 \times 10^{23}} = 8.17 \times 10^{-22} \text{ kJ} = 8.17 \times 10^{-19} \text{ J}$ to remove one electron

$$E = \frac{hc}{\lambda}, \quad \lambda = \frac{hc}{E} = \frac{6.626 \times 10^{-34} \text{ J s} \times 2.998 \times 10^8 \text{ m/s}}{8.17 \times 10^{-19} \text{ J}} = 2.43 \times 10^{-7} \text{ m} = 243 \text{ nm}$$

41.   Ionization energy = energy to remove an electron = 7.21×10^{-19} = E_{photon}

$E_{photon} = h\nu$ and $\lambda\nu = c$. So, $\nu = \frac{c}{\lambda}$ and $E = \frac{hc}{\lambda}$

$$\lambda = \frac{hc}{E_{photon}} = \frac{6.626 \times 10^{-34} \text{ J s} \times 2.998 \times 10^8 \text{ m/s}}{7.21 \times 10^{-19} \text{ J}} = 2.76 \times 10^{-7} \text{ m} = 276 \text{ nm}$$

42. $\frac{890.1 \text{ kJ}}{\text{mol}} \times \frac{1 \text{ mol}}{6.0221 \times 10^{23} \text{ atoms}} = \frac{1.478 \times 10^{-21} \text{ kJ}}{\text{atom}} = \frac{1.478 \times 10^{-18} \text{ J}}{\text{atom}}$ = ionization energy per atom

$$E = \frac{hc}{\lambda}, \quad \lambda = \frac{hc}{E} = \frac{6.6261 \times 10^{-34} \text{ J s} \times 2.9979 \times 10^8 \text{ m/s}}{1.478 \times 10^{-18} \text{ J}} = 1.344 \times 10^{-7} \text{ m} = 134.4 \text{ nm}$$

No, it will take light with a wavelength of 134.4 nm or less to ionize gold. A photon of light with a wavelength of 225 nm is longer wavelength and, thus, lower energy than 134.4 nm light.

43.   a.   5.0% of speed of light = $0.050 \times 3.00 \times 10^8$ m/s = 1.5×10^7 m/s

$$\lambda = \frac{h}{mv}, \quad \lambda = \frac{6.63 \times 10^{-34} \text{ J s}}{1.67 \times 10^{-27} \text{ kg} \times 1.5 \times 10^7 \text{ m/s}} = 2.6 \times 10^{-14} \text{ m} = 2.6 \times 10^{-5} \text{ nm}$$

Note: For units to come out, the mass must be in kg since $1 \text{ J} = \frac{1 \text{ kg m}^2}{\text{s}^2}$.

b.   mass = $5.2 \text{ oz} \times \frac{1 \text{ lb}}{16 \text{ oz}} \times \frac{1 \text{ kg}}{2.205 \text{ lb}} = 0.15 \text{ kg}$

velocity = $\frac{100.8 \text{ mi}}{\text{hr}} \times \frac{1 \text{ hr}}{3600 \text{ s}} \times \frac{1760 \text{ yd}}{\text{mi}} \times \frac{1 \text{ m}}{1.0936 \text{ yd}} = 45.06 \text{ m/s}$

$\lambda = \frac{h}{mv} = \frac{6.63 \times 10^{-34} \text{ J s}}{0.15 \text{ kg} \times 45.06 \text{ m/s}} = 9.8 \times 10^{-35} \text{ m} = 9.8 \times 10^{-26} \text{ nm}$

This number is so small that it is essentially zero. We cannot detect a wavelength this small. The meaning of this number is that we do not have to consider the wave properties of large objects.

44.   a. $\lambda = \frac{h}{mv} = \frac{6.626 \times 10^{-34} \text{ J s}}{1.675 \times 10^{-27} \text{ kg} \times (0.0100 \times 2.998 \times 10^8 \text{ m/s})} = 1.32 \times 10^{-13} \text{ m}$

b. $\lambda = \dfrac{h}{mv}$, $v = \dfrac{h}{\lambda m} = \dfrac{6.626 \times 10^{-34}\,\text{J s}}{75 \times 10^{-12}\,\text{m} \times 1.675 \times 10^{-27}\,\text{kg}} = 5.3 \times 10^3\,\text{m/s}$

45. $\lambda = \dfrac{h}{mv} = \dfrac{6.63 \times 10^{-34}\,\text{J s}}{9.11 \times 10^{-31}\,\text{kg} \times (1.0 \times 10^{-3} \times 3.00 \times 10^8\,\text{m/s})} = 2.4 \times 10^{-9}\,\text{m} = 2.4\,\text{nm}$

46. $\lambda = \dfrac{h}{mv}$, $v = \dfrac{h}{\lambda m}$;   For $\lambda = 1.0 \times 10^2\,\text{nm} = 1.0 \times 10^{-7}\,\text{m}$:

$$v = \dfrac{6.63 \times 10^{-34}\,\text{J s}}{9.11 \times 10^{-31}\,\text{kg} \times 1.0 \times 10^{-7}\,\text{m}} = 7.3 \times 10^3\,\text{m/s}$$

For $\lambda = 1.0\,\text{nm} = 1.0 \times 10^{-9}\,\text{m}$: $v = \dfrac{6.63 \times 10^{-34}\,\text{J s}}{9.11 \times 10^{-31}\,\text{kg} \times 1.0 \times 10^{-9}\,\text{m}} = 7.3 \times 10^5\,\text{m/s}$

Hydrogen Atom: The Bohr Model

47.   For the H-atom (Z = 1): $E_n = -2.178 \times 10^{-18}\,\text{J}/n^2$; For a spectral transition, $\Delta E = E_f - E_i$:

$$\Delta E = -2.178 \times 10^{-18}\,\text{J} \left(\dfrac{1}{n_f^2} - \dfrac{1}{n_i^2} \right)$$

where n_i and n_f are the levels of the initial and final states, respectively. A positive value of ΔE always corresponds to an absorption of light and a negative value of ΔE corresponds to an emission of light.

a. $\Delta E = -2.178 \times 10^{-18}\,\text{J} \left(\dfrac{1}{2^2} - \dfrac{1}{3^2} \right) = -2.178 \times 10^{-18}\,\text{J} \left(\dfrac{1}{4} - \dfrac{1}{9} \right)$

$\Delta E = -2.178 \times 10^{-18}\,\text{J} \times (0.2500 - 0.1111) = -3.025 \times 10^{-19}\,\text{J}$

The photon of light must have precisely this energy (3.025×10^{-19} J).

$|\Delta E| = E_{photon} = h\nu = \dfrac{hc}{\lambda}$ or $\lambda = \dfrac{hc}{|\Delta E|} = \dfrac{6.6261 \times 10^{-34}\,\text{J s} \times 2.9979 \times 10^8\,\text{m/s}}{3.025 \times 10^{-19}\,\text{J}}$

$$= 6.567 \times 10^{-7}\,\text{m} = 656.7\,\text{nm}$$

b. $\Delta E = -2.178 \times 10^{-18}\,\text{J} \left(\dfrac{1}{2^2} - \dfrac{1}{4^2} \right) = -4.084 \times 10^{-19}\,\text{J}$

$\lambda = \dfrac{hc}{|\Delta E|} = \dfrac{6.6261 \times 10^{-34}\,\text{J s} \times 2.9979 \times 10^8\,\text{m/s}}{4.084 \times 10^{-19}\,\text{J}} = 4.864 \times 10^{-7}\,\text{m} = 486.4\,\text{nm}$

c. $\Delta E = -2.178 \times 10^{-18}\,\text{J} \left(\dfrac{1}{1^2} - \dfrac{1}{2^2} \right) = -1.634 \times 10^{-18}\,\text{J}$

$\lambda = \dfrac{6.6261 \times 10^{-34}\,\text{J s} \times 2.9979 \times 10^8\,\text{m/s}}{1.634 \times 10^{-18}\,\text{J}} = 1.216 \times 10^{-7}\,\text{m} = 121.6\,\text{nm}$

48.   a. $\Delta E = -2.178 \times 10^{-18}\,\text{J}\left(\dfrac{1}{3^2} - \dfrac{1}{4^2}\right) = -1.059 \times 10^{-19}\,\text{J}$

$$\lambda = \frac{hc}{|\Delta E|} = \frac{6.6261 \times 10^{-34}\,\text{J s} \times 2.9979 \times 10^{8}\,\text{m/s}}{1.059 \times 10^{-19}\,\text{J}} = 1.876 \times 10^{-6}\,\text{m or } 1876\text{ nm}$$

b. $\Delta E = -2.178 \times 10^{-18}\,\text{J}\left(\dfrac{1}{4^2} - \dfrac{1}{5^2}\right) = -4.901 \times 10^{-20}\,\text{J}$

$$\lambda = \frac{hc}{|\Delta E|} = \frac{6.6261 \times 10^{-34}\,\text{J s} \times 2.9979 \times 10^{8}\,\text{m/s}}{4.901 \times 10^{-20}\,\text{J}} = \lambda = 4.053 \times 10^{-6}\,\text{m} = 4053\text{ nm}$$

c. $\Delta E = -2.178 \times 10^{-18}\,\text{J}\left(\dfrac{1}{3^2} - \dfrac{1}{5^2}\right) = -1.549 \times 10^{-19}\,\text{J}$

$$\lambda = \frac{hc}{|\Delta E|} = \frac{6.6261 \times 10^{-34}\,\text{J s} \times 2.9979 \times 10^{8}\,\text{m/s}}{1.549 \times 10^{-19}\,\text{J}} = 1.282 \times 10^{-6}\,\text{m} = 1282\text{ nm}$$

49.

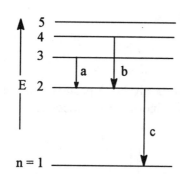

a. $3 \to 2$

b. $4 \to 2$

c. $2 \to 1$

Energy levels are not to scale.

50.

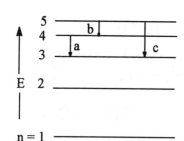

a. $4 \to 3$

b. $5 \to 4$

c. $5 \to 3$

51.   The longest wavelength light emitted will correspond to the transition with the lowest energy change. This is the transition from n = 6 to n = 5.

$$\Delta E = -2.178 \times 10^{-18}\,\text{J}\left(\frac{1}{5^2} - \frac{1}{6^2}\right) = -2.662 \times 10^{-20}\,\text{J}$$

$$\lambda = \frac{hc}{|\Delta E|} = \frac{6.6261 \times 10^{-34}\,\text{J s} \times 2.9979 \times 10^{8}\,\text{m/s}}{2.662 \times 10^{-20}\,\text{J}} = 7.462 \times 10^{-6}\,\text{m} = 7462\text{ nm}$$

The shortest wavelength emitted will correspond to the largest ΔE; this is $n = 6 \rightarrow n = 1$.

$$\Delta E = -2.178 \times 10^{-18} \text{ J} \left(\frac{1}{1^2} - \frac{1}{6^2} \right) = -2.118 \times 10^{-18} \text{ J}$$

$$\lambda = \frac{hc}{|\Delta E|} = \frac{6.6261 \times 10^{-34} \text{ J s} \times 2.9979 \times 10^8 \text{ m/s}}{2.118 \times 10^{-18} \text{ J}} = 9.379 \times 10^{-8} \text{ m} = 93.79 \text{ nm}$$

52. $\Delta E = h\nu = 6.6261 \times 10^{-34} \text{ J s} \times 1.141 \times 10^{14} \text{ s}^{-1} = 7.560 \times 10^{-20} \text{ J}$

$\Delta E = -7.560 \times 10^{-20} \text{ J}$ since light is emitted.

$$\Delta E = E_4 - E_n, \ -7.560 \times 10^{-20} \text{ J} = -2.178 \times 10^{-18} \text{ J} \left(\frac{1}{4^2} - \frac{1}{n^2} \right)$$

$$3.471 \times 10^{-2} = 6.250 \times 10^{-2} - \frac{1}{n^2}, \ \frac{1}{n^2} = 2.779 \times 10^{-2}, \ n^2 = 35.98, \ n = 6$$

53.   Ionization from $n = 1$ corresponds to the transition $n_i = 1 \rightarrow n_f = \infty$ where $E_\infty = 0$.

$$\Delta E = E_\infty - E_1 = -E_1 = 2.178 \times 10^{-18} \left(\frac{1}{1^2} \right) = 2.178 \times 10^{-18} \text{ J} = E_{photon}$$

$$\lambda = \frac{hc}{E} = \frac{6.6261 \times 10^{-34} \text{ J s} \times 2.9979 \times 10^8 \text{ m/s}}{2.178 \times 10^{-18} \text{ J}} = 9.120 \times 10^{-8} \text{ m} = 91.20 \text{ nm}$$

To ionize from $n = 2$, $\Delta E = E_\infty - E_2 = -E_2 = 2.178 \times 10^{-18} \left(\frac{1}{2^2} \right) = 5.445 \times 10^{-19} \text{ J}$

$$\lambda = \frac{6.6261 \times 10^{-34} \text{ J s} \times 2.9979 \times 10^8 \text{ m/s}}{5.445 \times 10^{-19} \text{ J}} = 3.648 \times 10^{-7} \text{ m} = 364.8 \text{ nm}$$

54.   To ionize from $n = 4$, $\Delta E = E_\infty - E_4 = 0 - E_4 = 2.178 \times 10^{-18} \left(\frac{1}{4^2} \right) = 1.361 \times 10^{-19} \text{ J} = E_{photon}$

$$\lambda = \frac{hc}{E} = \frac{6.6261 \times 10^{-34} \text{ J s} \times 2.9979 \times 10^8 \text{ m/s}}{1.361 \times 10^{-19} \text{ J}} = 1.460 \times 10^{-6} \text{ m} = 1460. \text{ nm}$$

To ionize from $n = 10$, $\Delta E = 2.178 \times 10^{-18} \left(\frac{1}{10^2} \right) = 2.178 \times 10^{-20} \text{ J}$

$$\lambda = \frac{hc}{\Delta E} = \frac{6.6261 \times 10^{-34} \text{ J s} \times 2.9979 \times 10^8 \text{ m/s}}{2.178 \times 10^{-20} \text{ J}} = 9.120 \times 10^{-6} \text{ m} = 9120. \text{ nm}$$

Quantum Mechanics, Quantum Numbers, and Orbitals

55. $\Delta mv = m\Delta v = 9.11 \times 10^{-31} \text{ kg} \times 0.100 \text{ m/s} = \dfrac{9.11 \times 10^{-32} \text{ kg m}}{\text{s}}$

$$\Delta x \Delta mv \geq \frac{h}{4\pi}, \ \Delta x = \frac{h}{4\pi \Delta mv} = \frac{6.626 \times 10^{-34} \text{ J s}}{4 \times 3.142 \times 9.11 \times 10^{-32} \text{ kg m/s}} = 5.79 \times 10^{-4} \text{ m}$$

$$\Delta x = 5.79 \times 10^{-4} \text{ m} = 5.79 \times 10^{-2} \text{ cm} = 5.79 \times 10^{5} \text{ nm}$$

Diameter of H atom is roughly 1.0×10^{-8} cm. The uncertainty in position is much larger than the size of the atom.

56. $$\Delta x = \frac{h}{4\pi \Delta mv} = \frac{6.626 \times 10^{-34} \text{ J s}}{4 \times 3.142 \times 0.145 \text{ kg} \times 0.100 \text{ m/s}} = 3.64 \times 10^{-33} \text{ m}$$

The uncertainty is insignificant compared to the size of a baseball.

57. $n = 1, 2, 3, \ldots ; \quad \ell = 0, 1, 2, \ldots (n - 1); \quad m_\ell = -\ell \ldots -2, -1, 0, 1, 2, \ldots +\ell$

58.   1p: $n = 1$, $\ell = 1$ is not possible; 3f: $n = 3$, $\ell = 3$ is not possible; 2d: $n = 2$, $\ell = 2$ is not possible; In all three incorrect cases, $n = \ell$. The maximum value ℓ can have is n - 1, not n.

59.   b. ℓ must be smaller than n.   d.   For $\ell = 0$, $m_\ell = 0$ is the only allowed value.

60.   b.   For $\ell = 3$, m_ℓ can range from -3 to +3; thus +4 is not allowed.

         c.   n cannot equal zero.   d. ℓ cannot be a negative number.

61. ψ^2 gives the probability of finding the electron at that point.

62.   The diagrams of the orbitals in the text only give 90% probabilities of where the electron may reside. We can never be 100% certain of the location of the electrons due to Heisenburg's uncertainty principle.

Polyelectronic Atoms

63.   5p: three orbitals; $3d_{z^2}$: one orbital;   4d: five orbitals

         $n = 5$: $\ell = 0$ (1 orbital), $\ell = 1$ (3 orbitals), $\ell = 2$ (5 orbitals), $\ell = 3$ (7 orbitals), $\ell = 4$ (9 orbitals)

              Total for n = 5 is 25 orbitals.

         $n = 4$: $\ell = 0$ (1), $\ell = 1$ (3), $\ell = 2$ (5), $\ell = 3$ (7); Total for n = 4 is 16 orbitals.

64.   1s, 2 electrons; 2p, 6 electrons; $3p_x$, 2 electrons (specifies one atomic orbital); 6f, 14 electrons; $2d_{xy}$, none (2d orbitals are not possible).

65.   a. $n = 4$: ℓ can be 0, 1, 2, or 3. Thus we have s (2 e⁻), p (6 e⁻), d (10 e⁻) and f (14 e⁻) orbitals present. Total number of electrons to fill these orbitals is 32.

         b. $n = 5$, $m_\ell = +1$: For $n = 5$, $\ell = 0, 1, 2, 3, 4$. For $\ell = 1, 2, 3, 4$, all can have $m_\ell = +1$. Four distinct orbitals, thus 8 electrons.

   c. $n = 5$, $m_s = +1/2$: For $n = 5$, $\ell = 0, 1, 2, 3, 4$. Number of orbitals $= 1, 3, 5, 7, 9$ for each value of ℓ, respectively. There are 25 orbitals with $n = 5$. They can hold 50 electrons and 25 of these electrons can have $m_s = +1/2$.

   d. $n = 3$, $\ell = 2$: These quantum numbers define a set of 3d orbitals. There are 5 degenerate 3d orbitals which can hold a total of 10 electrons.

   e. $n = 2$, $\ell = 1$: These define a set of 2p orbitals. There are 3 degenerate 2p orbitals which can hold a total of 6 electrons.

66.   a.   It is impossible for $n = 0$. Thus, no electrons can have this set of quantum numbers.

   b.   The four quantum numbers completely specify a single electron.

   c. $n = 3$: 3s, 3p and 3d orbitals all have $n = 3$. These orbitals can hold up to 18 electrons.

   d. $n = 2$, $\ell = 2$: This combination is not possible ($\ell \neq 2$ for $n = 2$). Zero electrons in an atom can have these quantum numbers.

   e. $n = 1$, $\ell = 0$, $m_\ell = 0$: These define a 1s orbital which can hold 2 electrons.

67.   Si: $1s^2 2s^2 2p^6 3s^2 3p^2$ or [Ne]$3s^2 3p^2$; Ga: $1s^2 2s^2 2p^6 3s^2 3p^6 4s^2 3d^{10} 4p^1$ or [Ar]$4s^2 3d^{10} 4p^1$

As: [Ar]$4s^2 3d^{10} 4p^3$; Ge: [Ar]$4s^2 3d^{10} 4p^2$; Al: [Ne]$3s^2 3p^1$; Cd: [Kr]$5s^2 4d^{10}$

S: [Ne]$3s^2 3p^4$; Se: [Ar]$4s^2 3d^{10} 4p^4$

68.   Cu: [Ar]$4s^2 3d^9$ (using periodic table), [Ar]$4s^1 3d^{10}$ (actual)

O: $1s^2 2s^2 2p^4$; La: [Xe]$6s^2 5d^1$; Y: [Kr]$5s^2 4d^1$; Ba: [Xe]$6s^2$

Tl: [Xe]$6s^2 4f^{14} 5d^{10} 6p^1$; Bi: [Xe]$6s^2 4f^{14} 5d^{10} 6p^3$

69.   The following are complete electron configurations. Noble gas shorthand notation could also be used.

Sc: $1s^2 2s^2 2p^6 3s^2\ 3p^6 4s^2 3d^1$; Fe: $1s^2 2s^2 2p^6 3s^2\ 3p^6 4s^2 3d^6$

P: $1s^2 2s^2 2p^6 3s^2\ 3p^3$; Cs: $1s^2 2s^2 2p^6 3s^2\ 3p^6 4s^2 3d^{10} 4p^6 5s^2 4d^{10} 5p^6 6s^1$

Eu: $1s^2 2s^2 2p^6 3s^2\ 3p^6 4s^2 3d^{10} 4p^6 5s^2 4d^{10} 5p^6 6s^2 4f^6 5d^1$*

Pt: $1s^2 2s^2 2p^6 3s^2\ 3p^6 4s^2 3d^{10} 4p^6 5s^2 4d^{10} 5p^6 6s^2 4f^{14} 5d^8$*

Xe: $1s^2 2s^2 2p^6 3s^2\ 3p^6 4s^2 3d^{10} 4p^6 5s^2 4d^{10} 5p^6$; Br: $1s^2 2s^2 2p^6 3s^2\ 3p^6 4s^2 3d^{10} 4p^5$

*Note: These electron configurations were written down using only the periodic table.

Actual electron configurations are: Eu: [Xe]$6s^2 4f^7$ and Pt: [Xe]$6s^1 4f^{14} 5d^9$

70.   K: $1s^2 2s^2 2p^6 3s^2 3p^6 4s^1$ or [Ar]$4s^1$

Fr: $1s^2 2s^2 2p^6 3s^2 3p^6 4s^2 3d^{10} 4p^6 5s^2 4d^{10} 5p^6 6s^2 4f^{14} 5d^{10} 6p^6 7s^1$ or [Rn]$7s^1$

Pu: [Rn]$7s^2 6d^1 5f^5$ (expected from periodic table)

Sb: $1s^2 2s^2 2p^6 3s^2 3p^6 4s^2 3d^{10} 4p^6 5s^2 4d^{10} 5p^3$ or [Kr] $5s^2 4d^{10} 5p^3$; Os: [Xe]$6s^2 4f^{14} 5d^6$

Pd: $1s^2 2s^2 2p^6 3s^2 3p^6 4s^2 3d^{10} 4p^6 5s^2 4d^8$ or [Kr] $5s^2 4d^8$ (expected from periodic table)

Pb: [Xe]$6s^2 4f^{14} 5d^{10} 6p^2$; I: $1s^2 2s^2 2p^6 3s^2 3p^6 4s^2 3d^{10} 4p^6 5s^2 4d^{10} 5p^5$ or [Kr]$5s^2 4d^{10} 5p^5$

71.   Exceptions: Cr, Cu, Nb, Mo, Tc, Ru, Rh, Pd, Ag, Pt, Au; Tc, Ru, Rh, Pd and Pt do not correspond to the supposed extra stability of half-filled and filled subshells.

72.   No, in the solid and liquid state the electrons may interact (bonding occurs) and the experimental configuration will not be valid. That is, the experimental electron configurations are for isolated atoms; in condensed phases, the atoms are no longer isolated.

73.   a.   The lightest halogen is fluorine: $1s^2 2s^2 2p^5$ b. K: $1s^2 2s^2 2p^6 3s^2 3p^6 4s^1$

c.   Be: $1s^2 2s^2$;   Mg: $1s^2 2s^2 2p^6 3s^2$;   Ca: $1s^2 2s^2 2p^6 3s^2 3p^6 4s^2$

d.   In: [Kr]$5s^2 4d^{10} 5p^1$ e. C: $1s^2 2s^2 2p^2$; Si: $1s^2 2s^2 2p^6 3s^2 3p^2$

f.   This will be element #118: [Rn]$7s^2 5f^{14} 6d^{10} 7p^6$

74.   a.   This atom has 10 electrons. Ne   b. S

c.   The ground state configuration is [Kr]$5s^2 4d^9$. The element is Ag.

d.   Bi: [Xe]$6s^2 4f^{14} 5d^{10} 6p^3$

75.   B : $1s^2 2s^2 2p^1$

	n	ℓ	m_ℓ	m_s
1s	1	0	0	+1/2
1s	1	0	0	-1/2
2s	2	0	0	+1/2
2s	2	0	0	-1/2
2p*	2	1	-1	+1/2

*This is only one of several possibilities for the 2p electron. The 2p electron in B could have m_ℓ = -1, 0 or +1, and m_s = +1/2 or -1/2, a total of six possibilities.

N : $1s^2 2s^2 2p^3$

	n	ℓ	m_ℓ	m_s
1s	1	0	0	+1/2
1s	1	0	0	-1/2
2s	2	0	0	+1/2
2s	2	0	0	-1/2
2p	2	1	-1	+1/2
2p	2	1	0	+1/2
2p	2	1	+1	+1/2

(Or all 2p electrons could have $m_s = -1/2$.)

76. Mg : $1s^2 2s^2 2p^6 \underline{3s^2}$

	n	ℓ	m_ℓ	m_s
3s	3	0	0	+1/2
3s	3	0	0	-1/2

As : $1s^2 2s^2 3s^2 3p^6 \underline{4s^2} 3d^{10} \underline{4p^3}$

	n	ℓ	m_ℓ	m_s
4s	4	0	0	+1/2
4s	4	0	0	-1/2
4p	4	1	-1	+1/2
4p	4	1	0	+1/2
4p	4	1	+1	+1/2

or all -1/2

Xe : $[Kr]\underline{5s^2} 4d^{10} \underline{5p^6}$

	n	ℓ	m_ℓ	m_s
5s	5	0	0	+1/2
5s	5	0	0	-1/2
5p	5	1	-1	+1/2
5p	5	1	-1	-1/2
5p	5	1	0	+1/2
5p	5	1	0	-1/2
5p	5	1	+1	+1/2
5p	5	1	+1	-1/2

77. O: $1s^2 2s^2 2p_x^2 2p_y^2$ ($\uparrow\downarrow$ $\uparrow\downarrow$ __); There are no unpaired electrons in this oxygen atom. This configuration would be an excited state and in going to the more stable ground state ($\uparrow\downarrow$ $\uparrow$ $\uparrow$), energy would be released.

78. The number of unpaired electrons are in parentheses.

a. excited state of boron (1)   b. ground state of neon (0)

B ground state: $1s^2 2s^2 2p^1$ (1)

c. exited state of fluorine (3)   d. excited state of iron (6)

F ground state: $1s^2 2s^2 2p^5$ (1)   Fe ground state: $[Ar]4s^2 3d^6$ (4)

79.   We get the number of unpaired electrons by looking at the incompletely filled subshells.

Sc: $[Ar]4s^23d^1$ $3d^1$:   ↑ __ __ __ __   one unpaired e⁻

Ti: $[Ar]4s^23d^2$ $3d^2$:   ↑ ↑ __ __ __   two unpaired e⁻

Al: $[Ne]3s^23p^1$ $3p^1$:   ↑ __ __   one unpaired e⁻

Sn: $[Kr]5s^24d^{10}5p^2$ $5p^2$:   ↑ ↑ __   two unpaired e⁻

Te: $[Ar]5s^24d^{10}5p^4$ $5p^4$:   ↑↓ ↑ ↑   two unpaired e⁻

Br: $[Ar]4s^23d^{10}4p^5$ $4p^5$:   ↑↓ ↑↓ ↑   one unpaired e⁻

80.   We get the number of unpaired electrons by looking at the incompletely filled subshells.

O: $[He]2s^22p^4$ $2p^4$:   ↑↓ ↑ ↑   two unpaired e⁻

O^+: $[He]2s^22p^3$ $2p^3$:   ↑ ↑ ↑   three unpaired e⁻

O^-: $[He]2s^22p^5$ $2p^5$:   ↑↓ ↑↓ ↑   one unpaired e⁻

Fe: $[Ar]4s^23d^6$ $3d^6$:   ↑↓ ↑ ↑ ↑ ↑   four unpaired e⁻

Mn: $[Ar]4s^23d^5$ $3d^5$:   ↑ ↑ ↑ ↑ ↑   five unpaired e⁻

S: $[Ne]3s^23p^4$ $3p^4$:   ↑↓ ↑ ↑   two unpaired e⁻

F: $[He]2s^22p^5$ $2p^5$:   ↑↓ ↑↓ ↑   one unpaired e⁻

Ar: $[Ne]3s^23p^6$ $3p^6$:   ↑↓ ↑↓ ↑↓   zero unpaired e⁻

The Periodic Table and Periodic Properties

81.   Size decreases left to right across the periodic table and size increases from top to bottom of the periodic table.

a.   Be < Mg < Ca   b.   Xe < I < Te   c.   Ge < Ga < In

82.   a.   F < N < As   b.   F < Cl < S   c.   Li < K < Cs

83.   The ionization energy trend is the opposite of the radii trend (see Exercise 7.81).

a.   Ca < Mg < Be   b.   Te < I < Xe   c.   In < Ga < Ge

84.   a.   As < N < F   b.   S < Cl < F   c.   Cs < K < Li

85.   Ge: $[Ar]4s^23d^{10}4p^2$; As: $[Ar]4s^23d^{10}4p^3$; Se: $[Ar]4s^23d^{10}4p^4$; There are extra electron-electron repulsions in Se because two electrons are in the same 4p orbital, resulting in a lower ionization energy. Se is an exception to the general ionization energy trend.

86.   Applying the general trends in radii and ionization energy allows matching the various values to the elements.

   Ar : $1s^22s^22p^63s^23p^6$: 1.527 MJ/mol : 0.98 Å

   Mg : $1s^22s^22p^63s^2$: 0.735 MJ/mol : 1.60 Å

   K   : $1s^22s^22p^63s^23p^64s^1$: 0.419 MJ/mol : 2.35 Å

   Size: Ar < Mg < K;   IE: K < Mg < Ar

87.   a. Li   b. P

   c. O^+. This ion has the fewest electrons as compared to the other oxygen species present. O^+ has the smallest amount of electron-electron repulsions which makes it the smallest ion with the largest ionization energy.

   d. From the radii trend, Ar < Cl < S and Kr > Ar. Since variation in size down a family is greater than the variation across a period, we would predict Cl to be the smallest of the three.

   e. Cu

88.   a. Cs   b. Ga   c. Tl   d. Tl

   e. O^{2-}; When comparing ions of the same element, the ion with the most electrons will have the largest amount of electron-electron repulsions. This makes it the largest ion with the smallest ionization energy.

89.   a. 106: $[Rn]7s^25f^{14}6d^4$ b. W   c. SgO_3 and SgO_4^{2-} (similar to Cr; Sg = 106)

90.   a. Uus will have 117 electrons. $[Rn]7s^25f^{14}6d^{10}7p^5$

   b. It will be in the halogen family and most similar to astatine, At.

   c. Uus should form -1 charged anions like the other halogens.

      NaUus, $Mg(Uus)_2$, $C(Uus)_4$, $O(Uus)_2$

   d. Assuming Uus is like the other halogens: $UusO^-$, $UusO_2^-$, $UusO_3^-$, $UusO_4^-$

91. $P(g) \rightarrow P^+(g) + e^-$; IE refers to atoms in the gas phase.

92. $P(g) + e^- \rightarrow P^-(g)$

93.   a.   More favorable EA: C and Br; The electron affinity trend is very erratic. Both N and Ar have positive EA values (unfavorable) due to their electron configuration (see text for detailed explanation).

      b.   Higher IE: N and Ar (follows the IE trend)

      c.   Larger size: C and Br (follows the radii trend)

94.   a.   More favorable EA: K and Cl; Mg has a positive EA value and F has a more positive EA value than expected from its position relative to Cl.

      b.   Higher IE: Mg and F   c.   Larger radius: K and Cl

95.   The electron affinity trend is very erratic. In general, EA decreases down the periodic table and the trend across the table is too erratic to be of much use.

      a.   Se < S; S is most exothermic.   b.   I < Br < F < Cl; Cl is most exothermic (F is an exception).

96.   a.   N < O < F, F is most exothermic.   b.   Al < P < Si; Si is most exothermic.

97.   Electron-electron repulsions are much greater in O^- than in S^- because the electron goes into a smaller 2p orbital vs. the larger 3p orbital in sulfur. This results in a more favorable (more exothermic) EA for sulfur.

98.   O; The electron-electron repulsions will be much more severe for $O^- + e^- \rightarrow O^{2-}$ than for $O + e^- \rightarrow O^-$.

99.   a.   The electron affinity of Mg^{2+} is ΔH for: $Mg^{2+}(g) + e^- \rightarrow Mg^+(g)$; This is just the reverse of the second ionization energy, or: $EA(Mg^{2+}) = -IE_2(Mg) = -1445$ kJ/mol (Table 7.5)

      b.   EA of Al^+ is ΔH for: $Al^+(g) + e^- \rightarrow Al(g)$; $EA(Al^+) = -IE_1(Al) = -580$ kJ/mol (Table 7.5)

100. a.   IE of Cl^- is ΔH for: $Cl^-(g) \rightarrow Cl(g) + e^-$; $IE(Cl^-) = -EA(Cl) = 348.7$ kJ/mol (Table 7.7)

      b. $Cl(g) \rightarrow Cl^+(g) + e^-$ IE = 1255 kJ/mol (Table 7.5)

      c. $Cl^+(g) + e^- \rightarrow Cl(g)$ $\Delta H = -IE_1 = -1255$ kJ/mol = EA (Cl^+)

Alkali and Alkaline Earth Elements

101. It should be potassium peroxide, K_2O_2, since K^+ ions are stable. K^{2+} ions are not stable; the second ionization energy of K is very large as compared to the first.

102. It should be magnesium(II) oxide since Mg^{2+} ions are stable. Mg^+ ions are not as stable as Mg^{2+} ions. The O_2^{2-} ion would oxidize Mg^+ to Mg^{2+} with O^{2-} as the other product.

103. $\nu = \dfrac{c}{\lambda} = \dfrac{2.9979 \times 10^8 \text{ m/s}}{455.5 \times 10^{-9} \text{ m}} = 6.582 \times 10^{14} \text{ s}^{-1}$

$E = h\nu = 6.6261 \times 10^{-34} \text{ J s} \times 6.582 \times 10^{14} \text{ s}^{-1} = 4.361 \times 10^{-19} \text{ J}$

104. For 589.0 nm: $\nu = \dfrac{c}{\lambda} = \dfrac{2.9979 \times 10^8 \text{ m/s}}{589.0 \times 10^{-9} \text{ m}} = 5.090 \times 10^{14} \text{ s}^{-1}$

$E = h\nu = 6.6261 \times 10^{-34} \text{ J s} \times 5.090 \times 10^{14} \text{ s}^{-1} = 3.373 \times 10^{-19} \text{ J}$

For 589.6 nm: $\nu = c/\lambda = 5.085 \times 10^{14} \text{ s}^{-1}$; $E = h\nu = 3.369 \times 10^{-19} \text{ J}$

The energies in kJ/mol are:

$$3.373 \times 10^{-19} \text{ J} \times \frac{1 \text{ kJ}}{1000 \text{ J}} \times \frac{6.0221 \times 10^{23}}{\text{mol}} = 203.1 \text{ kJ/mol}$$

$$3.369 \times 10^{-19} \text{ J} \times \frac{1 \text{ kJ}}{1000 \text{ J}} \times \frac{6.0221 \times 10^{23}}{\text{mol}} = 202.9 \text{ kJ/mol}$$

105. a. Li_3N; lithium nitride   b. NaBr; sodium bromide   c. K_2S; potassium sulfide

106. a. Li_3P; lithium phosphide   b. RbH; rubidium hydride

c. Na_2O or Na_2O_2; Sodium oxide and sodium peroxide can both form.

107. a. $4 \text{ Li(s)} + O_2(g) \rightarrow 2 \text{ Li}_2O(s)$ b. $2 \text{ K(s)} + \text{S(s)} \rightarrow K_2S(s)$

108. a. $2 \text{ Cs(s)} + 2 \text{ H}_2O(l) \rightarrow 2 \text{ CsOH(aq)} + H_2(g)$ b. $2 \text{ Na(s)} + Cl_2(g) \rightarrow 2 \text{ NaCl(s)}$

Additional Exercises

109. $E = \dfrac{310. \text{ kJ}}{\text{mol}} \times \dfrac{1 \text{ mol}}{6.022 \times 10^{23}} = 5.15 \times 10^{-22} \text{ kJ} = 5.15 \times 10^{-19} \text{ J}$

$E = \dfrac{hc}{\lambda}$, $\lambda = \dfrac{hc}{E} = \dfrac{6.626 \times 10^{-34} \text{ J s} \times 2.998 \times 10^8 \text{ m/s}}{5.15 \times 10^{-19}} = 3.86 \times 10^{-7} \text{ m} = 386 \text{ nm}$

110. a. False; It takes less energy to ionize an electron from $n = 3$ than from the ground state.

b. True

c. False; The energy difference from $n = 3 \rightarrow n = 2$ is less than the energy difference from $n = 3 \rightarrow n = 1$, thus, the wavelength is larger for $n = 3 \rightarrow n = 2$ than for $n = 3 \rightarrow n = 1$.

d. True

e. False; $n = 2$ is the first excited state and $n = 3$ is the second excited state.

111.   There are 4 possible transitions for an electron in the $n = 5$ level ($5 \rightarrow 4$, $5 \rightarrow 3$, $5 \rightarrow 2$ and $5 \rightarrow 1$). If an electron initially drops to the $n = 4$ level, three additional transitions can occur ($4 \rightarrow 3$, $4 \rightarrow 2$ and $4 \rightarrow 1$). Similarly, there are two more transitions from the $n = 3$ level ($3 \rightarrow 2$, $3 \rightarrow 1$) and one more transition for the $n = 2$ level ($2 \rightarrow 1$). There are a total of 10 possible transitions for an electron in the $n = 5$ level for a possible total of 10 different wavelength emissions.

112.   a.   n   b.   n and ℓ

113.   a. $n = 3$; We can have 3s, 3p, and 3d orbitals. Nine orbitals can hold 18 electrons.

         b. $n = 2$, $\ell = 0$; This is a 2s orbital. 2 electrons

         c. $n = 2$, $\ell = 2$; Not possible. No electrons can have this combination of quantum numbers.

         d.   These four quantum numbers completely specify a single electron.

114. $n = 5$; $m_\ell = -4, -3, -2, -1, 0, 1, 2, 3, 4$; 18 electrons

115.   b and f are the only possible sets of quantum numbers.

         a.   For $\ell = 0$, m_ℓ can only be 0.

         c. m_s can only be $+ 1/2$ or $-1/2$.

         d.   For $n = 1$, ℓ can only be 0.

         e.   For $\ell = 2$, m_ℓ cannot be -3. The lowest allowed m_ℓ value is -2.

116.   Ti : $[Ar]4s^2 3d^2$

	n	ℓ	m_ℓ	m_s
4s	4	0	0	+1/2
4s	4	0	0	-1/2
3d	3	2	-2	+1/2
3d	3	2	-1	+1/2

Only one of 10 possible combinations of m_ℓ and m_s for the first d electron. For the ground state, the second d electron should be in a different orbital with spin parallel; 4 possibilities.

117.   Sb: $1s^2 2s^2 2p^6 3s^2 3p^6 4s^2 3d^{10} 4p^6 5s^2 4d^{10} 5p^3$

         a. $\ell = 1$: Designates p orbitals. There are 21 electrons in p orbitals.

         b. $m_\ell = 0$: All s electrons, 2 out of each set of 2p, 3p, 4p electrons, 2 out of each set of 3d and 4d electrons, and one of the 5p electrons have $m_\ell = 0$. $10 + 6 + 4 + 1 = 21$ e$^-$ with $m_\ell = 0$.

         c. $m_\ell = 1$: 2 out of each set of 2p, 3p, and 4p electrons, 2 out of each set of 3d and 4d electrons, and one of the 5p electrons have $m_\ell = 1$. $6 + 4 + 1 = 11$ e$^-$ with $m_\ell = 1$.

118. He: $1s^2$; Ne: $1s^22s^22p^6$; Ar: $1s^22s^22p^63s^23p^6$; Each peak in the diagram corresponds to a subshell with different values of n. Corresponding subshells are closer to the nucleus for heavier elements because of the increased nuclear charge.

119. a. As: $1s^22s^22p^63s^23p^64s^23d^{10}4p^3$

   b. Element 116 will be below Po in the periodic table: $[Rn]7s^25f^{14}6d^{10}7p^4$

   c. Ta: $[Xe]6s^24f^{14}5d^3$ or Ir: $[Xe]6s^24f^{14}5d^7$

   d. Ti: $[Ar]4s^23d^2$; Ni: $[Ar]4s^23d^8$; Os: $[Xe]6s^24f^{14}5d^6$

120. It should be element #119 with the ground state electron configuration: $[Rn]7s^25f^{14}6d^{10}7p^68s^1$

121. a. The 4+ ion contains 20 electrons. Thus, the electrically neutral atom will contain 24 electrons. The atomic number is 24.

   b. The ground state electron configuration of the ion must be: $1s^22s^22p^63s^23p^64s^03d^2$; There are 6 electrons in s orbitals.

   c. 12   d. 2

   e. Because of the mass, this is an isotope of $^{50}_{24}Cr$. There are 26 neutrons in the nucleus.

   f. $1s^22s^22p^63s^23p^64s^13d^5$ is the ground state electron configuration for Cr. Cr is an exception to the normal filling order.

122. The valence electrons are strongly attracted to the nucleus for elements with large ionization energies. One would expect these species to readily accept another electron and have very exothermic electron affinities. The noble gases are an exception. The noble gases have a large IE but have an endothermic EA. Noble gases have a stable arrangement of electrons. Adding an electron disrupts this stable arrangement, resulting in unfavorable electron affinities.

123. Electron-electron repulsions become more important when we try to add electrons to an atom. From the standpoint of electron-electron repulsions, larger atoms would have more favorable (more exothermic) electron affinities. Considering only electron-nucleus attractions, smaller atoms would be expected to have the more favorable (more exothermic) EA's. These trends are exactly the opposite of each other. Thus, the overall variation in EA is not as great as ionization energy in which attractions to the nucleus dominate.

124. Al (-44), Si(-120), P (-74), S (-200.4), Cl (-348.7); Based on the increasing nuclear charge, we would expect the EA to become more exothermic as we go from left to right in the period. Phosphorus is out of line. The reaction for the EA of P is:

$$P(g) + e^- \rightarrow P^-(g)$$

$$[Ne]3s^23p^3 \qquad [Ne]3s^23p^4$$

The additional electron in P^- will have to go into an orbital that already has one electron. There will be greater repulsions between the paired electrons in P^-, causing the EA of P to be less favorable than predicted based solely on attractions to the nucleus.

125.   a.
$$Na(g) \rightarrow Na^+(g) + e^-$$ $I_1 = 495 \text{ kJ}$
$$Cl(g) + e^- \rightarrow Cl^-(g)$$ $EA = -348.7 \text{ kJ}$

$$Na(g) + Cl(g) \rightarrow Na^+(g) + Cl^-(g)$$ $\Delta H = 146 \text{ kJ}$

b.
$$Mg(g) \rightarrow Mg^+(g) + e^-$$ $I_1 = 735 \text{ kJ}$
$$F(g) + e^- \rightarrow F^-(g)$$ $EA = -327.8 \text{ kJ}$

$$Mg(g) + F(g) \rightarrow Mg^+(g) + F^-(g)$$ $\Delta H = 407 \text{ kJ}$

c.
$$Mg^+(g) \rightarrow Mg^{2+}(g) + e^-$$ $I_2 = 1445 \text{ kJ}$
$$F(g) + e^- \rightarrow F^-(g)$$ $EA = -327.8 \text{ kJ}$

$$Mg^+(g) + F(g) \rightarrow Mg^{2+}(g) + F^-(g)$$ $\Delta H = 1117 \text{ kJ}$

d.   From parts b and c we get:

$$Mg(g) + F(g) \rightarrow Mg^+(g) + F^-(g)$$ $\Delta H = 407 \text{ kJ}$
$$Mg^+(g) + F(g) \rightarrow Mg^{2+}(g) + F^-(g)$$ $\Delta H = 1117 \text{ kJ}$

$$Mg(g) + 2 F(g) \rightarrow Mg^{2+}(g) + 2 F^-(g)$$ $\Delta H = 1524 \text{ kJ}$

126.   a. $Se^{3+}(g) \rightarrow Se^{4+}(g) + e^-$ b. $S^-(g) + e^- \rightarrow S^{2-}(g)$

c. $Fe^{3+}(g) + e^- \rightarrow Fe^{2+}(g)$ d. $Mg(g) \rightarrow Mg^+(g) + e^-$

127.   The IE is for removal of the electron from the atom in the gas phase. The work function is for the removal of an electron from the solid.

$M(g) \rightarrow M^+(g) + e^-$ IE; $M(s) \rightarrow M^+(s) + e^-$ work function

128. Li^+ ions will be the smallest of the alkali metal cations and will be most strongly attracted to the water molecules.

Challenge Problems

129. $E_{photon} = \dfrac{hc}{\lambda} = \dfrac{6.6261 \times 10^{-34} \text{ J s} \times 2.9979 \times 10^8 \text{ m/s}}{253.4 \times 10^{-9} \text{ m}} = 7.839 \times 10^{-19} \text{ J}; \; \Delta E = -7.839 \times 10^{-19} \text{ J}$

$\Delta E = -2.178 \times 10^{-18} \text{ J } (Z)^2 \left(\dfrac{1}{n_f^2} - \dfrac{1}{n_i^2} \right)$, $Z = 4$ for Be^{3+}

$-7.839 \times 10^{-19} \text{ J} = -2.178 \times 10^{-18} (4)^2 \left(\dfrac{1}{n_f^2} - \dfrac{1}{5^2} \right)$

$$\frac{7.839 \times 10^{-19}}{2.178 \times 10^{-18} \times 16} + \frac{1}{25} = \frac{1}{n_f^2}, \quad \frac{1}{n_f^2} = 0.06249, \quad n_f = 4$$

This emission line corresponds to the $n = 5 \rightarrow n = 4$ electronic transition.

130.   a.   Each orbital could hold 3 electrons.

   b.   The first period corresponds to $n = 1$ which can only have 1s orbitals. The 1s orbital could hold 3 electrons hence the first period would have three elements. The second period corresponds to $n = 2$ which has 2s and 2p orbitals. These four orbitals can each hold three electrons. A total of 12 elements would be in the second period.

   c.   15   d.   21

131.   a.   1st period: $p = 1$, $q = 1$, $r = 0$, $s = \pm 1/2$ (2 elements)

   2nd period: $p = 2$, $q = 1$, $r = 0$, $s = \pm 1/2$ (2 elements)

   3rd period: $p = 3$, $q = 1$, $r = 0$, $s = \pm 1/2$ (2 elements)

   $p = 3$, $q = 3$, $r = -2$, $s = \pm 1/2$ (2 elements)

   $p = 3$, $q = 3$, $r = 0$, $s = \pm 1/2$ (2 elements)

   $p = 3$, $q = 3$, $r = +2$, $s = \pm 1/2$ (2 elements)

   4th period: $p = 4$; q and r values are the same as with $p = 3$ (8 total elements)

1							2
3							4
5	6	7	8	9	10	11	12
13	14	15	16	17	18	19	20

   b.   Elements 2, 4, 12 and 20 all have filled shells and will be least reactive.

   c.   Draw similarities to the modern periodic table.

   XY could be X^+Y^-, $X^{2+}Y^{2-}$ or $X^{3+}Y^{3-}$. Possible ions for each are:

   X^+ could be elements 1, 3, 5 or 13; Y^- could be 11 or 19.

   X^{2+} could be 6 or 14; Y^{2-} could be 10 or 18.

   X^{3+} could be 7 or 15; Y^{3-} could be 9 or 17.

Note: X^{4+} and Y^{4-} ions probably won't form.

XY_2 will be $X^{2+}(Y^-)_2$; See above for possible ions.

X_2Y will be $(X^+)_2Y^{2-}$; See above for possible ions.

XY_3 will be $X^{3+}(Y^-)_3$; See above for possible ions.

X_2Y_3 will be $(X^{3+})_2(Y^{2-})_3$; See above for possible ions.

d. $p = 4$, $q = 3$, $r = -2$, $s = \pm 1/2$ (2)

p = 4, q = 3, r = 0, s = ± 1/2 (2)

p = 4, q = 3, r = +2, s = ± 1/2 (2)

A total of 6 electrons can have $p = 4$ and $q = 3$.

e. $p = 3$, $q = 0$, $r = 0$: This is not allowed; q must be odd. Zero electrons can have these quantum numbers.

f. $p = 6$, $q = 1$, $r = 0$, $s = \pm 1/2$ (2)

p = 6, q = 3, r = -2, 0, +2; s = ± 1/2 (6)

p = 6, q = 5, r = -4, -2, 0, +2, 4; s = ± 1/2 (10)

Eighteen electrons can have $p = 6$.

132.   Size also decreases going across a period. Sc & Ti and Y & Zr are adjacent elements. There are 14 elements (the lanthanides) between La and Hf, making Hf considerable smaller.

133.   a.   As we remove succeeding electrons, the electron being removed is closer to the nucleus and there are fewer electrons left repelling it. The remaining electrons are more strongly attracted to the nucleus and it takes more energy to remove these electrons.

b.   Al : $1s^2 2s^2 2p^6 3s^2 3p^1$; For I_4, we begin removing an electron with $n = 2$. For I_3, we removed an electron with $n = 3$. In going from $n = 3$ to $n = 2$ there is a big jump in ionization energy because the $n = 2$ electrons are much closer to the nucleus on the average than $n = 3$ electrons. Since the $n = 2$ electrons are closer to the nucleus, then they are held more tightly and require a much larger amount of energy to remove them as compared to the $n = 3$ electrons.

c. Al^{4+}; The electron affinity for Al^{4+} is ΔH for the reaction:

$$Al^{4+}(g) + e^- \rightarrow Al^{3+}(g)   \Delta H = -I_4 = -11{,}600 \text{ kJ/mol}$$

d.   The greater the number of electrons, the greater the size.

Size trend: $Al^{4+} < Al^{3+} < Al^{2+} < Al^+ < Al$

134.   None of the noble gases and no subatomic particles had been discovered when Mendeleev published his periodic table. Thus, there was not an element out of place in terms of reactivity. There was no reason to predict an entire family of elements. Mendeleev ordered his table by mass; he had no way of knowing there were gaps in atomic numbers (they hadn't been invented yet).

CHAPTER EIGHT

BONDING: GENERAL CONCEPTS

Questions

10. a. Electronegativity: The ability of an atom <u>in a molecule</u> to attract electrons to itself.

Electron affinity: The energy change for $M(g) + e^- \rightarrow M^-(g)$. EA deals with isolated atoms in the gas phase.

b. Covalent bond: Sharing of electron pair(s); Polar covalent bond: Unequal sharing of electron pair(s).

c. Ionic bond: Electrons are no longer shared, i.e., complete transfer of electron(s) from one atom to another to form ions.

11. Isoelectronic: same number of electrons. There are two variables, number of protons and number of electrons, that will determine the size of an ion. Keeping the number of electrons constant we only have to consider the number of protons to predict trends in size.

12. Resonance occurs when more than one valid Lewis structure can be drawn for a particular molecule. A common characteristic in resonance structures is a multiple bond(s) that moves from one position to another. We say this multiple bond(s) is delocalized about the entire molecule which helps us rationalize why the bonds in a molecule that exhibits resonance are all equivalent in length and strength.

13. The two general requirements for a polar molecular are:

1. polar bonds

2. a structure such that the bond dipoles of the polar bonds do not cancel.

14. Nonmetals form covalent compounds. Nonmetals have valence electrons in the s and p orbitals. Since there are 4 total s and p orbitals, then there is room for only eight electrons (the octet rule).

Exercises

Chemical Bonds and Electronegativity

15. The general trend for electronegativity is:
     1) increase as we go from left to right across a period and
     2) decrease as we go down a group

     Using these trends, the expected orders are:

     a. C < N < O   b. Se < S < Cl   c. Sn < Ge < Si   d. Tl < Ge < S

16. a. Rb < K < Na   b. Ga < B < O   c. Br < Cl < F   d. S < O < F

17. The most polar bond will have the greatest difference in electronegativity between the two atoms.
     From positions in the periodic table, we would predict:

     a. Ge–F   b. P–Cl   c. S–F   d. Ti–Cl

18. a. Sn–H   b. Tl–Br   c. Si–O   d. O–F

19. The general trends in electronegativity used on Exercises 8.15 and 8.17 are only rules of thumb. In
     this exercise we use experimental values of electronegativities and can begin to see several
     exceptions. The order of EN from Figure 8.3 is:

     a. C (2.5) < N (3.0) < O (3.5)   same as predicted

     b. Se (2.4) < S (2.5) < Cl (3.0)   same

     c. Si = Ge = Sn (1.8)   different

     d. Tl (1.8) = Ge (1.8) < S (2.5)   different

     Most polar bonds using actual EN values:

     a. Si–F and Ge–F equal polarity (Ge–F predicted)

     b. P–Cl (same as predicted)

     c. S–F (same as predicted)   d. Ti–Cl (same as predicted)

20. a. Rb (0.8) = K (0.8) < Na (0.9), different   b. Ga (1.6) < B (2.0) < O (3.5), same

     c. Br (2.8) < Cl (3.0) < F (4.0), same   d. S (2.5) < O (3.5) < F (4.0), same

     Most polar bonds using actual EN values.

     a. C–H most polar (Sn–H predicted)

b. Al–Br most polar (Tl–Br predicted).   c. Si–O (same as predicted).

d. Each bond has the same polarity, but the bond dipoles point in opposite directions. Oxygen is the positive end in the O–F bond dipole and oxygen is the negative end in the O–Cl bond dipole. (O–F predicted.)

21. Electronegativity values increase from left to right across the periodic table. The order of electronegativities for the atoms from smallest to largest electronegativity will be H = P < C < N < O < F. The most polar bond will be F–H since it will have the largest difference in electronegativities and the least polar bond will be P–H since it will have the smallest difference in electronegativities (ΔEN = 0). The order of the bonds in decreasing polarity will be F–H > O–H > N–H > C–H > P–H.

22. Ionic character is proportional to the difference in electronegativity values between the two elements forming the bond. Using the trend in electronegativity, the order will be:

Br–Br < N–O < C–F < Ca–O < K–F
smallest   most
ionic character   ionic character

Note that Br–Br, N–O and C–F bonds are all covalent bonds since the elements are all nonmetals. The Ca–O and K–F bonds are ionic as is generally the case when a metal forms a bond with a nonmetal.

Ions and Ionic Compounds

23. Rb^+: $[Ar]4s^23d^{10}4p^6$; Ba^{2+}: $[Kr]5s^24d^{10}5p^6$; Se^{2-}: $[Ar]4s^23d^{10}4p^6$

I^-: $[Kr]5s^24d^{10}5p^6$

24. Te^{2-}: $[Kr]5s^24d^{10}5p^6$; Cl^-: $[Ne]3s^23p^6$; Sr^{2+}: $[Ar]4s^23d^{10}4p^6$

Li^+: $1s^2$

25. a. Mg^{2+}: $1s^22s^22p^6$; K^+: $1s^22s^22p^63s^23p^6$; Al^{3+}: $1s^22s^22p^6$

b. N^{3-}, O^{2-} and F^-: $1s^22s^22p^6$; Te^{2-}: $[Kr]5s^24d^{10}5p^6$

26. a. Sr^{2+}: $[Ar]4s^23d^{10}4p^6$; Cs^+: $[Kr]5s^24d^{10}5p^6$; In^+: $[Kr]5s^24d^{10}$

Pb^{2+}: $[Xe]6s^24f^{14}5d^{10}$

b. P^{3-} and S^{2-}: $[Ne]3s^23p^6$; Br^-: $[Ar]4s^23d^{10}4p^6$

27. a. Sc^{3+} b. Te^{2-} c. Ce^{4+} and Ti^{4+} d. Ba^{2+}

All of these have the number of electrons of a noble gas.

28. a. none   b. none   c. F^- d. Cs^+

29. There are many possible ions with 10 electrons. Some are: N^{3-}, O^{2-}, F^-, Na^+, Mg^{2+} and Al^{3+}. In terms of size, the ion with the most protons will hold the electrons the tightest and will be the smallest. The largest ion will be the ion with the fewest protons. The size trend is:

$$Al^{3+} < Mg^{2+} < Na^+ < F^- < O^{2-} < N^{3-}$$
smallest   largest

30. Some possibilities which contain 18 electrons are: P^{3-}, S^{2-}, Cl^-, K^+, Ca^{2+} and Sc^{3+}. The size trend is proportional to the number of protons in the nucleus. Size trend:

$$Sc^{3+} < Ca^{2+} < K^+ < Cl^- < S^{2-} < P^{3-}$$
smallest   largest

31. a. $Cu > Cu^+ > Cu^{2+}$ b. $Pt^{2+} > Pd^{2+} > Ni^{2+}$ c. $O^{2-} > O^- > O$

     d. $La^{3+} > Eu^{3+} > Gd^{3+} > Yb^{3+}$ e. $Te^{2-} > I^- > Cs^+ > Ba^{2+} > La^{3+}$

     For answer a, as electrons are removed from an atom, size decreases. Answers b and d follow the radii trend. For answer c, as electrons are added to an atom, size increases. Answer e follows the trend for an isoelectronic series, i.e., the smallest ion has the most protons.

32. a. $Co > Co^+ > Co^{2+} > Co^{3+}$ b. $N^{3-} > N^{2-} > N^- > N$ c. $Br^- > Cl^- > F^-$

     d. $S^{2-} > Cl^- > K^+ > Ca^{2+}$ e. $Ti^{2+} > Fe^{2+} > Ni^{2+} > Zn^{2+}$

33. a. Li^+ and N^{3-} are the expected ions. The formula of the compound would be Li_3N (lithium nitride).

     b. Ga^{3+} and O^{2-}; Ga_2O_3, gallium(III) oxide or gallium oxide

     c. Rb^+ and Cl^-; RbCl, rubidium chloride   d. Ba^{2+} and S^{2-}; BaS, barium sulfide

34. a. Al^{3+} and Cl^-; $AlCl_3$, aluminum chloride   b. Na^+ and O^{2-}; Na_2O, sodium oxide

     c. Sr^{2+} and F^-; SrF_2, strontium fluoride   d. Ca^{2+} and S^{2-}; CaS, calcium sulfide

35. Lattice energy is proportional to Q_1Q_2/r where Q is the charge of the ions and r is the distance between the ions. In general, charge effects on lattice energy are much greater than size effects.

     a. NaCl; Na^+ is smaller than K^+.   b. LiF; F^- is smaller than Cl^-.

     c. MgO; O^{2-} has a greater charge than OH^-.   d. $Fe(OH)_3$; Fe^{3+} has a greater charge than Fe^{2+}.

     e. Na_2O; O^{2-} has a greater charge than Cl^-.   f. MgO; The ions are smaller in MgO.

36.   a.   LiF; Li^+ is smaller than Cs^+.   b.   NaBr; Br^- is smaller than I^-.

        c.   BaO; O^{2-} has a greater charge than Cl^-.   d. $CaSO_4$; Ca^{2+} has a greater charge than Na^+.

        e. K_2O; O^{2-} has a greater charge than F^-.   f. Li_2O; The ions are smaller in Li_2O.

37. $Na(s) \rightarrow Na(g)$ $\Delta H = 109$ kJ   (sublimation)
                        $Na(g) \rightarrow Na^+(g) + e^-$ $\Delta H = 495$ kJ   (ionization energy)
                        $1/2\ Cl_2(g) \rightarrow Cl(g)$ $\Delta H = 239/2$ kJ (bond energy)
                        $Cl(g) + e^- \rightarrow Cl^-(g)$ $\Delta H = -349$ kJ (electron affinity)
                        $Na^+(g) + Cl^-(g) \rightarrow NaCl(s)$ $\Delta H = -786$ kJ (lattice energy)

        _____

                $Na(s) + 1/2\ Cl_2(g) \rightarrow NaCl(s)$ $\Delta H_f^\circ = -412$ kJ/mol

38. $Ba(s) \rightarrow Ba(g)$ $\Delta H = 178$ kJ   (sublimation)
                        $Ba(g) \rightarrow Ba^+(g) + e^-$ $\Delta H = 503$ kJ (IE_1)
                        $Ba^+(g) \rightarrow Ba^{2+}(g) + e^-$ $\Delta H = 965$ kJ (IE_2)
                        $Cl_2(g) \rightarrow 2\ Cl(g)$ $\Delta H = 239$ kJ   (BE)
                        $2\ Cl(g) + 2\ e^- \rightarrow 2\ Cl^-(g)$ $\Delta H = 2(-349)$ kJ (EA)
                $Ba^{2+}(g) + 2\ Cl^-(g) \rightarrow BaCl_2(s)$ $\Delta H = -2056$ kJ   (LE)

        _____

                $Ba(s) + Cl_2(g) \rightarrow BaCl_2(s)$ $\Delta H_f^\circ = -869$ kJ/mol

39.   From the data given, it costs less energy to produce $Mg^+(g) + O^-(g)$ than to produce $Mg^{2+}(g) + O^{2-}(g)$. However, the lattice energy for $Mg^{2+}O^{2-}$ will be much more exothermic than for Mg^+O^- (due to the greater charges in $Mg^{2+}O^{2-}$). The favorable lattice energy term will dominate and $Mg^{2+}O^{2-}$ forms.

40. $O(g) + e^- \rightarrow O^-(g)$ $\Delta H = -141$ kJ/mol
        $O^-(g) + e^- \rightarrow O^{2-}(g)$ $\Delta H = 878$ kJ/mol

        _____

        $O(g) + 2\ e^- \rightarrow O^{2-}(g)$ $\Delta H = 737$ kJ/mol

41. Ca^{2+} has greater charge than Na^+, and Se^{2-} is smaller than Te^{2-}. The effect of charge on the lattice energy is greater than the effect of size. We expect the trend from most exothermic to least exothermic to be:

                $CaSe\ >\ CaTe\ >\ Na_2Se\ >\ Na_2Te$
                (-2862)   (-2721)   (-2130)   (-2095)   This is what we observe.

42.   The compounds are FeO, Fe_2O_3, $FeCl_2$ and $FeCl_3$. Lattice energy is proportional to the charge of the cation times the charge of the anion, Q_1Q_2.

Compound	Q_1Q_2	Lattice Energy
$FeCl_2$	(+2)(-1)	-2631 kJ/mol
$FeCl_3$	(+3)(-1)	-3865 kJ/mol
FeO	(+2)(-2)	-5359 kJ/mol
Fe_2O_3	(+3)(-2)	-14,744 kJ/mol

Bond Energies

43.   a. $H - H + Cl - Cl \rightarrow 2\ H - Cl$

Bonds broken: Bonds formed:

    1 H – H (432 kJ/mol)   2 H – Cl (427 kJ/mol)
    1 Cl – Cl (239 kJ/mol)

$\Delta H = \Sigma D_{broken} - \Sigma D_{formed}$, $\Delta H = 432$ kJ + 239 kJ - 2(427) kJ = -183 kJ

b.

$$N \equiv N +\ 3\ H - H \longrightarrow 2\ \ H - \underset{\underset{H}{|}}{N} - H$$

Bonds broken: Bonds formed:

    1 N ≡ N (941 kJ/mol)   6 N – H (391 kJ/mol)
    3 H – H (432 kJ/mol)

$\Delta H = 941$ kJ + 3(432) kJ - 6(391) kJ = -109 kJ

44.   a.

Bonds broken: Bonds formed:

    1 C = C (614 kJ/mol)   1 C – C (347 kJ/mol)
    1 Br – Br (193 kJ/mol)   2 C – Br (276 kJ/mol)

$\Delta H = 614$ kJ + 193 kJ - [347 kJ + 2(276 kJ)] = -92 kJ

b.

Bonds broken: Bonds formed:

    1 C = C (614 kJ/mol)   1 C – C (347 kJ/mol)
    1 O – O (146 kJ/mol)   2 C – O (358 kJ/mol)

$\Delta H = 614 \text{ kJ} + 146 \text{ kJ} - [347 \text{ kJ} + 2(358 \text{ kJ})] = -303 \text{ kJ}$

Note: Sometimes some of the bonds remain the same between reactants and products. To save time, only break and form bonds that are involved in the reaction.

45.

Bonds broken: 1 C – N (305 kJ/mol)   Bonds formed: 1 C – C (347 kJ/mol)

$\Delta H = \Sigma D_{\text{broken}} - \Sigma D_{\text{formed}},\ \ \Delta H = 305 - 347 = -42 \text{ kJ}$

Note: Sometimes some of the bonds remain the same between reactants and products. To save time, only break and form bonds that are involved in the reaction.

46.

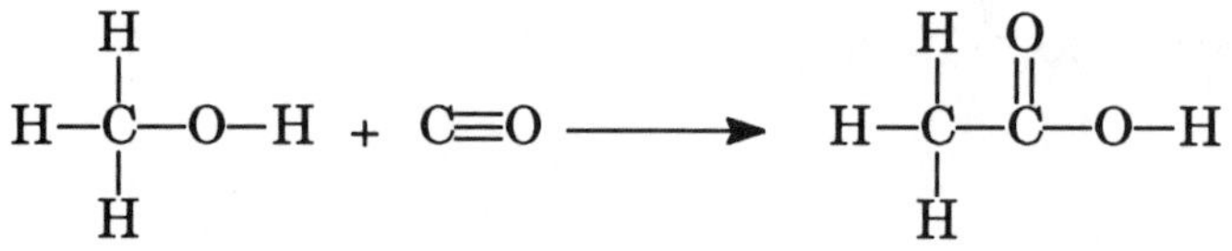

Bonds broken: Bonds formed:

    1 C ≡ O (1072 kJ/mol)   1 C – C (347 kJ/mol)
    1 C – O (358 kJ/mol)   1 C = O (745 kJ/mol)
                                               1 C – O (358 kJ/mol)

$\Delta H = 1072 + 358 - [347 + 745 + 358] = -20. \text{ kJ}$

47. $H - C \equiv C - H + 5/2\ O = O \rightarrow 2\ O = C = O + H - O - H$

Bonds broken: Bonds formed:

    1 C ≡ C (839 kJ/mol)   2 × 2 C = O (799 kJ/mol)
    2 C – H (413 kJ/mol)   2 O – H (467 kJ/mol)
    5/2 O = O (495 kJ/mol)

$\Delta H = 839 + 2(413) + 5/2\ (495) - [4(799) + 2(467)] = -1228 \text{ kJ}$

48.

Bonds broken: Bonds formed:

   1 C $=$ C (614 kJ/mol)   1 C $-$ C (347 kJ/mol)
   1 O $-$ O (146 kJ/mol)   1 C $=$ O (745 kJ/mol)
   1 C $-$ H (413 kJ/mol)   1 C $-$ H (413 kJ/mol)

ΔH = 614 + 146 + 413 - (347 + 745 + 413) = -332 kJ

49.

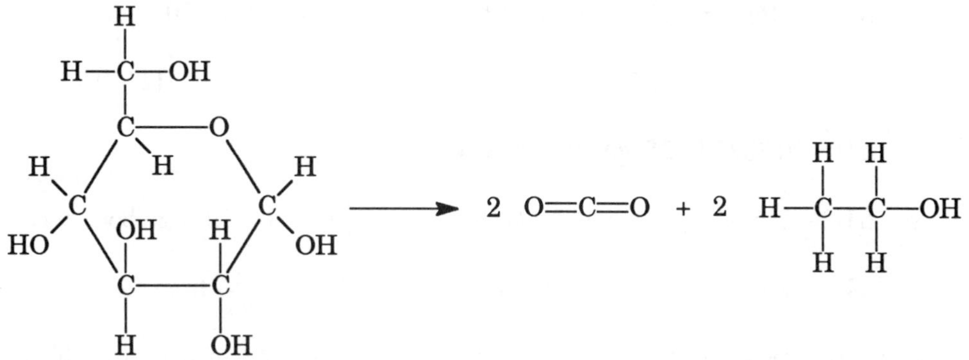

The molecules are complicated enough that it will be easier to break all bonds in glucose and make all the bonds in CO_2 and CH_3CH_2OH.

Bonds broken: Bonds formed:

   5 C $-$ C (347 kJ/mol)   2 × 2 C $=$ O (799 kJ/mol)
   7 C $-$ O (358 kJ/mol)   2 × 5 C $-$ H (413 kJ/mol)
   5 O $-$ H (467 kJ/mol)   2 C $-$ O (358 kJ/mol)
   7 C $-$ H (413 kJ/mol)   2 O $-$ H (467 kJ/mol)
                                          2 C $-$ C (347 kJ/mol)

ΔH = 5(347) + 7(358) + 5(467) + 7(413)

                  - [4(799) + 10(413) + 2(358) + 2(467) + 2(347)] = -203 kJ

50.

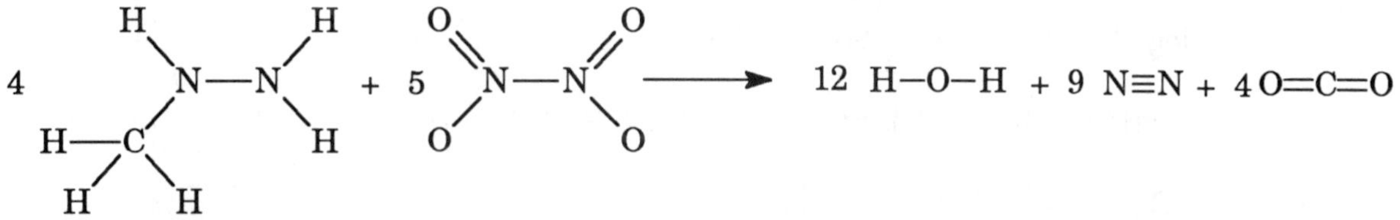

Bonds broken: Bonds made:

$$9 \; N-N \;\; (160. \; kJ/mol)$$
$$4 \; N-C \;\; (305 \; kJ/mol)$$
$$12 \; C-H \;\; (413 \; kJ/mol)$$
$$12 \; N-H \;\; (391 \; kJ/mol)$$
$$10 \; N=O \;\; (607 \; kJ/mol)$$
$$10 \; N-O \;\; (201 \; kJ/mol)$$

$$24 \; O-H \;\; (467 \; kJ/mol)$$
$$9 \; N \equiv N \;\; (941 \; kJ/mol)$$
$$8 \; C=O \;\; (799 \; kJ/mol)$$

$$\Delta H = 9(160.) + 4(305) + 12(413) + 12(391) + 10(607) + 10(201)$$

$$- [24(467) + 9(941) + 8(799)]$$

$$\Delta H = 20,388 \; kJ - 26,069 \; kJ = -5681 \; kJ$$

51.   a. $\Delta H° = 2 \; \Delta H°_{f, HCl} = 2 \; mol \; (-92 \; kJ/mol) = -184 \; kJ \; (= -183 \; kJ \; from \; bond \; energies)$

       b. $\Delta H° = 2 \; \Delta H°_{f, NH_3} = 2 \; mol \; (-46 \; kJ/mol) = -92 \; kJ \; (= -109 \; kJ \; from \; bond \; energies)$

Comparing the values for each reaction, bond energies seem to give a reasonably good estimate of the enthalpy change for a reaction. The estimate is especially good for gas phase reactions.

52. $CH_3OH(g) + CO(g) \rightarrow CH_3COOH(l)$

       $\Delta H° = -484 \; kJ - [(-201 \; kJ) + (-110.5 \; kJ)] = -173 \; kJ$

Using bond energies, $\Delta H = -20. \; kJ$. For this reaction, bond energies give a much poorer estimate for ΔH as compared to the gas phase reactions in Exercise 8.43. The reason is that not all species are gases in Exercise 8.46. Bond energies do not account for the energy changes that occur when liquids and solids form instead of gases. These energy changes are due to intermolecular forces and will be discussed in Chapter 10.

53.   a.   Using SF_4 data: $SF_4(g) \rightarrow S(g) + 4 \; F(g)$

             $\Delta H° = 4 \; D_{SF} = 278.8 + 4 \; (79.0) - (-775) = 1370. \; kJ$

             $$D_{SF} = \frac{1370. \; kJ}{4 \; mol \; SF \; bonds} = 342.5 \; kJ/mol$$

             Using SF_6 data: $SF_6(g) \rightarrow S(g) + 6 \; F(g)$

             $\Delta H° = 6 \; D_{SF} = 278.8 + 6 \; (79.0) - (-1209) = 1962 \; kJ$

             $$D_{SF} = \frac{1962 \; kJ}{6} = 327.0 \; kJ/mol$$

       b.   The $S-F$ bond energy in the table is 327 kJ/mol. The value in the table was based on the $S-F$ bond in SF_6.

c.   S(g) and F(g) are not the most stable form of the element at $25\,^{\circ}C$. The most stable forms are $S_8(s)$ and $F_2(g)$; $\Delta H_f^{\circ} = 0$ for these two species.

54. $NH_3(g) \rightarrow N(g) + 3\,H(g)$; $\Delta H^{\circ} = 3\,D_{NH} = 472.7 + 3(216.0) - (-46.1) = 1166.8$ kJ

$$D_{NH} = \frac{1166.8\ kJ}{3\ mol\ NH\ bonds} = 388.93\ kJ/mol \approx 389\ kJ/mol$$

$D_{calc} = 389$ kJ/mol as compared to 391 kJ/mol in the table. There is good agreement.

55. $N_2 + 3\,H_2 \rightarrow 2\,NH_3$; $\Delta H = D_{N_2} + 3\,D_{HN_2} - 6\,D_{NH}$; $\Delta H^{\circ} = 2(-46\ kJ) = -92$ kJ

-92 kJ = 941 kJ + 3(432 kJ) - (6 D_{N-H}), 6 D_{N-H} = 2329 kJ, D_{N-H} = 388.2 kJ/mol

Table in text: 391 kJ/mol;   There is good agreement.

56.

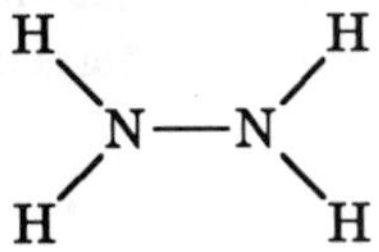

2 N(g) + 4 H(g) $\Delta H = D_{N-N} + 4\,D_{N-H} = D_{N-N} + 4(388.9)$

$\Delta H^{\circ} = 2\,\Delta H_{f,\,N}^{\circ} + 4\,\Delta H_{f,\,H}^{\circ} - \Delta H_{f,\,N_2H_4}^{\circ} = 2(472.7\ kJ) + 4(216.0\ kJ) - 95.4$ kJ

$\Delta H^{\circ} = 1714.0\ kJ = D_{N-N} + 4(388.9)$

D_{N-N} = 158.4 kJ/mol (160. kJ/mol in Table 8.5)

Lewis Structures and Resonance

57.   Drawing Lewis structures is mostly trial and error. However, the first two steps are always the same. These steps are 1) count the valence electrons available in the molecule and 2) attach all atoms to each other with single bonds (called the skeletal structure). Unless noted otherwise, the atom listed first is assumed to be the atom in the middle (called the central atom) and all other atoms in the formulas are attached to this atom. The most notable exceptions to the rule are formulas which begin with H, e.g., H_2O, H_2CO, etc. Hydrogen can never be a central atom since this would require H to have more than two electrons.

After counting valence electrons and drawing the skeletal structure, the rest is trial and error. We place the remaining electrons around the various atoms in an attempt to satisfy the octet rule (or duet rule for H). Keep in mind that practice makes perfect. After practicing you can (and will) become very adept at drawing Lewis structures.

a.   HCN has $1 + 4 + 5 = 10$ valence electrons.

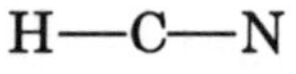

H—C—N   H—C≡N:

Skeletal   Lewis
structure   structure

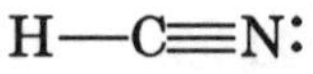

Skeletal structures uses 4 e⁻; 6 e⁻ remain

b. PH_3 has $5 + 3(1) = 8$ valence electrons.

H—P—H   H—P̈—H
    | |
    H   H

Skeletal   Lewis
structure   structure

Skeletal structure uses 6 e⁻; 2 e⁻ remain

c. $CHCl_3$ has $4 + 1 + 3(7) = 26$ valence electrons.

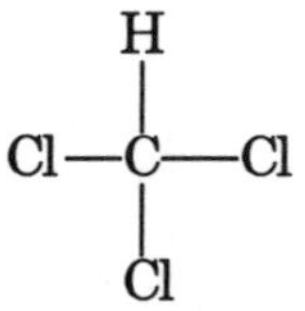

       H   H
       | |
Cl—C—Cl   :C̈l—C—C̈l:
       | |
       Cl   :C̈l:

Skeletal   Lewis
structure   structure

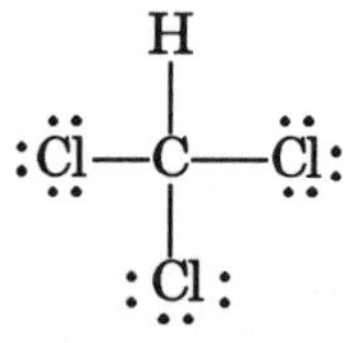

Skeletal structure uses 8 e⁻; 18 e⁻ remain

d. NH_4^+ has $5 + 4(1) - 1 = 8$ valence electrons.

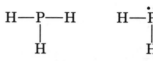

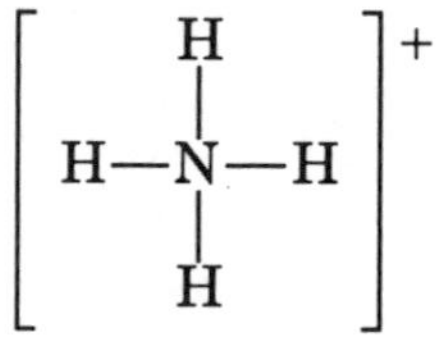

Lewis
structure

Note: Subtract
valence electrons for
positive charged ions.

e. H_2CO has $2(1) + 4 + 6 = 12$ valence electrons.

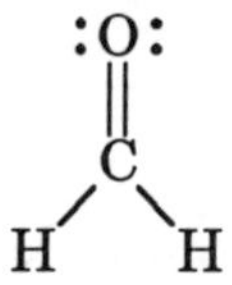

:O:
‖
C
H   H

f. SeF_2 has $6 + 2(7) = 20$ valence electrons.

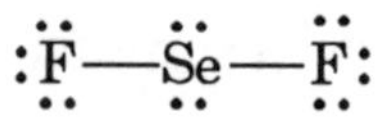

:F̈—S̈e—F̈:

g. CO_2 has $4 + 2(6) = 16$ valence electrons.

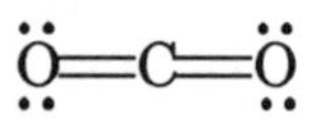

Ö=C=Ö

h. O_2 has $2(6) = 12$ valence electrons.

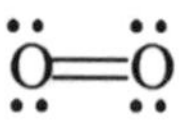

Ö=Ö

i.   HBr has $1 + 7 = 8$ valence electrons.

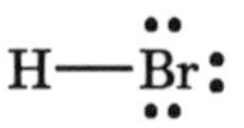

H—B̈r:

58.   a. $POCl_3$ has $5 + 6 + 3(7) = 32$ valence electrons.

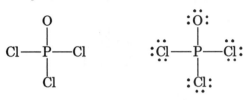

Skeletal   Lewis
structure   structure

This structure uses all 32 e⁻ while satisfying the octet rule for all atoms. This is a valid Lewis structure.

SO_4^{2-} has $6 + 4(6) + 2 = 32$ valence electrons.

Note: A negatively charged ion will have additional electrons to those that come from the valence shells of the atoms.

XeO_4, $8 + 4(6) = 32$ e⁻ PO_4^{3-}, $5 + 4(6) + 3 = 32$ e⁻

ClO_4^- has $7 + 4(6) + 1 = 32$ valence electrons.

Note: All of these species have the same number of atoms and the same number of valence electrons. They also have the same Lewis structure.

b. NF_3 has $5 + 3(7) = 26$ valence electrons. SO_3^{2-}, $6 + 3(6) + 2 = 26$ e⁻

Skeletal   Lewis
structure   structure

PO_3^{3-}, $5 + 3(6) + 3 = 26\ e^-$ ClO_3^-, $7 + 3(6) + 1 = 26\ e^-$

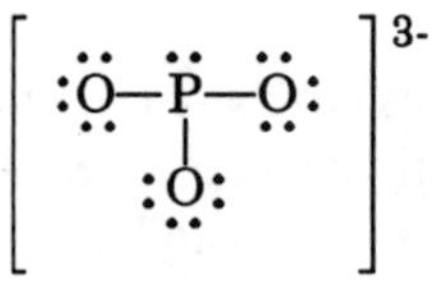

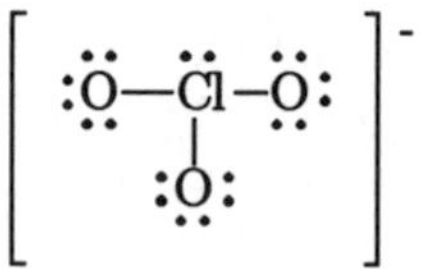

Note: Species with the same number of atoms and valence electrons have similar Lewis structures.

c. ClO_2^- has $7 + 2(6) + 1 = 20$ valence electrons.

Skeletal   Lewis
structure   structure

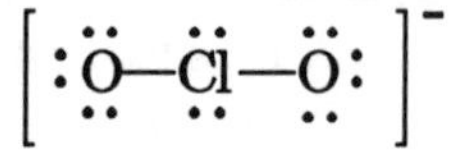

SCl_2, $6 + 2(7) = 20\ e^-$ PCl_2^-, $5 + 2(7) + 1 = 20\ e^-$

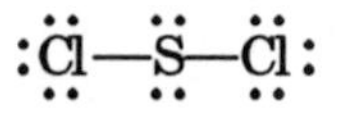

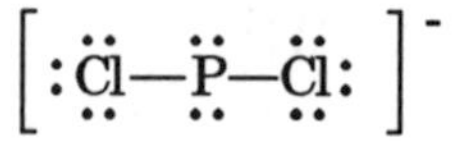

Note: Species with the same number of atoms and valence electrons have similar Lewis structures.

d.   Molecules/ions that have the same number of valence electrons and the same number of atoms will have similar Lewis structures.

59.   In each case in this problem, the octet rule cannot be satisfied for the central atom. BeH_2 and BH_3 all have too few electrons around the central atom and all the others have to many electrons around the central atom. Always try to satisfy the octet rule for every atom, but when it is impossible, the central atom is the species which will disobey the octet rule.

PF_5, $5 + 5(7) = 40\ e^-$ BeH_2, $2 + 2(1) = 4\ e^-$

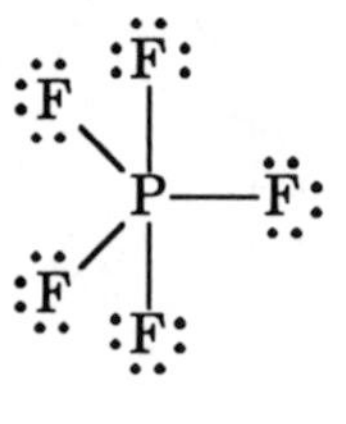

$H - Be - H$

BH$_3$, 3 + 3(1) = 6 e$^-$

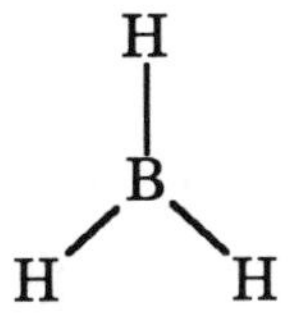

Br$_3^-$, 3(7) + 1 = 22 e$^-$

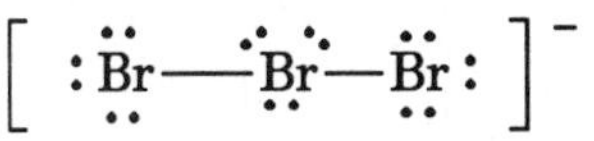

SF$_4$, 6 + 4(7) = 34 e$^-$

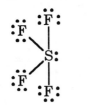

XeF$_4$, 8 + 4(7) = 36 e$^-$

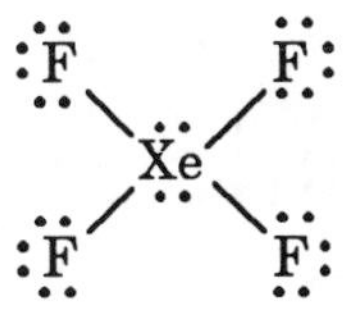

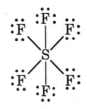

ClF$_5$, 7 + 5(7) = 42 e$^-$

SF$_6$, 6 + 6(7) = 48 e$^-$

60.   ClF$_3$ has 7 + 3(7) = 28 valence electrons.

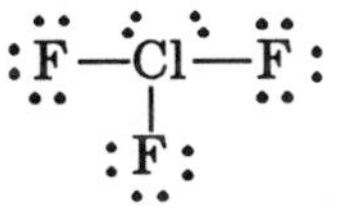

BrF$_3$ also has 28 valence electrons.

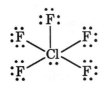

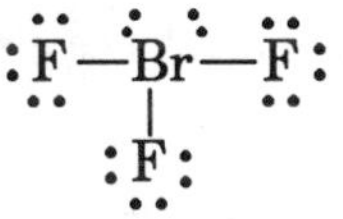

We expand the octet of the central Cl atom.

61.   a. NO$_2^-$ has 5 + 2(6) + 1 = 18 valence electrons. The skeletal structure is: O – N – O

To get an octet about the nitrogen and only use 18 e$^-$, we must form a double bond to one of the oxygen atoms.

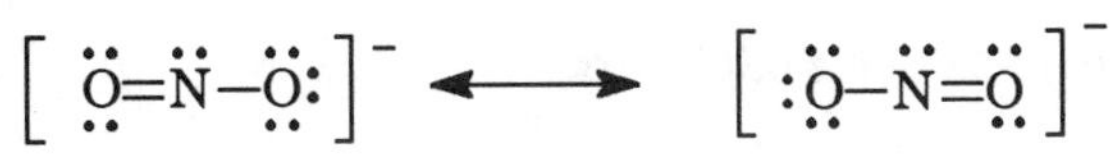

$$\left[\ \ddot{O}=\ddot{N}-\ddot{O}\colon\ \right]^- \longleftrightarrow \left[\ \colon\!\ddot{O}-\ddot{N}=\ddot{O}\ \right]^-$$

Since there is no reason to have the double bond to a particular oxygen atom, we can draw two resonance structures. Each Lewis structure uses the correct number of electrons and satisfies the octet rules so each is a valid Lewis structure. Resonance structures occur when you have multiple bonds that can be in various positions. We say the actual structure is an average of these two resonance structures.

NO_3^- has $5 + 3(6) + 1 = 24$ valence electrons. We can draw three resonance structures for NO_3^-, with the double bond rotating between the three oxygen atoms.

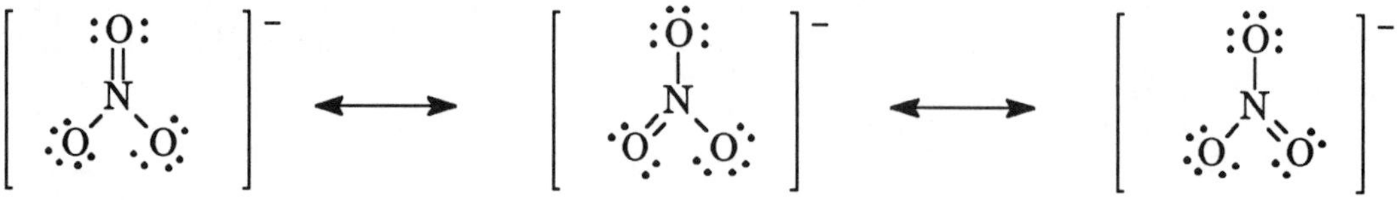

b. OCN^- has $6 + 4 + 5 + 1 = 16$ valence electrons. We can draw three resonance structures for OCN^-.

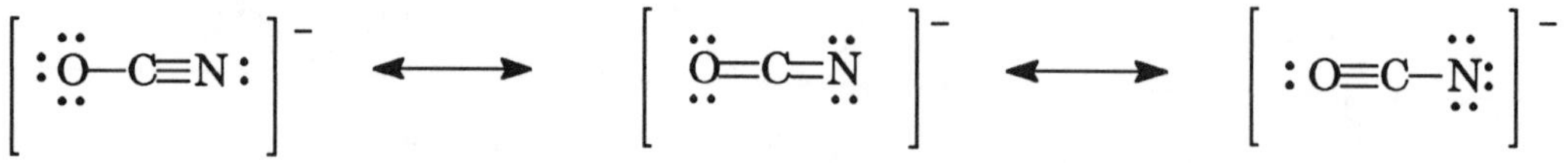

SCN^- has $6 + 4 + 5 + 1 = 16$ valence electrons. Three resonance structures can be drawn.

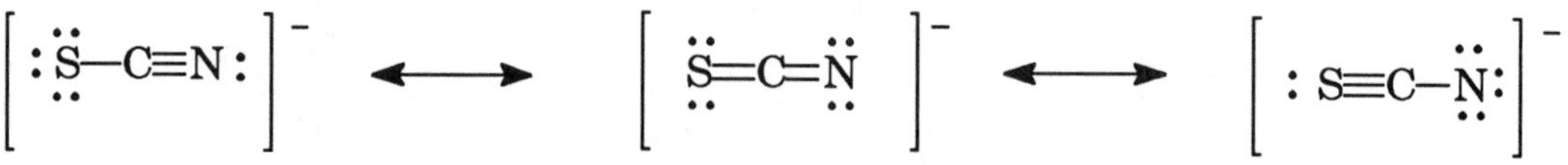

N_3^- has $3(5) + 1 = 16$ valence electrons. As with OCN^- and SCN^-, three different resonance structures can be drawn.

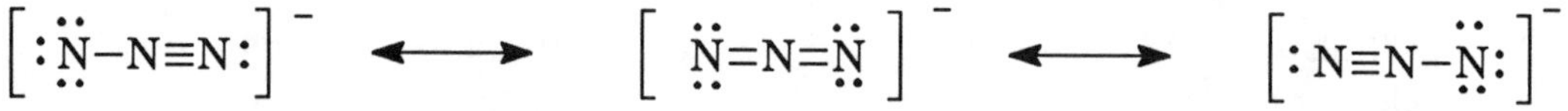

62. Ozone: O_3 has $3(6) = 18$ valence electrons.

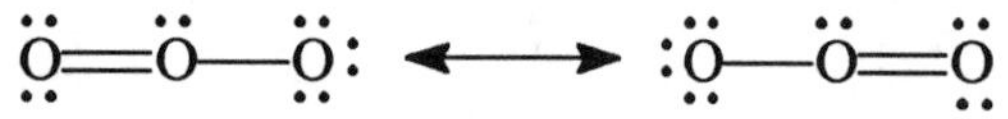

Sulfur dioxide: SO_2 has $6 + 2(6) = 18$ valence electrons.

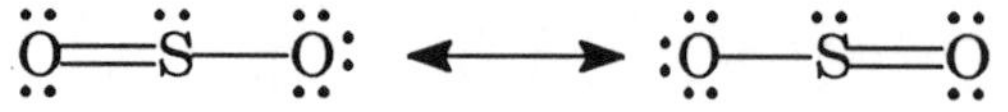

Sulfur trioxide: SO_3 has $6 + 3(6) = 24$ valence electrons.

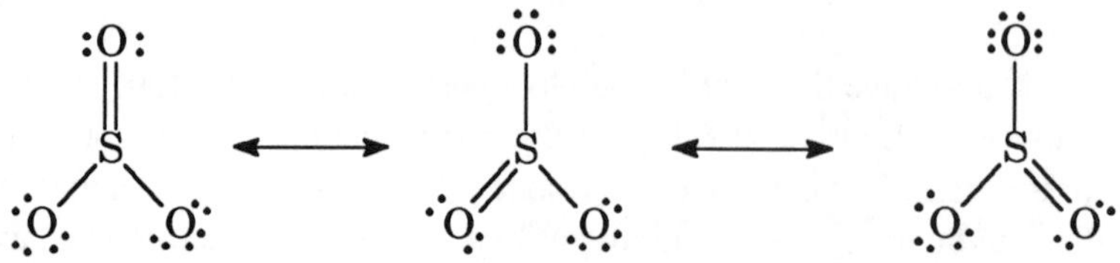

63.   Benzene has $6(4) + 6(1) = 30$ valence electrons. Two resonance structures can be drawn for benzene. The actual structure of benzene is an average of these two resonance structures, that is, all carbon-carbon bonds are equivalent with a bond length and bond strength somewhere between a single and a double bond.

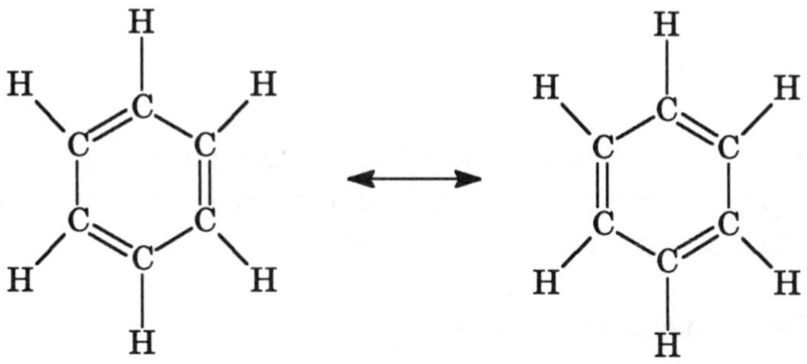

64.   Borazine $(B_3N_3H_6)$ has $3(3) + 3(5) + 6(1) = 30$ valence electrons. The possible resonance structures are similar to those of benzene in Exercise 8.63.

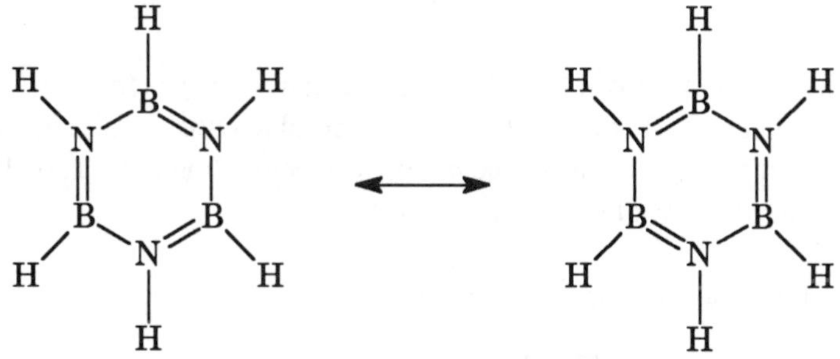

65.   We will use a hexagon to represent the six membered carbon ring and we will omit the 4 hydrogen atoms and the three lone pairs of electrons on each chlorine. If no resonance exists, we could draw 4 different molecules:

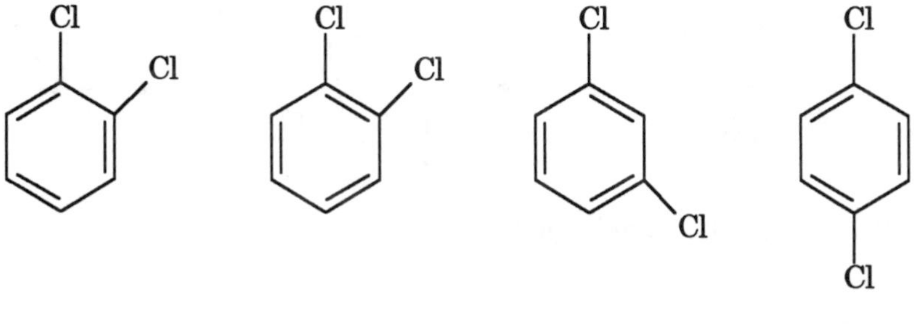

        If the double bonds in the benzene ring exhibit resonance, then we can draw only three different dichlorobenzenes. The circle in the hexagon represents the delocalization of the three double bonds in the benzene ring (see Exercise 8.63).

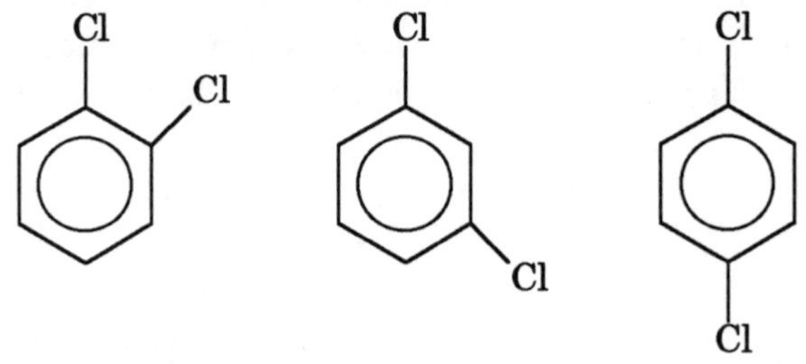

With resonance, all carbon-carbon bonds are equivalent. We can't distinguish between a single and double bond between adjacent carbons that have a chlorine attached. That only 3 isomers are observed provides evidence for the existence of resonance.

66. CO_3^{2-} has $4 + 3(6) + 2 = 24$ valence electrons.

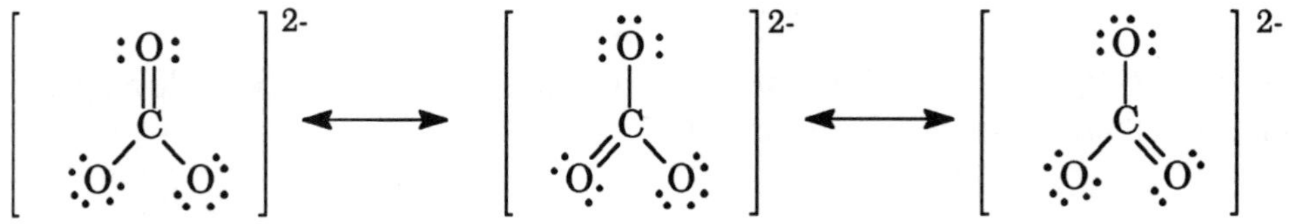

Three resonance structures can be drawn for CO_3^{2-}. The actual structure for CO_3^{2-} is an average of these three resonance structures. That is, the three C – O bond lengths are all equivalent, with a length somewhere between a single and a double bond. The actual bond length of 136 pm is consistent with this resonance view of CO_3^{2-}.

67.   The Lewis structures for the various species are below:

CO (10 e⁻):   :C≡≡O: Triple bond between C and O

CO_2 (16 e⁻):   Ö═══C═══Ö   Double bond between C and O

CO_3^{2-} (24 e⁻):

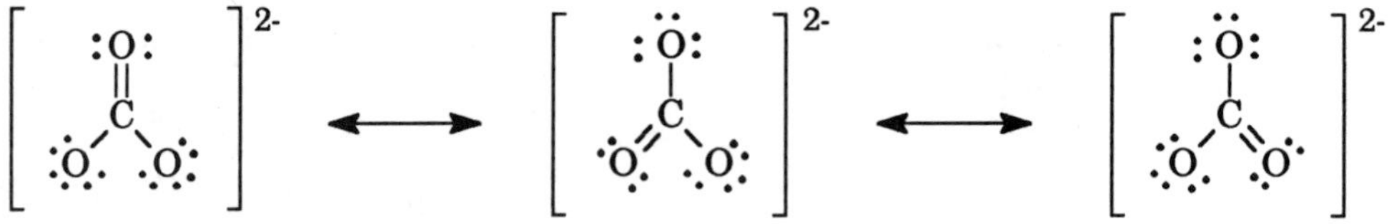

        Average of 1 1/3 bond between C and O

CH_3OH (14 e⁻):   H—C—Ö—H   Single bond between C and O

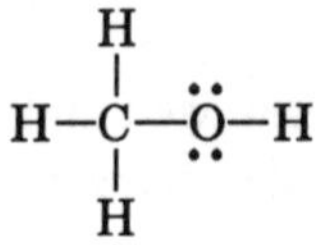

As the number of bonds increase between two atoms, bond length decreases and bond strength increases. With this in mind, then:

longest → shortest   C – O bond: $CH_3OH > CO_3^{2-} > CO_2 > CO$

weakest → strongest   C – O bond: $CH_3OH < CO_3^{2-} < CO_2 < CO$

68. H_2NOH (14 e⁻):   H—N̈—Ö—H   Single bond between N and O

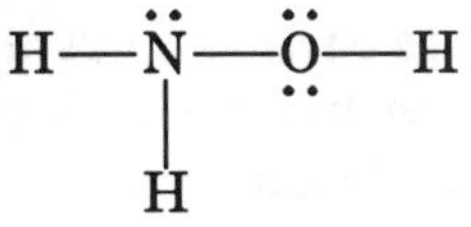

                                        |
                                        H

N_2O (16 e⁻):   N̈=N=Ö ⟷ :N≡N–Ö: ⟷ :N̈–N≡O:

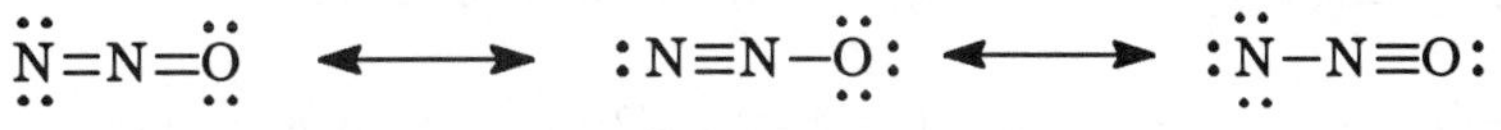

Average of a double bond between N and O

NO^+ (10 e⁻):   [:N≡O:]⁺   Triple bond between N and O

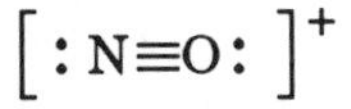

NO_2^- (18 e⁻):   [Ö=N̈–Ö:]⁻ ⟷ [:Ö–N̈=Ö]⁻

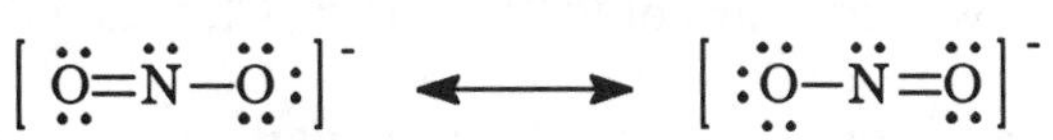

Average of 1 1/2 bond between N and O

NO_3^- (24 e⁻):   [structure]⁻ ⟷ [structure]⁻ ⟷ [structure]⁻

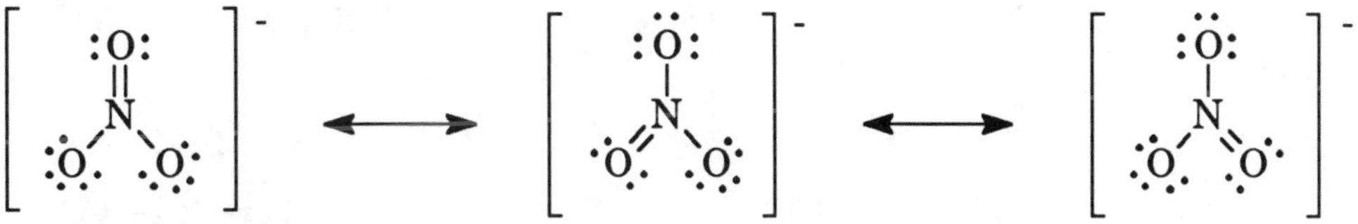

Average of 1 1/3 bond between N and O

From Lewis structures:

shortest → longest   N – O bond: $NO^+ < N_2O < NO_2^- < NO_3^- < H_2NOH$

Formal Charge

69. BF_3 has 3 + 3(7) = 24 valence electrons. The two Lewis structures to consider are:

The formal charge for the various atoms are assigned in the Lewis structures. Formal charge = number of valence electrons on free atom - number of lone pair electrons on atoms - 1/2 (number of shared electrons of atom). For B in the first Lewis structure, FC = 3 - 0 - 1/2(8) = -1. For F in the first structure with the double bond, FC = 7 - 4 - 1/2(4) = +1. The others all have a formal charge equal to zero.

The first Lewis structure obeys the octet rule but has a +1 formal charge on the most electronegative element there is, fluorine, and a negative formal charge on a much less electronegative element, boron. This is just the opposite of what we expect; negative formal charge on F and positive formal charge on B. The other Lewis structure does not obey the octet rule for B but has a zero formal charge on each element in BF_3. Since structures generally want to minimize formal charge, then BF_3 with only single bonds is best from a formal charge point of view.

70. $: C \equiv O :$ Carbon: FC = 4 - 2 - 1/2(6) = -1; Oxygen: FC = 6 - 2 - 1/2(6) = +1

Electronegativity predicts the opposite polarization. The two opposing effects seem to cancel to give a much less polar molecule than expected.

71.   See Exercise 8.58a for the Lewis structures of $POCl_3$, SO_4^{2-}, ClO_4^- and PO_4^{3-}.

a. $POCl_3$: P, FC = 5 - 1/2(8) = +1   b. SO_4^{2-}: S, FC = 6 - 1/2(8) = +2

c. ClO_4^-: Cl, FC = 7 - 1/2(8) = +3   d. PO_4^{3-}: P, FC = 5 - 1/2(8) = +1

e. SO_2Cl_2, 6 + 2(6) + 2(7) = 32 e⁻   f. XeO_4, 8 + 4(6) = 32 e⁻

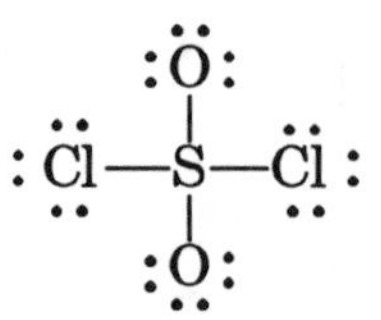

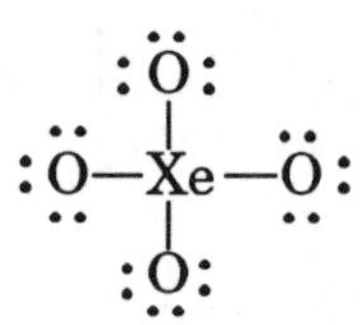

S, FC = 6 - 1/2(8) = +2   Xe, FC = 8 - 1/2(8) = +4

g. ClO_3^-, 7 + 3(6) + 1 = 26 e⁻   h. NO_4^{3-}, 5 + 4(6) + 3 = 32 e⁻

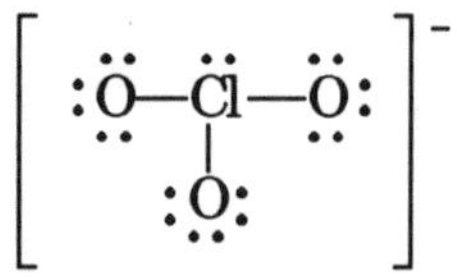

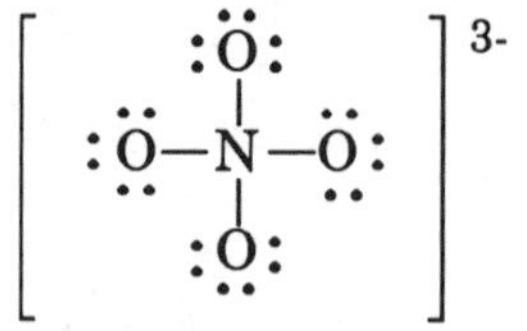

Cl, FC = 7 - 2 - 1/2(6) = +2   N, FC = 5 - 1/2(8) = +1

72.   For SO_4^{2-}, ClO_4^-, PO_4^{3-} and ClO_3^-, only one of the possible resonance structures is drawn.

a.   Must have five bonds to P to minimize formal charge of P. The best choice is to form a double bond to O since this will give O a formal charge of zero and single bonds to Cl for the same reason.

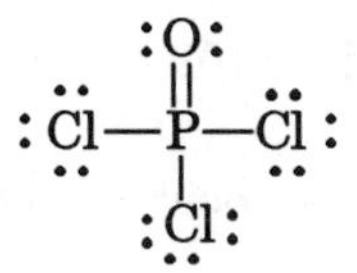

$$\begin{array}{c} :\!\ddot{O}\!: \\ \| \\ :\!\ddot{C}l\!-\!P\!-\!\ddot{C}l\!: \\ | \\ :\!\ddot{C}l\!: \end{array} \qquad \text{P, FC} = 0$$

b.   Must form six bonds to S to minimize formal charge of S.

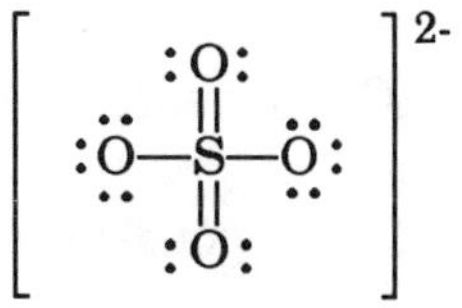

$$\left[\begin{array}{c} :\!\ddot{O}\!: \\ \| \\ :\!\ddot{O}\!-\!S\!-\!\ddot{O}\!: \\ | \\ :\!\ddot{O}\!: \end{array}\right]^{2-} \qquad \text{S, FC} = 0$$

c.   Must form seven bonds to Cl to minimize formal charge.

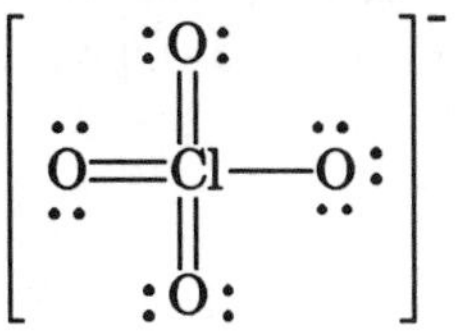

$$\left[\begin{array}{c} :\!\ddot{O}\!: \\ \| \\ \ddot{O}\!=\!Cl\!-\!\ddot{O}\!: \\ | \\ :\!\ddot{O}\!: \end{array}\right]^{-} \qquad \text{Cl, FC} = 0$$

d.   Must form five bonds to P to minimize formal charge.

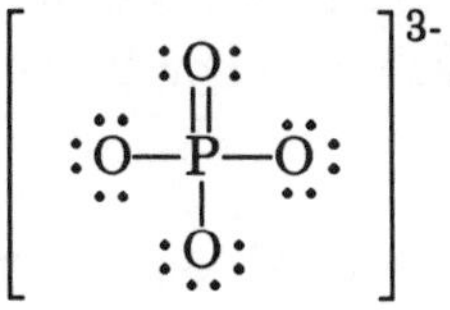

$$\left[\begin{array}{c} :\!\ddot{O}\!: \\ \| \\ :\!\ddot{O}\!-\!P\!-\!\ddot{O}\!: \\ | \\ :\!\ddot{O}\!: \end{array}\right]^{3-} \qquad \text{P, FC} = 0$$

e.

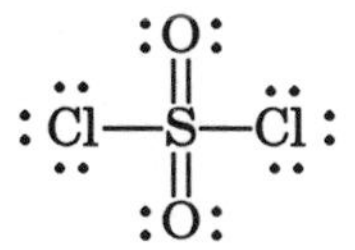

$$\begin{array}{c} :\!\ddot{O}\!: \\ \| \\ :\!\ddot{C}l\!-\!S\!-\!\ddot{C}l\!: \\ \| \\ :\!\ddot{O}\!: \end{array} \qquad \text{S, FC} = 0$$

f.

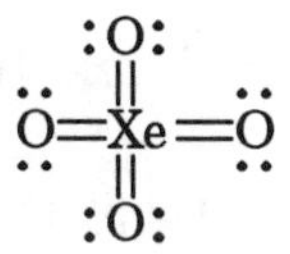

$$\begin{array}{c} :\!\ddot{O}\!: \\ \| \\ \ddot{O}\!=\!Xe\!=\!\ddot{O} \\ \| \\ :\!\ddot{O}\!: \end{array} \qquad \text{Xe, FC} = 0$$

g.

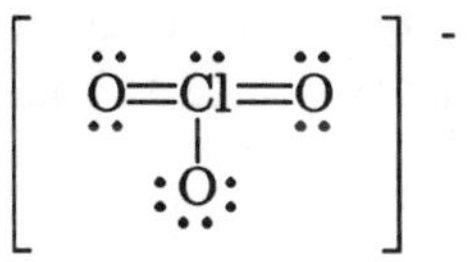

$$\left[\begin{array}{c} \ddot{O}\!=\!\ddot{C}l\!=\!\ddot{O} \\ | \\ :\!\ddot{O}\!: \end{array}\right]^{-} \qquad \text{Cl, FC} = 0$$

h.   We can't . The following structure has a zero formal charge for N:

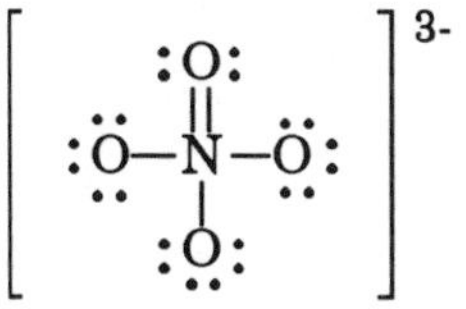

$$\left[\begin{array}{c} :\!\ddot{O}\!: \\ \| \\ :\!\ddot{O}\!-\!N\!-\!\ddot{O}\!: \\ | \\ :\!\ddot{O}\!: \end{array}\right]^{3-}$$

but N does not expand its octet. We wouldn't expect this resonance form to exist.

Molecular Geometry and Polarity

73.   The first step always is to draw a valid Lewis strucure when predicting molecular structure. When resonance is possible, only one of the possible resonance structures is necessary to predict the correct structure since all resonance structures give the same structure. The Lewis structures are in Exercises 8.57 and 8.61. The structures and bond angles for each follow.

8.57   a.   HCN: linear, 180°   b. PH_3: trigonal pyramid, < 109.5°

c. $CHCl_3$: tetrahedral, 109.5°   d. NH_4^+: tetrahedral, 109.5°

e. H_2CO: trigonal planar, 120°   f. SeF_2: V-shaped or bent, < 109.5°

g. CO_2: linear, 180°   h and i. O_2 and HBr are both linear, but there is no bond angle in either.

Note: PH_3 and SeF_2 both have lone pairs of electrons on the central atom which result in bond angles that are something less than predicted from a tetrahedral arrangement (109.5°). However, we cannot predict the exact number. For these cases, we will just insert a less than sign to show this phenomenon.

8.61   a. NO_2^-: V-shaped, < 120°; NO_3^-: trigonal planar, 120°

b. OCN^-, SCN^- and N_3^- are all linear with 180° bond angles.

74.   See Exercises 8.58 and 8.62 for the Lewis structures.

8.58   a.   All are tetrahedral; 109.5°

b.   All are trigonal pyramid; < 109.5°

c.   All are V-shaped; < 109.5°

8.62 O_3 and SO_2 are V-shaped (or bent) with a bond angle slightly less than 120° (due to the lone pair of electrons on the central atom). SO_3 is trigonal planar with a 120° bond angles.

75.   See Exercise 8.59 for the Lewis structures.

PF_5:   trigonal bipyramid, 120° and 90°
BeH_2:   linear, 180°
BH_3:   trigonal planar, 120°
Br_3^-:   linear, 180°
SF_4:   see-saw, ≈ 120° and ≈ 90°
XeF_4:   square planar, 90°
ClF_5:   square pyramid, ≈ 90°
SF_6:   octahedral, 90°

Note: Reference Figures 19.25 and 19.26 of the text for molecular structures based on the trigonal bipyramid and octahedral geometries. We will use the term see-saw to describe the molecular structure of SF_4 type molecules instead of distorted tetrahedron.

76. ClF_3 and BrF_3 each have similar Lewis structures. For ClF_3 the Lewis structure is:

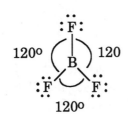

There are five pairs of electrons around the central chlorine atom so the electron arrangement is trigonal bipyramid. The positions of the atoms in ClF_3 and BrF_3 is called T-shaped with approximate bond angles of 90°. (See Figure 19.25 of the text.)

77.   a. BF_3, $3 + 3(7) = 24$ e⁻   b. $BeH_2{}^{2-}$, $2 + 2(1) + 2 = 6$ e⁻

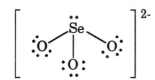

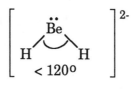

      Trigonal planar; All angles = 120°.   V-shaped; Angle is < 120°.

      Note: All of these structures have three effective pairs of electrons about the central atom. All of the structures are based on a trigonal planar geometry, but only BF_3 is described as having a trigonal planar structure. Molecular structure always describes the relative positions of the atoms.

78.   a. $SeO_3{}^{2-}$ has $6 + 3(6) + 2$ b. $SeCl_2$ has $6 + 2(7) =$
            $= 26$ valence electrons.   20 valence electrons.

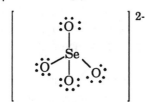

      Trigonal pyramid; All angles are < 109.5°.   V-shaped; Angle is < 109.5°.

      c. $SeO_4{}^{2-}$ has $6 + 4(6) + 2 = 32$ valence electrons.

                                                        Tetrahedral; All angles are 109.5°.

Note: There are 4 pairs of electrons about the Se atom in each case in this exercise. All of the structures are based on a tetrahedral geometry, but only SeO_4^{2-} has a tetrahedral structure. We consider only the relative positions of the atoms when describing the molecular structure.

79.   a. XeF_2 has $8 + 2(7) = 22$ valence electrons.

$$:\overset{\cdot\cdot}{\underset{\cdot\cdot}{F}} - \overset{\cdot\cdot}{Xe} - \overset{\cdot\cdot}{\underset{\cdot\cdot}{F}}:$$

180°

There are 5 pairs of electrons about the central Xe atom. The structure will be based on a trigonal bipyramid geometry. The most stable arrangement of the atoms in XeF_2 is linear with a 180° bond angle.

b. IF_3 has $7 + 3(7) = 28$ valence electrons.

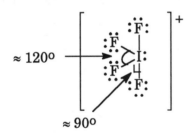

≈ 90°

≈ 90°

T-shaped; The FIF angles are ≈ 90°. Since the lone pairs will take up more space, the FIF bond angles will probably be slightly less than 90°.

c. IF_4^+ has $7 + 4(7) - 1 = 34$ valence electrons.

≈ 120°

≈ 90°

d. SF_5^+ has $6 + 5(7) - 1 = 40$ valence electrons.

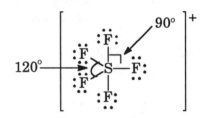

120°

90°

See-saw or teeter-totter
or distorted tetrahedron

Trigonal bipyramid

All of the species in this exercise have 5 pairs of electrons around the central atom. All of the structures are based on a trigonal bipyramid geometry, but only in SF_5^+ are all of the pairs bonding pairs. Thus, SF_5^+ is the only one we describe the molecular structure as trigonal bipyramid. Still, we had to begin with the trigonal bipyramid geometry to get to the structures of the others.

Note: Reference Figures 19.25 and 19.26 of the text for molecular structures based on the trigonal bipyramid geometry. We will use the term see-saw to describe the molecular structure of IF_4^+ type species instead of distorted tetrahedron.

80.   a. BrF_5, $7 + 5(7) = 42$ e⁻

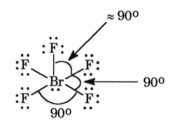

b. KrF_4, $8 + 4(7) = 36$ e⁻

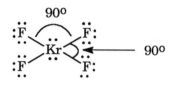

Square pyramid, ≈ 90°bond angles   Square planar, 90° bond angles

c. IF_6^+ has $7 + 6(7) - 1 = 48$ valence electrons.

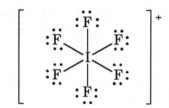

Octahedral, 90° bond angles

Note: All these species have 6 pairs of electrons around the central atom. All three structures are based on the octahedron, but only IF_6^+ has octahedral molecular structure. (See Figure 19.25 of the text.)

81. BF_3 and BeH_2^{2-} both have polar bonds but only BeH_2^{2-} has a dipole moment. The three bond dipoles from the three polar B – F bonds in BF_3 will all cancel when summed together. Hence, BF_3 is nonpolar since the overall molecule has no resulting dipole moment. In BeH_2^{2-}, the two Be – H bond dipoles do not cancel when summed together, hence BeH_2^{2-} has a dipole moment. Since H is more electronegative than Be, the negative end of the dipole moment is between the two H atoms and the positive end is around the Be atom. The arrow in the following illustration represents the overall dipole moment in BeH_2^{2-}.

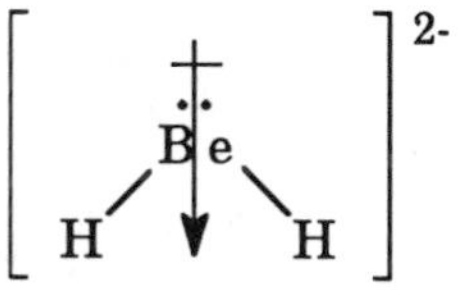

82.   All have polar bonds; in SeO_4^{2-} the individual bond dipoles cancel, and in SeO_3^{2-} and $SeCl_2$ the individual bond dipoles do not cancel. Therefore, SeO_4^{2-} has no dipole moment and SeO_3^{2-} and $SeCl_2$ have dipole moments. For SeO_3^{2-}, the negative end of the bond dipole is between the more electronegative oxygen atoms and the positive end is around Se. For $SeCl_2$, the negative end is between the more electronegative Cl atoms and the positive end of the dipole moment is around Se.

83.   All have polar bonds, but only IF_3 and IF_4^+ have dipole moments. The bond dipoles from the five S–F bonds in SF_5^+ cancel each other so SF_5^+ has no dipole moment. The bond dipoles in XeF_2 also cancel:

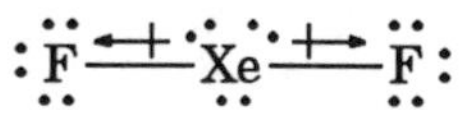

Since the bond dipoles from the two Xe – F bonds are equal in magnitude but point in opposite directions, then they cancel each other and XeF_2 has no dipole moment (is nonpolar). For IF_3 and IF_4^+, the arrangement of these molecules are such that the individual bond dipoles do <u>not</u> all cancel so each has an overall dipole moment.

84.   All have polar bonds, but only BrF_5 has an overall dipole moment. The six bond dipoles in IF_6^+ all cancel each other so IF_6^+ has no dipole moment. The same is true for KrF_4:

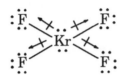

When the four bond dipoles are added together, they all cancel each other and KrF_4 has no overall dipole moment. BrF_5 has a structure where the individual bond dipoles do <u>not</u> all cancel, hence BrF_5 has a dipole moment.

85.   Molecules which have an overall dipole moment are called polar molecules and molecules which do not have an overall dipole moment are called nonpolar molecules.

a. OCl_2, $6 + 2(7) = 20\ e^-$ KrF_2, $8 + 2(7) = 22\ e^-$

V-shaped, polar; OCl_2 is polar because the two O – Cl bond dipoles don't cancel each other. The resultant dipole moment is shown in the drawing.

Linear, nonpolar; The molecule is nonpolar because the two Kr – F bond dipoles cancel each other.

BeH_2, $2 + 2(1) = 4\ e^-$ SO_2, $6 + 2(6) = 18\ e^-$

Linear, nonpolar; Be – H bond dipoles are equal and point in opposite directions. They cancel each other. BeH_2 is nonpolar.

V-shaped, polar; The S – O bond dipoles do not cancel so SO_2 is polar (has a dipole moment). Only one resonance structure is shown.

Note: All four species contain three atoms. They have different structures because the number of lone pairs of electrons around the central atom are different in each case.

b. SO_3, $6 + 3(6) = 24$ e⁻ NF_3, $5 + 3(7) = 26$ e⁻

Trigonal planar, nonpolar;   Trigonal pyramid, polar;
Bond dipoles cancel. Only one   Bond dipoles do not cancel.
resonance structure is shown.

ClF_3 has $7 + 3(7) = 28$ valence electrons.

T-shaped, polar; Bond dipoles do not cancel.

Note: Each molecule has the same number of atoms, but the structures are different because of
differing numbers of lone pairs around each central atom.

c. CF_4, $4 + 4(7) = 32$ e⁻ SeF_4, $6 + 4(7) = 34$ e⁻

Tetrahedral, nonpolar;   See-saw, polar;
Bond dipoles cancel.   Bond dipoles do not cancel.

XeF_4, $8 + 4(7) = 36$ valence electrons

Square planar, nonpolar;
Bond dipoles cancel.

Again, each molecule has the same number of atoms, but a different structure because of
differing numbers of lone pairs around the central atom.

d. IF_5, $7 + 5(7) = 42$ e⁻ AsF_5, $5 + 5(7) = 40$ e⁻

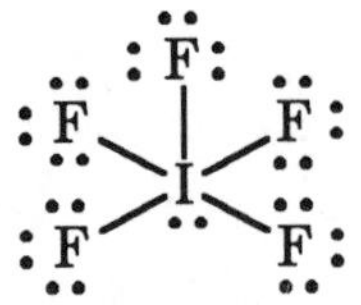

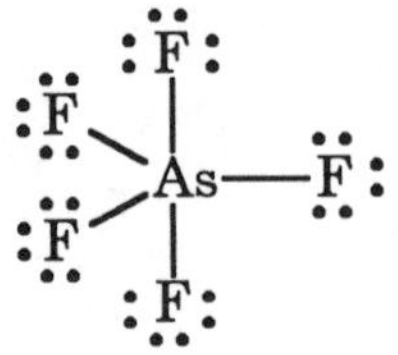

Square pyramid, polar;   Trigonal bipyramid, nonpolar;
Bond dipoles do not cancel.   Bond dipoles cancel.

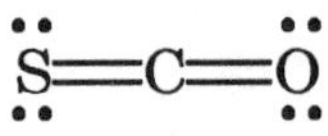

Yet again, the molecules have the same number of atoms, but different structures because of the
presence of differing numbers of lone pairs.

86.   a.   b.

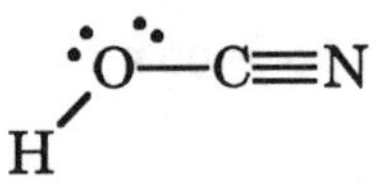

Polar; The bond dipoles do   Polar; The C – O bond is a more
not cancel.   polar bond than the C – S bond. So
                                            the two bond dipoles do not cancel
                                            each other.

c.   d.

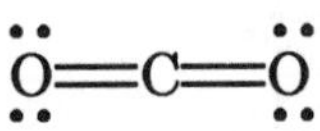

Nonpolar; The two C – O bond   Polar; All the bond dipoles are not
dipoles cancel each other.   equivalent, so they don't cancel each
                                            other.

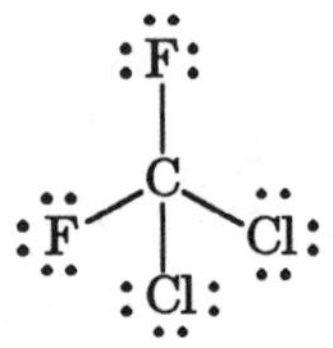

e.   f.

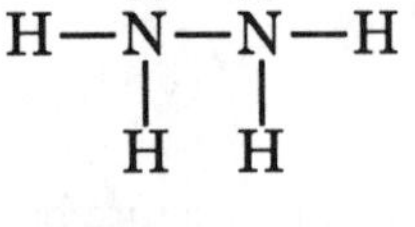

Polar; Bond dipoles do not cancel.   Polar; Bond dipoles are not equivalent
                                            so they don't cancel each other.

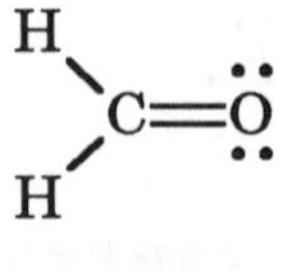

87.   SbF$_5$, 5 + 5(7) = 40 e$^-$ HF, 1 + 7 = 8 e$^-$

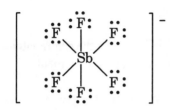

Trigonal bipyramid, nonpolar   Linear, polar

SbF$_6^-$, 5 + 6(7) + 1 = 48 e$^-$ H$_2$F$^+$, 2(1) + 7 - 1 = 8 e$^-$

Octahedral, nonpolar (no overall   V-shaped, polar (has an overall
dipole moment)   dipole moment)

88.   SF$_2$ has 6 + 2(7) = 20 valence electrons.   SF$_4$ has 6 + 4(7) = 34 valence electrons.

V-shaped, polar, < 109.5°   See-saw, polar
                                                                 a) ≈ 120°
                                                                 b) ≈ 90°

SF$_6$ has 6 + 6(7) = 48 valence electrons.   S$_2$F$_2$ has 2(6) + 2(7) = 26 e$^-$.

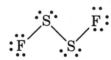

Octahedral, nonpolar, 90°   V-shaped about each sulfur, polar, ≈ 109.5°.
                                                                 Predicting polarity for this molecule is very
                                                                 difficult. Although it looks like the S − F bond
                                                                 dipoles could cancel in this drawing, in
                                                                 actuality, they do not and the molecule is polar.

S_2F_4 has $2(6) + 4(7) = 40$ valence electrons

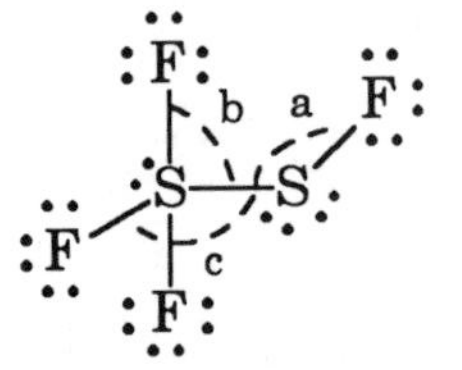

a) $\approx 109.5°$
b) $\approx 90°$
c) $\approx 120°$

There is no easy description for the S_2F_4 structure. There is a trigonal bipyramidal arrangement (3 F, 1 S, 1 lone pair) about one sulfur and a tetrahedral (1 F, 1 S, 2 lone pairs) arrangement of e⁻ pairs about the other sulfur. With this picture we can predict values for all bond angles. We also would predict S_2F_4 to be polar since the bond dipoles would not cancel.

89.   All these molecules have polar bonds that are symmetrically arranged about the central atoms. In each molecule, the individual bond dipoles cancel to give no net overall dipole moment.

90.   The lone pairs of electrons about each F are omitted in the following Lewis structures.

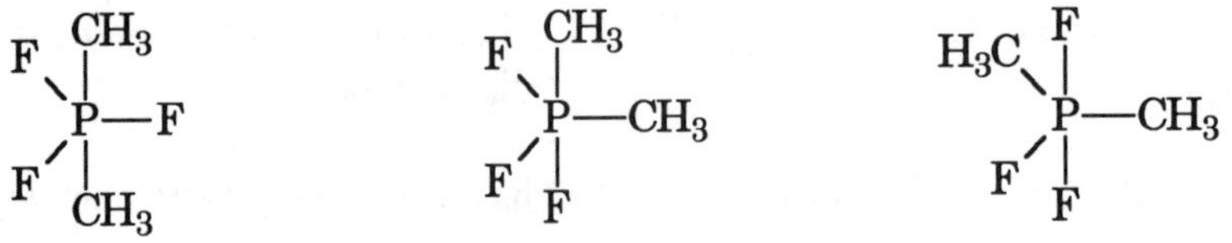

(i)   (ii)   (iii)

Structure i is nonpolar since the bond dipoles are arranged so that they cancel each other. Structure ii and iii are polar since the bonds are arranged so the bond dipoles do not cancel each other. Between structures ii and iii, ii is probably the most polar since the more polar P – F bond dipoles add together in structure ii, but are canceled somewhat in structure iii. Thus, dipole moments should help distinguish the three isomers; particularly in distinguishing i from the other two.

Additional Exercises

91.   All these series of ions are isoelectronic series. In each series, the smallest ion has the largest number of protons holding the constant number of electrons. The smallest ion will also have the largest ionization energy. The ion with the fewest protons will be largest and will have the smallest ionization energy.

a.   radius trend: $Mg^{2+} < Na^+ < F^- < O^{2-}$; IE trend: $O^{2-} < F^- < Na^+ < Mg^{2+}$

b.   radius trend: $Ca^{2+} < P^{3-}$; IE trend: $P^{3-} < Ca^{2+}$

c.   radius trend: $K^+ < Cl^- < S^{2-}$; IE trend: $S^{2-} < Cl^- < K^+$

92.   Ionic solids are held together by strong electrostatic forces which are omnidirectional.

    i.   For electrical conductivity, charged species must be free to move. In ionic solids the charged ions are held rigidly in place. Once the forces are disrupted (melting or dissolution) the ions can move about (conduct).

    ii.   Melting and boiling disrupts the attractions of the ions for each other. If the forces are strong, it will take a lot of energy (high temperature) to accomplish this.

    iii. If we try to bend a piece of material, the ions must slide across each other. For an ionic solid the following might happen:

          strong attraction   strong repulsion

        Just as the layers begin to slide, there will be very strong repulsions causing the solid to snap across a fairly clean plane.

    iv.   Polar molecules are attracted to ions and can break up the lattice.

These properties and their correlation to chemical forces will be discussed in detail in Chapters 10 and 11.

93.   a. $Na^+(g) + Cl^-(g) \rightarrow NaCl(s)$ b. $NH_4^+(g) + Br^-(g) \rightarrow NH_4Br(s)$

       c. $Mg^{2+}(g) + S^{2-}(g) \rightarrow MgS(s)$ d. $O_2(g) \rightarrow 2\ O(g)$

94. ΔH_f° for H(g) is ΔH° for the reaction $1/2\ H_2(g) \rightarrow H(g)$. The H_2 bond energy of 432 kJ/mol refers to ΔH for the reaction $H_2(g) \rightarrow 2\ H(g)$. Therefore, $\Delta H^\circ_{f,\,H} = D_{H\text{-}H}/2 = 432/2 = 216$ kJ/mol.

95. CO_3^{2-} has $4 + 3(6) + 2 = 24$ valence electrons.

HCO_3^- has $1 + 4 + 3(6) + 1 = 24$ valence electrons.

H_2CO_3 has $2(1) + 4 + 3(6) = 24$ valence electrons.

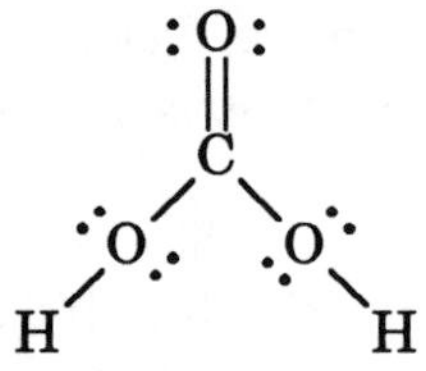

The Lewis structures for the reactants and products are:

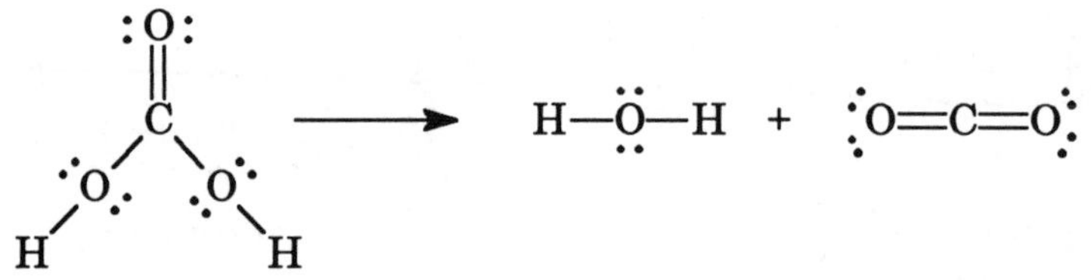

Bonds broken:	Bonds formed:
2 C – O (358 kJ/mol)	1 C = O (799 kJ/mol)
1 O – H (467 kJ/mol)	1 O – H (467 kJ/mol)

$\Delta H = 2(358) + 467 - [799 + 467] = -83$ kJ; The carbon-oxygen double bond is stronger than two carbon-oxygen single bonds, hence CO_2 and H_2O are more stable than H_2CO_3.

96.   The stable species are:

a. SO_4^{2-}: We can't draw a Lewis structure that obeys the octet rule for SO_4. The two extra electrons (from the -2 charge) complete octets.

b. PF_5:   N is too small and doesn't have low energy d-orbitals to expand its octet.

c. SF_6:   O is too small and doesn't have low energy d-orbitals to expand its octet.

d. BH_4^-: BH_3 doesn't obey octet rule.

e. MgO: In MgF we would have to have $Mg^+ F^-$ or $Mg^{2+} F^{2-}$. Neither Mg^+ nor F^{2-} are stable.

f. CsCl: Cs doesn't form stable Cs^{2+} ion.

g. KBr: Br doesn't form stable Br^{2-} ion.

97.   a. NO_2, $5 + 2(6) = 17 \ e^-$ N_2O_4, $2(5) + 4(6) = 34 \ e^-$

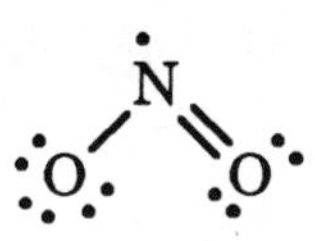

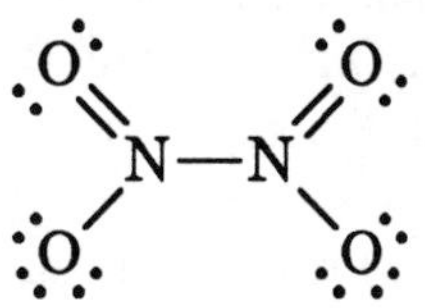

            plus other resonance structures   plus other resonance structures

        b. BF_3, $3 + 3(7) = 24 \ e^-$ NH_3, $5 + 3(1) = 8 \ e^-$

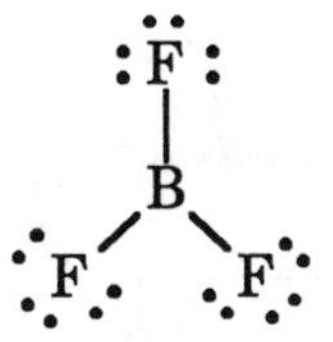

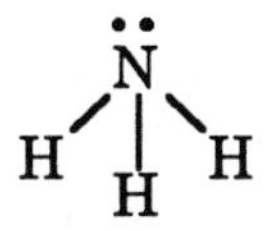

BF_3NH_3, $24 + 8 = 32 \ e^-$

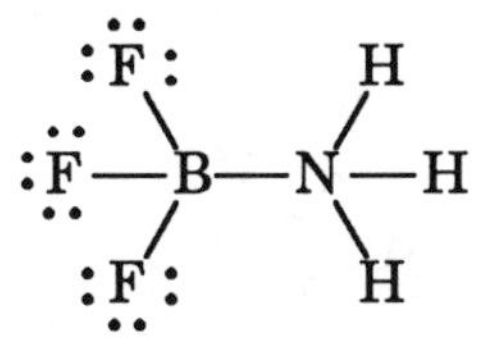

In reaction a, NO_2 has an odd number of electrons so it is impossible to satisfy the octet rule. By dimerizing to form N_2O_4, the odd electron on two NO_2 molecules can pair up giving a species whose Lewis structure can satisfy the octet rule. In general odd electron species and very reactive. In reaction b, BF_3 can be considered electron deficient. Boron only has six electrons around it. By forming BF_3NH_3, the boron atom satisfies the octet rule by accepting a lone pair of electrons from NH_3 to form a fourth bond.

98. S_2N_2 has $2(6) + 2(5) = 22$ valence electrons.

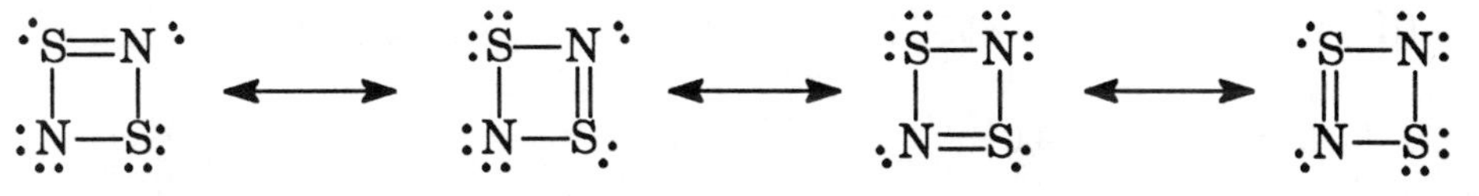

99.   The general structure of the trihalide ions is:

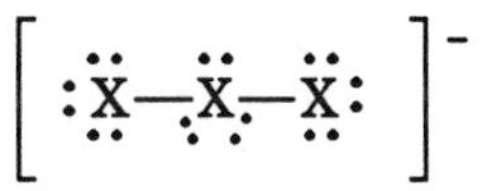

Bromine and iodine are large enough and have low energy, empty d-orbitals to accommodate the expanded octet. Fluorine is small and its valence shell contains only 2s and 2p orbitals (4 orbitals) and cannot expand its octet. The lowest energy d orbitals in F are 3d; they are too high in energy compared to 2s and 2p to be used in bonding.

100. H_2O and NH_3 have lone pair electrons on the central atoms. These lone pair electrons require more room than the bonding electrons which tends to compress the angles between the bonding pairs. The bond angles for H_2O is the smallest since oxygen has two lone pairs on the central atom and the bond angle is compressed more than in NH_3 where N has only one lone pair.

101.   Yes, each structure has the same number of effective pairs around the central atom. (A multiple bond is counted as a single group of electrons.)

102.   This is a very weak bond. The structure

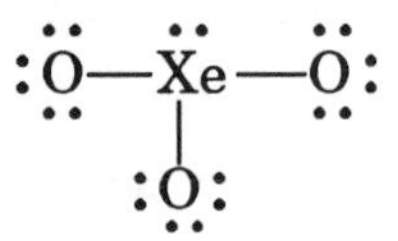

might be more accurate. It is difficult to think of a double bond as being so weak. This illustrates that no model is totally effective in describing the bonding in all molecules. We need to recognize that all models have limitations and that we can't apply all models to all molecules.

103.   a.

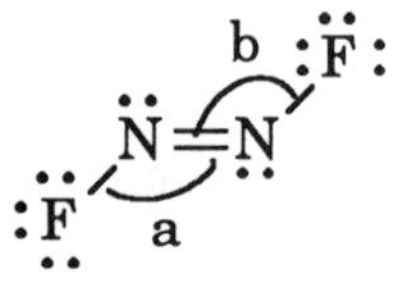

Angles a and b are both ≈ 120°. 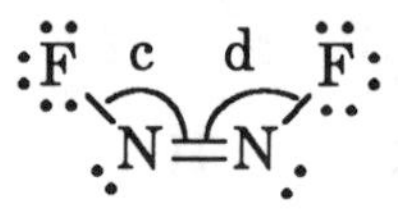   Angles c and d are both a little less than 120° (due to the lone pairs).

   b.   The N – F bond dipoles cancel in the first structure so it is nonpolar. The N – F bond dipoles do not cancel in the second structure so it is polar.

Challenge Problems

104.   (IE - EA)   (IE - EA)/502   EN (text)   2006/502 = 4.0

	(IE - EA)	(IE - EA)/502	EN (text)
F	2006 kJ/mol	4.0	4.0
Cl	1604	3.2	3.0
Br	1463	2.9	2.8
I	1302	2.6	2.5

The values calculated from IE and EA show the same trend (and agree fairly closely) to the values given in the text.

105.   C ≡ O (1072 kJ/mol) and N ≡ N (941 kJ/mol); CO is polar while N_2 is nonpolar. This may lead to a great reactivity for the CO bond.

106.   a.   I.

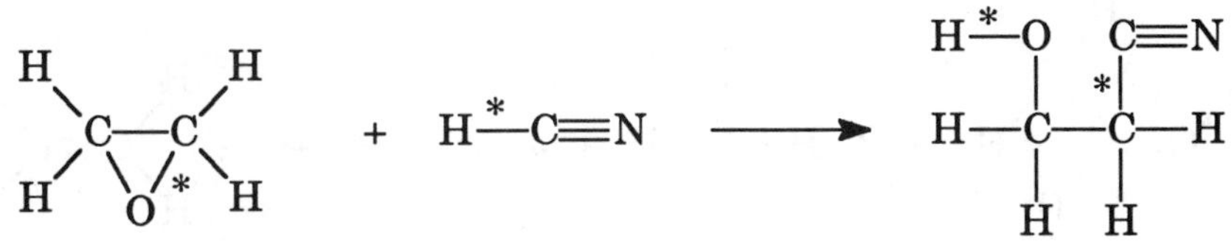

Bonds broken (*):   Bonds formed (*):

1 C – O (358 kJ)   1 O – H (467 kJ)
1 H – C (413 kJ)   1 C – C (347 kJ)

ΔH_I = 358 kJ + 413 kJ - [467 kJ + 347 kJ] = -43 kJ

II.

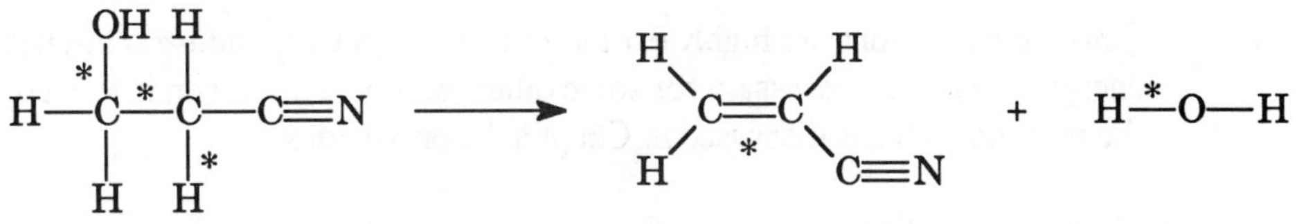

Bonds broken (*):   Bonds formed (*):

1 C – O (358 kJ/mol)   1 H – O (467 kJ/mol)
1 C – H (413 kJ/mol)   1 C = C (614 kJ/mol)
1 C – C (347 kJ/mol)

ΔH_{II} = 358 kJ + 413 kJ + 347 kJ - [467 kJ + 614 kJ] = +37 kJ

$\Delta H_{overall}$ = ΔH_I + ΔH_{II} = -43 kJ + 37 kJ = -6 kJ

b.

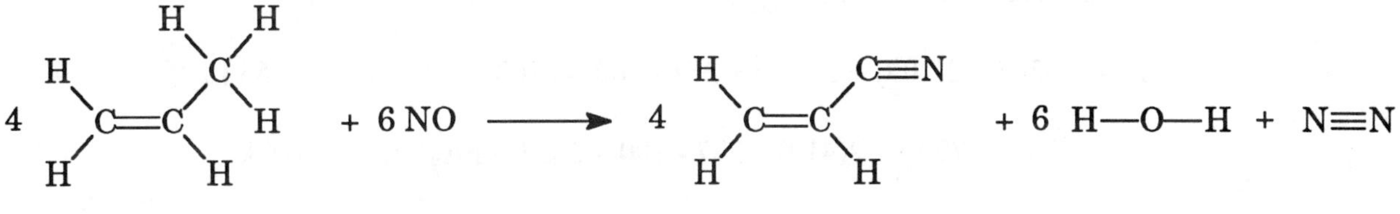

Bonds broken: Bonds formed:

4 × 3 C – H (413 kJ/mol)   4 C ≡ N (891 kJ/mol)
6 N = O (630. kJ/mol)   6 × 2 H – O (467 kJ/mol)
                                     1 N ≡ N (941 kJ/mol)

ΔH = 12(413) + 6(630.) - [4(891) + 12(467) + 941] = -1373 kJ

c.

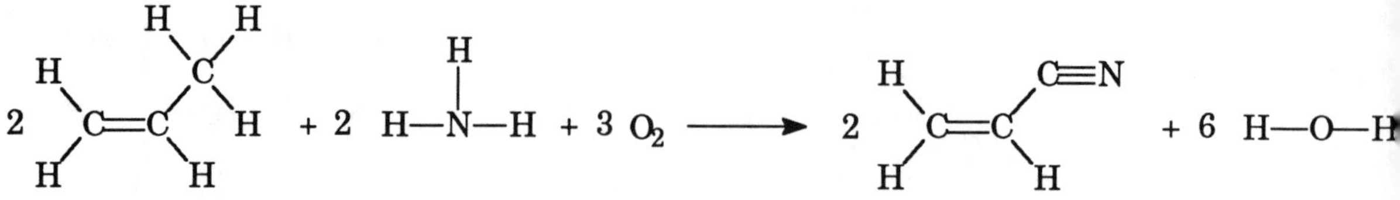

Bonds broken: Bonds formed:

2×3 C – H (413 kJ/mol)   2 C $\equiv$ N (891 kJ/mol)
2×3 N – H (391 kJ/mol) 6×2 O – H (467 kJ/mol)
3 O $=$ O (495 kJ/mol)

$\Delta H = 6(413) + 6(391) + 3(495) - [2(891) + 12(467)] = -1077$ kJ

d.   Since both reactions are highly exothermic, the high temperature is not needed to provide energy. It must be necessary for some other reason. The reason is to increase the speed of the reaction. This is discussed in Chapter 12 on kinetics.

107.   Let us look at the complete cycle for Li_2S.

$2 \, Li(s) \rightarrow 2 \, Li(g)$ $2 \, \Delta H_{sub, \, Li} = 2(161)$ kJ
$2 \, Li(g) \rightarrow 2 \, Li^+(g) + 2 \, e^-$ $2 \, IE = 2(520.)$ kJ
$S(s) \rightarrow S(g)$ $\Delta H_{sub, \, S} = 277$ kJ
$S(g) + e^- \rightarrow S^-(g)$ $EA_1 = -200.$ kJ
$S^-(g) + e^- \rightarrow S^{2-}(g)$ $EA_2 = ?$
$2 \, Li^+(g) + S^{2-}(g) \rightarrow Li_2S$ $LE = -2472$ kJ

_____

$2 \, Li(s) + S(s) \rightarrow Li_2S(s)$ $\Delta H_f^\circ = -500.$ kJ

$\Delta H_f^\circ = 2 \, \Delta H_{sub, \, Li} + 2 \, IE + \Delta H_{sub, \, S} + EA_1 + EA_2 + LE, \; -500. = -1033 + EA_2, \; EA_2 = 533$ kJ

For each salt: $\Delta H_f^\circ = 2 \, \Delta H_{sub, \, M} + 2 \, IE + 277 - 200. + LE + EA_2$

Na_2S: $-365 = 2(109) + 2(495) + 277 - 200. - 2203 + EA_2$; $EA_2 = 553$ kJ

K_2S: $-381 = 2(90.) + 2(419) + 277 - 200. - 2052 + EA_2$; $EA_2 = 576$ kJ

Rb_2S: $-361 = 2(82) + 2(409) + 277 - 200. - 1949 + EA_2$; $EA_2 = 529$ kJ

Cs_2S: $-360 = 2(78) + 2(382) + 277 - 200. - 1850 + EA_2$; $EA_2 = 493$ kJ

We get values from 493 to 576 kJ.

The mean value is: $\dfrac{533 + 553 + 576 + 529 + 493}{5} = 537$ kJ

We can represent the results as $EA_2 = 540 \pm 50$ kJ.

108.   If we can draw resonance forms for the anion after loss of H^+, we can argue that the extra stability of
the anion causes the proton to be more readily lost, i.e., makes the compound a better acid.

a.

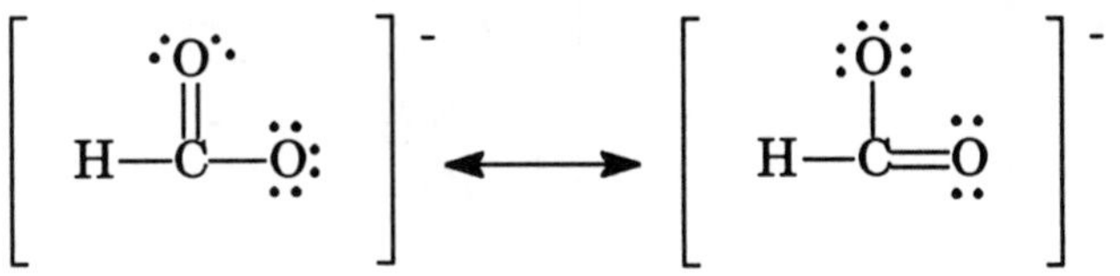

b.

c.

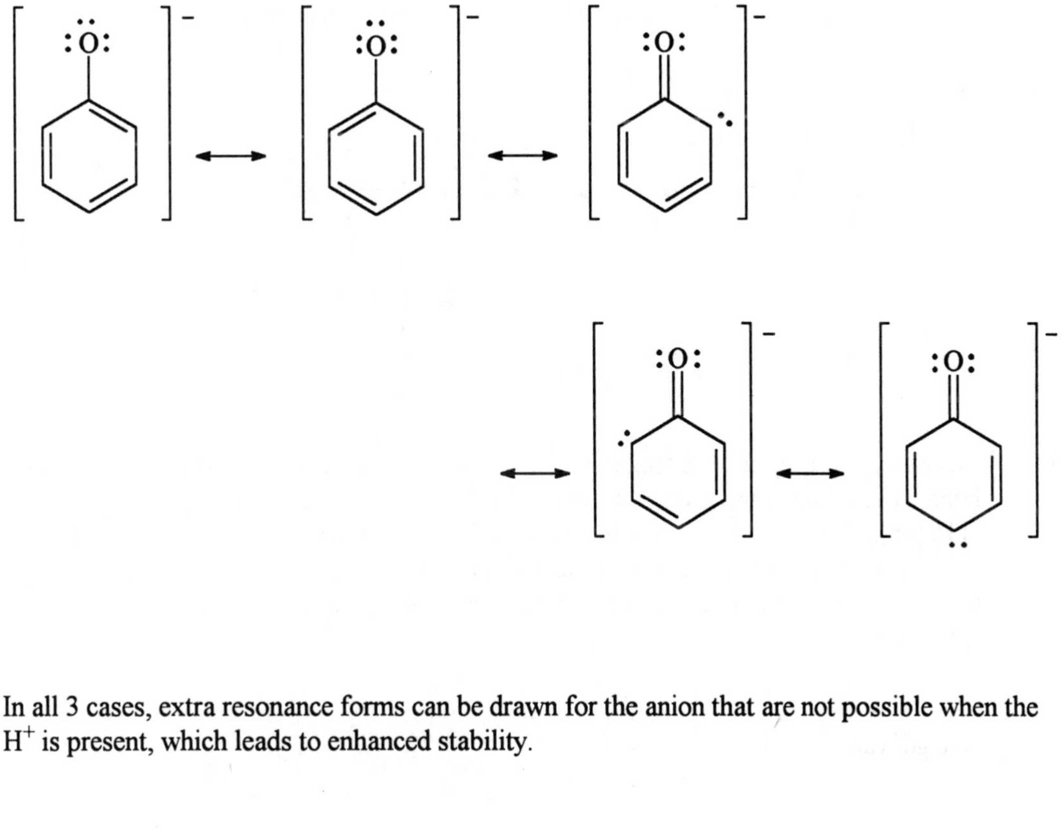

In all 3 cases, extra resonance forms can be drawn for the anion that are not possible when the
H^+ is present, which leads to enhanced stability.

109.   a. $N(NO_2)_2^-$ contains $5 + 2(5) + 4(6) + 1 = 40$ valence electrons.

The most likely structures are:

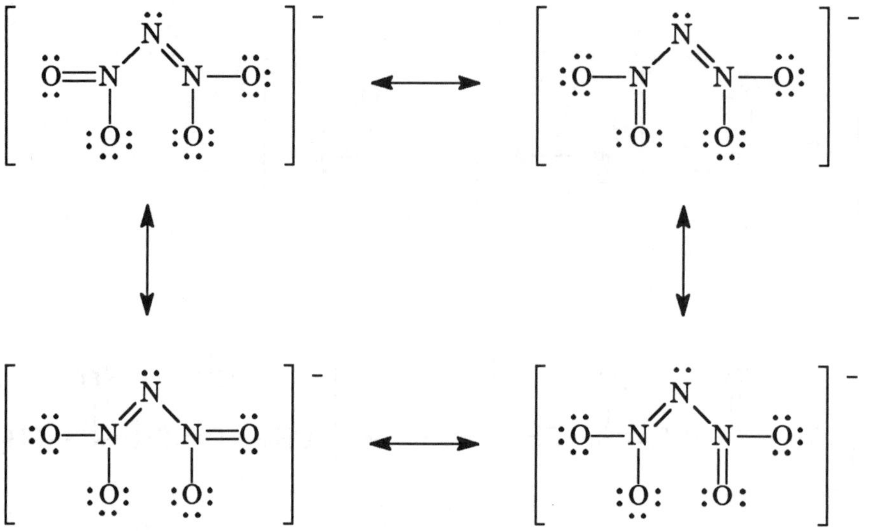

There are other possible resonance structures, but these are most likely.

b.   The NNN and all ONN and ONO bond angles should be about $120°$.

110.

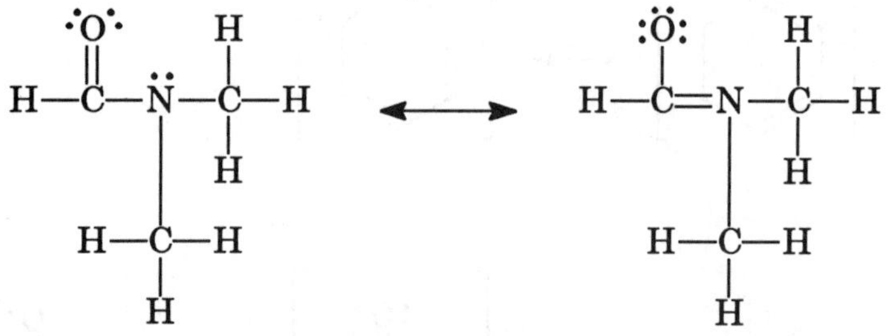

111.   The nitrogen-nitrogen bond length of 112 pm is between a double (120 pm) and a triple (110 pm) bond. The nitrogen-oxygen bond length of 119 pm is between a single (147 pm) and a double bond (115 pm). The last resonance structure doesn't appear to be as important as the other two since there is no evidence from bond lengths for a nitrogen-oxygen triple bond or a nitrogen-nitrogen single bond as in the third resonance form. We can adequately describe the structure of N_2O using the resonance forms:

Assigning formal charges for all 3 resonance forms:

$$\ddot{N}=N=\ddot{O} \quad \longleftrightarrow \quad :N\equiv N-\ddot{\ddot{O}}: \quad \longleftrightarrow \quad :\ddot{\ddot{N}}-N\equiv O:$$

$$-1 \quad +1 \quad 0 \qquad\qquad 0 \quad +1 \quad -1 \qquad\qquad -2 \quad +1 \quad +1$$

For:

$$\left(\ddot{\ddot{N}}=\right),\ \ FC = 5 - 4 - 1/2(4) = -1$$

$$\left(=N=\right),\ \ FC = 5 - 1/2(8) = +1,\ \ \text{Same for}\ \ \left(\equiv N-\right)\ \text{and}\ \left(-N\equiv\right)$$

$$\left(:\ddot{\ddot{N}}-\right),\ \ FC = 5 - 6 - 1/2(2) = -2;\ \ \left(:N\equiv\right),\ \ FC = 5 - 2 - 1/2(6) = 0$$

$$\left(=\ddot{\ddot{O}}\right),\ \ FC = 6 - 4 - 1/2(4) = 0;\ \ \left(-\ddot{\ddot{O}}:\right),\ \ FC = 6 - 6 - 1/2(2) = -1$$

$$\left(\equiv O:\right),\ \ FC = 6 - 2 - 1/2(6) = +1$$

We should eliminate N – N ≡ O since it has a formal charge of +1 on the most electronegative element (O). This is consistent with the observation that the N – N bond is between a double and triple bond and that the N – O bond is between a single and double bond.

112.   a.

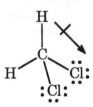

The C –H bonds are assumed nonpolar since the electronegativities of C and H are about equal.

δ+   δ-
C – Cl is the charge distribution for each C – Cl bond. The two individual C – Cl bond dipoles add together to give an overall dipole moment for the molecule. The overall dipole will point from C (positive end) to the midpoint of the two Cl atoms (negative end).

The C – H bond is essentially nonpolar. The three C – Cl bond dipoles $CHCl_3$ add together to give an overall dipole moment for the molecule. The overall dipole will have the negative end at the midpoint of the three chlorines and the positive end around the carbon.

CCl$_4$ is nonpolar. CCl$_4$ is a tetrahedral molecule where all four C – Cl bond dipoles cancel when added together. Let's consider just the C and two of the Cl atoms. There will be a net dipole pointing in the direction of the middle of the two Cl atoms.

There will be an equal and opposite dipole arising from the other two Cl atoms. Combining:

The two dipoles will cancel and CCl$_4$ is nonpolar.

b. CO$_2$ is nonpolar. CO$_2$ is a linear molecule with two equivalence bond dipoles that cancel. N$_2$O is polar since the bond dipoles do not cancel.

c. NH$_3$ is polar. The 3 N – H bond dipoles add together to give a net dipole in the direction of the lone pair. We would predict PH$_3$ to be nonpolar on the basis of electronegativitity, i.e., P – H bonds are nonpolar. However, the presence of the lone pair makes the PH$_3$ molecule slightly polar. The net dipole is in the direction of the lone pair and has a magnitude about one third that of the NH$_3$ dipole.

The As – H bonds are essentially nonpolar (ΔEN = 0.1). The lone pair on AsH$_3$ makes the molecule slightly polar with a dipole moment even less than the PH$_3$ dipole moment.

CHAPTER NINE

COVALENT BONDING: ORBITALS

Questions

7.   Bond energy is directly proportional to bond order. Bond length is inversely proportional to bond order. Bond energy and bond length can be measured.

8.   The electrons in sigma bonding molecular orbitals are attracted to two nuclei, which is a lower, more stable energy arrangement for the electrons than in the separated atoms. In sigma antibonding molecular orbitals, the electrons are mainly outside the space between the nuclei, which is a higher, less stable energy arrangement than in the separated atoms.

9.   Paramagnetic: Unpaired electrons are present. Measure the mass of a substance in the presence and absence of a magnetic field. A substance with unpaired electrons will be attracted by the magnetic field, giving an apparent increase in mass in the presence of the field. Greater number of unpaired electrons will give greater attraction and greater observed mass increase.

10.   Molecules that exhibit resonance have delocalized π bonding. In order to rationalize why the bond lengths are equal in molecules that exhibit resonance, we say that the π bonds are delocalized over the entire surface of the molecule.

Exercises

The Localized Electron Model and Hybrid Orbitals

11. H_2O has $2(1) + 6 = 8$ valence electrons.

H₂O has a tetrahedral arrangement of the electron pairs about the O atom which requires sp^3 hybridization. Two of the four sp^3 hybrid orbitals are used to form bonds to the two hydrogen atoms and the other two sp^3 hybrid orbitals hold the two lone pairs on oxygen. The two O – H bonds are formed from overlap of the sp^3 hybrid orbitals on oxygen with the 1s atomic orbitals on the hydrogen atoms.

12. CCl_4 has 4 + 4(7) = 32 valence electrons.

$$\ddot{\,:\!\overset{\displaystyle ..}{\underset{\displaystyle ..}{Cl}}\!:\,}$$

CCl_4 has a tetrahedral arrangement of the electron pairs about the carbon atom which requires sp^3 hybridization. The four sp^3 hybrid orbitals on carbon are used to form the four bonds to chlorine. The chlorine atoms also have a tetrahedral arrangement of electron pairs and we will assume that they are also sp^3 hybridized. The C – Cl sigma bonds are all formed from overlap of sp^3 hybrid orbitals on carbon with sp^3 hybrid orbitals on each chlorine atom.

13. H_2CO has 2(1) + 4 + 6 = 12 valence electrons.

The central carbon atom has a trigonal planar arrangement of the electron pairs which requires sp^2 hybridization. The two C – H sigma bonds are formed from overlap of the sp^2 hybrid orbitals on carbon with the hydrogen 1s atomic orbitals. The double bond between carbon and oxygen consists of one σ and one π bond. The oxygen atom, like the carbon atom, also has a trigonal planar arrangement of the electrons which requires sp^2 hybridization. The σ bond in the double bond is formed from overlap of a carbon sp^2 hybrid orbital with an oxygen sp^2 hybrid orbital. The π bond in the double bond is formed from overlap of the unhybridized p atomic orbitals. Carbon and oxygen each have one unhybridized p atomic orbital which are parallel to each other. When two parallel p atomic orbitals overlap, a π bond results.

14. C_2H_2 has 2(4) + 2(1) = 10 valence electrons.

$$H\!-\!C\!\equiv\!C\!-\!H$$

Each carbon atom in C_2H_2 is sp hybridized since each carbon atom is surrounded by two effective pairs of electrons, i.e., each carbon atom has a linear arrangement of the electrons. Since each carbon atom is sp hybridized, then each carbon atom has two unhybridized p atomic orbitals. The two C – H sigma bonds are formed from overlap of carbon sp hybrid orbitals with hydrogen 1s atomic orbitals. The triple bond is composed of one σ bond and two π bonds. The sigma bond between to the carbon atoms is formed from overlap of sp hybrid orbitals on each carbon atom. The two π bonds of the triple bond are formed from parallel overlap of the two unhybridized p atomic orbitals on each carbon.

15.   See Exercises 8.57 and 8.61 for the Lewis structures. To predict the hybridization, first determine the arrangement of electrons pairs about each central using the VSEPR model, then utilize the information in Figure 9.24 of the text to deduce the hybridization required for that arrangement of electron pairs.

8.57   a.   HCN; C is sp hybridized.   b. PH_3; P is sp^3 hybridized.

       c. $CHCl_3$; C is sp^3 hybridized.   d. NH_4^+; N is sp^3 hybridized.

       e. H_2CO; C is sp^2 hybridized.   f. SeF_2; Se is sp^3 hybridized.

       g. CO_2; C is sp hybridized.   h. O_2; Each O atom is sp^2 hybridized.

       i.   HBr; Br is sp^3 hybridized.

8.61   a.   In NO_2^-, N is sp^2 hybridized and in NO_3^-, N is also sp^2 hybridized.

       b.   In OCN^- and SCN^-, the central carbon atoms in each ion are sp hybridized and in N_3^-, the central N atom is also sp hybridized.

16.   See Exercises 8.58 and 8.62 for the Lewis structures.

    8.58   a.   All the central atoms are sp^3 hybridized.

        b.   All the central atoms are sp^3 hybridized.

        c.   All the central atoms are sp^3 hybridized.

    8.62   In O_3 and in SO_2, the central atoms are sp^2 hybridized and in SO_3, the central sulfur atom is also sp^2 hybridized.

17.   See Exercise 8.59 for the Lewis structures.

PF_5:	P is dsp^3 hybridized.	BeH_2:	Be is sp hybridized.
BH_3:	B is sp^2 hybridized.	Br_3^-:	Br is dsp^3 hybridized.
SF_4:	S is dsp^3 hybridized.	XeF_4:	Xe is d^2sp^3 hybridized.
ClF_5:	Cl is d^2sp^3 hybridized.	SF_6:	S is d^2sp^3 hybridized.

18.   In ClF_3, the central Cl atom is dsp^3 hybridized and in BrF_3, the central Br atom is also dsp^3 hybridized. See Exercise 8.60 for the Lewis structures.

19.   All have a trigonal planar arrangement of electron pairs about the central atom so all have central atoms with sp^2 hybridization. See Exercise 8.77 for the Lewis structures.

20.   All have a tetrahedral arrangement of electron pairs about the central atom so all have central atoms with sp^3 hybridization. See Exercise 8.78 for the Lewis structures.

21.   All have central atoms with dsp^3 hybridization since all are based on the trigonal bipyramid arrangement of electron pairs. See Exercise 8.79 for the Lewis structures.

22.   All have central atoms with d^2sp^3 hybridization since all are based on the octahedral arrangement of electron pairs. See Exercise 8.80 for the Lewis structures.

23.   a.   b.

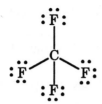

tetrahedral   sp^3 trigonal pyramid   sp^3
109.5°   nonpolar   ≈ 109.5°   polar

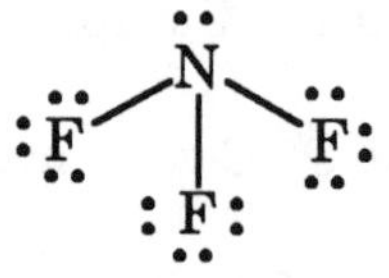

The angles in NF$_3$ should be slightly less than 109.5° because the lone pair requires more room than the bonding pairs. In Chapter 9, we will ignore this effect and say bond angles are ≈ 109.5°.

c.   d.

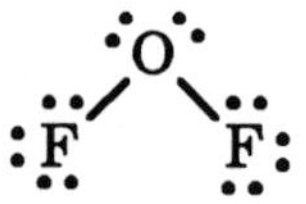

V-shaped   sp^3 trigonal planar   sp^2
≈ 109.5°   polar   120°   nonpolar

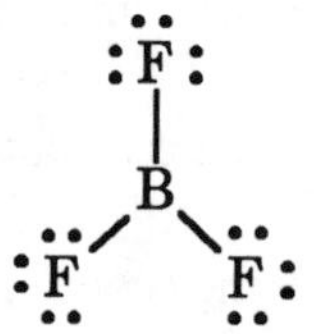

e.   f.

H——Be——H

linear   sp   see-saw
180°   nonpolar   a. ≈ 120°, b. ≈ 90°
                                              dsp^3 polar

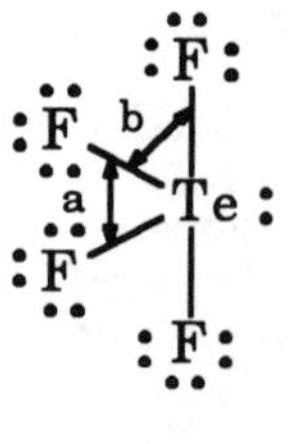

g.   h.

:F——Kr——F:

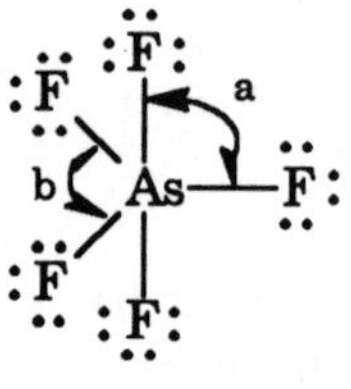

trigonal bipyramid   dsp^3 linear   dsp^3
a. 90°, b. 120°   nonpolar   180°   nonpolar

i.

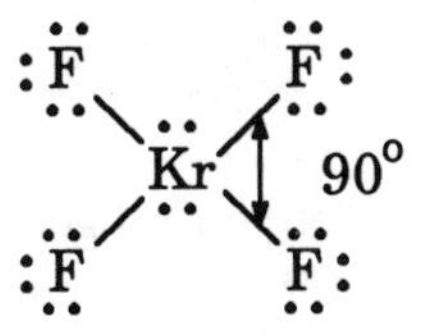

square planar d^2sp^3
90°   nonpolar

j.

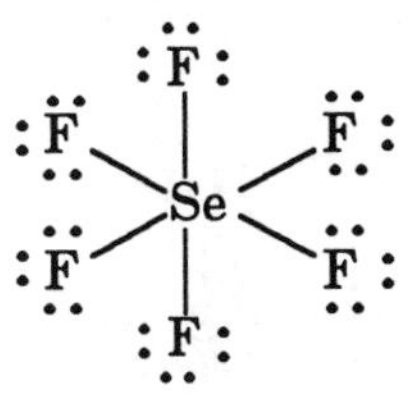

octahedral d^2sp^3
90°   nonpolar

k.

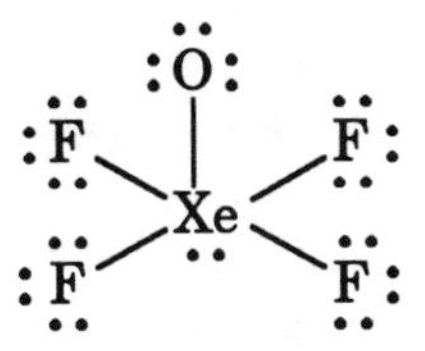

square pyramid d^2sp^3
≈ 90°   polar

l.

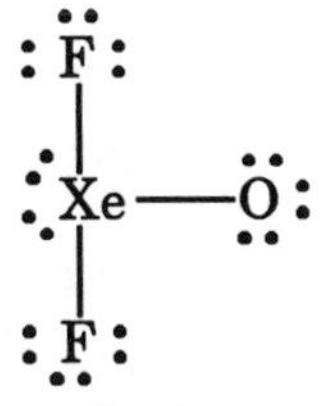

T-shaped dsp^3
≈ 90°   polar

m.

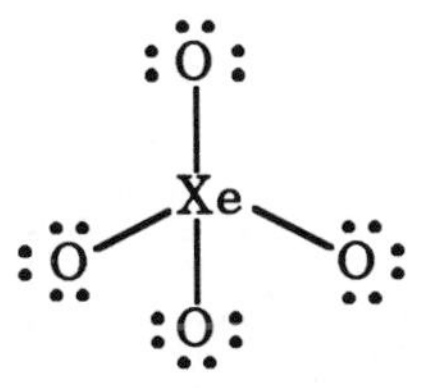

tetrahedral
109.5°
sp^3
nonpolar

24.   a.

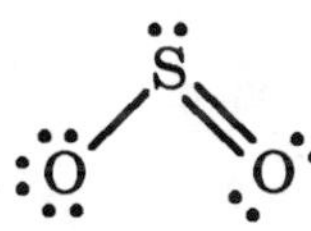

V-shaped
120°
sp^2

Only one resonance form is shown. Resonance does not change the position of the atoms. We can predict the geometry and hybridization from any one of the resonance structures.

b.

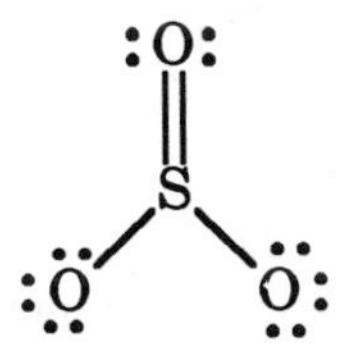

plus two other resonance structures

trigonal planar   120°
sp^2

c.

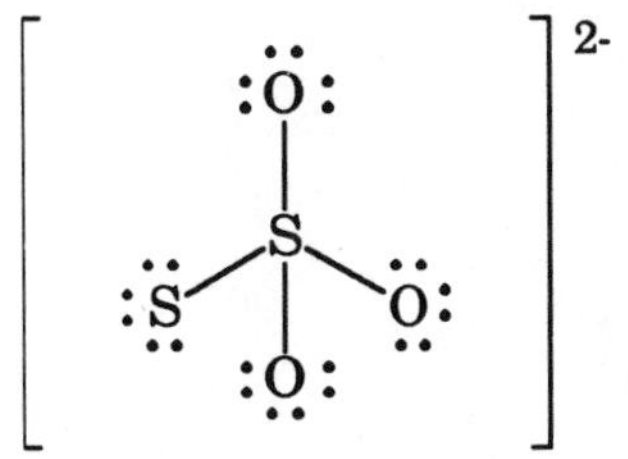

trigonal planar   120°
sp^2

tetradedral   109.5°
sp^3

d.

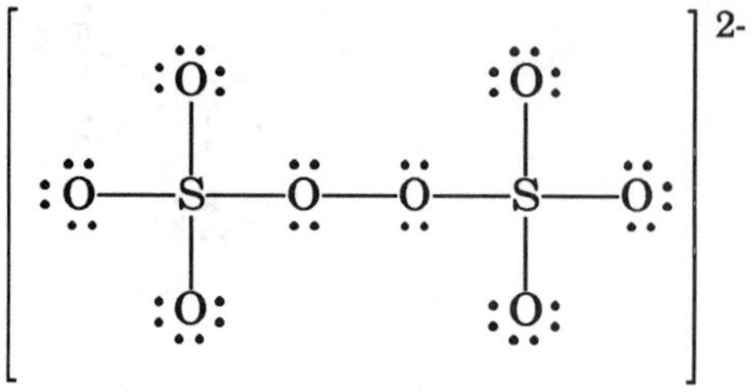

$$\left[\begin{array}{c} \ddot{O} \qquad \ddot{O} \\ \ddot{O}-S-O-O-S-\ddot{O} \\ \ddot{O} \qquad \ddot{O} \end{array} \right]^{2-}$$

Tetrahedral geometry about each S, 109.5°, sp³ hybrids; V-shaped arrangement about peroxide O's, ≈ 109.5°, sp³ hybrids

e.

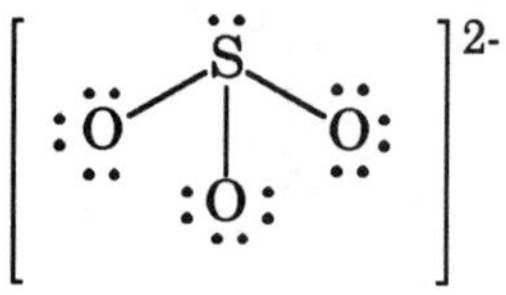

$$\left[\begin{array}{c} S \\ \ddot{O} \qquad \ddot{O} \\ \ddot{O} \end{array} \right]^{2-}$$

trigonal pyramid
≈ 109.5°
sp³

f.   g.

$$\left[\begin{array}{c} \ddot{O} \\ \ddot{O}-S-\ddot{O} \\ \ddot{O} \end{array} \right]^{2-}$$

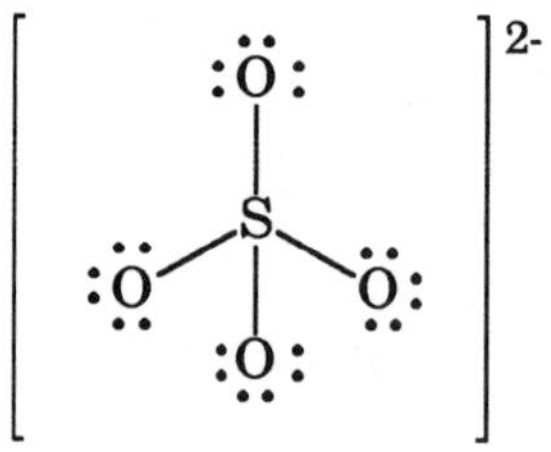

tetrahedral   109.5°   V-shaped   ≈ 109.5°
sp³   sp³

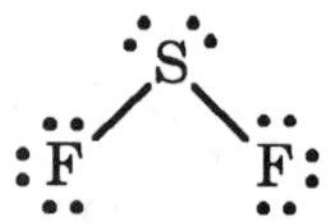

h.   i.

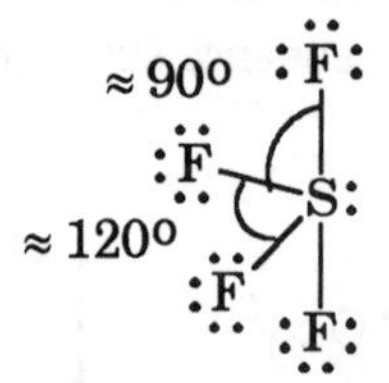

see-saw   ≈ 90°, ≈ 120°   octahedral   90°
dsp³   d²sp³

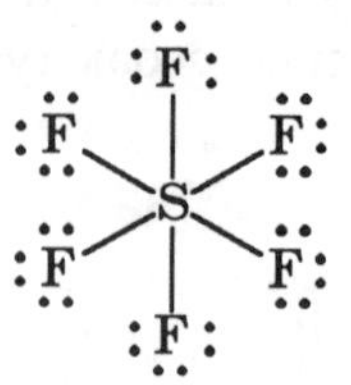

j.

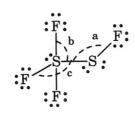

a)   ≈ 109.5°   b)   ≈ 90°   c)   ≈ 120°

See-saw about S atom with one lone pair (dsp^3);
bent about S atom with two lone pairs (sp^3)

k.

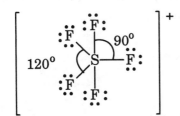

trigonal bipyramid
90° and 120°, dsp^3

25.

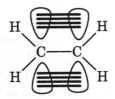

For the p-orbitals to properly line up to form the π bond, all six atoms are forced into the same plane. If the atoms are not in the same plane, then the π bond could not form since the p-orbitals would no longer be parallel to each other.

26.   No, the CH_2 planes are mutually perpendicular to each other. The center C atom is sp hybridized and is involved in two π-bonds. The p-orbitals used to form each π bond must be perpendicular to each other. This forces the two CH_2 planes to be perpendicular.

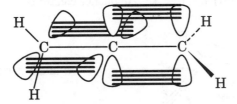

27.   To complete the Lewis structures, just add lone pairs of electrons to satisfy the octet rule for the atoms with fewer than eight electrons.

Biacetyl ($C_4H_6O_2$) has $4(4) + 6(1) + 2(6) = 34$ valence electrons.

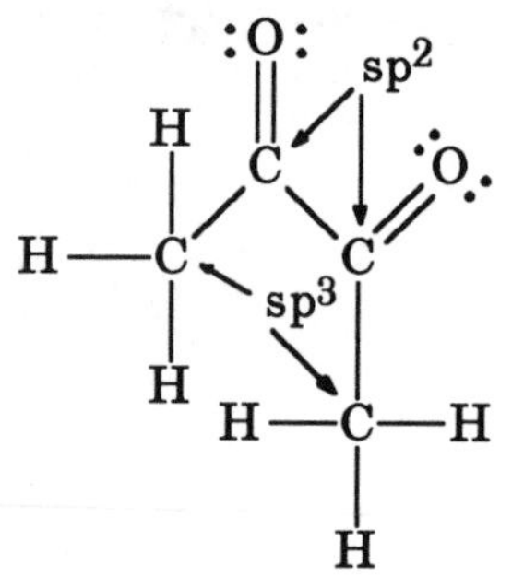

All CCO angles are 120°. The six atoms are not in the same plane because of free rotation about the carbon-carbon single (sigma) bonds. There are 11 σ and 2 π bonds in biacetyl.

Acetoin ($C_4H_8O_2$) has $4(4) + 8(1) + 2(6) = 36$ valence electrons.

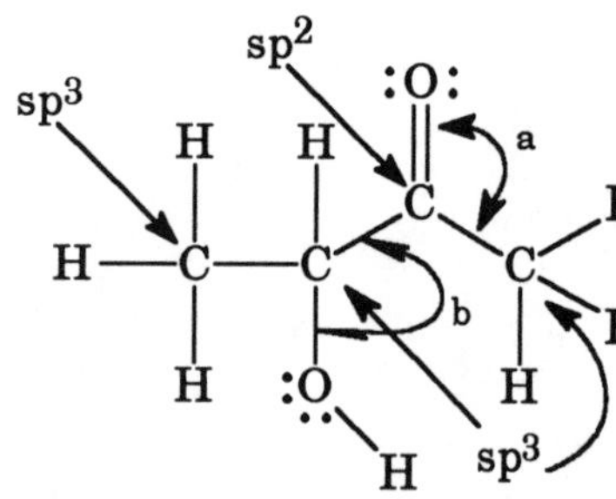

The carbon with the doubly bonded O is sp^2 hybridized. The other 3-C atoms are sp^3 hybridized. Angle a = 120° and angle b = 109.5°. There are 13 σ and 1 π bonds in acetoin.

Note: All single bonds are σ bonds, all double bonds are one σ and one π bond and all triple bonds are one σ and two π bonds.

28.   Acrylonitrile: C_3H_3N has $3(4) + 3(1) + 5 = 20$ valence electrons.

a.   120°
b.   120°
c.   180°

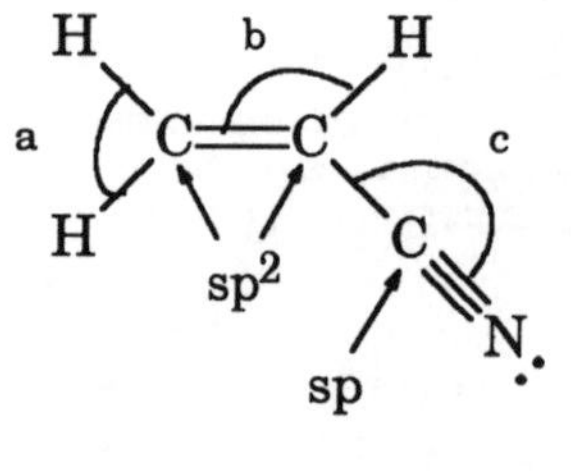

6 σ and 3 π bonds

All atoms of acrylonitrile line in the same plane. The π bond in the double bond dictates that the C and H atoms are all in the same plane and the triple bond dictates that N is in the same plane with the other atoms.

Methyl methacrylate ($C_5H_8O_2$) has $5(4) + 8(1) + 2(6) = 40$ valence electrons.

d. $120°$
e. $120°$
f. $\approx 109.5°$

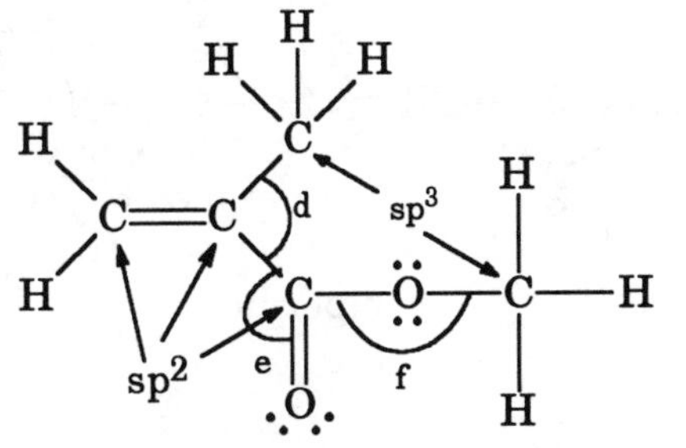

14 σ and 2 π bonds

29.   To complete the Lewis structure, just add lone pairs of electrons to satisfy the octet rule for the atoms that have fewer than eight electrons.

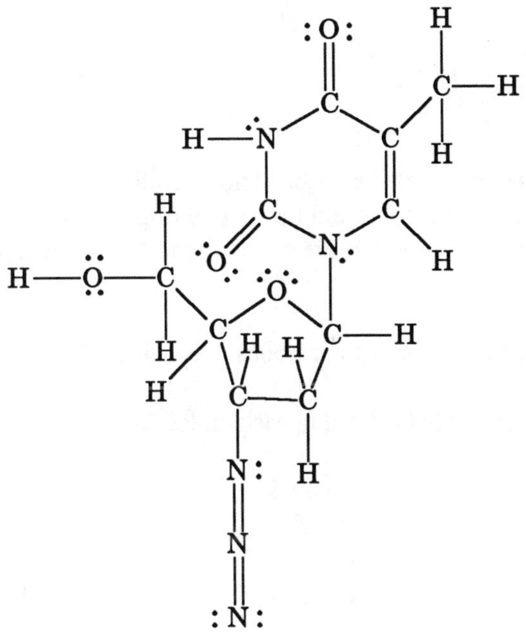

a.   6   b.   4   c.   The center N in $-N=N=N$ group

d.   33 σ   e.   5 π bonds   f. $180°$

g. $\approx 109.5°$ h. sp^3

30.   a.   Piperine and capsaicin are molecules classified as organic compounds, i.e., compounds based on carbon. The majority of Lewis structures for organic compounds have all atoms with zero formal charge. Therefore, carbon atoms in organic compounds will usually form four bonds, nitrogen atoms will form three bonds and complete the octet with one lone pair of electrons, and oxygen atoms will form two bonds and complete the octet with two lone pairs of electrons. Using these guidelines, the Lewis structures are:

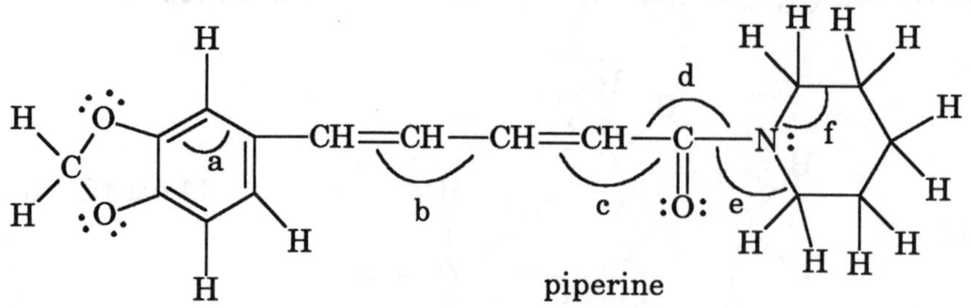

piperine

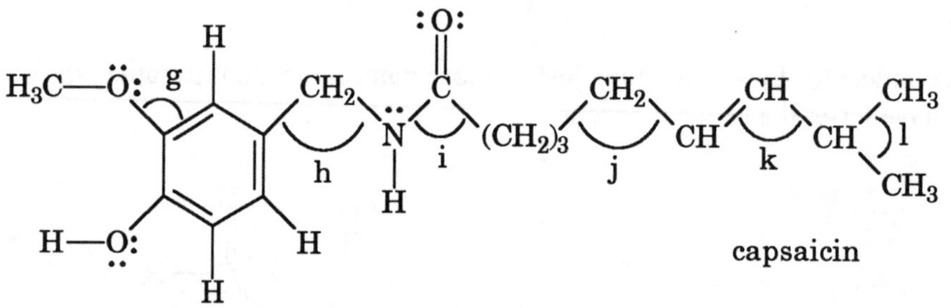

capsaicin

Note: The ring structures are all shorthand notation for rings of carbon atoms. In piperine, the first ring contains 6 carbon atoms and the second ring contains 5 carbon atoms (plus nitrogen). Also notice that CH_3, CH_2 and CH are shorthand for a carbon atoms singly bonded to hydrogen atoms.

b.   piperine: 0 sp, 11 sp^2 and 6 sp^3 carbons; capsaicin: 0 sp, 9 sp^2 and 9 sp^3 carbons

c.   The nitrogens are sp^3 hybridized in each molecule.

d. a.   120°   e.   ≈ 109.5°   i.   120°
    b.   120°   f.   109.5°   j.   109.5°
    c.   120°   g.   120°   k.   120°
    d.   120°   h.   109.5°   l.   109.5°

31.   a.   Add lone pairs to complete octets for each O and N.

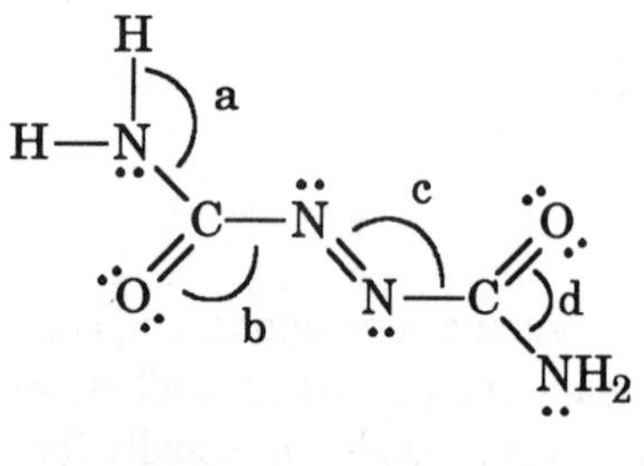

azidocarbonamide

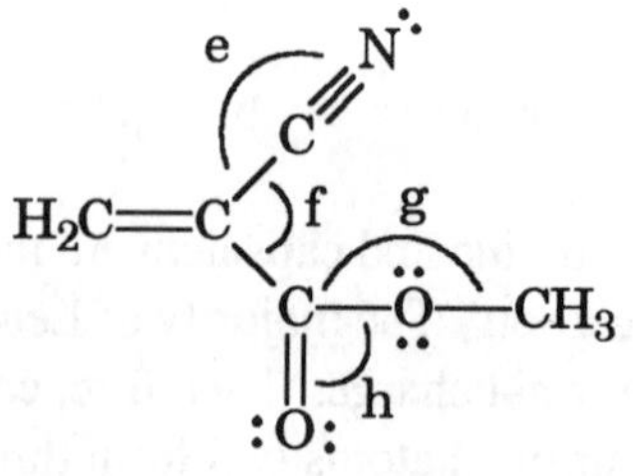

methyl cyanoacrylate

Note: NH_2, CH_2 and CH_3 are shorthand for carbon atoms singly bonded to hydrogen atoms.

b. In azidocarbonamide, the two carbon atoms are sp^2 hybridized. The two nitrogen atoms with hydrogens attached are sp^3 hybridized and the other two nitrogens are sp^2 hybridized. In methyl cyanoacrylate, the CH_3 carbon is sp^3 hybridized, the carbon with the triple bond is sp hybridized, and the other three carbons are sp^2 hybridized.

c. Azidocarbonamide contains 3π bonds and methyl cyanoacrylate contains 4π bonds.

d. a. $\approx 109.5°$ b. $120°$ c. $\approx 120°$ d. $120°$ e. $180°$

   f. $120°$ g. $\approx 109.5°$ h. $120°$

32. a. To complete the Lewis structure, add two lone pairs to each sulfur atom.

b. See Lewis structure. The four carbon atoms in the ring are all sp^2 hybridized and the two sulfur atoms are sp^3 hybridized.

c. 23σ and 9π bonds. Note: CH_3, CH_2 and CH are shorthand for carbon atoms singly bonded to hydrogen atoms.

The Molecular Orbital Model

33. If we calculate a non-zero bond order for a molecule, then we predict that it can exist (is stable).

a. H_2^+: $(\sigma_{1s})^1$ B.O. = (1-0)/2 = 1/2, stable
   H_2: $(\sigma_{1s})^2$ B.O. = (2-0)/2 = 1, stable
   H_2^-: $(\sigma_{1s})^2(\sigma_{1s}*)^1$ B.O. = (2-1)/2 = 1/2, stable
   H_2^{2-}: $(\sigma_{1s})^2(\sigma_{1s}*)^2$ B.O. = (2-2)/2 = 0, not stable

b. He_2^{2+}: $(\sigma_{1s})^2$ B.O. = (2-0)/2 = 1, stable
   He_2^+: $(\sigma_{1s})^2(\sigma_{1s}*)^1$ B.O. = (2-1)/2 = 1/2, stable
   He_2: $(\sigma_{1s})^2(\sigma_{1s}*)^2$ B.O. = (2-2)/2 = 0, not stable

34. a. N_2^{2-}: $(\sigma_{2s})^2(\sigma_{2s}*)^2(\pi_{2p})^4(\sigma_{2p})^2(\pi_{2p}*)^2$ B.O. = (8-4)/2 = 2, stable
   O_2^{2-}: $(\sigma_{2s})^2(\sigma_{2s}*)^2(\sigma_{2p})^2(\pi_{2p})^4(\pi_{2p}*)^4$ B.O. = (8-6)/2 = 1, stable
   F_2^{2-}: $(\sigma_{2s})^2(\sigma_{2s}*)^2(\sigma_{2p})^2(\pi_{2p})^4(\pi_{2p}*)^4(\sigma_{2p}*)^2$ B.O. = (8-8)/2 = 0, not stable

b. Be_2: $(\sigma_{2s})^2(\sigma_{2s}*)^2$ B.O. = (2-2)/2 = 0, not stable
   B_2: $(\sigma_{2s})^2(\sigma_{2s}*)^2(\pi_{2p})^2$ B.O. = (4-2)/2 = 1, stable
   Li_2: $(\sigma_{2s})^2$ B.O. = (2-0)/2 = 1, stable

35.   The electron configurations are:

a. H_2: $(\sigma_{1s})^2$ B.O. = (2-0)/2 = 1, diamagetic (0 unpaired e⁻)

b. B_2: $(\sigma_{2s})^2(\sigma_{2s}*)^2(\pi_{2p})^2$ B.O. = (4-2)/2 = 1, paramagnetic (2 unpaired e⁻)

c. F_2: $(\sigma_{2s})^2(\sigma_{2s}*)^2(\sigma_{2p})^2(\pi_{2p})^4(\pi_{2p}*)^4$ B.O. = (8-6)/2 = 1, diamagnetic (0 unpaired e⁻)

36.   The electron configurations are:

a. N_2: $(\sigma_{2s})^2(\sigma_{2s}*)^2(\pi_{2p})^4(\sigma_{2p})^2$ B.O. = 3, diamagnetic

b. N_2^+: $(\sigma_{2s})^2(\sigma_{2s}*)^2(\pi_{2p})^4(\sigma_{2p})^1$ B.O. = 2.5, paramagnetic

c. N_2^-: $(\sigma_{2s})^2(\sigma_{2s}*)^2(\pi_{2p})^4(\sigma_{2p})^2(\pi_{2p}*)^1$ B.O. = 2.5, paramagnetic

37.   The electron configurations are:

O_2^+: $(\sigma_{2s})^2(\sigma_{2s}*)^2(\sigma_{2p})^2(\pi_{2p})^4(\pi_{2p}*)^1$

O_2: $(\sigma_{2s})^2(\sigma_{2s}*)^2(\sigma_{2p})^2(\pi_{2p})^4(\pi_{2p}*)^2$

O_2^-: $(\sigma_{2s})^2(\sigma_{2s}*)^2(\sigma_{2p})^2(\pi_{2p})^4(\pi_{2p}*)^3$

O_2^{2-}: $(\sigma_{2s})^2(\sigma_{2s}*)^2(\sigma_{2p})^2(\pi_{2p})^4(\pi_{2p}*)^4$

	O_2^+	O_2	O_2^-	O_2^{2-}
Bond order	2.5	2	1.5	1
# of unpaired electrons	1	2	1	0

Bond energy: $O_2^{2-} < O_2^- < O_2 < O_2^+$; Bond length: $O_2^+ < O_2 < O_2^- < O_2^{2-}$

Bond energy is directly proportional to bond order and bond length is inversely proportional to bond order.

38.   Bond orders: N_2, 3; N_2^+, 2.5; N_2^-, 2.5

Bond length: $N_2 < N_2^+ \approx N_2^-$;   Bond energy: $N_2^- \approx N_2^+ < N_2$

We would expect the strengths and lengths of the bonds in N_2^+ and N_2^- to be close to each other since both have bond orders of 2.5. Since there are more electrons in N_2^- than N_2^+, electron repulsions will be greater in N_2^- than in N_2^+ and the bond in N_2^- probably is a little longer and a little weaker than the bond in N_2^+.

39. C_2^{2-} has 10 valence electrons. The Lewis structure predicts sp hybridization for each carbon with two unhybridized p orbitals on each carbon.

$$\left[:C{\equiv}C: \right]^{2-}$$

sp hybrids orbitals form the σ bond and the two unhybridized p atomic orbitals from each carbon form the two π bonds.

MO: $(\sigma_{2s})^2(\sigma_{2s}{}^*)^2(\pi_{2p})^4(\sigma_{2p})^2$, B.O. = (8 - 2)/2 = 3

Both give the same picture, a triple bond composed of a σ and two π-bonds. Both predict the ion to be diamagnetic. Lewis structures deal well with diamagnetic (all electrons paired) species. The Lewis model cannot really predict magnetic properties.

40.

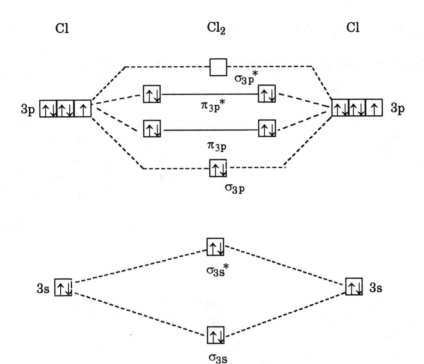

The bond order for Cl_2 is 1 and since all electrons are paired, we would expect Cl_2 to be diamagnetic.

41.   The electron configurations are:

a. CN^+: $(\sigma_{2s})^2(\sigma_{2s}{}^*)^2(\pi_{2p})^4$ B.O. = (6-2)/2 = 2, diamagnetic
b. CN: $(\sigma_{2s})^2(\sigma_{2s}{}^*)^2(\pi_{2p})^4(\sigma_{2p})^1$ B.O. = (7-2)/2 = 2.5, paramagnetic
c. CN^-: $(\sigma_{2s})^2(\sigma_{2s}{}^*)^2(\pi_{2p})^4(\sigma_{2p})^2$ B.O. = 3, diamagnetic

42.   The electron configurations are (assuming the same orbital order as that for N_2):

a. NO^+: $(\sigma_{2s})^2(\sigma_{2s}{}^*)^2(\pi_{2p})^4(\sigma_{2p})^2$ B.O. = (8-2)/2 = 3, diamagnetic
b. NO: $(\sigma_{2s})^2(\sigma_{2s}{}^*)^2(\pi_{2p})^4(\sigma_{2p})^2(\pi_{2p}{}^*)^1$ B.O. = (8-3)/2 = 2.5, paramagnetic
c. NO^-: $(\sigma_{2s})^2(\sigma_{2s}{}^*)^2(\pi_{2p})^4(\sigma_{2p})^2(\pi_{2p}{}^*)^2$ B.O. = (8-4)/2 = 2, paramagnetic

43.   The bond orders are: CN$^+$, 2; CN, 2.5; CN$^-$, 3; Since bond order is directly proportional to bond energy and inversely proportional to bond length, then:

shortest $\rightarrow$ longest bond length: CN$^-$ < CN < CN$^+$

lowest $\rightarrow$ highest bond energy: CN$^+$ < CN < CN$^-$

44.   The bond orders are: NO$^+$, 3; NO, 2.5; NO$^-$, 2

Shortest $\rightarrow$ longest bond length: NO$^+$ < NO < NO$^-$

Lowest $\rightarrow$ highest bond energy: NO$^-$ < NO < NO$^+$

45.   The two types of overlap that result in bond formation for p orbitals are side to side overlap (π bond) and head to head overlap (σ bond).

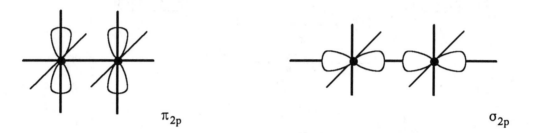

46.

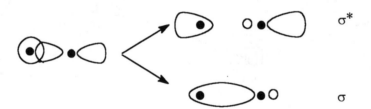

These molecular orbitals are sigma MOs since the electron density is cylindrically symmetric about the internuclear axis.

47.   a.   The electron density would be closer to F on the average. The F atom is more electronegative than the H atom and the 2p orbital of F is lower in energy than the 1s orbital of H.

b.   The bonding MO would have more fluorine 2p character since it is closer in energy to the fluorine 2p atomic orbital.

c.   The antibonding MO would place more electron density closer to H and would have a greater contribution from the higher energy hydrogen 1s atomic orbital.

48.   a.   The antibonding MO will have more hydrogen 1s character since the hydrogen 1s atomic orbital is closer in energy to the antibonding MO.

b.   No, the overall overlap is zero. The p_x orbital does not have proper symmetry to overlap with a 1s orbital. The $2p_x$ and $2p_y$ orbitals are called nonbonding orbitals.

c.

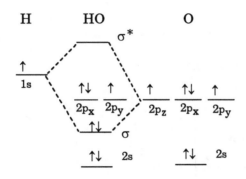

d.   Bond order $= \dfrac{2-0}{2} = 1$; Note: The 2s, $2p_x$, and $2p_y$ electrons have no effect on the bond order.

e.   To form OH$^+$ a nonbonding electron is removed from OH. Since the number of bonding electrons and antibonding electrons are unchanged, then the bond order is still equal to one.

49. O_3 and NO_2^- are isoelectronic, so we only need consider one of them since the same bonding ideas apply to both. The Lewis structures for O_3 are:

For each of the two resonance forms, the central O atom is sp^2 hybridized with one unhybridized p atomic orbital. The sp^2 hybrid orbitals are used to form the two sigma bonds to the central atom. The localized electron view of the π bond utilizes unhybridized p atomic orbitals. The π bond resonates between the two positions in the Lewis structures:

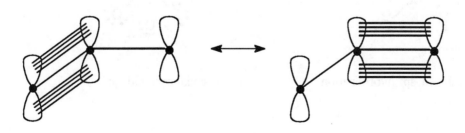

In the MO picture of the π bond, all three unhybridized p-orbitals overlap at the same time, resulting in π electrons that are delocalized over the entire surface of the molecule. This is represented as:

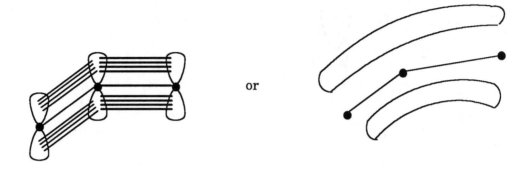

50.   The Lewis structures for CO_3^{2-} are (24 e⁻):

$$\left[\begin{array}{c} :O: \\ \| \\ C \\ :O: \quad :O: \end{array} \right]^{2-} \longleftrightarrow \left[\begin{array}{c} :O: \\ \| \\ C \\ \cdot O: \quad :O: \end{array} \right]^{2-} \longleftrightarrow \left[\begin{array}{c} :O: \\ \| \\ C \\ :O: \quad :O \cdot \end{array} \right]^{2-}$$

In the localized electron view, the central carbon atom is sp² hybridized which are used to form the three sigma bonds in CO_3^{2-}. The central C atom also has one unhybridized p atomic orbital which overlaps with another p atomic orbital from one of the oxygen atoms to form the π bond in each resonance structure. This localized π bond moves (resonates) from one position to another. In the molecular orbital model for CO_3^{2-}, all four atoms in CO_3^{2-} have a p atomic orbital which is perpendicular to the plane of the ion. All four of these p orbitals overlap at the same time to form a delocalized π bonding system where the π electrons can roam over the entire surface of the ion. The π molecular orbital system for CO_3^{2-} is analogous to that for NO_3^- which is shown in Figure 9.49 of the text.

Additional Exercises

51.   a.   FClO, $7 + 7 + 6 = 20$ e⁻   b. $FClO_2$, $7 + 7 + 2(6) = 26$ e⁻

V-shaped, sp³ hybridization   trigonal pyramid, sp³

c. $FClO_3$, $7 + 7 + 3(6) = 32$ e⁻   d. F_3ClO, $3(7) + 7 + 6 = 34$ e⁻

tetrahedral, sp^3 see-saw, dsp^3

e. F_3ClO_2, $3(7) + 7 + 2(6) = 40$ valence e⁻

trigonal bipyramid, dsp^3

Note: Two additional Lewis structures are possible, depending on the positions of the oxygen atoms.

52. $FClO_2 + F^- \rightarrow F_2ClO_2^-$ $F_3ClO + F^- \rightarrow F_4ClO^-$

$F_2ClO_2^-$, $2(7) + 7 + 2(6) + 1 = 34$ e⁻ F_4ClO^-, $4(7) + 7 + 6 + 1 = 42$ e⁻

see-saw, dsp^3 square pyramid, d^2sp^3

$F_3ClO \rightarrow F^- + F_2ClO^+$ $F_3ClO_2 \rightarrow F^- + F_2ClO_2^+$

F_2ClO^+, $2(7) + 7 + 6 - 1 = 26$ e⁻ $F_2ClO_2^+$, $2(7) + 7 + 2(6) - 1 = 32$ e⁻

trigonal pyramid, sp^3 tetrahedral, sp^3

53.   a.   There are 33 σ and 9 π bonds.

      b.   All C atoms are sp² hybridized since all have a trigonal planar arrangement of the electrons.

54.

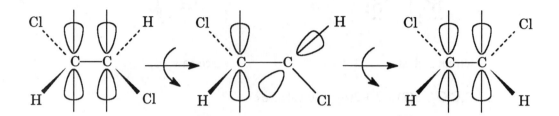

      In order to rotate about the double bond, the molecule must go through an intermediate stage where
      the π bond is broken while the sigma bond remains intact. Bond energies are 347 kJ/mol for C – C
      and 614 kJ/mol for C = C. If we take the single bond as the strength of the σ bond, then the strength
      of the π bond is (614 - 347 =) 267 kJ/mol. Thus, 267 kJ/mol must be supplied to rotate about a
      carbon-carbon double bond.

55.   a. $COCl_2$ has $4 + 6 + 2(7) = 24$ valence electrons.

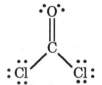

                                                      trigonal planar
                                                      polar
                                                      120°
                                                      sp²

      b. N_2F_2 has $2(5) + 2(7) = 24$ valence electrons.

                                                                                    Can also be:

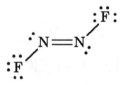

                          V-shaped about both Ns;
                          ≈ 120° about both Ns;
                          Both Ns: sp²

            polar   nonpolar

      These are distinctly different molecules.

      c.   COS has $4 + 6 + 6 = 16$ valence electrons.

            :Ö=C=S: linear, polar, 180°, sp

d. ICl_3 has $7 + 3(7) = 28$ valence electrons.

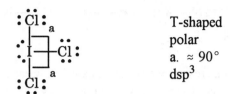

T-shaped
polar
a. $\approx 90°$
dsp^3

56.   a.   Yes, both have 4 sets of electrons about the P. We would predict a tetrahedral structure for both.
            See below for Lewis structures.

       b.   The hybridization is sp^3 for each P since both structures are tetrahedral.

       c.   P has to use one of its d orbitals to form the π bond since the p orbitals are all used to form the
            hybrid orbitals.

       d.   Formal charge = number of valence electrons of an atom - [(number of lone pair electrons) +
            1/2 (number of shared electrons)]. The formal charges calculated for the O and P atoms are next
            to the atoms in the following Lewis structures.

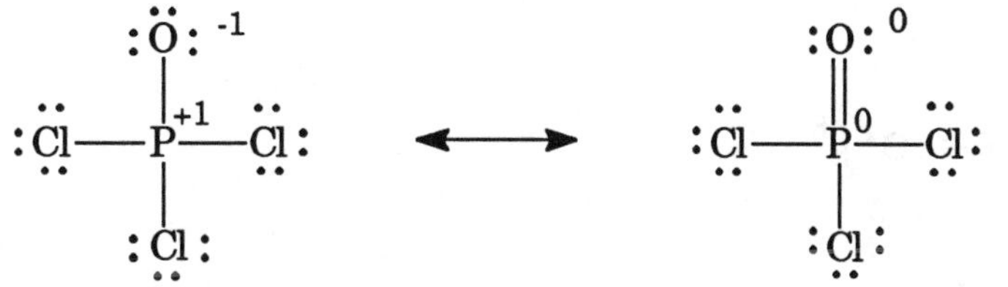

            In both structures, the formal charges of the Cl atoms are all zeros. The structure with the P $=$ O
            bond is favored on the basis of formal charge since it has a zero formal charge for all atoms.

57.   a.   The Lewis structures for NNO and NON are:

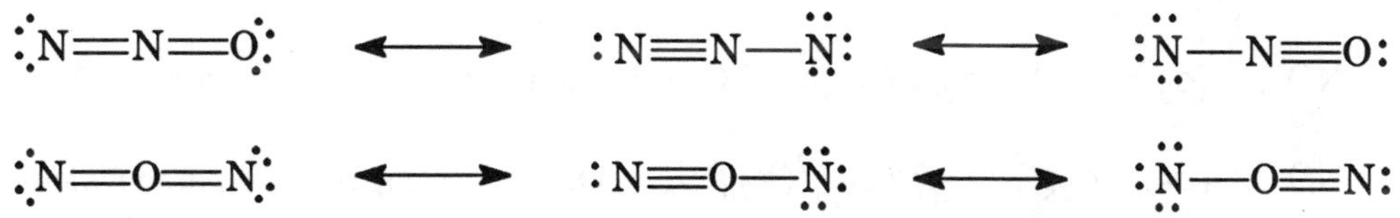

            The NNO structure is correct. From the Lewis structures we would predict both NNO and NON
            to be linear. However, we would predict NNO to be polar and NON to be nonpolar. Since
            experiments show N_2O to be polar, then NNO is the correct structure.

       b.   Formal charge = number of valence electrons of atoms - [(number of lone pair electrons) +
            1/2 (number of shared electrons)].

$$:N{=}N{=}O: \longleftrightarrow :N{\equiv}N{-}\ddot{O}: \longleftrightarrow :\ddot{N}{-}N{\equiv}O:$$

$$\;\;-1\;\;\;\;+1\;\;\;\;0\qquad\qquad 0\;\;\;\;+1\;\;\;\;-1\qquad\qquad -2\;\;\;\;+1\;\;\;\;+1$$

The formal charges for the atoms in the various resonance structures are below each atom. The central N is sp hybridized in all of the resonance structures. We can probably ignore the 3rd resonance structure on the basis of the relatively large formal charges as compared to the first two resonance structures.

c.   The sp hybrid orbitals on the center N overlap with atomic orbitals (or hybrid orbitals) on the other two atoms to form the two sigma bonds. The remaining two unhybridized p orbitals on the center N overlap with two p orbitals on the peripheral N to form the two π bonds.

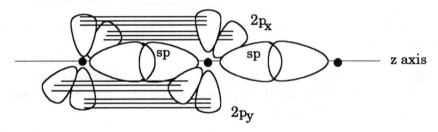

58.   Lewis structures:

NO^+: $\left[:N{\equiv}O: \right]^+$

NO: $\ddot{N}{=}\ddot{O} \longleftrightarrow \ddot{N}{=}\ddot{O} \longleftrightarrow \dot{N}{=}\dot{O}$

NO^-: $\left[\ddot{N}{=}\ddot{O} \right]^-$

M.O. model:

NO^+: $(\sigma_{2s})^2(\sigma_{2s}*)^2(\pi_{2p})^4(\sigma_{2p})^2$,   B.O. = 3, 0 unpaired e$^-$ (diamagnetic)

NO: $(\sigma_{2s})^2(\sigma_{2s}*)^2(\pi_{2p})^4(\sigma_{2p})^2(\pi_{2p}*)^1$,   B.O. = 2.5, 1 unpaired e$^-$ (paramagnetic)

NO^-: $(\sigma_{2s})^2(\sigma_{2s}*)^2(\pi_{2p})^4(\sigma_{2p})^2(\pi_{2p}*)^2$ B.O. = 2, 2 unpaired e$^-$ (paramagnetic)

The two models only give the same results for NO^+ (a triple bond with no unpaired electrons). Lewis structures are not adequate for NO and NO^-. The MO model gives a better representation for all three species. For NO, Lewis structures are poor for odd electron species. For NO^-, both models predict a double bond but only the MO model correctly predicts that NO^- is paramagnetic.

59. N_2 (ground state): $(\sigma_{2s})^2(\sigma_{2s}*)^2(\pi_{2p})^4(\sigma_{2p})^2$, B.O. = 3, diamagnetic (0 unpaired e$^-$)

       N_2 (1st excited state): $(\sigma_{2s})^2(\sigma_{2s}*)^2(\pi_{2p})^4(\sigma_{2p})^1(\pi_{2p}*)^1$

              B.O. = (7-3)/2 = 2, paramagnetic (2 unpaired e$^-$)

       The first excited state of N_2 should have a weaker bond and should be paramagnetic.

60.   Considering only the twelve valence electrons in O_2, the MO models would be:

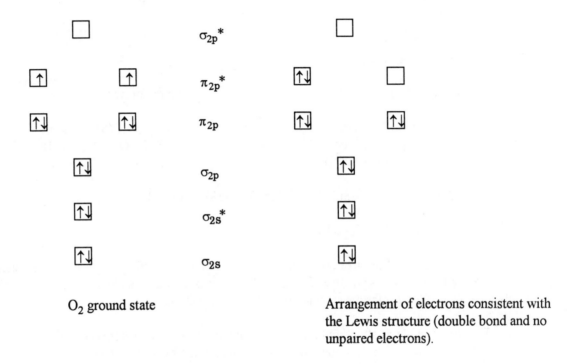

       O_2 ground state

Arrangement of electrons consistent with
the Lewis structure (double bond and no
unpaired electrons).

       It takes energy to pair electrons in the same orbital. Thus, the structure with no unpaired electrons is
       at a higher energy; it is an excited state.

Challenge Problems

61.   a.   No, some atoms are in different places. Thus, these are not resonance structures; they are
             different compounds.

       b.   For the first Lewis structure, all nitrogens are sp^3 hybridized and all carbons are sp^2 hybridized.
             In the second Lewis structure, all nitrogens and carbons are sp^2 hybridized.

c. For the reaction:

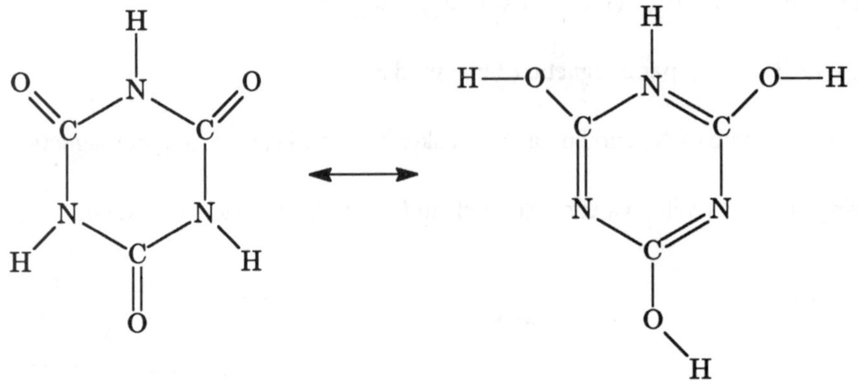

Bonds broken:

3 C = O (745 kJ/mol)
3 C – N (305 kJ/mol)
3 N – H (391 kJ/mol)

Bonds formed:

3 C = N (615 kJ/mol)
3 C – O (358 kJ/mol)
3 O – H (467 kJ/mol)

$\Delta H = 3(745) + 3(305) + 3(391) - [3(615) + 3(358) + 3(467)]$

$\Delta H = 4323$ kJ - 4320 kJ = 3 kJ

The bonds are slightly stronger in the first structure with the carbon-oxygen double bonds since ΔH for the reaction is positive. However, the value of ΔH is so small that the best conclusion is that the bond strengths are comparable in the two structures.

62. a. The complete Lewis structure follows. All but two of the carbon atoms are sp^3 hybridized. The two carbon atoms which contain the double bond are sp^2 hybridized (see *).

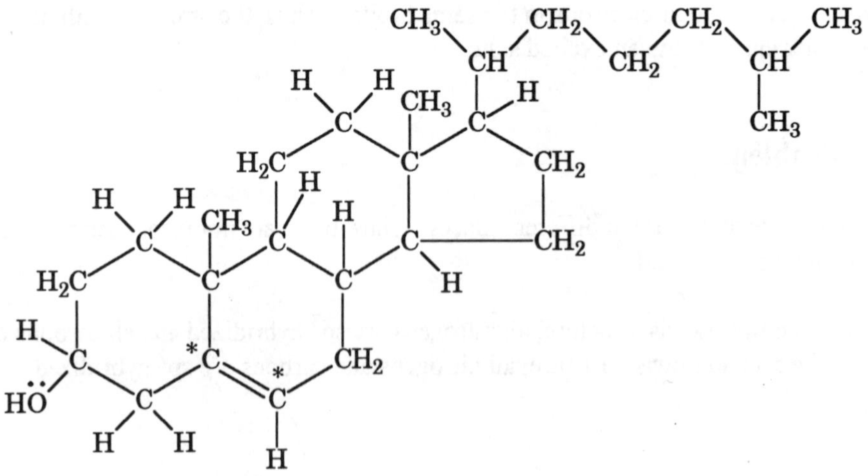

No; most of the carbons are not in the same plane since a majority of carbon atoms exhibit tetrahedral structure.

63.   a. NCN^{2-} has $5 + 4 + 5 + 2 = 16$ valence electrons.

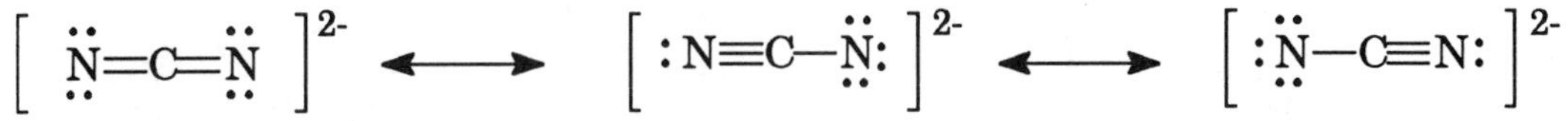

H_2NCN has $2(1) + 5 + 4 + 5 = 16$ valence electrons.

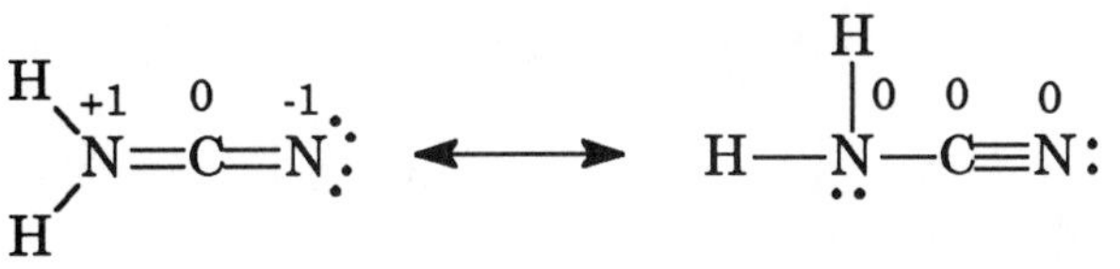

favored by formal charge

$NCNC(NH_2)_2$ has $5 + 4 + 5 + 4 + 2(5) + 4(1) = 32$ valence electrons.

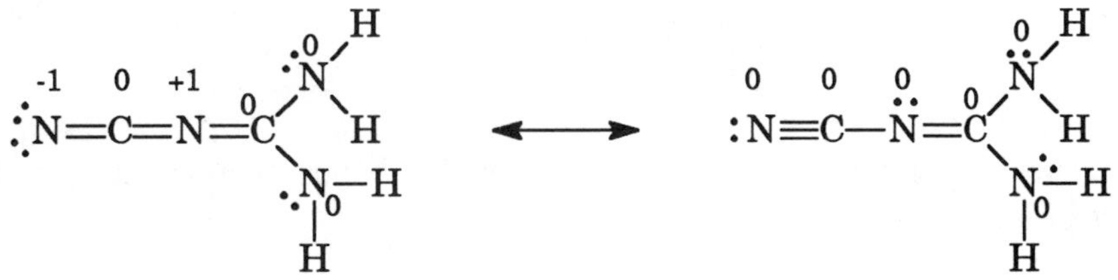

favored by formal charge

Melamine ($C_3N_6H_6$) has $3(4) + 6(5) + 6(1) = 48$ valence electrons.

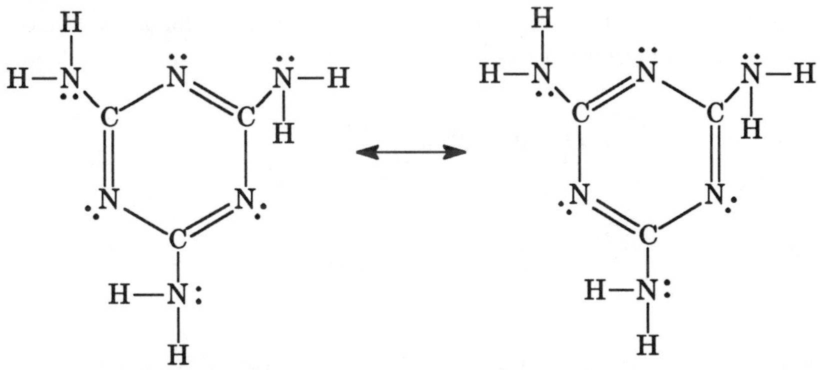

b. NCN^{2-}: C is sp hybridized. Depending on the resonance form, N can be sp, sp^2, or sp^3 hybridized. For the remaining compounds, we will give hybrids for the favored resonance structures as predicted from formal charge considerations.

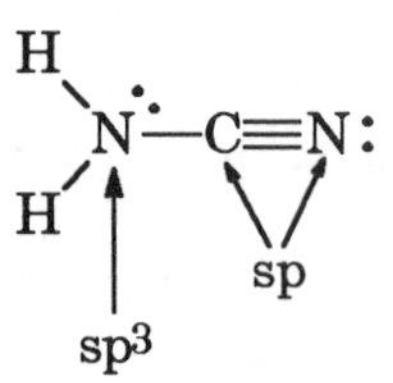

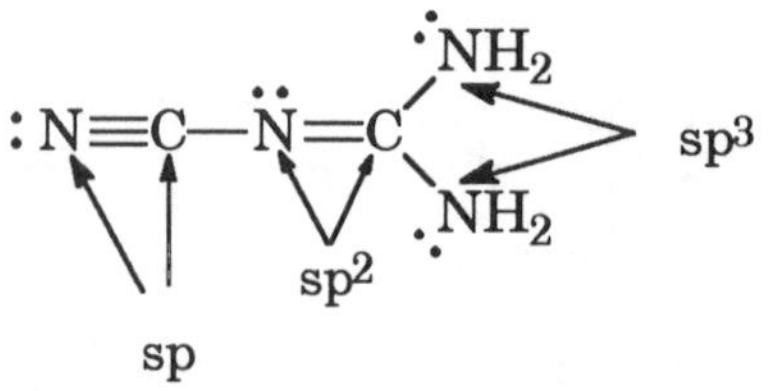

Melamine: N in NH_2 groups are all sp^3 hybridized. Atoms in ring are all sp^2 hybridized.

c. NCN^{2-}: 2 σ and 2 π bonds; H_2NCN: 4 σ and 2 π bonds; dicyandiamide: 9 σ and 3 π bonds; melamine: 15 σ and 3 π bonds

d.   The π-system forces the ring to be planar just as the benzene ring is planar.

e.   The structure:

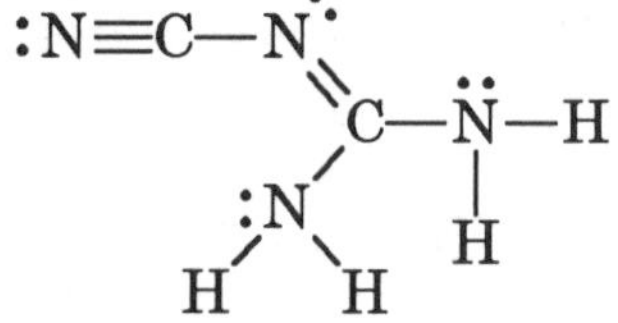

is the most important since it has three different CN bonds. This structure is also favored on the basis of formal charge.

64.   a.   The electron removed from N_2 is in the σ_{2p} molecular orbital which is lower in energy than the 2p atomic orbital from which the electron in atomic nitrogen is removed. Since the electron removed from N_2 is lower in energy than the electron in N, the ionization energy of N_2 should be greater than for N.

b. F_2 should have a lower ionization energy than F. The electron removed from F_2 is in a π_{2p}^* antibonding molecular orbital which is higher in energy than the 2p atomic orbitals. Thus, it is easier to remove an electron from F_2 than from F.

65. $$E = \frac{hc}{\lambda} = \frac{(6.626 \times 10^{-34} \text{ J s}) (2.998 \times 10^8 \text{ m/s})}{25 \times 10^{-9} \text{ m}} = 7.9 \times 16^{-18} \text{ J}$$

$$7.9 \times 10^{-18} \text{ J} \times \frac{6.022 \times 10^{23}}{\text{mol}} \times \frac{1 \text{ kJ}}{1000 \text{ J}} = 4800 \text{ kJ/mol}$$

Using Δ H values from the various reactions, 25 nm light has sufficient energy to ionize N_2 and N, and to break the triple bond. Thus, N_2, N_2^+, N, and N^+ will all be present, assuming excess N_2.

To produce atomic nitrogen but no ions, the range of energies of the light must be from 941 kJ/mol to just below 1402 kJ/mol.

$$\frac{941 \text{ kJ}}{\text{mol}} \times \frac{1 \text{ mol}}{6.022 \times 10^{23}} \times \frac{1000 \text{ J}}{\text{kJ}} = 1.56 \times 10^{-18} \text{ J/photon}$$

$$\lambda = \frac{hc}{E} = \frac{(6.626 \times 10^{-34} \text{ J s}) (2.998 \times 10^8 \text{ m/s})}{1.56 \times 10^{-18} \text{ J}} = 1.27 \times 10^{-7} \text{ m} = 127 \text{ nm}$$

$$\frac{1402 \text{ kJ}}{\text{mol}} \times \frac{1 \text{ mol}}{6.0221 \times 10^{23}} \times \frac{1000 \text{ J}}{\text{kJ}} = 2.328 \times 10^{-18} \text{ J/photon}$$

$$\lambda = \frac{hc}{E} = \frac{(6.6261 \times 10^{-34} \text{ J s}) (2.9979 \times 10^8 \text{ m/s})}{2.328 \times 10^{-18} \text{ J}} = 8.533 \times 10^{-8} \text{ m} = 85.33 \text{ nm}$$

Light with wavelengths in the range of 85.33 nm $< \lambda \leq$ 127 nm will produce N but no ions.

66.   The π bonds between S atoms and between C and S atoms are not as strong. The orbitals do not overlap with each other as well as the smaller atomic orbitals of C and O overlap.

67.   O=N–Cl: The bond order of the NO bond in NOCl is 2 (a double bond).

        NO: The bond order of this NO bond is 2.5 (see Exercise 9.42).

Both reactions apparently only involve the breaking of the N–Cl bond. However, in the reaction: ONCl → NO + Cl some energy is released in forming the stronger NO bond, lowering the value of ΔH. Therefore, the apparent N–Cl bond energy is artifically low for this reaction. The first reaction only involves the breaking of the N–Cl bond.

CHAPTER TEN

LIQUIDS AND SOLIDS

Questions

6. There is an electrostatic attraction between the permanent dipoles of polar molecules. The greater the polarity, the greater the attraction among molecules.

7. London dispersion (LD) < dipole-dipole < H-bonding < metallic bonding, covalent network, ionic.

    Yes, there is considerable overlap. Consider some of the examples in Exercise 10.86. Benzene (only LD forces) has a higher boiling point than acetone (dipole-dipole forces). Also, there is even more overlap between the stronger forces (metallic, covalent, and ionic).

8. As the size of the molecule increases, the strength of the London dispersion forces also increases. As the electron cloud gets larger, it is easier for the electrons to be drawn away from the nucleus (more polarizable).

9. As the strengths of interparticle forces increase: surface tension, viscosity, melting point and boiling point increase, while the vapor pressure decreases.

10. Dipole forces are generally weaker than hydrogen bonding. They are similar in that they arise from an unequal sharing of electrons. We can look at hydrogen bonding as a particularily strong dipole force.

11. a. Polarizability of an atom refers to the ease of distorting the electron cloud. It can also refer to distorting the electron clouds in molecules or ions. Polarity refers to the presence of a permanent dipole moment in a molecule.

    b. London dispersion forces are present in all substances. LD forces can be referred to as accidental dipole - induced dipole forces. Dipole - dipole forces involve the attraction of molecules with permanent dipoles for each other.

    c. inter: between; intra: within; For example, in Br_2 the covalent bond is an intramolecular force, holding the two Br-atoms together in the molecule. The much weaker London dispersion forces are the intermolecular forces of attraction which hold different molecules of Br_2 together in the liquid phase.

12.   Liquids and solids both have characteristic volume and are not very compressible. Liquids and gases flow and assume the shape of their container.

13.   Atoms have an approximately spherical shape (on the average). It is impossible to pack spheres together without some empty space between the spheres.

14.   Critical temperature: The temperature above which a liquid cannot exist, i.e., the gas cannot be liquified by increased pressure.

        Critical pressure: The pressure that must be applied to a substance at its critical temperature to produce a liquid.

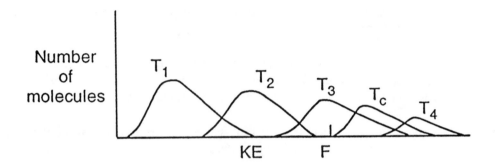

        The kinetic energy distribution changes as one raises the temperature ($T_4 > T_c > T_3 > T_2 > T_1$). At the critical temperature, T_c, all molecules have kinetic energies greater than the intermolecular forces, F, and a liquid can't form. Note: The distributions above are not to scale.

15.   As the intermolecular forces increase, the critical temperature increases.

16.   Evaporation takes place when some molecules at the surface of a liquid have enough energy to break the intermolecular forces holding them in the liquid phase. When a liquid evaporates, the molecules that escape have high kinetic energies. The average kinetic energy of the remaining molecules is lower, thus, the temperature of the liquid is lower.

17. a. Crystalline solid: Regular, repeating structure

         Amorphous solid: Irregular arrangement of atoms or molecules

     b. Ionic solid: Made up of ions held together by ionic bonding.

         Molecular solid: Made up of discrete covalently bonded molecules held together in the solid phase by weaker forces (LD, dipole or hydrogen bonds).

     c. Molecular solid: Discrete, individual molecules

         Covalent network solid: No discrete molecules; A covalent network solid is one large molecule. The interparticle forces are the covalent bonds between atoms.

d. Metallic solid: Completely delocalized electrons, conductor of electricity (ions in a sea of electrons)

Covalent network solid: Localized electrons; Insulator or semiconductor

18. A crystalline solid because a regular, repeating arrangement is necessary to produce planes of atoms that will diffract the x-rays in regular patterns. An amorphous solid does not have a regular repeating arrangement and will produce a complicated diffraction pattern.

19. No, an example is common glass which is primarily amorphous SiO_2 (a covalent network solid) as compared to ice (a crystalline solid held together by weaker H-bonds). The intermolecular forces in the amorphous solid in this case are stronger than those in the crystalline solid. Whether or not a solid is amorphous or crystalline depends on the long range order in the solid and not on the strengths of the intermolecular forces.

20. Conductor: The energy difference between the filled and unfilled molecular orbitals is minimal. We call this energy difference the band gap. Since the band gap is minimal, electrons can easily move into the conduction bands (the unfilled molecular orbitals).

Insulator: Large band gap; Electrons do not move from the filled molecular orbitals to the conduction bands since the energy difference is large.

Semiconductor: Small band gap; Since the energy difference between the filled and unfilled molecular orbitals is smaller than in insulators, some electrons can jump into the conduction bands. The band gap, however, is not as small as with conductors, so semiconductors have intermediate conductivity.

21. a. As the temperature is increased, more electrons in the filled molecular orbitals have sufficient kinetic energy to jump into the conduction bands (the unfilled molecular orbitals).

b. A photon of light is absorbed by an electron which then has sufficient energy to jump into the conduction bands.

c. An impurity either adds electrons at an energy near that of the conduction bands (n-type) or creates holes (empty energy levels) at energies in the previously filled molecular orbitals (p-type).

22. In conductors the electrical conductivity is inversely proportional to the temperature. Increases in temperature increases the motions of the atoms which gives rise to increased resistance (decreased conductivity). In a semiconductor the electrical conductivity is directly proportional to the temperature. An increase in temperature provides more electrons with enough kinetic energy to jump from the filled molecular orbitals to the conduction bands, increasing conductivity.

23. a. Condensation: vapor → liquid   b. Evaporation: liquid → vapor

c. Sublimation: solid → vapor

d. A supercooled liquid is a liquid which is at a temperature below its freezing point.

24.   Equilibrium: There is no change in composition; the vapor pressure is constant.

Dynamic: Two processes, vapor $\rightarrow$ liquid and liquid $\rightarrow$ vapor, are both occurring but with equal rates so the composition of the vapor is constant.

25.   a.   As the intermolecular forces increase, the rate of evaporation decreases.

b.   As temperature increases, the rate of evaporation increases.

c.   As surface area increases, the rate of evaporation increases.

26.   Sublimation will occur allowing water to escape as $H_2O(g)$.

27.   The phase change, $H_2O(g) \rightarrow H_2O(l)$, releases heat that can cause additional damage. Also steam can be at a temperature greater than $100°C$.

28.   Fusion refers to a solid converting to a liquid and vaporization refers to a liquid converting to a gas. Only a fraction of the hydrogen bonds are broken in going from the solid phase to the liquid phase. Most of the hydrogen bonds are still present in the liquid phase and must be broken during the liquid to gas phase transition. Thus, the enthalpy of vaporization is much larger than the enthalpy of fusion since more intermolecular forces are broken during the vaporization process.

Exercises

Intermolecular Forces and Physical Properties

29.   a.   ionic   b.   dipole, LD (LD = London dispersion)   c.   LD only

d.   LD only; For all practical purposes, we consider a C – H bond to be nonpolar.

e.   ionic   f.   LD only   g.   H-bonding, LD

30.   a.   ionic

b.   LD mostly; C – F bonds are polar, but polymers like teflon are so large the LD forces are the predominant intermolecular forces.

c.   LD   d.   dipole, LD   e.   H-bonding, LD

f.   dipole, LD   g.   LD

31.   a.   OCS; OCS is polar and has dipole-dipole forces in addition to London dispersion (LD) forces. All polar molecules have dipole forces. CO_2 is nonpolar and only has LD forces. In all of the following (b-d), only one molecule is polar and, in turn, has dipole-dipole forces. To predict polarity, draw the Lewis structure and deduce if the individual bond dipoles cancel.

b. PF_3; PF_3 is polar (PF_5 is nonpolar).   c. SF_2; SF_2 is polar (SF_6 is nonpolar).

    d. SO_2; SO_2 is polar (SO_3 is nonpolar).

32.   a. $H_2NCH_2CH_2NH_2$; More extensive hydrogen bonding is possible.

    b. H_2CO; H_2CO is polar while CH_3CH_3 is nonpolar. H_2CO_3 has dipole forces in addition to LD forces.

    c. CH_3OH; CH_3OH can form relatively strong H-bonding interactions, unlike H_2CO.

    d.   HF; HF is capable of forming H-bonding interactions, HBr is not.

33.   a.   Neopentane is more compact than n-pentane. There is less surface area contact between neopentane molecules. This leads to weaker LD forces and a lower boiling point.

    b.   Ethanol is capable of H-bonding, dimethyl ether is not.

    c.   HF is capable of H-bonding, HCl is not.

    d.   LiCl is ionic and HCl is a molecular solid with only dipole forces and LD forces. Ionic forces are much stronger than the forces for molecular solids.

    e.   n-pentane is a larger molecule so has stronger LD forces.

    f.   Dimethyl ether is polar so has dipole forces, in addition to LD forces, unlike n-propane which only has LD forces.

34. CH_3CO_2H: H-bonding + dipole forces + LD forces

    CH_2ClCO_2H: H-bonding + larger electronegative atom replacing H (greater dipole) + LD forces

    $CH_3CO_2CH_3$:   dipole forces (no H-bonding) + LD forces

From the intermolecular forces listed above, we predict $CH_3CO_2CH_3$ to have the weakest intermolecular forces and CH_2ClCO_2H to have the strongest. The boiling points are consistent with this view.

35.   See Question 10.9 to review the dependence of some physical properties on the strength of the intermolecular forces.

    a.   HCl; HCl is polar while Ar and F_2 are nonpolar. HCl has dipole forces unlike Ar and F_2.

    b.   NaCl; Ionic forces are much stronger than molecular forces.

    c. I_2; All are nonpolar so the largest molecule (I_2) will have the strongest LD forces and the lowest vapor pressure.

    d. N_2; Nonpolar and smallest, so has weakest intermolecular forces.

    e. CH_4; Smallest, nonpolar molecule so has weakest LD forces.

f.   HF; HF can form relatively strong H-bonding interactions unlike the others.

g. $CH_3CH_2CH_2OH$; H-bonding unlike the others so has strongest intermolecular forces.

36.   a.   NO; NO is polar while N_2 and O_2 are nonpolar. NO has dipole forces unlike N_2 and O_2.

b. CH_3CN; Polar, but no H-bonding so has weaker forces than H_2O and CH_3OH which can both H-bond.

c. H_2; Nonpolar like CH_4 but H_2 is smaller than CH_4, so H_2 has weaker LD forces.

d. H_2O; H_2O can H-bond unlike the others so has strongest intermolecular forces.

e. $HOCH_2CH_2OH$ (ethylene glycol); Greatest amount of H-bonding since two –OH groups are present.

f. NH_3; Need N – H, F – H, or O – H covalent bond for hydrogen bonding. NH_3 can form H-bonds.

g. H_2O; H_2O can H-bond unlike others.

h. CO_2; Nonpolar substance, unlike others, so has weakest intermolecular (LD) forces.

Properties of Liquids

37.   The attraction of H_2O for glass is stronger than the $H_2O – H_2O$ attraction. The miniscus is concave to increase the area of contact between glass and H_2O. The Hg – Hg attraction is greater than the Hg – glass attraction. The miniscus is convex to minimize the Hg – glass contact. Polyethylene is a nonpolar substance. The $H_2O – H_2O$ attraction is stronger than the H_2O – polyethylene attraction. Thus, the miniscus will have a convex shape.

38. H_2O will rise higher in a glass tube because of a greater attraction for glass than for polyethylene.

39.   The structure of H_2O_2 is H – O – O – H, which produces greater hydrogen bonding than water. Long chains of hydrogen bonded H_2O_2 molecules then get tangled together.

40. CO_2 is a gas at room temperature. As mp and bp increase, the strength of the intermolecular forces also increases. Therefore, the strength of forces is $CO_2 < CS_2 < CSe_2$. From a structural standpoint this is expected. All three are linear, nonpolar molecules. Thus, only London dispersion forces are present. Since the molecules increase in size from $CO_2 < CS_2 < CSe_2$, the strength of the intermolecular forces will increase in the same order.

Structures and Properties of Solids

41. $n\lambda = 2d \sin \theta$, $d = \dfrac{n\lambda}{2 \sin \theta} = \dfrac{1 \times 1.54 \text{ Å}}{2 \times \sin 14.22°} = 3.13 \text{ Å} = 3.13 \times 10^{-10} \text{ m} = 313 \text{ pm}$

42. $n\lambda = 2d \sin\theta$, $\sin\theta = \dfrac{n\lambda}{2d} = \dfrac{1 \times 0.712\,\text{Å}}{19.93\,\text{Å}} = 0.0357$, $\theta = 2.05°$

43.   A cubic closest packed structure has a face-centered cubic unit cell. In a face-centered cubic unit, there are:

$$8 \text{ corners} \times \frac{1/8 \text{ atom}}{\text{corner}} + 6 \text{ faces} \times \frac{1/2 \text{ atom}}{\text{face}} = 4 \text{ atoms}$$

The atoms in a face-centered cubic unit cell touch along the face diagonal of the cubic unit cell. Using the Pythagorean formula where l = length of the face diagonal and r = radius of the atom:

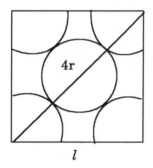

$$l^2 + l^2 = (4r)^2$$
$$2\,l^2 = 16\,r^2$$
$$l = r\sqrt{8}$$

$l = r\sqrt{8} = 125 \times 10^{-12}\,\text{m} \times \sqrt{8} = 3.54 \times 10^{-10}\,\text{m} = 3.54 \times 10^{-8}\,\text{cm}$

Volume of a unit cell $= l^3 = (3.54 \times 10^{-8}\,\text{cm})^3 = 4.44 \times 10^{-23}\,\text{cm}^3$

Mass of a unit cell $= 4 \text{ Co atoms} \times \dfrac{1 \text{ mol Co}}{6.022 \times 10^{23} \text{ atoms}} \times \dfrac{58.93 \text{ g Co}}{\text{mol Co}} = 3.914 \times 10^{-22}\,\text{g Co}$

density $= \dfrac{\text{mass}}{\text{volume}} = \dfrac{3.914 \times 10^{-22}\,\text{g}}{4.44 \times 10^{-23}\,\text{cm}^3} = 8.82 \text{ g/cm}^3$

44.   There are 4 Ni atoms in each unit cell: For a unit cell:

$$\text{density} = \frac{\text{mass}}{\text{volume}} = 6.84 \text{ g/cm}^3 = \frac{4 \text{ Ni atoms} \times \dfrac{1 \text{ mol Ni}}{6.022 \times 10^{23} \text{ atoms}} \times \dfrac{58.69 \text{ g Ni}}{\text{mol Ni}}}{l^3}$$

Solving: $l = 3.85 \times 10^{-8}\,\text{cm}$ = cube edge length

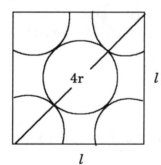

For a face centered cube:

$$(4r)^2 = l^2 + l^2 = 2\,l^2$$
$$r\sqrt{8} = l,\ r = l/\sqrt{8}$$
$$r = 3.85 \times 10^{-8}\,\text{cm}/\sqrt{8}$$
$$r = 1.36 \times 10^{-8}\,\text{cm} = 136 \text{ pm}$$

45.   The volume of a unit cell is:

$$V = l^3 = (383.3 \times 10^{-10} \text{ cm})^3 = 5.631 \times 10^{-23} \text{ cm}^3$$

There are 4 Ir atoms in the unit cell as is the case for all face-centered cubic unit cells. The mass of atoms in a unit cell is:

$$\text{mass} = 4 \text{ Ir atoms} \times \frac{1 \text{ mol Ir}}{6.022 \times 10^{23} \text{ atoms}} \times \frac{192.2 \text{ g Ir}}{\text{mol Ir}} = 1.277 \times 10^{-21} \text{ g}$$

$$\text{density} = \frac{\text{mass}}{\text{volume}} = \frac{1.277 \times 10^{-21} \text{ g}}{5.631 \times 10^{-23} \text{ cm}^3} = 22.68 \text{ g/cm}^3$$

46.   For a face-centered unit cell, the radius (r) of an atom is related to the length of a cube edge (l) by the equation $l = r\sqrt{8}$ (see Exercise 10.43).

$$\text{radius} = r = l/\sqrt{8} = 392 \times 10^{-12} \text{ m}/\sqrt{8} = 1.39 \times 10^{-10} \text{ m} = 1.39 \times 10^{-8} \text{ cm}$$

The volume of a unit cell is l^3, so the mass of the unknown metal (X) in a unit cell is:

$$\text{volume} \times \text{density} = (3.92 \times 10^{-8} \text{ cm})^3 \times \frac{21.45 \text{ g X}}{\text{cm}^3} = 1.29 \times 10^{-21} \text{ g X}$$

Since each face-centered unit cell contains 4 atoms of X, then:

$$\text{mol X in unit cell} = 4 \text{ atoms X} \times \frac{1 \text{ mol X}}{6.022 \times 10^{23} \text{ atoms X}} = 6.642 \times 10^{-24} \text{ mol X}$$

Therefore, each unit cell contains 1.29×10^{-21} g X which is equal to 6.642×10^{-24} mol X. The atomic mass of X is:

$$\frac{1.29 \times 10^{-21} \text{ g X}}{6.642 \times 10^{-24} \text{ mol X}} = 194 \text{ g/mol}$$

From the periodic table, the best choice for the metal is platinum.

47.   For a body-centered unit cell: $8 \text{ corners} \times \dfrac{1/8 \text{ Ti}}{\text{corner}} + \text{Ti at body center} = 2 \text{ Ti atoms}$

All body-centered unit cells have 2 atoms per unit cell. For a unit cell:

$$\text{density} = 4.50 \text{ g/cm}^3 = \frac{2 \text{ atoms Ti} \times \dfrac{1 \text{ mol Ti}}{6.022 \times 10^{23} \text{ atoms}} \times \dfrac{47.88 \text{ g Ti}}{\text{mol Ti}}}{l^3}, \; l = \text{cube edge length}$$

Solving: $l = $ edge length of unit cell $= 3.28 \times 10^{-8}$ cm $= 328$ pm

Assume Ti atoms just touch along the body diagonal of the cube, so body diagonal = 4 × radius of atoms = 4r.

The triangle we need to solve is:

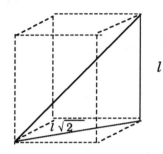

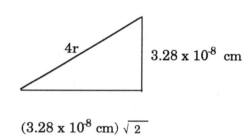

$(3.28 \times 10^{-8}$ cm$) \sqrt{2}$

$(4r)^2 = (3.28 \times 10^{-8}$ cm$)^2 + [(3.28 \times 10^{-8}$ cm$) \sqrt{2}\ ]^2$, r = 1.42 × 10⁻⁸ cm = 142 pm

For a body-centered unit cell, the radius of the atom is related to the cube edge length by $4r = l\sqrt{3}$ or $l = 4r/\sqrt{3}$.

48.   From Exercise 10.47:

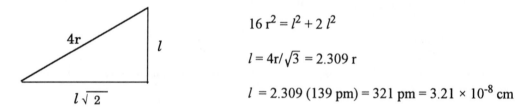

$16\ r^2 = l^2 + 2\ l^2$

$l = 4r/\sqrt{3} = 2.309\ r$

$l = 2.309\ (139\ \text{pm}) = 321\ \text{pm} = 3.21 \times 10^{-8}\ \text{cm}$

In bcc, there are 2 atoms/unit cell. For a unit cell:

$$\text{density} = \frac{\text{mass}}{\text{volume}} = \frac{2\ \text{atoms W}\ \times\ \dfrac{1\ \text{mol W}}{6.022 \times 10^{23}\ \text{atoms}}\ \times\ \dfrac{183.9\ \text{g W}}{\text{mol W}}}{(3.21 \times 10^{-8}\ \text{cm})^3} = \frac{18.5\ \text{g}}{\text{cm}^3}$$

49.   For the fcc unit cell:

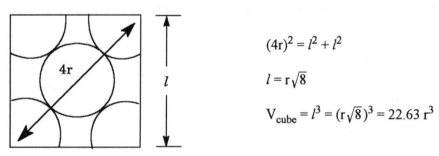

$(4r)^2 = l^2 + l^2$

$l = r\sqrt{8}$

$V_{\text{cube}} = l^3 = (r\sqrt{8})^3 = 22.63\ r^3$

There are four atoms in a face-centered cubic cell (see Exercise 10.43). Each atom has a volume of $4/3\ \pi r^3$.

$$V_{atoms} = 4 \times \frac{4}{3} \pi r^3 = 16.76 \, r^3$$

So, $\dfrac{V_{atom}}{V_{cube}} = \dfrac{16.76 \, r^3}{22.63 \, r^3} = 0.7406$ or 74.06% of the volume of each unit cell is occupied by atoms.

50.   From Exercise 10.47, a body-centered unit cell contains 2 atoms and the length of a cube edge (l) is related to the radius of the atom (r) by the equation $l = 4r/\sqrt{3}$.

Volume of unit cell $= l^3 = (4 \, r/\sqrt{3})^3 = 12.32 \, r^3$

Volume of atoms in unit cell $= 2 \times \dfrac{4}{3} \pi \, r^3 = 8.378 \, r^3$

So, $\dfrac{V_{atom}}{V_{cube}} = \dfrac{8.378 \, r^3}{12.32 \, r^3} = 0.6800 = 68.00\%$ occupied

To determine the radius of the Fe atoms, we need to determine the cube edge length (l).

$$\text{Volume of unit cell} = \left(2 \text{ Fe atoms} \times \frac{1 \text{ mol Fe}}{6.022 \times 10^{23} \text{ atoms}} \times \frac{55.85 \text{ g Fe}}{\text{mol Fe}} \right) \times \frac{1 \text{ cm}^3}{7.86 \text{ g}}$$

$$= 2.36 \times 10^{-23} \text{ cm}^3$$

Volume $= l^3 = 2.36 \times 10^{-23}$ cm^3, $l = 2.87 \times 10^{-8}$ cm

$l = 4r/\sqrt{3}$, $r = l\sqrt{3}/4 = 2.87 \times 10^{-8}$ cm $\times \sqrt{3}/4 = 1.24 \times 10^{-8}$ cm

51.   To produce a n-type semiconductor, dope Ge with a substance that has more than 4 valence electrons, e.g., a group 5A element. Phosphorus or arsenic are a couple substances which would produce n-type semiconductors when they are doped into germanium.

52.   To produce a p-type semiconductor, dope Ge with a substance that has fewer than 4 valence electrons, e.g., a group 3A element. Gallium or indium are a couple substances which would produce p-type semiconductors when they are doped into germanium.

53.   In has fewer valence electrons than Se, thus, Se doped with In would be a p-type semiconductor.

54.   To make a p-type semiconductor we need to dope the material with atoms that have fewer valence electrons. The average number of valence electrons is four when 50-50 mixtures of group 3A and group 5A elements are considered. We could dope with more of the Group 3A element or with atoms of Zn or Cd. Cadmium is the most common impurity used to produce p-type GaAs semiconductors. To make a n-type GaAs semiconductor, dope with an excess group 5A element or dope with a Group 6A element such as sulfur.

55. $E_{gap} = 2.5$ eV $\times 1.6 \times 10^{-19}$ J/eV $= 4.0 \times 10^{-19}$ J; We want $E_{gap} = E_{light}$, so:

$$E_{light} = \frac{hc}{\lambda}, \ \lambda = \frac{hc}{E} = \frac{(6.63 \times 10^{-34} \text{ J s}) (3.00 \times 10^8 \text{ m/s})}{4.0 \times 10^{-19} \text{ J}} = 5.0 \times 10^{-7} \text{ m} = 5.0 \times 10^2 \text{ nm}$$

56. $E = \dfrac{hc}{\lambda} = \dfrac{(6.626 \times 10^{-34}\,\text{J s})\,(2.998 \times 10^{8}\,\text{m/s})}{730.\times 10^{-9}\,\text{m}} = 2.72 \times 10^{-19}\,\text{J} = \text{energy of band gap}$

57.   a.   8 corners $\times \dfrac{1/8\ \text{Cl}}{\text{corner}} + 6$ faces $\times \dfrac{1/2\ \text{Cl}}{\text{face}} = 4$ Cl

  12 edges $\times \dfrac{1/4\ \text{Na}}{\text{edge}} + 1$ Na at body center = 4 Na;   NaCl is the formula.

  b.   1 Cs at body center;   8 corners $\times \dfrac{1/8\ \text{Cl}}{\text{corner}} = 1$ Cl;   CsCl is the formula.

  c.   There are 4 Zn inside the cube.

  8 corners $\times \dfrac{1/8\ \text{S}}{\text{corner}} + 6$ faces $\times \dfrac{1/2\ \text{S}}{\text{face}} = 4$ S;   ZnS is the formula.

  d.   8 corners $\times \dfrac{1/8\ \text{Ti}}{\text{corner}} + 1$ Ti at body center = 2 Ti

  4 faces $\times \dfrac{1/2\ \text{O}}{\text{face}} + 2$ O inside cube = 4 O; TiO_2 is the formula.

58.   Both As atoms are inside the unit cell. 8 corners $\times \dfrac{1/8\ \text{Ni}}{\text{corner}} + 4$ edges $\times \dfrac{1/4\ \text{Ni}}{\text{edge}} = 2$ Ni

  The unit cell contains 2 atoms of Ni and 2 atoms of As which gives a formula of NiAs.

59.   Re at 8 corners: 8(1/8) = 1 Re; O at 12 edges: 12(1/4) = 3 O

  Formula is ReO_3. If O has 2- charge, then charge on Re is +6.

60.   There are 2 tetrahedral holes per closest packed anion. Let f = fraction of tetrahedral holes filled by the cations.

  K_2O: cation to anion ratio $= \dfrac{2}{1} = \dfrac{2f}{1}$, f = 1; All of the tetrahedral holes are filled by K^+ cations.

  CuI: cation to anion ratio $= \dfrac{1}{1} = \dfrac{2f}{1}$, f $= \dfrac{1}{2}$; $\dfrac{1}{2}$ of the tetrahedral holes are filled by Cu^+ cations.

  ZrI_4: cation to anion ratio $= \dfrac{1}{4} = \dfrac{2f}{1}$, f $= \dfrac{1}{8}$; $\dfrac{1}{8}$ of the tetrahedral holes are filled by Zr^{4+} cations.

61.   Since magnesium oxide has the same structure as NaCl, each unit cell contains 4 Mg^{2+} ions and 4 O^{2-} ions. The mass of a unit cell is:

  4 MgO molecules $\left(\dfrac{1\ \text{mol MgO}}{6.022 \times 10^{23}\ \text{molecules}}\right)\left(\dfrac{40.31\ \text{g MgO}}{1\ \text{mol MgO}}\right) = 2.678 \times 10^{-22}$ g MgO

Volume of unit cell = 2.678×10^{-22} g MgO $\left(\dfrac{1\ cm^3}{3.58\ g} \right) = 7.48 \times 10^{-23}\ cm^3$

Volume of unit cell = l^3, l = cube edge length; $l = (7.48 \times 10^{-23}\ cm^3)^{1/3} = 4.21 \times 10^{-8}$ cm = 421 pm

From the NaCl structure in Figure 10.35 of the text, Mg^{2+} and O^{2-} ions should touch along the cube edge, l:

$$l = 2\ r_{Mg^{2+}} + 2\ r_{O^{2-}} = 2\ (65\ pm) + 2\ (140.\ pm) = 410.\ pm$$

The two values agree within 3%. In the actual crystals the Mg^{2+} and O^{2-} ions may not touch which is assumed in calculating the 410. pm value.

62.

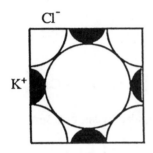

Assuming K^+ and Cl^- just touch along the cube edge, l:

$$l = 2(314\ pm) = 628\ pm = 6.28 \times 10^{-8}\ cm$$

Volume of unit cell = l^3

The unit cell contains 4 K^+ and 4 Cl^- ions. For a unit cell:

$$density = \dfrac{4\ KCl\ molecules \times \dfrac{1\ mol\ KCl}{6.022 \times 10^{23}\ molecules} \times \dfrac{74.55\ g\ KCl}{mol\ KCl}}{(6.28 \times 10^{-8}\ cm)^3} = 2.00\ g/cm^3$$

63. a. CO_2: molecular b. SiO_2: covalent network c. Si: atomic, covalent network

d. CH_4: molecular e. Ru: atomic, metallic f. I_2: molecular

g. KBr: ionic h. H_2O: molecular i. NaOH: ionic

j. U: atomic, metallic k. $CaCO_3$: ionic l. PH_3: molecular

64. a. C: atomic, covalent network b. CO: molecular c. P_4: molecular

d. S_8: molecular e. Mo: atomic, metallic f. Pt: atomic, metallic

g. NH_4Cl: ionic h. BaS: ionic i. Li_2O: ionic

j. Xe: atomic, Group 8A k. H_2S: molecular l. $NaHSO_4$: ionic

65.   Al: 8 corners $\times \dfrac{1/8 \text{ Al}}{\text{corner}} = 1$ Al; Ni: 6 face centers $\times \dfrac{1/2 \text{ Ni}}{\text{face center}} = 3$ Ni

Composition: $AlNi_3$

66.   a.   The unit cell consists of Ni at the cube corners and Ti at the body center, or Ti at the cube corners and Ni at the body center.

b. $8 \times 1/8 = 1$ atom from corners $+ 1$ atom at body center; Empirical formula = NiTi

c.   Both have coordination numbers of 8 (both are surrounded by 8 atoms).

67.   Structure 1   Structure 2

8 corners $\times \dfrac{1/8 \text{ Ca}}{\text{corner}} = 1$ Ca atom   8 corners $\times \dfrac{1/8 \text{ Ti}}{\text{corner}} = 1$ Ti atom

6 faces $\times \dfrac{1/2 \text{ O}}{\text{face}} = 3$ O atoms   12 edges $\times \dfrac{1/4 \text{ O}}{\text{edge}} = 3$ O atoms

1 Ti at body center. Formula = $CaTiO_3$ 1 Ca at body center. Formula = $CaTiO_3$

In the extended lattice of both structures, each Ti atom is surrounded by six O atoms.

68.   8 corners $\times \dfrac{1/8 \text{ Xe}}{\text{corner}} + 1$ Xe inside cell $= 2$ Xe; 8 edges $\times \dfrac{1/4 \text{ F}}{\text{edge}} + 2$ F inside cell $= 4$ F

Empirical formula is XeF_2. This is also the molecular formula.

69.   a.   Y: 1 Y in center; Ba: 2 Ba in center

Cu: 8 corners $\times \dfrac{1/8 \text{ Cu}}{\text{corner}} = 1$ Cu, 8 edges $\times \dfrac{1/4 \text{ Cu}}{\text{edge}} = 2$ Cu, total $= 3$ Cu atoms

O: 20 edges $\times \dfrac{1/4 \text{ O}}{\text{edge}} = 5$ oxygen, 8 faces $\times \dfrac{1/2 \text{ O}}{\text{face}} = 4$ oxygen, total $= 9$ O atoms

Formula: $YBa_2Cu_3O_9$

b.   The structure of this superconductor material follows the second perovskite structure described in Exercise 10.67. The $YBa_2Cu_3O_9$ structure is three of these cubic perovskite unit cells stacked on top of each other. The oxygen atoms are in the same places, Cu takes the place of Ti, two of the calcium atoms are replaced by two barium atoms and one Ca is replaced by Y.

c.   Y, Ba, and Cu are the same. Some oxygen atoms are missing.

12 edges $\times \dfrac{1/4 \text{ O}}{\text{edge}} = 3$ O, 8 faces $\times \dfrac{1/2 \text{ O}}{\text{face}} = 4$ O, total $= 7$ O atoms

Superconductor formula is $YBa_2Cu_3O_7$.

70.   a.   Structure i:

Ba: 2 Ba inside unit cell; Tl: 8 corners $\times \dfrac{1/8 \text{ Tl}}{\text{corner}} = 1$ Tl; Cu: 4 edges $\times \dfrac{1/4 \text{ Cu}}{\text{edge}} = 1$ Cu

O: 6 faces $\times \dfrac{1/2 \text{ O}}{\text{face}} + 8$ edges $\times \dfrac{1/4 \text{ O}}{\text{edge}} = 5$ O;   Formula = $TlBa_2CuO_5$

Structure ii:

Tl and Ba are the same as in structure i.

Ca: 1 Ca inside unit cell; Cu: 8 edges $\times \dfrac{1/4 \text{ Cu}}{\text{edge}} = 2$ Cu

O: 10 faces $\times \dfrac{1/2 \text{ O}}{\text{face}} + 8$ edges $\times \dfrac{1/4 \text{ O}}{\text{edge}} = 7$ O;   Formula = $TlBa_2CaCu_2O_7$

Structure iii:

Tl and Ba are the same and two Ca are located inside the unit cell.

Cu: 12 edges $\times \dfrac{1/4 \text{ Cu}}{\text{edge}} = 3$ Cu; O: 14 faces $\times \dfrac{1/2 \text{ O}}{\text{face}} + 8$ edges $\times \dfrac{1/4 \text{ O}}{\text{edge}} = 9$ O

Formula: $TlBa_2Ca_2Cu_3O_9$

Structure iv: Following similar calculations, formula = $TlBa_2Ca_3Cu_4O_{11}$

   b.   Structure i has one planar sheet of Cu and O atoms and the number increases by one for each of the remaining structures. The order of superconductivity temperature from lowest to highest temperature is: i < ii < iii < iv.

   c. $TlBa_2CuO_5$: $3 + 2(2) + x + 5(-2) = 0$, $x = +3$
        Only Cu^{3+} is present in each formula unit.

        $TlBa_2CaCu_2O_7$: $3 + 2(2) + 2 + 2(x) + 7(-2) = 0$, $x = +5/2$
        Each formula unit contains 1 Cu^{2+} and 1 Cu^{3+}.

        $TlBa_2Ca_2Cu_3O_9$: $3 + 2(2) + 2(2) + 3(x) + 9(-2) = 0$, $x = +7/3$
        Each formula unit contains 2 Cu^{2+} and 1 Cu^{3+}.

        $TlBa_2Ca_3Cu_4O_{11}$: $3 + 2(2) + 3(2) + 4(x) + 11(-2) = 0$, $x = +9/4$
        Each formula unit contains 3 Cu^{2+} and 1 Cu^{3+}.

   d.   This superconductor material achieves variable copper oxidation states by varying the numbers of Ca, Cu and O in each unit cell. The mixtures of copper oxidation states is discussed above. The superconductor material in Exercise 10.69 achieves variable copper oxidation states by omitting oxygen at various sites in the lattice.

Phase Changes and Phase Diagrams

71.   If we graph $\ln P_{vap}$ vs $1/T$, the slope of the resulting straight line will be $-\Delta H_{vap}/R$.

P_{vap}	$\ln P_{vap}$	T (Li)	1/T	T (Mg)	1/T
1 torr	0	1023 K	9.775×10^{-4} K^{-1}	893 K	11.2×10^{-4} K^{-1}
10.	2.3	1163	8.598×10^{-4}	1013	9.872×10^{-4}
100.	4.61	1353	7.391×10^{-4}	1173	8.525×10^{-4}
400.	5.99	1513	6.609×10^{-4}	1313	7.616×10^{-4}
760.	6.63	1583	6.317×10^{-4}	1383	7.231×10^{-4}

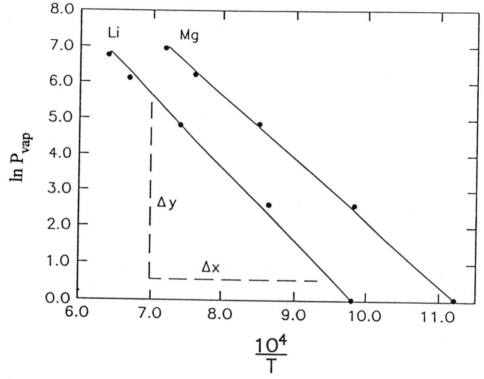

For Li:

We get the slope by taking two points (x, y) that are on the line we draw. For a line:

$$\text{slope} = \frac{\Delta y}{\Delta x} = \frac{y_2 - y_1}{x_2 - x_1}$$

or we can determine the straight line equation using a computer or calculator. The general straight line equation is $y = mx + b$ where m = slope and b = y-intercept.

The equation of the Li line is: $\ln P_{vap} = -1.90 \times 10^4 (1/T) + 18.6$, slope $= -1.90 \times 10^4$ K

Slope $= -\Delta H_{vap}/R$, $\Delta H_{vap} = -\text{slope} \times R = 1.90 \times 10^4$ K $\times$ 8.3145 J/K•mol

$\Delta H_{vap} = 1.58 \times 10^5$ J/mol = 158 kJ/mol

For Mg:

The equation of the line is: $\ln P_{vap} = -1.67 \times 10^4 (1/T) + 18.7$, slope $= -1.67 \times 10^4$ K

$\Delta H_{vap} = -\text{slope} \times R = 1.67 \times 10^4 \text{ K} \times 8.3145 \text{ J/K•mol}$, $\Delta H_{vap} = 1.39 \times 10^5$ J/mol $= 139$ kJ/mol

The bonding is stronger in Li since ΔH_{vap} is larger for Li.

72.   Again we graph $\ln P_{vap}$ vs $1/T$. The slope of the line equals $-\Delta H_{vap}/R$.

T(K)	$10^3/T$ (K^{-1})	P_{vap} (torr)	$\ln P_{vap}$
273	3.66	14.4	2.67
283	3.53	26.6	3.28
293	3.41	47.9	3.87
303	3.30	81.3	4.40
313	3.19	133	4.89
323	3.10	208	5.34
353	2.83	670.	6.51

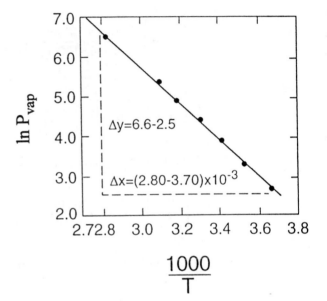

The slope of the line is -4600 K.

$$-4600 \text{ K} = \frac{-\Delta H_{vap}}{R} = \frac{-\Delta H_{vap}}{8.3145 \text{ J/K•mol}}$$

$\Delta H_{vap} = 38$ kJ/mol

To determine the normal boiling point, we can use the following formula:

$$\ln\left(\frac{P_1}{P_2}\right) = \frac{\Delta H_{vap}}{R}\left(\frac{1}{T_2} - \frac{1}{T_1}\right)$$

At the normal boiling point, the vapor pressure equals 1.00 atm or 760. torr. At 273 K, the vapor pressure is 14.4. torr (from data in the problem).

$$\ln\left(\frac{14.4}{760.}\right) = \frac{38,000 \text{ J/mol}}{8.3145 \text{ J/K•mol}}\left(\frac{1}{T_2} - \frac{1}{273 \text{ K}}\right), \quad -3.97 = 4.6 \times 10^3 \left(1/T_2 - 3.66 \times 10^{-3}\right)$$

$-8.6 \times 10^{-4} + 3.66 \times 10^{-3} = 1/T_2 = 2.80 \times 10^{-3}$, $T_2 = 357$ K $=$ normal boiling point

73.   At 100.°C (373 K), the vapor pressure of H_2O is 1.00 atm. For water, $\Delta H_{vap} = 40.7$ kJ/mol.

$$\ln\left(\frac{P_1}{P_2}\right) = \frac{\Delta H_{vap}}{R}\left(\frac{1}{T_2} - \frac{1}{T_1}\right) \text{ or } \ln\left(\frac{P_2}{P_1}\right) = \frac{\Delta H_{vap}}{R}\left(\frac{1}{T_1} - \frac{1}{T_2}\right)$$

$$\ln\left(\frac{P_2}{1.00 \text{ atm}}\right) = \frac{40.7 \times 10^3 \text{ J/mol}}{8.3145 \text{ J/K}\bullet\text{mol}}\left(\frac{1}{373 \text{ K}} - \frac{1}{388 \text{ K}}\right), \ \ln P_2 = 0.51, \ P_2 = e^{0.51} = 1.7 \text{ atm}$$

$$\ln\left(\frac{3.50}{1.00}\right) = \frac{40.7 \times 10^3 \text{ J/mol}}{8.3145 \text{ J/K}\bullet\text{mol}}\left(\frac{1}{373 \text{ K}} - \frac{1}{T_2}\right), \ 2.56 \times 10^{-4} = \left(\frac{1}{373} - \frac{1}{T_2}\right)$$

$$2.56 \times 10^{-4} = 2.68 \times 10^{-3} - \frac{1}{T_2}, \ \frac{1}{T_2} = 2.42 \times 10^{-3}, \ T_2 = \frac{1}{2.42 \times 10^{-3}} = 413 \text{ K or } 140.°\text{C}$$

74. $$\ln\left(\frac{P_2}{1.00}\right) = \frac{40.7 \times 10^3 \text{ J/mol}}{8.3145 \text{ J/K}\bullet\text{mol}}\left(\frac{1}{373 \text{ K}} - \frac{1}{623 \text{ K}}\right), \ \ln P_2 = 5.27, \ P_2 = e^{5.27} = 194 \text{ atm}$$

75. $$\ln\left(\frac{P_1}{P_2}\right) = \frac{\Delta H_{vap}}{R}\left(\frac{1}{T_2} - \frac{1}{T_1}\right)$$

At normal boiling point, $P_1 = 760.$ torr, $T_1 = 56.5°\text{C} = 329.7$ K; $T_2 = 25.0°\text{C} = 298.2$ K, $P_2 = ?$

$$\ln\left(\frac{760.}{P_2}\right) = \frac{32.0 \times 10^3 \text{ J/mol}}{8.3145 \text{ J/K}\bullet\text{mol}}\left(\frac{1}{298.2} - \frac{1}{329.7}\right), \ \ln 760. - \ln P_2 = 1.23$$

$\ln P_2 = 5.40, \ P_2 = e^{5.40} = 221$ torr

76. $$\ln\left(\frac{P_1}{P_2}\right) = \frac{\Delta H_{vap}}{R}\left(\frac{1}{T_2} - \frac{1}{T_1}\right)$$

$P_1 = 760.$ torr, $T_1 = 56.5°\text{C}$ (the normal bp from previous exercise); $P_2 = 630.$ torr, $T_2 = ?$

$$\ln\left(\frac{760.}{630.}\right) = \frac{32.0 \times 10^3 \text{ J/mol}}{8.3145 \text{ J/K}\bullet\text{mol}}\left(\frac{1}{T_2} - \frac{1}{329.7}\right), \ 0.188 = 3.85 \times 10^3\left(\frac{1}{T_2} - 3.033 \times 10^{-3}\right)$$

$$\frac{1}{T_2} - 3.033 \times 10^{-3} = 4.88 \times 10^{-5}, \ \frac{1}{T_2} = 3.082 \times 10^{-3}, \ T_2 = 324.5 \text{ K} = 51.3°\text{C}$$

$$\ln\left(\frac{630.}{P_2}\right) = \frac{32.0 \times 10^3}{8.3145}\left(\frac{1}{298.2} - \frac{1}{324.5}\right), \ \ln 630. - \ln P_2 = 1.05$$

$\ln P_2 = 5.40, \ P_2 = e^{5.40} = 221$ torr

77.

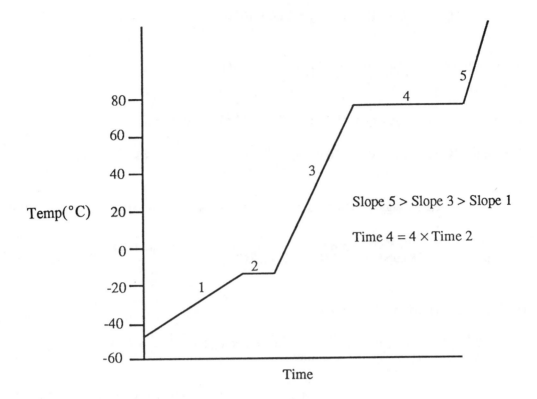

Slope 5 > Slope 3 > Slope 1

Time 4 = 4 × Time 2

78.   A typical heating curve looks like:

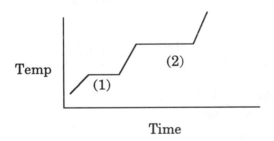

Plateau (1) gives a temperature at which the solid and liquid are in equilibrium. This temperature along with the pressure gives a point on the phase diagram. Plateau (2) will give a temperature and pressure point on the liquid-vapor line. There certainly is a lot more information needed to construct a detailed phase diagram. However, if several heating curves were determined as a function of pressure, then several more points on the phase diagram would be determined. This would eventually lead to a reasonable sketch of the phase diagram.

79. $H_2O(s, -20.\,°C) \rightarrow H_2O(s, 0°C),\ \Delta T = 20.\,°C$

$$q_1 = s_{ice} \times m \times \Delta T = \frac{2.1\,J}{g\,°C} \times 5.00 \times 10^2\,g \times 20.\,°C = 2.1 \times 10^4\,J = 21\,kJ$$

$H_2O(s, 0°C) \rightarrow H_2O(l, 0°C),\ q_2 = 5.00 \times 10^2\,g\,H_2O \times \dfrac{1\,mol}{18.02\,g} \times \dfrac{6.02\,kJ}{mol} = 167\,kJ$

$H_2O(l, 0°C) \rightarrow H_2O(l, 100.\,°C),\ q_3 = \dfrac{4.2\,J}{g\,°C} \times 5.00 \times 10^2\,g \times 100.\,°C = 2.1 \times 10^5 J = 210\,kJ$

$H_2O(l, 100.\,^\circ C) \rightarrow H_2O(g, 100.\,^\circ C)$, $q_4 = 5.00 \times 10^2 \text{ g} \times \dfrac{1 \text{ mol}}{18.02 \text{ g}} \times \dfrac{40.7 \text{ kJ}}{\text{mol}} = 1130 \text{ kJ}$

$H_2O(g, 100.\,^\circ C) \rightarrow H_2O(g, 250.\,^\circ C)$, $q_5 = \dfrac{2.0 \text{ J}}{\text{g }^\circ C} \times 5.00 \times 10^2 \text{ g} \times 150.\,^\circ C = 1.5 \times 10^5 \text{ J} = 150 \text{ kJ}$

$q_{total} = q_1 + q_2 + q_3 + q_4 + q_5 = 21 + 167 + 210 + 1130 + 150 = 1680 \text{ kJ}$

80.   To melt the 10.0 g of ice at $0^\circ C$: $10.0 \text{ g} \times \dfrac{1 \text{ mol}}{18.02 \text{ g}} \times \dfrac{6.02 \text{ kJ}}{\text{mol}} = 3.34 \text{ kJ} = 3340 \text{ J}$

Only 850 J of heat are added. Some, but not all of the ice will melt and the temperature will still be $0^\circ C$.

81.   Total mass $H_2O = 18 \text{ cubes} \times \dfrac{30.0 \text{ g}}{\text{cube}} = 540. \text{ g}$; $540. \text{ g } H_2O \times \dfrac{1 \text{ mol } H_2O}{18.02 \text{ g}} = 30.0 \text{ mol } H_2O$

Heat removed to produce ice at $-5.0^\circ C$:

$$\dfrac{4.18 \text{ J}}{\text{g }^\circ C} \times 540. \text{ g} \times 22.0\,^\circ C + \dfrac{6.02 \times 10^3 \text{ J}}{\text{mol}} \times 30.0 \text{ mol} + \dfrac{2.08 \text{ J}}{\text{g }^\circ C} \times 540. \text{ g} \times 5.0\,^\circ C$$

$$= 4.97 \times 10^4 \text{ J} + 1.81 \times 10^5 \text{ J} + 5.6 \times 10^3 \text{ J} = 2.36 \times 10^5 \text{ J}$$

$2.36 \times 10^5 \text{ J} \times \dfrac{1 \text{ g CF}_2\text{Cl}_2}{158 \text{ J}} = 1.49 \times 10^3 \text{ g CF}_2\text{Cl}_2$ must be vaporized.

82.   Energy added to ice cubes $= 5.00 \text{ min} \times \dfrac{60 \text{ s}}{\text{min}} \times \dfrac{750. \text{ J}}{\text{s}} = 2.25 \times 10^5 \text{ J} = 225 \text{ kJ}$

Energy to melt ice $= 475 \text{ g} \times \dfrac{1 \text{ mol}}{18.02 \text{ g}} \times \dfrac{6.02 \text{ kJ}}{\text{mol}} = 159 \text{ kJ}$

Energy to heat water $= 225 \text{ kJ} - 159 \text{ kJ} = 66 \text{ kJ}$; $66,000 \text{ J} = 475 \text{ g} \times \dfrac{4.18 \text{ J}}{\text{g }^\circ C} \times \Delta T$, $\Delta T = 33^\circ C$

The final temperature is $33^\circ C$. The water doesn't boil.

83.   A: solid;   B: liquid;   C: vapor

D: solid + vapor;   E: solid + liquid + vapor

F: liquid + vapor;   G: liquid + vapor;   H: vapor

triple point: E;   critical point: G

normal freezing point: temperature at which solid - liquid line is at 1.0 atm (see plot below).

normal boiling point: temperature at which liquid - vapor line is at 1.0 atm (see plot below).

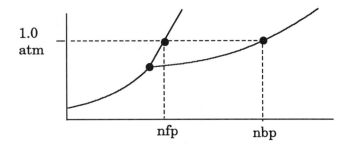

Since the solid-liquid equilibrium line has a positive slope, then the solid phase is denser than the liquid phase.

84.   a.   2

        b.   Triple point at 96°C: rhombic, monoclinic, vapor
             Triple point at 119°C: monoclinic, liquid, vapor

        c.   From the phase diagram, the rhombic phase is stable at T ≈ 20°C and P = 1.0 atm.

        d.   Yes, monoclinic sulfur and vapor (gas) share a common boundary line in the phase diagram
             (monoclinic sulfur can sublime at certain conditions).

        e.   The normal boiling point occurs at the temperature where the vapor pressure of the liquid equals
             760 torr. The vapor pressure at 119°C is 0.027 torr. Therefore, a temperature higher than
             119°C is needed to produce a vapor pressure of 760 torr.

Additional Exercises

85. $C_{25}H_{52}$ has the stronger intermolecular forces because it has the higher boiling point. Even though
        $C_{25}H_{52}$ is nonpolar, it is so large that its London dispersion forces are much stronger than the sum of
        the London dispersion and hydrogen bonding interactions found in H_2O.

86.   Benzene   Naphthalene

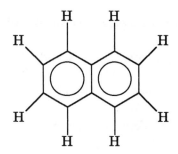

                LD forces only   LD forces only

Note: London dispersion forces in molecules like benzene and naphthalene are fairly large. The molecules are flat and there is efficient surface area contact between molecules. Large surface area contact leads to stronger London dispersion forces.

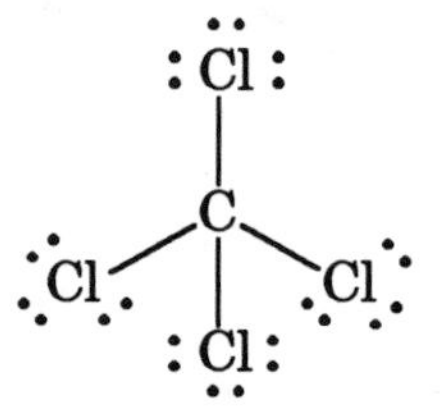

Carbon tetrachloride (CCl_4) has polar bonds but is a nnonpolar molecule. CCl_4 only has LD forces.

In terms of size and shape: $CCl_4 < C_6H_6 < C_{10}H_8$

The strengths of the LD forces are proportional to size and are related to shape. Although the size of CCl_4 is fairly large, the overall spherical shape gives rise to relatively weak LD forces as compared to flat molecules like benzene and and naphthalene. The physical properties given in the problem are consistent with the order listed above. Each of the physical properties will increase with an increase in intermolecular forces.

Acetone   Acetic Acid

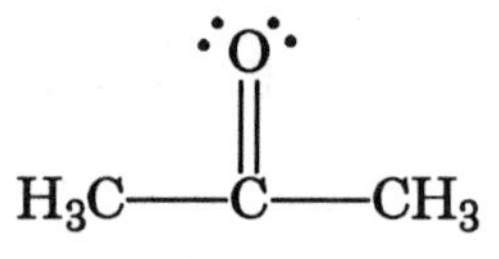

LD, dipole   LD, dipole, H-bonding

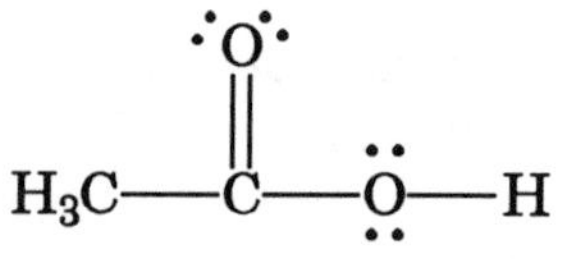

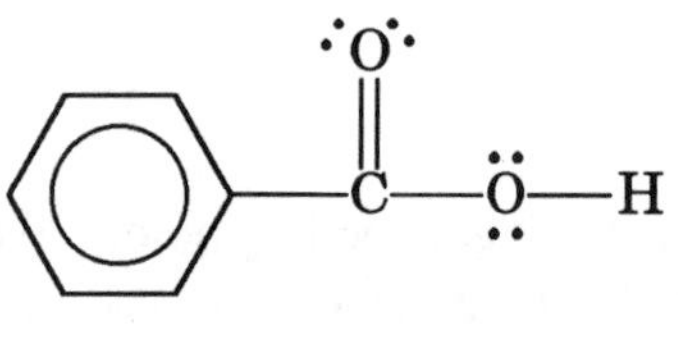

Benzoic Acid

LD, dipole, H-bonding

We would predict the strength of interparticle forces of the last three molecules to be:

    acetone   <   acetic acid   <   benzoic acid

    polar   H-bonding   H-bonding, but large LD forces because of greater size and shape.

This ordering is consistent with the values given for bp, mp, and ΔH_{vap}.

The overall order of the strengths of intermolecular forces based on physical properties are:

acetone $<$ CCl_4 $<$ C_6H_6 $<$ acetic acid $<$ naphthalene $<$ benzoic acid

The order seems reasonable except for acetone and naphthalene. Since acetone is polar, we would not expect it to boil at the lowest temperature. However, in terms of size and shape, acetone is the smallest molecule and the LD forces in acetone must be very small compared to the other molecules. Naphthalene must have very strong LD forces because of its size and flat shape.

87.

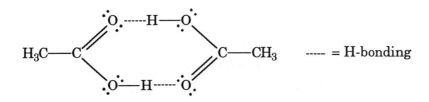

----- = H-bonding

88.   As the electronegativity of the atoms covalently bonded to H increases, the strength of the hydrogen bonding interaction increases.

$$N \cdots H - N < N \cdots H - O < O \cdots H - O < O \cdots H - F < F \cdots H - F$$

weakest   strongest

89. $n\lambda = 2d \sin\theta$, $\lambda = \dfrac{2d \sin\theta}{n} = \dfrac{2 \times 201 \text{ pm} \times \sin 34.68°}{1}$, $\lambda = 229$ pm $= 2.29 \times 10^{-10}$ m $= 0.229$ nm

90.   If face-centered cubic structure, then 4 atoms/unit cell and from Exercise 10.43:

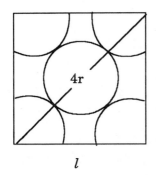

l

l

$2\,l^2 = 16\,r^2$

$l = r\sqrt{8} = 144$ pm $\sqrt{8} = 407$ pm

$l = 407 \times 10^{-12}$ m $= 407 \times 10^{-10}$ cm

density $= \dfrac{4 \text{ atoms Au} \times \dfrac{1 \text{ mol Au}}{6.022 \times 10^{23} \text{ atoms}} \times \dfrac{197.0 \text{ g Au}}{\text{mol Au}}}{(4.07 \times 10^{-8} \text{ cm})^3} = 19.4$ g/cm^3

If body-centered cubic structure, then 2 atoms/unit cell and from Exercise 10.47:

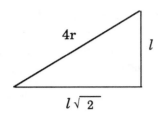

$$16\,r^2 = l^2 + 2\,l^2$$

$$l = 4r/\sqrt{3} = 333 \text{ pm} = 333 \times 10^{-12} \text{ m}$$

$$l = 333 \times 10^{-10} \text{ cm} = 3.33 \times 10^{-8} \text{ cm}$$

$$\text{density} = \frac{2 \text{ atoms Au} \times \dfrac{1 \text{ mol Au}}{6.022 \times 10^{23} \text{ atoms}} \times \dfrac{197.0 \text{ g Au}}{\text{mol Au}}}{(3.33 \times 10^{-8} \text{ cm})^3} = 17.7 \text{ g/cm}^3$$

The measured density is consistent with a face-centered cubic unit cell.

91.   If TiO_2 conducts electricity as a liquid then it is an ionic solid, if not then TiO_2 is a network solid.

92.   The student neglected to consider the enthalpy change that occurs at the melting point, called the enthalpy of fusion. To chill the hotter water (25°C) to 0°C requires:

$$q = \frac{4.2 \text{ J}}{g \bullet °C} \times 20.0 \text{ g} \times (-25°C) = -2100 \text{ J}$$

To heat ice from -10.0°C to 0°C requires:

$$q = \frac{2.1 \text{ J}}{g \bullet °C} \times 20.0 \text{ g} \times 10.°C = 420 \text{ J}$$

To convert ice to water at 0°C requires:

$$q = 20.0 \text{ g} \times \frac{1 \text{ mol}}{18.02} \times \frac{6.02 \text{ kJ}}{\text{mol}} = 6.68 \text{ kJ} = 6680 \text{ kJ}$$

Assuming no heat loss to the surroundings, the ice could easily chill the 25°C water to 0°C but in the process, some of the ice would be converted to water. At this point, a mixture of ice water remains as long as there is no heat loss to the surroundings. Therefore, the final temperature of the mixture would be 0°C.

93. $1.00 \text{ lb} \times \dfrac{454 \text{ g}}{\text{lb}} = 454 \text{ g } H_2O$; A change of 1.00°F is equal to a change of 5/9°C.

The amount of heat in J in 1 Btu is: $\dfrac{4.18 \text{ J}}{g \, °C} \times 454 \text{ g} \times \dfrac{5}{9}°C = 1.05 \times 10^3 \text{ J or } 1.05 \text{ kJ}$

It takes 40.7 kJ to vaporize 1 mol H_2O (ΔH_{vap}). Combining these:

$$\frac{1.00 \times 10^4 \text{ Btu}}{\text{hr}} \times \frac{1.05 \text{ kJ}}{\text{Btu}} \times \frac{1 \text{ mol } H_2O}{40.7 \text{ kJ}} = 258 \text{ mol/hr}$$

or: $\dfrac{258 \text{ mol}}{\text{hr}} \times \dfrac{18.02 \text{ g H}_2\text{O}}{\text{mol}} = 4650 \text{ g/hr} = 4.65 \text{ kg/hr}$

94.

T (°C)	T (K)	1/T (K^{-1})	P$_{vap}$ (torr)	ln P$_{vap}$
-6.0	267.2	3.743×10^{-3}	20.0	3.00
5.0	278.2	3.595×10^{-3}	40.0	3.69
12.1	285.3	3.505×10^{-3}	60.0	4.09
21.2	294.4	3.397×10^{-3}	100.0	4.605
49.9	323.1	3.095×10^{-3}	400.0	5.991

The straight line equation
(y = mx + b) from the graph is:
$\ln P_{vap} = -4620/T + 20.29$

Slope = m = $\dfrac{-\Delta H_{vap}}{R} = -4.62 \times 10^3$ K

$\Delta H_{vap} = 4.62 \times 10^3 \text{ K} \times \dfrac{8.3145 \text{ J}}{\text{mol K}}$

$\Delta H_{vap} = 38{,}400 \text{ J/mol} = 38.4 \text{ kJ/mol}$

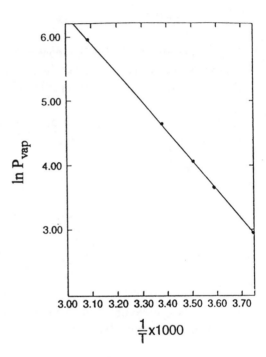

At normal boiling point, $P_{vap} = 760.$ torr. From the straight line equation:

$\ln 760. = -4620/T_b + 20.29, \quad T = \dfrac{-4620}{6.63 - 20.29} = 338 \text{ K} = 65°\text{C}$

95. $\ln\left(\dfrac{P_1}{P_2}\right) = \dfrac{\Delta H_{vap}}{R}\left(\dfrac{1}{T_2} - \dfrac{1}{T_1}\right)$; $P_1 = 760.$ torr, $T_1 = 630.$ K; $P_2 = ?$, $T_2 = 298$ K

$\ln\left(\dfrac{760.}{P_2}\right) = \dfrac{59.1 \times 10^3 \text{ J/mol}}{8.3145 \text{ J/K}\cdot\text{mol}}\left(\dfrac{1}{298 \text{ K}} - \dfrac{1}{630. \text{ K}}\right) = 12.6$

$760./ P_2 = e^{12.6}$, $P_2 = 760./2.97 \times 10^5 = 2.56 \times 10^{-3}$ torr

96.   a.   two

       b.   Higher pressure triple point: graphite, diamond and liquid; Lower pressure triple point:
             graphite, liquid and vapor

c.   It is converted to diamond (the more dense solid form).

d.   Diamond is more dense which is why graphite can be converted to diamond by applying pressure.

Challenge Problems

97.   A single hydrogen bond in H_2O has a strength of 21 kJ/mol. Each H_2O molecule forms two H-bonds. Thus, it should take 42 kJ/mol of energy to break all of the H-bonds in water. Consider the phase transitions:

$$\text{solid} \overset{6.0\,\text{kJ}}{\rightarrow} \text{liquid} \overset{40.7\,\text{kJ}}{\rightarrow} \text{vapor} \qquad \Delta H_{sub} = \Delta H_{fus} + \Delta H_{vap}$$

It takes a total of 46.7 kJ/mol to convert solid H_2O to vapor (ΔH_{sub}). This would be the amount of energy necessary to disrupt all of the intermolecular forces in ice. Thus, $(42 \div 46.7) \times 100 = 90\%$ of the attraction in ice can be attributed to H-bonding.

98.   Both molecules are capable of H-bonding. However, in oil of wintergreen the hydrogen bonding is <u>intra</u>molecular.

In methyl-4-hydroxybenzoate, the H-bonding is <u>inter</u>molecular, resulting in stronger intermolecular forces and a higher melting point.

99.   NaCl, $MgCl_2$, NaF, MgF_2 AlF_3 all have very high melting points indicative of strong intermolecular forces. They are all ionic solids. $SiCl_4$, SiF_4, F_2, Cl_2, PF_5 and SF_6 are nonpolar covalent molecules. Only LD forces are present. PCl_3 and SCl_2 are polar molecules. LD forces and dipole forces are present. In these 8 molecular substances, the intermolecular forces are weak and the melting points low. $AlCl_3$ doesn't seem to fit in as well. From the melting point, there are much stronger forces present than in the nonmetal halides, but they aren't as strong as we would expect for an ionic solid. $AlCl_3$ illustrates a gradual transition from ionic to covalent bonding; from an ionic solid to discrete molecules.

100.   One B atom and one N atom together have the same number of electrons as two C atoms. The description of physical properties sound a lot like the properties of graphite and diamond, the two solid forms of carbon. The two forms of BN have structures similar to graphite and diamond.

101. $n\lambda = 2d \sin \theta$;   100 pm = 10^{-10} m = 10^{-8} cm

$$d = \frac{n\lambda}{2 \sin \theta} = \frac{1 \times 71.2\,\text{pm}}{2 \times \sin 5.564} = 367\,\text{pm} = 3.67 \times 10^{-8}\,\text{cm} = \text{cube edge length} = l$$

Mass of Hf in unit cell = volume unit cell × density; volume = l^3 = $(3.67 \times 10^{-8} \text{ cm})^3$

Mass of Hf = $(3.67 \times 10^{-8} \text{ cm})^3 \times \dfrac{13.28 \text{ g Hf}}{\text{cm}^3}$ = 6.56×10^{-22} g Hf

Atoms Hf in unit cell = 6.56×10^{-22} g Hf × $\dfrac{1 \text{ mol Hf}}{178.5 \text{ g Hf}} \times \dfrac{6.022 \times 10^{23} \text{ atoms Hf}}{\text{mol Hf}}$ = 2.21

This is most consistent with a body-centered cubic unit cell which contains 2 atoms per unit cell. To determine the radius, the cube edge length (l) is related to the radius by the equation $l = 4r/\sqrt{3}$ (see Exercise 10.47).

radius = r = $l\sqrt{3}/4$ = 3.67×10^{-8} cm × $\sqrt{3}/4$ = 1.59×10^{-8} cm = 159 pm

102.   First we need to get the empirical formula of spinel. Assume 100.0 g of spinel.

37.9 g Al × $\dfrac{1 \text{ mol Al}}{26.98 \text{ g Al}}$ = 1.40 mol Al

The mole ratios are 2:1:4.

17.1 g Mg × $\dfrac{1 \text{ mol Mg}}{24.31 \text{ g Mg}}$ = 0.703 mol Mg

Empirical Formula = Al_2MgO_4

45.0 g O × $\dfrac{1 \text{ mol O}}{16.00 \text{ g O}}$ = 2.81 mol O

Assume each unit cell contains an integral value (n) of Al_2MgO_4 molecules. Each Al_2MgO_4 molecule has a mass of: 24.31 + 2(26.98) + 4(16.00) = 142.27 g/mol

density = $\dfrac{\text{n molecules} \times \dfrac{1 \text{ mol}}{6.022 \times 10^{23} \text{ molecules}} \times \dfrac{142.27 \text{ g}}{\text{mol}}}{(8.09 \times 10^{-8} \text{ cm})^3}$ = $\dfrac{3.57 \text{ g}}{\text{cm}^3}$, Solving: n = 8.00

Each unit cell has 8 molecules of Al_2MgO_4 or 16 Al, 8 Mg and 32 O atoms.

103.   A face-centered cubic unit cell contains 4 atoms. For a unit cell:

mass of X = volume × density = $(4.09 \times 10^{-8} \text{ cm})^3 \times 10.5 \text{ g/cm}^3$ = 7.18×10^{-22} g

mol X = 4 atoms X × $\dfrac{1 \text{ mol X}}{6.022 \times 10^{23} \text{ atoms}}$ = 6.642×10^{-24} mol X

Atomic mass = $\dfrac{7.18 \times 10^{-22} \text{ g X}}{6.642 \times 10^{-24} \text{ mol X}}$ = 108 g/mol; The metal is silver (Ag).

104.   a.   The arrangement of the layers are:

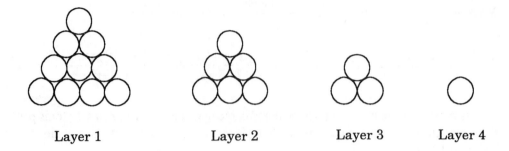

Layer 1   Layer 2   Layer 3   Layer 4

A total of 20 cannon balls will be needed.

b.   The layering alternates abcabc which is cubic closest packing.

c.   tetrahedron

105.

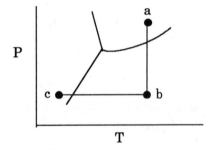

As P is lowered, we go from a to b on the phase diagram. The water boils. The evaporation of the water is endothermic and the water is cooled (b → c), forming some ice. If the pump is left on, the ice will sublime until none is left. This is the basis of freeze drying.

CHAPTER ELEVEN

PROPERTIES OF SOLUTIONS

Solution Review

9. $125 \text{ g sucrose} \times \dfrac{1 \text{ mol}}{342.30 \text{ g}} = 0.365 \text{ mol}; \ M = \dfrac{0.365 \text{ mol}}{1.00 \text{ L}} = \dfrac{0.365 \text{ mol sucrose}}{\text{L}}$

10. $0.250 \text{ L} \times \dfrac{0.100 \text{ mol}}{\text{L}} \times \dfrac{134.00 \text{ g}}{\text{mol}} = 3.35 \text{ g Na}_2\text{C}_2\text{O}_4$

11. $25.00 \times 10^{-3} \text{ L} \times \dfrac{0.308 \text{ mol}}{\text{L}} = 7.70 \times 10^{-3} \text{ mol}; \ \dfrac{7.70 \times 10^{-3} \text{ mol}}{0.500 \text{ L}} = \dfrac{1.54 \times 10^{-2} \text{ mol NiCl}_2}{\text{L}}$

$\text{NiCl}_2(s) \rightarrow \text{Ni}^{2+}(aq) + 2 \text{ Cl}^-(aq); \ M_{\text{Ni}^{2+}} = \dfrac{1.54 \times 10^{-2} \text{ mol}}{\text{L}}; \ M_{\text{Cl}^-} = \dfrac{3.08 \times 10^{-2} \text{ mol}}{\text{L}}$

12. a. $\text{Ca(NO}_3)_2(s) \rightarrow \text{Ca}^{2+}(aq) + 2 \text{ NO}_3^-(aq); \ M_{\text{Ca}^{2+}} = \dfrac{1.06 \times 10^{-3} \text{ mol}}{\text{L}}; \ M_{\text{NO}_3^-} = \dfrac{2.12 \times 10^{-3} \text{ mol}}{\text{L}}$

b. $1.0 \times 10^{-3} \text{ L} \times \dfrac{1.06 \times 10^{-3} \text{ mol Ca}^{2+}}{\text{L}} \times \dfrac{40.08 \text{ g Ca}^{2+}}{\text{mol}} = 4.2 \times 10^{-5} \text{ g Ca}^{2+}$

c. $1.0 \times 10^{-6} \text{ L} \times \dfrac{2.12 \times 10^{-3} \text{ mol NO}_3^-}{\text{L}} \times \dfrac{6.02 \times 10^{23} \text{ NO}_3^- \text{ ions}}{\text{mol NO}_3^-} = 1.3 \times 10^{15} \text{ NO}_3^- \text{ ions}$

13. $1.00 \text{ L} \times \dfrac{0.040 \text{ mol HCl}}{\text{L}} = 0.040 \text{ mol HCl}; \ 0.040 \text{ mol HCl} \times \dfrac{1 \text{ L}}{0.25 \text{ mol HCl}} = 0.16 \text{ L} = 160 \text{ mL}$

14. a. $\text{HNO}_3(l) \rightarrow \text{H}^+(aq) + \text{NO}_3^-(aq)$ b. $\text{Na}_2\text{SO}_4(s) \rightarrow 2 \text{ Na}^+(aq) + \text{SO}_4^{2-}(aq)$

c. $\text{AlCl}_3(s) \rightarrow \text{Al}^{3+}(aq) + 3 \text{ Cl}^-(aq)$ d. $\text{SrBr}_2(s) \rightarrow \text{Sr}^{2+}(aq) + 2 \text{ Br}^-(aq)$

e. $\text{KClO}_4(s) \rightarrow \text{K}^+(aq) + \text{ClO}_4^-(aq)$ f. $\text{NH}_4\text{Br}(s) \rightarrow \text{NH}_4^+(aq) + \text{Br}^-(aq)$

g. $\text{NH}_4\text{NO}_3(s) \rightarrow \text{NH}_4^+(aq) + \text{NO}_3^-(aq)$ h. $\text{CuSO}_4(s) \rightarrow \text{Cu}^{2+}(aq) + \text{SO}_4^{2-}(aq)$

i. $\text{NaOH}(s) \rightarrow \text{Na}^+(aq) + \text{OH}^-(aq)$

Questions

15. $\text{Molarity} = \dfrac{\text{moles solute}}{\text{L solution}}; \quad \text{Molality} = \dfrac{\text{moles solute}}{\text{kg solvent}}$

Since volume is temperature dependent and mass isn't, then molarity is temperature dependent and molality is temperature independent. In determining ΔT_f and ΔT_b, we are interested in how some temperature depends on composition. Thus, we don't want our expression of composition to also depend on temperature.

16.   The nature of the intermolecular forces. Polar solutes and ionic solutes dissolve in polar solvents and nonpolar solutes dissolve in nonpolar solvents.

17.   hydrophobic: water hating; hydrophilic: water loving

18.   As the temperature increases, the gas molecules will have a greater average kinetic energy. A greater fraction of the gas molecules in solution will have kinetic energy greater than the attractive forces between the gas molecules and the solvent molecules. More gas molecules escape to the vapor phase and the solubility of the gas decreases.

19.   If solute-solvent attractions > solvent-solvent and solute-solute attractions, then there is a negative deviation from Raoult's law. If solute-solvent attractions < solvent-solvent and solute-solute attractions, then there is a positive deviation from Raoult's law.

20.   A positive deviation from Raoult's law means the vapor pressure of the solution is greater than if the solution were ideal. At the boiling point, the vapor pressure equals the atmospheric pressure. For a solution with positive deviations, it will take a lower temperature to achieve a vapor pressure of one atmosphere. Therefore, the boiling point is lower than if the solution were ideal.

21.   No, the solution is not ideal. For an ideal solution, this strength of intermolecular forces in the solution are the same as in the pure solute and pure solvent. This results in $\Delta H_{soln} = 0$ for an ideal solution. ΔH_{soln} for methanol/water is not zero. Since $\Delta H_{soln} < 0$, then this solution shows negative deviation from Raoult's law.

22.   Ion pairing can occur, resulting in fewer particles than expected. This results in smaller freezing point depressions and smaller boiling point elevations ($\Delta T = km$). Ion pairing will increase as the concentration of electrolyte increases.

23.   With addition of salt or sugar, the osmotic pressure inside the fruit cells (and bacteria) is less than outside the cell. Water will leave the cells which will dehydrate bacteria present, causing them to die.

24.   A strong electrolyte dissociates completely into ions in solution. A weak electrolyte dissociates only partially into ions in solution. Colligative properties depend on the total number of particles in solution. By measuring a property such as freezing point depression, boiling point elevation or osmotic pressure, we can calculate the van't Hoff factor (i) to see if an electrolyte is strong or weak.

25.   Both solutions and colloids have suspended particles in some medium. The major difference between the two is the size of the particles. A colloid is a suspension of relatively large particles as compared to a solution. Because of this, colloids will scatter light while solutions will not. The scattering of light by a colloidal suspension is called the Tyndall effect.

26.   The micelles form so the ionic ends of the detergent molecules, the SO_4^- ends, are exposed to the polar water molecules on the outside, while the nonpolar hydrocarbon chains from the detergent molecules are hidden from the water by pointing toward the inside of the micelle. Dirt, which is basically nonpolar, is stabilized in the nonpolar interior of the micelle and is washed away.

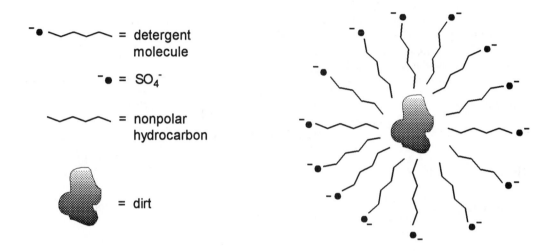

Exercises

Concentration of Solutions

27. $\text{mass \% CsCl} = \dfrac{\text{mass CsCl}}{\text{mass solution}} \times 100 = \dfrac{50.0 \text{ g CsCl}}{100.0 \text{ g solution}} \times 100 = 50.0\% \text{ CsCl by mass}$

$\text{molarity} = M = \dfrac{\text{mol solute}}{\text{L solution}} = \dfrac{50.0 \text{ g}}{63.3 \text{ mL}} \times \dfrac{1000 \text{ mL}}{\text{L}} \times \dfrac{1 \text{ mol}}{168.4 \text{ g}} = 4.69 \text{ mol/L}$

$\text{molality} = m = \dfrac{\text{mol solute}}{\text{kg solvent}} = \dfrac{50.0 \text{ g CsCl}}{50.0 \text{ g solvent}} \times \dfrac{1000 \text{ g}}{\text{kg}} \times \dfrac{1 \text{ mol CsCl}}{168.4 \text{ g}} = 5.94 \text{ mol/kg}$

$50.0 \text{ g CsCl} \times \dfrac{1 \text{ mol}}{168.4 \text{ g}} = 0.297 \text{ mol CsCl}; \ \ 50.0 \text{ g H}_2\text{O} \times \dfrac{1 \text{ mol}}{18.02 \text{ g}} = 2.77 \text{ mol H}_2\text{O}$

$\text{mole fraction CsCl} = \chi_{\text{CsCl}} = \dfrac{\text{mol CsCl}}{\text{total mol}} = \dfrac{0.297}{0.297 + 2.77} = 9.68 \times 10^{-2}$

28.   If we assume 100.0 g of solution, then the solution contains 5.0 g $HC_2H_3O_2$ and 95.0 g H_2O.

$$\text{mol } HC_2H_3O_2 = 5.0 \text{ g } HC_2H_3O_2 \times \frac{1 \text{ mol } HC_2H_3O_2}{60.05 \text{ g}} = 8.3 \times 10^{-2} \text{ mol } HC_2H_3O_2$$

$$\text{mol } H_2O = 95.0 \text{ g } H_2O \times \frac{1 \text{ mol } H_2O}{18.02 \text{ g}} = 5.27 \text{ mol } H_2O$$

$$\text{mol fraction } HC_2H_3O_2 = \chi_{HC_2H_3O_2} = \frac{8.3 \times 10^{-2} \text{ mol } HC_2H_3O_2}{(8.3 \times 10^{-2} + 5.27) \text{ mol total}} = 0.016$$

$$\text{molality} = m = \frac{8.3 \times 10^{-2} \text{ mol } HC_2H_3O_2}{95.0 \text{ g } H_2O \times \dfrac{1 \text{ kg}}{1000 \text{ g}}} = 0.87 \text{ mol/kg}$$

$$\text{molarity} = M = \frac{8.3 \times 10^{-2} \text{ mol } HC_2H_3O_2}{100.0 \text{ g solution} \times \dfrac{1 \text{ mL solution}}{1.01 \text{ g solution}} \times \dfrac{1 \text{ L}}{1000 \text{ mL}}} = 0.84 \text{ mol/L}$$

29.   Hydrochloric acid:

$$\text{molarity} = \frac{38 \text{ g HCl}}{100. \text{ g soln}} \times \frac{1.19 \text{ g soln}}{\text{cm}^3 \text{ soln}} \times \frac{1000 \text{ cm}^3}{L} \times \frac{1 \text{ mol HCl}}{36.46 \text{ g}} = 12 \text{ mol/L}$$

$$\text{molality} = \frac{38 \text{ g HCl}}{62 \text{ g solvent}} \times \frac{1000 \text{ g}}{\text{kg}} \times \frac{1 \text{ mol HCl}}{36.46 \text{ g}} = 17 \text{ mol/kg}$$

$$38 \text{ g HCl} \times \frac{1 \text{ mol}}{36.46 \text{ g}} = 1.0 \text{ mol HCl}; \quad 62 \text{ g } H_2O \times \frac{1 \text{ mol}}{18.02 \text{ g}} = 3.4 \text{ mol } H_2O$$

$$\text{mole fraction of HCl} = \chi_{HCl} = \frac{1.0}{3.4 + 1.0} = 0.23$$

Nitric acid:

$$\frac{70. \text{ g } HNO_3}{100. \text{ g soln}} \times \frac{1.42 \text{ g soln}}{\text{cm}^3 \text{ soln}} \times \frac{1000 \text{ cm}^3}{L} \times \frac{1 \text{ mol } HNO_3}{63.02 \text{ g}} = 16 \text{ mol/L}$$

$$\frac{70. \text{ g } HNO_3}{30. \text{ g solvent}} \times \frac{1000 \text{ g}}{\text{kg}} \times \frac{1 \text{ mol } HNO_3}{63.02 \text{ g}} = 37 \text{ mol/kg}$$

$$70. \text{ g } HNO_3 \times \frac{1 \text{ mol}}{63.02 \text{ g}} = 1.1 \text{ mol } HNO_3; \quad 30. \text{ g } H_2O \times \frac{1 \text{ mol}}{18.02 \text{ g}} = 1.7 \text{ mol } H_2O$$

$$\chi_{HNO_3} = \frac{1.1}{1.7 + 1.1} = 0.39$$

Sulfuric acid:

$$\frac{95 \text{ g H}_2\text{SO}_4}{100. \text{ g soln}} \times \frac{1.84 \text{ g soln}}{\text{cm}^3 \text{ soln}} \times \frac{1000 \text{ cm}^3}{\text{L}} \times \frac{1 \text{ mol H}_2\text{SO}_4}{98.09 \text{ g H}_2\text{SO}_4} = 18 \text{ mol/L}$$

$$\frac{95 \text{ g H}_2\text{SO}_4}{5 \text{ g H}_2\text{O}} \times \frac{1000 \text{ g}}{\text{kg}} \times \frac{1 \text{ mol}}{98.09 \text{ g}} = 194 \text{ mol/kg} \approx 200 \text{ mol/kg}$$

$$95 \text{ g H}_2\text{SO}_4 \times \frac{1 \text{ mol}}{98.09 \text{ g}} = 0.97 \text{ mol H}_2\text{SO}_4; \quad 5 \text{ g H}_2\text{O} \times \frac{1 \text{ mol}}{18.02 \text{ g}} = 0.3 \text{ mol H}_2\text{O}$$

$$\chi_{\text{H}_2\text{SO}_4} = \frac{0.97}{0.97 + 0.3} = 0.76$$

Acetic Acid:

$$\frac{99 \text{ g HC}_2\text{H}_3\text{O}_2}{100. \text{ g soln}} \times \frac{1.05 \text{ g soln}}{\text{cm}^3 \text{ soln}} \times \frac{1000 \text{ cm}^3}{\text{L}} \times \frac{1 \text{ mol}}{60.05 \text{ g}} = 17 \text{ mol/L}$$

$$\frac{99 \text{ g HC}_2\text{H}_3\text{O}_2}{1 \text{ g H}_2\text{O}} \times \frac{1000 \text{ g}}{\text{kg}} \times \frac{1 \text{ mol}}{60.05 \text{ g}} = 1600 \text{ mol/kg} \approx 2000 \text{ mol/kg}$$

$$99 \text{ g HC}_2\text{H}_3\text{O}_2 \times \frac{1 \text{ mol}}{60.05 \text{ g}} = 1.6 \text{ mol HC}_2\text{H}_3\text{O}_2; \quad 1 \text{ g H}_2\text{O} \times \frac{1 \text{ mol}}{18.02 \text{ g}} = 0.06 \text{ mol H}_2\text{O}$$

$$\chi_{\text{HC}_2\text{H}_3\text{O}_2} = \frac{1.6}{1.6 + 0.06} = 0.96$$

Ammonia:

$$\frac{28 \text{ g NH}_3}{100. \text{ g soln}} \times \frac{0.90 \text{ g}}{\text{cm}^3} \times \frac{1000 \text{ cm}^3}{\text{L}} \times \frac{1 \text{ mol}}{17.03 \text{ g}} = 15 \text{ mol/L}$$

$$\frac{28 \text{ g NH}_3}{72 \text{ g H}_2\text{O}} \times \frac{1000 \text{ g}}{\text{kg}} \times \frac{1 \text{ mol}}{17.03 \text{ g}} = 23 \text{ mol/kg}$$

$$28 \text{ g NH}_3 \times \frac{1 \text{ mol}}{17.03 \text{ g}} = 1.6 \text{ mol NH}_3; \quad 72 \text{ g H}_2\text{O} \times \frac{1 \text{ mol}}{18.02 \text{ g}} = 4.0 \text{ mol H}_2\text{O}$$

$$\chi_{\text{NH}_3} = \frac{1.6}{4.0 + 1.6} = 0.29$$

30.   a.   If we use 100. mL (100. g) of H_2O, we need:

$$0.100 \text{ kg H}_2\text{O} \times \frac{2.0 \text{ mol KCl}}{\text{kg}} \times \frac{74.55 \text{ g}}{\text{mol KCl}} = 14.9 \text{ g} = 15 \text{ g KCl}$$

Dissolve 15 g KCl in 100. mL H_2O to prepare a 2.0 m KCl solution. This will give us slightly more than 100 mL, but this will be the easiest way to make the solution. Since we don't know the density of the solution, we can't calculate the molarity and use a volumetric flask to make exactly 100 mL of solution.

b. If we took 15 g NaOH and 85 g H_2O, the volume would probably be less than 100 mL. To make sure we have enough solution, lets use 100. mL H_2O (100. g). Let x = mass of NaCl.

mass % = 15 = $\dfrac{x}{100. + x}$ × 100,   1500 + 15 x = 100. x, x = 17.6 g ≈ 18 g

Dissolve 18 g NaOH in 100. mL H_2O to make a 15% NaOH solution by mass.

c. In a fashion similar to 11.30b, let's use 100. mL CH_3OH. Let x = mass of NaOH.

100. mL CH_3OH × $\dfrac{0.79 \text{ g}}{\text{mL}}$ = 79 g CH_3OH

mass % = 25 = $\dfrac{x}{79 + x}$ × 100,   25(79) + 25 x = 100. x, x = 26.3 g ≈ 26 g

Dissolve 26 g NaOH in 100. mL CH_3OH.

d. To make sure we have enough solution, lets use 100. mL (100. g) of H_2O. Let x = mol $C_6H_{12}O_6$.

100. g H_2O × $\dfrac{1 \text{ mol } H_2O}{18.02 \text{ g}}$ = 5.55 mol H_2O

$\chi_{C_6H_{12}O_6}$ = 0.10 = $\dfrac{x}{x + 5.55}$;   0.10 x + 0.56 = x, x = 0.62 mol $C_6H_{12}O_6$

0.62 mol $C_6H_{12}O_6$ × $\dfrac{180.16 \text{ g}}{\text{mol}}$ = 110 g $C_6H_{12}O_6$

Dissolve 110 g $C_6H_{12}O_6$ in 100. mL of H_2O to prepare a solution with $\chi_{C_6H_{12}O_6}$ = 0.10.

31.   50.0 mL toluene × $\dfrac{0.867 \text{ g}}{\text{mL}}$ = 43.4 g toluene; 125 mL benzene × $\dfrac{0.874 \text{ g}}{\text{mL}}$ = 109 g benzene

mass % toluene = $\dfrac{\text{mass of toluene}}{\text{total mass}}$ × 100 = $\dfrac{43.4}{43.4 + 109}$ × 100 = 28.5%

molarity = $\dfrac{43.4 \text{ g toluene}}{175 \text{ mL soln}}$ × $\dfrac{1000 \text{ mL}}{\text{L}}$ × $\dfrac{1 \text{ mol toluene}}{92.13 \text{ g toluene}}$ = 2.69 mol/L

molality = $\dfrac{43.4 \text{ g toluene}}{109 \text{ g benzene}}$ × $\dfrac{1000 \text{ g}}{\text{kg}}$ × $\dfrac{1 \text{ mol toluene}}{92.13 \text{ g toluene}}$ = 4.32 mol/kg

43.4 g toluene × $\dfrac{1 \text{ mol}}{92.13 \text{ g}}$ = 0.471 mol toluene

$$109 \text{ g benzene} \times \frac{1 \text{ mol benzene}}{78.11 \text{ g benzene}} = 1.40 \text{ mol benzene}; \quad \chi_{\text{toluene}} = \frac{0.471}{0.471 + 1.40} = 0.252$$

32.   If there are 100.0 mL of wine:

$$12.5 \text{ mL } C_2H_5OH \times \frac{0.789 \text{ g}}{\text{mL}} = 9.86 \text{ g } C_2H_5OH \text{ and } 87.5 \text{ mL } H_2O \times \frac{1.00 \text{ g}}{\text{mL}} = 87.5 \text{ g } H_2O$$

$$\text{mass \% ethanol} = \frac{9.86}{87.5 + 9.86} \times 100 = 10.1\% \text{ by mass}$$

$$\text{molality} = \frac{9.86 \text{ g } C_2H_5OH}{0.0875 \text{ kg } H_2O} \times \frac{1 \text{ mol}}{46.07 \text{ g}} = 2.45 \text{ mol/kg}$$

33.   If we have 1.00 L of solution:

$$1.37 \text{ mol citric acid} \times \frac{192.12 \text{ g}}{\text{mol}} = 263 \text{ g citric acid}$$

$$1.00 \times 10^3 \text{ mL solution} \times \frac{1.10 \text{ g}}{\text{mL}} = 1.10 \times 10^3 \text{ g solution}$$

$$\text{mass \% of citric acid} = \frac{263 \text{ g}}{1.10 \times 10^3 \text{ g}} \times 100 = 23.9\%$$

In 1.00 L of solution, we have 263 g citric acid and $(1.10 \times 10^3 - 263) = 840$ g of H_2O.

$$\text{molality} = \frac{1.37 \text{ mol citric acid}}{0.84 \text{ kg } H_2O} = 1.6 \text{ mol/kg}$$

$$840 \text{ g } H_2O \times \frac{1 \text{ mol}}{18.02 \text{ g}} = 47 \text{ mol } H_2O; \quad \chi_{\text{citric acid}} = \frac{1.37}{47 + 1.37} = 0.028$$

Since citric acid is a triprotic acid, then the amount of protons citric acid can provide is three times the molarity. Therefore, normality = 3 × molarity:

$$\text{normality} = 3 \times 1.37 \, M = 4.11 \, N$$

34. $\dfrac{1.00 \text{ mol acetone}}{1.00 \text{ kg ethanol}} = 1.00 \text{ molal}; \quad 1.00 \times 10^3 \text{ g } C_2H_5OH \times \dfrac{1 \text{ mol}}{46.07 \text{ g}} = 21.7 \text{ mol } C_2H_5OH$

$$\chi_{\text{acetone}} = \frac{1.00}{1.00 + 21.7} = 0.0441$$

$$1 \text{ mol } CH_3COCH_3 \times \frac{58.08 \text{ g } CH_3COCH_3}{\text{mol } CH_3COCH_3} \times \frac{1 \text{ mL}}{0.788 \text{ g}} = 73.7 \text{ mL } CH_3COCH_3$$

$$1.00 \times 10^3 \text{ g ethanol} \times \frac{1 \text{ mL}}{0.789 \text{ g}} = 1270 \text{ mL}; \quad \text{Total volume} = 1270 + 73.7 = 1340 \text{ mL}$$

$$\text{molarity} = \frac{1.00 \text{ mol}}{1.34 \text{ L}} = 0.746 \, M$$

Energetics of Solutions and Solubility

35.   Using Hess's law:

$$KCl(s) \rightarrow K^+(g) + Cl^-(g) \qquad \Delta H = -\Delta H_{LE} = -(-715 \text{ kJ/mol})$$
$$K^+(g) + Cl^-(g) \rightarrow K^+(aq) + Cl^-(aq) \qquad \Delta H = \Delta H_{hyd} = -684 \text{ kJ/mol}$$

$$KCl(s) \rightarrow K^+(aq) + Cl^-(aq) \qquad \Delta H_{soln} = 31 \text{ kJ/mol}$$

ΔH_{soln} refers to the heat released or gained when a solute dissolves in a solvent. Here, an ionic compound dissolves in water.

36.   a. $$CsI(s) \rightarrow Cs^+(g) + I^-(g) \qquad \Delta H = -\Delta H_{LE} = -(-604 \text{ kJ})$$
$$Cs^+(g) + I^-(g) \rightarrow Cs^+(aq) + I^-(aq) \qquad \Delta H = \Delta H_{hyd}$$

$$CsI(s) \rightarrow Cs^+(aq) + I^-(aq) \qquad \Delta H_{soln} = 33 \text{ kJ}$$

$33 \text{ kJ} = 604 \text{ kJ} + \Delta H_{hyd}, \; \Delta H_{hyd} = -571 \text{ kJ}$

$$CsOH(s) \rightarrow Cs^+(g) + OH^-(g) \qquad \Delta H = -\Delta H_{LE} = -(-724 \text{ kJ})$$
$$Cs^+(g) + OH^-(g) \rightarrow Cs^+(aq) + OH^-(aq) \qquad \Delta H = \Delta H_{hyd}$$

$$CsOH(s) \rightarrow Cs^+(aq) + OH^-(aq) \qquad \Delta H_{soln} = -72 \text{ kJ}$$

$-72 \text{ kJ} = 724 \text{ kJ} + \Delta H_{hyd}, \; \Delta H_{hyd} = -796 \text{ kJ}$

b.   The enthalpy of hydration for CsOH is more exothermic than for CsI. Any differences must be due to differences in hydrations between OH$^-$ and I$^-$. Thus, the hydroxide ion is more strongly hydrated.

37.   Both Al(OH)$_3$ and NaOH are ionic compounds. Since the lattice energy is proportional to the charge of the ions, then the lattice energy of aluminum hydroxide is greater than that of sodium hydroxide. The attraction of water molecules for Al^{3+} and OH$^-$ cannot overcome the larger lattice energy and Al(OH)$_3$ is insoluble. For NaOH, the favorable hydration energy is large enough to overcome the smaller lattice energy and NaOH is soluble.

38.   The dissolving of an ionic solute in water can be thought of as taking place in two steps. The first step, called the lattice energy term, refers to breaking apart the ionic compound into gaseous ions. This step, as indicated in the problem requires a lot of energy and is unfavorable. The second step, called the hydration energy term, refers to the energy released when the separated gaseous ions are stabilized as water molecules surround the the ions. Since the interactions between water molecules and ions are strong, then a lot of energy is released when ions are hydrated. Thus, the dissolution process for ionic compounds can be thought of as consisting of an unfavorable and a favorable energy term. These two processes basically cancel each other out; so when ionic solids dissolve in water, the heat released or gained is minimal and the temperature change is minimal.

39.   Water is a polar solvent and dissolves polar solutes and ionic solutes. Carbon tetrachloride (CCl_4) is a nonpolar solvent and dissolves nonpolar solutes (like dissolves like).

   a.   Water; $Cu(NO_3)_2$ is an ionic solid.

   b. CCl_4; CS_2 is a nonpolar molecule.   c.   Water; CH_3CO_2H is polar.

   d. CCl_4; The long nonpolar hydrocarbon chain favors a nonpolar solvent (the molecule
            is mostly nonpolar).

   e.   Water; HCl is polar.   f. CCl_4; C_6H_6 is nonpolar.

40.   a.   water   b.   water   c.   hexane   d.   water

41.   Water is a polar molecule capable of hydrogen bonding. Polar molecules, especially molecules capable of hydrogen bonding, and ions are all attracted to water. For covalent compounds, as polarity increases, the attraction to water increases. For ionic compounds, as the charge of the ions increase and/or the size of the ions decrease, the attraction to water increases.

   a. CH_3CH_2OH; CH_3CH_2OH is polar while $CH_3CH_2CH_3$ is nonpolar.

   b. $CHCl_3$; $CHCl_3$ is polar while CCl_4 is nonpolar.

   c. CH_3CH_2OH; CH_3CH_2OH is much more polar than $CH_3(CH_2)_{14}CH_2OH$.

42.   a. NH_3; NH_3 is capable of H-bonding unlike PH_3.

   b. CH_3CN; CH_3CN is polar while CH_3CH_3 is nonpolar.

   c. CH_3CO_2H; CH_3CO_2H is capable of H-bonding, unlike the other compound.

43.   As the length of the hydrocarbon chain increases, the solubility decreases. The –OH end of the alcohols can hydrogen bond with water. The hydrocarbon chain, however, is basically nonpolar and interacts poorly with water. As the hydrocarbon chain gets longer, a greater portion of the molecule cannot interact with the water molecules and the solubility decreases, i.e., the effect of the –OH group decreases as the alcohols get larger.

44.   Benzoic acid is capable of hydrogen bonding, but a significant part of benzoic acid is the nonpolar benzene ring which is composed of only carbon and hydrogen. In benzene, a hydrogen bonded dimer forms:

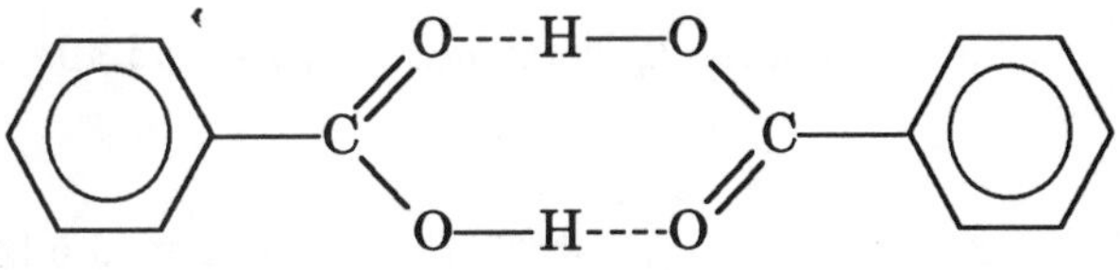

   The dimer is relatively nonpolar since the polar part of benzoic acid is hidden in the dimer formation. Thus, benzoic acid is more soluble in benzene than in water due to the dimer formation.

Benzoic acid would be more soluble in 0.1 M NaOH because of the reaction:

$$C_6H_5CO_2H + OH^- \rightarrow C_6H_5CO_2^- + H_2O$$

By removing the proton from benzoic acid, an anion forms, and like all anions, the species becomes more soluble in water.

45. $P_{gas} = kC$; 120 torr $\times \dfrac{1 \text{ atm}}{760 \text{ torr}} = \dfrac{780 \text{ atm L}}{\text{mol}} \times C$; $C = 2.0 \times 10^{-4}$ mol/L

46. $750.$ mL grape juice $\times \dfrac{12 \text{ mL } C_2H_5OH}{100. \text{ mL juice}} \times \dfrac{0.79 \text{ g } C_2H_5OH}{\text{mL}} \times \dfrac{1 \text{ mol } C_2H_5OH}{46.07 \text{ g}} \times \dfrac{2 \text{ mol } CO_2}{2 \text{ mol } C_2H_5OH}$

$$= 1.54 \text{ mol } CO_2 \quad \text{(carry extra significant figure)}$$

1.54 mol CO_2 = total mol CO_2 = mol $CO_2(g)$ + mol $CO_2(aq)$ = $n_g + n_{aq}$

$$P_{CO_2} = \dfrac{n_g RT}{V} = \dfrac{n_g\left(\dfrac{0.08206 \text{ L atm}}{\text{mol K}}\right)(298 \text{ K})}{75 \times 10^{-3} \text{ L}} = 326 \, n_g;$$

$$P_{CO_2} = kC = \dfrac{32 \text{ L atm}}{\text{mol}} \times \dfrac{n_{aq}}{0.750 \text{ L}} = 42.7 \, n_{aq}$$

$P_{CO_2} = 326 \, n_g = 42.7 \, n_{aq}$ and from above $n_{aq} = 1.54 - n_g$; Solving:

$326 \, n_g = 42.7(1.54 - n_g)$, $369 \, n_g = 65.8$, $n_g = 0.18$ mol

$P_{CO_2} = 326(0.18) = 59$ atm in gas phase; 59 atm $= \dfrac{32 \text{ L atm}}{\text{mol}} \times C$, $C = 1.8$ mol CO_2/L in wine

Vapor Pressures of Solutions

47. $P_{H_2O} = \chi_{H_2O} P_{H_2O}^{\circ}$; $\chi_{H_2O} = \dfrac{\text{mol } H_2O \text{ in solution}}{\text{total mol in solution}}$

50.0 g $C_6H_{12}O_6 \times \dfrac{1 \text{ mol } C_2H_{12}O_6}{180.16 \text{ g } C_6H_{12}O_6} = 0.278$ mol glucose

600.0 g $H_2O \times \dfrac{1 \text{ mol}}{18.02 \text{ g}} = 33.30$ mol H_2O; Total mol = $0.278 + 33.30 = 33.58$ mol

$\chi_{H_2O} = \dfrac{33.30}{33.58} = 0.9917$; $P_{H_2O} = \chi_{H_2O} P_{H_2O}^{\circ} = 0.9917 \times 23.8$ torr = 23.6 torr

48. $P = \chi P^{\circ}$; 710.0 torr = $\chi(760.0 \text{ torr})$; $\chi = 0.9342$ = mole fraction of methanol

49. $25.8 \text{ g CH}_4\text{N}_2\text{O} \times \dfrac{1 \text{ mol}}{60.06 \text{ g}} = 0.430 \text{ mol}; \ 275 \text{ g H}_2\text{O} \times \dfrac{1 \text{ mol}}{18.02 \text{ g}} = 15.3 \text{ mol}$

$\chi_{\text{H}_2\text{O}} = \dfrac{15.3}{15.3 + 0.430} = 0.973; \ P_{\text{H}_2\text{O}} = \chi_{\text{H}_2\text{O}} P_{\text{H}_2\text{O}}^{\circ} = 0.973 \,(23.8 \text{ torr}) = 23.2 \text{ torr at } 25\,^{\circ}\text{C}$

$P_{\text{H}_2\text{O}} = 0.973 \,(71.9 \text{ torr}) = 70.0 \text{ torr at } 45\,^{\circ}\text{C}$

50. $19.6 \text{ torr} = \chi_{\text{H}_2\text{O}} \,(23.8 \text{ torr}), \ \chi_{\text{H}_2\text{O}} = 0.824; \ \chi_{\text{solute}} = 1.000 - 0.824 = 0.176$

0.176 is the mol fraction of all the solute particles present. Since NaCl dissolves to produce two ions in solution (Na^+ and Cl^-), 0.176 is the mole fraction of Na^+ and Cl^- ions present. The mole fraction of NaCl is $1/2 \,(0.176) = 0.088 = \chi_{\text{NaCl}}$.

at $45\,^{\circ}\text{C}$, $P_{\text{H}_2\text{O}} = 0.824 \,(71.9 \text{ torr}) = 59.2 \text{ torr}$

51.   a. $25 \text{ mL C}_5\text{H}_{12} \times \dfrac{0.63 \text{ g}}{\text{mL}} \times \dfrac{1 \text{ mol}}{72.15 \text{ g}} = 0.22 \text{ mol C}_5\text{H}_{12}$

$45 \text{ mL C}_6\text{H}_{14} \times \dfrac{0.66 \text{ g}}{\text{mL}} \times \dfrac{1 \text{ mol}}{86.17 \text{ g}} = 0.34 \text{ mol C}_6\text{H}_{14}; \ \text{total mol} = 0.22 + 0.34 = 0.56 \text{ mol}$

$\chi_{\text{pen}}^{\text{L}} = \dfrac{\text{mol pentane in solution}}{\text{total mol in solution}} = \dfrac{0.22 \text{ mol}}{0.56 \text{ mol}} = 0.39, \ \chi_{\text{hex}}^{\text{L}} = 1.00 - 0.39 = 0.61$

$P_{\text{pen}} = \chi_{\text{pen}}^{\text{L}} P_{\text{pen}}^{\circ} = 0.39 \,(511 \text{ torr}) = 2.0 \times 10^2 \text{ torr}; \ P_{\text{hex}} = 0.61 \,(150.\text{ torr}) = 92 \text{ torr}$

$P_{\text{total}} = P_{\text{pen}} + P_{\text{hex}} = 2.0 \times 10^2 + 92 = 292 \text{ torr} = 290 \text{ torr}$

b.   From Chapter 5 on gases, the partial pressure of a gas is proportional to the number of moles of gas present. For the vapor phase:

$\chi_{\text{pen}}^{\text{V}} = \dfrac{\text{mol pentane in vapor}}{\text{total mol vapor}} = \dfrac{P_{\text{pen}}}{P_{\text{total}}} = \dfrac{2.0 \times 10^2 \text{ torr}}{290 \text{ torr}} = 0.69$

Note: In the Solutions Guide, we added V or L to the mole fraction symbol to emphasize which value we are solving. If the L or V is omitted, then the liquid phase is assumed.

52. $P_{\text{total}} = P_{\text{methanol}} + P_{\text{propanol}}; \ P = \chi^{\text{L}} P^{\circ}; \ \chi_{\text{methanol}}^{\text{L}} = \chi_{\text{propanol}}^{\text{L}} = \dfrac{1.00 \text{ mol}}{2.00 \text{ mol}} = 0.500$

$P_{\text{total}} = 0.500 \,(303 \text{ torr}) + 0.500 \,(44.6 \text{ torr}) = 152 + 22.3 = 174 \text{ torr}$

In the vapor: $\chi_{\text{methanol}}^{\text{V}} = \dfrac{P_{\text{methanol}}}{P_{\text{total}}} = \dfrac{152 \text{ torr}}{174 \text{ torr}} = 0.874; \ \chi_{\text{propanol}}^{\text{V}} = 1.000 - 0.874 = 0.126$

53. $P_{\text{total}} = P_{\text{pen}} + P_{\text{hex}}, \ 350.\text{ torr} = \chi_{\text{pen}}^{\text{L}} \,(511 \text{ torr}) + \chi_{\text{hex}}^{\text{L}} \,(150.\text{ torr}); \ \chi_{\text{hex}}^{\text{L}} = 1.000 - \chi_{\text{pen}}^{\text{L}}$

$350. = 511\,\chi_{\text{pen}}^{\text{L}} + (1.000 - \chi_{\text{pen}}^{\text{L}})150., \ \dfrac{200.}{361} = \chi_{\text{pen}}^{\text{L}} = 0.554; \ \chi_{\text{hex}}^{\text{L}} = 1.000 - 0.554 = 0.446$

For the vapor composition:

$$P_{pen} = \chi^L_{pen} P^\circ_{pen} = 0.554\,(511\text{ torr}) = 283\text{ torr}; \quad P_{hex} = 0.446\,(150.\text{ torr}) = 66.9\text{ torr}$$

Since partial pressures are proportional to the moles of gas present, then:

$$\chi^V_{pen} = \frac{P_{pen}}{P_{tot}} = \frac{283\text{ torr}}{350.\text{ torr}} = 0.809; \quad \chi^V_{hex} = 1.000 - 0.809 = 0.191$$

54. $P_{tol} = \chi^L_{tol} P^\circ_{tol}; \; P_{ben} = \chi^L_{ben} P^\circ_{ben}$; For the vapor, $\chi^V_A = P_A/P_{total}$. Since the mole fractions of benzene and toluene are equal in the vapor phase, then $P_{tol} = P_{ben}$.

$$\chi^L_{tol} P^\circ_{tol} = \chi^L_{ben} P^\circ_{ben} = (1.00 - \chi^L_{tol})P^\circ_{ben}, \quad \chi^L_{tol}(28\text{ torr}) = (1.00 - \chi^L_{tol})95\text{ torr}$$

$$123\,\chi^L_{tol} = 95, \quad \chi^L_{tol} = 0.77; \quad \chi^L_{ben} = 1.00 - 0.77 = 0.23$$

55.   Compared to H_2O, solution d (methanol/water) will have the highest vapor pressure since methanol is more volatile than water. Both solution b (glucose/water) and solution c (NaCl/water) will have a lower vapor pressure than water by Raoult's law. NaCl dissolves to give Na^+ ions and Cl^- ions; glucose is a nonelectrolyte. Since there are more solute particles in solution c, the vapor pressure of solution c will be the lowest.

56.   Solution d (methanol/water); Methanol is more volatile than water which will increase the total vapor pressure to a value greater than the vapor pressure of pure water at this temperature.

57. $$50.0\text{ g CH}_3\text{COCH}_3 \times \frac{1\text{ mol}}{58.08\text{ g}} = 0.861\text{ mol acetone}$$

$$50.0\text{ g CH}_3\text{OH} \times \frac{1\text{ mol}}{32.04\text{ g}} = 1.56\text{ mol methanol}$$

$$\chi^L_{acetone} = \frac{0.861}{0.861 + 1.56} = 0.356; \quad \chi^L_{methanol} = 1.000 - \chi^L_{acetone} = 0.644$$

$$P_{total} = P_{methanol} + P_{acetone} = 0.644(143\text{ torr}) + 0.356(271\text{ torr}) = 92.1\text{ torr} + 96.5\text{ torr} = 188.6\text{ torr}$$

Since partial pressures are proportional to the moles of gas present, then in the vapor phase:

$$\chi^V_{acetone} = \frac{P_{acetone}}{P_{total}} = \frac{96.5\text{ torr}}{188.6\text{ torr}} = 0.512; \quad \chi^V_{methanol} = 1.000 - 0.512 = 0.488$$

The actual vapor pressure of the solution (161 torr) is less than the calculated pressure assuming ideal behavior (188.6 torr). Therefore, the solution exhibits negative deviations from Raoult's law. This occurs when the solute-solvent interactions are stronger than in pure solute and pure solvent.

58.   a. CF_3CF_3 and $CF_3CF_2CF_3$; Solutions will be ideal when the intermolecular forces between the two substances are about equal. This usually occurs for two nonpolar substances of similar size since the strength of the London dispersion forces will be similar. CF_3CF_3 and $CF_3CF_2CF_3$ are both relatively nonpolar with similar size. Hydrogen bonding will be important in water/acetone solutions and will likely form solutions with a negative deviation from Raoult's law.

b. C_7H_{16} and C_6H_{14}; Both are nonpolar with similar size. The LD forces are approximately equal. Dipole forces are present in both CHF_3 and CH_3OCH_3 and it is unlikely that the dipole forces in solution are the same as in the two pure substances.

c. CCl_4 and CF_4 since both are nonpolar substances. There is hydrogen bonding in aqueous solutions of phosphoric acid and it is unlikely that the forces in solution are the same as in the pure substances.

Colligative Properties

59.   molality = m = $\dfrac{\text{mol solute}}{\text{kg solvent}}$ = $\dfrac{4.9 \text{ g sucrose}}{175 \text{ g solvent}}$ × $\dfrac{1000 \text{ g}}{\text{kg}}$ × $\dfrac{1 \text{ mol } C_{12}H_{22}O_{11}}{342.30 \text{ g } C_{12}H_{22}O_{11}}$ = 0.082 molal

$\Delta T_b = K_b m = \dfrac{0.51°C}{\text{molal}}$ × 0.082 molal = 0.042°C

The boiling point is raised from 100.000°C to 100.042°C. We assumed P = 1 atm and ample significant figures in the boiling point of pure water.

60. $m = \dfrac{\Delta T_b}{K_b} = \dfrac{0.55°C}{1.71 \text{ °C kg/mol}}$ = 0.32 mol/kg

mol hydrocarbon = 0.095 kg solvent × $\dfrac{0.32 \text{ mol hydrocarbon}}{\text{kg solvent}}$ = 0.030 mol hydrocarbon

From the problem, 3.75 g hydrocarbon was used which must contain 0.030 mol hydrocarbon. The molar mass of the hydrocarbon is:

$\dfrac{3.75 \text{ g}}{0.030 \text{ mol}}$ = 130 g/mol

61. $\Delta T_f = K_f m$, $\Delta T_f = 3.00°C = \dfrac{1.86°C}{\text{molal}}$ × m; m = 1.61 mol/kg

0.150 kg H_2O × $\dfrac{1.61 \text{ mol urea}}{\text{kg } H_2O}$ × $\dfrac{60.06 \text{ g urea}}{\text{mol urea}}$ = 14.5 g $(NH_2)_2CO$

62. ΔT = 25.50°C - 24.59°C = 0.91°C = $K_f m$, $m = \dfrac{0.91°C}{9.1°C/\text{molal}}$ = 0.10 mol/kg

mass H_2O = 0.0100 kg t-butanol $\left(\dfrac{0.10 \text{ mol } H_2O}{\text{kg t-butanol}} \right) \left(\dfrac{18.02 \text{ g } H_2O}{\text{mol } H_2O} \right)$ = 0.018 g H_2O

63.   molality = m = $\dfrac{40.0 \text{ g } C_2H_6O_2}{60.0 \text{ g } H_2O}$ × $\dfrac{1000 \text{ g}}{\text{kg}}$ × $\dfrac{1 \text{ mol}}{62.07 \text{ g}}$ = 10.7 mol/kg

$\Delta T_f = K_f m$ = 1.86°C/molal × 10.7 molal = 19.9°C; T_f = 0.0°C -19.9°C = -19.9°C

$\Delta T_b = K_b m$ = 0.51°C/molal × 10.7 molal = 5.5°C; T_b = 100.0°C + 5.5°C = 105.5°C

64. $m = \dfrac{\Delta T_f}{K_f} = \dfrac{30.0\,°C}{1.86\,°C \text{ kg/mol}} = 16.1 \text{ mol } C_2H_6O_2/kg$

Since the density of water is 1.00 g/cm^3, the moles of $C_2H_6O_2$ needed are:

$$15.0 \text{ L } H_2O \times \dfrac{1.00 \text{ kg } H_2O}{\text{L } H_2O} \times \dfrac{16.1 \text{ mol } C_2H_6O_2}{\text{kg } H_2O} = 242 \text{ mol } C_2H_6O_2$$

$$\text{Volume } C_2H_6O_2 = 242 \text{ mol } C_2H_6O_2 \times \dfrac{62.07 \text{ g}}{\text{mol } C_2H_6O_2} \times \dfrac{1 \text{ cm}^3}{1.11 \text{ g}} = 13{,}500 \text{ cm}^3 = 13.5 \text{ L}$$

$$\Delta T_b = K_b m = \dfrac{0.51\,°C}{\text{molal}} \times 16.1 \text{ molal} = 8.2\,°C; \quad T_b = 100.0\,°C + 8.2\,°C = 108.2\,°C$$

65. $\Delta T_f = K_f m, \quad m = \dfrac{\Delta T_f}{K_f} = \dfrac{2.63\,°C}{40.\,°C \text{ kg/mol}} = \dfrac{6.6 \times 10^{-2} \text{ mol reserpine}}{\text{kg solvent}}$

The mol of reserpine present is:

$$0.0250 \text{ kg solvent} \times \dfrac{6.6 \times 10^{-2} \text{ mol reserpine}}{\text{kg solvent}} = 1.7 \times 10^{-3} \text{ mol reserpine}$$

From the problem, 1.00 g reserpine was used which must contain 1.7×10^{-3} mol reserpine. The molar mass of reserpine is:

$$\text{molar mass} = \dfrac{1.00 \text{ g}}{1.7 \times 10^{-3} \text{ mol}} = 590 \text{ g/mol (610 g/mol if no rounding of numbers).}$$

66.   empirical formula mass $\approx 7(12) + 4(1) + 16 = 104$ g/mol

$$\Delta T_f = K_f m, \quad m = \dfrac{\Delta T_f}{K_f} = \dfrac{22.3\,°C}{40.\,°C/\text{molal}} = 0.56 \text{ molal}$$

$$\text{mol anthraquinone} = 0.0114 \text{ kg solvent} \times \dfrac{0.56 \text{ mol anthraquinone}}{\text{kg solvent}} = 6.4 \times 10^{-3} \text{ mol}$$

$$\text{molar mass} = \dfrac{1.32 \text{ g}}{6.4 \times 10^{-3} \text{ mol}} = 210 \text{ g/mol}$$

$$\dfrac{\text{molar mass}}{\text{empirical formula mass}} = \dfrac{210}{104} = 2.0; \quad \text{molecular formula} = C_{14}H_8O_2$$

67.   a. $M = \dfrac{1.0 \text{ g}}{\text{L}} \times \dfrac{1 \text{ mol}}{9.0 \times 10^4 \text{ g}} = 1.1 \times 10^{-5} \text{ mol/L}; \quad \pi = MRT$

At 298 K: $\pi = \dfrac{1.1 \times 10^{-5} \text{ mol}}{\text{L}} \times \dfrac{0.08206 \text{ L atm}}{\text{mol K}} \times 298 \text{ K} \times \dfrac{760 \text{ torr}}{\text{atm}}, \quad \pi = 0.20 \text{ torr}$

Since d = 1.0 g/cm^3, then 1.0 L solution has a mass of 1.0 kg. Since only 1.0 g of protein is present per liter solution, then 1.0 kg of H_2O is present and molality equals molarity.

$$\Delta T_f = K_f m = \frac{1.86\,°C}{molal} \times 1.1 \times 10^{-5}\ molal = 2.0 \times 10^{-5}\,°C$$

b. Osmotic pressure is better for determining the molar mass of large molecules. A temperature change of $10^{-5}\,°C$ is very difficult to measure. A change in height of a column of mercury by 0.2 mm (0.2 torr) is not as hard to measure precisely.

68. $\pi = MRT$, $M = \dfrac{\pi}{RT} = \dfrac{0.56\ torr \times \dfrac{1\ atm}{760\ torr}}{\dfrac{0.08206\ L\ atm}{mol\ K} \times 298\ K}$, $M = 3.0 \times 10^{-5}\ mol/L$

$$mol\ protein = 0.0250\ L \times \frac{3.0 \times 10^{-5}\ mol\ protein}{L} = 7.5 \times 10^{-7}\ mol$$

$$molar\ mass = \frac{0.0200\ g}{7.5 \times 10^{-7}\ mol} = 27,000\ g/mol$$

69. $\pi = MRT$, $M = \dfrac{\pi}{RT} = \dfrac{15\ atm}{0.08206 \times 295\ K} = 0.62\ M$; $\dfrac{0.62\ mol}{L} \times \dfrac{342.30\ g}{mol\ C_{12}H_{22}O_{11}} = 212\ g/L \approx 210\ g/L$

Dissolve 210 g of sucrose in some water and dilute to 1.0 L in a volumetric flask. To get 0.62 ± 0.01 mol/L, we need 212 ± 3 g sucrose.

70. $M = \dfrac{\pi}{RT} = \dfrac{15\ atm}{0.08206 \times 295\ K} = 0.62\ M$ solute particles

This represents the total molarity of solute particles. NaCl is a soluble ionic compound which breaks up into two ions, Na^+ and Cl^-. Therefore, the concentration of NaCl needed is $0.62/2 = 0.31\ M$.

$$1.0\ L \times \frac{0.31\ mol\ NaCl}{L} \times \frac{58.44\ g\ NaCl}{mol\ NaCl} = 18.1 \approx 18\ g\ NaCl$$

Dissolve 18 g of NaCl in some water and dilute to 1.0 L in a volumetric flask. To get 0.31 ± 0.01 mol/L, we need $18.1\ g \pm 0.6$ g NaCl.

Properties of Electrolyte Solutions

71. $MgCl_2$ and NaCl are strong electrolytes, HOCl is a weak electrolyte and glucose is a nonelectrolyte. The effective particle concentrations are $\sim$3.0 m $MgCl_2$, $\sim$2.0 m NaCl, $2.0 < m$ HOCl < 1.0, and 1.0 m glucose. The order of freezing point depressions ($\Delta T_f = K_f m$) from lowest to highest are: glucose $<$ HOCl $<$ NaCl $<$ $MgCl_2$ (a $<$ c $<$ b $<$ d).

72.   The solutions of glucose, NaCl and $CaCl_2$ will all have lower freezing points, higher boiling points and higher osmotic pressures than pure water. The solution with the largest particle concentration will have the lowest freezing point, the highest boiling point and the highest osmotic pressure. The $CaCl_2$ solution will have the largest effective particle concentration since it produces three ions per mol of compound.

a.   pure water   b. $CaCl_2$ solution   c. $CaCl_2$ solution

d.   pure water   e. $CaCl_2$ solution

73.   a. $m = \dfrac{5.0 \text{ g NaCl}}{0.025 \text{ kg}} \times \dfrac{1 \text{ mol}}{58.44 \text{ g}} = 3.4$ molal; $NaCl(aq) \rightarrow Na^+(aq) + Cl^-(aq)$, i = 2.0

$\Delta T_f = iK_f m = 2.0 \times 1.86°C/\text{molal} \times 3.4$ molal $= 13°C$; $T_f = -13°C$

$\Delta T_b = iK_b m = 2.0 \times 0.51°C/\text{molal} \times 3.4$ molal $= 3.5°C$; $T_b = 103.5°C$

b. $m = \dfrac{2.0 \text{ g Al(NO}_3)_3}{0.015 \text{ kg}} \times \dfrac{1 \text{ mol}}{213.01 \text{ g}} = 0.63$ mol/kg; $Al(NO_3)_3(aq) \rightarrow Al^{3+}(aq) + 3 \text{ NO}_3^-(aq)$, i = 4.0

$\Delta T_f = iK_f m = 4.0 \times 1.86°C/\text{molal} \times 0.63$ molal $= 4.7°C$; $T_f = -4.7°C$

$\Delta T_b = iK_b m = 4.0 \times 0.51°C/\text{molal} \times 0.63$ molal $= 1.3°C$; $T_b = 101.3°C$

74. $NaCl(s) \rightarrow Na^+(aq) + Cl^-(aq)$

$\pi = iMRT = 2.0 \times \dfrac{0.60 \text{ mol}}{L} \times \dfrac{0.08206 \text{ L atm}}{\text{mol K}} \times 298 \text{ K} = 29.3 \text{ atm} = 29 \text{ atm}$

A pressure greater than 30. atm should be applied to insure purification by reverse osmosis.

75. $\Delta T_f = iK_f m$, $i = \dfrac{\Delta T_f}{K_f m} = \dfrac{0.110°C}{1.86° \text{ C/molal} \times 0.0225 \text{ molal}} = 2.63$ for 0.0225 m $CaCl_2$

$i = \dfrac{0.440}{1.86 \times 0.0910} = 2.60$ for 0.0910 m $CaCl_2$; $i = \dfrac{1.330}{1.86 \times 0.278} = 2.57$ for 0.278 m $CaCl_2$

Note that i is less than the ideal value of 3.0 for $CaCl_2$. This is due to ion pairing in solution.

76.   For $CaCl_2$: $i = \dfrac{\Delta T_f}{K_f m} = \dfrac{0.440°C}{1.86° \text{ C/molal} \times 0.091 \text{ molal}} = 2.6$

$\% \text{ } CaCl_2 \text{ ionized} = \dfrac{2.6 - 1.0}{3.0 - 1.0} \times 100 = 80.\%$

For CsCl: $i = \dfrac{\Delta T_f}{K_f m} = \dfrac{0.302°C}{1.86°C/\text{molal} \times 0.091 \text{ molal}} = 1.8$

$\% \text{ CsCl ionized} = \dfrac{1.8 - 1.0}{2.0 - 1.0} \times 100 = 80.\%$

It appears that the i value for $CaCl_2$ is further from the ideal value. However, since both compounds are equally ionized, we can conclude that the extent of ion association is about the same in each solution.

77. $\pi = iMRT = 3.0 \times 0.50$ mol/L $\times 0.08206$ L atm/K•mol $\times 298$ K $= 37$ atm

Because of ion pairing in solution, we would expect i to be less than 3.0 which results in fewer solute particles in solution which results in a lower osmotic pressure than calculated above.

78.   a. $T_C = (T_F - 32)/9 = 5(-29 - 32)/9 = -34\,^{\circ}C$

Assuming the solubility of $CaCl_2$ is temperature independent, the molality of a saturated $CaCl_2$ solution is:

$$\frac{74.5 \text{ g } CaCl_2}{100.0 \text{ g } H_2O} \times \frac{1000 \text{ g}}{\text{kg}} \times \frac{1 \text{ mol } CaCl_2}{110.98 \text{ g } CaCl_2} = \frac{6.71 \text{ mol } CaCl_2}{\text{kg } H_2O}$$

$\Delta T_f = iK_f m = 3.00 \times 1.86\,^{\circ}C$ kg/mol $\times 6.71$ mol/kg $= 37.4\,^{\circ}C$

Assuming i = 3.00, a saturated solution of $CaCl_2$ can lower the freezing point of water to $-37.4\,^{\circ}C$. Assuming these conditions, a saturated $CaCl_2$ solution should melt ice at $-34\,^{\circ}C$ $(-29\,^{\circ}F)$.

b.   From Exercise 11.75, i ≈ 2.6; $\Delta T_f = iK_f m = 2.6 \times 1.86 \times 6.71 = 32\,^{\circ}C$; $T_f = -32\,^{\circ}C$

Assuming i = 2.6, a saturated $CaCl_2$ solution will not melt ice at $-34\,^{\circ}C(-29\,^{\circ}F)$.

Additional Exercises

79.   molality $= \dfrac{40.0 \text{ g EG}}{60.0 \text{ g } H_2O} \times \dfrac{1000 \text{ g}}{\text{kg}} \times \dfrac{1 \text{ mol EG}}{62.07 \text{ g}} = 10.7$ mol/kg

molarity $= \dfrac{40.0 \text{ g EG}}{100.0 \text{ g solution}} \times \dfrac{1.05 \text{ g}}{\text{cm}^3} \times \dfrac{1000 \text{ cm}^3}{\text{L}} \times \dfrac{1 \text{ mol}}{62.07 \text{ g}} = 6.77$ mol/L

40.0 g EG $\times \dfrac{1 \text{ mol}}{62.07 \text{ g}} = 0.644$ mol EG;   60.0 g $H_2O \times \dfrac{1 \text{ mol}}{18.02 \text{ g}} = 3.33$ mol H_2O

$\chi_{EG} = \dfrac{0.644}{3.33 + 0.644} = 0.162 =$ mole fraction ethylene glycol

80.   The main intermolecular forces are:

hexane (C_6H_{14}): London dispersion;   Chloroform, $(CHCl_3)$: dipole-dipole, London dispersion

methanol (CH_3OH): H-bonding; H_2O: H-bonding (two places)

There is a gradual change in the nature of the intermolecular forces (weaker to stronger). Each preceding solvent is miscible in its predecessor because there is not a great change in the strengths of the intermolecular forces from one solvent to the next.

81. NH_3 is capable of forming hydrogen bonding interactions with water so NH_3 is very soluble in water. O_2 is a nonpolar substance and is not soluble in water.

82. $CO_2 + OH^- \rightarrow HCO_3^-$; No, the reaction of CO_2 with OH^- greatly increases the solubility of CO_2 in water by forming the soluble bicarbonate anion.

83. $P_B = \chi_B P_B^{\circ}$, $\chi_B = P_B/P_B^{\circ} = 0.900 \text{ atm}/0.930 \text{ atm} = 0.968$

$0.968 = \dfrac{\text{mol benzene}}{\text{total mol}}$; mol benzene $= 78.11 \text{ g } C_6H_6 \times \dfrac{1 \text{ mol}}{78.11} = 1.000 \text{ mol}$

Let x = mol solute, then: $\chi_B = 0.968 = \dfrac{1.000 \text{ mol}}{1.000 + x}$, $0.968 + 0.968\, x = 1.000$, $x = 0.033 \text{ mol}$

molar mass $= \dfrac{10.0 \text{ g}}{0.033 \text{ mol}} = 303 \text{ g/mol} \approx 3.0 \times 10^2 \text{ g/mol}$

84. $m = \dfrac{24.0 \text{ g} \times \dfrac{1 \text{ mol}}{58.0 \text{ g}}}{0.600 \text{ kg}} = 0.690 \text{ mol/kg}$; $\Delta T_b = K_b m = 0.51\,°\text{C kg/mol} \times 0.690 \text{ mol/kg} = 0.35\,°\text{C}$

$T_b = 99.725\,°\text{C} + 0.35\,°\text{C} = 100.08\,°\text{C}$

85.   Out of 100.00 g, there are:

$31.57 \text{ g C} \times \dfrac{1 \text{ mol C}}{12.01 \text{ g}} = 2.629 \text{ mol C}$; $\dfrac{2.629}{2.629} = 1.000$

$5.30 \text{ g H} \times \dfrac{1 \text{ mol H}}{1.008 \text{ g}} = 5.26 \text{ mol H}$; $\dfrac{5.26}{2.629} = 2.00$

$63.13 \text{ g O} \times \dfrac{1 \text{ mol O}}{16.00 \text{ g}} = 3.946 \text{ mol O}$; $\dfrac{3.946}{2.629} = 1.501$

empirical formula: $C_2H_4O_3$; Use the freezing point data to determine the molar mass.

$m = \dfrac{\Delta T_f}{K_f} = \dfrac{5.20\,°\text{C}}{1.86\,°\text{C/molal}} = 2.80 \text{ molal}$

mol solute $= 0.0250 \text{ kg} \times \dfrac{2.80 \text{ mol solute}}{\text{kg}} = 0.0700 \text{ mol solute}$

molar mass $= \dfrac{10.56 \text{ g}}{0.0700 \text{ mol}} = 151 \text{ g/mol}$

The empirical formula mass of $C_2H_4O_3 = 76.05 \text{ g/mol}$. Since the molar mass is about twice the empirical mass, then the molecular formula is $C_4H_8O_6$ which has a molar mass of 152.10 g/mol.

Note: We use the experimental molar mass to determine the molecular formula. Knowing this, we calculate the molar mass precisely from the molecular formula using the periodic table.

86. $\pi = MRT$, $\pi = 18.6$ torr $\times \dfrac{1 \text{ atm}}{760 \text{ torr}} = M \times \dfrac{0.08206 \text{ L atm}}{\text{mol K}} \times 298$ K, $M = 1.00 \times 10^{-3}$ mol/L

mol protein $= 0.0020$ L $\times \dfrac{1.00 \times 10^{-3} \text{ mol protein}}{\text{L}} = 2.0 \times 10^{-6}$ mol protein

molar mass $= \dfrac{0.15 \text{ g}}{2.0 \times 10^{-6}} = 7.5 \times 10^4$ g/mol

87.   a.   As disccussed in Figure 11.18 of the text, the water would migrate from right to left. Initially, the level of liquid in the right arm would go down and the level in the left arm would go up. At some point, the rate of solvent transfer will be the same in both directions and the levels of the liquids in the two arms will stabilize. The height difference between the two arms will be a measure of the osmotic pressure of the NaCl solution.

       b.   Initially, H_2O molecules will have a net migration into the NaCl side. However, NaCl molecules can now migrate into the H_2O side. Because solute and solvent transfer are both possible, the levels of the liquids will be equal once the rate of solute and solvent transfer is equal in both directions. At this point, the concentration of NaCl will be equal in both chambers and the levels of liquid will be equal.

88.   If ideal, NaCl dissociates completely and i = 2.00. $\Delta T_f = iK_f m$; Assuming water freezes at $0.00°C$:

       $1.28°C = 2 \times 1.86°C$ kg/mol $\times m$, $m = 0.344$ mol NaCl/kg H_2O

       Assume an amount of solution which contains 1.00 kg of water (solvent).

       0.344 mol NaCl $\times \dfrac{58.44 \text{ g}}{\text{mol}} = 20.1$ g NaCl; mass % NaCl $= \dfrac{20.1 \text{ g}}{1.00 \times 10^3 \text{ g} + 20.1 \text{ g}} \times 100 = 1.97\%$

Challenge Problems

89. $\chi_{pen}^{V} = 0.15 = \dfrac{P_{pen}}{P_{total}}$; $P_{pen} = \chi_{pen}^{L} P_{pen}^{°} = \chi_{pen}^{L}(511 \text{ torr})$; $P_{total} = P_{pen} + P_{hex} = \chi_{pen}^{L}(511) + \chi_{hex}^{L}(150.)$

       Since $\chi_{hex}^{L} = 1.000 - \chi_{pen}^{L}$, then: $P_{total} = \chi_{pen}^{L}(511) + (1.000 - \chi_{pen}^{L})(150.) = 150. + 361 \chi_{pen}^{L}$

       $\chi_{pen}^{V} = \dfrac{P_{pen}}{P_{total}}$, $0.15 = \dfrac{\chi_{pen}^{L}(511)}{150. + 361 \chi_{pen}^{L}}$, $0.15 (150. + 361 \chi_{pen}^{L}) = 511 \chi_{pen}^{L}$

       $23 + 54 \chi_{pen}^{L} = 511 \chi_{pen}^{L}$, $\chi_{pen}^{L} = \dfrac{23}{457} = 0.050$

90.   a. $m = \dfrac{\Delta T_f}{K_f} = \dfrac{1.32°C}{5.12°C\ kg/mol} = 0.258\ mol/kg$

mol unknown $= 0.01560\ kg \times \dfrac{0.258\ mol\ unknown}{kg} = 4.02 \times 10^{-3}\ mol$

molar mass of unknown $= \dfrac{1.22\ g}{4.02 \times 10^{-3}\ mol} = 303\ g/mol$

Uncertainty in temperature $= \dfrac{0.04}{1.32} \times 100 = 3\%$; A 3% uncertainty in 303 g/mol = 9 g/mol.

So, molar mass = 303 ± 9 g/mol.

   b.   No, codeine could not be eliminated since its molar mass is in the possible range including the uncertainty.

   c.   We would really like the uncertainty to be ± 1 g/mol. We need the freezing point depression to be about 10 times what it was in this problem. Two possibilities are:

   1.   make the solution ten times more concentrated (may be solubility problem) or
   2.   use a solvent with a larger K_f value, e.g., camphor

91. $\Delta T_f = 5.51 - 2.81 = 2.70°C$; $m = \dfrac{\Delta T_f}{K_f} = \dfrac{2.7°C}{5.12°C/molal} = 0.527\ molal$

Let x = mass of naphthalene (molar mass: 128.2 g/mol). Then $1.60 - x$ = mass of anthracene (molar mass = 178.2 g/mol).

$\dfrac{x}{128.2}$ = moles naphthalene and $\dfrac{1.60 - x}{178.2}$ = moles anthracene

$\dfrac{0.527\ mol\ solute}{kg\ solvent} = \dfrac{\dfrac{x}{128.2} + \dfrac{1.60 - x}{178.2}}{0.0200\ kg\ solvent}$, $1.05 \times 10^{-2} = \dfrac{178.2\,x + 1.60\,(128.2) - 128.2\,x}{128.2\,(178.2)}$

$50.0\,x + 205 = 240.$, $50.0\,x = 35$, $x = 0.70$ g naphthalene

So mixture is: $\dfrac{0.70\ g}{1.60\ g} \times 100 = 44\%$ naphthalene by mass and 56% anthracene by mass

92. $\Delta T_f = K_f m$, $m = \dfrac{\Delta T}{K_f} = \dfrac{5.40°C}{1.86°C/molal} = 2.90\ molal$

$\dfrac{2.90\ mol\ solute}{kg\ solvent} = \dfrac{n}{0.0500\ kg}$, n = 0.145 mol of ions in solution

Since $NaNO_3$ and $Mg(NO_3)_2$ are strong electrolytes:

n = 2(x mol of $NaNO_3$) + 3[y mol $Mg(NO_3)_2$] = 0.145 mol ions

In addition: $6.50 \text{ g} = x \text{ mol NaNO}_3 \left(\dfrac{85.00 \text{ g}}{\text{mol}} \right) + y \text{ mol Mg(NO}_3)_2 \left(\dfrac{148.3 \text{ g}}{\text{mol}} \right)$

We have two equations: $2 x + 3 y = 0.145$ and $85.00 x + 148.3 y = 6.50$

Solving by simultaneous equations:

$-85.00 x - 127.5 y = -6.16$
$85.00 x + 148.3 y = 6.50$

$\overline{}$

$ 20.8 y = 0.34, \ y = 0.016 \text{ mol Mg(NO}_3)_2$

mass of Mg(NO$_3$)$_2$ = 0.016 mol × 148.3 g/mol = 2.4 g Mg(NO$_3$)$_2$

mass of NaNO$_3$ = 6.50 g - 2.4 g = 4.1 g NaNO$_3$

93. $HCO_2H \rightarrow H^+ + HCO_2^-$; Only 4.2% of HCO_2H ionizes. The amount of H^+ or HCO_2^- produced is:

$0.042 \times 0.10 \, M = 0.0042 \, M$

The amount of HCO_2H remaining in solution after ionization is:

$0.10 \, M - 0.0042 \, M = 0.10 \, M$

The total molarity of species present $= M_{HCO_2H} + M_{H^+} + M_{HCO_2^-} = 0.10 + 0.0042 + 0.0042 = 0.11 \, M$.

Assuming $0.11 \, M = 0.11$ molal and assuming ample significant figures in the freezing point and boiling point of water at P = 1 atm:

$\Delta T = K_f m = 1.86°C/\text{molal} \times 0.11 \text{ molal} = 0.20°C$; freezing point = -0.20°C

$\Delta T = K_b m = 0.51°C/\text{molal} \times 0.11 \text{ molal} = 0.056°C$;   boiling point = 100.056°C

94.   a.   Assuming MgCO$_3$(s) does not dissociate, the solute concentration in water is:

$$\frac{560 \, \mu g \, MgCO_3(s)}{mL} = \frac{560 \text{ mg}}{L} = \frac{560 \times 10^{-3} \text{ g}}{L} \times \frac{1 \text{ mol MgCO}_3}{84.32 \text{ g}} = 6.6 \times 10^{-3} \text{ mol MgCO}_3/L$$

An applied pressure of 8.0 atm will purify water up to a solute concentration of:

$$M = \frac{\pi}{RT} = \frac{8.0 \text{ atm}}{0.08206 \text{ L atm/mol} \cdot \text{K} \times 300. \text{ K}} = \frac{0.32 \text{ mol}}{L}$$

When the concentration of MgCO$_3$(s) reaches 0.32 mol/L, the reverse osmosis unit can no longer purify the water. Let V = volume (L) of water remaining after purifying 45 L of H$_2$O. When V + 45 L of water has been processed, the moles of solute particles will equal:

$6.6 \times 10^{-3} \text{ mol/L} \times (45 \text{ L} + V) = 0.32 \text{ mol/L} \times V$

Solving: $0.30 = (0.32 - 0.0066) \times V$, $V = 0.96$ L

The minimum total volume of water that must be processed is 45 L + 0.96 L = 46 L.

Note: If $MgCO_3$ does dissociate into Mg^{2+} and CO_3^{2-} ions, then the solute concentration increases to 1.3×10^{-2} M and at least 47 L of water must be processed.

b.   No; A reverse osmosis system that applies 8.0 atm can only purify water with a solute concentration less than 0.32 mol/L. Salt water has a solute concentration of 2(0.60 m) = 1.20 m ions. The solute concentration of salt water is much too high for this reverse osmosis unit to work.

CHAPTER TWELVE

CHEMICAL KINETICS

Questions

8.   The rate of a chemical reaction varies with time. Consider the general reaction:

$$A \rightarrow \text{Products where rate} = \frac{-\Delta[A]}{\Delta t}$$

If we graph [A] vs. t, it would roughly look like the dark line in the following plot.

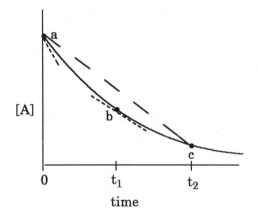

An instantaneous rate is the slope of a tangent line to the graph of [A] vs. t. We can determine the instantaneous rate at any time during the reaction. On the plot, tangent lines at $t = 0$ and $t = t_1$ are drawn. The slope of these tangent lines would be the instantaneous rates at $t \approx 0$ and $t = t_1$. We call the instantaneous rate at $t \approx 0$ the initial rate. The average rate is measured over a period of time. For example, the slope of the line connecting points a and c is the average rate of the reaction over the entire length of time 0 to t_2. The rate which is largest is the initial rate. At $t \approx 0$, the slope of the tangent line is greatest which means the rate is largest at $t \approx 0$.

9.   a.   An elementary step (reaction) is one in which the rate law can be written from the molecularity, i.e., from the coefficients in the balanced equation.

         b.   The mechanism of a reaction is the series of proposed elementary reactions that may occur to give the overall reaction. The sum of all the steps in the mechanism gives the balanced chemical reaction.

c.   The rate determining step is the slowest elementary reaction in any given mechanism.

10.   a.   The greater the frequency of collisions, the greater the opportunities for molecules to react, and, hence, the greater the rate.

b.   Chemical reactions involve the making and breaking of chemical bonds. The kinetic energy of the collisions can be used to break bonds. So, as the kinetic energy of the collisions increase, the rate increases.

c.   For a reaction to occur, it is the reactive portion of each molecule that must be involved in a collision. Only some of all the possible collisions have the correct orientation to convert from reactants to products.

11.   In a unimolecular reaction, a single reactant molecule decomposes to products. In a bimolecular reaction, two molecules collide to give products.

12.   The probability of the simultaneous collision of three molecules with the enough energy and orientation is very small.

13.   a.   A homogeneous catalyst is in the same phase as the reactants.

b.   A heterogeneous catalyst is in a different phase than the reactants. The catalyst is usually a solid, although a catalyst in a liquid phase can act as a heterogeneous catalyst for some gas phase reactions.

14.   A catalyst increases the rate of a reaction by providing reactants with an alternate pathway (mechanism) to convert to products. This alternate pathway has a lower activiation energy thus increasing the rate of the reaction.

15.   No, the catalyzed reaction has a different mechanism and hence, a different rate law.

16.   Some energy must be added to get the reaction started, that is, to overcome the activation energy barrier. Chemically what happens is:

$$Energy + H_2 \rightarrow 2\ H$$

The hydrogen atoms initiate a chain reaction that proceeds very rapidly. Collisions of H_2 and O_2 molecules at room temperature do not have sufficient kinetic energy to form hydrogen atoms and initiate the reaction.

17.   a.   Activation energy and ΔE are independent of each other. Activation energy depends on the path reactants to take to convert to products. The overall energy change, ΔE, only depends on the initial and final energy states of the reactants and products. ΔE is path independent.

b.   The rate law can only be determined from experiment, not from the overall balanced reaction.

c.   Most reactions occur by a series of steps. The rate of the reaction is determined by the rate of the slowest step in the mechanism.

18.   All of these choices would affect the rate of the reaction, but only b and c affect the rate by affecting the value of the rate constant k. The value of the rate constant is dependent on temperature. The value of the rate constant also depends on the activation energy. A catalyst will change the value of k because the activation energy changes. Increasing the concentration of either H_2 or NO does not affect the value of k, but they do increase the rate of the reaction because both concentrations appear in the rate law.

Exercises

Reaction Rates

19.   The coefficients in the balanced equation tell us that the rate of consumption of $S_2O_3^{2-}$ will equal the rate of production of I^-, as well as telling us that the rate of consumption of I_2 and the rate of production of $S_4O_6^{2-}$ will be one-half the rate of consumption of $S_2O_3^{2-}$ (due to the 1:2 mol ratios).

$$\text{Rate} = \frac{-\Delta[S_2O_3^{2-}]}{\Delta t} = \frac{0.0080 \text{ mol}}{\text{L s}}; \quad I_2 \text{ is consumed at half this rate.} \quad \frac{-\Delta[I_2]}{\Delta t} = 0.0040 \text{ mol/L} \cdot \text{s}$$

$$\frac{\Delta[S_4O_6^{2-}]}{\Delta t} = 0.0040 \text{ mol/L} \cdot \text{s}; \quad \frac{\Delta[I^-]}{\Delta t} = 0.0080 \text{ mol/L} \cdot \text{s}$$

20. $$\frac{\Delta[H_2]}{\Delta t} = 3\frac{\Delta[N_2]}{\Delta t} \text{ and } \frac{\Delta[NH_3]}{\Delta t} = -2\frac{\Delta[N_2]}{\Delta t}; \quad \text{So,} \quad -\frac{1}{3}\frac{\Delta[H_2]}{\Delta t} = \frac{1}{2}\frac{\Delta[NH_3]}{\Delta t}$$

$$\text{or} \quad \frac{\Delta[NH_3]}{\Delta t} = -\frac{2}{3}\frac{\Delta[H_2]}{\Delta t}$$

Ammonia is produced at a rate equal to 2/3 of the rate of consumption of hydrogen.

21.   a.   The units for rate are always mol/L·s.   b.   Rate = k; k must have units of mol/L·s

c.   Rate = k[A], $\quad \dfrac{\text{mol}}{\text{L s}} = k\left(\dfrac{\text{mol}}{\text{L}}\right)$ d.   Rate = k[A]^2, $\quad \dfrac{\text{mol}}{\text{L s}} = k\left(\dfrac{\text{mol}}{\text{L}}\right)^2$

k must have units of s^{-1}.   k must have units of L/mol·s.

e. $L^2/\text{mol}^2 \cdot \text{s}$

22.   Rate = k[Cl]$^{1/2}$[CHCl$_3$], $\quad \dfrac{\text{mol}}{\text{L s}} = k\left(\dfrac{\text{mol}}{\text{L}}\right)^{1/2}\left(\dfrac{\text{mol}}{\text{L}}\right)$, k must have units of $L^{1/2}/\text{mol}^{1/2} \cdot \text{s}$.

Rate Laws from Experimental Data: Initial Rates Method

23.   a.   In the first two experiments, [NO] is held constant and [Cl$_2$] is doubled. The rate also doubled. Thus, the reaction is first order with respect to Cl$_2$. Or mathematically: Rate = k[NO]^x[Cl$_2$]^y

$$\frac{0.36}{0.18} = \frac{k(0.10)^x(0.20)^y}{k(0.10)^x(0.10)^y} = \frac{(0.20)^y}{(0.10)^y}, \; 2.0 = 2.0^y, \; y = 1$$

We can get the dependence on NO from the second and third experiments. Here, as the NO concentration doubles (Cl_2 concentration is constant), the rate increases by a factor of four. Thus, the reaction is second order with respect to NO. Or mathematically:

$$\frac{1.45}{0.36} = \frac{k(0.20)^x(0.20)}{k(0.10)^x(0.20)} = \frac{(0.20)^x}{(0.10)^x}, \; 4.0 = 2.0^x, \; x = 2; \; \text{So, Rate} = k[NO]^2[Cl_2]$$

Try to examine experiments where only one concentration changes at a time. The more variables that change, the harder it is to determine the orders. Also, these types of problems can usually be solved by inspection. In general, we will solve using a mathematical approach, but keep in mind, you probably can solve for the orders by simple inspection of the data.

b. The rate constant k can be determined from the experiments. From experiment 1:

$$\frac{0.18 \; \text{mol}}{L \, \text{min}} = k\left(\frac{0.10 \; \text{mol}}{L}\right)^2\left(\frac{0.10 \; \text{mol}}{L}\right), \; k = 180 \; L^2/\text{mol}^2\bullet\text{min}$$

From the other experiments:

$$k = 180 \; L^2/\text{mol}^2\bullet\text{min (2nd exp.);} \; k = 180 \; L^2/\text{mol}^2\bullet\text{min (3rd exp.)}$$

The average rate constant is $k_{mean} = 1.8 \times 10^2 \; L^2/\text{mol}^2\bullet\text{min}$.

24.   a.   Rate = $k[I^-]^x[S_2O_8^{2-}]^y$; $\; \dfrac{12.5 \times 10^{-6}}{6.25 \times 10^{-6}} = \dfrac{k(0.080)^x(0.040)^y}{k(0.040)^x(0.040)^y}, \; 2.00 = 2.0^x, \; x = 1$

$$\frac{12.5 \times 10^{-6}}{6.25 \times 10^{-6}} = \frac{k(0.080)(0.040)^y}{k(0.080)(0.020)^y}, \; 2.00 = 2.0^y, \; y = 1; \; \text{Rate} = k[I^-][S_2O_8^{2-}]$$

b. For the first experiment:

$$\frac{12.5 \times 10^{-6} \; \text{mol}}{L \, s} = k\left(\frac{0.080 \; \text{mol}}{L}\right)\left(\frac{0.040 \; \text{mol}}{L}\right), \; k = 3.9 \times 10^{-3} \; L/\text{mol}\bullet s$$

The other values are:

Initial Rate mol/L•s	k L/mol•s
12.5×10^{-6}	3.9×10^{-3}
6.25×10^{-6}	3.9×10^{-3}
6.25×10^{-6}	3.9×10^{-3}
5.00×10^{-6}	3.9×10^{-3}
7.00×10^{-6}	3.9×10^{-3}

$k_{mean} = 3.9 \times 10^{-3} \; L/\text{mol}\bullet s$

25.   a.   Rate = $k[NOCl]^n$; Using experiments two and three:

$$\frac{2.66 \times 10^4}{6.64 \times 10^3} = \frac{k(2.0 \times 10^{16})^n}{k(1.0 \times 10^{16})^n}, \quad 4.01 = 2.0^n, \ n = 2; \ \text{Rate} = k[NOCl]^2$$

       b. $\dfrac{5.98 \times 10^4 \text{ molecules}}{cm^3 \ s} = k\left(\dfrac{3.0 \times 10^{16} \text{ molecules}}{cm^3}\right)^2, \ k = 6.6 \times 10^{-29} \ cm^3/\text{molecules}\bullet s$

             The other three experiments give $(6.7, 6.6 \text{ and } 6.6) \times 10^{-29} \ cm^3/\text{molecules}\bullet s$, respectively.

             The mean value for k is $6.6 \times 10^{-29} \ cm^3/\text{molecules}\bullet s$.

       c. $\dfrac{6.6 \times 10^{-29} \ cm^3}{\text{molecules } s} \times \dfrac{1 \ L}{1000 \ cm^3} \times \dfrac{6.022 \times 10^{23} \text{molecules}}{mol} = \dfrac{4.0 \times 10^{-8} \ L}{mol \ s}$

26.   a.   Rate = $k[I^-]^x[OCl^-]^y$; $\dfrac{7.91 \times 10^{-2}}{3.95 \times 10^{-2}} = \dfrac{k(0.12)^x(0.18)^y}{k(0.060)^x(0.18)^y} = 2.0^x, \ 2.00 = 2.0^x, \ x = 1$

             $\dfrac{3.95 \times 10^{-2}}{9.88 \times 10^{-3}} = \dfrac{k(0.060)(0.18)^y}{k(0.030)(0.090)^y}, \ 4.00 = 2.0 \times 2.0^y, \ 2.0 = 2.0^y, \ y = 1$

             Rate = $k[I^-][OCl^-]$

       b.   From the first experiment: $\dfrac{7.91 \times 10^{-2} \ mol}{L \ s} = k\left(\dfrac{0.12 \ mol}{L}\right)\left(\dfrac{0.18 \ mol}{L}\right), \ k = 3.7 \ L/\text{mol}\bullet s$

             All four experiments give the same value of k to two significant figures.

27.   a.   Rate = $k[Hb]^x[CO]^y$; Comparing the first two experiments, [CO] is unchanged, [Hb] doubles, and the rate doubles. Therefore, the reaction is first order in Hb. Comparing the second and third experiments, [Hb] is unchanged, [CO] triples and the rate triples. Therefore, $y = 1$ and the reaction is first order in CO.

       b.   Rate = $k[Hb][CO]$

       c.   From the first experiment:

             $0.619 \ \mu mol/L\bullet s = k \ (2.21 \ \mu mol/L)(1.00 \ \mu mol/L), \ k = 0.280 \ L/\mu mol\bullet s$

             The second and third experiments give similar k values, so $k_{mean} = 0.280 \ L/\mu mol\bullet s$.

       d.   Rate = $k[Hb][CO] = \dfrac{0.280 \ L}{\mu mol \ s} \times \dfrac{3.36 \ \mu mol}{L} \times \dfrac{2.40 \ \mu mol}{L} = 2.26 \ \mu mol/L\bullet s$

28.   a.   Rate = $k[ClO_2]^x[OH^-]^y$; From the first two experiments:

             $2.30 \times 10^{-1} = k(0.100)^x(0.100)^y \text{ and } 5.75 \times 10^{-2} = k(0.0500)^x(0.100)^y$

Dividing the two rate laws: $4.00 = \dfrac{(0.100)^x}{(0.0500)^x} = 2.00^x$, $x = 2$

Comparing the second and third experiments:

$2.30 \times 10^{-1} = k(0.100)(0.100)^y$ and $1.15 \times 10^{-1} = k(0.100)(0.050)^y$

Dividing: $2.00 = \dfrac{(0.100)^y}{(0.050)^y} = 2.0^y$, $y = 1$

The rate law is: Rate $= k[ClO_2]^2[OH^-]$

2.30×10^{-1} mol/L•s $= k(0.100 \text{ mol/L})^2(0.100 \text{ mol/L})$, $k = 2.30 \times 10^2$ L^2/mol^2•s $= k_{mean}$

b.   Rate $= k[ClO_2]^2[OH^-] = \dfrac{2.30 \times 10^2 \, L^2}{mol^2 \, s} \times \left(\dfrac{0.175 \text{ mol}}{L}\right)^2 \times \dfrac{0.0844 \text{ mol}}{L} = 0.594$ mol/L•s

Integrated Rate Laws

29.   The first assumption to make is that the reaction is first order. For a first order reaction, a graph of ln [H$_2$O$_2$] vs time will yield a straight line. If this plot is not linear, then the reaction is not first order and we make another assumption. The data and plot for the first order assumption is below.

Time (s)	[H$_2$O$_2$] (mol/L)	ln [H$_2$O$_2$]
0	1.00	0.000
120.	0.91	-0.094
300.	0.78	-0.25
600.	0.59	-0.53
1200.	0.37	-0.99
1800.	0.22	-1.51
2400.	0.13	-2.04
3000.	0.082	-2.50
3600.	0.050	-3.00

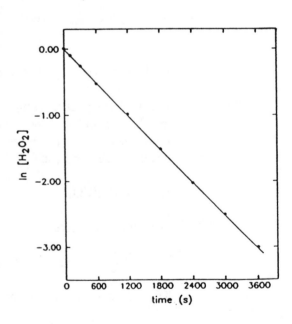

Note: We carried extra significant figures in some of the ln values in order to reduce round off error. For the plots, we will do this most of the time when the ln function is involved.

The plot of ln [H$_2$O$_2$] vs. time is linear. Thus, the reaction is first order. The rate law and integrated rate law are: Rate $= k[H_2O_2]$ and ln [H$_2$O$_2$] $= -kt + $ ln [H$_2$O$_2$]$_o$.

We determine the rate constant k by determining the slope of the ln [H$_2$O$_2$] vs time plot (slope = -k). Using two points on the curve gives:

$$\text{slope} = -k = \frac{\Delta y}{\Delta x} = \frac{0 - (3.00)}{0 - 3600.} = -8.3 \times 10^{-4} \text{ s}^{-1}, \ k = 8.3 \times 10^{-4} \text{ s}^{-1}$$

To determine [H$_2$O$_2$] at 4000. s, use the integrated rate law where at t = 0, [H$_2$O$_2$]$_0$ = 1.00 M.

$$\ln [\text{H}_2\text{O}_2] = -kt + \ln [\text{H}_2\text{O}_2]_0 \ \text{ or } \ \ln\left(\frac{[\text{H}_2\text{O}_2]}{[\text{H}_2\text{O}_2]_0}\right) = -kt$$

$$\ln\left(\frac{[\text{H}_2\text{O}_2]}{1.00}\right) = -8.3 \times 10^{-4} \text{ s}^{-1} \times 4000. \text{ s}, \ \ln [\text{H}_2\text{O}_2] = -3.3, \ [\text{H}_2\text{O}_2] = e^{-3.3} = 0.037 \ M$$

30. a. Since the ln[A] vs time plot was linear, then the reaction is first order in A. The slope of the ln[A] vs time plot equals -k. Therefore, the rate law, the integrated rate law and the rate constant value are:

$$\text{Rate} = k[\text{A}]; \ \ln[\text{A}] = -kt + \ln[\text{A}]_0; \ k = 6.90 \times 10^{-2} \text{ s}^{-1}$$

b. The half-life expression for a first order rate law is:

$$t_{1/2} = \frac{\ln 2}{k} = \frac{0.6931}{k}, \ t_{1/2} = \frac{0.6931}{6.90 \times 10^{-2} \text{ s}^{-1}} = 10.0 \text{ s}$$

c. When a first order reaction is 87.5% complete (or 12.5% remains), then the reaction has gone through 3 half-lives:

$$\begin{array}{ccccccc}
100\% & \to & 50.0\% & \to & 25\% & \to & 12.5\%; \ \ t = 3 \times t_{1/2} = 3 \times 10.0 \text{ s} = 30.0 \text{ s} \\
 & t_{1/2} & & t_{1/2} & & t_{1/2} &
\end{array}$$

Or we can use the integrated rate law where [A] = 0.125 [A]$_0$.

$$\ln\left(\frac{[\text{A}]}{[\text{A}]_0}\right) = -kt, \ \ln\left(\frac{0.125 [\text{A}]_0}{[\text{A}]_0}\right) = -(6.90 \times 10^{-3} \text{ s}^{-1})\,t, \ t = \frac{\ln (0.125)}{-6.90 \times 10^{-3} \text{ s}^{-1}} = 30.1 \text{ s}$$

31. Assume the reaction is first order and see if the plot of ln [NO$_2$] vs. time is linear. If this isn't linear, try the second order plot of 1/[NO$_2$] vs. time. The data and plots follow.

Time (s)	[NO$_2$] (M)	ln [NO$_2$]	1/[NO$_2$] (M^{-1})
0	0.500	-0.693	2.00
1.20×10^3	0.444	-0.812	2.25
3.00×10^3	0.381	-0.965	2.62
4.50×10^3	0.340	-1.079	2.94
9.00×10^3	0.250	-1.386	4.00
1.80×10^4	0.174	-1.749	5.75

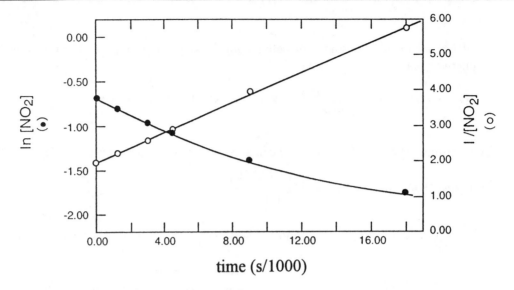

time (s/1000)

The plot of $1/[NO_2]$ vs. time is linear. The reaction is second order in NO_2. The rate law and integrated rate law are: Rate = $k[NO_2]^2$ and $\dfrac{1}{[NO_2]} = kt + \dfrac{1}{[NO_2]_o}$.

The slope of the plot $1/[NO_2]$ vs. t gives the value of k. Using a couple points on the plot:

$$\text{slope} = k = \frac{\Delta y}{\Delta x} = \frac{(5.75 - 2.00)\,M^{-1}}{(1.80 \times 10^{-4} - 0)\,\text{s}} = 2.08 \times 10^{-4}\ \text{L/mol}\bullet\text{s}$$

To determine $[NO_2]$ at 2.70×10^4 s, use the integrated rate law where $1/[NO_2]_o = 1/0.500\,M = 2.00\ M^{-1}$

$$\frac{1}{[NO_2]} = kt + \frac{1}{[NO_2]_o},\quad \frac{1}{[NO_2]} = \frac{2.08 \times 10^{-4}\,\text{L}}{\text{mol s}} \times 2.70 \times 10^4\,\text{s} + 2.00\,M^{-1}$$

$$\frac{1}{[NO_2]} = 7.62,\quad [NO_2] = 0.131\,M$$

32.   a.   Since the $1/[A]$ vs time plot was linear, then the reaction is second order in A. The slope of the $1/[A]$ vs. time plot equals the rate constant k. Therefore, the rate law, the integrated rate law and the rate constant value are:

$$\text{Rate} = k[A]^2;\quad \frac{1}{[A]} = kt + \frac{1}{[A]_o};\quad k = 4.15 \times 10^{-3}\ \text{L/mol}\bullet\text{s}$$

b.   The half-life expression for a second order reaction is: $t_{1/2} = \dfrac{1}{k[A]_o}$

For this reaction: $t_{1/2} = \dfrac{1}{4.15 \times 10^{-3}\ \text{L/mol}\bullet\text{s} \times 0.100\ \text{mol/L}} = 2.41 \times 10^3\ \text{s}$

c.   Since the half-life for a second order reaction depends on concentration, then we must use the integrated rate law to solve. If a reaction is 75.0% complete, this means that 25.0% of A remains.

$[A]_o = 0.100\ M;\ \ [A] = 0.250 \times 0.100\ M = 2.50 \times 10^{-2}\ M$

$$\frac{1}{[A]} = kt + \frac{1}{[A]_o}, \quad \frac{1}{2.50 \times 10^{-2}\,M} = \frac{4.15 \times 10^{-3}\,L}{mol \bullet s} \times t + \frac{1}{0.100\,M}$$

$40.0 - 10.0 = 4.15 \times 10^{-3}\ t, \ \ t = 7.23 \times 10^3\ s$

33.   a.   The integrated rate law for this zero order reaction is: $[HI] = -kt + [HI]_o$. This equation is in the form of the generic straight line equation, $y = mx + b$. A plot of $[HI]$ vs time will give a straight line with a negative slope equal to $-k$.

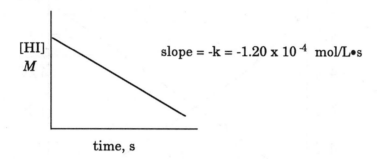

slope = $-k = -1.20 \times 10^{-4}$ mol/L$\bullet$s

time, s

b. $[HI] = -kt + [HI]_o, \ \ [HI] = -\left(\dfrac{1.20 \times 10^{-4}\,mol}{L\,s}\right) \times \left(25\,min \times \dfrac{60\,s}{min}\right) + \dfrac{0.250\,mol}{L}$

$[HI] = -0.18\ mol/L + 0.250\ mol/L = 0.07\ M$

c. $[HI] = 0 = -kt + [HI]_o, \ \ kt = [HI]_o, \ \ t = \dfrac{[HI]_o}{k}$

$t = \dfrac{0.250\ mol/L}{1.20 \times 10^{-4}\ mol/L \bullet s} = 2080\ s = 34.7\ min$

34.   a. $[A] = -kt + [A]_o, \ \ [A] = -(5.0 \times 10^{-2}\ mol/L \bullet s)\,t + 1.0 \times 10^{-3}\ mol/L$

b.   The half-life expression for a zero order reaction is: $t_{1/2} = \dfrac{[A]_o}{2\,k}$

$t_{1/2} = \dfrac{1.0 \times 10^{-3}\ mol/L}{2 \times 5.0 \times 10^{-2}\ mol/L \bullet s} = 1.0 \times 10^{-2}\ s$

c. $[A] = -5.0 \times 10^{-2}\ mol/L \bullet s \times 5.0 \times 10^{-3}\ s + 1.0 \times 10^{-3}\ mol/L = 7.5 \times 10^{-4}\ mol/L$

Since $7.5 \times 10^{-4}\ M$ A remains, then $2.5 \times 10^{-4}\ M$ A reacted which means that $2.5 \times 10^{-4}\ M$ B has been produced.

35.   The first assumption to make is that the reaction is first order. For a first order reaction, a graph of ln $[C_4H_6]$ vs. t should yield a straight line. If this isn't linear, then try the second order plot of $1/[C_4H_6]$ vs. t. The data and the plots follow.

Time	195	604	1246	2180	6210 s
$[C_4H_6]$	1.6×10^{-2}	1.5×10^{-2}	1.3×10^{-2}	1.1×10^{-2}	$0.68 \times 10^{-2} M$
$\ln [C_4H_6]$	-4.14	-4.20	-4.34	-4.51	-4.99
$1/[C_4H_6]$	62.5	66.7	76.9	90.9	$147 M^{-1}$

Note: To reduce round off error, we carried extra sig. figs. in the data points.

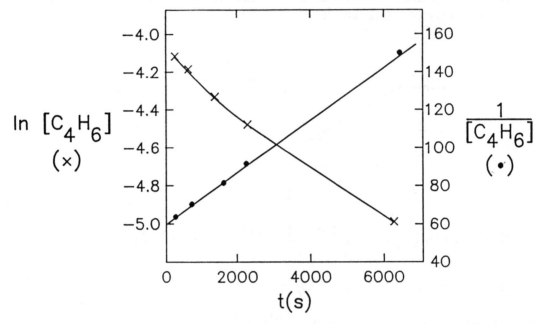

The natural log plot is not linear, so the reaction is not first order. Since the second order plot of $1/[C_4H_6]$ vs. t is linear, then we can conclude that the reaction is second order in butadiene. The rate law is:

$$Rate = k[C_4H_6]^2$$

For a second order reaction, the integrated rate law is: $\dfrac{1}{[C_4H_6]} = kt + \dfrac{1}{[C_4H_6]_o}$

The slope of the straight line equals the value of the rate constant. Using the points on the line at 1000. and 6000. s:

$$k = slope = \dfrac{144 \text{ L/mol} - 73 \text{ L/mol}}{6000. \text{ s} - 1000. \text{ s}} = 1.4 \times 10^{-2} \text{ L/mol} \cdot s$$

36. a.   First, assume the reaction to be first order with respect to O. Hence, a graph of ln [O] vs. t should be linear if the reaction is first order.

t(s)	[O] (atoms/cm^3)	ln[O]
0	5.0×10^9	22.33
$10. \times 10^{-3}$	1.9×10^9	21.37
$20. \times 10^{-3}$	6.8×10^8	20.34
$30. \times 10^{-3}$	2.5×10^8	19.34

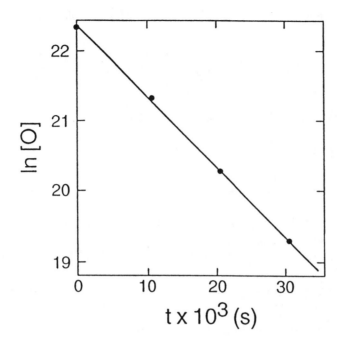

Since the graph is linear, we can conclude that the reaction is first order with respect to O.

b.   The overall rate law is: Rate = $k[NO_2][O]$

Since NO_2 was in excess, its concentration is constant. So for this experiment, the rate law is: Rate = $k'[O]$ where $k' = k[NO_2]$. In a typical first order plot, the slope equals -k. For this experiment, the slope equals $-k' = -k[NO_2]$. From the graph:

$$\text{slope} = -\frac{19.34 - 22.33}{(30. \times 10^{-3} - 0)\text{ s}} = -1.0 \times 10^2 \text{ s}^{-1}, \ k' = -\text{slope} = 1.0 \times 10^2 \text{ s}^{-1}$$

To determine k, the actual rate constant:

$$k' = k[NO_2], \ 1.0 \times 10^2 \text{ s}^{-1} = k(1.0 \times 10^{13} \text{ molecules/cm}^3), \ k = 1.0 \times 10^{-11} \text{ cm}^3/\text{molecules} \cdot \text{s}$$

37.   If $[A]_o = 100.0$, then after 65 s, 45.0% of A has reacted or $[A] = 55.0$. For first order reactions:

$$\ln\left(\frac{[A]}{[A]_o}\right) = -kt, \ \ln\left(\frac{55.0}{100.0}\right) = -k(65 \text{ s}), \ k = 9.2 \times 10^{-3} \text{ s}^{-1}; \ t_{1/2} = \frac{\ln 2}{k} = \frac{0.693}{k} = 75 \text{ s}$$

38. $\ln\left(\dfrac{[A]}{[A]_o}\right) = -kt; \ \ k = \dfrac{\ln 2}{t_{1/2}} = \dfrac{0.6931}{14.3 \text{ d}} = 4.85 \times 10^{-2} \text{ d}^{-1}$

If $[A]_o = 100.0$, then after 95.0% completion, $[A] = 5.0$.

$\ln\left(\dfrac{5.0}{100.0}\right) = -4.85 \times 10^{-2} \text{ d}^{-1} \times t, \ t = 62 \text{ days}$

39.   a.   If the reaction is 38.5% complete, then 38.5% of the original concentration is consumed, leaving 61.5%.

$[A] = 61.5\%$ of $[A]_o$ or $[A] = 0.615\,[A]_o$; $\ln\left(\dfrac{[A]}{[A]_o}\right) = -kt$, $\ln\left(\dfrac{0.615\,[A]_o}{[A]_o}\right) = -k(480\text{ s})$

$\ln(0.615) = -k(480.\text{ s})$, $-0.486 = -k(480.\text{ s})$, $k = 1.01 \times 10^{-3}\text{ s}^{-1}$

b. $t_{1/2} = (\ln 2)/k = 0.6931/1.01 \times 10^{-3}\text{ s}^{-1} = 686\text{ s}$

c.   25% complete: $[A] = 0.75\,[A]_o$; $\ln(0.75) = -1.01 \times 10^{-3}\,(t)$, $t = 280\text{ s}$

75% complete: $[A] = 0.25\,[A]_o$; $\ln(0.25) = -1.01 \times 10^{-3}\,(t)$, $t = 1.4 \times 10^3\text{ s}$

Or, we know it takes $2 \times t_{1/2}$ for reaction to be 75% complete. $t = 2 \times 686\text{ s} = 1370\text{ s}$

95% complete: $[A] = 0.05\,[A]_o$; $\ln(0.05) = -1.01 \times 10^{-3}\,(t)$, $t = 3 \times 10^3\text{ s}$

40.   Since the reaction is 50.% complete in 3.5 hours, then $t_{1/2} = 3.5$ hours.

$k = \dfrac{\ln 2}{t_{1/2}} = \dfrac{0.693}{3.5\text{ hr}} = 0.20\text{ hr}^{-1}$

When 88% complete, then $[A] = 0.12\,[A]_o$:

$\ln\left(\dfrac{[A]}{[A]_o}\right) = -kt$, $\ln(0.12) = -0.20\text{ hr}^{-1}\,(t)$, $t = 11\text{ hr}$

41.   For a second order reaction: $t_{1/2} = \dfrac{1}{k[A]_o}$ or $k = \dfrac{1}{t_{1/2}[A]_o}$

$k = \dfrac{1}{143\text{ s}(0.060\text{ mol/L})} = 0.12\text{ L/mol}\bullet\text{s}$

42. $\dfrac{1}{[A]} = kt + \dfrac{1}{[A]_o}$, $\dfrac{1}{0.020\text{ mol/L}} = 0.40\text{ L/mol}\bullet\text{min} \times t + \dfrac{1}{0.10\text{ mol/L}}$, $t = \dfrac{50.-10.}{0.40} = 1.0 \times 10^2\text{ min}$

43. $100\% \to 50\% \to 25\% \to 12.5\%$; This process is 3 half-lives $= 3(14\text{ h}) = 42$ hours.

44.   Successive half-lives increase in time for a second order reaction. Therefore, assume reaction is second order in A.

$t_{1/2} = \dfrac{1}{k[A]_o}$, $k = \dfrac{1}{t_{1/2}[A]_o} = \dfrac{1}{10.0\text{ min}\,(0.10\,M)} = 1.0\text{ L/mol}\bullet\text{min}$

a. $\dfrac{1}{[A]} = kt + \dfrac{1}{[A]_o} = \dfrac{1.0\text{ L}}{\text{mol min}} \times 80.0\text{ min} + \dfrac{1}{0.10\,M} = 90.\,M^{-1}$, $[A] = 1.1 \times 10^{-2}\,M$

b.   30.0 min = 2 half-lives, so 25% of original A is remaining.

[A] = 0.25(0.10 M) = 0.025 M

Reaction Mechanisms

45.   For elementary reactions, the rate law can be written using the coefficients in the balanced equation to determine orders.

a.   Rate = $k[CH_3NC]$

b.   Rate = $k[O_3][NO]$

46.   a.   Rate = $k[O_3]$

b.   Rate = $k[O_3][O]$

47.   From experiment (Exercise 12.29), we know the rate law is: Rate = $k[H_2O_2]$. A mechanism consists of a series of elementary reactions where the rate law for each step can be determined using the coefficients in the balanced equations. For a plausible mechanism, the rate law derived from a mechanism must agree with the rate law determined from experiment. To derive the rate law from the mechanism, the rate of the reaction is assumed to equal the rate of the slowest step in the mechanism.

This mechanism will agree with the experimentally determined rate law only if step 1 is the slow step (called the rate determining step). If step 1 is slow, then Rate = $k[H_2O_2]$ which agrees with experiment.

Another important property of a mechanism is that the sum of all steps must give the overall balanced equation. Summing all steps gives:

$$H_2O_2 \rightarrow 2\ OH$$
$$H_2O_2 + OH \rightarrow H_2O + HO_2$$
$$HO_2 + OH \rightarrow H_2O + O_2$$

$$\overline{}$$

$$2\ H_2O_2 \rightarrow 2\ H_2O + O_2$$

48.   Since the rate of the slowest elementary step equals the rate of a reaction, then:

Rate = rate of step 1 = $k[NO_2]^2$

The sum of all steps in a plausible mechanism must give the overall balanced reaction. Summing all steps gives:

$$NO_2 + NO_2 \rightarrow NO_3 + NO$$
$$NO_3 + CO \rightarrow NO_2 + CO_2$$

$$\overline{}$$

$$NO_2 + CO \rightarrow NO + CO_2$$

Temperature Dependence of Rate Constants and the Collision Model

49.   In the following plot, R = reactants, P = products, E_a = activation energy and RC = reaction
        coordinate which is the same as reaction progress. Note for this reaction that ΔE is positive since the
        products are at a higher energy than the reactants.

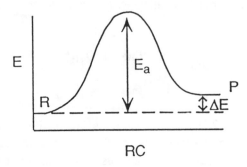

50.   When ΔE is positive, the products are at a higher energy relative to reactants and when ΔE is
        negative, the products are at a lower energy relative to reactants.

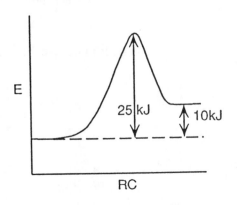

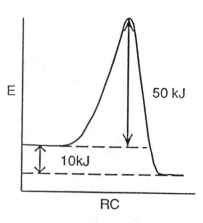

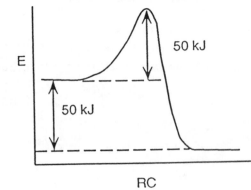

51.

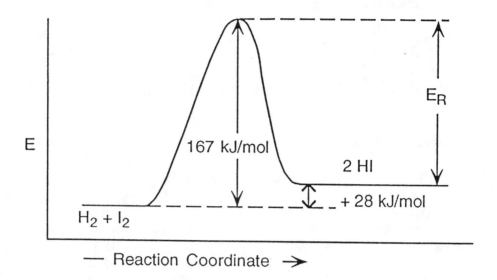

The activation energy for the reverse reaction is E_R in the diagram. $E_R = 167 - 28 = 139$ kJ/mol

52.   An example of this type of reaction is the reaction in Exercise 12.51. From the energy profile in
Exercise 51, when $E_{a, \text{ forward}} > E_{a, \text{ reverse}}$, then ΔE is positive. When ΔE is negative, then
$E_{a, \text{ reverse}} > E_{a, \text{ forward}}$.

53.   The Arrhenius equation is: $k = A \exp(-E_a/RT)$ or in logarithmic form, $\ln k = -E_a/RT + \ln A$.
Hence, a graph of $\ln k$ vs. $1/T$ should yield a straight line with a slope equal to $-E_a/R$ since the
logarithmic form of the Arrhenius equation is in the form of a straight line equation, $y = mx + b$.
Note: We carried one extra significant figure in the following $\ln k$ values in order to reduce round off
error.

T (K)	1/T (K^{-1})	k (L/mol•s)	ln k
195	5.13×10^{-3}	1.08×10^{9}	20.80
230.	4.35×10^{-3}	2.95×10^{9}	21.81
260.	3.85×10^{-3}	5.42×10^{9}	22.41
298	3.36×10^{-3}	12.0×10^{9}	23.21
369	2.71×10^{-3}	35.5×10^{9}	24.29

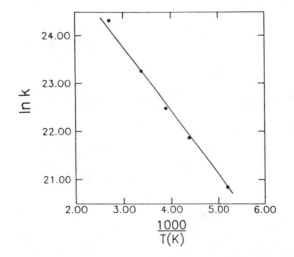

From the "eyeball" line on the graph:

$$\text{slope} = \frac{20.95 - 23.65}{5.00 \times 10^{-3} - 3.00 \times 10^{-3}} = \frac{-2.70}{2.00 \times 10^{-3}} = -1.35 \times 10^3 \text{ K} = \frac{-E_a}{R}$$

$$E_a = 1.35 \times 10^3 \text{ K} \times \frac{8.3145 \text{ J}}{\text{K mol}} = 1.12 \times 10^4 \text{ J/mol} = 11.2 \text{ kJ/mol}$$

From the best straight line (by calculator): slope = -1.43×10^3 K and $E_a = 11.9$ kJ/mol

54.   A graph of ln k vs. 1/T should be linear with slope = $-E_a/R$.

T (K)	1/T (K^{-1})	k (s^{-1})	ln k
338	2.96×10^{-3}	4.9×10^{-3}	-5.32
318	3.14×10^{-3}	5.0×10^{-4}	-7.60
298	3.36×10^{-3}	3.5×10^{-5}	-10.26

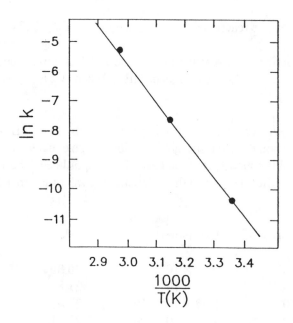

$$\text{Slope} = \frac{-10.76 - (-5.85)}{3.40 \times 10^{-3} - 3.00 \times 10^{-3}} = -1.2 \times 10^4 \text{ K} = -E_a/R$$

$$E_a = -\text{slope} \times R = 1.2 \times 10^4 \text{ K} \times \frac{8.3145 \text{ J}}{\text{K mol}}, \quad E_a = 1.0 \times 10^5 \text{ J/mol} = 1.0 \times 10^2 \text{ kJ/mol}$$

55. $k = A \exp(-E_a/RT)$ or $\ln k = \dfrac{-E_a}{RT} + \ln A$ (the Arrhenius equation)

For two conditions: $\ln\left(\dfrac{k_2}{k_1}\right) = \dfrac{E_a}{R}\left(\dfrac{1}{T_1} - \dfrac{1}{T_2}\right)$ (Assuming A is temperature independent.)

Let $k_1 = 2.0 \times 10^3 \text{ s}^{-1}$, $T_1 = 298$ K; $k_2 = ?$, $T_2 = 348$ K; $E_a = 15.0 \times 10^3$ J/mol

$$\ln\left(\frac{k_2}{2.0 \times 10^3 \text{ s}^{-1}}\right) = \frac{15.0 \times 10^3 \text{ J/mol}}{8.3145 \text{ J/mol}\bullet\text{K}}\left(\frac{1}{298 \text{ K}} - \frac{1}{348 \text{ K}}\right) = 0.87$$

$$\ln\left(\frac{k_2}{2.0 \times 10^3}\right) = 0.87, \quad \frac{k_2}{2.0 \times 10^3} = e^{0.87} = 2.4, \quad k_2 = 2.4(2.0 \times 10^3) = 4.8 \times 10^3 \text{ s}^{-1}$$

56.   For two conditions: $\ln\left(\dfrac{k_2}{k_1}\right) = \dfrac{E_a}{R}\left(\dfrac{1}{T_1} - \dfrac{1}{T_2}\right)$ (Assuming A factor is T independent.)

$$\ln\left(\frac{8.1 \times 10^{-2} \text{ s}^{-1}}{4.6 \times 10^{-2} \text{ s}^{-1}}\right) = \frac{E_a}{8.3145 \text{ J/mol}\bullet\text{K}}\left(\frac{1}{273 \text{ K}} - \frac{1}{293 \text{ K}}\right)$$

$$0.57 = \frac{E_a}{8.3145}(2.5 \times 10^{-4}), \quad E_a = 1.9 \times 10^4 \text{ J/mol} = 19 \text{ kJ/mol}$$

57.   a. $\ln\left(\dfrac{k_2}{k_1}\right) = \dfrac{E_a}{R}\left(\dfrac{1}{T_1} - \dfrac{1}{T_2}\right)$, $\ln\left(\dfrac{0.950}{2.45 \times 10^{-4}}\right) = \dfrac{E_a}{8.3145}\left(\dfrac{1}{575} - \dfrac{1}{781}\right)$

$$8.26 = \frac{E_a}{8.3145}(4.6 \times 10^{-4}), \quad E_a = 1.5 \times 10^5 \text{ J/mol} = 150 \text{ kJ/mol}$$

$$\ln k_1 = -E_a/RT + \ln A, \quad \ln(2.45 \times 10^{-4}) = \left(\frac{-1.5 \times 10^5 \text{ J/mol}}{8.3145 \text{ J/mol}\bullet\text{K} \times 575 \text{ K}}\right) + \ln A$$

$$-8.314 = -31.38 + \ln A, \quad \ln A = 23.07, \quad A = e^{23.07} = 1.0 \times 10^{10} \text{ L/mol}\bullet\text{s}$$

Note: We carried extra significant figures in order to avoid large round off error.

b. $k = A \exp(-E_a/RT) = 1.0 \times 10^{10} \times \exp\left(\dfrac{-1.5 \times 10^5}{8.3145 \times 648}\right) = 8.1 \times 10^{-3}$ L/mol$\bullet$s

58. $\ln\left(\dfrac{k_2}{k_1}\right) = \dfrac{E_a}{R}\left(\dfrac{1}{T_1} - \dfrac{1}{T_2}\right)$; Since the rate doubles, then $k_2 = 2 k_1$.

$$\ln 2.00 = \frac{E_a}{8.3145 \text{ J/mol}\bullet\text{K}}\left(\frac{1}{298 \text{ K}} - \frac{1}{308 \text{ K}}\right), \quad E_a = 5.3 \times 10^4 \text{ J/mol} = 53 \text{ kJ/mol}$$

Catalysts

59.   a.   NO is the catalyst. NO is present in the first step of the mechanism on the reactant side, but it is not a reactant since it is regenerated in the second step.

b. NO_2 is an intermediate. Intermediates also never appear in the overall balanced equation. In a mechanism, intermediates always appear first on the product side while catalysts always appear first on the reactant side. Intermediates are substances we make up in an attempt to explain the steps reactants take to get to products while catalysts are actual compounds used to speed up the reaction.

c. $k = A \exp(-E_a/RT)$; $\dfrac{k_{cat}}{k_{un}} = \dfrac{A \exp[-E_a(cat)/RT]}{A \exp[-E_a(un)/RT]} = \exp\left(\dfrac{E_a(un) - E_a(cat)}{RT}\right)$

$$\frac{k_{cat}}{k_{un}} = \exp\left(\frac{2100 \text{ J/mol}}{8.3145 \text{ J/mol} \cdot \text{K} \times 298 \text{ K}}\right) = e^{0.85} = 2.3$$

The catalyzed reaction is 2.3 times faster than the uncatalyzed reaction at 25°C.

60.   The mechanism for the chlorine catalyzed destruction of ozone is:

$$O_3 + Cl \rightarrow O_2 + ClO \quad \text{(slow)}$$
$$ClO + O \rightarrow O_2 + Cl \quad \text{(fast)}$$
$$\overline{}$$
$$O_3 + O \rightarrow 2 O_2$$

Since the chlorine atom catalyzed reaction has a lower activation energy, then the Cl catalyzed rate is faster. Hence, Cl is a more effective catalyst. Using the activation energy, we can estimate the efficiency that Cl atoms destroy ozone as compared to NO molecules (see Exercise 12.59c).

At 25°C: $\dfrac{k_{Cl}}{k_{NO}} = \exp\left(\dfrac{-E_a(Cl)}{RT} + \dfrac{E_a(NO)}{RT}\right) = \exp\left(\dfrac{(-2100 + 11{,}900) \text{ J/mol}}{(8.3145 \times 298) \text{ J/mol}}\right) = e^{3.96} = 52$

At 25°C, the Cl catalyzed reaction is roughly 52 times faster than the NO catalyzed reaction, assuming the frequency factor A is the same for each reaction.

61.   The reaction at the surface of the catalyst follows the steps:

Thus, $CH_2D–CH_2D$ should be the product.

62.   If the mechanism is possible, then the reaction must be:

$$C_2H_4 + D_2 \rightarrow CH_2DCH_2D$$

If we got this product, then we could conclude that this is a possible mechanism. If we got some other product, e.g., CH_3CHD_2, then we would conclude that the mechanism is wrong. Even though this mechanism correctly predicts the products of the reaction, we cannot say conclusively that this is the correct mechanism; we might be able to conceive of other mechanisms that would give the same products as our proposed one.

Additional Exercises

63.   Rate = $k[H_2SeO_3]^x[H^+]^y[I^-]^z$; Comparing the first and second experiments:

$$\frac{3.33 \times 10^{-7}}{1.66 \times 10^{-7}} = \frac{k(2.0 \times 10^{-4})^x (2.0 \times 10^{-2})^y (2.0 \times 10^{-2})^z}{k(1.0 \times 10^{-4})^x (2.0 \times 10^{-2})^y (2.0 \times 10^{-2})^z},\ \ 2.01 = 2.0^x,\ x = 1$$

Comparing the first and fourth experiments:

$$\frac{6.66 \times 10^{-7}}{1.66 \times 10^{-7}} = \frac{k(1.0 \times 10^{-4}) (4.0 \times 10^{-2})^y (2.0 \times 10^{-2})^z}{k(1.0 \times 10^{-4}) (2.0 \times 10^{-2})^y (2.0 \times 10^{-2})^z},\ \ 4.01 = 2.0^y,\ y = 2$$

Comparing the first and sixth experiments:

$$\frac{13.2 \times 10^{-7}}{1.66 \times 10^{-7}} = \frac{k(1.0 \times 10^{-4}) (2.0 \times 10^{-2})^2 (4.0 \times 10^{-2})^z}{k(1.0 \times 10^{-4}) (2.0 \times 10^{-2})^2 (2.0 \times 10^{-2})^z}$$

$$7.95 = 2.0^z,\ \log 7.95 = z \log 2.0,\ z = \frac{\log 7.95}{\log 2.0} = 2.99 \approx 3$$

Rate = $k[H_2SeO_3][H^+]^2[I^-]^3$

Experiment #1:

$$\frac{1.66 \times 10^{-7}\,\text{mol}}{L\,s} = k\left(\frac{1.0 \times 10^{-4}\,\text{mol}}{L}\right)\left(\frac{2.0 \times 10^{-2}\,\text{mol}}{L}\right)^2\left(\frac{2.0 \times 10^{-2}\,\text{mol}}{L}\right)^3$$

$$k = 5.19 \times 10^5\ L^5/\text{mol}^5{\cdot}s = 5.2 \times 10^5\ L^5/\text{mol}^5{\cdot}s = k_{mean}$$

64.   The integrated rate law for each reaction is:

$$\ln[A] = -4.50 \times 10^{-4}\ s^{-1}(t) + \ln[A]_o\ \text{ and }\ \ln[B] = -3.70 \times 10^{-3}\ s^{-1}(t) + \ln[B]_o$$

Subtracting the second equation from the first equation ($\ln[A]_o = \ln[B]_o$):

$$\ln[A] - \ln[B] = -4.50 \times 10^{-4}\,(t) + 3.70 \times 10^{-3}\,(t),\ \ln\left(\frac{[A]}{[B]}\right) = 3.25 \times 10^{-3}\ s^{-1}\,(t)$$

When [A] = 4.00 [B], $\ln 4.00 = 3.25 \times 10^{-3}\,(t)$, t = 427 s

65.   The pressure of a gas is proportional to concentration. Therefore, we will use the pressure data to solve the problem since Rate = $-\Delta[SO_2Cl_2]/\Delta t = -\Delta P_{SO_2Cl_2}/\Delta t$.

Assuming a first order equation, the data and plot follow.

Time (hour)	0.00	1.00	2.00	4.00	8.00	16.00
$P_{SO_2Cl_2}$ (atm)	4.93	4.26	3.52	2.53	1.30	0.34
$\ln P_{SO_2Cl_2}$	1.595	1.449	1.258	0.928	0.262	-1.08

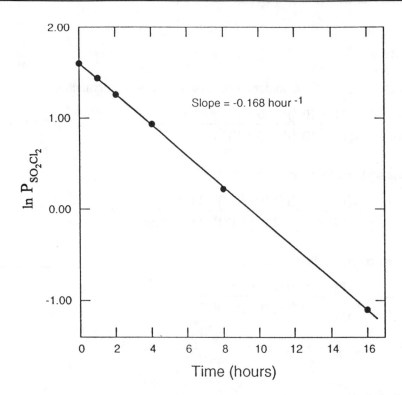

Since the ln $P_{SO_2Cl_2}$ vs. time plot is linear, then the reaction is first order in SO_2Cl_2.

a. Slope of ln(P) vs. t plot is -0.168 hour^{-1} = -k, k = 0.168 hour^{-1} = 4.67×10^{-5} s^{-1}

Since concentration units don't appear in first order rate constants, this value of k determined from the pressure data will be the same as if concentration data in molarity units were used.

b. $t_{1/2} = \dfrac{\ln 2}{k} = \dfrac{0.6931}{k} = \dfrac{0.6931}{0.168 \text{ h}^{-1}} = 4.13$ hour

c. $\ln\left(\dfrac{P_{SO_2Cl_2}}{P_O}\right) = -kt = -0.168 \text{ h}^{-1} (20.0 \text{ hr}) = -3.36$, $\left(\dfrac{P_{SO_2Cl_2}}{P_O}\right) = e^{-3.36} = 3.47 \times 10^{-2}$

Fraction left = 0.0347 = 3.47%

66. a.

t (s)	$[C_4H_6]$ (M)	$\ln[C_4H_6]$	$1/[C_4H_6]$ (M^{-1})
0	0.01000	-4.6052	1.000×10^2
1000.	0.00629	-5.069	1.59×10^2
2000.	0.00459	-5.384	2.18×10^2
3000.	0.00361	-5.624	2.77×10^2

The plot of $1/[C_4H_6]$ vs. t is linear, thus the reaction is second order in butadiene. From the plot (not included), the integrated rate law is:

$$\dfrac{1}{[C_4H_6]} = 5.90 \times 10^{-2} \, t + 100.0$$

b. From Exercise 12.35, k = 1.4×10^{-2} L/mol•s at 500. K. From this problem, k = 5.90×10^{-2} L/mol•s at 620. K.

$$\ln\left(\frac{k_2}{k_1}\right) = \frac{E_a}{R}\left(\frac{1}{T_1} - \frac{1}{T_2}\right), \ \ln\left(\frac{5.90 \times 10^{-2}}{1.4 \times 10^{-2}}\right) = \frac{E_a}{8.3145 \ \text{J/mol•K}}\left(\frac{1}{500.\ \text{K}} - \frac{1}{620.\ \text{K}}\right)$$

$12 = E_a \ (3.9 \times 10^{-4})$, $E_a = 3.1 \times 10^4$ J/mol = 31 kJ/mol

67. $\ln\left(\frac{k_2}{k_1}\right) = \frac{E_a}{R}\left(\frac{1}{T_1} - \frac{1}{T_2}\right)$; $\frac{k_2}{k_1} = 7.00$; $T_1 = 275$ K; $T_2 = 300.$ K

$$\ln 7.00 = \frac{E_a}{8.3145 \ \text{J/mol•K}}\left(\frac{1}{275\ \text{K}} - \frac{1}{300.\ \text{K}}\right)$$

$$E_a = \frac{(8.3145 \ \text{J/mol•K}) \ (\ln 7.00)}{3.0 \times 10^{-4} \ \text{K}^{-1}} = 5.4 \times 10^4 \ \text{J/mol} = 54 \ \text{kJ/mol}$$

68.   The observed rate law for this reaction is: Rate = $k[NO]^2[H_2]$. For a mechanism to be plausible, the sum of all steps must give the overall balanced equation (true for all the proposed mechanisms in this problem) and the rate law derived from the mechanism must agree with the observed mechanism. In each mechanism (I - III), the first elementary step is the rate determining step (the slow step), so the derived rate law for each mechanism will be the rate of the first step. The derived rate laws follow:

Mechanism I: Rate = $k[H_2]^2[NO]^2$
Mechanism II: Rate = $k[H_2][NO]$
Mechanism III: Rate = $k[H_2][NO]^2$

Only in Mechanism III does the derived rate law agree with the observed rate law. Thus, only Mechanism III is a plausible mechanism for this reaction.

69.   The rate depends on the number of reactant molecules adsorbed on the surface of the catalyst. This quantity is proportional to the concentration of reactant. However, when all of the catalyst surface sites are occupied, the rate becomes independent of the concentration of reactant.

70.   a. W since it has a lower activation energy than the Os catalyst.

b. $k_w = A_w \exp[-E_a(W)/RT]$; $k_{uncat} = A_{uncat} \exp[-E_a(uncat)/RT]$; Assume $A_w = A_{uncat}$

$$\frac{k_w}{k_{uncat}} = \exp\left(\frac{-E_a(W)}{RT} + \frac{E_a(uncat)}{RT}\right)$$

$$\frac{k_w}{k_{uncat}} = \exp\left(\frac{-163,000 \ \text{J/mol} + 335,000 \ \text{J/mol}}{8.3145 \ \text{J/mol•K} \times 298 \ \text{K}}\right) = 1.41 \times 10^{30}$$

The W catalyzed reaction is approximately 10^{30} times faster than the uncatalyzed reaction.

c.   Since $[H_2]$ is in the denominator of the rate law, then H_2 decreases the rate of the reaction. For the decomposition to occur, NH_3 molecules must be adsorbed on the surface of the catalyst. If H_2 is also adsorbed on the catalyst surface, then there are fewer sites for NH_3 molecules to be adsorbed and the rate decreases.

Challenge Problems

71.   Rate = $k[I^-]^x[OCl^-]^y[OH^-]^z$; Comparing the first and second experiments:

$$\frac{18.7 \times 10^{-3}}{9.4 \times 10^{-3}} = \frac{k(0.0026)^x (0.012)^y (0.10)^z}{k(0.0013)^x (0.012)^y (0.10)^z}, \ 2.0 = 2.0^x, \ x = 1$$

Comparing the first and third experiments:

$$\frac{9.4 \times 10^{-3}}{4.7 \times 10^{-3}} = \frac{k(0.0013) (0.012)^y (0.10)^z}{k(0.0013) (0.006)^y (0.10)^z}, \ 2.0 = 2^y, \ y = 1$$

Comparing the first and sixth experiments:

$$\frac{4.7 \times 10^{-3}}{9.4 \times 10^{-3}} = \frac{k(0.0013) (0.012) (0.20)^z}{k(0.0013) (0.012) (0.10)^z}, \ 1/2 = 2.0^z, \ z = -1$$

Rate = $\dfrac{k[I^-][OCl^-]}{[OH^-]}$; The presence of OH^- decreases the rate of the reaction.

For the first experiment:

$$\frac{9.4 \times 10^{-3} \, \text{mol}}{L \, s} = k \, \frac{(0.0013 \, \text{mol/L}) (0.012 \, \text{mol/L})}{(0.10 \, \text{mol/L})}, \ k = 60.3 \, s^{-1} = 60. \, s^{-1}$$

For all experiments, $k_{mean} = 60. \, s^{-1}$.

72.   For second order kinetics: $\dfrac{1}{[A]} - \dfrac{1}{[A]_o} = kt$ and $t_{1/2} = \dfrac{1}{k[A]_o}$

a. $\dfrac{1}{[A]} = (0.250 \, \text{L/mol} \cdot s)t + \dfrac{1}{[A]_o}, \ \dfrac{1}{[A]} = 0.250 \times 180. \, s + \dfrac{1}{1.00 \times 10^{-2} M}$

$\dfrac{1}{[A]} = 145 \, M^{-1}, \ [A] = 6.90 \times 10^{-3} \, M$

Amount of A that reacted = $0.0100 - 0.00690 = 0.0031 \, M$

$[A_2] = \dfrac{1}{2}(3.1 \times 10^{-3} \, M) = 1.6 \times 10^{-3} \, M$

b.   After 3.00 minutes (180. s): $[A] = 3.00 \, [B], \ 6.90 \times 10^{-3} \, M = 3.00 \, [B], \ [B] = 2.30 \times 10^{-3} \, M$

$\dfrac{1}{[B]} = k_2 t + \dfrac{1}{[B]_o}, \ \dfrac{1}{2.30 \times 10^{-3} M} = k_2(180. \, s) + \dfrac{1}{2.50 \times 10^{-2} M}, \ k_2 = 2.19 \, \text{L/mol} \cdot s$

c. $t_{1/2} = \dfrac{1}{k[A]_o} = \dfrac{1}{0.250\ \text{L/mol}\cdot\text{s} \times 1.00 \times 10^{-2}\ \text{mol/L}} = 4.00 \times 10^2\ \text{s}$

73.   a. We check for first order dependence by graphing ln [concentration] vs. time for each set of data. The rate dependence on NO is determined from the first set of data since the ozone concentration is relatively large compared to the NO concentration, so it is effectively constant.

time (ms)	[NO] (molecules/cm^3)	ln [NO]
0	6.0×10^8	20.21
100.	5.0×10^8	20.03
500.	2.4×10^8	19.30
700.	1.7×10^8	18.95
1000.	9.9×10^7	18.41

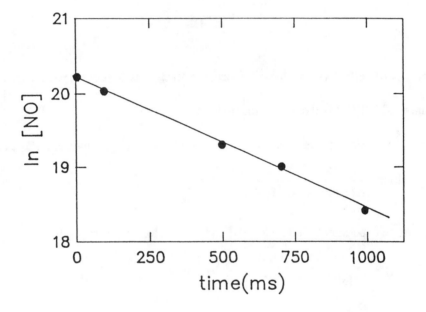

Since ln [NO] vs. t is linear, the reaction is first order with respect to NO.

We follow the same procedure for ozone using the second set of data. The data and plot are:

time (ms)	[O$_3$] (molecules/cm^3)	ln [O$_3$]
0	1.0×10^{10}	23.03
50.	8.4×10^9	22.85
100.	7.0×10^9	22.67
200.	4.9×10^9	22.31
300.	3.4×10^9	21.95

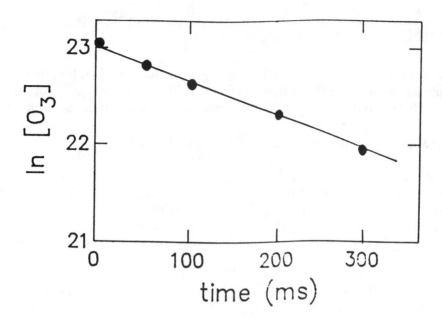

The plot of ln [O$_3$] vs. t is linear. Hence, the reaction is first order with respect to ozone.

b.   Rate = k[NO][O$_3$] is the overall rate law.

c.   For NO experiment, Rate = k′[NO] and k′ = -(slope from graph of ln [NO] vs. t).

$$k' = \text{-slope} = -\frac{18.41 - 20.21}{(1000. - 0) \times 10^{-3}\,s} = 1.8\ s^{-1}$$

For ozone experiment, Rate = k″[O$_3$] and k″ = -(slope from ln [O$_3$] vs. t).

$$k'' = \text{-slope} = -\frac{(21.95 - 23.03)}{(300. - 0) \times 10^{-3}\,s} = 3.6\ s^{-1}$$

d.   From NO experiment, Rate = k[NO][O$_3$] = k′[NO] where k′ = k[O$_3$].

k′ = 1.8 s^{-1} = k(1.0 × 10^{14} molecules/cm^3), k = 1.8 × 10^{-14} cm^3/molecules•s

We can check this from the ozone data. Rate = k″[O$_3$] = k[NO][O$_3$] where k″ = k[NO].

k″ = 3.6 s^{-1} = k(2.0 × 10^{14} molecules/cm^3), k = 1.8 × 10^{-14} cm^3/molecules•s

Both values of k agree.

74.   On the following energy profile, R = reactants, P = products, E_a = activation energy, ΔE = overall energy change for the reaction and I = intermediate.

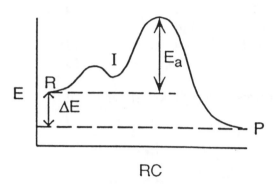

a - d.   See above plot

e.   This is a two step reaction since an intermediate plateau appears between the reactant and the products. This plateau represents the energy of the intermediate. The general reaction mechanism for this reaction is:

$$R \rightarrow I$$
$$\underline{I \rightarrow P}$$
$$R \rightarrow P$$

In a mechanism, the rate of the slowest step determines the rate of the reaction. The activation energy for the slowest step will be the largest energy barrier that the reaction must overcome. Since the second hump in the diagram is at the highest energy, then the second step has the largest activation energy and will be the rate determining step (the slow step).

75.   a.

T (K)	1/T (K^{-1})	k (min^{-1})	ln k
298.2	3.353×10^{-3}	178	5.182
293.5	3.407×10^{-3}	126	4.836
290.5	3.442×10^{-3}	100.	4.605

The plot of ln k vs. 1/T gives a straight line. The equation for the straight line is:

ln k = -6.48 × 10^3 (1/T) + 26.9

For a ln k vs. 1/T plot, the slope = $-E_a/R$ = -6.48 × 10^3 K

-6.48 × 10^3 K = $-E_a$/8.3145 J/mol•K, E_a = 5.39 × 10^4 J/mol = 53.9 kJ/mol

b.   ln k = -6.48 × 10^3(1/288.2) + 26.9 = 4.42, k = $e^{4.42}$ = 83 min^{-1}

About 83 chirps per minute per insect. Note: We carried extra sig. figs.

c. k gives the number of chirps per minute. The number or chirps in 15 sec is k/4.

T (C°)	T (°F)	k (min^{-1})	42 + 0.80 (k/4)
25.0	77.0	178	78° F
20.3	68.5	126	67°F
17.3	63.1	100.	62°F
15.0	59.0	83	59°F

The rule of thumb appears to be fairly accurate, about ± 1°F.

CHAPTER THIRTEEN

CHEMICAL EQUILIBRIUM

Questions

9. a. The rates of the forward and reverse reactions are equal at equilibrium.

    b. There is no net change in the composition (as long as temperature is constant).

10. False. Equilibrium and rates of reaction (kinetics) are independent of each other. A reaction with a large equilibrium constant value may be a fast reaction or a slow reaction. The same is true for a reaction with a small equilibrium constant value. Kinetics is discussed in detail in Chapter 12 of the text.

11. The equilibrium constant is a number that tells us the relative concentrations (pressures) of reactants and products at equilibrium. An equilibrium position is a set of concentrations that satisfy the equilibrium constant expression. More than one equilibrium position can satisfy the same equilibrium constant expression.

12. For the gas phase reaction $a A + b B \rightleftharpoons c C + d D$:

    the equilibrium constant expression is: $K = \dfrac{[C]^c [D]^d}{[A]^a [B]^b}$

    and the reaction quotient has the same form: $Q = \dfrac{[C]^c [D]^d}{[A]^a [B]^b}$

    The difference is that in the expression for K we use equilibrium concentrations, i.e., [A], [B], [C], and [D] are all in equilibrium with each other. Any set of concentrations can be plugged into the reaction quotient expression. Typically, we plug in initial concentrations into the Q expression then compare the value of Q to K to see how far we are from equilibrium. If $Q = K$, then the reaction is at equilibrium with these concentrations. If $Q \neq K$, then the reaction will have to shift either to products or to reactants to reach equilibrium.

13. Table 13.1 of the text illustrates this nicely. Each of the three experiments in Table 13.1 have different equilibrium positions, that is, each experiment has different equilibrium concentrations. However, when these equilibrium concentrations are inserted into the equilibrium constant expression, each experiment gives the same value for K. The equilibrium position depends on the initial concentrations that one starts with. Since there are an infinite number of initial conditions,

then there are an infinite number of equilibrium positions. However, each one of these infinite equilibrium positions will always give the same value for the equilibrium constant (assuming temperature is constant).

14.   The size of the equilibrium constant value can tell us whether a reaction contains mostly products or mostly reactants at equilibrium. When a reaction contains mostly products at equilibrium, the forward reaction is dominant. This situation is indicated by value of K which is much larger than 1. When a reaction contains mostly reactants at equilibrium, the reverse reaction is dominant, i.e., the forward reaction does not occur to any great extent. This situation is indicated by a value of K which is much smaller than 1. When K is close to 1, both the forward and reverse reaction occur significantly (neither reaction is dominant).

15.   When we change the pressure by adding an unreactive gas, we do not change the partial pressures of any of the substances in equilibrium with each other. In this case the equilibrium will not shift. If we change the pressure by changing the volume, we will change the partial pressures of all the substances in equilibrium by the same factor. If there are unequal number of gaseous particles on the two sides of the equation, then the reaction is no longer at equilibrium and must shift to either products or reactants to return to equilibrium.

16.   A change in volume will change the partial pressure of all reactants and products by the same factor. The shift in equilibrium depends on the number of gaseous particles on each side. An increase in volume will shift the equilibrium to the side with the greater number of particles in the gas phase. A decrease in volume will favor the side with the fewer number of gas phase particles. If there are the same number of gas phase particles on each side of the reaction, then a change in volume will not shift the equilibrium.

Exercises

Characteristics of Chemical Equilibrium

17.   No, equilibrium is a dynamic process. Both reactions:

$$H_2O + CO \rightarrow H_2 + CO_2 \text{ and } H_2 + CO_2 \rightarrow H_2O + CO$$

are occurring, but at equal rates. Thus, ^{14}C atoms will be distributed between CO and CO_2.

18.   No, it doesn't matter which direction the equilibrium position is reached. Both experiments will give the same equilibrium position since both experiments started with stoichiometric amounts of reactants or products.

The Equilibrium Constant

19.   a. $K = \dfrac{[NO_2][O_2]}{[NO][O_3]}$ b. $K = \dfrac{[O_2][O]}{[O_3]}$

         c. $K = \dfrac{[ClO][O_2]}{[Cl][O_3]}$ d. $K = \dfrac{[O_2]^3}{[O_3]^2}$

20. a. $K_p = \dfrac{P_{NO_2} \times P_{O_2}}{P_{NO} \times P_{O_3}}$

b. $K_p = \dfrac{P_{O_2} \times P_O}{P_{O_3}}$

c. $K_p = \dfrac{P_{ClO} \times P_{O_2}}{P_{Cl} \times P_{O_3}}$

d. $K_p = \dfrac{P_{O_2}^3}{P_{O_3}^2}$

21. $K = 278 = \dfrac{[SO_3]^2}{[SO_2]^2[O_2]}$ for $2\,SO_2(g) + O_2(g) \rightleftharpoons 2\,SO_3(g)$

When a reaction is reversed, then $K_{new} = 1/K_{original}$. When a reaction is multiplied through by a value of n, then $K_{new} = (K_{original})^n$.

a. $SO_2(g) + 1/2\,O_2(g) \rightleftharpoons SO_3(g)$, $K' = \dfrac{[SO_3]}{[SO_2][O_2]^{1/2}} = K^{1/2} = (278)^{1/2} = 16.7$

b. $2\,SO_3(g) \rightleftharpoons 2\,SO_2(g) + O_2(g)$, $K'' = \dfrac{[SO_2]^2[O_2]}{[SO_3]^2} = \dfrac{1}{K} = \dfrac{1}{278} = 3.60 \times 10^{-3}$

c. $SO_3(g) \rightleftharpoons SO_2(g) + 1/2\,O_2(g)$, $K''' = \dfrac{[SO_2][O_2]^{1/2}}{[SO_3]} = \left(\dfrac{1}{K}\right)^{1/2} = 6.00 \times 10^{-2}$

d. $4\,SO_2(g) + 2\,O_2(g) \rightleftharpoons 4\,SO_3(g)$, $K'''' = \dfrac{[SO_3]^4}{[SO_2]^4[O_2]^2} = K^2 = 7.73 \times 10^4$

22. $H_2(g) + Br_2(g) \rightleftharpoons 2\,HBr(g)$ $K_p = \dfrac{P_{HBr}^2}{(P_{H_2})(P_{Br_2})} = 3.5 \times 10^4$

a. $HBr \rightleftharpoons 1/2\,H_2 + 1/2\,Br_2$ $K_p' = \dfrac{(P_{H_2})^{1/2}(P_{Br_2})^{1/2}}{P_{HBr}} = \left(\dfrac{1}{K_p}\right)^{1/2} = \left(\dfrac{1}{3.5 \times 10^4}\right)^{1/2} = 5.3 \times 10^{-3}$

b. $2\,HBr \rightleftharpoons H_2 + Br_2$ $K_p'' = \dfrac{(P_{H_2})(P_{Br_2})}{P_{HBr}^2} = \dfrac{1}{K_p} = \dfrac{1}{3.5 \times 10^4} = 2.9 \times 10^{-5}$

c. $1/2\,H_2 + 1/2\,Br_2 \rightleftharpoons HBr$ $K_p''' = \dfrac{P_{HBr}}{(P_{H_2})^{1/2}(P_{Br_2})^{1/2}} = (K_p)^{1/2} = 190$

23. $K = \dfrac{[NO]^2}{[N_2][O_2]} = \dfrac{(4.7 \times 10^{-4}\,M)^2}{(0.041\,M)(0.0078\,M)} = 6.9 \times 10^{-4}$

24. $K = \dfrac{[NCl_3]^2}{[N_2][Cl_2]^3} = \dfrac{(0.19\,M)^2}{(1.4 \times 10^{-3}\,M)(4.3 \times 10^{-4}\,M)^3} = 3.2 \times 10^{11}\,L^2/mol^2$

25. $H_2(g) + I_2(g) \rightleftharpoons 2\,HI(g)$

$K = \dfrac{[HI]^2}{[H_2][I_2]} = \dfrac{\left(\dfrac{3.50\,mol}{3.00\,L}\right)^2}{\left(\dfrac{4.10\,mol}{3.00\,L}\right)\left(\dfrac{0.30\,mol}{3.00\,L}\right)} = 10.$

26. $K = \dfrac{[H_2S]^2}{[H_2]^2[S_2]} = \dfrac{\left(\dfrac{2.0\ mol}{2.0\ L}\right)^2}{\left(\dfrac{0.40\ mol}{2.0\ L}\right)^2\left(\dfrac{0.80\ mol}{2.0\ L}\right)} = 62.5 = 63\ L/mol$

27. $K_p = \dfrac{P_{NO}^2 \times P_{Cl_2}}{P_{NOCl}^2} = \dfrac{(1.25 \times 10^{-2}\ atm)^2(0.300\ atm)}{(1.20\ atm)^2} = 3.26 \times 10^{-5}\ atm$

28. $K_p = \dfrac{P_{NH_3}^2}{P_{N_2} \times P_{H_2}^3} = \dfrac{(3.1 \times 10^{-2}\ atm)^2}{(0.85\ atm)(3.1 \times 10^{-3}\ atm)^3} = 3.8 \times 10^4\ atm^{-2}$

29. $K_p = K(RT)^{\Delta n}$ where Δn = sum of gaseous product coefficients - sum of gaseous reactant coefficients. For this reaction, $\Delta n = 4 - 2 = 2$.

$K_p = \dfrac{2.6 \times 10^{-5}\ mol^2}{L^2} \times \left(\dfrac{0.08206\ L\ atm}{mol\ K} \times 400.\ K\right)^2 = 2.8 \times 10^{-2}\ atm^2$

30. $K_p = K(RT)^{\Delta n}$, $K = \dfrac{K_p}{(RT)^{\Delta n}}$; $\Delta n = 2 - 3 = -1$; $K = \dfrac{0.25\ atm^{-1}}{\left(\dfrac{0.08206\ L\ atm}{mol\bullet K} \times 1100\ K\right)^{-1}} = 23\ L/mol$

31.   Solids and liquids do not appear in the equilibrium expression. Only gases and dissolved solutes appear in the equilibrium expression.

a. $K = \dfrac{1}{[O_2]^5}$ b. $K = [N_2O][H_2O]^2$

c. $K = \dfrac{1}{[CO_2]}$ d. $K = \dfrac{[SO_2]^8}{[O_2]^8}$

32.   a. $K_p = \dfrac{1}{(P_{O_2})^{3/2}}$ b. $K_p = \dfrac{1}{P_{CO_2}}$

c. $K_p = \dfrac{P_{CO} \times P_{H_2}}{P_{H_2O}}$ d. $K_p = \dfrac{P_{O_2}^3}{P_{H_2O}^2}$

33.   Since solids do not appear in the equilibrium constant expression, then $K = [CO_2] = 2.1 \times 10^{-3}$ mol/L.

34. $C(s) + CO_2(g) \rightleftharpoons 2\ CO(g)$ $K_p = \dfrac{P_{CO}^2}{P_{CO_2}} = \dfrac{(2.6\ atm)^2}{2.9\ atm} = 2.3\ atm$

35. $K_p = K(RT)^{\Delta n}$ where Δn equals the difference in the sum of the coefficients between gaseous products and gaseous reactants (Δn = mol gaseous products - mol gaseous reactants). When $\Delta n = 0$, then $K_p = K$. In Exercise 13.19, only reactions a and c have $\Delta n = 0$, so only these reactions have K_p = K. Reaction b has $\Delta n = 2 - 1 = 1$ and reaction d has $\Delta n = 3 - 2 = 1$.

36.   When $\Delta n = 0$, then $K_p = K$. In Exercise 13.31, only reaction d has $\Delta n = 0$ so only reaction d has K_p $= K$. In Exercise 13.32, none of the reactions have $K_p = K$ since none of the reactions have $\Delta n = 0$.

Equilibrium Calculations

37. $H_2O(g) + Cl_2O(g) \rightarrow 2\ HOCl(g)$ $K = \dfrac{[HOCl]}{[H_2O][Cl_2O]} = 0.0900$

Use the reaction quotient Q to determine which way the reaction shifts to reach equilibrium. If $Q <$ K, then the reaction shifts right to reach equilibrium. If $Q > K$, then the reaction shifts left to reach equilibrium. If $Q = K$, then the reaction does not have a shift in either direction since the reaction is at equilibrium.

a. $Q = \dfrac{[HOCl]_o^2}{[H_2O]_o[Cl_2O]_o} = \dfrac{\left(\dfrac{1.0\ mol}{1.0\ L}\right)^2}{\left(\dfrac{0.10\ mol}{1.0\ L}\right)\left(\dfrac{0.10\ mol}{1.0\ L}\right)} = 1.0 \times 10^2$

$Q > K$ so the reaction shifts left to produce more reactants at equilibrium.

b. $Q = \dfrac{\left(\dfrac{0.084\ mol}{2.0\ L}\right)^2}{\left(\dfrac{0.98\ mol}{2.0\ L}\right)\left(\dfrac{0.080\ mol}{2.0\ L}\right)} = 0.090 = K;$ at equilibrium

c. $Q = \dfrac{\left(\dfrac{0.25\ mol}{3.0\ L}\right)^2}{\left(\dfrac{0.56\ mol}{3.0\ L}\right)\left(\dfrac{0.0010\ mol}{3.0\ L}\right)} = 110 > K$

Reaction will proceed to the left to reach equilibrium.

38.   As in Exercise 13.37, determine Q for each reaction and compare this value to K_p ($= 0.0900$) to determine which direction the reaction shifts to reach equilibrium.

a. $Q = \dfrac{P_{HOCl}^2}{P_{H_2O} \times P_{Cl_2O}} = \dfrac{(1.00\ atm)^3}{(1.00\ atm)(1.00\ atm)} = 1.00$

$Q > K_p$ so the reaction shifts left to reach equilibrium.

b. $Q = \dfrac{(21.0\ torr)^2}{(200.\ torr)(49.8\ torr)} = 4.43 \times 10^{-2} < K_p$

Reaction will proceed to the right to reach equilibrium. Note: Since Q and K_p are unitless, we can use any pressure units to determine Q.

c. $Q = \dfrac{(20.0 \text{ torr})^2}{(296 \text{ torr})(15.0 \text{ torr})} = 0.0901 \approx K_p$; at equilibrium

39. $CaCO_3(s) \rightleftharpoons CaO(s) + CO_2(g)$ $K_p = P_{CO_2} = 1.04 \text{ atm}$

a. $Q = P_{CO_2}$; We only need the partial pressure of CO_2 to determine Q since solids do not appear in equilibrium expressions (or Q expressions). At this temperature all CO_2 will be in the gas phase. Q = 2.55 atm so Q > K_p; Reaction will shift to the left to reach equilibrium; the mass of CaO will decrease.

b.   Q = 1.04 atm = K_p so the reaction is at equilibrium; mass of CaO will not change.

c.   Q = 1.04 atm = K_p so the reaction is at equilibrium; mass of CaO will not change.

d.   Q = 0.211 atm < K_p; The reaction will shift to the right to reach equilibrium; the mass of CaO will increase.

40. $CH_3CO_2H + C_2H_5OH \rightleftharpoons CH_3CO_2C_2H_5 + H_2O$ $K = \dfrac{[CH_3CO_2C_2H_5][H_2O]}{[CH_3CO_2H][C_2H_5OH]} = 2.2$

a. $Q = \dfrac{(0.22)(0.10)}{(0.010)(0.010)} = 220 > K$; Reaction will shift left to reach equilibrium so the concentration of water will decrease.

b. $Q = \dfrac{(0.22)(0.0020)}{(0.0020)(0.10)} = 2.2 = K$; Reaction is at equilibrium so the concentration of water will remain the same.

c. $Q = \dfrac{(0.88)(0.12)}{(0.044)(6.0)} = 0.40 < K$; Since Q < K, the concentration of water will increase since the reaction shifts right to reach equilibrium.

d. $Q = \dfrac{(4.4)(4.4)}{(0.88)(10.0)} = 2.2 = K$;   At equilibrium so the water concentration is unchanged.

e. $K = 2.2 = \dfrac{(2.0\,M)[H_2O]}{(0.10\,M)(5.0\,M)}$, $[H_2O] = 0.55\,M$

f.   Water is a product of the reaction, but it is not the solvent. Thus, the concentration of water must be included in the equilibrium expression since it is a solute in the reaction. Only when water is the solvent do we not include it in the equilibrium expression.

41. $K = \dfrac{[HF]^2}{[H_2][F_2]} = 2.1 \times 10^3$; $2.1 \times 10^3 = \dfrac{[HF]^2}{(0.0021\,M)(0.0021\,M)}$, $[HF] = 0.096\,M$

42. $K_p = \dfrac{P_{HI}^2}{P_{H_2} \times P_{I_2}} = 45.9$; $45.9 = \dfrac{(4.00 \text{ atm})^2}{0.200 \text{ atm} \times P_{I_2}}$, $P_{I_2} = 1.74 \text{ atm}$

43. $SO_2(g) + NO_2(g) \rightleftharpoons SO_3(g) + NO(g)$ $K = \dfrac{[SO_3][NO]}{[SO_2][NO_2]}$

To determine K, we must calculate the equilibrium concentrations. The initial concentrations are:

$$[SO_3]_o = [NO]_o = 0; \quad [SO_2]_o = [NO_2]_o = \frac{2.00 \text{ mol}}{1.00 \text{ L}} = 2.00 \, M$$

Next, we determine the change required to reach equilibrium. At equilibrium, [NO] = 1.30 mol/1.00 L = 1.30 M. Since there was zero amount of NO present initially, then 1.30 M of SO_2 and 1.30 M NO_2 must have reacted to produce 1.30 M NO as well as 1.30 M SO_3, all required by the balanced reaction. The equilibrium concentration for each substance is the sum of the initial concentration plus the change concentration necessary to reach equilibrium. The equilibrium concentrations are:

$$[SO_3] = [NO] = 0 + 1.30 \, M = 1.30 \, M; \quad [SO_2] = [NO_2] = 2.00 \, M - 1.30 \, M = 0.70 \, M$$

We now use these equilibrium concentrations to calculate K:

$$K = \frac{[SO_3][NO]}{[SO_2][NO_2]} = \frac{(1.30 \, M)(1.30 \, M)}{(0.70 \, M)(0.70 \, M)} = 3.4$$

44. $PCl_5(g) \rightleftharpoons PCl_3(g) + Cl_2(g) \quad K_p = \dfrac{P_{PCl_3} \times P_{Cl_2}}{P_{PCl_5}}$

Initially: P_{PCl_5} = 0.50 atm and $P_{PCl_3} = P_{Cl_2} = 0$

Change: Since 0.16 atm of PCl_5 remains at equilibrium then 0.50 - 0.16 = 0.34 atm of PCl_5 must have reacted to reach equilibrium. If 0.34 atm of PCl_5 reacted, then 0.34 atm of PCl_3 and 0.34 atm of Cl_2 must have been produced (moles and pressure are directly related at constant T and V).

Equilibrium: P_{PCl_5} = 0.16 atm, $P_{PCl_3} = P_{Cl_2}$ = 0.34 atm

$$K_p = \frac{P_{PCl_3} \times P_{Cl_2}}{P_{PCl_5}} = \frac{(0.34 \text{ atm})(0.34 \text{ atm})}{0.16 \text{ atm}} = 0.72 \text{ atm}$$

45.   When solving equilibrium problems, a common method to summarize all the information in the problem is to set up a table. We commonly call this table the ICE table since it summarizes initial concentrations, changes that must occur to reach equilibrium and equilibrium concentrations (the sum of the initial and change columns). For the change column, we will generally use the variable x which will be defined as the amount of reactant (or product) that must react to reach equilibrium. In this problem, the reaction must shift right since there are no products present initially. The general ICE table for this problem is:

	2 NO_2(g)	$\rightleftharpoons$	2 NO(g)	+	O_2(g)	$K = \dfrac{[NO]^2[O_2]}{[NO_2]^2}$
Initial	8.0 mol/1.0 L		0		0	
	Let x mol/L of NO_2 react to reach equilibrium					
Change	-x	$\rightarrow$	+x		+x/2	
Equil.	8.0 - x		x		x/2	

Note that we must use the coefficients in the balanced equation to determine how much products are produced letting x mol/L of NO_2 react to reach equilibrium. In the problem, we are told that $[NO]_e = 2.0\,M$. From the set-up, $[NO]_e = x = 2.0\,M$. Solving for the other concentrations: $[NO]_e = 8.0 - x = 8.0 - 2.0 = 6.0\,M$; $[O_2]_e = x/2 = 2.0/2 = 1.0\,M$. Calculating K:

$$K = \frac{[NO]^2[O_2]}{[NO_2]^2} = \frac{(2.0\,M)^2(1.0\,M)}{(6.0\,M)^2} = 0.11 \text{ mol/L}$$

Alternate Method: Fractions in the change column can be avoided (if you want) be defining x differently. If we were to let $2x$ mol/L of NO_2 react to reach equilibrium then the ICE table set-up is:

$$2\,NO_2(g) \quad \rightleftharpoons \quad 2\,NO(g) \quad + \quad O_2(g) \qquad K = \frac{[NO]^2[O_2]}{[NO_2]^2}$$

Initial	8.0 M	0	0
	Let $2x$ mol/L of NO_2 react to reach equilibrium		
Change	-2x	$\rightarrow$ +2x	+x
Equil.	8.0 - 2x	2x	x

Solving: $2x = [NO]_e = 2.0\,M$, $x = 1.0\,M$; $[NO_2]_e = 8.0 - 2(1.0) = 6.0\,M$; $[O_2]_e = x = 1.0\,M$

These are exactly the same equilibrium concentrations as solved for previously, thus K will be the same (as it must be). The moral of the story is define x in a manner that is most comfortable for you. Your final answer is independent of how you define x initially.

46.
$$2\,NH_3(g) \quad \rightleftharpoons \quad N_2(g) \quad + \quad 3\,H_2(g) \qquad K = \frac{[N_2][H_2]^3}{[NH_3]^2}$$

Initial	4.0 mol/2.0 L	0	0
	Let $2x$ mol/L of NH_3 react to reach equilibrium		
Change	-2x	$\rightarrow$ +x	+3x
Equil.	2.0 - 2x	x	3x

From the problem: $[NH_3]_e = 2.0$ mol/2.0 L $= 1.0\,M = 2.0 - 2x$, $x = 0.50\,M$

$[N_2] = x = 0.50\,M$; $[H_2] = 3x = 3(0.50\,M) = 1.5\,M$

$$K = \frac{[N_2][H_2]^3}{[NH_3]^2} = \frac{(0.50\,M)(1.5\,M)^3}{(1.0\,M)^2} = 1.7 \text{ mol}^2/\text{L}^2$$

47. $Q = 1.00$ which is less than K. Reaction shifts to the right to reach equilibrium. Summarizing the equilibrium problem in a table:

$$SO_2(g) \quad + \quad NO_2(g) \quad \rightleftharpoons \quad SO_3(g) \quad + \quad NO(g) \quad K = 2.50$$

Initial	1.00 M	1.00 M	1.00 M	1.00 M
	x mol/L of SO_2 reacts to reach equilibrium			
Change	-x	-x	$\rightarrow$ +x	+x
Equil.	1.00 - x	1.00 - x	1.00 + x	1.00 + x

Plug the equilibrium concentrations into the equilibrium constant expression:

$$K = \frac{[SO_3][NO]}{[SO_2][NO_2]} = 2.50 = \frac{(1.00+x)^2}{(1.00-x)^2};$$ Take the square root of both sides and solve for x:

$$\frac{1.00+x}{1.00-x} = 1.58, \quad 1.00+x = 1.58 - 1.58\,x, \quad 2.58\,x = 0.58, \quad x = 0.22\,M$$

The equilibrium concentrations are:

$$[SO_3] = [NO] = 1.00 + x = 1.00 + 0.22 = 1.22\,M; \quad [SO_2] = [NO_2] = 1.00 - x = 0.78\,M$$

48. $Q = 1.00$ which is less than K. Reaction shifts right to reach equilibrium.

$$H_2(g) \quad + \quad I_2(g) \quad \rightleftharpoons \quad 2\,HI(g) \qquad K = \frac{[HI]^2}{[H_2][I_2]} = 100.$$

Initial	1.00 M	1.00 M	1.00 M
	x mol/L of H_2 reacts to reach equilibrium		
Change	$-x$	$-x$ $\rightarrow$	$+2x$
Equil.	1.00 - x	1.00 - x	1.00 + $2x$

$$K = 100. = \frac{(1.00+2x)^2}{(1.00-x)^2}; \quad \text{Taking the square root of both sides:}$$

$$10.0 = \frac{1.00+2x}{1.00-x}, \quad 10.0 - 10.0\,x = 1.00 + 2x, \quad 12.0\,x = 9.0, \quad x = 0.75\,M$$

$$[H_2] = [I_2] = 1.00 - 0.75 = 0.25\,M; \quad [HI] = 1.00 + 2(0.75) = 2.50\,M$$

49.   Since only reactants are present initially, the reaction must proceed to the right to reach equilibrium. Summarizing the problem in a table:

$$N_2(g) \quad + \quad O_2(g) \quad \rightleftharpoons \quad 2\,NO(g) \qquad K_p = 0.050$$

Initial	0.80 atm	0.20 atm	0
	x atm of N_2 reacts to reach equilibrium		
Change	$-x$	$-x$ $\rightarrow$	$+2x$
Equil.	0.80 - x	0.20 - x	$2x$

$$K = 0.050 = \frac{P_{NO}^2}{P_{N_2} \times P_{O_2}} = \frac{(2x)^2}{(0.80-x)(0.20-x)}, \quad 0.050(0.16 - 1.00\,x + x^2) = 4\,x^2$$

$$4\,x^2 = 8.0 \times 10^{-3} - 0.050\,x + 0.050\,x^2, \quad 3.95\,x^2 + 0.050\,x - 8.0 \times 10^{-3} = 0$$

Solving using the quadratic equation (see Appenxix 1.4 of the text):

$$x = \frac{-b \pm (b^2 - 4ac)^{1/2}}{2a} = \frac{-0.050 \pm [(0.050)^2 - 4(3.95)(-8.0 \times 10^{-3})]^{1/2}}{2(3.95)}$$

$x = 3.9 \times 10^{-2}$ atm or $x = -5.2 \times 10^{-2}$ atm; Only $x = 3.9 \times 10^{-2}$ atm makes sense (x cannot be negative), so the equilibrium NO concentration is:

$P_{NO} = 2x = 2(3.9 \times 10^{-2}$ atm$) = 7.8 \times 10^{-2}$ atm

50. $2 SO_2(g)$ + $O_2(g)$ $\rightleftharpoons$ $2 SO_3(g)$ $K_p = 0.25$

Initial   0.50 atm   0.50 atm   0
                   2x atm of SO_2 reacts to reach equilibrium
Change   -2x   -x $\rightarrow$ +2x
Equil.   0.50 - 2x   0.50 - x   2x

$$K_p = 0.25 = \frac{P_{SO_3}^2}{P_{SO_2}^2 \times P_{O_2}} = \frac{(2x)^2}{(0.50 - 2x)^2(0.50 - x)}$$

This will give a cubic equation which are very difficult to solve. To solve this problem, we will use the method of successive approximations (see Appendix 1.4 of the text). The first step is to guess a value for x. Since the value of K is small (K < 1), then not much of the forward reaction will occur to reach equilibrium. This tells us that x is small. Lets guess that $x = 0.050$ atm. Now we take this guessed value for x and substitute it into the equation everywhere that x appears except for one. For equilibrium problems, we will substitute the guessed value for x into the denominator, then solve for the numerator value of x. We continue this process until the value of x guessed and the calculated value of x converge on the same number. This is the same answer that we would get if we were to solve the cubic equation exactly. Applying the method of successive approximations and carrying extra significant figures:

$$\frac{4x^2}{[0.50 - 2(0.050)]^2 [0.50 - (0.050)]} = \frac{4x^2}{(0.40)^2(0.45)} = 0.25, \ x = 0.067$$

$$\frac{4x^2}{[0.50 - 2(0.067)]^2 [0.50 - (0.067)]} = \frac{4x^2}{(0.366)^2(0.433)} = 0.25, \ x = 0.060$$

$$\frac{4x^2}{(0.38)^2(0.44)} = 0.25, \ x = 0.063; \quad \frac{4x^2}{(0.374)^2(0.437)} = 0.25, \ x = 0.062$$

The next trial gives the same value for $x = 0.062$ atm. We are done except for determining the equilibrium concentrations. They are:

$P_{SO_2} = 0.50 - 2x = 0.50 - 2(0.062) = 0.376 = 0.38$ atm

$P_{O_2} = 0.50 - x = 0.438 = 0.44$ atm; $P_{SO_3} = 2x = 0.124 = 0.12$ atm

51.   a.   The reaction must shift to reactants to reach equilibrium. Summarize the problem in a table:

                         $N_2O_4(g)$ $\rightleftharpoons$ $2 NO_2(g)$ $K_p = 0.25$

Initial   0   0.050 atm
                   2x atm of NO_2 reacts to reach equilibrium
Change   +x $\leftarrow$ -2x
Equil.   x   0.050 - 2x

$$K_p = \frac{P^2_{NO_2}}{P_{N_2O_4}}, \quad 0.25 = \frac{(0.050 - 2x)^2}{x}, \quad 0.25\,x = 2.5 \times 10^{-3} - 0.20\,x + 4\,x^2$$

$4\,x^2 - 0.45\,x + 2.5 \times 10^{-3} = 0$;   Solving using the quadratic equation:

$$x = \frac{-(-0.45) \pm [(-0.45)^2 - 4(4)(2.5 \times 10^{-3})]^{1/2}}{2(4)} = 5.9 \times 10^{-3} \text{ (other value, 0.11,}$$
makes no sense)

$P_{N_2O_4} = x = 5.9 \times 10^{-3}$ atm; $P_{NO_2} = 0.050 - 2(0.0059) = 3.8 \times 10^{-2}$ atm

b.   This reaction must proceed to products to reach equilibrium.

$$N_2O_4(g) \quad \rightleftharpoons \quad 2\,NO_2(g) \quad K_p = 0.25$$

Initial	0.040 atm	0
	x atm of N_2O_4 reacts to reach equilibrium	
Change	$-x$ $\rightarrow$	$+2x$
Equil.	0.040 - x	$2x$

$$K_p = \frac{(2x)^2}{0.040 - x} = 0.25, \quad 4\,x^2 = 0.010 - 0.25\,x, \quad 4\,x^2 + 0.25\,x - 0.010 = 0$$

$$x = \frac{-0.25 \pm [(0.25)^2 - 4(4)(-0.010)]^{1/2}}{2(4)} = 0.028 \text{ (other value is negative)}$$

$P_{N_2O_4} = 0.040 - x = 0.012$ atm; $P_{NO_2} = 2x = 0.056$ atm

c. $Q = 1.00$, so some NO_2 must react to form N_2O_4 $(Q > K_p)$.

$$N_2O_4(g) \quad \rightleftharpoons \quad 2\,NO_2(g)$$

Initial	1.00 atm	1.00 atm
	$2x$ atm of NO_2 reacts to reach equilibrium	
Change	$+x$ $\leftarrow$	$-2x$
Equil.	1.00 + x	1.00 - $2x$

$$K_p = \frac{(1.00 - 2x)^2}{1.00 + x} = 0.25, \quad 1.00 - 4.00\,x + 4\,x^2 = 0.25 + 0.25\,x, \quad 4\,x^2 - 4.25\,x + 0.75 = 0$$

$$x = \frac{-(-4.25) \pm [(-4.25)^2 - 4(4)(0.75)]^{1/2}}{2(4)} = 0.22 \quad \text{(other value, 0.84, makes no sense)}$$

$P_{N_2O_4} = 1.00 + x = 1.22$ atm; $P_{NO_2} = 1.00 - 2x = 0.56$ atm

52.   a. $\qquad H_2(g) \quad + \quad F_2(g) \quad \rightleftharpoons \quad 2\,HF(g) \quad K = 1.00 \times 10^2$

Initial	2.00 M	2.00 M	0
	x mol/L of H_2 reacts to reach equilibrium		
Change	$-x$	$-x$ $\rightarrow$	$+2x$
Equil.	2.00 - x	2.00 - x	$2x$

$$K = \frac{[HF]^2}{[H_2][F_2]} = \frac{(2x)^2}{(2.00 - x)^2} = 1.00 \times 10^2; \quad \text{This is a perfect square. Solving:}$$

$$\frac{2x}{2.00 - x} = 10.0, \ 2x = 20.0 - 10.0\,x, \ 12.0\,x = 20.0, \ x = 1.67$$

$[HF] = 2x = 2(1.67) = 3.34\,M; \ [H_2] = [F_2] = 2.00 - 1.67 = 0.33\,M$

b. Reaction will shift right to reach equilibrium after 0.50 mol of H_2 is added (Q < K).

$$H_2 \quad + \quad F_2 \quad \rightleftharpoons \quad 2\,HF$$

Initial	0.33 M	0.33 M	3.34 M (From part a.)
	Add 0.50 mol H_2 to the 1.00 L flask.		
New Initial	0.83	0.33	3.34
	x mol/L of H_2 reacts to reach equilibrium		
Change	$-x$	$-x$ →	$+2x$
Equil.	0.83 - x	0.33 - x	3.34 + 2x

$$K = \frac{(3.34 + 2x)^2}{(0.83 - x)(0.33 - x)} = 1.00 \times 10^2, \ 11.2 + 13.4\,x + 4\,x^2 = 27.4 - 116\,x + 100.\,x^2$$

$96\,x^2 - 129.4\,x + 16.2 = 0;$ Solving using the quadratic formula (Appendix 1.4 of text):

$$x = \frac{-(-129.4) \pm [(-129.4)^2 - 4(96)(16.2)]^{1/2}}{2(96)} = \frac{129.4 \pm 102.5}{192} = 0.14 \ (x \neq 1.2)$$

Note: When solving using the quadratic equation, we will carry extra significant figures until x is calculated, then we will round-off.

$[H_2] = 0.83 - x = 0.69\,M; \ [F_2] = 0.33 - x = 0.19\,M; \ [HF] = 3.34 + 2x = 3.62\,M$

53.   a. The reaction must proceed to products to reach equilibrium since only reactants are present initially. Summarizing the problem in a table:

$$2\,NOCl(g) \quad \rightleftharpoons \quad 2\,NO(g) \quad + \quad Cl_2(g) \qquad K = 1.6 \times 10^{-5}$$

Initial	$\dfrac{2.0\ \text{mol}}{2.0\ \text{L}} = 1.0\,M$	0	0
	$2x$ mol/L of NOCl reacts to reach equilibrium		
Change	$-2x$ →	$+2x$	$+x$
Equil.	1.0 - 2x	2x	x

$$K = 1.6 \times 10^{-5} = \frac{[NO]^2[Cl_2]}{[NOCl]^2} = \frac{(2x)^2\,(x)}{(1.0 - 2x)^2}$$

If we assume that $1.0 - 2x \approx 1.0$ (from the size of K, we know that not much reaction will occur so x is small), then:

$$1.6 \times 10^{-5} = \frac{4x^3}{1.0^2}, \quad x = 1.6 \times 10^{-2}; \quad \text{Now we must check the assumption.}$$

$$1.0 - 2x = 1.0 - 2(0.016) = 0.97 = 1.0 \text{ (to proper significant figures)}$$

Our error is about 3%, i.e., $2x$ is 3.2% of 1.0 M. Generally, if the error we introduce by making simplifying assumptions is less than 5%, we go no further, the assumption is said to be valid. We call this the 5% rule. Solving for the equilibrium concentrations:

$$[NO] = 2x = 0.032 \; M; \quad [Cl_2] = x = 0.016 \; M; \quad [NOCl] = 1.0 - 2x = 0.97 \; M \approx 1.0 \; M$$

Note: If we were to solve this cubic equation exactly (a long and tedious process), we get $x = 0.016$. This is the exact same answer we determined by making a simplifying assumption. We saved time and energy. Whenever K is a very small value, always make the assumption that x is small. If the assumption introduces an error of less than 5%, then the answer you calculated making the assumption will be considered the correct answer.

b.   2 NOCl(g) $\rightleftharpoons$ 2 NO(g)   + Cl_2(g)

Initial	1.0 M	1.0 M	0
	$2x$ mol/L of NOCl reacts to reach equilibrium		
Change	$-2x$ $\rightarrow$	$+2x$	$+x$
Equil.	1.0 - $2x$	1.0 + $2x$	x

$$1.6 \times 10^{-5} = \frac{(1.0 + 2x)^2(x)}{(1.0 - 2x)^2} \approx \frac{(1.0)^2(x)}{(1.0)^2} \quad \text{(assuming } 2x \ll 1.0)$$

$x = 1.6 \times 10^{-5}$; Assumptions are great ($2x$ is 3.2×10^{-3}% of 1.0).

$[Cl_2] = 1.6 \times 10^{-5} \; M$ and $[NOCl] = [NO] = 1.0 \; M$

c.   2 NOCl(g) $\rightleftharpoons$ 2 NO(g)   + Cl_2(g)

Initial	2.0 M	0	1.0 M
	$2x$ mol/L of NOCl reacts to reach equilibrium		
Change	$-2x$ $\rightarrow$	$+2x$	$+x$
Equil.	2.0 - $2x$	$2x$	1.0 + x

$$1.6 \times 10^{-5} = \frac{(2x)^2(1.0 + x)}{(2.0 - 2x)^2} \approx \frac{4x^2}{4.0} \quad \text{(assuming } x \ll 1.0)$$

Solving: $x = 4.0 \times 10^{-3}$; Assumptions good (x is 0.4% of 1.0 and $2x$ is 0.4% of 2.0).

$[Cl_2] = 1.0 + x = 1.0 \; M; \quad [NO] = 2(4.0 \times 10^{-3}) = 8.0 \times 10^{-3}; \quad [NOCl] = 2.0 \; M$

54. $N_2O_4(g)$ $\rightleftharpoons$ $2 NO_2(g)$ $K = \dfrac{[NO_2]^2}{[N_2O_4]} = 4.0 \times 10^{-7}$

Initial   1.0 mol/10.0 L   0
                 x mol/L of N_2O_4 reacts to reach equilibrium
Change $-x$ $\rightarrow$ $+2x$
Equil. $0.10 - x$ $2x$

$K = \dfrac{[NO_2]^2}{[N_2O_4]} = \dfrac{(2x)^2}{0.10 - x} = 4.0 \times 10^{-7}$; Since K has a small value, then assume that x is small
                                                                  compared to 0.10 so that $0.10 - x \approx 0.10$. Solving:

$4.0 \times 10^{-7} \approx \dfrac{4x^2}{0.10}$, $4x^2 = 4.0 \times 10^{-8}$, $x = 1.0 \times 10^{-4} M$

Checking assumption by the 5% rule: $\dfrac{x}{0.10} \times 100 = \dfrac{1.0 \times 10^{-4}}{0.10} \times 100 = 0.10\%$

Since this number is less than 5%, then we will say that the assumption is valid.

$[N_2O_4] = 0.10 - 1.0 \times 10^{-4} = 0.10\,M$; $[NO_2] = 2x = 2(1.0 \times 10^{-4}) = 2.0 \times 10^{-4}\,M$

55. $2 CO_2(g)$ $\rightleftharpoons$ $2 CO(g)$ $+$ $O_2(g)$ $K = \dfrac{[CO]^2[O_2]}{[CO_2]^2} = 2.0 \times 10^{-6}$

Initial   2.0 mol/5.0 L   0   0
                 $2x$ mol/L of CO_2 reacts to reach equilibrium
Change $-2x$ $\rightarrow$ $+2x$ x
Equil. $0.40 - 2x$ $2x$ x

$K = 2.0 \times 10^{-6} = \dfrac{[CO]^2[O_2]}{[CO_2]^2} = \dfrac{(2x)^2(x)}{(0.40 - 2x)^2}$; Assuming $2x << 0.40$:

$2.0 \times 10^{-6} \approx \dfrac{4x^3}{(0.40)^2}$, $2.0 \times 10^{-6} = \dfrac{4x^3}{0.16}$, $x = 4.3 \times 10^{-3}\,M$

Checking assumption: $\dfrac{2(4.3 \times 10^{-3})}{0.40} \times 100 = 2.2\%$; Assumption valid by the 5% rule.

$[CO_2] = 0.40 - 2x = 0.40 - 2(4.3 \times 10^{-3}) = 0.39\,M$

$[CO] = 2x = 2(4.3 \times 10^{-3}) = 8.6 \times 10^{-3}\,M$; $[O_2] = x = 4.3 \times 10^{-3}\,M$

56. $COCl_2(g)$ $\rightleftharpoons$ $CO(g)$ $+$ $Cl_2(g)$ $K_p = \dfrac{P_{CO} \times P_{Cl_2}}{P_{COCl_2}} = 6.8 \times 10^{-9}$

Initial   1.0 atm   0   0
                 x atm of $COCl_2$ reacts to reach equilibrium
Change $-x$ $\rightarrow$ $+x$ $+x$
Equil. $1.0 - x$ x x

$6.8 \times 10^{-9} = \dfrac{P_{CO} \times P_{Cl_2}}{P_{COCl_2}} = \dfrac{x^2}{1.0 - x} \approx \dfrac{x^2}{1.0}$ (Assuming $1.0 - x \approx 1.0$.)

$x = 8.2 \times 10^{-5}$ atm; Assumption good (x is 8.2×10^{-3}% of 1.0).

$P_{COCl_2} = 1.0 - x = 1.0 - 8.2 \times 10^{-5} = 1.0$ atm; $P_{CO} = P_{Cl_2} = x = 8.2 \times 10^{-5}$ atm

57.   This is a typical equilibrium problem except that the reaction contains a solid. Whenever solids and liquids are present, we basically ignore them in the equilibrium problem.

$$NH_4Cl(s) \quad \rightleftharpoons \quad NH_3(g) \quad + \quad HCl(g)$$

Initial	-	0	0

$NH_4Cl(s)$ decomposes to produce x atm each of NH_3 and HCl at equilibrium

Change	-	$\rightarrow$ +x	+x
Equil.	-	x	x

$K_p = P_{NH_3} \times P_{HCl} = (x)(x)$; From the problem, $P_{NH_3} = x = 2.2$ atm. Solving for K_p:

$\quad K_p = x^2 = (2.2 \text{ atm})^2 = 4.8 \text{ atm}^2$

58. $$H_2(g) \quad + \quad S(s) \quad \rightleftharpoons \quad H_2S(g)$$

Initial	0.15 M	-	0

x mol/L of H_2 reacts to reach equilibrium

Change	-x	-	$\rightarrow$ +x
Equil.	0.15 - x	-	x

$K = 6.8 \times 10^{-2} = \dfrac{[H_2S]}{[H_2]} = \dfrac{x}{0.15 - x}$, $0.010 - 0.068 \, x = x$

$x = \dfrac{0.010}{1.068} = 9.4 \times 10^{-3} \, M$; $[H_2S] = x = 9.4 \times 10^{-3} \, M$

Le Chatelier's Principle

59.   a.   No effect; Adding more of a pure solid or pure liquid has no effect on the equilibrium position.

b.   Shifts left; HF(g) will be removed by reaction with the glass. As HF(g) is removed, the reaction will shift left to produce more HF(g).

c.   Shifts right; As $H_2O(g)$ is removed, the reaction will shift right to produce more $H_2O(g)$.

60.   When the volume of a reaction container is increased, the reaction itself will want to increase its own volume by shifting to the side of the reaction which contains the most molecules of gas. When the molecules of gas are equal on both sides of the reaction, then the reaction will remain at equilibrium no matter what happens to the volume of the container.

a.   Reaction shifts left (to reactants) since the reactants contain 4 molecules of gas compared to 2 molecules of gas on the product side.

b.   Reaction shifts right (to products) since there are more product molecules of gas (2) as compared to reactant molecules (1).

c. No change since there are equal reactant and product molecules of gas.

d. Reaction shifts right.

e. Reaction shifts right to produce more $CO_2(g)$. One can ignore the solids and only concentrate on the gases since gases occupy a relatively large volume as compared to solids. We make the same assumption when liquids are present (only worry about the gas molecules).

61.   a. right   b. right   c. no effect; $He(g)$ is neither a reactant nor a product.

d. left; Since the reaction is exothermic, heat is a product:

$$CO(g) + H_2O(g) \rightarrow H_2(g) + CO_2(g) + Heat$$

Increasing T will add heat. The equilibrium shifts to the left to use up the added heat.

e. no effect; Since there are equal numbers of gas molecules on both sides of the reaction, then a change in volume has no effect on the equilibrium position.

62.   a. The moles of SO_3 will increase since the reaction will shift left to use up the added $O_2(g)$.

b. Increase; Since there are fewer reactant gas molecules than product gas molecules, then the reaction shifts left with a decrease in volume.

c. No effect; The partial pressures of sulfur trioxide, sulfur dioxide, and oxygen are unchanged.

d. Increase; Heat $+ 2\,SO_3 \rightleftharpoons 2\,SO_2 + O_2$; Decreasing T will remove heat, shifting this endothermic reaction to the left.

e. Decrease

63.   a. left   b. right   c. left

d. no effect (reactant and product concentrations are unchanged)

e. no effect; Since there are equal numbers of product and reactant gas molecules, then a change in volume has no effect on the equilibrium position.

f. right; A decrease in temperature will shift the equilibrium to the right since heat is a product in this reaction (as is true in all exothermic reactions).

64.   a. shift to left

b. shift to right; Since the reaction is endothermic (heat is a reactant), then an increase in temperature will shift the equilibrium to the right.

c. no effect   d. shift to right

e. shift to right; Since there are more gaseous product molecules than gaseous reactant molecules, then the equilibrium will shift right with an increase in volume.

65.   In an exothermic reaction, heat is a product. To maximize yield of products, one would want as low a temperature as possible since high temperatures would shift the reaction left (away from products). Since temperature changes also change the value of K, then at low temperatures the value of K will be largest which maximizes yield of products.

66.   As temperature increases, the value of K decreases. This is consistent with an exothermic reaction. In an exothermic reaction, heat is a product and an increase in temperature shifts the equilibrium to the reactant side (as well as lowering the value of K).

Additional Exercises

67.

$$O(g) + NO(g) \rightleftharpoons NO_2(g) \qquad K = 1/6.8 \times 10^{-49} = 1.5 \times 10^{48}$$
$$NO_2(g) + O_2(g) \rightleftharpoons NO(g) + O_3(g) \qquad K = 1/5.8 \times 10^{-34} = 1.7 \times 10^{33}$$

$$\overline{O_2(g) + O(g) \rightleftharpoons O_3(g) \qquad K = (1.5 \times 10^{48})(1.7 \times 10^{33}) = 2.6 \times 10^{81} \text{ L/mol}}$$

68.   a. $N_2(g) + O_2(g) \rightleftharpoons 2\,NO(g)$ $K_p = 1 \times 10^{-31} = \dfrac{P_{NO}^2}{P_{N_2} \times P_{O_2}} = \dfrac{P_{NO}^2}{(0.8)(0.2)}$

$$P_{NO} = 1 \times 10^{-16} \text{ atm}; \quad n_{NO} = \dfrac{PV}{RT} = \dfrac{(1 \times 10^{-16} \text{ atm})\,(1.0 \times 10^{-3} \text{ L})}{\left(\dfrac{0.08206 \text{ L atm}}{\text{mol K}}\right)(298 \text{ K})} = 4 \times 10^{-21} \text{ mol NO}$$

$$\dfrac{4 \times 10^{-21} \text{ mol NO}}{\text{cm}^3} \times \dfrac{6.02 \times 10^{23} \text{ molecules}}{\text{mol NO}} = \dfrac{2 \times 10^3 \text{ molecules NO}}{\text{cm}^3}$$

b.   There is more NO in the atmosphere than we would expect from the value of K. The answer must lie in the rates of the reaction. At 25°C the rates of both reactions:

$$N_2 + O_2 \rightarrow 2\,NO \text{ and } 2\,NO \rightarrow N_2 + O_2$$

are so slow that they are essentially zero. Very strong bonds must be broken; the activation energy is very high. Nitric oxide is produced in high energy or high temperature environments. In nature some NO is produced by lightning and the primary manmade source is from automobiles. The production of NO is endothermic ($\Delta H = +90$ kJ/mol). At high temperatures, K will increase and the rates of the reaction will also increase, resulting in a higher production of NO. Once the NO gets into a more normal temperature environment, it doesn't go back to N_2 and O_2 because of the slow rate.

69. $H_2O(g) + Cl_2O(g) \rightleftharpoons 2\,HOCl(g) \qquad K = 0.090 = \dfrac{[HOCl]^2}{[H_2O]\,[Cl_2O]}$

a.   The initial concentrations of H_2O and Cl_2O are:

$$\dfrac{1.0 \text{ g } H_2O}{1.0 \text{ L}} \times \dfrac{1 \text{ mol}}{18.02 \text{ g}} = 5.5 \times 10^{-2} \text{ mol/L}; \quad \dfrac{2.0 \text{ g } Cl_2O}{1.0 \text{ L}} \times \dfrac{1 \text{ mol}}{86.90 \text{ g}} = 2.3 \times 10^{-2} \text{ mol/L}$$

$$H_2O(g) \quad + \quad Cl_2O(g) \quad \rightleftharpoons \quad 2\, HOCl(g)$$

Initial	$5.5 \times 10^{-2}\, M$	$2.3 \times 10^{-2}\, M$	0
	x mol/L of H_2O reacts to reach equilibrium		
Change	$-x$	$-x$ $\rightarrow$	$+2x$
Equil.	$5.5 \times 10^{-2} - x$	$2.3 \times 10^{-2} - x$	$2x$

$$K = 0.090 = \frac{(2x)^2}{(5.5 \times 10^{-2} - x)(2.3 \times 10^{-2} - x)}, \quad 1.14 \times 10^{-4} - 7.02 \times 10^{-3}\, x + 0.090\, x^2 = 4\, x^2$$

$3.91\, x^2 + 7.02 \times 10^{-3}\, x - 1.14 \times 10^{-4} = 0$ (We carried extra significant figures.)

Solving using the quadratic equation:

$$x = \frac{-7.02 \times 10^{-3} \pm (4.93 \times 10^{-5} + 1.78 \times 10^{-3})^{1/2}}{7.82} = 4.6 \times 10^{-3} \text{ or } -6.4 \times 10^{-3}$$

A negative answer makes no physical sense; we can't have less than nothing.
So $x = 4.6 \times 10^{-3}\, M$.

$[HOCl] = 2x = 9.2 \times 10^{-3}\, M$; $[Cl_2O] = 2.3 \times 10^{-2} - x = 0.023 - 0.0046 = 1.8 \times 10^{-2}\, M$

$[H_2O] = 5.5 \times 10^{-2} - x = 0.055 - 0.0046 = 5.0 \times 10^{-2}\, M$

b.

$$H_2O(g) \quad + \quad Cl_2O(g) \quad \rightleftharpoons \quad 2\, HOCl(g)$$

Initial	0	0	1.0 mol/2.0 L $= 0.50\, M$
	$2x$ mol/L of HOCl reacts to reach equilibrium		
Change	$+x$	$+x$ $\leftarrow$	$-2x$
Equil.	x	x	$0.50 - 2x$

$$K = 0.090 = \frac{[HOCl]^2}{[H_2O]\,[Cl_2O]} = \frac{(0.50 - 2x)^2}{x^2}$$

The expression is a perfect square, so we can take the square root of each side:

$$0.30 = \frac{0.50 - 2x}{x}, \quad 0.30\, x = 0.50 - 2x, \quad 2.30\, x = 0.50$$

$x = 0.217$ (We carried extra significant figures.)

$x = [H_2O] = [Cl_2O] = 0.217 = 0.22\, M$; $[HOCl] = 0.50 - 2x = 0.50 - 0.434 = 0.07\, M$

70.

$$FeSCN^{2+}(aq) \quad \rightleftharpoons \quad Fe^{3+}(aq) \quad + \quad SCN^-(aq) \qquad K = 9.1 \times 10^{-4}$$

Initial	$2.0\, M$	0	0
	x mol/L of $FeSCN^{2+}$ reacts to reach equilibrium		
Change	$-x$ $\rightarrow$	$+x$	$+x$
Equil.	$2.0 - x$	x	x

$$9.1 \times 10^{-4} = \frac{[Fe^{3+}][SCN^-]}{[FeSCN^{2+}]} = \frac{x^2}{2.0 - x} \approx \frac{x^2}{2.0} \quad \text{(Assuming } 2.0 - x \approx 2.0.\text{)}$$

$x = 4.3 \times 10^{-2} \, M$; Assumption good by the 5% rule (x is 2.2% of 2.0).

$[FeSCN^{2+}] = 2.0 - x = 2.0 - 4.3 \times 10^{-2} = 2.0 \, M$; $[Fe^{3+}] = [SCN^-] = x = 4.3 \times 10^{-2} \, M$

71. a. $P_{PCl_5} = \dfrac{nRT}{V} = \dfrac{\dfrac{2.450 \text{ g } PCl_5}{208.22 \text{ g/mol}} \times \dfrac{0.08206 \text{ L atm}}{\text{mol K}} \times 600. \text{ K}}{0.500 \text{ L}} = 1.16 \text{ atm}$

b. $PCl_5(g) \quad \rightleftharpoons \quad PCl_3(g) \quad + \quad Cl_2(g) \qquad K_p = \dfrac{P_{PCl_3} \times P_{Cl_2}}{P_{PCl_5}} = 11.5$

Initial   1.16 atm   0   0
          x atm of PCl_5 reacts to reach equilibrium
Change $-x$ $\rightarrow$ $+x$ $+x$
Equil. $1.16 - x$ x x

$K_p = \dfrac{x^2}{1.16 - x} = 11.5$, $x^2 + 11.5\,x - 13.3 = 0$;   Using the quadratic formula: $x = 1.06$ atm

$P_{PCl_5} = 1.16 - 1.06 = 0.10$ atm

c. $P_{PCl_3} = P_{Cl_2} = 1.06$ atm; $P_{PCl_5} = 0.10$ atm

$P_{tot} = P_{PCl_5} + P_{PCl_3} + P_{Cl_2} = 0.10 + 1.06 + 1.06 = 2.22$ atm

d. Percent dissociation $= \dfrac{x}{1.16} \times 100 = \dfrac{1.06}{1.16} \times 100 = 91.4\%$

72. $PCl_5(g) \quad \rightleftharpoons \quad PCl_3(g) \quad + \quad Cl_2(g)$

Initial   189.2 torr   0   0
              x torr of PCl_5 reacts to reach equilibrium
Change $-x$ $\rightarrow$ $+x$ $+x$
Equil. $189.2 - x$ x x

$P_{total} = P_{PCl_5} + P_{PCl_3} + P_{Cl_2} = 189.2 - x + x + x$, 358.7 torr $= 189.2$ torr $+ x$, $x = 169.5$ torr

$P_{PCl_3} = P_{Cl_2} = x = 169.5$ torr $= 0.2230$ atm

$P_{PCl_5} = 189.2 - 169.5 = 19.7$ torr $= 0.0259$ atm

$K_p = \dfrac{P_{PCl_3} \times P_{Cl_2}}{P_{PCl_5}} = \dfrac{(0.2230 \text{ atm})^2}{0.0259 \text{ atm}} = 1.92$ atm

73.   a. $N_2O_4(g)$ $\rightleftharpoons$ $2NO_2(g)$

Initial	4.5 atm	0

x atm of N_2O_4 reacts to reach equilibrium

Change	-x	$\rightarrow$	+2x
Equil.	4.5 - x		2x

$$K_p = \frac{P_{NO_2}^2}{P_{N_2O_4}} = \frac{(2x)^2}{4.5-x} = 0.25$$

Solving using successive approximations or the quadratic formula: $x = 0.50$ atm

$P_{NO_2} = 2x = 1.0$ atm; $P_{N_2O_4} = 4.5 - x = 4.0$ atm

b. $N_2O_4(g)$ $\rightleftharpoons$ $2\,NO_2(g)$

Initial	0	9.0 atm

$2x$ atm of NO_2 reacts to reach equilibrium

Change	+x	$\leftarrow$	-2x
Equil.	x		9.0 - 2x

$\dfrac{(9.0 - 2x)^2}{x} = 0.25$; Solving using quadratic formula: $x = 4.0$ (the other value makes no sense).

$P_{N_2O_4} = x = 4.0$ atm; $P_{NO_2} = 9.0 - 2x = 1.0$ atm

c.   No, we get the same equilibrium position starting with either pure N_2O_4 or pure NO_2 in stoichiometric amounts.

74. $CoCl_2(s) + 6\,H_2O(g) \rightleftharpoons CoCl_2 \cdot 6\,H_2O(s)$; If rain is imminent, there would be a lot of water vapor in the air. The reaction would shift to the right and would take on the color of $CoCl_2 \cdot 6\,H_2O$, pink.

75. $H^+ + OH^- \rightarrow H_2O$; Sodium hydroxide (NaOH) will react with the H^+ on the product side of the reaction. This effectively removes H^+ from the equilibrium which will shift the reaction to the right to produce more H^+ and CrO_4^{2-}. Since more CrO_4^{2-} is produced, the solution turns yellow.

76. $N_2(g) + 3\,H_2(g) \rightleftharpoons 2\,NH_3(g) + $ heat

a.   This reaction is exothermic so an increase in temperature will decrease the value of K (see Table 13.3 of text.) This has the effect of lowering the amount of $NH_3(g)$ produced at equilibrium. The temperature increase, therefore, must be for kinetics reasons. As the temperature increases, the rate that the reaction reaches equilibrium will increase . At low temperatures, this reaction is very slow, too slow to be of any use.

b.   As $NH_3(g)$ is removed, this shifts the reaction to the right to produce more $NH_3(g)$.

c.   A catalyst has no effect on the equilibrium position. The purpose of a catalyst is to speed up a reaction so it reaches equilibrium quickly.

d.   When the pressure of reactants and products is high, the reaction shifts to the side that has fewer gas molecules. Since the product side contains 2 molecules of gas as compared to 4 molecules of gas on the reactant side, then the reaction shifts right to products at high pressures of reactants and products.

Challenge Problems

77.   There is a little trick we can use to solve this problem in order to avoid solving a cubic equation. Since K for this reaction is very small, then the dominant reaction is the reverse reaction. We will let the products react to completion by the reverse reaction, then we will solve the forward equilibrium problem to determine the equilibrium concentrations. Summarizing these steps to solve in a table:

$$2\ NOCl(g) \;\rightleftharpoons\; 2\ NO(g) \;+\; Cl_2(g) \qquad K = 1.6 \times 10^{-5}$$

Before	0		2.0 M	1.0 M
	Let 1.0 mol/L Cl_2 react completely.			(K is small, reactants dominate.)
Change	+2.0	$\leftarrow$	-2.0	-1.0	React completely
After	2.0		0	0	New initial conditions
	2x mol/L of NOCl reacts to reach equilibrium			
Change	-2x	$\rightarrow$	+2x	+x
Equil.	2.0 - 2x		2x	x

$$K = 1.6 \times 10^{-5} = \frac{(2x)^2\,(x)}{(2.0 - 2x)^2} \approx \frac{4x^3}{2.0^2} \quad \text{(assuming } 2.0 - 2x \approx 2.0)$$

$x^3 = 1.6 \times 10^{-5}$, $x = 2.5 \times 10^{-2}$ Assumption good by the 5% rule (2x is 2.5% of 2.0).

[NOCl] = 2.0 - 0.050 = 1.95 M = 2.0 M; [NO] = 0.050 M; [Cl$_2$] = 0.025 M

Note: If we do not break this problem into two parts (a stoichiometric part and an equilibrium part), then we are faced with solving a cubic equation. The set-up would be:

$$2\ NOCl \;\rightleftharpoons\; 2\ NO \;+\; Cl_2$$

Initial	0		2.0 M	1.0 M
Change	+2y	$\leftarrow$	-2y	-y
Equil.	2y		2.0 - 2y	1.0 - y

$1.6 \times 10^{-5} = \dfrac{(2.0 - 2y)^2\,(1.0 - y)}{(2y)^2}$;   If we say that y is small to simplify the problem, then:

$1.6 \times 10^{-5} = \dfrac{2.0^2}{4y^2}$;   We get $y = 250$.   This is impossible!

To solve this equation we cannot make any simplifying assumptions; we have to find a way to solve a cubic equation. Or, we can use some chemical common sense and solve the problem the easier way.

78.   a. $2 NO(g)$ + $Br_2(g)$ $\rightleftharpoons$ $2 NOBr(g)$

Initial   98.4 torr   41.3 torr   0
                   $2x$ torr of NO reacts to reach equilibrium
Change $-2x$ $-x$ $\rightarrow$ $+2x$
Equil.   98.4 - $2x$ 41.3 - x $2x$

$P_{total} = P_{NO} + P_{Br_2} + P_{NOBr} = (98.4 - 2x) + (41.3 - x) + 2x = 139.7 - x$

$P_{total} = 110.5 = 139.7 - x,\ x = 29.2$ torr; $P_{NO} = 98.4 - 2(29.2) = 40.0$ torr $= 0.0526$ atm

$P_{Br_2} = 41.3 - 29.2 = 12.1$ torr $= 0.0159$ atm; $P_{NOBr} = 2(29.2) = 58.4$ torr $= 0.0768$ atm

$K_p = \dfrac{P_{NOBr}^2}{P_{NO}^2 \times P_{Br_2}} = \dfrac{(0.0768\ atm)^2}{(0.0526\ atm)^2\,(0.0159\ atm)} = 134\ atm^{-1}$

   b. $2 NO(g)$ + $Br_2(g)$ $\rightleftharpoons$ $2 NOBr(g)$

Initial   0.30 atm   0.30 atm   0
               $2x$ atm of NO reacts to reach equilibrium
Change $-2x$ $-x$ $\rightarrow$ $+2x$
Equil.   0.30 - $2x$ 0.30 - x $2x$

This would yield a cubic equation. There is no way we can simplify the task. K_p is pretty large, so let us approach equilibrium in two steps; assume the reaction goes to completion then solve the back equilibrium problem.

                         $2 NO$ + Br_2 $\rightleftharpoons$ $2 NOBr$

Before   0.30 atm   0.30 atm   0
               Let 0.30 atm NO react completely.
Change   -0.30   -0.15 $\rightarrow$ +0.30   React completely
After   0   0.15   0.30   New initial
               $2y$ atm of NOBr reacts to reach equilibrium
Change $+2y$ $+y$ $\leftarrow$ $-2y$
Equil. $2y$ 0.15 + y 0.30 - $2y$

$\dfrac{(0.30 - 2y)^2}{(2y)^2\,(0.15 + y)} = 134,\quad \dfrac{(0.30 - 2y)^2}{(0.15 + y)} = 134 \times 4\,y^2 = 536\,y^2$

If $y \ll 0.15$: $\dfrac{(0.30)^2}{0.15} \approx 536\,y^2$ and $y = 0.034$;   Assumptions are poor (y is 23% of 0.15).

Use 0.034 as approximation for y and solve by successive approximations:

$\dfrac{(0.30 - 0.068)^2}{0.15 + 0.034} = 536\,y^2,\ y = 0.023;\quad \dfrac{(0.30 - 0.046)^2}{0.15 + 0.023} = 536\,y^2,\ y = 0.026$

$\dfrac{(0.30 - 0.052)^2}{0.15 + 0.026} = 536\,y^2,\ y = 0.026$

So: $P_{NO} = 2y = 0.052$ atm; $P_{Br_2} = 0.15 + y = 0.18$ atm; $P_{NOBr} = 0.30 - 2y = 0.25$ atm

79. $N_2(g)$ + $3 H_2(g)$ $\rightleftharpoons$ $2 NH_3(g)$

Initial   0   0 P_0 P_0 = initial pressure of NH_3
                          $2x$ atm of NH_3 reacts to reach equilibrium
Change $+x$ $+3x$ $\leftarrow$ $-2x$
Equil. x $3x$ $P_0 - 2x$

From problem, $P_0 - 2x = \dfrac{P_0}{2.00}$, so $P_0 = 4.00 x$

$$K_p = \frac{(4.00 x - 2x)^2}{(x)(3x)^3} = \frac{(2.00 x)^2}{(x)(3x)^3} = \frac{4.00 x^2}{27 x^4} = \frac{4.00}{27 x^2} = 5.3 \times 10^5,\ x = 5.3 \times 10^{-4}\ \text{atm}$$

$P_0 = 4.00 x = 4.00 \times (5.3 \times 10^{-4})$ atm $= 2.1 \times 10^{-3}$ atm

80. $P_4(g) \rightleftharpoons 2 P_2(g)$ $K_p = 0.100 = \dfrac{P_{P_2}^2}{P_{P_4}}$; $P_{P_4} + P_{P_2} = P_{total} = 1.00$ atm, $P_{P_4} = 1.00 - P_{P_2}$

Let $y = P_{P_2}$ at equilibrium, then $\dfrac{y^2}{1.00 - y} = 0.100$

Solving: $y = 0.270$ atm $= P_{P_2}$; $P_{P_4} = 1.00 - 0.270 = 0.73$ atm

To solve for the fraction dissociated, we need the initial pressure of P_4.

                     $P_4(g)$ $\rightleftharpoons$ $2 P_2(g)$

Initial P_0 0 P_0 = initial pressure of P_4
                          x atm of P_4 reacts to reach equilibrium
Change $-x$ $\rightarrow$ $+2x$
Equil. $P_0 - x$ $2x$

$P_{total} = P_0 - x + 2x = 1.00$ atm $= P_0 + x$

Solving: 0.270 atm $= P_{P_2} = 2x$, $x = 0.135$ atm; $P_0 = 1.00 - 0.135 = 0.87$ atm

Fraction dissociation $= \dfrac{x}{P_0} = \dfrac{0.135}{0.87} = 0.16$ or 16% of P_4 is dissociated to reach equilibrium.

81. $N_2O_4(g) \rightleftharpoons 2 NO_2(g)$ $K_p = \dfrac{P_{NO_2}^2}{P_{N_2O_4}} = \dfrac{(1.2\ \text{atm})^2}{0.33\ \text{atm}} = 4.4$ atm

Doubling the volume decreases each partial pressure by a factor of 2 (P = nRT/V).

$P_{NO_2} = 0.60$ atm and $P_{N_2O_4} = 0.17$ atm are the new partial pressures.

$Q = \dfrac{(0.60)^2}{0.17} = 2.1$, so $Q < K$;   Equilibrium will shift to the right.

$$N_2O_4(g) \quad \rightleftharpoons \quad 2\,NO_2(g)$$

Initial   0.17 atm   0.60 atm

$\quad\quad\quad\quad\quad$ x atm of N_2O_4 reacts to reach equilibrium

Change $-x$ $\rightarrow$ $+2x$

Equil. $0.17 - x$ $0.60 + 2x$

$K_p = 4.4 = \dfrac{(0.60 + 2x)^2}{(0.17 - x)}$, $4x^2 + 6.8\,x - 0.39 = 0$

Solving using the quadratic formula: $x = 0.056$

$P_{NO_2} = 0.60 + 2(0.056) = 0.71$ atm; $P_{N_2O_4} = 0.17 - 0.056 = 0.11$ atm

82.   a. $2\,NaHCO_3(s) \quad \rightleftharpoons \quad Na_2CO_3(s) \quad + \quad CO_2(g) \quad + \quad H_2O(g)$

Initial   -   -   0   0

$\quad\quad\quad\quad$ $NaHCO_3(s)$ decomposes to form x atm each of $CO_2(g)$ and $H_2O(g)$ at equilibrium.

Change   - $\rightarrow$ - $+x$ $+x$

Equil.   -   - x x

$0.25 = K_p = P_{CO_2} \times P_{H_2O}$, $0.25 = x^2$, $x = P_{CO_2} = P_{H_2O} = 0.50$ atm

b. $n_{CO_2} = \dfrac{PV}{RT} = \dfrac{(0.50\ \text{atm})\,(1.00\ \text{L})}{(0.08206\ \text{L atm/mol•K})\,(398\ \text{K})} = 1.5 \times 10^{-2}$ mol CO_2

Mass of Na_2CO_3 produced:

$$1.5 \times 10^{-2}\ \text{mol}\ CO_2 \times \frac{1\ \text{mol}\ Na_2CO_3}{\text{mol}\ CO_2} \times \frac{105.99\ \text{g}\ Na_2CO_3}{\text{mol}\ Na_2CO_3} = 1.6\ \text{g}\ Na_2CO_3$$

Mass of $NaHCO_3$ reacted:

$$1.5 \times 10^{-2}\ \text{mol}\ CO_2 \times \frac{2\ \text{mol}\ NaHCO_3}{1\ \text{mol}\ CO_2} \times \frac{84.01\ \text{g}\ NaHCO_3}{\text{mol}} = 2.5\ \text{g}\ NaHCO_3$$

Mass of $NaHCO_3$ remaining = 10.0 - 2.5 = 7.5 g

c. $10.0\ \text{g}\ NaHCO_3 \times \dfrac{1\ \text{mol}\ NaHCO_3}{84.01\ \text{g}\ NaHCO_3} \times \dfrac{1\ \text{mol}\ CO_2}{2\ \text{mol}\ NaHCO_3} = 5.95 \times 10^{-2}$ mol CO_2

When all of the $NaHCO_3$ has just been consumed, we will have 5.95×10^{-2} mol CO_2 gas at a pressure of 0.50 atm (from a).

$$V = \frac{nRT}{P} = \frac{(5.95 \times 10^{-2}\ \text{mol})\,(0.08206\ \text{L atm/mol•K})\,(398\ \text{K})}{0.50\ \text{atm}} = 3.9\ \text{L}$$

CHAPTER FOURTEEN

ACIDS AND BASES

Questions

13. The Arrhenius definitions are: acids produce H^+ in water and bases produce OH^- in water. The difference between strong and weak acids and bases is the amount of H^+ and OH^- produced.

   a. A strong acid is 100% dissociated in water.
   b. A strong base is 100% dissociated in water.
   c. A weak acid is much less than 100% dissociated in water (typically 1-10% dissociated).
   d. A weak base is one that only a small percentage of the molecules react with water to form OH^-.

14. When a strong acid (HX) is added to water, the reaction $HX + H_2O \rightarrow H_3O^+ + X^-$ basically goes to completion. All strong acids in water are completely converted into H_3O^+ and X^-. Thus, no acid stronger than H_3O^+ will remain undissociated in water. Similarly, when a strong base (B) is added to water, the reaction $B + H_2O \rightarrow BH^+ + OH^-$ basically goes to completion. All bases stronger than OH^- are completely converted into OH^- and BH^+. Even though there are acids and bases stronger than H_3O^+ and OH^-, in water these acids and bases are completely converted into H_3O^+ and OH^-.

15. $H_2O \rightleftharpoons H^+ + OH^- \qquad K_w = [H^+] [OH^-] = 1.0 \times 10^{-14}$

   Neutral solution: $[H^+] = [OH^-]$; $[H^+] = 1.0 \times 10^{-7} M$; $pH = -\log (1.0 \times 10^{-7}) = 7.00$

16. 10.78 (4 S.F.); 6.78 (3 S.F.); 0.78 (2 S.F.); A pH value is a logarithm. The numbers to the left of the decimal place identifies the power of ten that $[H^+]$ is expressed in scientific notation, e.g., 10^{-11}, 10^{-7}, 10^{-1}. The number of decimal places in a pH value identifies the number of significant figures in $[H^+]$. In all three pH values, the $[H^+]$ should only be expressed to two significant figures since all pH values only have two decimal places.

17. a. Arrhenius acid: produce H^+ in water
    b. Brönsted-Lowry acid: proton (H^+) donor
    c. Lewis acid: electron pair acceptor

18. The Lewis definition is most general. The Lewis definition can apply to all Arrhenius and Brönsted-Lowry acids; H^+ has an empty 1s orbital and forms bonds to all bases by accepting a pair of electrons from the base. In addition the Lewis definition incorporates other reactions not typically considered acid-base reactions, e.g., $BF_3(g) + NH_3(g) \rightarrow F_3B-NH_3(s)$. NH_3 is something we usually consider a base and it is in this reaction using the Lewis definition; NH_3 donates a pair of electrons to form the N–B bond.

19.   An organic compound containing nitrogen with a lone pair of electrons will most likely produce a basic solution.

20.   a.   The weaker the X – H bond, the stronger the acid.
      b.   As the electronegativity of neighboring atoms increase, the strength of the acid increases.
      c.   As the number of oxygen atoms increase, the strength of the acid increases.

21.   In general, the weaker the acid, the stronger the conjugate base and vice versa.

      a.   Since acid strength increases as the X – H bond strength decreases, then conjugate base strength will increase as the strength of the X – H bond increases.
      b.   Since acid strength increases as the electronegativity of neighboring atoms increase, then conjugate base strength will decrease as the electronegativity of neighboring atoms increase.
      c.   Since acid strength increases as the number of oxygen atoms increase, then conjugate base strength decreases as the number of oxygen atoms increase.

22.   A Lewis acid must have an empty orbital to accept an electron pair and a Lewis base must have an unshared pair of electrons.

23.   a. H_2O and $CH_3CO_2^-$

      b.   An acid-base reaction can be thought of as a competition between two opposing bases. Since this equilibrium lies far to the left ($K_a < 1$), then $CH_3CO_2^-$ is a stronger base than H_2O.

      c.   The acetate ion is a better base than water and produces basic solutions in water. When we put acetate ion into solution as the only major basic species, the reaction is:

$$CH_3CO_2^- + H_2O \rightleftharpoons CH_3CO_2H + OH^-$$

      Now the competition is between $CH_3CO_2^-$ and OH^- for the proton. Hydroxide ion is the strongest base possible in water. The above equilibrium lies far to the left resulting in a K_b value less than one. Those species we specifically call weak bases ($10^{-14} < K_b < 1$) lie between H_2O and OH^- in base strength. Weak bases are stronger bases than water but are weaker bases than OH^-.

24.   The NH_4^+ ion is a weak acid because it lies between H_2O and H_3O^+ (H^+) in terms of acid strength. Weak acids are better acids than water, thus their aqueous solutions are acidic. They are weak acids because they are not as strong as H_3O^+ (H^+). Weak acids only partially dissociate in water and have K_a values between 10^{-14} and 1.

Exercises

Nature of Acids and Bases

25.   a. $HClO_4(aq) + H_2O(l) \rightarrow H_3O^+(aq) + ClO_4^-(aq)$. Only the forward reaction is indicated since $HClO_4$ is a strong acid and is basically 100% dissociated in water. For acids, the dissociation reaction is commonly written without water as a reactant. The common abbreviation for this reaction is: $HClO_4(aq) \rightarrow H^+(aq) + ClO_4^-(aq)$. This reaction is also called the K_a reaction as the equilibrium constant for this reaction is called K_a.

b.   Propanoic acid is a weak acid, so it is only partially dissociated in water. The dissociation reaction is: $CH_3CH_2CO_2H(aq) + H_2O(l) \rightleftharpoons H_3O^+(aq) + CH_3CH_2CO_2^-(aq)$ or $CH_3CH_2CO_2H(aq) \rightleftharpoons H^+(aq) + CH_3CH_2CO_2^-(aq)$.

c. NH_4^+ is a weak acid. Similar to propanoic acid, the dissociation reaction is:

$$NH_4^+(aq) + H_2O(l) \rightleftharpoons H_3O^+(aq) + NH_3(aq) \text{ or } NH_4^+(aq) \rightleftharpoons H^+(aq) + NH_3(aq)$$

26.   The dissociation reaction (the K_a reaction) of an acid in water commonly omits water as a reactant. We will follow this practice. All dissociation reactions produce H^+ and the conjugate base of the acid that is dissociated.

a. $HNO_2(aq) \rightleftharpoons H^+(aq) + NO_2^-(aq)$ $K_a = \dfrac{[H^+][NO_2^-]}{[HNO_2]}$

b. $Ti(H_2O)_6^{4+}(aq) \rightleftharpoons H^+(aq) + Ti(H_2O)_5(OH)^{3+}(aq)$ $K_a = \dfrac{[H^+][Ti(H_2O)_5(OH)^{3+}]}{[Ti(H_2O)_6^{4+}]}$

c. $HCN(aq) \rightleftharpoons H^+(aq) + CN^-(aq)$ $K_a = \dfrac{[H^+][CN^-]}{[HCN]}$

27.   An acid is a proton (H^+) donor and the base is a proton acceptor. A conjugate acid-base pair only differs by a proton (H^+).

	Acid	Base	Conjugate Base of Acid	Conjugate Acid of Base
a.	HF	H_2O	F^-	H_3O^+
b.	H_2SO_4	H_2O	HSO_4^-	H_3O^+
c.	HSO_4^-	H_2O	SO_4^{2-}	H_3O^+

28.

	Acid	Base	Conjugate Base of Acid	Conjugate Acid of Base
a.	$Fe(H_2O)_6^{3+}$	H_2O	$Fe(H_2O)_5(OH)^{2+}$	H_3O^+
b.	$C_2H_5NH^+$	H_2O	C_2H_5N	H_3O^+
c.	HF	NH_3	F^-	NH_4^+

29.   Strong acids have a $K_a \gg 1$ and weak acids have $K_a < 1$. Table 14.2 in the text lists some K_a values for weak acids. K_a values for strong acids are hard to determine so they are not listed in the text. However, there are only a few common strong acids so if you memorize the strong acids, then all other acids will be weak acids. The strong acids to memorize are HCl, HBr, HI, HNO_3, $HClO_4$ and H_2SO_4.

a. HNO_2 is a weak acid ($K_a = 4.0 \times 10^{-4}$).
b. HNO_3 is a strong acid.
c. HCl is a strong acid.
d. HF is a weak acid ($K_a = 7.2 \times 10^{-4}$).

30.   a. $HClO_4$ is a strong acid (one to memorize).
       b.   HOCl is a weak acid ($K_a = 3.5 \times 10^{-8}$).
       c. H_2SO_4 is a strong acid (one to memorize).
       d. HSO_4^- is a weak acid ($K_a = 1.2 \times 10^{-2}$).

31.   The K_a value is directly related to acid strength. As K_a increases, acid strength increases. For water, use K_w when comparing the acid strength of water to other species. The K_a values are:

       HNO_3: strong acid ($K_a \gg 1$); HOCl: $K_a = 3.5 \times 10^{-8}$

       NH_4^+: $K_a = 5.6 \times 10^{-10}$; H_2O: $K_a = K_w = 1.0 \times 10^{-14}$

       From the K_a values, the ordering is: $HNO_3 > HOCl > NH_4^+ > H_2O$.

32.   Except for water, these are the conjugate bases of the acids in the previous exercise. In general, the weaker the acid, the stronger the conjugate base. NO_3^- is the conjugate base of a strong acid. It is a terrible base (worse than water). The ordering is: $NH_3 > OCl^- > H_2O > NO_3^-$

33.   a. $HClO_4$ is a strong acid and water is a weak acid with $K_a = K_w = 1.0 \times 10^{-14}$. $HClO_4$ is a stronger acid than H_2O.

       b. H_2O, $K_a = K_w = 1.0 \times 10^{-14}$; $HClO_2$, $K_a = 1.2 \times 10^{-2}$; $HClO_2$ is a stronger acid than H_2O since K_a for $HClO_2 > K_a$ for H_2O.

       c.   HF, $K_a = 7.2 \times 10^{-4}$; HCN, $K_a = 6.2 \times 10^{-10}$; HF is a stronger acid than HCN since K_a for HF $> K_a$ for HCN.

34.   a. H_2O; The conjugate bases of strong acids are terrible bases ($K_b < 10^{-14}$).

       b. ClO_2^-; The conjugate bases of weak acids are weak bases ($10^{-14} < K_b < 1$).

       c. CN^-; For a conjugate acid-base pair, $K_a \times K_b = K_w$. From this relationship, the stronger the acid the weaker the conjugate base (K_b decreases as K_a increases). Since HF is a stronger acid than HCN (K_a for HF $> K_a$ for HCN), then CN^- will be a stronger base than F^-.

Autoionization of Water and the pH Scale

35.   At $25\,^{\circ}C$, the relationship: $[H^+][OH^-] = K_w = 1.0 \times 10^{-14}$ always holds for aqueous solutions. When $[H^+]$ is greater than $1.0 \times 10^{-7}\ M$, then the solution is acidic; when $[H^+]$ is less than $1.0 \times 10^{-7}\ M$, then the solution is basic; when $[H^+] = 1.0 \times 10^{-7}\ M$, then the solution is neutral. In terms of $[OH^-]$, an acidic solution has $[OH^-] < 1.0 \times 10^{-7}\ M$, a basic solution has $[OH^-] > 1.0 \times 10^{-7}\ M$ and a neutral solution has $[OH^-] = 1.0 \times 10^{-7}\ M$.

       a. $[OH^-] = \dfrac{K_w}{[H^+]} = \dfrac{1.0 \times 10^{-14}}{1.0 \times 10^{-7}} = 1.0 \times 10^{-7}\ M$; The solution is neutral.

       b. $[OH^-] = \dfrac{1.0 \times 10^{-14}}{1.4 \times 10^{-3}} = 7.1 \times 10^{-12}\ M$; The solution is acidic.

c. $[OH^-] = \dfrac{1.0 \times 10^{-14}}{2.5 \times 10^{-10}} = 4.0 \times 10^{-5}\,M$; The solution is basic.

d. $[OH^-] = \dfrac{1.0 \times 10^{-14}}{6.1} = 1.6 \times 10^{-15}\,M$; The solution is acidic.

36.   a. $[H^+] = \dfrac{K_w}{[OH^-]} = \dfrac{1.0 \times 10^{-14}}{3.5 \times 10^{-2}} = 2.9 \times 10^{-13}\,M$; basic

b. $[H^+] = \dfrac{1.0 \times 10^{-14}}{8.0 \times 10^{-11}} = 1.3 \times 10^{-4}\,M$; acidic

c. $[H^+] = \dfrac{1.0 \times 10^{-14}}{1.0 \times 10^{-7}} = 1.0 \times 10^{-7}\,M$; neutral

d. $[H^+] = \dfrac{1.0 \times 10^{-14}}{5.0} = 2.0 \times 10^{-15}\,M$; basic

37.   a.   Since the value of the equilibrium constant increases as the temperature increases, then the reaction is endothermic. In endothermic reactions, heat is a reactant so an increase in temperature (heat) shifts the reaction to produce more products and increases K in the process.

b. $H_2O(l) \rightleftharpoons H^+(aq) + OH^-(aq)$ $K_w = 5.47 \times 10^{-14} = [H^+][OH^-]$;

In pure water $[H^+] = [OH^-]$, so $5.47 \times 10^{-14} = [H^+]^2$, $[H^+] = 2.34 \times 10^{-7}\,M = [OH^-]$

38.   a. $H_2O(l) \rightleftharpoons H^+(aq) + OH^-(aq)$ $K_w = 1.14 \times 10^{-15} = [H^+][OH^-]$

In pure water $[H^+] = [OH^-]$, so $1.14 \times 10^{-15} = [H^+]^2$, $[H^+] = 3.38 \times 10^{-8}\,M = [OH^-]$

b.   0°C: $pH = -\log[H^+] = -\log(3.38 \times 10^{-8}) = 7.471$

50°C: From Exercise 14.37b, $[H^+] = [OH^-] = 2.34 \times 10^{-7}\,M$; $pH = -\log(2.34 \times 10^{-7}) = 6.631$

39. $pH = -\log[H^+]$; $pOH = -\log[OH^-]$; At 25°C, $pH + pOH = 14.00$

a. $pH = -\log[H^+] = -\log(1.0 \times 10^{-7}) = 7.00$; $pOH = 14.00 - pH = 14.00 - 7.00 = 7.00$

b. $pH = -\log(1.4 \times 10^{-3}) = 2.85$; $pOH = 14.00 - 2.85 = 11.15$

c. $pH = -\log(2.5 \times 10^{-10}) = 9.60$; $pOH = 14.00 - 9.60 = 4.40$

d. $pH = -\log(6.1) = -0.79$; $pOH = 14.00 - (-0.79) = 14.79$

Note that pH is less than zero when $[H^+]$ is greater than $1.0\,M$ (an extremely acidic solution).

40.   There are several ways to determine pH and pOH. In this problem we will determine the pOH from the $[OH^-]$ given in the problem ($pOH = -\log[OH^-]$), then determine the pH from the relationship $pH + pOH = 14.00$, which always holds at 25°C for an aqueous solution.

a.   pOH = -log [OH⁻] = -log (3.5 × 10⁻²) = 1.46; pH = 14.00 - pOH = 14.00 - 1.46 = 12.54

b.   pOH = -log (8.0 × 10⁻¹¹) = 10.10; pH = 14.00 - 10.10 = 3.90

c.   pOH = -log (1.0 × 10⁻⁷) = 7.00; pH = 14.00 - 7.00 = 7.00

d.   pOH = -log (5.0) = -0.70; pH = 14.00 - (-0.70) = 14.70

Note that pH is greater than 14.00 when [OH⁻] is greater than 1.0 M (an extremely basic solution).

41.   a. $[H^+] = 10^{-pH}$, $[H^+] = 10^{-7.41} = 3.9 \times 10^{-8}\,M$

pOH = 14.00 - pH = 14.00 - 7.41 = 6.59; $[OH^-] = 10^{-pOH} = 10^{-6.59} = 2.6 \times 10^{-7}\,M$

or $[OH^-] = \dfrac{K_w}{[H^+]} = \dfrac{1.0 \times 10^{-14}}{3.9 \times 10^{-8}} = 2.6 \times 10^{-7}\,M$

b. $[H^+] = 10^{-15.3} = 5 \times 10^{-16}\,M$; pOH = 14.00 - 15.3 = -1.3; $[OH^-] = 10^{-(-1.3)} = 20\,M$

c. $[H^+] = 10^{-(-1.0)} = 10\,M$; pOH = 14.0 - (-1.0) = 15.0; $[OH^-] = 10^{-15.0} = 1 \times 10^{-15}\,M$

d. $[H^+] = 10^{-3.2} = 6 \times 10^{-4}\,M$; pOH = 14.0 - 3.2 = 10.8; $[OH^-] = 10^{-10.8} = 2 \times 10^{-11}\,M$

e. $[OH^-] = 10^{-5.0} = 1 \times 10^{-5}\,M$; pH = 14.0 - pOH = 14.0 - 5.0 = 9.0; $[H^+] = 10^{-9.0} = 1 \times 10^{-9}\,M$

f. $[OH^-] = 10^{-9.6} = 3 \times 10^{-10}\,M$; pH = 14.0 - 9.6 = 4.4; $[H^+] = 10^{-4.4} = 4 \times 10^{-5}\,M$

42.   a. $[H^+] = 10^{-pH} = 10^{-2.66} = 2.2 \times 10^{-3}\,M$

pOH = 14.00 - pH = 14.00 - 2.66 = 11.34; $[OH^-] = 10^{-pOH} = 10^{-11.34} = 4.6 \times 10^{-12}\,M$

b. $[H^+] = 10^{-8.66} = 2.2 \times 10^{-9}\,M$; pOH = 14.00 - 8.66 = 5.34; $[OH^-] = 10^{-5.34} = 4.6 \times 10^{-6}\,M$

c. $[H^+] = 10^{-(-0.34)} = 2.2\,M$; pOH = 14.00 - (-0.34) = 14.34; $[OH^-] = 10^{-14.34} = 4.6 \times 10^{-15}\,M$

d. $[OH^-] = 10^{-2.48} = 3.3 \times 10^{-3}\,M$

pH = 14.00 - pOH = 14.00 - 2.48 = 11.52; $[H^+] = 10^{-pH} = 10^{-11.52} = 3.0 \times 10^{-12}\,M$

e. $[OH^-] = 10^{-8.48} = 3.3 \times 10^{-9}\,M$; pH = 14.00 - 8.48 = 5.52; $[H^+] = 10^{-5.52} = 3.0 \times 10^{-6}\,M$

f. $[OH^-] = 10^{-15.00} = 1.0 \times 10^{-15}$; pH = 14.00 - 15.00 = -1.00, $[H^+] = 10^{-(-1.00)} = 10.\,M$

43.   pOH = 14.00 - pH = 14.00 - 6.77 = 7.23; $[H^+] = 10^{-pH} = 10^{-6.77} = 1.7 \times 10^{-7}\,M$

$[OH^-] = \dfrac{K_w}{[H^+]} = \dfrac{1.0 \times 10^{-14}}{1.7 \times 10^{-7}} = 5.9 \times 10^{-8}\,M$ or $[OH^-] = 10^{-pOH} = 10^{-7.23} = 5.9 \times 10^{-8}\,M$

Milk is slightly acidic since the pH is less than 7.00 at 25°C.

44. $pH = 14.00 - pOH = 14.00 - 5.74 = 8.26$; $[H^+] = 10^{-pH} = 10^{-8.26} = 5.5 \times 10^{-9} M$

$$[OH^-] = \frac{K_w}{[H^+]} = \frac{1.0 \times 10^{-14}}{5.5 \times 10^{-9}} = 1.8 \times 10^{-6} M \text{ or } [OH^-] = 10^{-pOH} = 10^{-5.74} = 1.8 \times 10^{-6} M$$

A solution of baking soda is basic since the pH is greater than 7.00 at 25°C.

Solutions of Acids

45.   All the acids in this problem are strong acids which are always assumed to completely dissociate in water. The general dissociation reaction for a strong acid is: $HA(aq) \rightarrow H^+(aq) + A^-(aq)$ where A^- is the conjugate base of the strong acid HA. For $0.250 M$ solutions of these strong acids, $0.250 M H^+$ and $0.250 M A^-$ are present when the acids completely dissociate. The amount of H^+ donated from water will be insignificant in this problem since H_2O is a very weak acid.

   a.   Major species present after dissociation = $H^+(aq)$, $Cl^-(aq)$ and H_2O (always present),

        $pH = -\log [H^+] = -\log (0.250) = 0.602$

   b.   Major species = $H^+(aq)$, $Br^-(aq)$ and H_2O; $pH = 0.602$

46. $HClO_4$ and HNO_3 are strong acids to memorize. As in Exercise 14.45, $0.250 M$ solutions of these strong acids give $0.250 M H^+$ and $0.250 M A^-$ where A^- is the conjugate base of the strong acid.

   a.   Major species = H^+, ClO_4^- and H_2O; $pH = -\log [H^+] = -\log (0.250) = 0.602$

   b.   Major species = H^+, NO_3^- and H_2O; $pH = 0.602$

47.   Strong acids are assumed to completely dissociate in water: $HCl(aq) \rightarrow H^+(aq) + Cl^-(aq)$

   a.   A $0.10 M$ HCl solution gives $0.10 M H^+$ and $0.10 M Cl^-$ since HCl completely dissociates. The amount of H^+ from H_2O will be insignificant.

        $pH = -\log [H^+] = -\log (0.10) = 1.00$

   b. $5.0 M H^+$ is produced when $5.0 M$ HCl completely dissociates. The amount of H^+ from H_2O will be insignificant. $pH = -\log (5.0) = -0.70$ (Negative pH values just indicate very concentrated acid solutions).

   c. $1.0 \times 10^{-11} M H^+$ is produced when $1.0 \times 10^{-11} M$ HCl completely dissociates. If you take the negative log of 1.0×10^{-11} this gives $pH = 11.00$. This is impossible! We dissolved an acid in water and got a basic pH. What we must consider in this problem is that water by itself donates $1.0 \times 10^{-7} M H^+$. We can normally ignore the small amount of H^+ from H_2O except when we have a very dilute solution of an acid (as in the case here). Therefore, the pH is that of neutral water (pH = 7.00) since the amount of HCl present is insignificant.

48. $HNO_3(aq) \rightarrow H^+(aq) + NO_3^-(aq)$; HNO_3 is a strong acid which means it is assumed to completely dissociate in water. The initial concentration of HNO_3 will equal the $[H^+]$ donated by the strong acid.

   a. $pH = -\log [H^+] = -\log (2.0 \times 10^{-2}) = 1.70$

   b. $pH = -\log (4.0) = -0.60$

   c.   Since the concentration of HNO_3 is so dilute, the pH will be that of neutral water (pH = 7.00). In this problem, water is the major H^+ producer present.

49.   Both are strong acids.

   $0.0500\ L \times 0.050\ mol/L = 2.5 \times 10^{-3}\ mol\ HCl = 2.5 \times 10^{-3}\ mol\ H^+ + 2.5 \times 10^{-3}\ mol\ Cl^-$

   $0.1500\ L \times 0.10\ mol/L = 1.5 \times 10^{-2}\ mol\ HNO_3 = 1.5 \times 10^{-2}\ mol\ H^+ + 1.5 \times 10^{-2}\ mol\ NO_3^-$

   $[H^+] = \dfrac{(2.5 \times 10^{-3} + 1.5 \times 10^{-2})\ mol}{0.2000\ L} = 0.088\ M$; $[OH^-] = \dfrac{K_w}{[H^+]} = 1.1 \times 10^{-13}\ M$

   $[Cl^-] = \dfrac{2.5 \times 10^{-3}\ mol}{0.2000\ L} = 0.013\ M$; $[NO_3^-] = \dfrac{1.5 \times 10^{-2}\ mol}{0.2000\ L} = 0.075\ M$

50. $90.0 \times 10^{-3}\ L \times \dfrac{5.00\ mol}{L} = 0.450\ mol\ H^+$ from HCl

   $30.0 \times 10^{-3}\ L \times \dfrac{8.00\ mol}{L} = 0.240\ mol\ H^+$ from HNO_3

   $[H^+] = \dfrac{0.450\ mol + 0.240\ mol}{1.00\ L} = 0.690\ M$; $pH = -\log (0.690) = 0.161$

   $pOH = 14.000 - 0.161 = 13.839$; $[OH^-] = 10^{-13.839} = 1.45 \times 10^{-14}\ M$

51. $[H^+] = 10^{-pH} = 10^{-2.50} = 3.2 \times 10^{-3}\ M$. Since HCl is a strong acid, then a $3.2 \times 10^{-3}\ M$ HCl solution will produce $3.2 \times 10^{-3}\ M\ H^+$ giving a pH = 2.50.

52. $[H^+] = 10^{-pH} = 10^{-4.25} = 5.6 \times 10^{-5}\ M$. Since HNO_3 is a strong acid, then a $5.6 \times 10^{-5}\ M\ HNO_3$ solution is necessary to produce a pH = 4.25 solution.

53.   a. HNO_2 ($K_a = 4.0 \times 10^{-4}$) and H_2O ($K_a = K_w = 1.0 \times 10^{-14}$) are the major species. HNO_2 is much stronger acid than H_2O so it is the major source of H^+. However, HNO_2 is a weak acid ($K_a < 1$) so it only partially dissociates in water. We must solve an equilibrium problem to determine $[H^+]$. In the Solutions Guide, we will summarize the initial, change and equilibrium concentrations into one table called the ICE table. Solving the weak acid problem:

$$HNO_2 \quad \rightleftharpoons \quad H^+ \quad + \quad NO_2^-$$

Initial   0.250 M ~0   0

          x mol/L HNO_2 dissociates to reach equilibrium

Change   -x $\rightarrow$ +x +x

Equil.   0.250 -x x x

$$K_a = \frac{[H^+][NO_2^-]}{[HNO_2]} = 4.0 \times 10^{-4} = \frac{x^2}{0.250 - x}; \text{ If we assume } x \ll 0.250, \text{ then:}$$

$$4.0 \times 10^{-4} \approx \frac{x^2}{0.250}, x = \sqrt{4.0 \times 10^{-4}(0.250)} = 0.010\,M$$

We must check the assumption: $\dfrac{x}{0.250} \times 100 = \dfrac{0.010}{0.250} \times 100 = 4.0\%$

All the assumptions are good. The H^+ contribution from water ($10^{-7}\,M$) is negligible and x is small compared to 0.250 (percent error = 4.0%). If the percent error is less than 5% for an assumption, we will consider it a valid assumption (called the 5% rule). Finishing the problem: $x = 0.010\,M = [H^+]$; pH = -log(0.010) = 2.00

b. CH_3CO_2H ($K_a = 1.8 \times 10^{-5}$) and H_2O ($K_a = K_w = 1.0 \times 10^{-14}$) are the major species. CH_3CO_2H is the major source of H^+. Solving the weak acid problem:

$$CH_3CO_2H \quad \rightleftharpoons \quad H^+ \quad + \quad CH_3CO_2^-$$

Initial   0.250 M ~0   0

          x mol/L CH_3CO_2H dissociates to reach equilibrium

Change   -x $\rightarrow$ +x +x

Equil.   0.250 - x x x

$$K_a = \frac{[H^+][CH_3CO_2^-]}{[CH_3CO_2H]} = 1.8 \times 10^{-5} = \frac{x^2}{0.250 - x} \approx \frac{x^2}{0.250} \quad (\text{assuming } x \ll 0.250)$$

$x = 2.1 \times 10^{-3}\,M$; Checking assumption: $\dfrac{2.1 \times 10^{-3}}{0.250} \times 100 = 0.84\%$. Assumptions good.

$[H^+] = x = 2.1 \times 10^{-3}\,M$; pH = -log (2.1×10^{-3}) = 2.68

54. a. HOC_6H_5 ($K_a = 1.6 \times 10^{-10}$) and H_2O ($K_a = K_w = 1.0 \times 10^{-14}$) are the major species. The major equilibrium is the dissociation of HOC_6H_5. Solving the weak acid problem:

$$HOC_6H_5 \quad \rightleftharpoons \quad H^+ \quad + \quad OC_6H_5^-$$

Initial   0.250 M ~0   0

          x mol/L HOC_6H_5 dissociates to reach equilibrium

Change   -x $\rightarrow$ +x +x

Equil.   0.250 - x x x

$$K_a = 1.6 \times 10^{-10} = \frac{[H^+][OC_6H_5^-]}{[HOC_6H_5]} = \frac{x^2}{0.250 - x} \approx \frac{x^2}{0.250} \quad \text{(assuming } x \ll 0.250\text{)}$$

$x = [H^+] = 6.3 \times 10^{-6} M$; Checking assumption: x is $2.5 \times 10^{-3}\%$ of 0.250 so assumption is valid by the 5% rule.

$pH = -\log(6.3 \times 10^{-6}) = 5.20$

b. HCN ($K_a = 6.2 \times 10^{-10}$) and H_2O are the major species. HCN is the major source of H^+.

$$\text{HCN} \quad \rightleftharpoons \quad \text{H}^+ \quad + \quad \text{CN}^-$$

Initial	0.250 M	~ 0	0

x mol/L HCN dissociates to reach equilibrium

Change	$-x$	$\rightarrow$	$+x$	$+x$
Equil.	0.250 $- x$		x	x

$$K_a = 6.2 \times 10^{-10} = \frac{[H^+][CN^-]}{[HCN]} = \frac{x^2}{0.250 - x} \approx \frac{x^2}{0.250} \quad \text{(assuming } x \ll 0.250\text{)}$$

$x = [H^+] = 1.2 \times 10^{-5} M$; Checking assumption: x is $4.8 \times 10^{-3}\%$ of 0.250

Assumptions good. $pH = -\log(1.2 \times 10^{-5}) = 4.92$

55. Major species: $HC_2H_3O_2$ ($K_a = 1.8 \times 10^{-5}$) and water; Major source of $H^+ = HC_2H_3O_2$. Since K_a for $HC_2H_3O_2$ is less than one, then $HC_2H_3O_2$ is a weak acid and we must solve an equilibrium problem to determine $[H^+]$. The set-up is:

$$HC_2H_3O_2 \quad \rightleftharpoons \quad H^+ + \quad C_3H_3O_2^-$$

Initial	0.20 M	~ 0	0

x mol/L $HC_2H_3O_2$ dissociates to reach equilibrium

Change	$-x$	$\rightarrow$	$+x$	$+x$
Equl.	0.20 $- x$		x	x

$$K_a = 1.8 \times 10^{-5} = \frac{[H^+][C_2H_3O_2^-]}{[HC_2H_3O_2]} = \frac{x^2}{0.20 - x} \approx \frac{x^2}{0.20} \quad \text{(assuming } x \ll 0.20\text{)}$$

$x = [H^+] = 1.9 \times 10^{-3} M$

We have made two assumptions which we must check.

1. $0.20 - x \approx 0.20$

$(x/0.20) \times 100 = (1.9 \times 10^{-3}/0.20) \times 100 = 0.95\%$. Good assumption (1% error). If the percent error in the assumption is $< 5\%$, then the assumption is valid.

2. Acetic acid is the major source of H^+, i.e., we can ignore $10^{-7}\ M$ H^+ already present in neutral H_2O.

$[H^+]$ from $HC_2H_3O_2 = 1.9 \times 10^{-3} >> 10^{-7}$;   This assumption is valid.

In future problems we will always begin the problem solving process by making these assumptions and we will always check them. However, we may not explicitly state that the assumptions are valid. We will <u>always</u> state when the assumptions are <u>not</u> valid and we have to use other techniques to solve the problem. Remember, anytime we make an assumption, we must check its validity before the solution to the problem is complete. Answering the question:

$$[H^+] = [C_2H_3O_2^-] = 1.9 \times 10^{-3}\ M;\ \ [OH^-] = 5.3 \times 10^{-12}\ M$$

$$[HC_2H_3O_2] = 0.20 - x = 0.198 \approx 0.20\ M;\ \ pH = -\log(1.9 \times 10^{-3}) = 2.72$$

56. HNO_2 ($K_a = 4.0 \times 10^{-4}$) is the dominant producer of H^+. Solving the weak acid problem:

$$HNO_2 \rightleftharpoons H^+ + NO_2^- \qquad K_a = 4.0 \times 10^{-4}$$

Initial	1.5 M	~0	0
	x mol/L HNO_2 dissociates to reach equilibrium		
Change	$-x$ $\rightarrow$	$+x$	$+x$
Equl.	1.5 - x	x	x

$$K_a = 4.0 \times 10^{-4} = \frac{[H^+][NO_2^-]}{[HNO_2]} = \frac{x^2}{1.5 - x} \approx \frac{x^2}{1.5}\ \ (\text{assuming } x << 1.5)$$

$x = [H^+] = 2.4 \times 10^{-2}\ M;$ Assumptions good: $10^{-7} << 2.4 \times 10^{-2} << 1.5$

$[H^+] = [NO_2^-] = 2.4 \times 10^{-2}\ M;\ [OH^-] = 4.2 \times 10^{-13}\ M$

$[HNO_2] = 1.5 - x = 1.48 \approx 1.5\ M;\ pH = -\log(2.4 \times 10^{-2}) = 1.62$

57. Boric acid is a weak acid; it is the major source of H^+.

$$B(OH)_3 + H_2O \rightleftharpoons B(OH)_4^- + H^+$$

Initial	0.50 M	0	~0
	x mol/L $B(OH)_3$ reacts to reach equilibrium		
Change	$-x$ $\rightarrow$	$+x$	$+x$
Equil.	0.50 - x	x	x

$$K_a = 5.8 \times 10^{-10} = \frac{[B(OH)_4^-][H^+]}{[B(OH)_3]} = \frac{x^2}{0.50 - x} \approx \frac{x^2}{0.50}\ \ (\text{assuming } x << 0.50)$$

$x = [H^+] = 1.7 \times 10^{-5}\ M;\ pH = 4.77$ Assumptions good (x is 3.4×10^{-3}% of 0.50).

58.   This is a weak acid in water. We must solve the weak acid problem.

$$12.2 \text{ g } C_6H_5CO_2H \text{ (HBz)} \times \frac{1 \text{ mol HBz}}{122.12 \text{ g}} = 9.99 \times 10^{-2} \text{ mol}; \ [HBz]_o = 9.99 \times 10^{-2} M$$

$$\text{HBz} \quad \rightleftharpoons \quad H^+ \ + \ Bz^- \quad Bz^- = C_6H_5CO_2^-$$

	HBz		H⁺		Bz⁻

Initial $9.99 \times 10^{-2} M$ ~0   0

x mol/L HBz dissociates to reach equilibrium

Change $-x$ $\rightarrow$ $+x$ $+x$

Equil. $9.99 \times 10^{-2} - x$ x x

$$K_a = 6.4 \times 10^{-5} = \frac{[H^+][Bz^-]}{[HBz]} = \frac{x^2}{9.99 \times 10^{-2} - x} \approx \frac{x^2}{9.99 \times 10^{-2}} \quad \text{(assuming } x << 9.99 \times 10^{-2})$$

$x = [H^+] = 2.5 \times 10^{-3} M$; Assumptions good ($x$ is 2.5% of 9.99×10^{-2}).

So: $x = [H^+] = [Bz^-] = 2.5 \times 10^{-3} M$; $[HBz] = 9.99 \times 10^{-2} - x = 9.74 \times 10^{-2} M$

pH = 2.60; pOH = 11.40; $[OH^-] = 10^{-11.40} = 4.0 \times 10^{-12} M$

59.   This is a weak acid in water. Solving the weak acid problem:

$$\text{HF} \quad \rightleftharpoons \quad H^+ \ + \ F^- \quad K_a = 7.2 \times 10^{-4}$$

Initial $0.020 M$ ~0   0

x mol/L HF dissociates to reach equilibrium

Change $-x$ $\rightarrow$ $+x$ $+x$

Equl. $0.020 - x$ x x

$$K_a = 7.2 \times 10^{-4} = \frac{[H^+][F^-]}{[HF]} = \frac{x^2}{0.020 - x} \approx \frac{x^2}{0.020} \quad \text{(assuming } x << 0.020)$$

$x = [H^+] = 3.8 \times 10^{-3} M$; Check assumptions: $\dfrac{x}{0.020} \times 100 = \dfrac{3.8 \times 10^{-3}}{0.020} \times 100 = 19\%$

The assumption $x << 0.020$ is not good (x is more than 5% of 0.020). We must solve $x^2/0.20 - x = 7.2 \times 10^{-4}$ exactly by using either the quadratic formula or by the method of successive approximations (see Appendix 1.4 of text). Using successive approximations, we let $0.016 M$ be a new approximation for [HF]. That is, in the denominator try $x = 0.0038$ (the value of x we calculated making the normal assumption), so $0.020 - 0.0038 = 0.016$, then solve for a new value of x in the numerator.

$$\frac{x^2}{0.020 - x} \approx \frac{x^2}{0.016} = 7.2 \times 10^{-4}, \ x = 3.4 \times 10^{-3}$$

We use this new value of x to further refine our estimate of [HF], i.e., $0.020 - x = 0.020 - 0.0034 = 0.0166$ (carry extra significant figure).

$$\frac{x^2}{0.020 - x} \approx \frac{x^2}{0.0166} = 7.2 \times 10^{-4}, \ x = 3.5 \times 10^{-3}$$

We repeat, until we get a self-consistent answer. This would be the same answer we would get solving exactly using the quadratic equation. In this case it is: $x = 3.5 \times 10^{-3}$

So: $[H^+] = [F^-] = x = 3.5 \times 10^{-3} M$; $[OH^-] = K_w/[H^+] = 2.9 \times 10^{-12} M$

$[HF] = 0.020 - x = 0.020 - 0.0035 = 0.017 M$; $pH = 2.46$

Note: When the 5% assumption fails, use whichever method you are most comfortable with to solve exactly. The method of successive approximations is probably fastest when the percent error is less than ~25%.

60.   Major species: HIO_3, H_2O; Major source of H^+: HIO_3 (a weak acid, $K_a = 0.17$)

$$HIO_3 \quad \rightleftharpoons \quad H^+ \quad + \quad IO_3^-$$

Initial	0.20 M	~0	0

x mol/L HIO_3 dissociates to reach equilibrium

Change	-x	$\rightarrow$ +x	+x
Equil.	0.20 - x	x	x

$$K_a = 0.17 = \frac{[H^+][IO_3^-]}{[HIO_3]} = \frac{x^2}{0.20 - x} \approx \frac{x^2}{0.20}, \quad x = 0.18; \quad \text{Check assumption.}$$

Assumption is horrible (x is 90% of 0.20). When the assumption is this poor, it is generally quickest to solve exactly using the quadratic formula (see Appendix 1.4 in text). The method of successive approximations will require many trials to finally converge on the answer. For this problem, 5 trials were required. Using the quadratic formula and carrying extra significant figures:

$$0.17 = \frac{x^2}{0.20 - x}, \quad x^2 = 0.17(0.20 - x), \quad x^2 + 0.17x - 0.034 = 0$$

$$x = \frac{-0.17 \pm [(0.17)^2 - 4(1)(-0.034)]^{1/2}}{2(1)} = \frac{-0.17 \pm 0.406}{2}, \quad x = 0.12 \text{ or } -0.29$$

Only $x = 0.12$ makes sense. $x = 0.12 M = [H^+]$; $pH = -\log(0.12) = 0.92$

61.   This is a weak acid in water. Solving the weak acid problem:

$$HCO_2H \quad \rightleftharpoons \quad H^+ \quad + \quad HCO_2^- \quad K_a = 1.8 \times 10^{-4}$$

Initial	0.025 M	~0	0

x mol/L HCO_2H dissociates to reach equilibrium

Change	-x	$\rightarrow$ +x	+x
Equil.	0.025 - x	x	x

$$K_a = 1.8 \times 10^{-4} = \frac{[H^+][HCO_2^-]}{[HCO_2H]} = \frac{x^2}{0.025 - x} \approx \frac{x^2}{0.025}$$

$x = [H^+] = 2.1 \times 10^{-3} M$; Check assumptions: $\dfrac{2.1 \times 10^{-3}}{0.025} \times 100 = 8.4\%$

Assumption that $x \ll 0.025$ is not good (fails the 5% rule). Solving using the method of successive approximations (see Appendix 1.4 in text):

$$\frac{x^2}{0.025 - x} = \frac{x^2}{0.025 - 0.0021} = \frac{x^2}{0.023} = 1.8 \times 10^{-4}, \ x = 2.0 \times 10^{-3} \text{ which we get}$$
$$\text{consistently.}$$

$x = [H^+] = 2.0 \times 10^{-3} \, M; \ \text{pH} = 2.70$

62.   Major species: $HC_2H_2ClO_2$ ($K_a = 1.35 \times 10^{-3}$) and H_2O; Major source of H^+: $HC_2H_2ClO_2$

$$HC_2H_2ClO_2 \ \rightleftharpoons \ H^+ \ + \ C_2H_2ClO_2$$

Initial   0.10 M ~0   0
                 x mol/L $HC_2H_2ClO_2$ dissociates to reach equilibrium
Change   -x $\rightarrow$ +x +x
Equil.   0.10 - x x x

$K_a = 1.35 \times 10^{-3} = \dfrac{x^2}{0.10 - x} \approx \dfrac{x^2}{0.10}, \ x = 1.2 \times 10^{-2} \, M$

Checking the assumptions finds that x is 12% of 0.10 which fails the 5% rule. We must solve $1.35 \times 10^{-3} = x^2/0.10 - x$ exactly using either the method of successive approximations or the quadratic equation. Using either method gives $x = [H^+] = 1.1 \times 10^{-2} \, M$. pH = -log $[H^+]$ = -log $(1.1 \times 10^{-2}) = 1.96$.

63.   HCl is a strong acid. It will produce 0.10 M H^+. HOCl is a weak acid. Let's consider the equilibrium:

$$HOCl \ \rightleftharpoons \ H^+ \ + \ OCl^-$$

Initial   0.10 M 0.10 M 0
                 x mol/L HOCl dissociates to reach equilibrium
Change   -x $\rightarrow$ +x +x
Equil.   0.10 - x 0.10 + x x

$K_a = 3.5 \times 10^{-8} = \dfrac{[H^+][OCl^-]}{[HOCl]} = \dfrac{(0.10 + x)(x)}{0.10 - x} \approx x, x = 3.5 \times 10^{-8}$

Assumptions are great (x is 3.5×10^{-5}% of 0.10). We are really assuming that HCl is the only important source of H^+, which it is. The $[H^+]$ contribution from the HOCl, x, is negligible. Therefore, $[H^+] = 0.10 \, M$ and pH = 1.00

64.   HF and HOC_6H_5 are both weak acids with K_a values of 7.2×10^{-4} and 1.6×10^{-10}, respectively. Since the K_a value for HF is much greater than the K_a value for HOC_6H_5, then HF will be the dominant producer of H^+ (we can ignore the amount of H^+ produced from HOC_6H_5 since it will be insignificant).

$$HF \quad \rightleftharpoons \quad H^+ \quad + \quad F^-$$

Initial   0.10 M ~0   0

x mol/L HF dissociates to reach equilibrium

Change $-x$ $\rightarrow$ $+x$ $+x$

Equil.   0.10 - x x x

$$K_a = 7.2 \times 10^{-4} = \frac{[H^+][F^-]}{[HF]} = \frac{x^2}{1.0 - x} \approx \frac{x^2}{1.0}$$

$x = [H^+] = 2.7 \times 10^{-2}\,M;$ pH = -log (2.7×10^{-2}) = 1.57   Assumptions good.

Solving for $[OC_6H_5^-]$ using $HOC_6H_5 \rightleftharpoons H^+ + OC_6H_5^-$ equilibrium:

$$K_a = 1.6 \times 10^{-10} = \frac{[H^+][OC_6H_5^-]}{[HOC_6H_5]} = \frac{(2.7 \times 10^{-2})[OC_6H_5^-]}{1.0}, \quad [OC_6H_5^-] = 5.9 \times 10^{-9}\,M$$

Note that this answer indicates that only $5.9 \times 10^{-9}\,M$ HOC_6H_5 dissociates which indicates that HF is truly the only significant producer of H^+ in this solution.

65.   In all parts of this problem, acetic acid ($HC_2H_3O_2$) is the best weak acid present. We must solve a weak acid problem.

a. $$HC_2H_3O_2 \quad \rightleftharpoons \quad H^+ \quad + \quad C_2H_3O_2^-$$

Initial   0.50 M ~0   0

x mol/L $HC_2H_3O_2$ dissociates to reach equilibrium

Change $-x$ $\rightarrow$ $+x$ $+x$

Equil.   0.50 - x x x

$$K_a = 1.8 \times 10^{-5} = \frac{[H^+][C_2H_3O_2^-]}{[HC_2H_3O_2]} = \frac{x^2}{0.50 - x} \approx \frac{x^2}{0.50}$$

$x = [H^+] = [C_2H_3O_2^-] = 3.0 \times 10^{-3}\,M$ Assumptions good.

$$\text{Perecent dissociation} = \frac{[H^+]}{[HC_2H_3O_2]_o} \times 100 = \frac{3.0 \times 10^{-3}}{0.50} \times 100 = 0.60\%$$

b.   The set-up for solutions b and c are similar to solution a except the final equation is slightly different, reflecting the new concentration of $HC_2H_3O_2$.

$$K_a = 1.8 \times 10^{-5} = \frac{x^2}{0.050 - x} \approx \frac{x^2}{0.050}$$

$x = [H^+] = [C_2H_3O_2^-] = 9.5 \times 10^{-4}\,M$ Assumptions good.

$$\% \text{ dissociation} = \frac{9.5 \times 10^{-4}}{0.050} \times 100 = 1.9\%$$

c. $K_a = 1.8 \times 10^{-5} = \dfrac{x^2}{0.0050 - x} \approx \dfrac{x^2}{0.0050}$

$x = [H^+] = [C_2H_3O_2^-] = 3.0 \times 10^{-4}\, M$; Check assumptions.

Assumption that x is negligible is borderline (6.0% error). We should solve exactly. Using the method of successive approximations (see Appendix 1.4 of text):

$1.8 \times 10^{-5} = \dfrac{x^2}{0.0050 - 3.0 \times 10^{-4}} = \dfrac{x^2}{0.0047}, \; x = 2.9 \times 10^{-4}$

Next trial also gives $x = 2.9 \times 10^{-4}$.

% dissociation $= \dfrac{2.9 \times 10^{-4}}{5.0 \times 10^{-3}} \times 100 = 5.8\%$

d. As we dilute a solution, all concentrations are decreased. Dilution will shift the equilibrium to the side with the greater number of particles. For example, suppose we double the volume of an equilibrium mixture of a weak acid by adding water, then:

$$Q = \dfrac{\left(\dfrac{[H^+]_{eq}}{2}\right)\left(\dfrac{[X^-]_{eq}}{2}\right)}{\left(\dfrac{[HX]_{eq}}{2}\right)} = \dfrac{1}{2}K_a$$

$Q < K_a$, so the equilibrium shifts to the right or towards a greater percent dissociation.

e. $[H^+]$ depends on the initial concentration of weak acid and on how much weak acid dissociates. For solutions a-c the initial concentration of acid decreases more rapidly than the percent dissociation increases. Thus, $[H^+]$ decreases.

66. a. $\quad\quad\quad\quad\quad\quad$ HOCl $\quad\rightleftharpoons\quad$ H$^+$ $\quad+\quad$ OCl$^-$ $\quad\quad$ $K_a = 3.5 \times 10^{-8}$

Initial $\quad\quad$ 0.100 M $\quad\quad\quad\quad$ ~0 $\quad\quad\quad$ 0
$\quad\quad\quad\quad\quad$ x mol/L HOCl dissociates to reach equilibrium
Change $\quad\quad$ $-x$ $\quad\quad\rightarrow\quad$ $+x$ $\quad\quad$ $+x$
Equil. $\quad\quad$ 0.100 $- x$ $\quad\quad\quad$ x $\quad\quad$ x

$K_a = 3.5 \times 10^{-8} = \dfrac{[H^+][OCl^-]}{[HOCl]} = \dfrac{x^2}{0.100 - x} \approx \dfrac{x^2}{0.100}$

$x = [H^+] = [OCl^-] = 5.9 \times 10^{-5}\, M$; Assumptions good.

% dissociation $= \dfrac{[H^+]}{[HOCl]_o} \times 100 = \dfrac{5.9 \times 10^{-5}}{0.100} \times 100 = 0.059\%$

b.

$$HCN \rightleftharpoons H^+ + CN^- \qquad K_a = 6.2 \times 10^{-10}$$

Initial	0.100 M		~0	0
	x mol/L HCN dissociates to reach equilibrium			
Change	-x	$\rightarrow$	+x	+x
Equil.	0.100 - x		x	x

$$K_a = 6.2 \times 10^{-10} = \frac{[H^+][CN^-]}{[HCN]} = \frac{x^2}{0.100 - x} \approx \frac{x^2}{0.100}$$

$$x = [H^+] = [CN^-] = 7.9 \times 10^{-6} M; \quad \text{Assumptions good.}$$

$$\% \text{ dissociation} = \frac{7.9 \times 10^{-6}}{0.100} \times 100 = 7.9 \times 10^{-3}\%$$

c. HCl is a strong acid; it is 100% dissociated in solution.

d. For the same initial concentration, the percent dissociation increases as the strength of the acid increases (as K_a increases).

67. $HC_3H_5O_2$ ($K_a = 1.3 \times 10^{-5}$) and H_2O ($K_a = K_w = 1.0 \times 10^{-14}$) are the major species present. $HC_3H_5O_2$ will be the dominant producer of H^+ since $HC_3H_5O_2$ is a stronger acid than H_2O. Solving the weak acid problem:

$$HC_3H_5O_2 \rightleftharpoons H^+ + C_3H_5O_2^-$$

Initial	0.10 M		~0	0
	x mol/L $HC_3H_5O_2$ dissociates to reach equilibrium			
Change	-x	$\rightarrow$	+x	+x
Equil.	0.10 - x		x	x

$$K_a = 1.3 \times 10^{-5} = \frac{[H^+][C_3H_5O_2^-]}{[HC_3H_5O_2]} = \frac{x^2}{0.10 - x} \approx \frac{x^2}{0.10}$$

$$x = [H^+] = 1.1 \times 10^{-3} M; \quad pH = -\log(1.1 \times 10^{-3}) = 2.96$$

Assumption follows the 5% rule (x is 1.1% of 0.10).

$$\text{Percent dissociation} = \frac{[H^+]}{[HC_3H_5O_2]_o} \times 100 = \frac{1.1 \times 10^{-3}}{0.10} \times 100 = 1.1\%$$

68.

$$HClO_2 \rightleftharpoons H^+ + ClO_2^- \qquad K_a = 1.2 \times 10^{-2}$$

Initial	0.22 M		~0	0
	x mol/L $HClO_2$ dissociates to reach equilibrium			
Change	-x	$\rightarrow$	+x	+x
Equil.	0.22 - x		x	x

$$K_a = 1.2 \times 10^{-2} = \frac{[H^+][ClO^-]}{[HClO_2]} = \frac{x^2}{0.22 - x} \approx \frac{x^2}{0.22}, \quad x = 5.1 \times 10^{-2}$$

The assumption that x is small is not good (x is 23% of 0.22). Using the method of successive approximations and carrying extra significant figures:

$$\frac{x^2}{0.22 - 0.051} = \frac{x^2}{0.169} = 1.2 \times 10^{-2}, \ x = 4.5 \times 10^{-2}$$

$$\frac{x^2}{0.175} = 1.2 \times 10^{-2}, \ x = 4.6 \times 10^{-2}; \ x = 4.6 \times 10^{-2} \text{ repeats}$$

$[H^+] = [ClO_2^-] = x = 4.6 \times 10^{-2} \, M; \ \%$ dissociation $= \dfrac{4.6 \times 10^{-2}}{0.22} \times 100 = 21\%$

69.   Let HX symbolize the weak acid. Set-up the problem like a typical weak acid equilibrium problem.

$$HX \rightleftharpoons H^+ + X^-$$

Initial	0.15	~0	0
	x mol/L HX dissociates to reach equilibrium		
Change	$-x$ $\rightarrow$	$+x$	$+x$
Equil.	$0.15 - x$	x	x

If the acid is 3.0% dissociated, then $x = [H^+]$ is 3.0% of 0.15: $x = 0.030 \times (0.15 \, M) = 4.5 \times 10^{-3} \, M$
Now that we know the value of x, we can solve for K_a.

$$K_a = \frac{[H^+][X^-]}{[HX]} = \frac{x^2}{0.15 - x} = \frac{(4.5 \times 10^{-3})^2}{0.15 - 4.5 \times 10^{-3}} = 1.4 \times 10^{-4}$$

70. $$HF \rightleftharpoons H^+ + F^-$$

Initial	$0.100 \, M$	~0	0
	x mol/L HF dissociates to reach equilibrium		
Change	$-x$ $\rightarrow$	$+x$	$+x$
Equil.	$0.100 - x$	x	x

$$K_a = \frac{[H^+][F^-]}{[HF]} = \frac{x^2}{0.100 - x}; \ x = [H^+] = [F^-] = 0.081 \times (0.100 \, M) = 8.1 \times 10^{-3} \, M$$

$[HF] = 0.100 - 8.1 \times 10^{-3} = 0.092 \, M; \ K_a = \dfrac{(8.1 \times 10^{-3})^2}{0.092} = 7.1 \times 10^{-4}$

71.   Set-up the problem using the K_a equilibrium reaction for HOBr.

$$HOBr \rightleftharpoons H^+ + OBr^-$$

Initial	$0.063 \, M$	~0	0
	x mol/L HOBr dissociates to reach equilibrium		
Change	$-x$ $\rightarrow$	$+x$	$+x$
Equil.	$0.063 - x$	x	x

$$K_a = \frac{[H^+][OBr^-]}{[HOBr]} = \frac{x^2}{0.063 - x}; \quad \text{Since pH} = 4.95, \text{ then: } x = [H^+] = 10^{-pH} = 10^{-4.95} = 1.1 \times 10^{-5} M$$

$$K_a = \frac{(1.1 \times 10^{-5})^2}{0.063 - 1.1 \times 10^{-5}} = 1.9 \times 10^{-9}$$

72. Set-up the problem using the K_a equilibrium reaction for CCl_3CO_2H.

$$CCl_3CO_2H \quad \rightleftharpoons \quad H^+ \quad + \quad CCl_3CO_2^-$$

Initial	$0.050\,M$	~0	0
Equil.	$0.050 - x$	x	x

$$K_a = \frac{[H^+][CCl_3CO_2^-]}{[CCl_3CO_2H]} = \frac{x^2}{0.050 - x}; \quad x = [H^+] = 10^{-1.40} = 4.0 \times 10^{-2} M$$

$$K_a = \frac{(4.0 \times 10^{-2})^2}{0.050 - 4.0 \times 10^{-2}} = 0.16$$

73. Major species: $HC_2H_3O_2$ (acetic acid) and H_2O; Major source of H^+: $HC_2H_3O_2$

$$HC_2H_3O_2 \quad \rightleftharpoons \quad H^+ \quad + \quad C_2H_3O_2^-$$

Initial	C	~0	0	where C = $[HC_2H_3O_2]_o$

x mol/L $HC_2H_3O_2$ dissociates to reach equilibrium

Change	$-x$	$\rightarrow$	$+x$	$+x$
Equil.	$C - x$		x	x

$$K_a = 1.8 \times 10^{-5} = \frac{[H^+][C_2H_3O_2^-]}{[HC_2H_3O_2]} = \frac{x^2}{C - x} \text{ where } x = [H^+]$$

$$1.8 \times 10^{-5} = \frac{[H^+]^2}{C - [H^+]}; \quad \text{Since pH} = 3.0, \text{ then: } [H^+] = 10^{-3.0} = 1 \times 10^{-3} M$$

$$1.8 \times 10^{-5} = \frac{(1 \times 10^{-3})^2}{C - 1 \times 10^{-3}}, \quad C - 1 \times 10^{-3} = \frac{1 \times 10^{-6}}{1.8 \times 10^{-5}}, \quad C = 5.7 \times 10^{-2} \approx 6 \times 10^{-2} M$$

A $6 \times 10^{-2} M$ acetic acid solution will have pH = 3.0.

74. $[HA]_o = \dfrac{1.0 \text{ mol}}{2.0 \text{ L}} = 0.50 \text{ mol/L}; \quad$ Solve using the K_a equilibrium reaction.

$$HA \quad \rightleftharpoons \quad H^+ \quad + \quad A^-$$

Initial	$0.50\,M$	~0	0
Equil.	$0.50 - x$	x	x

$$K_a = \frac{[H^+][A^-]}{[HA]} = \frac{x^2}{0.50 - x}; \quad \text{In this problem, [HA]} = 0.45\,M \text{ so:}$$

$$[HA] = 0.45\,M = 0.50\,M - x,\ \ x = 0.05\,M;\ \ K_a = \frac{(0.05)^2}{0.45} = 6 \times 10^{-3}$$

Solutions of Bases

75. a. $NH_3(aq) + H_2O(l) \rightleftharpoons NH_4^+(aq) + OH^-(aq)$ $K_b = \dfrac{[NH_4^+]\,[OH^-]}{[NH_3]}$

b. $C_5H_5N(aq) + H_2O(l) \rightleftharpoons C_5H_5NH^+(aq) + OH^-(aq)$ $K_b = \dfrac{[C_5H_5NH^+]\,[OH^-]}{[C_5H_5N]}$

76. a. $C_6H_5NH_2(aq) + H_2O(l) \rightleftharpoons C_6H_5NH_3^+(aq) + OH^-(aq)$ $K_b = \dfrac{[C_6H_5NH_3^+]\,[OH^-]}{[C_6H_5NH_2]}$

b. $(CH_3)_2NH(aq) + H_2O(l) \rightleftharpoons (CH_3)_2NH_2^+(aq) + OH^-(aq)$ $K_b = \dfrac{[(CH_3)_2NH_2^+]\,[OH^-]}{[(CH)_3)_2NH]}$

77. NO_3^-: $K_b \ll K_w$ since HNO_3 is a strong acid. All conjugate bases of strong acids have no base strength. H_2O: $K_b = K_w = 1.0 \times 10^{-14}$; NH_3: $K_b = 1.8 \times 10^{-5}$; CH_3NH_2: $K_b = 4.38 \times 10^{-4}$

$CH_3NH_2 > NH_3 > H_2O > NO_3^-$ (As K_b increases, base strength increases.)

78. Excluding water these are the conjugate acids of the bases in the previous exercise. In general, the stronger the base, the weaker the conjugate acid. Note: Even though NH_4^+ and $CH_3NH_3^+$ are conjugate acids of weak bases, they are still weak acids with K_a values between K_w and 1. Prove this to yourself by calculating the K_a values for NH_4^+ and $CH_3NH_3^+$ ($K_a = K_w/K_b$).

$HNO_3 > NH_4^+ > CH_3NH_3^+ > H_2O$

79. a. NH_3 b. NH_3 c. OH^- d. CH_3NH_2

The base with the largest K_b value is the strongest base. OH^- is the strongest base possible in water.

80. a. HNO_3 b. NH_4^+ c. NH_4^+

The acid with the largest K_a value is the strongest acid. To calculate K_a values for NH_4^+ and $CH_3NH_3^+$, use $K_a = K_w/K_b$ where K_b refers to the bases NH_3 or CH_3NH_2.

81. $NaOH(aq) \rightarrow Na^+(aq) + OH^-(aq)$; NaOH is a strong base which completely dissociates into Na^+ and OH^-. The initial concentration of NaOH will equal the concentration of OH^- donated by NaOH.

a. $[OH^-] = 0.10\,M$; pOH = $-\log[OH^-]$ = $-\log(0.10)$ = 1.00

pH = 14.00 - pOH = 14.00 - 1.00 = 13.00

Note that H_2O is also present but the amount of OH^- produced by H_2O will be insignificant as compared to 0.10 M OH^- produced from the NaOH.

b. The [OH$^-$] concentration donated by the NaOH is $1.0 \times 10^{-10}\,M$. Water by itself donates $1.0 \times 10^{-7}\,M$. In this problem, water is the major OH$^-$ contributor and [OH$^-$] = $1.0 \times 10^{-7}\,M$.

pOH = -log (1.0×10^{-7}) = 7.00; pH = 14.00 - 7.00 = 7.00

c. [OH$^-$] = 2.0 M; pOH = -log (2.0) = -0.30; pH = 14.00 - (-0.30) = 14.30

82.   Ca(OH)$_2$(aq) → Ca^{2+}(aq) + 2 OH$^-$(aq); The OH$^-$ donated by the strong base Ca(OH)$_2$ is twice the initial concentration of Ca(OH)$_2$.

a. [OH$^-$] = 2(0.00040) = $8.0 \times 10^{-4}\,M$; pOH = -log [OH$^-$] = -log (8.0×10^{-4}) = 3.10

pH = 14.00 - pOH = 14.00 - 3.10 = 10.90; The OH$^-$ donated from H$_2$O will be insignificant.

b. [OH$^-$] = 2(1.0) = 2.0 M; pOH = -log(2.0) = -0.30; pH = 14.00 - (-0.30) = 14.30

c. This is an extremely dilute solution of Ca(OH)$_2$. The $1.0 \times 10^{-7}\,M$ OH$^-$ donated from H$_2$O will be the dominant OH$^-$ source. pOH = 7.00 = pH

83.   a. Major species: K$^+$, OH$^-$, H$_2$O (KOH is a strong base.)

[OH$^-$] = 0.150 M, pOH = -log(0.150) = 0.824; pH = 14.000 - pOH = 13.176

b. Major species: Sr^{2+}, OH$^-$, H$_2$O; Sr(OH)$_2$(aq) → Sr^{2+}(aq) + 2 OH$^-$(aq); Since the strong base Sr(OH)$_2$ dissolves in water to produce two mol OH$^-$, then [OH$^-$] = 2(0.150 M) = 0.300 M.

pOH = -log(0.300) = 0.523; pH = 14.000 - 0.523 = 13.477

84.   a. Major species: Na$^+$, Li$^+$, OH$^-$, H$_2$O (NaOH and LiOH are both strong bases.)

[OH$^-$] = 0.050 + 0.050 = 0.100 M; pOH = 1.000; pH = 13.000

b. Major species: Ba^{2+}, Rb$^+$, OH$^-$, H$_2$O; Both Ba(OH)$_2$ and RbOH are strong bases and Ba(OH)$_2$ donates 2 mol OH$^-$ per mol Ba(OH)$_2$.

[OH$^-$] = 2(0.0010) + 0.020 = 0.022 M; pOH = -log(0.022) = 1.66; pH = 12.34

85.   pH = 10.50; pOH = 14.00 - 10.50 = 3.50; [OH$^-$] = $10^{-3.50}$ = $3.2 \times 10^{-4}\,M$

KOH(aq) ⇌ K$^+$(aq) + OH$^-$(aq); Since KOH is a strong base, a $3.2 \times 10^{-4}\,M$ KOH solution will produce a pH = 10.50 solution.

86.   pH = 13.00; pOH = 14.00 - 13.00 = 1.00; [OH$^-$] = $10^{-1.00}$ = 0.10 M

Ca(OH)$_2$(aq) → Ca^{2+}(aq) + 2 OH$^-$(aq); Ca(OH)$_2$ donates two mol OH$^-$ per mol Ca(OH)$_2$.

[Ca(OH)$_2$] = 0.10 M OH$^-$ $\times \left(\dfrac{1\,M\,\text{Ca(OH)}_2}{2\,M\,\text{OH}^-} \right)$ = 0.050 M Ca(OH)$_2$

A 0.050 M $Ca(OH)_2$ solution will produce a pH = 13.00 solution.

87. NH_3 is a weak base with $K_b = 1.8 \times 10^{-5}$. The major species present will be NH_3 and H_2O ($K_b = K_w$ = 1.0×10^{-14}). Since NH_3 has a much larger K_b value as compared to H_2O, then NH_3 is the stronger base present and will be the major producer of OH^-. To determine the amount of OH^- produced from NH_3, we must perform an equilibrium calculation.

$$NH_3(aq) \;+\; H_2O(l) \;\rightleftharpoons\; NH_4^+(aq) \;+\; OH^-(aq)$$

Initial	0.150 M	0	~0

x mol/L NH_3 reacts with H_2O to reach equilibrium

Change	-x	$\rightarrow$	+x	+x
Equil.	0.150 - x		x	x

$$K_b = 1.8 \times 10^{-5} = \frac{[NH_4^+][OH^-]}{[NH_3]} = \frac{x^2}{0.150 - x} \approx \frac{x^2}{0.150} \quad \text{(assuming } x \ll 0.150)$$

$x = [OH^-] = 1.6 \times 10^{-3} \, M$; Check assumptions: x is 1.1% of 0.150 so the assumption 0.150 - $x \approx$ 0.150 is valid by the 5% rule. Also, the contribution of OH^- from water will be insignificant (which will usually be the case). Finishing the problem, pOH = -log [OH⁻] = -log ($1.6 \times 10^{-3} \, M$) = 2.80; pH = 14.00 - pOH = 14.00 - 2.80 = 11.20.

88.   Major species: H_2NNH_2 ($K_b = 3.0 \times 10^{-6}$) and H_2O ($K_b = K_w = 1.0 \times 10^{-14}$); The weak base H_2NNH_2 will dominate OH^- production. We must perform a weak base equilibrium calculation.

$$H_2NNH_2 \;+\; H_2O \;\rightleftharpoons\; H_2NNH_3^+ \;+\; OH^- \qquad K_b = 3.0 \times 10^{-6}$$

Initial	2.0 M	0	~0

x mol/L H_2NNH_2 reacts with H_2O to reach equilibrium

Change	-x	$\rightarrow$	+x	+x
Equil.	2.0 - x		x	x

$$K_b = 3.0 \times 10^{-6} = \frac{[H_2NNH_3^+][OH^-]}{[H_2NNH_2]} = \frac{x^2}{2.0 - x} \approx \frac{x^2}{2.0} \quad \text{(assuming } x \ll 2.0)$$

$x = [OH^-] = 2.4 \times 10^{-3} \, M$; pOH = 2.62; pH = 11.38   Assumptions good (x is 0.12% of 2.0).

89.   These are solutions of weak bases in water. We must solve the equilibrium weak base problem.

a. $$(C_2H_5)_3N \;+\; H_2O \;\rightleftharpoons\; (C_2H_5)_3NH^+ \;+\; OH^- \qquad K_b = 4.0 \times 10^{-4}$$

Initial	0.20 M	0	~0

x mol/L of $(C_2H_5)_3N$ reacts with H_2O to reach equilibrium

Change	-x	$\rightarrow$	+x	+x
Equil.	0.20 - x		x	x

$$K_b = 4.0 \times 10^{-4} = \frac{[(C_2H_5)_3NH^+][OH^-]}{[(C_2H_5)_3N]} = \frac{x^2}{0.20 - x} \approx \frac{x^2}{0.20}, \; x = [OH^-] = 8.9 \times 10^{-3} \, M$$

Assumptions good (x is 4.5% of 0.20). $[OH^-] = 8.9 \times 10^{-3}\ M$

$$[H^+] = \frac{K_w}{[OH^-]} = \frac{1.0 \times 10^{-14}}{8.9 \times 10^{-3}} = 1.1 \times 10^{-12}\ M;\ pH = 11.96$$

b. $\qquad\qquad HONH_2 + H_2O \rightleftharpoons HONH_3^+ + OH^- \qquad K_b = 1.1 \times 10^{-8}$

Initial	0.20 M	0	~0
Equil.	0.20 - x	x	x

$$K_b = 1.1 \times 10^{-8} = \frac{x^2}{0.20 - x} \approx \frac{x^2}{0.20},\ x = [OH^-] = 4.7 \times 10^{-5}\ M;\ \text{Assumptions good.}$$

$[H^+] = 2.1 \times 10^{-10}\ M;\ pH = 9.68$

90.   These are solutions of weak bases in water.

a. $\qquad\qquad C_6H_5NH_2 + H_2O \rightleftharpoons C_6H_5NH_3^+ + OH^- \qquad K_b = 3.8 \times 10^{-10}$

Initial	0.20 M	0	~0
	x mol/L of $C_6H_5NH_2$ reacts with H_2O to reach equilibrium		
Change	-x $\rightarrow$	+x	+x
Equil.	0.20 - x	x	x

$$3.8 \times 10^{-10} = \frac{x^2}{0.20 - x} \approx \frac{x^2}{0.20},\ x = [OH^-] = 8.7 \times 10^{-6}\ M;\ \text{Assumptions good.}$$

$[H^+] = K_w/[OH^-] = 1.1 \times 10^{-9}\ M;\ pH = 8.96$

b. $\qquad\qquad C_5H_5N + H_2O \rightleftharpoons C_5H_5NH^+ + OH^- \qquad K_b = 1.7 \times 10^{-9}$

Initial	0.20 M	0	~0
Equil.	0.20 - x	x	x

$$K_b = 1.7 \times 10^{-9} = \frac{x^2}{0.20 - x} \approx \frac{x^2}{0.20},\ x = 1.8 \times 10^{-5}\ M;\ \text{Assumptions good.}$$

$[OH^-] = 1.8 \times 10^{-5}\ M;\ [H^+] = 5.6 \times 10^{-10}\ M;\ pH = 9.25$

91.   This is a solution of a weak base in water. We must solve the weak base equilibrium problem.

$$C_2H_5NH_2 + H_2O \rightleftharpoons C_2H_5NH_3^+ + OH^- \qquad K_b = 5.6 \times 10^{-4}$$

Initial	0.20 M	0	~0
	x mol/L $C_2H_5NH_2$ reacts with H_2O to reach equilibrium		
Change	-x $\rightarrow$	+x	+x
Equil.	0.20 - x	x	x

$$K_b = \frac{[C_2H_5NH_3^+][OH^-]}{[C_2H_5NH_2]} = \frac{x^2}{0.20 - x} \approx \frac{x^2}{0.20} \qquad \text{(assuming } x \ll 0.20)$$

$x = 1.1 \times 10^{-2}$;   Checking assumption: $\dfrac{1.1 \times 10^{-2}}{0.20} \times 100 = 5.5\%$

Assumption fails the 5% rule. We must solve exactly using either the quadratic equation or the method of successive approximations (see Appendix 1.4 of the text). Using successive approximations and carrying extra significant figures:

$$\frac{x^2}{0.20 - 0.011} = \frac{x^2}{0.189} = 5.6 \times 10^{-4}, \; x = 1.0 \times 10^{-2} \, M \; \text{(consistent answer)}$$

$$x = [OH^-] = 1.0 \times 10^{-2} \, M; \; [H^+] = \frac{K_w}{[OH^-]} = \frac{1.0 \times 10^{-14}}{1.0 \times 10^{-2}} = 1.0 \times 10^{-12} \, M; \; pH = 12.00$$

92. $$(C_2H_5)_2NH + H_2O \; \rightleftharpoons \; (C_2H_5)_2NH_2^+ + OH^- \qquad K_b = 1.3 \times 10^{-3}$$

Initial	0.20 M	0	~0
	x mol/L $(C_2H_5)_2NH$ reacts with H_2O to reach equilibrium		
Change	$-x$ $\rightarrow$	$+x$	$+x$
Equil.	0.20 - x	x	x

$$K_b = 1.3 \times 10^{-3} = \frac{[(C_2H_5)_2NH_2^+][OH^-]}{[(C_2H_5)_2NH)]} = \frac{x^2}{0.20 - x} \approx \frac{x^2}{0.20} \quad \text{(assuming } x \ll 0.20)$$

$x = 1.6 \times 10^{-2}$;   Assumption is bad (x is 8.0% of 0.20).

Using successive approximations:

$$\frac{x^2}{0.20 - 0.016} = \frac{x^2}{0.184} = 1.3 \times 10^{-3}, \; x = 1.55 \times 10^{-2} \; \text{(carry extra significant figure)}$$

$$\frac{x^2}{0.185} = 1.3 \times 10^{-3}, \; x = 1.55 \times 10^{-2} \; \text{(consistent answer)}$$

$[OH^-] = x = 1.55 \times 10^{-2} \, M; \; [H^+] = 6.45 \times 10^{-13} \, M;$ To correct significant figures:

$[OH^-] = 1.6 \times 10^{-2} \, M; \; [H^+] = 6.5 \times 10^{-13} \, M; \; pH = 12.19$

93. To solve for percent ionization, just solve the weak base equilibrium problem.

a. $$NH_3 + H_2O \; \rightleftharpoons \; NH_4^+ + OH^- \qquad K_b = 1.8 \times 10^{-5}$$

Initial	0.10 M	0	~0
Equil.	0.10 - x	x	x

$$K_b = 1.8 \times 10^{-5} = \frac{x^2}{0.10 - x} \approx \frac{x^2}{0.10}, \; x = [OH^-] = 1.3 \times 10^{-3} \, M; \; \text{Assumptions good.}$$

$$\text{Percent ionization} = \frac{[OH^-]}{[NH_3]_o} \times 100 = \frac{1.3 \times 10^{-3} \, M}{0.10 \, M} \times 100 = 1.3\%$$

b.
$$NH_3 \; + \; H_2O \; \rightleftharpoons \; NH_4^+ \; + \; OH^-$$

Initial	$0.010\,M$	0	~ 0
Equil.	$0.010 - x$	x	x

$$1.8 \times 10^{-5} = \frac{x^2}{0.010 - x} \approx \frac{x^2}{0.010}, \; x = [OH^-] = 4.2 \times 10^{-4}\,M; \; \text{Assumptions good.}$$

$$\text{Percent ionization} = \frac{4.2 \times 10^{-4}}{0.010} \times 100 = 4.2\%$$

Note: For the same base, the percent ionization increases as the initial concentration of base decreases.

94.   a.
$$HONH_2 \; + \; H_2O \; \rightleftharpoons \; HONH_3^+ \; + \; OH^- \qquad K_b = 1.1 \times 10^{-8}$$

Initial	$0.10\,M$	0	~ 0
Equil.	$0.10 - x$	x	x

$$1.1 \times 10^{-8} = \frac{x^2}{0.10 - x} \approx \frac{x^2}{0.10}, \; x = [OH^-] = 3.3 \times 10^{-5}\,M; \; \text{Assumptions good.}$$

$$\text{Percent ionization} = \frac{[OH^-]}{[HONH_2]_o} \times 100 = \frac{3.3 \times 10^{-5}}{0.10} \times 100 = 0.033\%$$

b.
$$CH_3NH_2 \; + \; H_2O \; \rightleftharpoons \; CH_3NH_3^+ \; + \; OH^- \qquad K_b = 4.38 \times 10^{-4}$$

Initial	$0.10\,M$	0	~ 0
Equil.	$0.10 - x$	x	x

$$4.38 \times 10^{-4} = \frac{x^2}{0.10 - x} \approx \frac{x^2}{0.10}, \; x = 6.6 \times 10^{-3}; \; \text{Assumptions fails the 5\% rule (x is 6.6\%}$$
of 0.10).

Using successive approximations and carrying extra significant figures:

$$\frac{x^2}{0.10 - 0.0066} = \frac{x^2}{0.093} = 4.38 \times 10^{-4}; \; x = 6.4 \times 10^{-3} \quad \text{(consistent answer)}$$

$$\text{Percent ionization} = \frac{6.4 \times 10^{-3}}{0.10} \times 100 = 6.4\%$$

95.   Using the K_b reaction to solve where PT = p-toluidine, $CH_3C_6H_4NH_2$:

$$PT \; + \; H_2O \; \rightleftharpoons \; PTH^+ \; + \; OH^-$$

Initial	$0.016\,M$	0	~ 0
	\multicolumn{3}{l}{x mol/L of PT reacts with H_2O to reach equilibrium}	
Change	$-x$	$\rightarrow$ $+x$	$+x$
Equil.	$0.016 - x$	x	x

$$K_b = \frac{[PTH^+][OH^-]}{[PT]} = \frac{x^2}{0.016 - x}$$

Since pH = 8.60: pOH = 14.00 - 8.60 = 5.40 and $[OH^-] = x = 10^{-5.40} = 4.0 \times 10^{-6}\,M$

$$K_b = \frac{(4.0 \times 10^{-6})^2}{0.016 - 4.0 \times 10^{-6}} = 1.0 \times 10^{-9}$$

96. Using the K_b reaction to solve where PY = pyrrolidine, C_4H_8NH:

$$PY + H_2O \rightleftharpoons PYH^+ + OH^-$$

Initial $1.00 \times 10^{-3}\,M$ 0   ~0
Equil. $1.00 \times 10^{-3} - x$ x x

$$K_b = \frac{x^2}{1.00 \times 10^{-3} - x};\ \text{Since pH = 10.82: pOH = 3.18 and } [OH^-] = x = 10^{-3.18} = 6.6 \times 10^{-4}\,M$$

$$K_b = \frac{(6.6 \times 10^{-4})^2}{1.00 \times 10^{-3} - 6.6 \times 10^{-4}} = 1.3 \times 10^{-3}$$

Polyprotic Acids

97. $H_2SO_3(aq) \rightleftharpoons HSO_3^-(aq) + H^+(aq)$ $K_{a_1} = \dfrac{[HSO_3^-][H^+]}{[H_2SO_3]}$

$HSO_3^-(aq) \rightleftharpoons SO_3^{2-}(aq) + H^+(aq)$ $K_{a_2} = \dfrac{[SO_3^{2-}][H^+]}{[HSO_3^-]}$

98. $H_3AsO_4(aq) \rightleftharpoons H_2AsO_4^-(aq) + H^+(aq)$ $K_{a_1} = \dfrac{[H_2AsO_4^-][H^+]}{[H_3AsO_4]}$

$H_2AsO_4^-(aq) \rightleftharpoons HAsO_4^{2-}(aq) + H^+(aq)$ $K_{a_2} = \dfrac{[HAsO_4^{2-}][H^+]}{[H_2AsO_4^-]}$

$HAsO_4^{2-}(aq) \rightleftharpoons AsO_4^{3-}(aq) + H^+(aq)$ $K_{a_3} = \dfrac{[AsO_4^{3-}][H^+]}{[HAsO_4^{2-}]}$

99.   In both these polyprotic acid problems, the dominate equilibrium is the K_{a_1} reaction. The amount of H^+ produced from the subsequent K_a reactions will be minimal since they are all have much smaller K_a values.

a. $H_3AsO_4 \rightleftharpoons H^+ + H_2AsO_4^-$ $K_{a_1} = 5 \times 10^{-3}$

Initial $0.10\,M$ ~0   0
             x mol/L H_3AsO_4 dissociates to reach equilibrium
Change $-x$ $\rightarrow$ $+x$ $+x$
Equil. $0.10 - x$ x x

$$K_{a_1} = 5 \times 10^{-3} = \frac{[H^+][H_2AsO_4^-]}{[H_3AsO_4]} = \frac{x^2}{0.10 - x} \approx \frac{x^2}{0.10}, \ x = 2 \times 10^{-2}$$

Assumption is bad (x is 20% of 0.10). Using successive approximations:

$$\frac{x^2}{0.10 - 0.02} = \frac{x^2}{0.08} = 5 \times 10^{-3}, \ x = 2 \times 10^{-2} \quad \text{(consistent answer)}$$

$$x = [H^+] = 2 \times 10^{-2} \, M; \ pH = -\log(2 \times 10^{-2}) = 1.7$$

b. $\qquad\qquad H_2CO_3 \ \rightleftharpoons \ H^+ \ + \ HCO_3^- \qquad K_{a_1} = 4.3 \times 10^{-7}$

Initial	0.10 M	~0	0
Equil.	0.10 - x	x	x

$$K_{a_1} = 4.3 \times 10^{-7} = \frac{[H^+][HCO_3^-]}{[H_2CO_3]} = \frac{x^2}{(0.10 - x)} \approx \frac{x^2}{0.10}$$

$$x = [H^+] = 2.1 \times 10^{-4} \, M; \ pH = 3.68; \ \text{Assumptions good.}$$

100.   The reactions are:

$$H_3PO_4 \rightleftharpoons H^+ + H_2PO_4^- \quad K_{a_1} = 7.5 \times 10^{-3}$$

$$H_2PO_4^- \rightleftharpoons H^+ + HPO_4^{2-} \quad K_{a_2} = 6.2 \times 10^{-8}$$

$$HPO_4^{2-} \rightleftharpoons H^+ + PO_4^{3-} \quad K_{a_3} = 4.8 \times 10^{-13}$$

We will deal with the reactions in order of importance, beginning with the largest K_a, K_{a_1}.

$$H_3PO_4 \ \rightleftharpoons \ H^+ \ + \ H_2PO_4^- \quad K_{a_1} = 7.5 \times 10^{-3} = \frac{[H^+][H_2PO_4^-]}{[H_3PO_4]}$$

Initial	0.100 M	~0	0
Equil.	0.100 - x	x	x

$$7.5 \times 10^{-3} = \frac{x^2}{0.100 - x}, \ x = 2.4 \times 10^{-2} \, M \qquad \begin{array}{l}\text{(By using successive approximations} \\ \text{or the quadratic formula.)}\end{array}$$

$$[H^+] = [H_2PO_4^-] = 2.4 \times 10^{-2} \, M; \ [H_3PO_4] = 0.100 - 0.024 = 0.076 \, M$$

Since $K_{a_2} = \dfrac{[H^+][HPO_4^{2-}]}{[H_2PO_4^-]} = 6.2 \times 10^{-8}$ is much smaller than the K_{a_1} value, very little of the second (and third) reactions occur as compared to the first reaction. Therefore, $[H^+]$ and $[H_2PO_4^-]$ will not change significantly by the second reaction. Using the above concentrations of H^+ and $H_2PO_4^-$ to calculate the concentration of HPO_4^{2-}:

$$6.2 \times 10^{-8} = \frac{(2.4 \times 10^{-2}) [HPO_4^{2-}]}{2.4 \times 10^{-2}}, \quad [HPO_4^{2-}] = 6.2 \times 10^{-8} M$$

Assumption that second reaction does not change $[H^+]$ and $[H_2PO_4^-]$ is good. We repeat the process using K_{a_3} to get $[PO_4^{3-}]$.

$$K_{a_3} = 4.8 \times 10^{-13} = \frac{[H^+][PO_4^{3-}]}{[HPO_4^{2-}]} = \frac{(2.4 \times 10^{-2})[PO_4^{3-}]}{(6.2 \times 10^{-8})}$$

$[PO_4^{3-}] = 1.2 \times 10^{-18} M$ Assumptions good.

So in 0.100 M analytical concentration of H_3PO_4:

$$[H_3PO_4] = 7.6 \times 10^{-2} M; \quad [H^+] = [H_2PO_4^-] = 2.4 \times 10^{-2} M$$

$$[HPO_4^{2-}] = 6.2 \times 10^{-8} M; \quad [PO_4^{3-}] = 1.2 \times 10^{-18} M$$

$$[OH^-] = K_w/[H^+] = 4.2 \times 10^{-13} M$$

101.   The dominant H^+ producer is the strong acid H_2SO_4. A 2.0 M H_2SO_4 solution produces 2.0 M HSO_4^- and 2.0 M H^+. However, HSO_4^- is a weak acid which could also add H^+ to the solution.

	HSO_4^-	$\rightleftharpoons$	H^+	+	SO_4^{2-}
Initial	2.0 M		2.0 M		0
	x mol/L HSO_4^- dissociates to reach equilibrium				
Change	$-x$	$\rightarrow$	$+x$		$+x$
Equil.	2.0 - x		2.0 + x		x

$$K_{a_2} = 1.2 \times 10^{-2} = \frac{[H^+][SO_4^{2-}]}{[HSO_4^-]} = \frac{(2.0 + x)(x)}{2.0 - x} \approx \frac{2.0(x)}{2.0}, \quad x = 1.2 \times 10^{-2}$$

Since x is 0.60% of 2.0, then the assumption is valid by the 5% rule. The amount of additional H^+ from HSO_4^- is 1.2×10^{-2}. The total amount of H^+ present is:

$$[H^+] = 2.0 + 1.2 \times 10^{-2} = 2.0 M; \quad pH = -\log(2.0) = -0.30$$

Note: In this problem, H^+ from HSO_4^- could have been ignored. However, this is not usually the case, especially in more dilute solutions of H_2SO_4.

102.   For H_2SO_4, the first dissociation occurs to completion. The hydrogen sulfate ion, HSO_4^-, is a weak acid with $K_{a_2} = 1.2 \times 10^{-2}$. We will consider this equilibrium for additional H^+ production:

$$HSO_4^- \quad \rightleftharpoons \quad H^+ \quad + \quad SO_4^{2-}$$

Initial	$0.0010\,M$	$0.0010\,M$	0

x mol/L HSO_4^- dissociates to reach equilibrium

Change	$-x$ $\longrightarrow$	$+x$	$+x$
Equil.	$0.0010 - x$	$0.0010 + x$	x

$$K_{a_2} = 0.012 = \frac{(0.0010 + x)\,(x)}{(0.0010 - x)} \approx x,\ x = 0.012;\quad \text{Assumption is horrible (1200\% error).}$$

Using the quadratic formula and carrying extra significant figures:

$$1.2 \times 10^{-5} - 0.012\,x = x^2 + 0.0010\,x,\ x^2 + 0.013\,x - 1.2 \times 10^{-5} = 0$$

$$x = \frac{-0.013 \pm (1.69 \times 10^{-4} + 4.80 \times 10^{-5})^{1/2}}{2} = \frac{-0.013 \pm 0.0147}{2},\ x = 8.5 \times 10^{-4}$$

$$[H^+] = 0.0010 + x = 0.0010 + 0.00085 = 0.0019\,M;\ pH = 2.72$$

Note: We had to consider both H_2SO_4 and HSO_4^- for H^+ production in this problem.

Acid-Base Properties of Salts

103.   One difficult aspect of acid-base chemistry is recognizing what types of species are present in solution, i.e., whether a species is a strong acid, strong base, weak acid, weak base or a neutral species. Below are some ideas and generalizations to keep in mind that will help in recognizing types of species present.

a.   Memorize the following strong acids: HCl, HBr, HI, HNO_3, $HClO_4$ and H_2SO_4

b.   Memorize the following strong bases: LiOH, NaOH, KOH, RbOH, $Ca(OH)_2$, $Sr(OH)_2$ and $Ba(OH)_2$

c.   All weak acids have a K_a value less than 1 but greater than K_w. Some weak acids are in Table 14.2 of the text. All weak bases have a K_b value less than 1 but greater than K_w. Some weak bases are in Table 14.3 of the text.

d.   All conjugate bases of weak acids are weak bases, i.e., all have a K_b value less than 1 but greater than K_w. Some examples of these are the conjugate bases of the weak acids in Table 14.2 of the text.

e.   All conjugate acids of weak bases are weak acids, i.e., all have a K_a value less than 1 but greater than K_w. Some examples of these are the conjugate acids of the weak bases in Table 14.3 of the text.

f.   Alkali metal ions (Li^+, Na^+, K^+, Rb^+, Cs^+) and heavier alkaline earth metal ions (Ca^{2+}, Sr^{2+}, Ba^{2+}) have no acidic or basic properties in water.

g.   All conjugate bases of strong acids (Cl^-, Br^-, I^-, NO_3^-, ClO_4^-, HSO_4^-) have no basic properties in water ($K_b \ll K_w$) and only HSO_4^- has any acidic properties in water.

Lets apply these ideas to this problem to see what type of species are present. The letters in parenthesis is/are the generalization(s) above which identifies that species.

KOH: strong base (b)
KCl: neutral; K^+ and Cl^- have no acidic/basic properties (f and g).
KCN: CN^- is a weak base, $K_b = 1.0 \times 10^{-14}/6.2 \times 10^{-10} = 1.6 \times 10^{-5}$ (c and d). Ignore K^+(f).
NH_4Cl: NH_4^+ is a weak acid, $K_a = 5.6 \times 10^{-10}$ (c and e). Ignore Cl^-(g).
HCl: strong acid (a)

The most acidic solution will be the strong acid followed by the weak acid. The most basic solution will be the strong base followed by the weak base. The KCl solution will be between the acidic and basic solutions at pH = 7.00.

Most acidic → most basic; $HCl > NH_4Cl > KCl > KCN > KOH$

104.   See Exercise 14.103 for some generalizations to keep in mind. The letters in parenthesis is/are the generalization(s) listed in Exercise 14.103 which identifies that species.

$Ca(NO_3)_2$: neutral; Ca^{2+} and NO_3^- have no acidic/basic properties (f and g).
$NaNO_2$: NO_2^- is a weak base, $K_b = 1.0 \times 10^{-14}/4.0 \times 10^{-4} = 2.5 \times 10^{-11}$ (c and d). Ignore Na^+(f).
HNO_3:   strong acid (a)
NH_4NO_3: NH_4^+ is a weak acid, $K_a = 5.6 \times 10^{-10}$ (c and e). Ignore NO_3^- (g).
$Ca(OH)_2$: strong base (b)

Most acidic → most basic: $HNO_3 > NH_4NO_3 > Ca(NO_3)_2 > NaNO_2 > Ca(OH)_2$

105.   From the K_a values, acetic acid is a stronger acid than hypochlorous acid. Conversely, the conjugate base of acetic acid, $C_2H_3O_2^-$, will be a weaker base than the conjugate base of hypochlorous acid, OCl^-. Thus, the hypochlorite ion, OCl^-, is a stronger base than the acetate ion, $C_2H_3O_2^-$. In general, the stronger the acid, the weaker the conjugate base. This statement comes from the relationship $K_w = K_a \times K_b$ which holds for all conjugate acid-base pairs.

106.   Since NH_3 is a weaker base (smaller K_b value) than CH_3NH_2, then the conjugate acid of NH_3 will be a stronger acid than the conjugate acid of CH_3NH_2. Thus, NH_4^+ is a stronger acid than $CH_3NH_3^+$.

107. $NaN_3 \rightarrow Na^+ + N_3^-$; Azide, N_3^-, is a weak base since it is the conjugate base of a weak acid. All conjugate bases of weak acids are weak bases ($K_w < K_b < 1$). Ignore Na^+.

$$N_3^- + H_2O \rightleftharpoons HN_3 + OH^- \qquad K_b = \frac{K_w}{K_a} = \frac{1.0 \times 10^{-14}}{1.9 \times 10^{-5}} = 5.3 \times 10^{-10}$$

Initial   0.010 M 0   ~0
              x mol/L of N_3^- reacts with H_2O to reach equilibrium
Change $-x$ → $+x$ $+x$
Equil. $0.010 - x$ x x

$$K_b = \frac{[HN_3][OH^-]}{[N_3^-]} = 5.3 \times 10^{-10} = \frac{x^2}{0.010 - x} \approx \frac{x^2}{0.010} \quad \text{(assuming } x \ll 0.010)$$

$$x = [OH^-] = 2.3 \times 10^{-6}\,M;\ [H^+] = \frac{1.0 \times 10^{-14}}{2.3 \times 10^{-6}} = 4.3 \times 10^{-9}\,M \quad \text{Assumptions good.}$$

$[HN_3] = [OH^-] = 2.3 \times 10^{-6}\,M;\ [Na^+] = 0.010\,M;\ [N_3^-] = 0.010 - 2.3 \times 10^{-6} = 0.010\,M$

108. $C_2H_5NH_3Cl \rightarrow C_2H_5NH_3^+ + Cl^-$; $C_2H_5NH_3^+$ is the conjugate acid of the weak base $C_2H_5NH_2$ ($K_b = 5.6 \times 10^{-4}$). As is true for all conjugate acids of weak bases, $C_2H_5NH_3^+$ is a weak acid. Cl^- has no basic (or acidic) properties. Ignore Cl^-. Solving the weak acid problem:

$$C_2H_5NH_3^+ \quad \rightleftharpoons \quad C_2H_5NH_2 \ + \ H^+ \qquad K_a = K_w/5.6 \times 10^{-4} = 1.8 \times 10^{-11}$$

Initial	0.25 M	0	~0

Initial   0.25 M   0   ~0

x mol/L $C_2H_5NH_3^+$ dissociates to reach equilibrium

Change   -*x* $\rightarrow$ +*x*   +*x*
Equil.   0.25 - *x*   *x*   *x*

$$K_a = 1.8 \times 10^{-11} = \frac{[C_2H_5NH_2]\,[H^+]}{[C_2H_5NH_3^+]} = \frac{x^2}{0.25 - x} \approx \frac{x^2}{0.25} \quad \text{(assuming } x \ll 0.25\text{)}$$

$x = [H^+] = 2.1 \times 10^{-6}\,M;\ pH = 5.68;\ \text{Assumptions good.}$

$[C_2H_5NH_2] = [H^+] = 2.1 \times 10^{-6}\,M;\ [C_2H_5NH_3^+] = 0.25\,M;\ [Cl^-] = 0.25\,M$

$[OH^-] = K_w/[H^+] = 4.8 \times 10^{-9}\,M$

109. a. $CH_3NH_3Cl \rightarrow CH_3NH_3^+ + Cl^-$: $CH_3NH_3^+$ is a weak acid. Cl^- is the conjugate base of a strong acid. Cl^- has no basic (or acidic) properties.

$$CH_3NH_3^+ \rightleftharpoons CH_3NH_2 + H^+ \quad K_a = \frac{[CH_3NH_2]\,[H^+]}{[CH_3NH_3^+]} = \frac{K_w}{K_b} = \frac{1.00 \times 10^{-14}}{4.38 \times 10^{-4}} = 2.28 \times 10^{-11}$$

$$CH_3NH_3^+ \quad \rightleftharpoons \quad CH_3NH_2 \ + \ H^+$$

Initial   0.10 M   0   ~0

x mol/L $CH_3NH_3^+$ dissociates to reach equilibrium

Change   -*x* $\rightarrow$ +*x*   +*x*
Equil.   0.10 - *x*   *x*   *x*

$$K_a = 2.28 \times 10^{-11} = \frac{x^2}{0.10 - x} \approx \frac{x^2}{0.10} \quad \text{(assuming } x \ll 0.10\text{)}$$

$x = [H^+] = 1.5 \times 10^{-6}\,M;\ pH = 5.82 \qquad \text{Assumptions good.}$

b. $NaCN \rightarrow Na^+ + CN^-$: CN^- is a weak base. Na^+ has no acidic (or basic) properties.

$$CN^- + H_2O \rightleftharpoons HCN + OH^- \quad K_b = \frac{K_w}{K_a} = \frac{1.0 \times 10^{-14}}{6.2 \times 10^{-10}} = 1.6 \times 10^{-5}$$

Initial $0.050\,M$ 0   ~0

x mol/L CN^- reacts with H_2O to reach equilibrium

Change $-x$ $\rightarrow$ $+x$ $+x$

Equil. $0.050 - x$ x x

$$K_b = 1.6 \times 10^{-5} = \frac{[HCN][OH^-]}{[CN^-]} = \frac{x^2}{0.050 - x} \approx \frac{x^2}{0.050}$$

$x = [OH^-] = 8.9 \times 10^{-4}\,M$; pOH = 3.05; pH = 10.95   Assumptions good.

110.   a. $NaNO_2 \rightarrow Na^+ + NO_2^-$: NO_2^- is a weak base. Ignore Na^+.

$$NO_2^- + H_2O \rightleftharpoons HNO_2 + OH^- \quad K_b = \frac{K_w}{K_a} = \frac{1.0 \times 10^{-14}}{4.0 \times 10^{-4}} = 2.5 \times 10^{-11}$$

Initial $0.12\,M$ 0   ~0

Equil. $0.12 - x$ x x

$$K_b = 2.5 \times 10^{-11} = \frac{[OH^-][HNO_2]}{[NO_2^-]} = \frac{x^2}{0.12 - x} \approx \frac{x^2}{0.12}$$

$x = [OH^-] = 1.7 \times 10^{-6}\,M$; pOH = 5.77; pH = 8.23   Assumptions good.

b. $NaOCl \rightarrow Na^+ + OCl^-$: OCl^- is a weak base. Ignore Na^+.

$$OCl^- + H_2O \rightleftharpoons HOCl + OH^- \quad K_b = \frac{K_w}{K_a} = \frac{1.0 \times 10^{-14}}{3.5 \times 10^{-8}} = 2.9 \times 10^{-7}$$

Initial $0.45\,M$ 0   ~0

Equil. $0.45 - x$ x x

$$K_b = 2.9 \times 10^{-7} = \frac{[HOCl][OH^-]}{[OCl^-]} = \frac{x^2}{0.45 - x} \approx \frac{x^2}{0.45}$$

$x = [OH^-] = 3.6 \times 10^{-4}\,M$; pOH = 3.44; pH = 10.56   Assumptions good.

c. $NH_4ClO_4 \rightarrow NH_4^+ + ClO_4^-$: NH_4^+ is a weak acid. ClO_4^- is the conjugate base of a strong acid. ClO_4^- has no basic (or acidic) properties.

$$NH_4^+ \rightleftharpoons NH_3 + H^+ \quad K_a = \frac{K_w}{K_b} = \frac{1.0 \times 10^{-14}}{1.8 \times 10^{-5}} = 5.6 \times 10^{-10}$$

Initial $0.40\,M$ 0   ~0

Equil. $0.40 - x$ x x

$$K_a = 5.6 \times 10^{-10} = \frac{[NH_3][H^+]}{[NH_4^+]} = \frac{x^2}{0.40 - x} \approx \frac{x^2}{0.40}$$

$x = [H^+] = 1.5 \times 10^{-5}\,M$; pH = 4.82;   Assumptions good.

111.   All these salts contain Na^+ which has no acidic/basic properties and a conjugate base of a weak acid (except for NaCl where Cl^- is a neutral species.). All conjugate bases of weak acids are weak bases since K_b for these species are between K_w and 1. To identify the species, we will use the data given to determine the K_b value for the weak conjugate base. From the K_b value and data in Table 14.2 of the text, we can identify the conjugate base present by calculating the K_a value for the weak acid. We will use A^- as an abbreviation for the weak conjugate base.

$$A^- + H_2O \rightleftharpoons HA + OH^-$$

Initial	0.100 mol/1.00 L		0	~0
	x mol/L A^- reacts with H_2O to reach equilibrium			
Change	$-x$	$\rightarrow$	$+x$	$+x$
Equil.	$0.100 - x$		x	x

$K_b = \dfrac{[HA][OH^-]}{[A^-]} = \dfrac{x^2}{0.100 - x}$; From the problem, pH = 8.07:

pOH = 14.00 - 8.07 = 5.93; $[OH^-] = x = 10^{-5.93} = 1.2 \times 10^{-6} M$

$K_b = \dfrac{(1.2 \times 10^{-6})^2}{0.100 - 1.2 \times 10^{-6}} = 1.4 \times 10^{-11} = K_b$ value for the conjugate base of a weak acid.

The K_a value for the weak acid equals K_w/K_b: $K_a = \dfrac{1.0 \times 10^{-14}}{1.4 \times 10^{-11}} = 7.1 \times 10^{-4}$

From Table 14.2 of the text, this K_a value is closest to HF. Therefore, the unknown salt is NaF.

112. $BHCl \rightarrow BH^+ + Cl^-$; Cl^- is the conjugate base of the strong acid HCl, so Cl^- has no acidic/basic properties. BH^+ is a weak acid since it is the conjugate acid of a weak base, B. Determining the K_a value for BH^+:

$$BH^+ \rightleftharpoons B + H^+$$

Initial	0.10 M		0	~0
	x mol/L BH^+ dissociates to reach equilibrium			
Change	$-x$	$\rightarrow$	$+x$	$+x$
Equil.	$0.10 - x$		x	x

$K_a = \dfrac{[B][H^+]}{[BH^+]} = \dfrac{x^2}{0.10 - x}$; From the problem, pH = 5.82:

$[H^+] = x = 10^{-5.82} = 1.5 \times 10^{-6} M$; $K_a = \dfrac{(1.5 \times 10^{-6})^2}{0.10 - 1.5 \times 10^{-6}} = 2.3 \times 10^{-11}$

K_b for the base, $B = K_w\backslash K_a = 1.0 \times 10^{-14}/2.3 \times 10^{-11} = 4.3 \times 10^{-4}$.

From Table 14.3 of the text, this K_b value is closest to CH_3NH_2 so the unknown salt is CH_3NH_3Cl.

113. Major species present: $Al(H_2O)_6^{3+}$ ($K_a = 1.4 \times 10^{-5}$), NO_3^- (neutral) and H_2O ($K_w = 1.0 \times 10^{-14}$); $Al(H_2O)_6^{3+}$ is a stronger acid than water so it will be the dominant H^+ producer.

$$Al(H_2O)_6^{3+} \quad \rightleftharpoons \quad Al(H_2O)_5(OH)^{2+} \quad + \quad H^+$$

Initial	$0.050\,M$	0	~ 0
	x mol/L $Al(H_2O)_6^{3+}$ dissociates to reach equilibrium		
Change	$-x$ $\rightarrow$	$+x$	$+x$
Equil.	$0.050 - x$	x	x

$$K_a = 1.4 \times 10^{-5} = \frac{[Al(H_2O)_5(OH)^{2+}]\,[H^+]}{[Al(H_2O)_6]^{3+}} = \frac{x^2}{0.050 - x} \approx \frac{x^2}{0.050}$$

$x = 8.4 \times 10^{-4}\,M = [H^+]$; $pH = -\log(8.4 \times 10^{-4}) = 3.08$; Assumptions good.

114. Major species: $Co(H_2O)_6^{3+}$ ($K_a = 1.0 \times 10^{-5}$), Cl^- (neutral) and H_2O ($K_w = 1.0 \times 10^{-14}$); $Co(H_2O)_6^{3+}$ will determine the pH since it is a stronger acid than water. Solving the weak acid problem in the usual manner:

$$Co(H_2O)_6^{3+} \quad \rightleftharpoons \quad Co(H_2O)_5(OH)^{2+} \quad + \quad H^+ \qquad K_a = 1.0 \times 10^{-5}$$

Initial	$0.10\,M$	0	~ 0
Equil.	$0.10 - x$	x	x

$$K_a = 1.0 \times 10^{-5} = \frac{x^2}{0.10 - x} \approx \frac{x^2}{0.10}, \quad x = [H^+] = 1.0 \times 10^{-3}\,M$$

$pH = -\log(1.0 \times 10^{-3}) = 3.00$; Assumptions good.

115. Reference Table 14.6 of the text and the solution to Exercise 14.103 for some generalizations on acid-base properties of salts.

a. $NaNO_3 \rightarrow Na^+ + NO_3^-$ neutral; Neither species has any acidic/basic properties.

b. $NaNO_2 \rightarrow Na^+ + NO_2^-$ basic; NO_2^- is a weak base and Na^+ has no effect on pH.

$$NO_2^- + H_2O \rightleftharpoons HNO_2 + OH^- \qquad K_b = \frac{K_w}{K_{a,\,HNO_2}} = \frac{1.0 \times 10^{-14}}{4.0 \times 10^{-4}} = 2.5 \times 10^{-11}$$

c. $NH_4NO_3 \rightarrow NH_4^+ + NO_3^-$ acidic; NH_4^+ is a weak acid and NO_3^- has no effect on pH.

$$NH_4^+ \rightleftharpoons H^+ + NH_3 \qquad K_a = \frac{K_w}{K_{b,\,NH_3}} = \frac{1.0 \times 10^{-14}}{1.8 \times 10^{-5}} = 5.6 \times 10^{-10}$$

d. $NH_4NO_2 \rightarrow NH_4^+ + NO_2^-$ acidic; NH_4^+ is a weak acid ($K_a = 5.6 \times 10^{-10}$) and NO_2^- is a weak base ($K_b = 2.5 \times 10^{-11}$). Since $K_{a,\,NH_4^+} > K_{b,\,NO_2^-}$, then the solution is acidic.

$$NH_4^+ \rightleftharpoons H^+ + NH_3 \quad K_a = 5.6 \times 10^{-10}; \quad NO_2^- + H_2O \rightleftharpoons HNO_2 + OH^- \quad K_b = 2.5 \times 10^{-11}$$

e. $NaOCl \rightarrow Na^+ + OCl^-$ basic; OCl^- is a weak base and Na^+ has no effect on pH.

$$OCl^- + H_2O \rightleftharpoons HOCl + OH^-   \quad K_b = \frac{K_w}{K_{a,\,HOCl}} = \frac{1.0 \times 10^{-14}}{3.5 \times 10^{-8}} = 2.9 \times 10^{-7}$$

f. $NH_4OCl \rightarrow NH_4^+ + OCl^-$ basic; NH_4^+ is a weak acid and OCl^- is a weak base. Since $K_{b,\,OCl^-} > K_{a,\,NH_4^+}$, then the solution is basic.

$$NH_4^+ \rightleftharpoons NH_3 + H^+   \quad K_a = 5.6 \times 10^{-10}; \quad OCl^- + H_2O \rightleftharpoons HOCl + OH^-   \quad K_b = 2.9 \times 10^{-7}$$

116.   a. $KCl \rightarrow K^+ + Cl^-$ neutral; K^+ and Cl^- have no effect on pH.

b. $NH_4C_2H_3O_2 \rightarrow NH_4^+ + C_2H_3O_2^-$ neutral; NH_4^+ is a weak acid and $C_2H_3O_2^-$ is a weak base. Since $K_{a,\,NH_4^+} = K_{b,\,C_2H_3O_2^-}$, then pH = 7.00.

$$NH_4^+ \rightleftharpoons NH_3 + H^+   \quad K_a = \frac{K_w}{K_{b,\,NH_3}} = \frac{1.0 \times 10^{-14}}{1.8 \times 10^{-5}} = 5.6 \times 10^{-10}$$

$$C_2H_3O_2^- + H_2O \rightleftharpoons HC_2H_3O_2 + OH^-   \quad K_b = \frac{K_w}{K_{a,\,HC_2H_3O_2}} = \frac{1.0 \times 10^{-14}}{1.8 \times 10^{-5}} = 5.6 \times 10^{-10}$$

c. $CH_3NH_3Cl \rightarrow CH_3NH_3^+ + Cl^-$ acidic; $CH_3NH_3^+$ is a weak acid and Cl^- has no effect on pH.

$$CH_3NH_3^+ \rightleftharpoons H^+ + CH_3NH_2   \quad K_a = \frac{K_w}{K_{b,\,CH_3NH_2}} = \frac{1.00 \times 10^{-14}}{4.38 \times 10^{-4}} = 2.28 \times 10^{-11}$$

d. $KF \rightarrow K^+ + F^-$ basic; F^- is a weak base and K^+ has no effect on pH.

$$F^- + H_2O \rightleftharpoons HF + OH^-   \quad K_b = \frac{K_w}{K_{a,\,HF}} = \frac{1.0 \times 10^{-14}}{7.2 \times 10^{-4}} = 1.4 \times 10^{-11}$$

e. $NH_4F \rightarrow NH_4^+ + F^-$ acidic; NH_4^+ is a weak acid and F^- is a weak base. Since $K_{a,\,NH_4^+} > K_{b,\,F^-}$, then the solution is acidic.

$$NH_4^+ \rightleftharpoons H^+ + NH_3   \quad K_a = 5.6 \times 10^{-10}; \quad F^- + H_2O \rightleftharpoons HF + OH^-   \quad K_b = 1.4 \times 10^{-11}$$

f. $CH_3NH_3CN \rightarrow CH_3NH_3^+ + CN^-$ basic; $CH_3NH_3^+$ is a weak acid and CN^- is a weak base. Since $K_{b,\,CN^-} > K_{a,\,CH_3NH_3^+}$, then the solution is basic.

$$CH_3NH_3^+ \rightleftharpoons H^+ + CH_3NH_2   \quad K_a = 2.28 \times 10^{-11}$$

$$CN^- + H_2O \rightleftharpoons HCN + OH^-   \quad K_b = \frac{K_w}{K_{a,\,HCN}} = \frac{1.0 \times 10^{-14}}{6.2 \times 10^{-10}} = 1.6 \times 10^{-5}$$

Relationships Between Structure and Strengths of Acids and Bases

117.   a. $HBrO < HBrO_2 < HBrO_3$; As the number of oxygen atoms attached to the bromine atom increases, acid strength increases.

    b. $HIO_2 < HBrO_2 < HClO_2$; As the electronegativity of the central atom increases, acid strength increases.

    c. $HBrO_3 < HClO_3$; Same reasoning as in b.

    d. $H_2SO_3 < H_2SO_4$; Same reasoning as in a.

118.   a. $BrO_3^- < BrO_2^- < BrO^-$; These are the conjugate bases of the acids in Exercise 14.117a. Since $HBrO_3$ is the strongest acid, then the conjugate base of $HBrO_3$ (BrO_3^-) will be the weakest base. BrO^- will be the strongest base since $HBrO$ is the weakest acid.

    b. $ClO_2^- < BrO_2^- < IO_2^-$; These are the conjugate bases of the acids in Exercise 14.117b. Conjugate base strength is inversely related to acid strength.

119.   a. $H_2O < H_2S < H_2Se$; As the strength of the H – X bond decreases, acid strength increases.

    b. $CH_3CO_2H < FCH_2CO_2H < F_2CHCO_2H < F_3CCO_2H$; As the electronegativity of neighboring atoms increase, acid strength increases.

    c. $NH_4^+ < HONH_3^+$; Same reason as in b.

    d. $NH_4^+ < PH_4^+$; Same reason as in a.

120.   In general, the stronger the acid, the weaker the conjugate base.

    a. $SeH^- < SH^- < OH^-$; These are the conjugate bases of the acids in Exercise 14.119a. The ordering of the base strength is the opposite of the acids.

    b. $PH_3 < NH_3$ (See Exercise 14.119d.)

    c. $HONH_2 < NH_3$ (See Exercise 14.119c.)

121.   In general, metal oxides form basic solutions in water and nonmetal oxides form acidic solutions in water.

    a.   basic; $CaO(s) + H_2O(l) \rightarrow Ca(OH)_2(aq)$, $Ca(OH)_2$ is a strong base.

    b.   acidic; $SO_2(g) + H_2O(l) \rightarrow H_2SO_3(aq)$, H_2SO_3 is a weak diprotic acid.

    c.   acidic; $Cl_2O(g) + H_2O(l) \rightarrow 2\ HOCl(aq)$, HOCl is a weak acid.

122.   a.   basic; $Na_2O(s) + H_2O(l) \rightarrow 2\ NaOH(aq)$, NaOH is a strong base.

    b.   acidic; $P_4O_{10}(s) + 6\ H_2O(l) \rightarrow 4\ H_3PO_4(aq)$, H_3PO_4 is a weak triprotic acid.

    c.   acidic; $2\ NO_2(g) + H_2O(l) \rightarrow HNO_3(aq) + HNO_2(aq)$, HNO_3 is a strong acid and HNO_2 is a weak acid.

Lewis Acids and Bases

123.   A Lewis base is an electron pair donor and a Lewis acid is an electron pair acceptor.

   a. $B(OH)_3$, acid; H_2O, base   b. Ag^+, acid; NH_3, base   c. BF_3, acid; NH_3, base

124.   a. I_2, acid; I^-, base   b. $Zn(OH)_2$, acid; OH^-, base   c. Fe^{3+}, acid; SCN^-, base

125. $Al(OH)_3(s) + 3\ H^+(aq) \rightarrow Al^{3+}(aq) + 3\ H_2O(l)$ (Bronsted-Lowry base, H^+ acceptor)

   $Al(OH)_3(s) + OH^-(aq) \rightarrow Al(OH)_4^-(aq)$ (Lewis acid, electron pair acceptor)

126. $Zn(OH)_2(s) + 2\ H^+(aq) \rightarrow Zn^{2+}(aq) + 2\ H_2O(l)$ (Bronsted-Lowry base)

   $Zn(OH)_2(s) + 2\ OH^-(aq) \rightarrow Zn(OH)_4^{2-}(aq)$ (Lewis acid)

127. Fe^{3+} should be the stronger Lewis acid. Fe^{3+} is smaller and has a greater positive charge. Because of this, Fe^{3+} will be more strongly attracted to lone pairs of electrons as compared to Fe^{2+}.

128.   The Lewis structures for the reactants and products are:

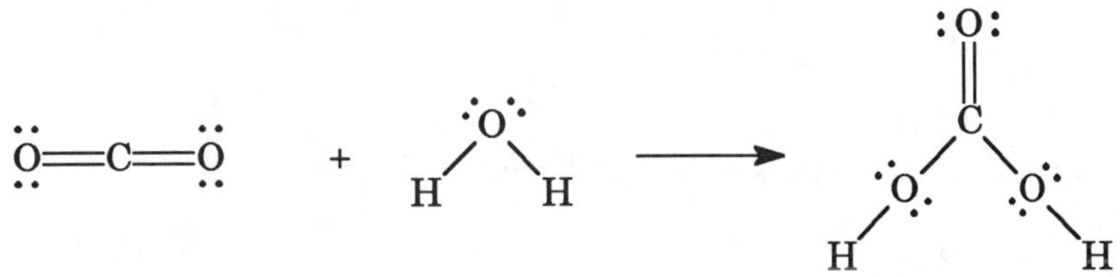

   In this reaction, H_2O donates a pair of electrons to carbon in CO_2 which is followed by a proton shift to form H_2CO_3. H_2O is the Lewis base and CO_2 is the Lewis acid.

Additional Exercises

129. $2.48\ g\ TlOH \times \dfrac{1\ mol}{221.4\ g} = 1.12 \times 10^{-2}\ mol$; TlOH is a strong base, so $[OH^-] = 1.12 \times 10^{-2}\ M$.

   $pOH = -\log[OH^-] = 1.951$; $pH = 14.000 - pOH = 12.049$

130. $K_a \times K_b = K_w$, $-\log(K_a \times K_b) = -\log K_w$

   $-\log K_a - \log K_b = -\log K_w$, $pK_a + pK_b = pK_w = 14.00$ (at $25\,°C$)

131.   Let HBz = benzoic acid. A saturated solution is one where the maximum amount of solute (benzoic acid) has dissolved. Let C = equilibrium concentration of HBz in the saturated solution. Since benzoic acid is a weak acid ($K_a = 6.4 \times 10^{-5}$), then at equilibrium we also have H^+ and the conjugate base of benzoic acid (Bz^-) present. H^+ and Bz^- are both produced from the dissociation of HBz so their concentrations must be equal. Let $x = [H^+] = [Bz^-]$ at equilibrium. The general set-up for this problem is:

$$HBz \quad \rightleftharpoons \quad H^+ \quad + \quad Bz^-$$

Equil.   C x x

Solve for C, the equilibrium solubility of HBz, using the K_a expression.

$$K_a = \frac{[H^+][Bz^-]}{[HBz]} = 6.4 \times 10^{-5} = \frac{x^2}{C}; \quad \text{Since pH} = 2.80: \ x = [H^+] = [Bz^-] = 10^{-2.80} = 1.6 \times 10^{-3} \, M$$

$$C = \frac{(1.6 \times 10^{-3})^2}{6.4 \times 10^{-5}} = 4.0 \times 10^{-2} \, M$$

The equilibrium molar solubility of benzoic acid is 4.0×10^{-2} mol/L. The initial amount of benzoic acid that will dissolve in water is $4.0 \times 10^{-2} \, M + 1.6 \times 10^{-3} \, M = 4.2 \times 10^{-2}$ mol/L (this is the equilibrium concentration of benzoic acid plus the amount of benzoic acid that dissociated, x).

132. $\dfrac{5.0 \times 10^{-3} \, g}{0.0100 \, L} \times \dfrac{1 \, mol}{299.36 \, g} = 1.7 \times 10^{-3} \, M = [codeine]_o; \ $ Let cod = codeine, $C_{18}H_{21}NO_3$

Solving the weak base equilibrium problem:

$$cod \ + \ H_2O \quad \rightleftharpoons \quad codH^+ \quad + \quad OH^- \qquad K_b = 10^{-6.05} = 8.9 \times 10^{-7}$$

Initial $1.7 \times 10^{-3} \, M$ 0 ~ 0

            x mol/L codeine reacts with H_2O to reach equilibrium

Change $-x$ $\rightarrow$ $+x$ $+x$
Equil. $1.7 \times 10^{-3} - x$ x x

$$K_b = 8.9 \times 10^{-7} = \frac{x^2}{1.7 \times 10^{-3} - x} \approx \frac{x^2}{1.7 \times 10^{-3}}, \ x = 3.9 \times 10^{-5} \qquad \text{Assumptions good.}$$

$[OH^-] = 3.9 \times 10^{-5} \, M; \ [H^+] = K_w/[OH^-] = 2.6 \times 10^{-10} \, M; \ pH = 9.59$

133.   For this problem we will abbreviate $CH_2=CHCO_2H$ as Hacr and $CH_2=CHCO_2^-$ as acr$^-$.

a.   Solving the weak acid problem:

$$Hacr \quad \rightleftharpoons \quad H^+ \quad + \quad acr^- \qquad K_a = 5.6 \times 10^{-5}$$

Initial   0.10 M ~ 0 0
Equil.   0.10 - x x x

$$\frac{x^2}{0.10 - x} = 5.6 \times 10^{-5} \approx \frac{x^2}{0.10}, \ x = [H^+] = 2.4 \times 10^{-3} \, M; \ pH = 2.62; \ \text{Assumptions good.}$$

b.   % dissociation $= \dfrac{[H^+]}{[Hacr]_o} \times 100 = \dfrac{2.4 \times 10^{-3}}{0.10} \times 100 = 2.4\%$

c.   acr$^-$ is a weak base and the major source of OH$^-$ in this solution.

$$\text{acr}^- \;+\; H_2O \;\rightleftharpoons\; Hacr \;+\; OH^- \qquad K_b = \frac{K_w}{K_a} = \frac{1.0 \times 10^{-14}}{5.6 \times 10^{-5}}$$

Initial	$0.050\,M$	0	~ 0
Equil.	$0.050 - x$	x	x

$K_b = 1.8 \times 10^{-10}$

$$K_b = \frac{[Hacr]\,[OH^-]}{[acr^-]} = 1.8 \times 10^{-10} = \frac{x^2}{0.050 - x} \approx \frac{x^2}{0.050}$$

$x = [OH^-] = 3.0 \times 10^{-6}\,M;\;\; pOH = 5.52;\;\; pH = 8.48$ Assumptions good.

134. B^- is a weak base. Use the weak base data to determine K_b for B^-.

$$B^- \;+\; H_2O \;\rightleftharpoons\; HB \;+\; OH^-$$

Initial	$0.050\,M$	0	~ 0
Equil.	$0.050 - x$	x	x

Since $pH = 9.00$, then $pOH = 5.00$ and $[OH^-] = 1.0 \times 10^{-5}\,M = x$.

$$K_b = \frac{[HB]\,[OH^-]}{[B^-]} = \frac{x^2}{0.050 - x} = \frac{(1.0 \times 10^{-5})^2}{0.050 - 1.0 \times 10^{-5}} = 2.0 \times 10^{-9}$$

Since B^- is a weak base, then HB is a weak acid. Solving the weak acid problem:

$$HB \;\rightleftharpoons\; H^+ \;+\; B^-$$

Initial	$0.010\,M$	~ 0	0
Equil.	$0.010 - x$	x	x

$$K_a = \frac{K_w}{K_b} = \frac{1.0 \times 10^{-14}}{2.0 \times 10^{-9}} = 5.0 \times 10^{-6} = \frac{x^2}{0.010 - x} \approx \frac{x^2}{0.010}$$

$x = [H^+] = 2.2 \times 10^{-4}\,M;\;\; pH = 3.66$ Assumptions good.

135.   a. $$Fe(H_2O)_6^{3+} + H_2O \;\rightleftharpoons\; Fe(H_2O)_5(OH)^{2+} \;+\; H_3O^+$$

Initial	$0.10\,M$	0	~ 0
Equil.	$0.10 - x$	x	x

$$K_a = \frac{[Fe(H_2O)_5(OH)^{2+}]\,[H_3O^+]}{[Fe(H_2O)_6^{3+}]} = 6.0 \times 10^{-3} = \frac{x^2}{0.10 - x} \approx \frac{x^2}{0.10}$$

$x = 2.4 \times 10^{-2}$; Assumption is poor (x is 24% of 0.10). Using successive approximations:

$$\frac{x^2}{0.10 - 0.024} = 6.0 \times 10^{-3},\; x = 0.021$$

$$\frac{x^2}{0.10 - 0.021} = 6.0 \times 10^{-3},\; x = 0.022; \qquad \frac{x^2}{0.10 - 0.022} = 6.0 \times 10^{-3},\; x = 0.022$$

$x = [H^+] = 0.022 \, M; \; pH = 1.66$

b. Because of the lower charge, $Fe^{2+}(aq)$ will not be as strong an acid as $Fe^{3+}(aq)$. A solution of iron(II) nitrate will be less acidic (have a higher pH) than a solution with the same concentration of iron(III) nitrate.

136.   See generalizations in Exercise 14.103.

a. HI: strong acid; HF: weak acid ($K_a = 7.2 \times 10^{-4}$)

    NaF: F^- is the conjugate base of the weak acid HF so F^- is a weak base. The K_b value for F^- = $K_w/K_{a, \, HF}$ = 1.4×10^{-11}. Na^+ has no acidic or basic properties.

    NaI: neutral (pH = 7.0); Na^+ and I^- have no acidic/basic properties.

                To place in order of increasing pH, we place the compounds from most acidic (lowest pH) to most basic (highest pH). Increasing pH: HI < HF < NaI < NaF.

b. NH_4Br: NH_4^+ is a weak acid ($K_a = 5.6 \times 10^{-10}$) and Br^- is a neutral species.
    HBr: strong acid
    KBr: neutral; K^+ and Br^- have no acidic/basic properties
    NH_3:   weak base, $K_b = 1.8 \times 10^{-5}$

    Increasing pH: HBr < NH_4Br < KBr < NH_3
                     most   most
                     acidic   basic

c. NH_4NO_3: NH_4^+ is a weak acid ($K_a = 5.6 \times 10^{-10}$) and NO_3^- is a neutral species.
    $NaNO_3$:   neutral; Na^+ and NO_3^- have no acidic/basic properties.
    NaOH: strong base
    HF: weak acid ($K_a = 7.2 \times 10^{-4}$)
    KF: F^- is a weak base ($K_b = 1.4 \times 10^{-11}$) and K^+ is a neutral species.
    NH_3:   weak base ($K_b = 1.8 \times 10^{-5}$)
    HNO_3:   strong acid

    This is a little more difficult than the previous parts of this problem because two weak acids and two weak bases are present. Between the weak acids, HF is a stronger weak acid than NH_4^+ since the K_a value for HF is larger than the K_a value for NH_4^+. Between the two weak bases, since the K_b value for NH_3 is larger than the K_b value for F^-, then NH_3 is a stronger weak base than F^-.

    Increasing pH: HNO_3 < HF < NH_4NO_3 < $NaNO_3$ < KF < NH_3 < NaOH
                     most   most
                     acidic   basic

137.   Solution is acidic from $HSO_4^- \rightleftharpoons H^+ + SO_4^{2-}$. Solving the weak acid problem:

$$HSO_4^- \quad \rightleftharpoons \quad H^+ \quad + \quad SO_4^{2-} \qquad K_a = 1.2 \times 10^{-2}$$

Initial	0.10 M	~0	0
Equil.	0.10 - x	x	x

$$1.2 \times 10^{-2} = \frac{[H^+][SO_4^{2-}]}{[HSO_4^-]} = \frac{x^2}{0.10 - x} \approx \frac{x^2}{0.10}, \quad x = 0.035$$

Assumption is not good (x is 35% of 0.10). Using successive approximations:

$$\frac{x^2}{0.10 - x} \approx \frac{x^2}{0.10 - 0.035} = 1.2 \times 10^{-2}, \quad x = 0.028$$

$$\frac{x^2}{0.10 - 0.028} = 1.2 \times 10^{-2}, \quad x = 0.029; \qquad \frac{x^2}{0.10 - 0.029} = 1.2 \times 10^{-2}, \quad x = 0.029$$

$x = [H^+] = 0.029\ M;\ pH = 1.54$

138.   The relevant reactions are:

$$H_2CO_3 \rightleftharpoons H^+ + HCO_3^- \quad K_{a_1} = 4.3 \times 10^{-7}; \quad HCO_3^- \rightleftharpoons H^+ + CO_3^{2-} \quad K_{a_2} = 5.6 \times 10^{-11}$$

Initially, we deal only with the first reaction (since $K_{a_1} \gg K_{a_2}$) and then let these results control values of concentrations in the second reaction.

$$H_2CO_3 \quad \rightleftharpoons \quad H^+ \quad + \quad HCO_3^-$$

Initial	0.010 M	~0	0
Equil.	0.010 - x	x	x

$$K_{a_1} = 4.3 \times 10^{-7} = \frac{[H^+][HCO_3^-]}{[H_2CO_3]} = \frac{x^2}{0.010 - x} \approx \frac{x^2}{0.010}$$

$x = 6.6 \times 10^{-5}\ M = [H^+] = [HCO_3^-]$ Assumptions good.

$$HCO_3^- \quad \rightleftharpoons \quad H^+ \quad + \quad CO_3^{2-}$$

Initial	$6.6 \times 10^{-5}\ M$	$6.6 \times 10^{-5}\ M$	0
Equil.	6.6×10^{-5} - y	6.6×10^{-5} + y	y

If y is small, then $[H^+] = [HCO_3^-]$ and $K_{a_2} = 5.6 \times 10^{-11} = \dfrac{[H^+][CO_3^{2-}]}{[HCO_3^-]} \approx y$

$y = [CO_3^{2-}] = 5.6 \times 10^{-11}\ M$ Assumptions good.

The amount of H^+ from the second dissociation is $5.6 \times 10^{-11}\ M$ or:

$$\frac{5.6 \times 10^{-11}}{6.6 \times 10^{-5}} \times 100 = 8.5 \times 10^{-5}\ \%$$

This result justifies our treating the equilibria separately. If the second dissociation contributed a significant amount of H^+, then we would have to treat both equilibria simultaneously. The reaction that occurs when acid is added to a solution of HCO_3^- is:

$$HCO_3^-(aq) + H^+(aq) \rightarrow H_2CO_3(aq) \rightarrow H_2O(l) + CO_2(g)$$

The bubbles are $CO_2(g)$ and are formed by the breakdown of unstable H_2CO_3 molecules. We should write $H_2O(l) + CO_2(aq)$ or $CO_2(aq)$ for what we call carbonic acid. It is for convenience, however, that we write $H_2CO_3(aq)$.

139.   a.   In the lungs, there is a lot of O_2 and the equilibrium favors $Hb(O_2)_4$. In the cells there is a deficiency of O_2 and the equilibrium favors HbH_4^{4+}.

       b. CO_2 is a weak acid, $CO_2 + H_2O \rightleftharpoons HCO_3^- + H^+$. Removing CO_2 essentially decreases H^+. $Hb(O_2)_4$ is then favored and O_2 is not released by hemoglobin in the cells. Breathing into a paper bag increases $[CO_2]$ in the blood, thus increasing $[H^+]$ which shifts the reaction left.

       c. CO_2 builds up in the blood and it becomes too acidic, driving the equilibrium to the left. Hemoglobin can't bind O_2 as strongly in the lungs. Bicarbonate ion acts as a base in water and neutralizes the excess acidity.

140.   a. $NH_3 + H_3O^+ \rightleftharpoons NH_4^+ + H_2O$

$$K_{eq} = \frac{[NH_4^+]}{[NH_3][H^+]} = \frac{1}{K_a \text{ for } NH_4^+} = \frac{K_b}{K_w} = \frac{1.8 \times 10^{-5}}{1.0 \times 10^{-14}} = 1.8 \times 10^9$$

       b. $NO_2^- + H_3O^+ \rightleftharpoons H_2O + HNO_2$ $K_{eq} = \dfrac{[HNO_2]}{[NO_2^-][H^+]} = \dfrac{1}{K_a \text{ for } HNO_2} = \dfrac{1}{4.0 \times 10^{-4}} = 2.5 \times 10^3$

       c. $NH_4^+ + OH^- \rightleftharpoons NH_3 + H_2O$ $K_{eq} = \dfrac{1}{K_b \text{ for } NH_3} = \dfrac{1}{1.8 \times 10^{-5}} = 5.6 \times 10^4$

       d. $HNO_2 + OH^- \rightleftharpoons H_2O + NO_2^-$

$$K_{eq} = \frac{[NO_2^-]}{[HNO_2][OH^-]} \times \frac{[H^+]}{[H^+]} = \frac{K_a \text{ for } HNO_2}{K_w} = \frac{4.0 \times 10^{-4}}{1.0 \times 10^{-14}} = 4.0 \times 10^{10}$$

141.   a. H_2SO_3 b. $HClO_3$ c. H_3PO_3

NaOH and KOH are soluble ionic compounds composed of Na^+ and K^+ cations and OH^- anions. All soluble ionic compounds dissolve to form the ions from which they are formed. In oxyacids, the compounds are all covalent compounds in which electrons are shared to form bonds (unlike ionic compounds). When these compounds are dissolved in water, the covalent bond between oxygen and hydrogen breaks to form H^+ ions.

Challenge Problems

142.   The pH of this solution is not 8.00 because water will donate a significant amount of H^+ from the autoionization of water. The pertinent reactions are:

$$H_2O \rightleftharpoons H^+ + OH^-   \quad K_w = [H^+][OH^-] = 1.0 \times 10^{-14}$$

$HCl \rightarrow H^+ + Cl^-$ K_a is very large, so we assume that only the forward reaction occurs.

In any solution, the overall net positive charge must equal the overall net negative charge (called the charge balance). For this problem:

[positive charge] = [negative charge], so $[H^+] = [OH^-] + [Cl^-]$

From K_w, $[OH^-] = K_w/[H^+]$, and from $1.0 \times 10^{-8} \, M$ HCl, $[Cl^-] = 1.0 \times 10^{-8} \, M$. Substituting into the charge balance equation:

$$[H^+] = \frac{1.0 \times 10^{-14}}{[H^+]} + 1.0 \times 10^{-8}, \, [H^+]^2 - 1.0 \times 10^{-8} \, [H^+] - 1.0 \times 10^{-14} = 0$$

Using the quadratic formula to solve:

$$[H^+] = \frac{-(-1.0 \times 10^{-8}) \pm [(-1.0 \times 10^{-8})^2 - 4(1)(-1.0 \times 10^{-14})]^{1/2}}{2(1)}, \, [H^+] = 1.1 \times 10^{-7} \, M$$

$$pH = -\log(1.1 \times 10^{-7}) = 6.96$$

143.   Since this is a very dilute solution of NaOH, then we must worry about the amount of OH^- donated from the autoionization of water.

$$NaOH \rightarrow Na^+ + OH^-$$

$$H_2O \rightleftharpoons H^+ + OH^-   \quad K_w = [H^+][OH^-] = 1.0 \times 10^{-14}$$

This solution, like all solutions, must be charged balance, that is [positive charge] = [negative charge]. For this problem, the charge balance equation is:

$$[Na^+] + [H^+] = [OH^-], \text{ where } [Na^+] = 1.0 \times 10^{-7} \, M \text{ and } [H^+] = \frac{K_w}{[OH^-]}$$

Substituting into the charge balance equation:

$$1.0 \times 10^{-7} + \frac{1.0 \times 10^{-14}}{[OH^-]} = [OH^-], \, [OH^-]^2 - 1.0 \times 10^{-7} \, [OH^-] - 1.0 \times 10^{-14} = 0$$

Using the quadratic formula to solve:

$$[OH^-] = \frac{-(-1.0 \times 10^{-7}) \pm [(-1.0 \times 10^{-7})^2 - 4(1)(-1.0 \times 10^{-14})]^{1/2}}{2(1)}$$

$$[OH^-] = 1.6 \times 10^{-7} \, M; \, pOH = -\log(1.6 \times 10^{-7}) = 6.80; \, pH = 7.20$$

144.   HA $\rightleftharpoons$ H^+ + A^- $K_a = 1.00 \times 10^{-6}$

Initial   C   ~0   0   C = $[HA]_o$ for pH = 4.000 solution
Equil.   C - 1.00×10^{-4} 1.00×10^{-4} 1.00×10^{-4} $x = [H^+] = 1.00 \times 10^{-4}\,M$

$$K_a = \frac{(1.00 \times 10^{-4})^2}{C - 1.00 \times 10^{-4}} = 1.00 \times 10^{-6};\ \text{Solving: } C = 0.0101\,M$$

The solution initially contains $50.0 \times 10^{-3}\,L \times 0.0101\,mol/L = 5.05 \times 10^{-4}$ mol HA. We then dilute to a total volume V in liters. The resulting pH = 5.000, so $[H^+] = 1.00 \times 10^{-5}$. In the typical weak acid problem, $x = [H^+]$, so:

                    HA $\rightleftharpoons$ H^+ + A^-

Initial 5.05×10^{-4} mol/V   ~0   0
Equil. $5.05 \times 10^{-4}/V - 1.00 \times 10^{-5}$ 1.00×10^{-5} 1.00×10^{-5}

$$K_a = \frac{(1.00 \times 10^{-5})^2}{5.05 \times 10^{-4}/V - 1.00 \times 10^{-5}} = 1.00 \times 10^{-6},\ 1.00 \times 10^{-4} = 5.05 \times 10^{-4}/V - 1.00 \times 10^{-5}$$

V = 4.59 L; 50.0 mL are present initially, so we need to add 4540 mL of water.

145.   HBrO $\rightleftharpoons$ H^+ + BrO^- $K_a = 2 \times 10^{-9}$

Initial $1.0 \times 10^{-6}\,M$ ~0   0
             x mol/L HBrO dissociates to reach equilibrium
Change   -x $\rightarrow$ +x   +x
Equil. $1.0 \times 10^{-6} - x$ x   x

$$K_a = 2 \times 10^{-9} = \frac{x^2}{1.0 \times 10^{-6} - x} \approx \frac{x^2}{1.0 \times 10^{-6}};\ x = [H^+] = 4 \times 10^{-8}\,M;\ \text{pH} = 7.4$$

Lets check the assumptions. This answer is impossible! We can't add a small amount of an acid to a neutral solution and get a basic solution. The highest pH possible for an acid in water is 7.0. In the correct solution, we would have to take into account the autoionization of water.

146. $HIO_3 \rightleftharpoons H^+ + IO_3^-$ $K_a = \dfrac{[H^+][IO_3^-]}{[HIO_3]} = 0.17$

When pH = 1.20, then $[H^+] = 10^{-1.20} = 0.063\,M$. The initial concentration of HIO_3 was $0.10\,M$; if x amount of HIO_3 dissociates, then at equilibrium $[HIO_3] = 0.10 - x$ and $[IO_3^-] = x$. Adding these two equilibrium concentrations together gives: $[HIO_3] + [IO_3^-] = 0.10 - x + x = 0.10\,M$.

$$K_a = \frac{[H^+][IO_3^-]}{[HIO_3]} = 0.17 = 0.063 \times \frac{[IO_3^-]}{[HIO_3]},\ [IO_3^-] = 2.7\,[HIO_3]$$

$0.10\ M = [HIO_3] + [IO_3^-] = [HIO_3] + 2.7\ [HIO_3],\ \ [HIO_3] = 0.027\ M$

$[IO_3^-] = 2.7 \times 0.027\ M = 0.073\ M$

147.   Since NH_3 is so concentrated, we need to calculate the OH^- contribution from the weak base NH_3.

$$NH_3\ +\ H_2O\ \ \rightleftharpoons\ \ NH_4^+\ +\ OH^-\ \ \ \ K_b = 1.8 \times 10^{-5}$$

Initial $15.0\ M$ 0 $0.0100\ M$ (Assume no volume change.)
Equil. $15.0 - x$ x $0.0100 + x$

$K_b = 1.8 \times 10^{-5} = \dfrac{x(0.0100 + x)}{15.0 - x} \approx \dfrac{x(0.0100)}{15.0},\ x = 0.027;$ Assumption is horrible (x is 270% of 0.0100).

Using the quadratic formula:

$1.8 \times 10^{-5}(15.0 - x) = 0.0100\ x + x^2,\ x^2 + 0.0100\ x - 2.7 \times 10^{-4} = 0$

$x = 1.2 \times 10^{-2},\ \ [OH^-] = 1.2 \times 10^{-2} + 0.0100 = 0.022\ M$

148.   For H_3PO_4, $K_{a_1} = 7.5 \times 10^{-3}$, $K_{a_2} = 6.2 \times 10^{-8}$ and $K_{a_3} = 4.8 \times 10^{-13}$. Since K_{a_1} is much larger than K_{a_2} and K_{a_3}, then the dominant H^+ producer is H_3PO_4 and the H^+ from $H_2PO_4^-$ and HPO_4^{2-} can be ignored. Solving the weak acid problem in the typical manner:

$$H_3PO_4\ \ \rightleftharpoons\ \ H_2PO_4^-\ +\ H^+\ \ \ \ K_{a_1} = 7.5 \times 10^{-3}$$

Initial $0.007\ M$ 0 ~ 0
Equil. $0.007 - x$ x x

$K_{a_1} = 7.5 \times 10^{-3} = \dfrac{[H_2PO_4^-]\,[H^+]}{[H_3PO_4]} = \dfrac{x^2}{0.007 - x} \approx \dfrac{x^2}{0.007}$

$x = 7 \times 10^{-3};$ Assumption is horrible since x is 100% of 0.007. We will use the quadratic equation to solve exactly.

$7.5 \times 10^{-3} = \dfrac{x^2}{0.007 - x},\ x^2 = 5 \times 10^{-5} - 7.5 \times 10^{-3}\ x,\ x^2 + 7.5 \times 10^{-3}\ x - 5 \times 10^{-5} = 0$

$x = [H^+] = \dfrac{-7.5 \times 10^{-3} \pm [(7.5 \times 10^{-3})^2 - 4(1)(-5 \times 10^{-5})]^{1/2}}{2(1)} = 4 \times 10^{-3}\ M$

$pH = -\log(4 \times 10^{-3}) = 2.4$

149. PO_4^{3-} is the conjugate base of HPO_4^{2-}. The K_a value for HPO_4^{2-} is $K_{a_3} = 4.8 \times 10^{-13}$.

$$PO_4^{3-}(aq) + H_2O(l) \rightleftharpoons HPO_4^{2-}(aq) + OH^-(aq)\ \ \ K_b = \dfrac{K_w}{K_{a_3}} = \dfrac{1.0 \times 10^{-14}}{4.8 \times 10^{-13}} = 0.021$$

HPO_4^{2-} is the conjugate base of $H_2PO_4^-$ ($K_{a_2} = 6.2 \times 10^{-8}$).

$$HPO_4^{2-} + H_2O \rightleftharpoons H_2PO_4^- + OH^- \quad K_b = \frac{K_w}{K_{a_2}} = \frac{1.0 \times 10^{-14}}{6.2 \times 10^{-8}} = 1.6 \times 10^{-7}$$

$H_2PO_4^-$ is the conjugate base of H_3PO_4 ($K_{a_1} = 7.5 \times 10^{-3}$).

$$H_2PO_4^- + H_2O \rightleftharpoons H_3PO_4 + OH^- \quad K_b = \frac{K_w}{K_{a_1}} = \frac{1.0 \times 10^{-14}}{7.5 \times 10^{-3}} = 1.3 \times 10^{-12}$$

From the K_b values, PO_4^{3-} is the strongest base. This is expected since PO_4^{3-} is the conjugate base of the weakest acid (HPO_4^{2-}).

150.   a. $HCO_3^- + HCO_3^- \rightleftharpoons H_2CO_3 + CO_3^{2-}$

$$K_{eq} = \frac{[H_2CO_3]\,[CO_3^{2-}]}{[HCO_3^-]\,[HCO_3^-]} \times \frac{[H^+]}{[H^+]} = \frac{K_{a_2}}{K_{a_1}} = \frac{5.6 \times 10^{-11}}{4.3 \times 10^{-7}} = 1.3 \times 10^{-4}$$

   b. $[H_2CO_3] = [CO_3^{2-}]$ since the reaction in part a is the principle equilibrium reaction.

   c. $H_2CO_3 \rightleftharpoons 2\,H^+ + CO_3^{2-} \quad K_{eq} = \dfrac{[H^+]^2\,[CO_3^{2-}]}{[H_2CO_3]} = K_{a_1} \times K_{a_2}$

Since, $[H_2CO_3] = [CO_3^{2-}]$ from part b, $[H^+]^2 = K_{a_1} \times K_{a_2}$

$[H^+] = (K_{a_1} \times K_{a_2})^{1/2}$ or pH $= \dfrac{pK_{a_1} + pK_{a_2}}{2}$

   d. $[H^+] = [(4.3 \times 10^{-7}) \times (5.6 \times 10^{-11})]^{1/2}$, $[H^+] = 4.9 \times 10^{-9}\,M$; pH $= 8.31$

CHAPTER FIFTEEN

APPLICATIONS OF AQUEOUS EQUILIBRIA

Questions

12. A common ion is an ion that appears in an equilibrium reaction but came from a source other than that reaction. Addition of a common ion (H^+ or NO_2^-) to the reaction $HNO_2 \rightleftharpoons H^+ + NO_2^-$, will drive the equilibrium to the left as predicted by Le Chatelier's principle.

13. A buffered solution must contain both a weak acid and a weak base. Most buffered solutions are prepared using a weak acid plus the conjugate base of the weak acid (which is a weak base). Buffered solutions are useful for controlling the pH of a solution since they resist pH change.

14. The capacity of a buffer is a measure of how much strong acid or strong base the buffer can neutralize. All the buffers listed have the same pH ($= pK_a = 4.74$) since they all have a 1:1 concentration ratio between the weak acid and the conjugate base. The $1.0\ M$ buffer has the greatest capacity; the $0.01\ M$ buffer the least capacity. In general, the larger the concentrations of weak acid and conjugate base, the greater the buffer capacity, i.e., the greater the ability to neutralize added strong acid or strong base.

15. No, as long as there is both a weak acid and a weak base present, the solution will be buffered. If the concentrations are the same, the buffer will have the same capacity towards both added H^+ and OH^-. In addition, buffers with equal concentrations of weak acid and conjugate base have $pH = pK_a$.

16. Between the starting point of the titration and the equivalence point, we are dealing with a buffer solution. The Henderson-Hasselbalch equation can be used to determine pH:

$$pH = pK_a + \log \frac{[\text{Base}]}{[\text{Acid}]}$$

At the halfway point to equivalence: $[\text{Base}] = [\text{Acid}]$ so $pH = pK_a + \log 1 = pK_a$

The K_a value can be determined at any point in a titration, from the initial point to the equivalence point. In Chapter 14 we calculated the K_a from a solution of only the weak acid. In the buffer region, we can calculate the ratio of the base to acid form of the weak acid and use the Henderson-Hasselbalch equation to determine the K_a value. The equivalence point data can be used to calculate the K_b value for the conjugate base which is related to K_a by the equation $K_a = K_w/K_b$.

17.   No, since there are three colored forms, there must be two proton transfer reactions. Thus, there must be at least two acidic protons in the acid (orange) form of thymol blue.

18.   Equivalence point: moles acid = moles base. End point: indicator changes color. We want the indicator to tell us when we have reached the equivalence point. We can detect the end point visually and assume it is the equivalence point for doing stoichiometric calculations. They don't have to be as close as 0.01 pH units since at the equivalence point the pH is changing very rapidly with added titrant. The range over which an indicator changes color only needs to be close to the pH of the equivalence point.

19.   The two forms of an indicator are different colors. The HIn form has one color and the In⁻ form has another color. To see only one color, that form must be in an approximately ten fold excess or greater over the other form. When the ratio of the two forms is less than 10, both colors are present. To go from $[HIn]/[In^-] = 10$ to $[HIn]/[In^-] = 0.1$ requires a change of 2 pH units (a 100 fold decrease in $[H^+]$) as the indicator changes from the HIn color to the In⁻ color.

20.   If the number of ions in the two salts are the same, then the K_{sp}'s can be compared, i.e., 1:1 electrolytes (1 cation:1 anion) can be compared to each other; 2:1 electrolytes can be compared to each other, etc. If the ions are the same, then the salt with the largest K_{sp} value has the largest molar solubility.

Exercises

Buffers

21.   When strong acid or strong base is added to an acetic acid/sodium acetate mixture, the strong acid/base is neutralized. The reaction goes to completion resulting in the strong acid/base being replaced with a weak acid/base which results in a new buffer solution. The reactions are:

$$H^+(aq) + CH_3CO_2^-(aq) \rightarrow CH_3CO_2H(aq); \quad OH^- + CH_3CO_2H(aq) \rightarrow CH_3CO_2^-(aq) + H_2O(l)$$

22. $NH_3(aq) + H^+(aq) \rightarrow NH_4^+; \quad NH_4^+(aq) + OH^-(aq) \rightarrow NH_3(aq) + H_2O(l)$

23.   a.   This is a weak acid problem. Let $HC_3H_5O_2 = HOPr$ and $C_3H_5O_2^- = OPr^-$.

$$HOPr \quad \rightleftharpoons \quad H^+ \quad + \quad OPr^- \qquad K_a = 1.3 \times 10^{-5}$$

Initial	0.100 M	~0	0

x mol/L HOPr dissociates to reach equilibrium

Change	-x	$\rightarrow$	+x	+x
Equil.	0.100 - x		x	x

$$K_a = 1.3 \times 10^{-5} = \frac{[H^+][OPr^-]}{[HOPr]} = \frac{x^2}{0.100 - x} \approx \frac{x^2}{0.100}$$

$x = [H^+] = 1.1 \times 10^{-3} M$; pH = 2.96   Assumptions good by the 5% rule.

b. This is a weak base problem.

$$OPr^- \ + \ H_2O \ \rightleftharpoons \ HOPr \ + \ OH^- \qquad K_b = \frac{K_w}{K_a} = 7.7 \times 10^{-10}$$

Initial   0.100 M 0   ~0
            x mol/L OPr$^-$ reacts with H_2O to reach equilibrium
Change   -x $\rightarrow$ +x +x
Equil.   0.100 - x x x

$$K_b = 7.7 \times 10^{-10} = \frac{[HOPr][OH^-]}{[OPr^-]} = \frac{x^2}{0.100-x} \approx \frac{x^2}{0.100}$$

$x = [OH^-] = 8.8 \times 10^{-6} \, M$; pOH = 5.06; pH = 8.94 $\qquad$ Assumptions good.

c. pure H_2O, $[H^+] = [OH^-] = 1.0 \times 10^{-7} \, M$; pH = 7.00

d. This solution contains a weak acid and its conjugate base. This is a buffer solution. We will solve for the pH through the weak acid equilibrium reaction.

$$HOPr \ \rightleftharpoons \ H^+ \ + \ OPr^- \qquad K_a = 1.3 \times 10^{-5}$$

Initial   0.100 M ~0   0.100 M
            x mol/L HOPr dissociates to reach equilibrium
Change   -x $\rightarrow$ +x +x
Equil.   0.100 - x x 0.100 + x

$$1.3 \times 10^{-5} = \frac{(0.100+x)(x)}{0.100-x} \approx \frac{(0.100)(x)}{0.100} = x = [H^+]$$

$[H^+] = 1.3 \times 10^{-5} \, M$; pH = 4.89 $\quad$ Assumptions good.

Alternately, we can use the Henderson-Hasselbalch equation to calculate the pH of buffer solutions.

$$pH = pK_a + \log \frac{[Base]}{[Acid]} = pK_a + \log \frac{(0.100)}{(0.100)} = pK_a = -\log(1.3 \times 10^{-5}) = 4.89$$

The Henderson-Hasselbalch equation will be valid when an assumption of the type, $0.1 + x \approx 0.1$, that we just made in this problem is valid. From a practical standpoint, this will almost always be true for useful buffer solutions. If the assumption is not valid, the solution will have such a low buffering capacity it will not be of any use to control the pH. Note: The Henderson-Hasselbalch equation can only be used to solve for the pH of buffer solutions.

24.   a.   Weak base problem:

$$NH_3 + H_2O \rightleftharpoons NH_4^+ + OH^- \qquad K_b = 1.8 \times 10^{-5}$$

Initial	0.100 M	0	~0

x mol/L NH_3 reacts with H_2O to reach equilibrium

Change	$-x$	$\rightarrow$ $+x$	$+x$
Equil.	0.100 - x	x	x

$$K_b = 1.8 \times 10^{-5} = \frac{x^2}{(0.100 - x)} \approx \frac{x^2}{0.100}$$

$x = [OH^-] = 1.3 \times 10^{-3}\,M$, pOH = 2.89; pH = 11.11   Assumptions good.

b.   Weak acid problem:

$$NH_4^+ \rightleftharpoons NH_3 + H^+$$

Initial	0.100 M	0	~0

x mol/L NH_4^+ dissociates to reach equilibrium

Change	$-x$	$\rightarrow$ $+x$	$+x$
Equil.	0.100 - x	x	x

$$K_a = \frac{K_w}{K_b} = 5.6 \times 10^{-10} = \frac{[NH_3][H^+]}{[NH_4^+]} = \frac{x^2}{0.100 - x} \approx \frac{x^2}{0.100}$$

$x = [H^+] = 7.5 \times 10^{-6}\,M$; pH = 5.12   Assumptions good.

c.   Pure H_2O, pH = 7.00

d.   Buffer solution where $pK_a = -\log(5.6 \times 10^{-10}) = 9.25$. Using the Henderson-Hasselbalch equation:

$$pH = pK_a + \log \frac{[Base]}{[Acid]} = 9.25 + \log \frac{[NH_3]}{[NH_4^+]} = 9.25 + \log \frac{(0.100)}{(0.100)} = 9.25$$

25.   0.100 M $HC_3H_5O_2$: percent dissociation $= \dfrac{[H^+]}{[HC_3H_5O_2]_o} \times 100 = \dfrac{1.1 \times 10^{-3}\,M}{0.100\,M} \times 100 = 1.1\%$

0.100 M $HC_3H_5O_2$ + 0.100 M $NaC_3H_5O_2$: % dissociation $= \dfrac{1.3 \times 10^{-5}}{0.100} \times 100 = 1.3 \times 10^{-2}\%$

The percent dissociation of the acid decreases from 1.1% to 1.3×10^{-2} % when $C_3H_5O_2^-$ is present. This is known as the common ion effect. The presence of the conjugate base of the weak acid inhibits the acid dissociation reaction.

26.   0.100 M NH_3: percent ionization $= \dfrac{[OH^-]}{[NH_3]_o} \times 100 = \dfrac{1.3 \times 10^{-3}\,M}{0.100\,M} \times 100 = 1.3\%$

0.100 M NH$_3$ + 0.100 M NH$_4$Cl: % ionization = $\dfrac{1.8 \times 10^{-5}}{0.100} \times 100 = 1.8 \times 10^{-2}$%

The percent ionization decreases by about a factor of ~100. The presence of the conjugate acid of the weak base inhibits the weak base reaction with water. This is known as the common ion effect.

27.   a.   We have a weak acid (HOPr = HC$_3$H$_5$O$_2$) and a strong acid (HCl) present. The amount of H$^+$ donated by the weak acid will be negligible. To prove it lets consider the weak acid equilibrium reaction:

$$\text{HOPr} \;\rightleftharpoons\; \text{H}^+ \;+\; \text{OPr}^- \qquad K_a = 1.3 \times 10^{-5}$$

	HOPr		H$^+$	OPr$^-$
Initial	0.100 M		0.020 M	0

Initial   0.100 M 0.020 M 0
                 x mol/L HOPr dissociates to reach equilibrium
Change   -x →   +x +x
Equil.   0.100 - x 0.020 + x x

[H$^+$] = 0.020 + x ≈ 0.020 M; pH = 1.70   Assumption good ($x = 6.5 \times 10^{-5}$ which is << 0.020).

Note: The H$^+$ contribution from the weak acid HOPr was negligible. The pH of the solution can be determined by only considering the amount of strong acid present.

   b.   Added H$^+$ reacts completely with the best base present, OPr$^-$.

$$\text{OPr}^- \;+\; \text{H}^+ \;\rightarrow\; \text{HOPr}$$

	OPr$^-$	H$^+$		HOPr	
Before	0.100 M	0.020 M		0	
Change	-0.020	-0.020	→	+0.020	Reacts completely
After	0.080	0		0.020 M	

After reaction, a weak acid, HOPr , and its conjugate base, OPr$^-$, are present. This is a buffer solution. Using the Henderson-Hasselbalch equation where pK_a = -log (1.3 × 10^{-5}) = 4.89:

$$\text{pH} = \text{p}K_a + \log \frac{[\text{Base}]}{[\text{Acid}]} = 4.89 + \log \frac{(0.080)}{(0.020)} = 5.49 \qquad \text{Assumptions good.}$$

   c.   This is a strong acid problem. [H$^+$] = 0.020 M; pH = 1.70

   d.   Added H$^+$ reacts completely with the best base present, OPr$^-$.

$$\text{OPr}^- \;+\; \text{H}^+ \;\rightarrow\; \text{HOPr}$$

	OPr$^-$	H$^+$		HOPr	
Before	0.100 M	0.020 M		0.100 M	
Change	-0.020	-0.020	→	+0.020	Reacts completely
After	0.080	0		0.120	

A buffer solution results (weak acid + conjugate base). Using the Henderson-Hasselbalch equation:

$$pH = pK_a + \log \frac{[Base]}{[Acid]} = 4.89 + \log \frac{(0.080)}{(0.120)} = 4.71$$

28.   a.   Added H^+ reacts completely with NH_3 (the best base present) to form NH_4^+.

	NH_3	+	H^+	$\rightarrow$	NH_4^+	
Before	0.100 M		0.020 M		0	
Change	-0.020		-0.020	$\rightarrow$	+0.020	Reacts completely
After	0.080		0		0.020	

After this reaction, a buffer solution exists, i.e., a weak acid (NH_4^+) and its conjugate base (NH_3) present at the same time. Using the Henderson-Hasselbalch equation to solve for the pH:

$$pH = pK_a + \log \frac{[base]}{[acid]} = 9.25 + \log \frac{(0.080)}{(0.020)} = 9.25 + 0.60 = 9.85$$

   b.   We have a weak acid and a strong acid present at the same time. The H^+ contribution from the weak acid, NH_4^+, will be negligible. So, we only have to consider the H^+ from HCl. $[H^+]$ = 0.020 M; pH = 1.70

   c.   This is a strong acid in water. $[H^+]$ = 0.020 M; pH = 1.70

   d.   Major species: H_2O, Cl^-, NH_3, NH_4^+, H^+
                                        0.100   0.100   0.020

   H^+ will react completely with NH_3, the best base present.

	NH_3	+	H^+	$\rightarrow$	NH_4^+	
Before	0.100 M		0.020 M		0.100 M	
Change	-0.020		-0.020	$\rightarrow$	+0.020	Reacts completely
After	0.080		0		0.120	

A buffer solution results after reaction. Using the Henderson-Hasselbalch equation:

$$pH = 9.25 + \log \frac{[NH_3]}{[NH_4^+]} = 9.25 + \log \frac{(0.080)}{(0.120)} = 9.25 - 0.18 = 9.07$$

29.   a. OH^- will react completely with the best acid present, HOPr.

	HOPr	+	OH^-	$\rightarrow$	OPr^-	+	H_2O	
Before	0.100 M		0.020 M		0			
Change	-0.020		-0.020	$\rightarrow$	+0.020			Reacts completely
After	0.080		0		0.020			

A buffer solution results after the reaction. Using the Henderson-Hasselbalch equation:

$$pH = pK_a + \log \frac{[Base]}{[Acid]} = 4.89 + \log \frac{(0.020)}{(0.080)} = 4.29$$

b.   We have a weak base and a strong base present at the same time. The amount of OH⁻ added by
the weak base will be negligible. To prove it, lets consider the weak base equilibrium:

$$OPr^- \ + \ H_2O \ \rightleftharpoons \ HOPr \ + \ OH^- \qquad K_b = 7.7 \times 10^{-10}$$

Initial	0.100 M	0	0.020 M

x mol/L OPr⁻ reacts with H_2O to reach equilibrium

Change	$-x$	$\rightarrow$	$+x$	$+x$
Equil.	0.100 - x		x	0.020 + x

$[OH^-] = 0.020 + x \approx 0.020\,M$; pOH = 1.70; pH = 12.30   Assumption good.

Note: The OH⁻ contribution from the weak base OPr⁻ was negligible ($x = 3.9 \times 10^{-9}\,M$ as
compared to $0.020\,M$ OH⁻ from the strong base). The pH can be determined by only considering
the amount of strong base present.

c.   This is a strong base in water. $[OH^-] = 0.020\,M$; pOH = 1.70; pH = 12.30

d.   OH⁻ will react completely with HOPr, the best acid present.

$$HOPr \ + \ OH^- \ \rightarrow \ OPr^- \ + \ H_2O$$

Before	0.100 M	0.020 M	0.100 M	
Change	-0.020	-0.020	$\rightarrow$	+0.020	Reacts completely
After	0.080	0	0.120	

Using the Henderson-Hasselbalch equation to solve for the pH of the resulting buffer solution:

$$pH = pK_a + \log \ \frac{[Base]}{[Acid]} = 4.89 + \log \frac{(0.120)}{(0.080)} = 5.07$$

30.   a.   We have a weak base and a strong base present at the same time. The OH⁻ contribution from the
weak base, NH_3, will be negligible. Consider only the added strong base as the primary source
of OH⁻.

$[OH^-] = 0.020\,M$; pOH = 1.70; pH = 12.30

b.   Added strong base will react to completion with the best acid present, NH_4^+.

$$OH^- \ + \ NH_4^+ \ \rightarrow \ NH_3 \ + \ H_2O$$

Before	0.020 M	0.100 M	0	
Change	-0.020	-0.020	$\rightarrow$	+0.020	Reacts completely
After	0	0.080	0.020	

The resulting solution is a buffer (a weak acid and its conjugate base). Using the Henderson-
Hasselbalch equation:

$$pH = 9.25 + \log \left(\frac{0.020}{0.080} \right) = 9.25 - 0.60 = 8.65$$

c.   This is a strong base in water. $[OH^-] = 0.020\,M$; $pOH = 1.70$; $pH = 12.30$

d.   Major species: H_2O, Cl^-, Na^+, NH_3, NH_4^+, OH^-
                                        0.100   0.100   0.020

Again, the added strong base reacts completely with the best acid present, NH_4^+.

$$NH_4^+ \quad + \quad OH^- \quad \rightarrow \quad NH_3 \quad + \quad H_2O$$

Before	$0.100\,M$	$0.020\,M$	$0.100\,M$	
Change	-0.020	-0.020	$\rightarrow$	$+0.020$	Reacts completely
After	0.080	0	0.120	

A buffer solution results. Using the Henderson-Hasselbalch equation:

$$pH = 9.25 + \log \frac{[NH_3]}{[NH_4^+]} = 9.25 + \log\left(\frac{0.12}{0.080}\right) = 9.25 + 0.18 = 9.43$$

31.   Consider all of the results to Exercises 15.23, 15.27, and 15.29:

Solution	Initial pH	after added acid	after added base
a	2.96	1.70	4.29
b	8.94	5.49	12.30
c	7.00	1.70	12.30
d	4.89	4.71	5.07

The solution in Exercise 15.23d is a buffer; it contains both a weak acid ($HC_3H_5O_2$) and a weak base ($C_3H_5O_2^-$). Solution d shows the greatest resistance to changes in pH when either strong acid or strong base is added, which is the primary property of buffers.

32.   Consider all of the results to Exercises 15.24, 15.28, and 15.30.

Solution	Initial pH	after added acid	after added base
a	11.11	9.85	12.30
b	5.12	1.70	8.65
c	7.00	1.70	12.30
d	9.25	9.07	9.43

The solution in Exercise 15.24d is a buffer; it shows the greatest resistance to a change in pH when strong acid or base is added. The solution in Exercise 15.24d contains a weak acid (NH_4^+) and a weak base (NH_3), which constitutes a buffer solution.

33.   Major species: HNO_2, NO_2^- and Na^+. Na^+ has no acidic or basic properties. The appropriate equilibrium reaction to use is the K_a reaction of HNO_2 which contains both HNO_2 and NO_2^-. Solving the equilibrium problem (called a buffer problem):

$$HNO_2 \rightleftharpoons NO_2^- + H^+$$

Initial	1.00 M	1.00 M	~0

x mol/L HNO_2 dissociates to reach equilibrium

Change	-x	$\rightarrow$	+x	+x
Equil.	1.00 - x		1.00 + x	x

$$K_a = 4.0 \times 10^{-4} = \frac{[NO_2^-][H^+]}{[HNO_2]} = \frac{(1.00 + x)(x)}{(1.00 - x)} \approx \frac{1.00(x)}{1.00} \quad \text{(assuming } x \ll 1.00\text{)}$$

$x = 4.0 \times 10^{-4} \, M = [H^+]$;   Assumptions good ($x$ is 4.0×10^{-2}% of 1.00).

$pH = -\log(4.0 \times 10^{-4}) = 3.40$

Note: We would get the same answer using the Henderson-Hasselbalch equation. Use whichever method is easiest for you.

34.   Major species: HF, F^- and K^+ (no acidic/basic properties). The appropriate equilibrium reaction to use is the K_a reaction of HF which contains both HF and F^-.

$$HF \rightleftharpoons F^- + H^+$$

Initial	0.60 M	1.00 M	~0

x mol/L HF dissociates to reach equilibrium

Change	-x	$\rightarrow$	+x	+x
Equil.	0.60 - x		1.00 + x	x

$$K_a = 7.2 \times 10^{-4} = \frac{[F^-][H^+]}{[HF]} = \frac{(1.00 + x)(x)}{(0.60 - x)} \approx \frac{1.00(x)}{0.60} \quad \text{(assuming } x \ll 0.60\text{)}$$

$x = [H^+] = 0.60 \times (7.2 \times 10^{-4}) = 4.3 \times 10^{-4} \, M$; Assumptions good ($x$ is 7.2×10^{-2}% of 0.60).

$pH = -\log(4.3 \times 10^{-4}) = 3.37$

35.   Major species: HNO_2, NO_2^-, Na^+ and OH^-. The OH^- from the strong base will react with the best acid present (HNO_2). Any reaction involving a strong base is assumed to go to completion. The stoichiometry problem is:

$$OH^- + HNO_2 \rightarrow NO_2^- + H_2O$$

Before	0.10 mol/1.00 L	1.00 M		1.00 M	
Change	-0.10 M	-0.10 M	$\rightarrow$	+0.10 M	Reacts completely
After	0	0.90		1.10	

After all the OH^- reacts, we are left with a solution containing a weak acid (HNO_2) and its conjugate base (NO_2^-). This is what we call a buffer problem. We will solve this buffer problem using the K_a equilibrium reaction.

$$HNO_2 \quad \rightleftharpoons \quad NO_2^- \quad + \quad H^+$$

Initial	$0.90\,M$	$1.10\,M$	~ 0
	x mol/L HNO_2 dissociates to reach equilibrium		
Change	$-x$ $\rightarrow$	$+x$	$+x$
Equil.	$0.90 - x$	$1.10 + x$	x

$$K_a = 4.0 \times 10^{-4} = \frac{(1.10 + x)(x)}{(0.90 - x)} \approx \frac{1.10(x)}{0.90}, \quad x = [H^+] = 3.3 \times 10^{-4}\,M; \quad pH = 3.48; \quad \text{Assumptions good.}$$

Note: The added NaOH to this buffer solution only changes the pH from 3.40 to 3.48. If the NaOH were added to 1.0 L of pure water, the pH would change from 7.00 to 13.00.

36.   Added OH^- reacts completely with the best acid present, HF.

$$OH^- \quad + \quad HF \quad \rightarrow \quad F^- \quad + \quad H_2O$$

Before	$\dfrac{0.10\ \text{mol}}{1.00\ \text{L}}$	$0.60\,M$	$1.00\,M$	
Change	$-0.10\,M$	$-0.10\,M$ $\rightarrow$	$+0.10\,M$	Reacts completely
After	0	0.50	1.10	

We now have a new buffer problem. Solving the equilibrium part of the problem:

$$HF \quad \rightleftharpoons \quad F^- \quad + \quad H^+$$

Initial	$0.50\,M$	$1.10\,M$	~ 0
Equil.	$0.50 - x$	$1.10 + x$	x

$$K_a = 7.2 \times 10^{-4} = \frac{(1.10 + x)(x)}{(0.50 - x)} \approx \frac{1.10(x)}{(0.50)}, \quad x = [H^+] = 3.3 \times 10^{-4}\,M$$

pH = 3.48;   Assumptions good.

37.   Major species: HNO_2, NO_2^-, H^+, Na^+, Cl^-; The added H^+ from the strong acid will react completely with the best base present (NO_2^-).

$$H^+ \quad + \quad NO_2^- \quad \rightarrow \quad HNO_2$$

Before	$\dfrac{0.20\ \text{mol}}{1.00\ \text{L}}$	$1.00\,M$	$1.00\,M$
Change	$-0.20\,M$	$-0.20\,M$ $\rightarrow$	$+0.20\,M$ Reacts completely
After	0	0.80	1.20

After all the H^+ has reacted, we have a buffer solution (a solution containing a weak acid and its conjugate base). Solving the buffer problem:

$$HNO_2 \rightleftharpoons NO_2^- + H^+$$

Initial	1.20 M	0.80 M	0
Equil.	1.20 - x	0.80 + x	+x

$K_a = 4.0 \times 10^{-4} = \dfrac{(0.80 + x)(x)}{1.20 - x} \approx \dfrac{0.80(x)}{1.20}$, $x = [H^+] = 6.0 \times 10^{-4} M$;   pH = 3.22;   Assumptions good.

Note: The added HCl to this buffer solution only changes the pH from 3.40 to 3.22. If the HCl were added to 1.0 L of pure water, the pH would change from 7.00 to 0.70.

38.   Added H^+ reacts to completion with the best base (F^-) present.

$$H^+ + F^- \rightarrow HF$$

Before	0.20 M	1.00 M	0.60 M
After	0	0.80 M	0.80 M

After reaction, we have a buffer problem:

$$HF \rightleftharpoons F^- + H^+$$

Initial	0.80 M	0.80 M	0
Equil.	0.80 - x	0.80 + x	x

$K_a = 7.2 \times 10^{-4} = \dfrac{(0.80 + x)(x)}{0.80 - x} \approx x$, $x = [H^+] = 7.2 \times 10^{-4} M$;   pH = 3.14;   Assumptions good.

39.   75.0 g $NaC_2H_3O_2 \times \dfrac{1\ mol}{82.03\ g} = 0.914$ mol; $[C_2H_3O_2^-] = \dfrac{0.914\ mol}{0.5000\ L} = 1.83\ M$

We will solve this buffer problem using the Henderson-Hasselbalch equation.

$pH = pK_a + log \dfrac{[C_2H_3O_2^-]}{[HC_2H_3O_2]} = -log\,(1.8 \times 10^{-5}) + log\left(\dfrac{1.83}{0.64}\right) = 4.74 + 0.46 = 5.20$

40.   50.0 g $NH_4Cl \times \dfrac{1\ mol\ NH_4Cl}{53.49\ g\ NH_4Cl} = 0.935$ mol NH_4Cl added to 1.00 L; $[NH_4^+] = 0.935\ M$

$pH = pK_a + log \dfrac{[NH_3]}{[NH_4^+]} = -log\,(5.6 \times 10^{-10}) + log\left(\dfrac{0.75}{0.935}\right) = 9.25 - 0.096 = 9.15$

41.   a. $HC_2H_3O_2 \rightleftharpoons H^+ + C_2H_3O_2^-$ $K_a = 1.8 \times 10^{-5}$

Initial	0.10 M	~0	0.25 M
	x mol/L $HC_2H_3O_2$ dissociates to reach equilibrium		
Change	-x $\rightarrow$	+x	+x
Equil.	0.10 - x	x	0.25 + x

$$1.8 \times 10^{-5} = \frac{x(0.25 + x)}{(0.10 - x)} \approx \frac{x(0.25)}{0.10} \quad \text{(assuming } 0.25 + x \approx 0.25 \text{ and } 0.10 - x \approx 0.10\text{)}$$

$x = [H^+] = 7.2 \times 10^{-6} M; \quad pH = 5.14$ Assumptions good by the 5% rule.

Alternatively, we can use the Henderson-Hasselbalch equation:

$$pH = pK_a + \log \frac{[Base]}{[Acid]} \quad \text{where } pK_a = -\log (1.8 \times 10^{-5}) = 4.74$$

$$pH = 4.74 + \log \frac{(0.25)}{(0.10)} = 4.74 + 0.40 = 5.14$$

The Henderson-Hasselbalch equation will be valid when assumptions of the type, $0.10 - x \approx 0.10$, that we just made are valid. From a practical standpoint, this will almost always be true for useful buffer solutions. Note: The Henderson-Hasselbalch equation can <u>only</u> be used to solve for the pH of buffer solutions.

b. $pH = 4.74 + \log \dfrac{(0.10)}{(0.25)} = 4.74 + (-0.40) = 4.34$

c. $pH = 4.74 + \log \dfrac{(0.20)}{(0.080)} = 4.74 + 0.40 = 5.14$

d. $pH = 4.74 + \log \dfrac{(0.080)}{(0.20)} = 4.74 + (-0.40) = 4.34$

42.   We will use the Henderson-Hasselbalch equation to solve for the pH of these buffer solutions.

a. $pH = pK_a + \log \dfrac{[base]}{[acid]}$; $[base] = [C_2H_5NH_2] = 0.50\,M$; $[acid] = [C_2H_5NH_3^+] = 0.25\,M$

$$K_a = \frac{K_w}{K_b} = \frac{1.0 \times 10^{-14}}{5.6 \times 10^{-4}} = 1.8 \times 10^{-11}$$

$$pH = -\log (1.8 \times 10^{-11}) + \log \left(\frac{0.50\,M}{0.25\,M} \right) = 10.74 + 0.30 = 11.04$$

b. $pH = 10.74 + \log \left(\dfrac{0.25\,M}{0.50\,M} \right) = 10.74 + (-0.30) = 10.44$

c. $pH = 10.74 + \log \left(\dfrac{0.50\,M}{0.50\,M} \right) = 10.74 + 0 = 10.74$

43.   This is a buffer solution since a weak base (CH_3NH_2) and its conjugate acid ($CH_3NH_3^+$) are present at the same time. Using the Henderson Hasselbalch equation:

$$pH = pK_a + \log \frac{[base]}{[acid]}; \quad K_a = \frac{K_w}{K_b} = \frac{1.00 \times 10^{-14}}{4.38 \times 10^{-4}} = 2.28 \times 10^{-11}$$

$$\text{pH} = -\log (2.28 \times 10^{-11}) + \log \left(\frac{0.50\,M}{0.70\,M} \right) = 10.642 + (-0.15) = 10.49$$

44.   Added NaOH will react completely with the best acid present, $CH_3NH_3^+$.

$$CH_3NH_3^+ \quad + \quad OH^- \quad \rightarrow \quad CH_3NH_2 \quad + \quad H_2O$$

Before	$0.70\,M$	$\dfrac{0.10\ \text{mol}}{1.0\ \text{L}}$	$0.50\,M$	
Change	$-0.10\,M$	$-0.10\,M$	$\rightarrow$ $+0.10\,M$	Reacts completely
After	$0.60\,M$	0	$0.60\,M$	

After reaction, a buffer solution remains. Using the Henderson-Hasselbalch equation:

$$\text{pH} = \text{p}K_a + \log \frac{[CH_3NH_2]}{[CH_3NH_3^+]} = 10.642 + \log \left(\frac{0.60}{0.60} \right) = 10.642 + 0 = 10.642$$

Added HCl will react completely the best base present (CH_3NH_2). Since 0.10 mol of HCl is present, then 0.10 mol of CH_3NH_2 will be reacted, and 0.10 mol of $CH_3NH_3^+$ will be produced. After this reaction: $[CH_3NH_2] = 0.50 - 0.10 = 0.40\,M$ and $[CH_3NH_3^+] = 0.70 + 0.10 = 0.80\,M$. Solving the buffer problem:

$$\text{pH} = 10.642 + \log \left(\frac{0.40}{0.80} \right) = 10.642 + (-0.30) = 10.34$$

45. $[H^+]$ added $= \dfrac{0.010\ \text{mol}}{0.25\ \text{L}} = 0.040\,M$; The added H^+ reacts completely with NH_3 to form NH_4^+.

a. $NH_3 \quad + \quad H^+ \quad \rightarrow \quad NH_4^+$

Before	$0.050\,M$	$0.040\,M$	$0.15\,M$	
Change	-0.040	-0.040	$\rightarrow$ $+0.040$	Reacts completely
After	0.010	0	0.19	

A buffer solution still exists after H^+ reacts completely. Using the Henderson-Hasselbalch equation:

$$\text{pH} = \text{p}K_a + \log \frac{[NH_3]}{[NH_4^+]} = -\log (5.6 \times 10^{-10}) + \log \left(\frac{0.010}{0.19} \right) = 9.25 + (-1.28) = 7.97$$

b. $NH_3 \quad + \quad H^+ \quad \rightarrow \quad NH_4^+$

Before	$0.50\,M$	$0.040\,M$	$1.50\,M$	
Change	-0.040	-0.040	$\rightarrow$ $+0.040$	Reacts completely
After	0.46	0	1.54	

A buffer solution still exists. $\text{pH} = \text{p}K_a + \log \dfrac{[NH_3]}{[NH_4^+]} = 9.25 + \log \left(\dfrac{0.46}{1.54} \right) = 8.73$

The two buffers differ in their capacity and not the pH (both buffers had an initial pH = 8.77). Solution b has the greatest capacity since it has the largest concentrations of weak acid and conjugate base. Buffers with greater capacities will be able to absorb more H^+ or OH^- added.

46.   a.   Major species: H_2O, Na^+, OH^-,   HOPr, OPr^- where $HOPr = HC_3H_5O_2$
                                                   $0.15\,M$ $0.050\,M$ $0.080\,M$ and $OPr^- = C_3H_5O_2^-$

OH^- will react completely with HOPr. Here, HOPr is the limiting reagent.

	HOPr	+	OH^-	$\rightarrow$	OPr^-	+	H_2O	
Before	$0.050\,M$		$0.15\,M$		$0.080\,M$			
Change	-0.050		-0.050	$\rightarrow$	+0.050			Reacts completely
After	0		0.10		0.130			

OH^- from the strong base is in excess. The OH^- contribution from the weak base, OPr^-, will be negligible. Consider only the excess strong base to determine the pH.

$[OH^-] = 0.10\,M$; pOH = 1.00; pH = 13.00

Note: Original pH of buffer is: $4.89 + \log\left(\dfrac{0.080}{0.050}\right) = 5.09$

b. OH^- will react completely with HOPr. Here, OH^- is the limiting reagent.

	HOPr	+	OH^-	$\rightarrow$	OPr^-	+	H_2O	
Before	$0.50\,M$		$0.15\,M$		$0.80\,M$			
Change	-0.15		-0.15	$\rightarrow$	+0.15			Reacts completely
After	0.35		0		0.95			

A buffer solution results since after reaction HOPr and OPr^- are present.

$$pH = pK_a + \log\frac{[OPr^-]}{[HOPr]} = 4.89 + \log\left(\frac{0.95}{0.35}\right) = 4.89 + 0.43 = 5.32$$

c.   Although solutions a and b both started out as buffers with the same pH (= 5.09), solution a is no longer a buffer after 0.15 mol NaOH was added. We added more strong base than there was weak acid present in the buffer; we overloaded the buffer system in solution a. We say that solution b has the greater buffer capacity since it has larger initial concentrations of weak acid and conjugate base.

47. $NH_4^+ \rightleftharpoons H^+ + NH_3$ $K_a = \dfrac{K_w}{K_b} = 5.6 \times 10^{-10}$; $pK_a = -\log(5.6 \times 10^{-10}) = 9.25$; We will use the

Henderson-Hasselbalch equation to calculate the concentration ratio necessary for each buffer.

$$pH = pK_a + \log\frac{[base]}{[acid]}, \quad pH = 9.25 + \log\frac{[NH_3]}{[NH_4^+]}$$

a. $9.00 = 9.25 + \log \dfrac{[NH_3]}{[NH_4^+]}$ b. $8.80 = 9.25 + \log \dfrac{[NH_3]}{[NH_4^+]}$

$\log \dfrac{[NH_3]}{[NH_4^+]} = -0.25$ $\log \dfrac{[NH_3]}{[NH_4^+]} = -0.45$

$\dfrac{[NH_3]}{[NH_4^+]} = 10^{-0.25} = 0.56$ $\dfrac{[NH_3]}{[NH_4^+]} = 10^{-0.45} = 0.35$

c. $10.00 = 9.25 + \log \dfrac{[NH_3]}{[NH_4^+]}$ d. $9.60 = 9.25 + \log \dfrac{[NH_3]}{[NH_4^+]}$

$\dfrac{[NH_3]}{[NH_4^+]} = 10^{0.75} = 5.6$ $\dfrac{[NH_3]}{[NH_4^+]} = 10^{0.35} = 2.2$

48. $pH = pK_a + \log \dfrac{[Base]}{[Acid]}$, $7.41 = -\log (4.3 \times 10^{-7}) + \log \dfrac{[HCO_3^-]}{[H_2CO_3]} = 6.37 + \log \dfrac{[HCO_3^-]}{[H_2CO_3]}$

$\dfrac{[HCO_3^-]}{[H_2CO_3]} = 10^{1.04} = 11$ or $\dfrac{[CO_2]}{[HCO_3^-]} = \dfrac{1}{11} = 0.091$

49.   A best buffer has large and equal quantities of weak acid and conjugate base. Since [acid] = [base] for a best buffer, then $pH = pK_a + \log \dfrac{[base]}{[acid]} = pK_a + 0 = pK_a$.

The best acid choice for a pH = 7.00 buffer would be the weak acid with a pK_a close to 7 or $K_a \approx 1 \times 10^{-7}$. HOCl is the best choice in Table 14.2 ($K_a = 3.5 \times 10^{-8}$; $pK_a = 7.46$). To make this buffer, we need to calculate the [base]/[acid] ratio.

$7.00 = 7.46 + \log \dfrac{[base]}{[acid]}$, $\dfrac{[OCl^-]}{[HOCl]} = 10^{-0.46} = 0.35$

Any OCl$^-$/HOCl buffer in a concentration ratio of 0.35:1 will have a pH = 7.00. One possibility is [NaOCl] = 0.35 M and [HOCl] = 1.0 M.

50.   For a pH = 5.00 buffer, we want an acid with pK_a close to 5.00. For a conjugate acid-base pair, $14.00 = pK_a + pK_b$. So for a pH = 5.00 buffer, we want the base to have a pK_b close to 14 - 5 = 9 or a K_b close to 1×10^{-9}. The best choice in Table 14.3 is pyridine, C_5H_5N, with $K_b = 1.7 \times 10^{-9}$.

$pH = pK_a + \log \dfrac{[base]}{[acid]}$; $pK_a = \dfrac{K_w}{K_b} = \dfrac{1.0 \times 10^{-14}}{1.7 \times 10^{-9}} = 5.9 \times 10^{-6}$

$5.00 = -\log (5.9 \times 10^{-6}) + \log \dfrac{[base]}{[acid]}$, $\dfrac{[C_5H_5N]}{[C_5H_5NH^+]} = 10^{-0.23} = 0.59$

There are several possibilities to make this buffer. One possibility is a solution of [C_5H_5N] = 0.59 M and [C_5H_5NHCl] = 1.0 M. The pH of this solution will be 5.00 since the base to acid concentration ratio is 0.59:1.

51.   The reaction $H^+ + NH_3 \rightarrow NH_4^+$ goes to completion for solutions b-d. After this reaction occurs, there must be both NH_3 and NH_4^+ in solution for it to be a buffer. The important components of each solution (after any reaction) are:

a.   0.05 M NH_4^+ and 0.05 M H^+ (two acids present, no buffer)

b.   0.05 M NH_4^+ (no NH_3 remains, no buffer)

c.   0.05 M NH_4^+ and 0.05 M H^+ (too much H^+ added)

d.   0.05 M NH_4^+ and 0.05 M NH_3 (a buffer solution results)

Only the combination in mixture d results in a buffer. Note that the concentrations are halved from the initial values. This is because equal volumes of two solutions were added together which halves the initial concentrations.

52.   a.   No; A strong acid and neutral salt do not constitute a buffer solution. A buffer solution always consists of a weak acid and a weak base (usually the conjugate base of the weak acid).

b.   Yes; HNO_2 = weak acid, NO_2^- = weak base; Since a weak acid and a weak base are both present at the same time, then a buffer solution exists.

c. OH^- reacts completely with HNO_2.

$$\text{mol } OH^- = 0.0250 \text{ L} \times \frac{0.10 \text{ mol } OH^-}{L} = 2.5 \times 10^{-3} \text{ mol } OH^-$$

$$\text{mol } HNO_2 = 0.0500 \text{ L} \times \frac{0.10 \text{ mol } HNO_2}{L} = 5.0 \times 10^{-3} \text{ mol } HNO_2$$

	OH^-	+	HNO_2	$\rightarrow$	NO_2^-	+	H_2O	
Before	2.5×10^{-3} mol		5.0×10^{-3} mol		0			
Change	-2.5×10^{-3} mol		-2.5×10^{-3} mol	$\rightarrow$	$+2.5 \times 10^{-3}$ mol			Reacts completely
After	0		2.5×10^{-3} mol		2.5×10^{-3} mol			

After the initial stoichiometry reaction, a buffer solution results since HNO_2 and NO_2^- are both present at the same time. So yes this is a buffer solution and if asked, the pH would equal the pK_a value for HNO_2 since the moles (and concentrations) of HNO_2 and NO_2^- are equal.

53.   When OH^- is added, it converts $HC_2H_3O_2$ into $C_2H_3O_2^-$: $HC_2H_3O_2 + OH^- \rightarrow C_2H_3O_2^- + H_2O$

From this reaction, the moles of $C_2H_3O_2^-$ produced <u>equals</u> the moles of OH^- added. Also, the total concentration of acetic acid plus acetate ion must equal 2.0 M (assuming no volume change on addition of NaOH). Summarizing for each solution:

$[C_2H_3O_2^-] + [HC_2H_3O] = 2.0$ M and $[C_2H_3O_2^-]$ produced = $[OH^-]$ added

a. $pH = pK_a + \log \dfrac{[C_2H_3O_2^-]}{[HC_2H_3O_2]}$; For $pH = pK_a$, $\log \dfrac{[C_2H_3O_2^-]}{[HC_2H_3O_2]} = 0$

Therefore, $\dfrac{[C_2H_3O_2^-]}{[HC_2H_3O_2]} = 1.0$ and $[C_2H_3O_2^-] = [HC_2H_3O_2]$

Since $[C_2H_3O_2^-] + [HC_2H_3O_2] = 2.0\,M$, then $[C_2H_3O_2^-] = [HC_2H_3O_2] = 1.0\,M = [OH^-]$ added

To produce a $1.0\,M$ $C_2H_3O_2^-$ solution we need to add 1.0 mol of NaOH to 1.0 L of the $2.0\,M$ $HC_2H_3O_2$ solution. The resultant solution will have $pH = pK_a = 4.74$.

b. $4.00 = 4.74 + \log \dfrac{[C_2H_3O_2^-]}{[HC_2H_3O_2]}$, $\dfrac{[C_2H_3O_2^-]}{[HC_2H_3O_2]} = 10^{-0.74} = 0.18$

$[C_2H_3O_2^-] = 0.18\,[HC_2H_3O_2]$ or $[HC_2H_3O_2] = 5.6\,[C_2H_3O_2^-]$; Since $[C_2H_3O_2^-] + [HC_2H_3O_2] = 2.0\,M$, then:

$[C_2H_3O_2^-] + 5.6\,[C_2H_3O_2^-] = 2.0\,M$, $[C_2H_3O_2^-] = \dfrac{2.0}{6.6} = 0.30\,M = [OH^-]$ added

We need to add 0.30 mol of NaOH to 1.0 L of $2.0\,M\,HC_2H_3O_2$ solution to produce $0.30\,M$ $C_2H_3O_2^-$. The resultant solution will have $pH = 4.00$.

c. $5.00 = 4.74 + \log \dfrac{[C_2H_3O_2^-]}{[HC_2H_3O_2]}$, $\dfrac{[C_2H_3O_2^-]}{[HC_2H_3O_2]} = 10^{0.26} = 1.8$

$1.8\,[HC_2H_3O_2] = [C_2H_3O_2^-]$ or $[HC_2H_3O_2] = 0.56\,[C_2H_3O_2^-]$; Since $[HC_2H_3O_2] + [C_2H_3O_2^-] = 2.0\,M$, then:

$1.56\,[C_2H_3O_2^-] = 2.0\,M$, $[C_2H_3O_2^-] = 1.3\,M = [OH^-]$ added

We need to add 1.3 mol of NaOH to 1.0 L of $2.0\,M\,HC_2H_3O_2$ to produce a solution with $pH = 5.00$.

54. When H^+ is added, it converts $C_2H_3O_2^-$ into $HC_2H_3O_2$: $C_2H_3O_2^- + H^+ \rightarrow HC_2H_3O_2$. From this reaction, the moles of $HC_2H_3O_2$ produced must equal the moles of H^+ added and the total concentration of acetate ion + acetic acid must equal $1.0\,M$ (assuming no volume change). Summarizing for each solution:

$[C_2H_3O_2^-] + [HC_2H_3O_2] = 1.0\,M$ and $[HC_2H_3O_2] = [H^+]$ added

a. $pH = pK_a + \log \dfrac{[C_2H_3O_2^-]}{[HC_2H_3O_2]}$; For $pH = pK_a$, $[C_2H_3O_2^-] = [HC_2H_3O_2]$.

For this to be true, $[C_2H_3O_2^-] = [HC_2H_3O_2] = 0.50\,M = [H^+]$ added, which means that 0.50 mol of HCl must have been added to 1.0 L of the initial solution to produce a solution with $pH = pK_a$.

b. $4.20 = 4.74 + \log \dfrac{[C_2H_3O_2^-]}{[HC_2H_3O_2]}, \quad \dfrac{[C_2H_3O_2^-]}{[HC_2H_3O_2]} = 10^{-0.54} = 0.29$

$[C_2H_3O_2^-] = 0.29 \, [HC_2H_3O_2]; \ \ 0.29 \, [HC_2H_3O_2] + [HC_2H_3O_2] = 1.0 \, M$

$[HC_2H_3O_2] = 0.78 \, M = [H^+]$ added

0.78 mol of HCl must be added to produce a solution with pH = 4.20.

c. $5.00 = 4.74 + \log \dfrac{[C_2H_3O_2^-]}{[HC_2H_3O_2]}, \quad \dfrac{[C_2H_3O_2^-]}{[HC_2H_3O_2]} = 10^{0.26} = 1.8$

$[C_2H_3O_2^-] = 1.8 \, [HC_2H_3O_2]; \ \ 1.8 \, [HC_2H_3O_2] + [HC_2H_3O_2] = 1.0 \, M$

$[HC_2H_3O_2] = 0.36 \, M = [H^+]$ added

0.36 mol of HCl must be added to produce a solution with pH = 5.00.

Acid-Base Titrations

55.

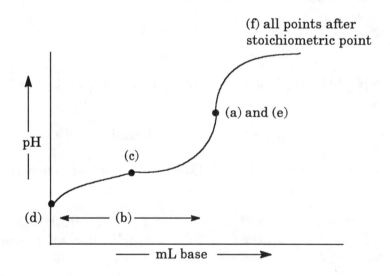

$HA + OH^- \rightarrow A^- + H_2O$; Added OH^- from the strong base converts the weak acid, HA, into its conjugate base, A^-. Initially before any OH^- is added (point d), HA is the dominant species present. After OH^- is added, both HA and A^- are present and a buffer solution results (region b). At the equivalence point (points a and e), exactly enough OH^- has been added to convert all of the weak acid, HA, into its conjugate base, A^-. Past the equivalence point (region f), excess OH^- is present. For the answer to part b, we included almost the entire buffer region. The maximum buffer region (or the region which is the best buffer solution) is around the halfway point to equivalence (point c). At this point, enough OH^- has been added to convert exactly one-half of the weak acid present initially into its conjugate base so [HA] = [A^-] and pH = pK_a. A best buffer has about equal concentrations of weak acid and conjugate base present.

56.

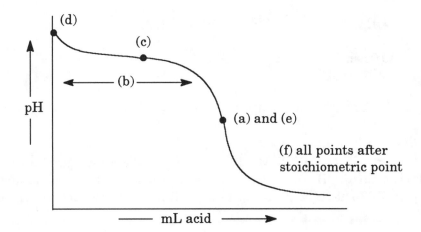

$B + H^+ \rightarrow BH^+$; Added H^+ from the strong acid converts the weak base, B, into its conjugate acid, BH^+. Initially before any H^+ is added (point d), B is the dominant species present. After H^+ is added, both B and BH^+ are present and a buffer solution results (region b). At the equivalence point (points a and e), exactly enough H^+ has been added to convert all of the weak base present initially into its conjugate acid, BH^+. Past the equivalence point (region f), excess H^+ is present. For the answer to b, we included almost the entire buffer region. The maximum buffer region is around the halfway point to equivalence (point c) where $[B] = [BH^+]$. Here, $pH = pK_a$ which is a characteristic of a best buffer.

57.   This is a strong acid (HCl) titrated by a strong base (NaOH). Added OH^- from the strong base will react completely with the H^+ present from the strong acid.

a.   Only strong acid present. $[H^+] = 0.100\ M$; $pH = 1.000$

b.   mmol OH^- added $= 20.0\ mL \times \dfrac{0.20\ mmol\ OH^-}{mL} = 4.00\ mmol\ OH^-$

mmol H^+ present $= 100.0\ mL \times \dfrac{0.100\ mmol\ H^+}{mL} = 10.0\ mmol\ H^+$

Note: The units mmoles are usually easier numbers to work with. The units for molarity are moles/L but are also equal to mmoles/mL.

$$H^+ \quad + \quad OH^- \quad \rightarrow \quad H_2O$$

	H^+	OH^-	
Before	10.0 mmol	4.00 mmol	
Change	-4.00 mmol	-4.00 mmol	Reacts completely
After	6.0 mmol	0	

The excess H^+ determines pH. $[H^+]_{excess} = \dfrac{6.0\ mmol\ H^+}{100.0\ mL + 20.0\ mL} = 0.050\ M$; $pH = 1.30$

c. mmol OH⁻ added = 30.0 mL × 0.20 M = 6.00 mmol OH⁻

$$H^+ \quad + \quad OH^- \quad \rightarrow \quad H_2O$$

Before   10.0 mmol   6.00 mmol
After   4.0 mmol   0

$[H^+]_{excess} = \dfrac{4.0 \text{ mmol}}{(100.0 + 30.0) \text{ mL}} = 0.031\, M;\ pH = 1.51$

d. mmol OH⁻ added = 50.0 mL × 0.20 M = 10.0 mmol OH⁻; This is the equivalence point since we have added just enough OH⁻ to react with all the acid present. For a strong acid-strong base titration, pH = 7.00 at the equivalence point since only neutral species are present (Na⁺, Cl⁻, H_2O).

e. mmol OH⁻ added = 75.0 mL × 0.200 M = 15.0 mmol OH⁻

$$H^+ \quad + \quad OH^- \quad \rightarrow \quad H_2O$$

Before   10.0 mmol   15.0 mmol
After   0   5.0 mmol

Past the equivalence point, the pH is determined by the excess OH⁻ present.

$[OH^-]_{excess} = \dfrac{5.0 \text{ mmol}}{(100.0 + 75.0) \text{ mL}} = 0.029\, M;\ pOH = 1.54;\ pH = 12.46$

58. This is a strong base, $Ca(OH)_2$, titrated by a strong acid, $HClO_4$. The added strong acid will neutralize the OH⁻ from the strong base. As is always the case when a strong acid and/or strong base reacts, the reaction is assumed to go to completion.

a. Only a strong base is present, but it breaks up into two mol of OH⁻ ions for every mol of $Ca(OH)_2$. [OH⁻] = 2 × 0.150 M = 0.300 M; pOH = 0.523; pH = 13.477

b. mmol OH⁻ present = 50.0 mL × $\dfrac{0.150 \text{ mmol } Ca(OH)_2}{mL}$ × $\dfrac{2 \text{ mmol OH}^-}{\text{mmol } Ca(OH)_2}$ = 15.0 mmol OH⁻

mmol H⁺ added = 50.0 mL × $\dfrac{0.100 \text{ mmol H}^+}{mL}$ = 5.00 mmol H⁺

$$OH^- \quad + \quad H^+ \quad \rightarrow \quad H_2O$$

Before   15.0 mmol   5.00 mmol
Change   -5.00 mmol   -5.00 mmol   Reacts completely
After   10.0 mmol   0

$[OH^-]_{excess} = \dfrac{10.0 \text{ mmol OH}^-}{50.0 \text{ mL} + 50.0 \text{ mL}} = 0.100\, M;\ pOH = 1.000,\ pH = 13.000$

c.   mmol H^+ added = 100.0 mL $\times$ 0.100 M = 10.0 mmol H^+

$$OH^- \quad + \quad H^+ \quad \rightarrow \quad H_2O$$

Before   15.0 mmol   10.0 mmol
After   5.0 mmol   0

$[OH^-]_{excess} = \dfrac{5.0 \text{ mmol } OH^-}{(50.0 + 100.0) \text{ mL}} = 0.033\ M$; pOH = 1.48; pH = 12.52

d.   mmol H^+ added = 150.0 mL $\times$ 0.100 M = 15.0 mmol H^+; This is the equivalence point. Since the H^+ will exactly neutralize the OH^- from the strong base, then all we have in solution is Ca^{2+}, ClO_4^- and H_2O. All are neutral species so pH = 7.00.

e.   mmol H^+ added = 200.0 mL $\times$ 0.100 M = 20.0 mmol H^+

$$OH^- \quad + \quad H^+ \quad \rightarrow \quad H_2O$$

Before   15.0 mmol   20.0 mmol
After   0   5.0 mmol

$[H^+]_{excess} = \dfrac{5.0 \text{ mmol}}{(50.0 + 200.0) \text{ mL}} = 0.020\ M$; pH = 1.70

59.   This is a weak acid ($HC_2H_3O_2$) titrated by a strong base (NaOH).

a.   Only weak acid is present. Solving the weak acid problem:

$$HC_2H_3O_2 \quad \rightleftharpoons \quad H^+ \quad + \quad C_2H_3O_2^-$$

Initial   0.200 M ~0   0
               x mol/L $HC_2H_3O_2$ dissociates to reach equilibrium
Change   -x $\rightarrow$ +x +x
Equil.   0.200 - x x x

$K_a = 1.8 \times 10^{-5} = \dfrac{x^2}{0.200 - x} = \dfrac{x^2}{0.200}$, $x = [H^+] = 1.9 \times 10^{-3}\ M$

pH = 2.72; Assumptions good.

b.   The added OH^- will react completely with the best acid present, $HC_2H_3O_2$.

mmol $HC_2H_3O_2$ present = 100.0 mL $\times \dfrac{0.200 \text{ mmol } HC_2H_3O_2}{\text{mL}} = 20.0$ mmol $HC_2H_3O_2$

mmol OH^- added = 50.0 mL $\times \dfrac{0.100 \text{ mmol } OH^-}{\text{mL}} = 5.00$ mmol OH^-

$$HC_2H_3O_2 \quad + \quad OH^- \quad \rightarrow \quad C_2H_3O_2^- \quad + \quad H_2O$$

Before	20.0 mmol	5.00 mmol	0	
Change	-5.00 mmol	-5.00 mmol $\rightarrow$	+5.00 mmol	Reacts completely
After	15.0 mmol	0	5.00 mmol	

After reaction of all the strong base, we have a buffer solution containing a weak acid ($HC_2H_3O_2$) and its conjugate base ($C_2H_3O_2^-$). We will use the Henderson-Hasselbalch equation to solve for the pH.

$$pH = pK_a + \log \frac{[C_2H_3O_2^-]}{[HC_2H_3O_2]} = -\log (1.8 \times 10^{-5}) + \log \left(\frac{5.00 \text{ mmol}/V_T}{15.0 \text{ mmol}/V_T} \right) \text{ where } V_T = \text{total volume}$$

$$pH = 4.74 + \log \left(\frac{5.00}{15.0} \right) = 4.74 + (-0.477) = 4.26$$

Note that the total volume cancels in the Henderson-Hasselbalch equation. For the [base]/[acid] term, the mole ratio equals the concentration ratio since the components of the buffer are always in the same volume of solution.

c.   mmol OH^- added = 100.0 mL $\times$ 0.100 mmol OH^-/mL = 10.0 mmol OH^-; The same amount (20.0 mmol) of $HC_2H_3O_2$ is present as before (it never changes). As before, let the OH^- react to completion, then see what is remaining in solution after this reaction.

$$HC_2H_3O_2 \quad + \quad OH^- \quad \rightarrow \quad C_2H_3O_2^- \quad + \quad H_2O$$

Before	20.0 mmol	10.0 mmol	0	
After	10.0 mmol	0	10.0 mmol	

A buffer solution results after reaction. Since $[C_2H_3O_2^-] = [HC_2H_3O_2] = 10.0$ mmol/total volume, then $pH = pK_a$. This is always true at the halfway point to equivalence for a weak acid/strong base titration, $pH = pK_a$.

$$pH = -\log (1.8 \times 10^{-5}) = 4.74$$

d.   mmol OH^- added = 150.0 mL $\times$ 0.100 M = 15.0 mmol OH^-. Added OH^- reacts completely with the weak acid.

$$HC_2H_3O_2 \quad + \quad OH^- \quad \rightarrow \quad C_2H_3O_2^- \quad + \quad H_2O$$

Before	20.0 mmol	15.0 mmol	0	
After	5.0 mmol	0	15.0 mmol	

We have a buffer solution after all the OH^- reacts to completion. Using the Henderson-Hasselbalch equation:

$$pH = 4.74 + \log \frac{[C_2H_3O_2^-]}{[HC_2H_3O_2]} = 4.74 + \log \left(\frac{15.0 \text{ mmol}}{5.0 \text{ mmol}} \right) \quad \text{(Total volume cancels, so we can use mol ratios.)}$$

pH = 4.74 + 0.48 = 5.22

e.   mmol OH⁻ added = 200.00 mL × 0.100 M = 20.0 mmol OH⁻; As before, let the added OH⁻ react
to completion with the weak acid, then see what is in solution after this reaction.

$$HC_2H_3O_2 \quad + \quad OH^- \quad \rightarrow \quad C_2H_3O_2^- \quad + \quad H_2O$$

Before	20.0 mmol	20.0 mmol	0
After	0	0	20.0 mmol

This is the equivalence point. Enough OH⁻ has been added to exactly neutralize all the weak acid
present initially. All that remains that affects the pH at the equivalence point is the conjugate
base of the weak acid, $C_2H_3O_2^-$. This is a weak base equilibrium problem.

$$C_2H_3O_2^- + H_2O \rightleftharpoons HC_2H_3O_2 + OH^- \quad K_b = \frac{K_w}{K_a} = \frac{1.0 \times 10^{-14}}{1.8 \times 10^{-5}} = 5.6 \times 10^{-10}$$

Initial	20.0 mmol/300.0 mL	0	0
	x mol/L $C_2H_3O_2^-$ reacts with H_2O to reach equilibrium		
Change	$-x$ $\rightarrow$	$+x$	$+x$
Equil.	0.0667 - x	x	x

$$K_b = 5.6 \times 10^{-10} = \frac{x^2}{0.0667 - x} \approx \frac{x^2}{0.0667}, \quad x = [OH^-] = 6.1 \times 10^{-6}\, M$$

pOH = 5.21; pH = 8.79; Assumptions good.

f.   mmol OH⁻ added = 250.0 mL × 0.100 M = 25.0 mmol OH⁻

$$HC_2H_3O_2 \quad + \quad OH^- \quad \rightarrow \quad C_2H_3O_2^- \quad + \quad H_2O$$

Before	20.0 mmol	25.0 mmol	0
After	0	5.0 mmol	20.0 mmol

After the titration reaction, we have a solution containing excess OH⁻ and a weak base, $C_2H_3O_2^-$.
When a strong base and a weak base are both present, assume the amount of OH⁻ added from the
weak base will be minimal, i.e., the pH past the equivalence point is determined by the amount of
excess base.

$$[OH^-]_{excess} = \frac{5.0\ mmol}{100.0\ mL + 250.0\ mL} = 0.014\ M;\ pOH = 1.85;\ pH = 12.15$$

60.   This is a weak base ($HONH_2$) titrated by a strong acid (HNO_3). To calculate the pH at the various
points, let the strong acid react completely with the weak base present, then see what is in solution.

a.   Only a weak base is present. Solve the weak base equilibrium problem.

$$HONH_2 \;+\; H_2O \;\rightleftharpoons\; HONH_3^+ \;+\; OH^-$$

Initial	0.200 M	0	~0
Equil.	0.200 - x	x	x

$$K_b = 1.1 \times 10^{-8} = \frac{x^2}{0.200 - x} \approx \frac{x^2}{0.200}, \; x = [OH^-] = 4.7 \times 10^{-5} \, M$$

pOH = 4.33; pH = 9.67; Assumptions good.

b.   mmol $HONH_2$ present = 50.0 mL $\times \dfrac{0.200 \text{ mmol } HONH_2}{mL}$ = 10.0 mmol $HONH_2$

mmol H^+ added = 25.0 mL $\times \dfrac{0.100 \text{ mmol } H^+}{mL}$ = 2.50 mmol H^+

$$HONH_2 \quad + \quad H^+ \quad \rightarrow \quad HONH_3^+$$

	$HONH_2$	H^+		$HONH_3^+$	
Before	10.0 mmol	2.50 mmol		0	
Change	-2.50 mmol	-2.50 mmol	$\rightarrow$	+2.50 mmol	Reacts completely
After	7.5 mmol	0		2.50 mmol	

A buffer solution results after the titration reaction. Solving using the Henderson-Hasselbalch equation:

$$pH = pK_a + \log \frac{[base]}{[acid]}, \quad K_a = \frac{K_w}{K_b} = \frac{1.0 \times 10^{-14}}{1.1 \times 10^{-8}} = 9.1 \times 10^{-7}$$

$$pH = -\log (9.1 \times 10^{-7}) + \log \left(\frac{7.5 \text{ mmol}/V_T}{2.50 \text{ mmol}/V_T} \right) \text{ where } V_T = \text{total volume}$$

$$pH = 6.04 + \log (3.0) = 6.04 + 0.48 = 6.52$$

c.   mmol H^+ added = 50.0 mL $\times$ 0.100 M = 5.00 mmol H^+

$$HONH_2 \quad + \quad H^+ \quad \rightarrow \quad HONH_3^+$$

	$HONH_2$	H^+	$HONH_3^+$
Before	10.0 mmol	5.00 mmol	0
After	5.0 mmol	0	5.00 mmol

This is the halfway point to equivalence since $[HONH_3^+] = [HONH_2]$. At this point, pH = pK_a (which is characteristic of the halfway point for any weak base/strong acid titration).

$$pH = -\log (9.1 \times 10^{-7}) = 6.04$$

d.   mmol H^+ added = 75.0 mL × 0.100 M = 7.50 mmol H^+

	$HONH_2$	+	H^+	→	$HONH_3^+$
Before	10.0 mmol		7.50 mmol		0
After	2.5 mmol		0		7.50 mmol

A buffer solution results.

$$pH = pK_a + \log \frac{[base]}{[acid]} = 6.04 + \log \left(\frac{2.5 \text{ mmol}/V_T}{7.50 \text{ mmol}/V_T} \right) = 6.04 + (-0.48) = 5.56$$

e.   mmol H^+ added = 100.0 mL × 0.100 M = 10.0 mmol H^+

	$HONH_2$	+	H^+	→	$HONH_3^+$
Before	10.0 mmol		10.0 mmol		0
After	0		0		10.0 mmol

As is always the case in a weak base/strong acid titration, the pH at the equivalence point is acidic because only a weak acid ($HONH_3^+$) is present. Solving the weak acid equilibrium problem:

	$HONH_3^+$	⇌	H^+	+	$HONH_2$
Initial	10.0 mmol/150.0 mL		0		0
Equil.	0.0667 - x		x		x

$$K_a = 9.1 \times 10^{-7} = \frac{x^2}{0.0667 - x} \approx \frac{x^2}{0.0667}, \quad x = [H^+] = 2.5 \times 10^{-4} M$$

pH = 3.60; Assumptions good.

f.   mmol H^+ added = 125.0 mL × 0.100 M = 12.5 mmol H^+

	$HONH_2$	+	H^+	→	$HONH_3^+$
Before	10.0 mmol		12.5 mmol		0
After	0		2.5 mmol		10.0 mmol

Two acids are present past the equivalence point, but the excess H^+ will determine the pH of the solution since $HONH_3^+$ is a weak acid.

$$[H^+]_{excess} = \frac{2.5 \text{ mmol}}{50.0 \text{ mL} + 125.0 \text{ mL}} = 0.014 \, M; \quad pH = 1.85$$

61.   We will do sample calculations for the various parts of the titration. All results are summarized in
       Table 15.1 at the end of Exercise 15.64.

       At the beginning of the titration, only the weak acid $HC_3H_5O_3$ is present.

$$HLac \rightleftharpoons H^+ + Lac^- \quad K_a = 10^{-3.86} = 1.4 \times 10^{-4} \quad HLac = HC_3H_5O_3$$
$$Lac^- = C_3H_5O_3^-$$

Initial	$0.100\,M$	~ 0	0

       x mol/L HLac dissociates to reach equilibrium

| Change | $-x$ | $\rightarrow$ $+x$ | $+x$ |
| Equil. | $0.100 - x$ | x | x |

$$1.4 \times 10^{-4} = \frac{x^2}{0.100 - x} = \frac{x^2}{0.100}, \quad x = [H^+] = 3.7 \times 10^{-3}\,M;\ pH = 2.43 \quad \text{Assumptions good.}$$

Up to the stoichiometric point, we calculate the pH using the Henderson-Hasselbalch equation. This
is the buffer region. For example, at 4.0 mL of NaOH added:

$$\text{initial mmol HLac present} = 25.0\ mL \times \frac{0.100\ mmol}{mL} = 2.50\ mmol\ HLac$$

$$\text{mmol } OH^- \text{ added} = 4.0\ mL \times \frac{0.100\ mmol}{mL} = 0.40\ mmol\ OH^-$$

Note: The units mmol are usually easier numbers to work with. The units for molarity are moles/L
but are also equal to mmoles/mL.

The 0.40 mmol added OH^- converts 0.40 mmoles HLac to 0.40 mmoles Lac^- according to the
equation:

$$HLac + OH^- \rightarrow Lac^- + H_2O \qquad \text{Reacts completely}$$

mmol HLac remaining = 2.50 - 0.40 = 2.10 mmol; mmol Lac^- produced = 0.40 mmol

We have a buffer solution. Using the Henderson-Hasselbalch equation where $pK_a = 3.86$:

$$pH = pK_a + \log \frac{[Lac^-]}{[HLac]} = 3.86 + \log \frac{(0.40)}{(2.10)} \qquad \text{(Total volume cancels, so we can use}$$
$$\text{use the ratio of moles or mmoles.)}$$

$$pH = 3.86 - 0.72 = 3.14$$

Other points in the buffer region are calculated in a similar fashion. Perform a stoichiometry
problem first, followed by a buffer problem. The buffer region includes all points up to 24.9 mL
OH^- added.

At the stoichiometric point (25.0 mL OH^- added), we have added enough OH^- to convert all of the
HLac (2.50 mmol) into its conjugate base, Lac^-. All that is present is a weak base. To determine the
pH, we perform a weak base calculation.

$$[Lac^-]_o = \frac{2.50\ mmol}{25.0\ mL + 25.0\ mL} = 0.0500\,M$$

$$Lac^- + H_2O \; \rightleftharpoons \; HLac + OH^- \qquad K_b = \frac{1.0 \times 10^{-14}}{1.4 \times 10^{-4}} = 7.1 \times 10^{-11}$$

Initial	$0.0500\,M$		0	0

x mol/L Lac$^-$ reacts with H_2O to reach equilibrium

Change	$-x$	$\rightarrow$	$+x$	$+x$
Equil.	$0.0500 - x$		x	x

$$K_b = \frac{x^2}{0.0500 - x} = \frac{x^2}{0.0500} = 7.1 \times 10^{-11}$$

$x = [OH^-] = 1.9 \times 10^{-6}\,M$; pOH = 5.72; pH = 8.28   Assumptions good.

Past the stoichiometric point, we have added more than 2.50 mmol of NaOH. The pH will be determined by the excess OH^- ion present. An example of this calculation follows.

At 25.1 mL: OH^- added = $25.1\,\text{mL} \times \dfrac{0.100\,\text{mmol}}{\text{mL}} = 2.51$ mmol OH^-; 2.50 mmol OH^- neutralizes all the weak acid present. The remainder is excess OH^-. excess $OH^- = 2.51 - 2.50 = 0.01$ mmol

$$[OH^-]_{excess} = \frac{0.01\,\text{mmol}}{(25.0 + 25.1)\,\text{mL}} = 2 \times 10^{-4}\,M; \text{ pOH} = 3.7; \text{ pH} = 10.3$$

All results are listed in Table 15.1 at the end of the solution to Exercise 15.64.

62.   Results for all points are summarized in Table 15.1 at the end of the solution to Exercise 15.64. At the beginning of the titration, we have a weak acid problem:

$$HOPr \; \rightleftharpoons \; H^+ + OPr^- \qquad\qquad HOPr = HC_3H_5O_2$$
$$OPr^- = C_3H_5O_2^-$$

Initial	$0.100\,M$		~ 0	0

x mol/L HOPr acid dissociates to reach equilibrium

Change	$-x$	$\rightarrow$	$+x$	$+x$
Equil.	$0.100 - x$		x	x

$$K_a = \frac{[H^+][OPr^-]}{[HOPr]} = 1.3 \times 10^{-5} = \frac{x^2}{0.100 - x} \approx \frac{x^2}{0.100}$$

$x = [H^+] = 1.1 \times 10^{-3}\,M$; pH = 2.96   Assumptions good.

The buffer region is from 4.0 - 24.9 mL of OH^- added. We will do a sample calculation at 24.0 mL OH^- added.

$$\text{initial mmol HOPr present} = 25.0\,\text{mL} \times \frac{0.100\,\text{mmol}}{\text{mL}} = 2.50\text{ mmol HOPr}$$

$$\text{mmol } OH^- \text{ added} = 24.0\,\text{mL} \times \frac{0.100\,\text{mmol}}{\text{mL}} = 2.40\text{ mmol } OH^-$$

The added strong base converts HOPr into OPr⁻.

$$HOPr \quad + \quad OH^- \quad \rightarrow \quad OPr^- + H_2O$$

	HOPr	OH⁻		OPr⁻	
Before	2.50 mmol	2.40 mmol		0	
Change	-2.40	-2.40	$\rightarrow$	+2.40	Reacts completely
After	0.10 mmol	0		2.40 mmol	

A buffer solution results. Using the Henderson-Hasselbalch equation where $pK_a = -\log(1.3 \times 10^{-5}) = 4.89$:

$$pH = pK_a + \log \frac{[Base]}{[Acid]} = 4.89 + \log \frac{[OPr^-]}{[HOPr]}$$

$$pH = 4.89 + \log \left(\frac{2.40}{0.10} \right) = 4.89 + 1.38 = 6.27 \quad \text{(Volume cancels, so we can use the mol ratio.)}$$

All points in the buffer region, 4.0 mL to 24.9 mL, are calculated this way. See Table 15.1 at the end of Exercise 15.64 for all the results.

At the stoichiometric point, only a weak base (OPr⁻) present:

$$OPr^- \quad + \quad H_2O \quad \rightleftharpoons \quad OH^- \quad + \quad HOPr$$

	OPr⁻	OH⁻	HOPr
Initial	$\frac{2.50 \text{ mmol}}{50.0 \text{ mL}} = 0.0500\,M$	0	0

x mol/L OPr⁻ reacts with H_2O to reach equilibrium

	OPr⁻		OH⁻	HOPr
Change	-x	$\rightarrow$	+x	+x
Equil.	0.0500 - x		x	x

$$K_b = \frac{[OH^-][HOPr]}{[OPr^-]} = \frac{K_w}{K_a} = 7.7 \times 10^{-10} = \frac{x^2}{0.0500 - x} \approx \frac{x^2}{0.0500}$$

$$x = 6.2 \times 10^{-6}\,M = [OH^-], \quad pOH = 5.21, \quad pH = 8.79 \qquad \text{Assumptions good.}$$

Beyond the stoichiometric point, the pH is determined by the excess strong base added. The results are the same as those in Exercise 15.61 (see Table 15.1).

For example at 26.0 mL NaOH added:

$$[OH^-] = \frac{2.60 \text{ mmol} - 2.50 \text{ mmol}}{(25.0 + 26.0) \text{ mL}} = 2.0 \times 10^{-3}\,M; \quad pOH = 2.70; \quad pH = 11.30$$

63.   At beginning of the titration, only the weak base NH_3 is present. As always, solve for the pH using the K_b reaction for NH_3.

$$NH_3 + H_2O \rightleftharpoons NH_4^+ + OH^- \quad K_b = 1.8 \times 10^{-5}$$

Initial	$0.100\,M$	0	~ 0
Equil.	$0.100 - x$	x	x

$$K_b = \frac{x^2}{0.100 - x} = \frac{x^2}{0.100} = 1.8 \times 10^{-5}$$

$x = [OH^-] = 1.3 \times 10^{-3}\,M;\ pOH = 2.89;\ pH = 11.11$ Assumptions good.

In the buffer region (4.0 - 24.9 mL), we can use the Henderson-Hasselbalch equation:

$$K_a = \frac{1.0 \times 10^{-14}}{1.8 \times 10^{-5}} = 5.6 \times 10^{-10};\ pK_a = 9.25;\ pH = 9.25 + \log \frac{[NH_3]}{[NH_4^+]}$$

We must determine the amounts of NH_3 and NH_4^+ present after the added H^+ reacts completely with the NH_3. For example, after 8.0 mL HCl added:

$$\text{initial mmol } NH_3 \text{ present} = 25.0 \text{ mL} \times \frac{0.100 \text{ mmol}}{\text{mL}} = 2.50 \text{ mmol } NH_3$$

$$\text{mmol } H^+ \text{ added} = 8.0 \text{ mL} \times \frac{0.100 \text{ mmol}}{\text{mL}} = 0.80 \text{ mmol } H^+$$

Added H^+ reacts with NH_3 to completion: $NH_3 + H^+ \rightarrow NH_4^+$

mmol NH_3 remaining = 2.50 - 0.80 = 1.70 mmol; mmol NH_4^+ produced = 0.80 mmol

$$pH = 9.25 + \log \frac{1.70}{0.80} = 9.58 \quad \text{(Mole ratios can be used since the total volume cancels.)}$$

Other points in the buffer region are calculated in similar fashion. Results are summarized in Table 15.1 at the end of Exercise 15.64.

At the stoichiometric point (25.0 mL H^+ added), just enough HCl has been added to convert all of the weak base (NH_3) into its conjugate acid (NH_4^+). Perform a weak acid calculation. $[NH_4^+]_o = 2.50$ mmol/50.0 mL = $0.0500\,M$

$$NH_4^+ \rightleftharpoons H^+ + NH_3 \quad K_a = 5.6 \times 10^{-10}$$

Initial	$0.0500\,M$	0	0
Equil.	$0.0500 - x$	x	x

$$5.6 \times 10^{-10} = \frac{x^2}{0.0500 - x} = \frac{x^2}{0.0500}, \quad x = [H^+] = 5.3 \times 10^{-6}\,M;\ pH = 5.28 \quad \text{Assumptions good.}$$

Beyond the stoichiometric point, the pH is determined by the excess H^+. For example, at 28.0 mL of H^+ added:

$$H^+ \text{ added} = 28.0 \text{ mL} \times \frac{0.100 \text{ mmol}}{\text{mL}} = 2.80 \text{ mmol } H^+$$

$$\text{Excess } H^+ = 2.80 \text{ mmol} - 2.50 \text{ mmol} = 0.30 \text{ mmol excess } H^+$$

$$[H^+]_{excess} = \frac{0.30 \text{ mmol}}{(25.0 + 28.0) \text{ mL}} = 5.7 \times 10^{-3} M; \text{ pH} = 2.24$$

All results are summarized in Table 15.1.

64.   Initially, a weak base problem:

$$\text{py} \quad + \quad H_2O \quad \rightleftharpoons \quad Hpy^+ \quad + \quad OH^- \qquad \text{py is pyridine}$$

Initial	0.100 *M*		0	~0
Equil.	0.100 - *x*		*x*	*x*

$$K_b = \frac{[Hpy^+][OH^-]}{[py]} = \frac{x^2}{0.100 - x} = \frac{x^2}{0.100} = 1.7 \times 10^{-9}$$

$x = [OH^-] = 1.3 \times 10^{-5} M$; pOH = 4.89; pH = 9.11   Assumptions good.

Buffer region (4.0 - 24.5 mL): Added H^+ reacts completely with py: $py + H^+ \rightarrow Hpy^+$. Determine the moles (or mmoles) of py and Hpy^+ after reaction and use the Henderson-Hasselbalch equation to solve for the pH.

$$K_a = \frac{K_w}{K_b} = \frac{1.0 \times 10^{-14}}{1.7 \times 10^{-9}} = 5.9 \times 10^{-6}; \text{ pK}_a = 5.23; \text{ pH} = 5.23 + \log \frac{[py]}{[Hpy^+]}$$

Results in the buffer region are summarized in Table 15.1 that follows this problem. See Exercise 15.63 for a similar sample calculation.

At the stoichiometric point (25.0 mL H^+ added), this is a weak acid problem since just enough H^+ has been added to convert all of the weak base into its conjugate acid. The initial concentration of $[Hpy^+] = 0.0500 M$.

$$Hpy^+ \quad \rightleftharpoons \quad \text{py} \quad + \quad H^+ \qquad K_a = 5.9 \times 10^{-6}$$

Initial	0.0500 *M*		0	0
Equil.	0.0500 - *x*		*x*	*x*

$$5.9 \times 10^{-6} = \frac{x^2}{0.0500 - x} = \frac{x^2}{0.0500}, \; x = [H^+] = 5.4 \times 10^{-4} M; \text{ pH} = 3.27 \quad \text{Assumptions good.}$$

Beyond the equivalence point, the pH determination is made by calculating the concentration of excess H^+. See Exercise 15.63 for an example. All results are summarized in Table 15.1.

Table 15.1: Summary of pH Results for Exercises 15.61 - 15.64   (Graph follows)

titrant mL	Exercise 15.61	Exercise 15.62	Exercise 15.63	Exercise 15.64
0.0	2.43	2.96	11.11	9.11
4.0	3.14	4.17	9.97	5.95
8.0	3.53	4.56	9.58	5.56
12.5	3.86	4.89	9.25	5.23
20.0	4.46	5.49	8.65	4.63
24.0	5.24	6.27	7.87	3.85
24.5	5.6	6.6	7.6	3.5
24.9	6.3	7.3	6.9	-
25.0	8.28	8.79	5.28	3.27
25.1	10.3	10.3	3.7	-
26.0	11.29	11.30	2.71	2.71
28.0	11.75	11.75	2.24	2.25
30.0	11.96	11.96	2.04	2.04

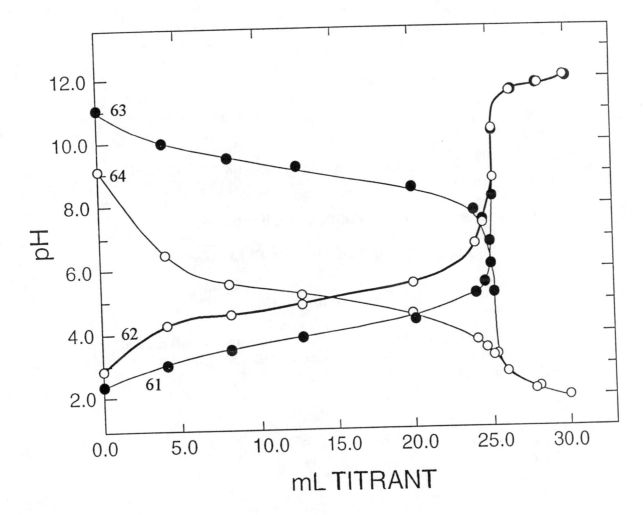

65.   a.   This is a weak acid/strong base titration. At the halfway point to equivalence, [weak acid] = [conjugate base], so pH = pK_a (always for a weak acid/strong base titration).

pH = -log (6.4×10^{-5}) = 4.19

mmol $HC_7H_5O_2$ present = 100. mL $\times$ 0.10 M = 10. mmol $HC_7H_5O_2$. For the equivalence point, 10. mmol of OH^- must be added. The volume of OH^- added to reach the equivalence point is:

$$10. \text{ mmol } OH^- \times \frac{1 \text{ mL}}{0.10 \text{ mmol } OH^-} = 1.0 \times 10^2 \text{ mL } OH^-$$

At the equivalence point, 10. mmol of $HC_7H_5O_2$ is neutralized by 10. mmol of OH^- to produce 10. mmol of $C_7H_5O_2^-$. This is a weak base. The total volume of the solution is 100.0 mL + 1.0 $\times 10^2$ mL = 2.0 $\times 10^2$ mL. Solving the weak base equilibrium problem:

$$C_7H_5O_2^- + H_2O \rightleftharpoons HC_7H_5O_2 + OH^- \quad K_b = \frac{K_w}{K_a} = \frac{1.0 \times 10^{-14}}{6.4 \times 10^{-5}} = 1.6 \times 10^{-10}$$

Initial   10. mmol/2.0 $\times 10^2$ mL   0   0
Equil.   0.050 - x x x

$$K_b = 1.6 \times 10^{-10} = \frac{x^2}{0.050 - x} \approx \frac{x^2}{0.050}, \quad x = [OH^-] = 2.8 \times 10^{-6} \, M$$

pOH = 5.55; pH = 8.45   Assumptions good.

b.   At the halfway point to equivalence for a weak base/strong acid titration, pH = pK_a since [weak base] = [conjugate acid].

$$K_a = \frac{K_w}{K_b} = \frac{1.0 \times 10^{-14}}{5.6 \times 10^{-4}} = 1.8 \times 10^{-11}; \quad pH = pK_a = -\log(1.8 \times 10^{-11}) = 10.74$$

For the equivalence point (mmol acid added = mmol base present):

mmol $C_2H_5NH_2$ present = 100.0 mL $\times$ 0.10 M = 10. mmol $C_2H_5NH_2$

$$\text{mL } HNO_3 \text{ added} = 10. \text{ mmol } H^+ \times \frac{1 \text{ mL}}{0.20 \text{ mmol } H^+} = 50. \text{ mL } H^+$$

The strong acid added completely converts the weak base into its conjugate acid. Therefore, at the equivalence point, $[C_2H_5NH_3^+]_o$ = 10. mmol/(100.0 + 50.) mL = 0.067 M. Solving the weak acid equilibrium problem:

$$C_2H_5NH_3^+ \rightleftharpoons H^+ + C_2H_5NH_2$$

Initial   0.067 M 0   0
Equil.   0.067 - x x x

$$K_a = 1.8 \times 10^{-11} = \frac{x^2}{0.067 - x} \approx \frac{x^2}{0.067}, \quad x = [H^+] = 1.1 \times 10^{-6} \, M$$

pH = 5.96; Assumptions good.

c.   In a strong acid/strong base titration, the halfway point has no special significance other than exactly one-half of the original amount of acid present has been neutralized.

mmol H^+ present = 100.0 mL × 0.50 M = 50. mmol H^+

mL OH^- added = 25. mmol OH^- × $\dfrac{1\ mL}{0.25\ mmol}$ = 1.0 × 10² mL OH^-

$$H^+ \quad + \quad OH^- \quad \rightarrow \quad H_2O$$

	50. mmol	25 mmol
Before	50. mmol	25 mmol
After	25 mmol	0

$[H^+]_{excess} = \dfrac{25\ mmol}{(100.0 + 1.0 \times 10^2)\ mL}$ = 0.13 M; pH = 0.89

At the equivalence point of a strong acid/strong base titration, only neutral speices are present (Na^+, Cl^-, H_2O) so the pH = 7.00.

66.   50.0 mL × 1.0 M = 50. mmol CH_3NH_2 present initially

a.   50.0 mL × 0.50 M = 25. mmol HCl added. This is the halfway point to equivalence where $[CH_3NH_2]$ = $[CH_3NH_3^+]$.

pH = pK_a + log $\dfrac{[CH_3NH_2]}{[CH_3NH_3^+]}$ = pK_a; K_a = $\dfrac{1.0 \times 10^{-14}}{4.4 \times 10^{-4}}$ = 2.3 × 10⁻¹¹

pH = pK_a = -log (2.3 × 10⁻¹¹) = 10.64

b.   It will take 100. mL of HCl solution to reach the stoichiometric point.

$[CH_3NH_3^+]_o$ = $\dfrac{50.\ mmol}{150.\ mL}$ = 0.33 M

$$CH_3NH_3^+ \quad \rightleftharpoons \quad H^+ \quad + \quad CH_3NH_2 \qquad K_a = \dfrac{K_w}{K_b} = 2.3 \times 10^{-11}$$

	0.33 M	0	0
Initial	0.33 M	0	0
Equil.	0.33 - x	x	x

2.3 × 10⁻¹¹ = $\dfrac{x^2}{0.33 - x}$ ≈ $\dfrac{x^2}{0.33}$, x = $[H^+]$ = 2.8 × 10⁻⁶ M; pH = 5.55   Assumptions good.

67.   mmol OH^- added = 40.0 mL × $\dfrac{0.100\ mmol\ OH^-}{mL}$ = 4.00 mmol OH^-

The added OH^- will react completely with the weak acid, HA.

$$HA \quad + \quad OH^- \quad \rightarrow \quad A^- \quad + \quad H_2O$$

Before	10.0 mmol	4.00 mmol	0
After	6.0 mmol	0	4.00 mmol

After reaction, a buffer solution exists.

$$pH = pK_a + \log \frac{[base]}{[acid]}, \quad 4.00 = pK_a + \log \left(\frac{4.00 \ mmol/V_T}{6.0 \ mmol/V_T} \right)$$

$$pK_a = 4.00 - \log (4.00/6.0) = 4.00 - (-0.18) = 4.18; \quad K_a = 10^{-4.18} = 6.6 \times 10^{-5}$$

68. a. 500.0 mL of HCl added represents the halfway point to equivalence. At the halfway point to equivalence in a weak base/strong acid titration, $pH = pK_a$. So, $pH = pK_a = 5.00$ and $K_a = 1.0 \times 10^{-5}$. Since $K_a = 1.0 \times 10^{-5}$ for the acid, then K_b for the conjugate base, A^-, equals 1.0×10^{-9} ($K_a \times K_b = K_w$).

   b. Added H^+ converts A^- into HA. At the equivalence point, only HA is present. The moles of HA present equals the mol of H^+ added (= 1.00 L $\times$ 0.100 mol/L = 0.100 mol). The concentration of HA at the equivalence point is:

$$[HA]_o = \frac{0.100 \ mol}{1.10 \ L} = 0.0909 \ M$$

Solving the weak acid equilibrium problem:

$$HA \quad \rightleftharpoons \quad H^+ \quad + \quad A^- \qquad K_a = 1.0 \times 10^{-5}$$

Initial	0.0909 M	0	0
Equil.	0.0909 - x	x	x

$$1.0 \times 10^{-5} = \frac{x^2}{0.0909 - x} \approx \frac{x^2}{0.0909}$$

$$x = 9.5 \times 10^{-4} M = [H^+]; \quad pH = 3.02 \quad \text{Assumptions good.}$$

Indicators

69. $HIN \rightleftharpoons In^- + H^+ \quad K_a = \dfrac{[In^-][H^+]}{[HIn]} = 1.0 \times 10^{-9}$

   a. In a very acid solution, the HIn form dominates so the solution will be yellow.

   b. The color change occurs when the concentration of the more dominate form is approximately ten times as great as the less dominant form of the indicator.

$$\frac{[HIn]}{[In^-]} = \frac{10}{1}; \quad K_a = 1.0 \times 10^{-9} = \left(\frac{1}{10} \right) [H^+], \quad [H^+] = 1 \times 10^{-8} M; \quad pH = 8.0 \ \text{at color change}$$

c. This is way past the equivalence point (100.0 mL OH⁻ added) so the solution is very basic and the In⁻ form of the indicator dominates. The solution will be blue.

70. The color of the indicator will change over the approximate range of pH = pK$_a$ ± 1 = 5.3 ± 1. Therefore, the useful pH range of methyl red where it changes color would be about: 4.3 (red) - 6.3 (yellow). Note that at pH < 4.3, the HIn form of the indicator dominates and the color of the solution is the color of HIn (red). At pH > 6.3, the In⁻ form of the indicator dominates and the color of the solution is the color of In⁻ (yellow). In titrating a weak acid with base, we start off with an acidic solution with pH < 4.3 so the color would change from red to orange at pH ~ 4.3. In titrating a weak base with acid, the color change would be from yellow to orange at pH ~ 6.3. Only a weak base/strong acid titration would have an acidic pH at the equivalence point so only in this type of titration would the color change of methyl red indicate the approximate endpoint.

71. When choosing an indicator, we want the color change of the indicator to occur approximately at the pH of the equivalence point. Since the pH generally changes very rapidly at the equivalence point, we don't have to be exact. This is especially true for strong acid/strong base titrations. Some choices where color change occurs at about the pH of the equivalence point are:

Exercise	pH at eq. pt.	Indicator
15.57	7.00	bromthymol blue or phenol red
15.59	8.79	o-cresolphthalein or phenolphthalein

72.

Exercise	pH at eq. pt.	Indicator
15.58	7.00	bromthymol blue or phenol red
15.60	3.60	bromphenol blue or methyl orange

73.

Exercise	pH at eq. pt.	Indicator
15.61	8.28	phenolphthalein
15.63	5.28	bromcresol green

74.

Exercise	pH at eq. pt.	Indicator
15.62	8.79	phenolphthalein
15.64	3.27	2,4-dinitrophenol

The titration in 15.64 is not feasible. The pH break at the equivalence point is too small.

75. The pH will be less than about 0.5 since crystal violet is yellow at a pH less than about 0.5. The methyl orange result only tells us that the pH is less than about 3.5.

76. pH > 5 for bromcresol green to be blue. pH < 8 for thymol blue to be yellow. The pH is between 5 and 8.

77. a. yellow   b. orange   c. blue   d. bluish-green

78.   a.   yellow   b.   yellow

        c.   green (Both yellow and blue forms are present.)   d.   colorless

Solubility Equilibria

79.   a. $AgC_2H_3O_2(s) \rightleftharpoons Ag^+(aq) + C_2H_3O_2^-(aq)$ $K_{sp} = [Ag^+][C_2H_3O_2^-]$

        b. $Al(OH)_3(s) \rightleftharpoons Al^{3+}(aq) + 3\ OH^-(aq)$ $K_{sp} = [Al^{3+}][OH^-]^3$

        c. $Ca_3(PO_4)_2(s) \rightleftharpoons 3\ Ca^{2+}(aq) + 2\ PO_4^{3-}(aq)$ $K_{sp} = [Ca^{2+}]^3[PO_4^{3-}]^2$

80.   a. $PbI_2(s) \rightleftharpoons Pb^{2+}(aq) + 2\ I^-(aq)$ $K_{sp} = [Pb^{2+}][I^-]^2$

        b. $Cu_2S(s) \rightleftharpoons 2\ Cu^+(aq) + S^{2-}(aq)$ $K_{sp} = [Cu^+]^2[S^{2-}]$

        c. $Ni(OH)_2(s) \rightleftharpoons Ni^{2+}(aq) + 2\ OH^-(aq)$ $K_{sp} = [Ni^{2+}][OH^-]^2$

81.   In our set-up, s = solubility of the ionic solid in mol/L. This is defined as the maximum amount of a
        salt which can dissolve. Since solids do not appear in the K_{sp} expression, we do not need to worry
        about their initial and equilibrium amounts.

        a. $CaC_2O_4(s)$ $\rightleftharpoons$ $Ca^{2+}(aq)$ + $C_2O_4^{2-}(aq)$

        Initial   0   0
                          s mol/L of $CaC_2O_4(s)$ dissolves to reach equilibrium
        Change   -s $\rightarrow$ +s   +s
        Equil.   s   s

        From the problem, s = 4.8×10^{-5} mol/L.

        $K_{sp} = [Ca^{2+}][C_2O_4^{2-}] = (s)(s) = s^2$, $K_{sp} = (4.8 \times 10^{-5})^2 = 2.3 \times 10^{-9}$

        b. $BiI_3(s)$ $\rightleftharpoons$ $Bi^{3+}(aq)$ + $3\ I^-(aq)$

        Initial   0   0
                          s mol/L of $BiI_3(s)$ dissolves to reach equilibrium
        Change   -s $\rightarrow$ +s   +3s
        Equil.   s   3s

        $K_{sp} = [Bi^{3+}][I^-]^3 = (s)(3s)^3 = 27\ s^4$, $K_{sp} = 27(1.32 \times 10^{-5})^4 = 8.20 \times 10^{-19}$

82.   a. $Ca_3(PO_4)_2(s)$ $\rightleftharpoons$ $3\ Ca^{2+}(aq)$ + $2\ PO_4^{3-}(aq)$

        Initial   0   0
                          s mol/L of $Ca_3(PO_4)_2(s)$ dissolves to reach equilibrium = molar solubility
        Change   -s $\rightarrow$ +3s   +2s
        Equil.   3s   2s

$$K_{sp} = [Ca^{2+}]^3 [PO_4^{3-}]^2 = (3s)^3(2s)^2 = 108\,s^5, \quad K_{sp} = 108(1.6 \times 10^{-7})^5 = 1.1 \times 10^{-32}$$

b.
$$PbCl_2(s) \rightleftharpoons Pb^{2+}(aq) + 2\,Cl^-(aq)$$

Initial   s = solubility (mol/L)   0   0
Equil.   s   2s

$$K_{sp} = [Pb^{2+}][Cl^-]^2 = s(2s)^2 = 4s^3, \quad K_{sp} = 4(1.6 \times 10^{-2})^3 = 1.6 \times 10^{-5}$$

83.
$$PbBr_2(s) \rightleftharpoons Pb^{2+}(aq) + 2\,Br^-(aq)$$

Initial   0   0
                s mol/L of PbBr$_2$(s) dissolves to reach equilibrium
Change   -s   →   +s   +2s
Equil.   s   2s

From the problem, s = [Pb^{2+}] = 2.14 × 10^{-2} M. So:

$$K_{sp} = [Pb^{2+}][Br^-]^2 = s(2s)^2 = 4s^3, \quad K_{sp} = 4(2.14 \times 10^{-2})^3 = 3.92 \times 10^{-5}$$

84.
$$Ce(IO_3)_3(s) \rightleftharpoons Ce^{3+}(aq) + 3\,IO_3^-(aq)$$

Initial   s = solubility (mol/L)   0   0
Equil.   s   3s

From problem, [IO$_3^-$] = 3s = 5.6 × 10^{-3}, s = 1.9 × 10^{-3} M

$$K_{sp} = [Ce^{3+}][IO_3^-]^3 = s(3s)^3 = 27s^4 = 27(1.9 \times 10^{-3})^4 = 3.5 \times 10^{-10}$$

85.   In our set-ups, s = solubility in mol/L. Since solids do not appear in the K$_{sp}$ expression, we do not need to worry about their initial or equilibrium amounts.

a.
$$Ag_3PO_4(s) \rightleftharpoons 3\,Ag^+(aq) + PO_4^{3-}(aq)$$

Initial   0   0
                s mol/L of Ag$_3$PO$_4$(s) dissolves to reach equilibrium
Change   -s   →   +3s   +s
Equil.   3s   s

$$K_{sp} = 1.8 \times 10^{-18} = [Ag^+]^3[PO_4^{3-}] = (3s)^3(s) = 27\,s^4$$

$$27s^4 = 1.8 \times 10^{-18}, \quad s = (6.7 \times 10^{-20})^{1/4} = 1.6 \times 10^{-5}\ mol/L = molar\ solubility$$

b.
$$CaCO_3(s) \rightleftharpoons Ca^{2+} + CO_3^{2-}$$

Initial   s = solubility (mol/L)   0   0
Equil.   s   s

$$K_{sp} = 8.7 \times 10^{-9} = [Ca^{2+}][CO_3^{2-}] = s^2, \quad s = 9.3 \times 10^{-5}\ mol/L$$

c. $Hg_2Cl_2(s)$ $\rightleftharpoons$ Hg_2^{2+} + $2\ Cl^-$

Initial   s = solubility (mol/L)   0   0
Equil.   s   2s

$K_{sp} = 1.1 \times 10^{-18} = [Hg_2^{2+}][Cl^-]^2 = (s)(2s)^2 = 4s^3,\ s = 6.5 \times 10^{-7}\ mol/L$

86.   a. $Pb_3(PO_4)_2(s)$ $\rightleftharpoons$ $3\ Pb^{2+}(aq)$ + $2\ PO_4^{3-}(aq)$

Initial   s = solubility (mol/L)   0   0
Equil.   3s   2s

$K_{sp} = 1 \times 10^{-54} = [Pb^{2+}]^3[PO_4^{3-}]^2 = (3s)^3(2s)^2 = 108\ s^5$

$s = (1 \times 10^{-54}/108)^{1/5} = 6 \times 10^{-12}\ mol/L$ = molar solubility

b. $SrSO_4(s)$ $\rightleftharpoons$ Sr^{2+} + SO_4^{2-}

Initial   s = solubility (mol/L)   0   0
Equil.   s   s

$K_{sp} = 3.2 \times 10^{-7} = [Sr^{2+}][SO_4^{2-}] = s^2,\ s = 5.7 \times 10^{-4}\ mol/L$

c. $Ag_2CO_3(s)$ $\rightleftharpoons$ $2\ Ag^+$ + CO_3^{2-}

Initial   s = solubility (mol/L)   0   0
Equil.   2s   s

$K_{sp} = 8.1 \times 10^{-12} = [Ag^+]^2[CO_3^{2-}] = (2s)^2(s) = 4s^3,\ s = 1.3 \times 10^{-4}\ mol/L$

87.   Let s = solubility of $Al(OH)_3$ in mol/L. Note: Since solids do not appear in the K_{sp} expression, we do not need to worry about their initial or equilibrium amounts.

$Al(OH)_3(s)$ $\rightleftharpoons$ $Al^{3+}(aq)$ + $3\ OH^-(aq)$

Initial   0 $1.0 \times 10^{-7}\ M$ from water
            s mol/L of $Al(OH)_3(s)$ dissolves to reach equilibrium = molar solubility
Change   -s $\rightarrow$ +s   +3s
Equil.   s $1.0 \times 10^{-7} + 3s$

$K_{sp} = 2 \times 10^{-32} = [Al^{3+}][OH^-]^3 = (s)(1.0 \times 10^{-7} + 3s)^3 \approx s(1.0 \times 10^{-7})^3$

$s = \dfrac{2 \times 10^{-32}}{1.0 \times 10^{-21}} = 2 \times 10^{-11}\ mol/L$ Assumption good $(1.0 \times 10^{-7} + 3s \approx 1.0 \times 10^{-7})$.

88. $Mg(OH)_2(s)$ $\rightleftharpoons$ $Mg^{2+}(aq)$ + $2\ OH^-(aq)$

Initial   s = solubility (mol/L)   0 $1.0 \times 10^{-7}\ M$
Equil.   s $1.0 \times 10^{-7} + 2s$

$K_{sp} = [Mg^{2+}] [OH^-]^2 = s(1.0 \times 10^{-7} + 2s)^2$; Assume that $1.0 \times 10^{-7} + 2s \approx 2s$, then:

$$K_{sp} = 8.9 \times 10^{-12} = s(2s)^2 = 4s^3, \quad s = 1.3 \times 10^{-4} \text{ mol/L}$$

Assumption is good (1.0×10^{-7} is 0.04% of 2s). Molar solubility = 1.3×10^{-4} mol/L

89.   a.   Since both solids dissolve to produce 3 ions in solution, then we can compare values of K_{sp} to determine relative solubility. Since the K_{sp} for CaF_2 is the smallest, then $CaF_2(s)$ is the least soluble (in mol/L).

b.   We must calculate solubilities since each salt yields a different number of ions when it dissolves.

$$Ca_3(PO_4)_2(s) \quad \rightleftharpoons \quad 3\,Ca^{2+}(aq) \; + \; 2\,PO_4^{3-}(aq) \qquad K_{sp} = 1.3 \times 10^{-32}$$

Initial   s = solubility (mol/L)   0   0
Equil.   3s   2s

$K_{sp} = [Ca^{2+}]^3 [PO_4^{3-}]^2 = (3s)^3(2s)^2 = 108s^5, \quad s = (1.3 \times 10^{-32}/108)^{1/5} = 1.6 \times 10^{-7}$ mol/L

$$FePO_4(s) \quad \rightleftharpoons \quad Fe^{3+}(aq) \; + \; PO_4^{3-}(aq) \qquad K_{sp} = 1.0 \times 10^{-22}$$

Initial   s = solubility (mol/L)   0   0
Equil.   s   s

$K_{sp} = [Fe^{3+}] [PO_4^{3-}] = s^2, \quad s = \sqrt{1.0 \times 10^{-22}} = 1.0 \times 10^{-11}$ mol/L

Since the molar solubility of $FePO_4$ is the smallest, then $FePO_4$ is least soluble (in mol/L).

90.   a. $$Ag_3PO_4(s) \quad \rightleftharpoons \quad 3\,Ag^+(aq) \; + \; PO_4^{3-}(aq)$$

Equil.   s = solubility (mol/L)   3s   s

$K_{sp} = 1.8 \times 10^{-18} = [Ag^+]^3 [PO_4^{3-}] = (3s)^3 s = 27s^4$

$s = (1.8 \times 10^{-18}/27)^{1/4} = 1.6 \times 10^{-5}$ mol/L

$$Hg_2Cl_2(s) \quad \rightleftharpoons \quad Hg_2^{2+}(aq) \; + \; 2\,Cl^-(aq)$$

Equil.   s   2s

$K_{sp} = 1.1 \times 10^{-18} = [Hg_2^{2+}] [Cl^-]^2 = s(2s)^2 = 4s^3, \quad s = (1.1 \times 10^{-18}/4)^{1/3} = 6.5 \times 10^{-7}$ mol/L

By comparing calculated molar solubilities, $Hg_2Cl_2(s)$ is least soluble (in mol/L).

b.   Since each salt produces 4 ions in solution, then we can compare K_{sp} values to determine relative molar solubilities. Therefore, $Co(OH)_3(s)$ will be the least soluble salt (in mol/L) since it has a smaller K_{sp} value.

91.   a. $Fe(OH)_3(s)$ $\rightleftharpoons$ Fe^{3+} + $3\ OH^-$ pH = 7.0 so $[OH^-] = 1 \times 10^{-7}\ M$

Initial   0 $1 \times 10^{-7}\ M$

s mol/L of $Fe(OH)_3(s)$ dissolves to reach equilibrium = molar solubility

Change   -s $\rightarrow$ +s   +3s
Equil.   s $3s + 1 \times 10^{-7}$

$K_{sp} = 4 \times 10^{-38} = [Fe^{3+}][OH^-]^3 = (s)(3s + 1 \times 10^{-7})^3 \approx s(1 \times 10^{-7})^3$

$s = 4 \times 10^{-17}$ mol/L   Assumption good $(3s << 1 \times 10^{-7})$.

b. $Fe(OH)_3(s)$ $\rightleftharpoons$ Fe^{3+} + $3\ OH^-$ pH = 5.0 so $[OH^-] = 1 \times 10^{-9}\ M$

Initial   0 $1 \times 10^{-9}\ M$ (buffered)

s mol/L dissolves to reach equilibrium

Change   -s $\rightarrow$ +s   -----   (assume no pH change in buffer)
Equil.   s 1×10^{-9}

$K_{sp} = 4 \times 10^{-38} = [Fe^{3+}][OH^-]^3 = (s)(1 \times 10^{-9})^3$, $s = 4 \times 10^{-11}$ mol/L = molar solubility

c. $Fe(OH)_3(s)$ $\rightleftharpoons$ Fe^{3+} + $3\ OH^-$ pH = 11.0 so $[OH^-] = 1 \times 10^{-3}\ M$

Initial   0 $0.001\ M$ (buffered)

s mol/L dissolves to reach equilibrium

Change   -s $\rightarrow$ +s   -----   (assume no pH change)
Equil.   s   0.001

$K_{sp} = 4 \times 10^{-38} = [Fe^{3+}][OH^-]^3 = (s)(0.001)^3$, $s = 4 \times 10^{-29}$ mol/L = molar solubility

Note: As $[OH^-]$ increases, solubility decreases. This is the common ion effect.

92.   a. $PbI_2(s)$ $\rightleftharpoons$ Pb^{2+} + $2\ I^-$

Initial   s = solubility (mol/L)   0   0
Equil.   s   2s

$K_{sp} = 1.4 \times 10^{-8} = [Pb^{2+}][I^-]^2 = 4s^3$, $s = 1.5 \times 10^{-3}$ mol/L

b. $PbI_2(s)$ $\rightleftharpoons$ Pb^{2+} + $2\ I^-$

Initial   s = solubility (mol/L)   0.10 M   0
Equil.   0.10 + s   2s

$1.4 \times 10^{-8} = (0.10 + s)(2s)^2 \approx (0.10)(2s)^2 = 0.40\ s^2$, $s = 1.9 \times 10^{-4}$ mol/L   Assumption good.

c. $PbI_2(s)$ $\rightleftharpoons$ Pb^{2+} + $2\ I^-$

Initial   s = solubility (mol/L)   0   0.010 M
Equil.   s   0.010 + 2s

$1.4 \times 10^{-8} = (s)(0.010 + 2s)^2 \approx (s)(0.010)^2$, $s = 1.4 \times 10^{-4}$ mol/L   Assumption good.

The presence of a common ion decreases the solubility.

93. $Ca_3(PO_4)_2(s)$ $\rightleftharpoons$ $3\ Ca^{2+}(aq)$ + $2\ PO_4^{3-}(aq)$

Initial		0	0.20 M
	s mol/L of $Ca_3(PO_4)_2(s)$ dissolves to reach equilibrium		
Change	-s $\rightarrow$	+3s	+2s
Equil.		3s	0.20 + 2s

$K_{sp} = 1.3 \times 10^{-32} = [Ca^{2+}]^3\ [PO_4^{3-}]^2 = (3s)^3(0.20 + 2s)^2$

Assuming $0.20 + 2s \approx 0.20$: $1.3 \times 10^{-32} = (3s)^2(0.20)^2 = 27s^3(0.040)$

s = molar solubility = 2.3×10^{-11} mol/L; Assumptions good.

94. $Pb(IO_3)_2(s)$ $\rightleftharpoons$ $Pb^{2+}(aq)$ + $2\ IO_3^-(aq)$

Initial	s = solubility (mol/L)	0	0.10 M
Equil.		s	0.10 + 2s

$K_{sp} = [Pb^{2+}]\ [IO_3^-]^2 = (s)(0.10 + 2s)^2$

From the problem, s = 2.6×10^{-11} mol/L; Solving for K_{sp}:

$K_{sp} = (2.6 \times 10^{-11}) \times [0.10 + 2(2.6 \times 10^{-11})]^2 = 2.6 \times 10^{-13}$

95.   If the anion in the salt can act as a base in water, then the solubility of the salt will increase as the solution becomes more acidic. Added H^+ will react with the base forming the conjugate acid. As the basic anion is removed, more of the salt will dissolve to replenish the basic anion. The salts with basic anions are Ag_3PO_4, $CaCO_3$, $Pb_3(PO_4)_2$, $SrSO_4$ and Ag_2CO_3. Only Hg_2Cl_2 does not have any pH dependence since Cl^- is a terrible base (the conjugate base of a strong acid).

$Ag_3PO_4(s) + H^+(aq) \rightarrow 3\ Ag^+(aq) + HPO_4^{2-}(aq)$ $\xrightarrow{\text{excess } H^+}$ $3\ Ag^+(aq) + H_3PO_4(aq)$

$CaCO_3(s) + H^+ \rightarrow Ca^{2+} + HCO_3^-$ $\xrightarrow{\text{excess } H^+}$ $Ca^{2+} + H_2CO_3\ [H_2O(l) + CO_2(g)]$

$Pb_3(PO_4)_2(s) + 2\ H^+ \rightarrow 3\ Pb^{2+} + 2\ HPO_4^{2-}$ $\xrightarrow{\text{excess } H^+}$ $3\ Pb^{2+} + 2\ H_3PO_4$

$SrSO_4(s) + H^+ \rightarrow Sr^{2+} + HSO_4^-$

$Ag_2CO_3(s) + H^+ \rightarrow 2\ Ag^+ + HCO_3^-$ $\xrightarrow{\text{excess } H^+}$ $2\ Ag^+ + H_2CO_3\ [H_2O(l) + CO_2(g)]$

96.   a.   AgF   b. $Pb(OH)_2$ c. $Sr(NO_2)_2$ d. $Ni(CN)_2$

All the above salts have anions that are bases. The anions of the other choices are conjugate bases of strong acids. They have no basic properties in water and, therefore, do not have solubilities which depend on pH.

97.   Potentially, $BaSO_4(s)$ could form if Q is greater than K_{sp}.

$$BaSO_4(s) \rightleftharpoons Ba^{2+}(aq) + SO_4^{2-}(aq)   K_{sp} = 1.5 \times 10^{-9}$$

To calculate Q, we need the initial concentrations of Ba^{2+} and SO_4^{2-}.

$$[Ba^{2+}]_o = \frac{mmoles\ Ba^{2+}}{total\ mL\ solution} = \frac{75.0\ mL \times \dfrac{0.020\ mmol\ Ba^{2+}}{mL}}{75.0\ mL \times 125\ mL} = 0.0075\ M$$

$$[SO_4^{2-}]_o = \frac{mmoles\ SO_4^{2-}}{total\ mL\ solution} = \frac{125\ mL \times \dfrac{0.040\ mmol\ SO_4^{2-}}{mL}}{200.\ mL} = 0.025\ M$$

$$Q = [Ba^{2+}]_o[SO_4^{2-}]_o = (0.0075\ M)(0.025\ M) = 1.9 \times 10^{-4}$$

$Q > K_{sp}$ (1.5×10^{-9}) so $BaSO_4(s)$ will form.

98.   Only $PbCl_2(s)$ is a possible precipitate. After mixing: $[Pb^{2+}]_o = [Cl^-]_o = 0.010\ M$. The ion concentrations were halved from the initial concentrations since the volume doubled.

$$Q = [Pb^{2+}]_o[Cl^-]_o^2 = (0.010)(0.010)^2 = 1.0 \times 10^{-6} < K_{sp}\ (1.6 \times 10^{-5});\ \ No\ precipitate\ will\ form.$$

99.   The concentrations of ions is large so Q will be greater than K_{sp} and $BaC_2O_4(s)$ will form. To solve this problem, we will assume that the precipitation reaction goes to completion, then we will solve an equilibrium problem to get the actual ion concentrations.

$$100.\ mL \times \frac{0.200\ mmol\ K_2C_2O_4}{mL} = 20.0\ mmol\ K_2C_2O_4$$

$$150.\ mL \times \frac{0.250\ mmol\ BaBr_2}{mL} = 37.5\ mmol\ BaBr_2$$

	$Ba^{2+}(aq)$	+	$C_2O_4^{2-}(aq)$	$\rightarrow$	$BaC_2O_4(s)$	$K = 1/K_{sp} \gg 1$
Before	37.5 mmol		20.0 mmol		0	
Change	-20.0		-20.0	$\rightarrow$	+20.0	Reacts completely (K is large)
After	17.5		0		20.0	

New initial concentrations (after complete precipitation) are: $[Ba^{2+}] = \dfrac{17.5\ mmol}{250.\ mL} = 7.00 \times 10^{-2}\ M$

$$[K^+] = \frac{2(20.0 \text{ mmol})}{250. \text{ mL}} = 0.160 \, M; \quad [Br^-] = \frac{2(37.5 \text{ mmol})}{250. \text{ mL}} = 0.300 \, M$$

For K^+ and Br^-, these are also the final concentrations. For Ba^{2+} and $C_2O_4^{2-}$, we need to perform an equilibrium calculation.

$$BaC_2O_4(s) \quad \rightleftharpoons \quad Ba^{2+}(aq) + C_2O_4^{2-}(aq) \quad K_{sp} = 2.3 \times 10^{-8}$$

Initial   0.0700 M 0
            s mol/L $BaC_2O_4(s)$ dissolves to reach equilibrium
Equil.   0.0700 + s   s

$$K_{sp} = 2.3 \times 10^{-8} = [Ba^{2+}][C_2O_4^{2-}] = (0.0700 + s)(s) \approx 0.0700 \, s$$

$s = [C_2O_4^{2-}] = 3.3 \times 10^{-7}$ mol/L; $[Ba^{2+}] = 0.0700 \, M$ Assumption good (s << 0.0700).

100.   50.0 mL $\times$ 0.00200 M = 0.100 mmol Ag^+; 50.0 mL $\times$ 0.0100 M = 0.500 mmol IO_3^-

Assume $AgIO_3(s)$ precipitates completely. After reaction, 0.400 mmol IO_3^- is remaining. Now, let some $AgIO_3(s)$ dissolve in solution with excess IO_3^- to reach equilibrium.

$$AgIO_3 \quad \rightleftharpoons \quad Ag^+ + IO_3^-$$

Initial   0 $\dfrac{0.400 \text{ mmol}}{100.0 \text{ mL}} = 4.00 \times 10^{-3} \, M$

            s mol/L $AgIO_3(s)$ dissolves to reach equilibrium
Equil.   s $4.00 \times 10^{-3} + s$

$$K_{sp} = [Ag^+][IO_3^-] = 3.0 \times 10^{-8} = s(4.00 \times 10^{-3} + s) \approx s(4.00 \times 10^{-3})$$

$s = 7.5 \times 10^{-6}$ mol/L = $[Ag^+]$ Assumptions good.

101. $Ag_3PO_4(s) \rightleftharpoons 3 \, Ag^+(aq) + PO_4^{3-}(aq)$; When Q is greater than K_{sp}, then precipitation will occur. We will calculate the $[Ag^+]_o$ necessary for $Q = K_{sp}$. Any $[Ag^+]_o$ greater than this calculated number will cause precipitation of $Ag_3PO_4(s)$. In this problem, $[PO_4^{3-}]_o = [Na_3PO_4]_o = 1.0 \times 10^{-5} \, M$

$$K_{sp} = 1.8 \times 10^{-18}; \quad Q = 1.8 \times 10^{-18} = [Ag^+]_o^3 [PO_4^{3-}]_o = [Ag^+]_o^3 (1.0 \times 10^{-5} \, M)$$

$$[Ag^+]_o = \left(\frac{1.8 \times 10^{-18}}{1.0 \times 10^{-5}} \right)^{1/3}, \quad [Ag^+]_o = 5.6 \times 10^{-5} \, M$$

When $[Ag^+]_o = [AgNO_3]_o$ is greater than $5.6 \times 10^{-5} \, M$, then precipitation of $Ag_3PO_4(s)$ will occur.

102.   For each lead salt, we will calculate the $[Pb^{2+}]_o$ necessary for $Q = K_{sp}$. Any $[Pb^{2+}]_o$ greater than this value will cause precipitation of the salt ($Q > K_{sp}$).

$$PbF_2(s) \rightleftharpoons Pb^{2+}(aq) + 2 \, F^-(aq) \quad K_{sp} = 4 \times 10^{-8}; \quad Q = 4.0 \times 10^{-8} = [Pb^{2+}]_o [F^-]_o^2$$

$$[Pb^{2+}]_o = \frac{4.0 \times 10^{-8}}{(1 \times 10^{-4})^2} = 4 \, M$$

$$PbS(s) \rightleftharpoons Pb^{2+}(aq) + S^{2-}(aq) \qquad K_{sp} = 7 \times 10^{-29}; \; Q = 7 \times 10^{-29} = [Pb^{2+}]_o \, [S^{2-}]_o$$

$$[Pb^{2+}]_o = \frac{7 \times 10^{-29}}{1 \times 10^{-4}} = 7 \times 10^{-25} \, M$$

$$Pb_3(PO_4)_2(s) \rightleftharpoons 3 \, Pb^{2+}(aq) + 2 \, PO_4^{3-}(aq) \quad K_{sp} = 1 \times 10^{-54}; \; Q = 1 \times 10^{-54} = [Pb^{2+}]_o^3 \, [PO_4^{3-}]_o^2$$

$$[Pb^{2+}]_o = \left(\frac{1 \times 10^{-54}}{(1 \times 10^{-4})^2} \right)^{1/3} = 5 \times 10^{-16} \, M$$

From the calculated $[Pb^{2+}]_o$, the least soluble salt is PbS(s) and will form first. $Pb_3(PO_4)_2(s)$ will form second and $PbF_2(s)$ will form last since it requires the largest $[Pb^{2+}]_o$ in order for precipitation to occur.

Complex Ion Equilibria

103. a.
$$\begin{array}{ll}
Co^{2+} + NH_3 \rightleftharpoons CoNH_3^{2+} & K_1 \\
CoNH_3^{2+} + NH_3 \rightleftharpoons Co(NH_3)_2^{2+} & K_2 \\
Co(NH_3)_2^{2+} + NH_3 \rightleftharpoons Co(NH_3)_3^{2+} & K_3 \\
Co(NH_3)_3^{2+} + NH_3 \rightleftharpoons Co(NH_3)_4^{2+} & K_4 \\
Co(NH_3)_4^{2+} + NH_3 \rightleftharpoons Co(NH_3)_5^{2+} & K_5 \\
Co(NH_3)_5^{2+} + NH_3 \rightleftharpoons Co(NH_3)_6^{2+} & K_6 \\
\hline
Co^{2+} + 6 \, NH_3 \rightleftharpoons Co(NH_3)_6^{2+} & K_f = K_1 K_2 K_3 K_4 K_5 K_6
\end{array}$$

Note: The various K's included are for your information. Each NH_3 adds with a corresponding K value associated with that reaction. The overall formation constant, K_f, for the overall reaction is equal to the product of all the stepwise K values.

b.
$$\begin{array}{ll}
Ag^+ + NH_3 \rightleftharpoons AgNH_3^+ & K_1 \\
AgNH_3^+ + NH_3 \rightleftharpoons Ag(NH_3)_2^+ & K_2 \\
\hline
Ag^+ + 2 \, NH_3 \rightleftharpoons Ag(NH_3)_2^+ & K_f = K_1 K_2
\end{array}$$

104. a.
$$\begin{array}{ll}
Ni^{2+} + CN^- \rightleftharpoons NiCN^+ & K_1 \\
NiCN^+ + CN^- \rightleftharpoons Ni(CN)_2 & K_2 \\
Ni(CN)_2 + CN^- \rightleftharpoons Ni(CN)_3^- & K_3 \\
Ni(CN)_3^- + CN^- \rightleftharpoons Ni(CN)_4^{2-} & K_4 \\
\hline
Ni^{2+} + 4 \, CN^- \rightleftharpoons Ni(CN)_4^{2-} & K_f = K_1 K_2 K_3 K_4
\end{array}$$

b.
$$\begin{array}{ll}
Mn^{2+} + C_2O_4^{2-} \rightleftharpoons MnC_2O_4 & K_1 \\
MnC_2O_4 + C_2O_4^{2-} \rightleftharpoons Mn(C_2O_4)_2^{2-} & K_2 \\
\hline
Mn^{2+} + 2 \, C_2O_4^{2-} \rightleftharpoons Mn(C_2O_4)_2^{2-} & K_f = K_1 K_2
\end{array}$$

105.   An ammonia solution is basic. Initially the reaction that occurs is $Cu^{2+}(aq) + 2\ OH^-(aq) \rightarrow$
$Cu(OH)_2(s)$. As the concentration of NH_3 increases, the soluble complex ion $Cu(NH_3)_4^{2+}$ forms:
$Cu(OH)_2(s) + 4\ NH_3(aq) \rightleftharpoons Cu(NH_3)_4^{2+}(aq) + 2\ OH^-(aq)$

106. $Ag^+(aq) + Cl^-(aq) \rightleftharpoons AgCl(s)$, white ppt.; $AgCl(s) + 2\ NH_3(aq) \rightleftharpoons Ag(NH_3)_2^+(aq) + Cl^-(aq)$

$Ag(NH_3)_2^+(aq) + Br^-(aq) \rightleftharpoons AgBr(s) + 2\ NH_3(aq)$, pale yellow ppt.

$AgBr(s) + 2\ S_2O_3^{2-}(aq) \rightleftharpoons Ag(S_2O_3)_2^{3-}(aq) + Br^-(aq)$

$Ag(S_2O_3)_2^{3-}(aq) + I^-(aq) \rightleftharpoons AgI(s) + 2\ S_2O_3^{2-}(aq)$, yellow ppt.

The least soluble salt (smallest K_{sp} value) must be AgI since it forms in the presence of Cl^- and Br^-.
The most soluble salt (largest K_{sp} value) must be AgCl since it forms initially, but never reforms.
The order of K_{sp} values are: $K_{sp}\ (AgCl) > K_{sp}\ (AgBr) > K_{sp}\ (AgI)$

107.   The formation constant for HgI_4^{2-} is an extremely large number. Because of this, we will let the
Hg^{2+} and I^- ions present initially react to completion, and then solve an equilibrium problem to
determine the Hg^{2+} concentration.

$$Hg^{2+}\quad +\quad 4\ I^-\quad \rightleftharpoons\quad HgI_4^{2-}\qquad K = 1.0 \times 10^{30}$$

Before	$0.010\ M$	$0.78\ M$	0	
Change	-0.010	-0.040 $\rightarrow$	+0.010	Reacts completely (K large)
After	0	0.74	0.010	New initial

x mol/L HgI_4^{2-} dissociates to reach equilibrium

Change	$+x$	$+4x$ $\leftarrow$	$-x$	
Equil.	x	$0.74 + 4x$	$0.010 - x$	

$$K = 1.0 \times 10^{30} = \frac{[HgI_4^{2-}]}{[Hg^{2+}]\,[I^-]^4} = \frac{(0.010 - x)}{(x)\,(0.74 + 4x)^4};\quad \text{Making normal assumptions:}$$

$$1.0 \times 10^{30} \approx \frac{(0.010)}{(x)\,(0.74)^4},\quad x = [Hg^{2+}] = 3.3 \times 10^{-32}\ M \quad \text{Assumptions good.}$$

Note: 3.3×10^{-32} mol/L corresponds to one Hg^{2+} ion per 5×10^7 L. It is very reasonable to
approach the equilibrium in two steps. The reaction does essentially go to completion.

108. $$Ni^{2+}\quad +\quad 6\ NH_3\quad \rightleftharpoons\quad Ni(NH_3)_6^{2+}\qquad K = 5.5 \times 10^8$$

Initial	0	$3.0\ M$	0.10 mol/0.50 L = 0.20 M	

x mol/L $Ni(NH_3)_6^{2+}$ dissociates to reach equilibrium

Change	$+x$	$+6x$ $\leftarrow$	$-x$	
Equil.	x	$3.0 + 6x$	$0.20 - x$	

$$K = 5.5 \times 10^8 = \frac{[Ni(NH_3)_6^{2+}]}{[Ni^{2+}]\,[NH_3]^6} = \frac{(0.20 - x)}{(x)\,(3.0 + 6x)^6},\quad 5.5 \times 10^8 \approx \frac{(0.20)}{(x)\,(3.0)^6}$$

$x = [Ni^{2+}] = 5.0 \times 10^{-13}\ M;\ [Ni(NH_3)_6^{2+}] = 0.20\ M - x = 0.20\ M$ Assumptions good.

109. a.

$$CuCl(s) \rightleftharpoons Cu^+ + Cl^-$$

Initial	s = solubility (mol/L)	0	0
Equil.		s	s

$K_{sp} = 1.2 \times 10^{-6} = [Cu^+][Cl^-] = s^2,\ s = 1.1 \times 10^{-3}$ mol/L

b. Cu^+ forms the complex ion $CuCl_2^-$ in the presence of Cl^-. We will consider both the K_{sp} reaction and the complex ion reaction at the same time.

$CuCl(s) \rightleftharpoons Cu^+ (aq) + Cl^- (aq)$	$K_{sp} = 1.2 \times 10^{-6}$
$Cu^+ (aq) + 2\ Cl^- (aq) \rightleftharpoons CuCl_2^-(aq)$	$K_f = 8.7 \times 10^4$

$$CuCl\ (s) + Cl^-(aq) \rightleftharpoons CuCl_2^-(aq) \qquad K = K_{sp} \times K_f = 0.10$$

Initial	0.10 M	0
Equil.	0.10 - s	s	where s = solubility of CuCl(s) in mol/L

$K = 0.10 = \dfrac{[CuCl_2^-]}{[Cl^-]} = \dfrac{s}{0.10 - s},\quad 1.0 \times 10^{-2} - 0.10\ s = s,\ 1.10\ s = 1.0 \times 10^{-2},\ s = 9.1 \times 10^{-3}$ mol/L

110.

$AgBr(s) \rightleftharpoons Ag^+ + Br^-$	$K_{sp} = 5.0 \times 10^{-13}$
$Ag^+ + 2\ NH_3 \rightleftharpoons Ag(NH_3)_2^+$	$K_f = K_1K_2 = (2.1 \times 10^3)(8.2 \times 10^3) = 1.7 \times 10^7$

$$AgBr(s) + 2\ NH_3 \rightleftharpoons Ag(NH_3)_2^+ + Br^- \qquad K = K_{sp} \times K_f = 8.5 \times 10^{-6}$$

$$AgBr(s) + 2\ NH_3 \rightleftharpoons Ag(NH_3)_2^+ + Br^-$$

Initial	0.200 M	0	0
	s mol/L of AgBr(s) dissolves to reach equilibrium = molar solubility		
Equil.	0.200 - 2s	s	s

$K = \dfrac{s^2}{(0.200 - 2s)^2} = 8.5 \times 10^{-6} \approx \dfrac{s^2}{(0.200)^2},\ s = 5.8 \times 10^{-4}$ mol/L   Assumption good.

111. In 2.0 M NH_3, the soluble complex ion $Ag(NH_3)_2^+$ forms which increases the solubility of AgCl(s). The reaction is: $AgCl(s) + 2\ NH_3 \rightleftharpoons Ag(NH_3)_2^+ + Cl^-$. In 2.0 M NH_4NO_3, NH_3 is only formed by the dissociation of the weak acid NH_4^+. There is not enough NH_3 produced by this reaction to dissolve AgCl(s) by the formation of the complex ion.

112. In NH_3, Cu^{2+} forms the soluble complex ion, $Cu(NH_3)_4^{2+}$. This increases the solubility of $Cu(OH)_2(s)$ since added NH_3 removes Cu^{2+} from the equilibrium which causes more $Cu(OH)_2(s)$ to dissolve. In HNO_3, H^+ removes OH^- from the K_{sp} equilibrium causing more $Cu(OH)_2(s)$ to dissolve. Any salt with basic anions will be more soluble in an acid solution.

$AgC_2H_3O_2(s)$ will be more soluble in either NH_3 or HNO_3. This is because Ag^+ forms the complex ion $Ag(NH_3)_2^+$ and $C_2H_3O_2^-$ is a weak base, so it will react with added H^+. AgCl(s) will only be more soluble in NH_3 due to $Ag(NH_3)_2^+$ formation. In acid, Cl^- is a horrible base so it doesn't react with added H^+. $AgCl(s)$ will not be more soluble in HNO_3.

Additional Exercises

113. $NH_3 + H_2O \rightleftharpoons NH_4^+ + OH^-$ $K_b = \dfrac{[NH_4^+][OH^-]}{[NH_3]}$; Taking the -log of the K_b expression:

$$-\log K_b = -\log[OH^-] - \log\frac{[NH_4^+]}{[NH_3]}, \quad -\log[OH^-] = -\log K_b + \log\frac{[NH_4^+]}{[NH_3]}$$

$$pOH = pK_b + \log\frac{[NH_4^+]}{[NH_3]} \quad\text{or}\quad pOH = pK_b + \log\frac{[Acid]}{[Base]}$$

114.   a. $pH = pK_a = -\log(6.4 \times 10^{-5}) = 4.19$ since $[HBz] = [Bz^-]$ where $HBz = C_6H_5CO_2H$ and $[Bz^-] = C_6H_5CO_2^-$.

b. $[Bz^-]$ will increase to $0.120\ M$ and $[HBz]$ will decrease to $0.080\ M$ after OH^- reacts completely with HBz.

$$pH = pK_a + \log\frac{[Bz^-]}{[HBz]}, \quad pH = 4.19 + \log\frac{(0.120)}{(0.080)} = 4.37$$

c.

	Bz^-	+	H_2O	$\rightleftharpoons$	HBz	+	OH^-
Initial	$0.120\ M$				$0.080\ M$		0
Equil.	$0.120 - x$				$0.080 + x$		x

$$K_b = \frac{K_w}{K_a} = \frac{1.0 \times 10^{-14}}{6.4 \times 10^{-5}} = \frac{(0.080 + x)(x)}{(0.120 - x)} \approx \frac{(0.080)(x)}{0.120}$$

$x = [OH^-] = 2.34 \times 10^{-10}\ M$ (carrying extra sig. figs.); Assumptions good.

$pOH = 9.63$; $pH = 4.37$

d.   We get the same answer. Both equilibria involve the two major species, benzoic acid and benzoate anion. Both equilibria must hold true. K_b is related to K_a by K_w and $[OH^-]$ is related to $[H^+]$ by K_w, so all constants are interrelated.

115.   A best buffer is when $pH \approx pK_a$ since these solutions have about equal concentrations of weak acid and conjugate base. Therefore, choose combinations that yield a buffer where $pH \approx pK_a$, i.e., look at the acids available and choose the one whose pK_a is closest to the pH.

a.   Potassium fluoride + HCl will yield a buffer consisting of HF ($pK_a = 3.14$) and F^-.

b.   Benzoic acid + NaOH will yield a buffer consisting of benzoic acid ($pK_a = 4.19$) and benzoate anion.

c.   Sodium acetate + acetic acid ($pK_a = 4.74$) is the best choice for pH = 5.0 buffer since acetic acid has a pK_a value closest to 5.0.

d.   HOCl and NaOH: This is the best choice to produce a conjugate acid/base pair with pH = 7.0. This mixture would yield a buffer consisting of HOCl ($pK_a = 7.46$) and OCl^-. Actually the best choice for a pH = 7.0 buffer is an equimolar mixture of ammonium chloride and sodium acetate. NH_4^+ is a weak acid ($K_a = 5.6 \times 10^{-10}$) and $C_2H_3O_2^-$ is a weak base ($K_b = 5.6 \times 10^{-10}$). A mixture of the two will give a buffer at pH = 7 since the weak acid and weak base are the same strengths. $NH_4C_2H_3O_2$ is commercially available and its solutions are used for pH = 7.0 buffers.

e.   Ammonium chloride + NaOH will yield a buffer consisting of NH_4^+ ($pK_a = 9.26$) and NH_3.

116.   a.   The optimum pH for a buffer is when pH = pK_a. At this pH a buffer will have equal neutralization capacity for both added acid and base. As shown below, since the pK_a for $TRISH^+$ is about 8, then the optimal buffer pH is about 8.

$$K_b = 1.19 \times 10^{-6}; \ K_a = K_w/K_b = 8.40 \times 10^{-9}; \ pK_a = -\log(8.40 \times 10^{-9}) = 8.076$$

b. $pH = pK_a + \log \dfrac{[TRIS]}{[TRISH^+]}, \ \ 7.00 = 8.076 + \log \dfrac{[TRIS]}{[TRISH^+]}$

$\dfrac{[TRIS]}{[TRISH^+]} = 10^{-1.08} = 0.083 \ \ (\text{at pH} = 7.00)$

$9.00 = 8.076 + \log \dfrac{[TRIS]}{[TRISH^+]}, \ \dfrac{[TRIS]}{[TRISH^+]} = 10^{0.92} = 8.3 \ \ (\text{at pH} = 9.00)$

c. $\dfrac{50.0 \text{ g TRIS}}{2.0 \text{ L}} \times \dfrac{1 \text{ mol}}{121.14 \text{ g}} = 0.206 \, M = 0.21 \, M = [TRIS]$

$\dfrac{65.0 \text{ g TRISHCl}}{2.0 \text{ L}} \times \dfrac{1 \text{ mol}}{157.60 \text{ g}} = 0.206 \, M = 0.21 \, M = [TRISHCl] = [TRISH^+]$

$pH = pK_a + \log \dfrac{[TRIS]}{[TRISH^+]} = 8.076 + \log \dfrac{(0.21)}{(0.21)} = 8.08$

The amount of H^+ added from HCl is: $0.50 \times 10^{-3} \text{ L} \times \dfrac{12 \text{ mol}}{L} = 6.0 \times 10^{-3} \text{ mol } H^+$

The H^+ from HCl will convert TRIS into $TRISH^+$. The reaction is:

	TRIS	+	H^+	$\rightarrow$	$TRISH^+$	
Before	0.21 M		$\dfrac{6.0 \times 10^{-3}}{0.2005} = 0.030 \, M$		0.21 M	
Change	-0.030		-0.030	$\rightarrow$	+0.030	Reacts completely
After	0.18		0		0.24	

Now use the Henderson-Hasselbalch equation to solve the buffer problem.

$$pH = 8.076 + \log\left(\frac{0.18}{0.24}\right) = 7.95$$

117.   a. $NH_3 + H_2O \rightleftharpoons OH^- + NH_4^+$; $pH = 8.95$; $pOH = 5.05$

$$K_b = 1.8 \times 10^{-5} = \frac{[OH^-][NH_4^+]}{[NH_3]} = \frac{(10^{-5.05})[NH_4^+]}{(0.500)}, \quad [NH_4^+] = 1.0\,M$$

   b. $4.00\text{ g NaOH} \times \dfrac{1\text{ mol}}{40.00\text{ g}} = 0.100\text{ mol}$; OH^- converts 0.100 mol NH_4^+ into 0.100 mol NH_3.

   $NH_4^+ + OH^- \rightarrow NH_3 + H_2O$; After this reaction goes to completion, a buffer solution still exists where mol $NH_4^+ = 1.0 - 0.100 = 0.9$ mol and mol $NH_3 = 0.500 + 0.100 = 0.600$ mol. The pH of this solution is:

$$pH = 9.26 + \log\left(\frac{0.600}{0.9}\right) = 9.26 + (-0.2) = 9.1$$

118.   a. $HC_2H_3O_2 + OH^- \rightleftharpoons C_2H_3O_2^- + H_2O$

$$K_{eq} = \frac{[C_2H_3O_2^-]}{[HC_2H_3O_2][OH^-]} \times \frac{[H^+]}{[H^+]} = \frac{K_{a,\,HC_2H_3O_2}}{K_w} = \frac{1.8 \times 10^{-5}}{1.0 \times 10^{-14}} = 1.8 \times 10^9$$

   b. $C_2H_3O_2^- + H^+ \rightleftharpoons HC_2H_3O_2$ $K_{eq} = \dfrac{[HC_2H_3O_2]}{[H^+][C_2H_3O_2^-]} = \dfrac{1}{K_{a,\,HC_2H_3O_2}} = 5.6 \times 10^4$

   c. $HCl + NaOH \rightarrow NaCl + H_2O$

   Net ionic equation is: $H^+ + OH^- \rightleftharpoons H_2O$; $K_{eq} = \dfrac{1}{K_w} = 1.0 \times 10^{14}$

119.   At a pH = 0.00, the $[H^+] = 10^{-0.00} = 1.0\,M$. Begin with 1.0 L $\times$ 2.0 mol/L NaOH = 2.0 mol OH^-. We will need 2.0 mol HCl to neutralize the OH^- plus an additional 1.0 mol excess H^+ to reduce the pH to 0.00. We need 3.0 mol HCl total assuming 1.0 L of solution.

120.   For a titration of a strong acid with a strong base, the added OH^- reacts completely with the H^+ present. To determine the pH, we calculate the concentration of excess H^+ or OH^- after the neutralization reaction and then calculate the pH.

   0 mL: $[H^+] = 0.100\,M$ from HNO_3; pH = 1.000

   4.0 mL: initial mmol H^+ present = 25.0 mL $\times \dfrac{0.100\text{ mmol }H^+}{mL} = 2.50$ mmol H^+

   mmol OH^- added = 4.0 mL $\times \dfrac{0.100\text{ mmol }OH^-}{mL} = 0.40$ mmol OH^-

   0.40 mmol OH^- reacts completely with 0.40 mmol H^+: $OH^- + H^+ \rightarrow H_2O$

   $[H^+]_{excess} = \dfrac{(2.50 - 0.40)\text{ mmol}}{(25.0 + 4.0)\text{ mL}} = 7.24 \times 10^{-2}\,M$; pH = 1.140

We follow the same procedure for the remaining calculations.

8.0 mL: $[H^+]_{excess} = \dfrac{(2.50 - 0.80)\ mmol}{33.0\ mL} = 5.15 \times 10^{-2}\ M;\ pH = 1.288$

12.5 mL: $[H^+]_{excess} = \dfrac{(2.50 - 1.25)\ mmol}{37.5\ mL} = 3.33 \times 10^{-2}\ M;\ pH = 1.478$

20.0 mL: $[H^+]_{excess} = \dfrac{(2.50 - 2.00)\ mmol}{45.0\ mL} = 1.1 \times 10^{-2}\ M;\ pH = 1.96$

24.0 mL: $[H^+]_{excess} = \dfrac{(2.50 - 2.40)\ mmol}{49.0\ mL} = 2.0 \times 10^{-3}\ M;\ pH = 2.70$

24.5 mL: $[H^+]_{excess} = \dfrac{(2.50 - 2.45)\ mmol}{49.5\ mL} = 1 \times 10^{-3}\ M;\ pH = 3.0$

24.9 mL: $[H^+]_{excess} = \dfrac{(2.50 - 2.49)\ mmol}{49.9\ mL} = 2 \times 10^{-4}\ M;\ pH = 3.7$

25.0 mL: Equivalence point; We have a neutral solution since there is no excess H^+ or OH^- remaining after the neutralization reaction. pH = 7.00

25.1 mL: base in excess, $[OH^-]_{excess} = \dfrac{(2.51 - 2.50)\ mmol}{50.1\ mL} = 2 \times 10^{-4}\ M;\ pOH = 3.7;\ pH = 10.3$

26.0 mL: $[OH^-]_{excess} = \dfrac{(2.60 - 2.50)\ mmol}{51.0\ mL} = 2.0 \times 10^{-3}\ M;\ pOH = 2.70;\ pH = 11.30$

28.0 mL: $[OH^-]_{excess} = \dfrac{(2.80 - 2.50)\ mmol}{53.0\ mL} = 5.7 \times 10^{-3}\ M;\ pOH = 2.24;\ pH = 11.76$

30.0 mL: $[OH^-]_{excess} = \dfrac{(3.00 - 2.50)\ mmol}{55.0\ mL} = 9.1 \times 10^{-3}\ M;\ pOH = 2.04;\ pH = 11.96$

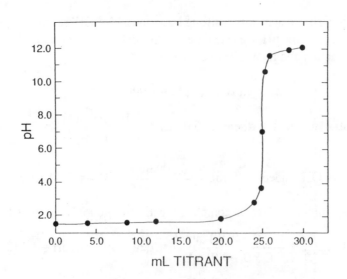

121. $HA + OH^- \rightarrow A^- + H_2O$ where HA = acetylsalicylic acid (assuming a monoprotic acid)

mmol HA present = $27.36 \text{ mL } OH^- \times \dfrac{0.5106 \text{ mmol } OH^-}{\text{mL } OH^-} \times \dfrac{1 \text{ mmol } HA}{\text{mmol } OH^-} = 13.97$ mmol HA

Molar mass of HA = $\dfrac{\text{grams}}{\text{moles}} = \dfrac{2.51 \text{ g HA}}{13.97 \times 10^{-3} \text{ mol HA}} = 180.$ g/mol

To determine the K_a value, use the pH data. After complete neutralization of acetylsalicylic acid by OH^-, we have 13.97 mmol of A^- produced from the neutralization reaction. A^- will react completely with the added H^+ and reform acetylsalicylic acid, HA.

mmol H^+ added = $13.68 \text{ mL} \times \dfrac{0.5106 \text{ mmol } H^+}{\text{mL}} = 6.985$ mmol H^+

	A^-	+	H^+	$\rightarrow$	HA	
Before	13.97 mmol		6.985 mmol		0	
Change	-6.985		-6.985	$\rightarrow$	+6.985	Reacts completely
After	6.985 mmol		0		6.985 mmol	

We have back titrated this solution to the halfway point to equivalence where pH = pK_a (assuming HA is a weak acid). We know this because after H^+ reacts completely, equal mmol of HA and A^- are present which only occurs at the halfway point to equivalence. Assuming acetylsalicylic acid is a weak acid, then pH = pK_a = 3.48. $K_a = 10^{-3.48} = 3.3 \times 10^{-4}$

122.   At the equivalence point, P^{2-} is the major species. It is a weak base in water since it is the conjugate base of a weak acid.

	P^{2-}	+	H_2O	$\rightleftharpoons$	HP^-	+	OH^-	
Initial	$\dfrac{0.5 \text{ g}}{0.1 \text{ L}} \times \dfrac{1 \text{ mol}}{204.22 \text{ g}} = 0.024 \, M$				0		~0	(carry extra sig. fig.)
Equil.	0.024 - x				x		x	

$K_b = \dfrac{[HP^-][OH^-]}{[P^{2-}]} = \dfrac{K_w}{K_a} = \dfrac{1.0 \times 10^{-14}}{10^{-5.51}} = 3.2 \times 10^{-9} = \dfrac{x^2}{0.024 - x} \approx \dfrac{x^2}{0.024}$

$x = [OH^-] = 8.8 \times 10^{-6} \, M$; pOH = 5.1; pH = 8.9   Assumptions good.

Phenolphthalein would be the best indicator for this titration since it changes color at pH ~ 9.

123. $Ca_5(PO_4)_3OH(s) \quad \rightleftharpoons \quad 5 \, Ca^{2+} + 3 \, PO_4^{3-} + OH^-$

		Ca^{2+}	PO_4^{3-}	OH^-
Initial	s = solubility (mol/L)	0	0	1.0×10^{-7}
Equil.		5s	3s	$s + 1.0 \times 10^{-7} \approx s$

$K_{sp} = 6.8 \times 10^{-37} = [Ca^{2+}]^5 [PO_4^{3-}]^3 [OH^-] = (5s)^5 (3s)^3 (s)$

$6.8 \times 10^{-37} = (3125)(27)s^9$, $s = 2.7 \times 10^{-5}$ mol/L   Assumption good.

The solubility of hydroxyapatite will increase as the solution gets more acidic since both phosphate and hydroxide can react with H^+.

$$Ca_5(PO_4)_3F(s) \quad \rightleftharpoons \quad 5\,Ca^{2+} \; + \; 3\,PO_4^{3-} \; + \; F^-$$

Initial	s = solubility (mol/L)	0	0	0
Equil.		5s	3s	s

$K_{sp} = 1 \times 10^{-60} = (5s)^5(3s)^3(s) = (3125)(27)s^9$, $s = 6 \times 10^{-8}$ mol/L

The hydroxyapatite in the tooth enamel is converted to the less soluble fluorapatite by fluoride treated water. The less soluble fluorapatite is more difficult to remove, making teeth less susceptible to decay.

124. $K_{sp} = 6.4 \times 10^{-9} = [Mg^{2+}]\,[F^-]^2$, $6.4 \times 10^{-9} = (0.00375 - y)(0.0625 - 2y)^2$

This is a cubic equation. No simplifying assumptions can be made since y is relatively large. Solving cubic equations is difficult. We could use a computer or a calculator to solve the equation by numerical methods. By the time we've done all that, we could have solved the problem several times over using the approximations based on our "chemical common sense."

125.   a. $$Pb(OH)_2(s) \quad \rightleftharpoons \quad Pb^{2+} \; + \; 2\,OH^-$$

Initial	s = solubility (mol/L)	0	$1.0 \times 10^{-7}\,M$ from water
Equil.		s	$1.0 \times 10^{-7} + 2s$

$K_{sp} = 1.2 \times 10^{-15} = [Pb^{2+}]\,[OH^-]^2 = s(1.0 \times 10^{-7} + 2s)^2 \approx s(2s^2) = 4s^3$

$s = [Pb^{2+}] = 6.7 \times 10^{-6}\,M$;   Assumption to ignore OH^- from water is good by the 5% rule.

b. $$Pb(OH)_2(s) \; \rightleftharpoons \; Pb^{2+} \; + \; 2\,OH^-$$

Initial		0	$0.10\,M$ pH = 13.00, $[OH^-] = 0.10\,M$
	s mol/L $Pb(OH)_2(s)$ dissolves to reach equilibrium		
Equil.		s	0.10 (buffered solution)

$1.2 \times 10^{-15} = (s)(0.10)^2$, $s = [Pb^{2+}] = 1.2 \times 10^{-13}\,M$

c.   We need to calculate the Pb^{2+} concentration in equilibrium with $EDTA^{4-}$. Since K is large for the formation of $PbEDTA^{2-}$, let the reaction go to completion then solve an equilibrium problem to get the Pb^{2+} concentration.

$$Pb^{2+} \quad + \quad EDTA^{4-} \quad \rightleftharpoons \quad PbEDTA^{2-} \quad K = 1.1 \times 10^{18}$$

Before	$0.010\,M$	$0.050\,M$	0

0.010 mol/L Pb^{2+} reacts completely (large K)

Change	-0.010	-0.010	$\rightarrow$	+0.010	Reacts completely
After	0	0.040		0.010	New initial

x mol/L $PbEDTA^{2-}$ dissociates to reach equilibrium

Equil.	x	$0.040 + x$	$0.010 - x$

$$1.1 \times 10^{18} = \frac{(0.010 - x)}{(x)\,(0.040 + x)} \approx \frac{(0.010)}{x(0.040)}, \quad x = [Pb^{2+}] = 2.3 \times 10^{-19}\,M \quad \text{Assumptions good.}$$

Now calculate the solubility quotient for $Pb(OH)_2$ to see if precipitation occurs. The concentration of OH^- is $0.10\,M$ since we have a solution buffered at pH = 13.00.

$$Q = [Pb^{2+}]_o\,[OH^-]_o^2 = (2.3 \times 10^{-19})(0.10)^2 = 2.3 \times 10^{-21} < K_{sp} \ (1.2 \times 10^{-15})$$

$Pb(OH)_2(s)$ will not form since Q is less than K_{sp}.

Challenge Problems

126.   At 4.0 mL NaOH added: $\left|\dfrac{\Delta pH}{\Delta mL}\right| = \left|\dfrac{2.43 - 3.14}{0 - 4.0}\right| = 0.18$

The other points are calculated in a similar fashion. The results are summarized and plotted below. As can be seen from the plot, the advantage of this approach is that it is much easier to accurately determine the location of the equivalence point.

mL	pH	$\lvert \Delta pH/\Delta mL \rvert$
0	2.43	-
4.0	3.14	0.18
8.0	3.53	0.098
12.5	3.86	0.073
20.0	4.46	0.080
24.0	5.24	0.20
24.5	5.6	0.7
24.9	6.3	2
25.0	8.28	20
25.1	10.3	20
26.0	11.29	1
28.0	11.75	0.23
30.0	11.96	0.11

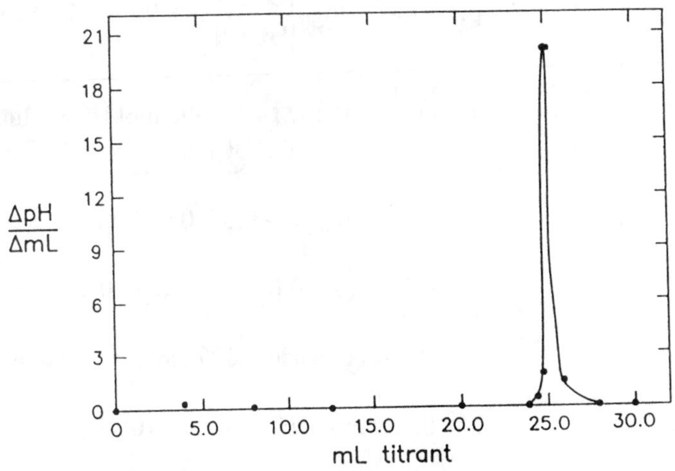

127.   At equivalence point: $16.00 \text{ mL} \times 0.125 \text{ mmol/mL} = 2.00 \text{ mmol OH}^-$ added; There must be 2.00 mmol HX present initially.

2.00 mL NaOH added = $2.00 \text{ mL} \times 0.125 \text{ mmol/mL} = 0.250 \text{ mmol OH}^-$; 0.250 mmol of OH^- added will convert 0.250 mmol HX into 0.250 mmol X^-. Remaining HX = 2.00 - 0.250 = 1.75 mmol HX; This is a buffer solution where $[H^+] = 10^{-6.912} = 1.22 \times 10^{-7} M$. Since total volume cancels:

$$K_a = \frac{[H^+][X^-]}{[HX]} = \frac{1.22 \times 10^{-7}(0.250)}{1.75} = 1.74 \times 10^{-8}$$

Note: We could solve for K_a using the Henderson-Hasselbalch equation.

128. $0.400 \text{ mol/L} \times V_{NH_3} = \text{mol NH}_3 = \text{mol NH}_4^+$ after reaction with HCl at the equivalence point.

At the equivalence point: $[NH_4^+]_o = \dfrac{\text{mol NH}_4^+}{\text{total volume}} = \dfrac{0.400 \times V_{NH_3}}{1.50 \times V_{NH_3}} = 0.267 \, M$

$$NH_4^+ \quad \rightleftharpoons \quad H^+ \quad + \quad NH_3$$

Initial	$0.267 \, M$	0	0
Equil.	$0.267 - x$	x	x

$$K_a = \frac{K_w}{K_b} = \frac{1.0 \times 10^{-14}}{1.8 \times 10^{-5}} = 5.6 \times 10^{-10} = \frac{x^2}{0.267 - x} \approx \frac{x^2}{0.267}$$

$x = [H^+] = 1.2 \times 10^{-5} M$; pH = 4.92   Assumptions good.

129.   For HOCl, $K_a = 3.5 \times 10^{-8}$ and $pK_a = -\log(3.5 \times 10^{-8}) = 7.46$; This will be a buffer solution since the pH is close to the pK_a value.

$$pH = pK_a + \log \frac{[OCl^-]}{[HOCl]}, \quad 8.00 = 7.46 + \log \frac{[OCl^-]}{[HOCl]}, \quad \frac{[OCl^-]}{[HOCl]} = 10^{0.54} = 3.5$$

1.00 L $\times$ 0.0500 M = 0.0500 mol HOCl initially. Added OH^- converts HOCl into OCl^-. The total moles of OCl^- and HOCl must equal 0.0500 mol. Solving where n = moles:

$n_{OCl^-} + n_{HOCl} = 0.0500$ and $n_{OCl^-} = 3.5 \, n_{HOCl}$

$4.5 \, n_{HOCl} = 0.0500$, $n_{HOCl} = 0.011$ mol; $n_{OCl^-} = 0.039$ mol

We need to add 0.039 mol NaOH to produce 0.039 mol OCl^-.

$0.039 \text{ mol OH}^- = V \times 0.0100 \, M$, V = 3.9 L NaOH

130. $50.0 \text{ mL} \times 0.100 \, M = 5.00 \text{ mmol H}_2SO_4$; $30.0 \text{ mL} \times 0.10 \, M = 3.0 \text{ mmol HOCl}$

$25.0 \text{ mL} \times 0.20 \, M = 5.0 \text{ mmol NaOH}$; $25.0 \text{ mL} \times 0.10 \, M = 2.5 \text{ mmol Ca(OH)}_2 = 5.00 \text{ mmol OH}^-$

10.0 mL $\times$ 0.15 M = 1.5 mmol KOH; We've added 11.5 mmol OH⁻ total.

Let the OH⁻ react with the best acid present. This is H_2SO_4 which is a diprotic acid. For H_2SO_4, $K_{a_1} \gg 1$ and $K_{a_2} = 1.2 \times 10^{-2}$. The reaction is:

10.0 mmol OH⁻ + 5.00 mmol H_2SO_4 → 10.0 mmol H_2O + 5.0 mmol SO_4^{2-}

Now we have 1.5 mmol of OH⁻ remaining which will react with the next best acid available, 3 mmol of HOCl ($K_a = 3.5 \times 10^{-8}$). The remaining 1.5 mmol OH⁻ will convert 1.5 mmol HOCl to OCl⁻, resulting in a solution containing 1.5 mmol OCl⁻ and 1.5 mmol HOCl. Major species at this point: HOCl, OCl⁻, SO_4^{2-}, H_2O plus cations that don't affect pH. SO_4^{2-} is an extremely weak base ($K_b = 8.3 \times 10^{-13}$). Major equilibrium affecting pH: HOCl $\rightleftharpoons$ H⁺ + OCl⁻. Since [HOCl] = [OCl⁻], then:

$$[H^+] = K_a = 3.5 \times 10^{-8}\, M; \; pH = 7.46$$

131. Phenolphthalein will change color at pH ~ 9. Phenolphthalein will mark the second end point. Therefore, at the phenolphthalein end point we will have titrated both protons on malonic acid.

$H_2Mal + 2\, OH^- \rightarrow 2\, H_2O + Mal^{2-}$ where H_2Mal = malonic acid

$$31.50 \text{ mL} \times \frac{0.0984 \text{ mmol NaOH}}{\text{mL}} \times \frac{1 \text{ mmol } H_2Mal}{2 \text{ mmol NaOH}} = 1.55 \text{ mmol } H_2Mal$$

$$[H_2Mal] = \frac{1.55 \text{ mmol}}{25.00 \text{ mL}} = 0.0620\, M$$

132. An indicator changes color at pH $\approx$ $pK_a \pm 1$. The results from each indicator tells us something about the pH. The conclusions are summarized below:

Results from	pH
bromphenol blue	$\geq$ ~ 5.0
bromcresol purple	$\leq$ ~ 5.0
bromcresol green *	pH ~ pK_a ~ 4.8 ~ 5.0
alizarin	$\leq$ ~ 5.5

*For bromcresol green, the resultant color is green.
 This is a combination of the extremes (yellow and blue).
 This occurs when pH ~ pK_a of the indicator.

From the indicator results, the pH of the solution is about 5.0. We solve for K_a by setting up the typical weak acid problem.

	HX	$\rightleftharpoons$	H⁺	+	X⁻
Initial	1.0 M		~0		0
Equil.	1.0 - x		x		x

$$K_a = \frac{[H^+][X^-]}{[HX]} = \frac{x^2}{1.0 - x}; \quad \text{Since pH} \sim 5.0, \text{ then } [H^+] = x \approx 1 \times 10^{-5} \, M.$$

$$K_a \approx \frac{(1 \times 10^{-5})^2}{1.0 - 1 \times 10^{-5}} \approx 1 \times 10^{-10}$$

133.

$$AgBr(s) \rightleftharpoons Ag^+ + Br^- \qquad K_{sp} = 5.0 \times 10^{-13}$$
$$Ag^+ + 2\,S_2O_3^{2-} \rightleftharpoons Ag(S_2O_3)_2^{3-} \qquad K_f = 2.9 \times 10^{13}$$

$$\overline{AgBr(s) + 2\,S_2O_3^{2-} \rightleftharpoons Ag(S_2O_3)_2^{3-} + Br^-} \qquad K = K_{sp} \times K_f = 14.5 \qquad \text{(Carry extra sig. figs.)}$$

$$AgBr(s) \quad + \quad 2\,S_2O_3^{3-} \quad \rightleftharpoons \quad Ag(S_2O_3)_2^{3-} \quad + \quad Br^-$$

Initial		$0.500 \, M$	0	0
		s mol/L AgBr(s) dissolves to reach equilibrium		
Change	$-s$	$-2s$ $\longrightarrow$	$+s$	$+s$
Equil.		$0.500 - 2s$	s	s

$$K = \frac{s^2}{0.500 - 2s} = 14.5; \quad \text{Using the quadratic equation since s is not small:}$$

$$s^2 = 7.25 - 29.0\,s, \quad s^2 + 29.0\,s - 7.25 = 0; \quad s = \frac{-29.0 + \sqrt{(29.0)^2 + 4(7.25)}}{2} = 0.248 \text{ mol/L}$$

$$1.00 \text{ L} \times \frac{0.248 \text{ mol AgBr}}{L} \times \frac{187.8 \text{ g AgBr}}{\text{mol AgBr}} = 46.6 \text{ g AgBr} = 47 \text{ g AgBr}$$

CHAPTER SIXTEEN

SPONTANEITY, ENTROPY, AND FREE ENERGY

Questions

7.   A spontaneous process is one that occurs without any outside intervention.

8.   Entropy is a measure of disorder or randomness.

9.   a.   Entropy increases; there is a greater volume accessible to the randomly moving gas molecules which increases disorder.

      b.   The positional entropy doesn't change. There is no change in volume and thus, no change in the numbers of positions of the molecules. The total entropy (ΔS_{univ}) increases because the increase in temperature increases the energy disorder (ΔS_{surr}).

      c.   Entropy decreases because the volume decreases (P and V are inversely related).

10.   a.   The system is the portion of the universe in which we are interested.

      b.   The surroundings are everything else in the universe besides the system.

11.   Living organisms need an external source of energy to carry out these processes. Green plants use the energy from sunlight to produce glucose from carbon dioxide and water by photosynthesis. In the human body, the energy released from the metabolism of glucose helps drive the synthesis of proteins. For all processes combined, ΔS_{univ} must be greater than zero (2nd law).

12.   No; When using ΔG_f° values in Appendix 4, we have specified a temperature of $25\,^\circ C$. Further, if gases or solutions are involved, we have specified partial pressures of 1 atm and solute concentrations of 1 molar. At other temperatures and compositions, the reaction may not be spontaneous. A negative ΔG° value means the reaction is spontaneous under <u>standard conditions</u>.

13.   Dispersion increases the entropy of the universe since the more widely something is dispersed, the greater the disorder. We must do work to overcome this disorder. In terms of the 2nd law, it would be more advantageous to prevent contamination of the environment rather than to clean it up later. As a substance disperses, we have a much larger area that must be decontaminated.

14. All thermodynamic functions depend on temperature. However, ΔH and ΔS are the least dependent on temperature and are commonly assumed to be temperature independent. ΔG has the strongest temperature dependence. To determine ΔG, the temperature must be known in order to use the equation $\Delta G = \Delta H - T\Delta S$.

15. $w_{max} = \Delta G$; When ΔG is negative, the magnitude of ΔG is equal to the maximum possible useful work obtainable from the process (at constant T and P). When ΔG is positive, the magnitude of ΔG is equal to the minimum amount of work that must be expended to make the process spontaneous. Due to waste energy (heat) in any real process, the amount of useful work obtainable from a spontaneous process is always less than w_{max} and for a nonspontaneous reaction, an amount of work greater than w_{max} must be applied to make the process spontaneous.

16. The rate of a reaction is directly related to temperature. As temperature increases, the rate of a reaction increases. Spontaneity, however, does not necessarily have a direct relationship to temperature. The temperature dependence of spontaneity depends on the signs to the values for ΔH and ΔS (see Table 16.5 of the text). For example, when ΔH and ΔS are both negative, the reaction becomes more favorable thermodynamically (ΔG becomes more negative) with decreasing temperature. This is just the opposite of the kinetics dependence on temperature. Other sign combinations of ΔH and ΔS have different spontaneity temperature dependence.

Exercises

Spontaneity, Entropy, and the Second Law of Thermodynamics: Free Energy

17. a, b and c; From our own experiences, salt water, colored water and rust form without any outside intervention. A bedroom, however, spontaneously gets cluttered. It takes an outside energy source to clean a bedroom.

18. c and d; It takes an outside energy source to build a house and to launch and keep a satellite in orbit.

19. We draw all of the possible arrangements of the two particles in the three levels.

2 kJ	___	___	x	___	x	xx
1 kJ	___	x	___	xx	x	___
0 kJ	xx	x	x	___	___	___
Total E =	0 kJ	1 kJ	2 kJ	2 kJ	3 kJ	4 kJ

The most likely total energy is 2 kJ.

20. There are more ways to roll a seven. We can consider all of the possible throws by constructing a table.

one die	1	2	3	4	5	6	
1	2	3	4	5	6	7	
2	3	4	5	6	7	8	
3	4	5	6	7	8	9	sum of the two dice
4	5	6	7	8	9	10	
5	6	7	8	9	10	11	
6	7	8	9	10	11	12	

There are six ways to get a seven, more than any other number. The seven is not favored by energy; rather it is favored by probability. To change the probability we would have to expend energy (do work).

21.   a. H_2 at 100°C and 0.5 atm; Higher temperature and lower pressure means greater volume and hence, greater positional entropy.

b. N_2 at STP has the greater volume.   c. $H_2O(l)$ is more disordered than $H_2O(s)$.

22.   a.   He at 1 atm; Greater volume of gas, hence, a greater positional disorder.

b.   He at STP; Higher T and lower P leads to a greater volume.

c.   He(g) at 5 K; A gas is much more disordered than a solid and at 0 K, all substances have no motion so all have S = 0 (assuming a perfect crystal).

23.   Of the three phases (solid, liquid, and gas), solids are most ordered and gases are most disordered. Thus, a and b (melting and sublimation) involve an increase in the entropy of the system since going from a solid to a liquid or a solid to a gas increases disorder. For freezing (process c), a substance goes from the more disordered liquid state to the more ordered solid state, hence, entropy decreases.

24.   Processes a and c (mixing and boiling) involve an increase in disorder (entropy). Separation increases order (decreases the entropy of the system).

25.   a.   To boil a liquid requires heat. Hence, this is an endothermic process. All endothermic processes decrease the entropy of the surroundings (ΔS_{surr} is negative).

b.   This is an exothermic process. Heat is released when gas molecules slow down enough to form the solid. In exothermic processes, the entropy of the surroundings increases (ΔS_{surr} is positive).

26.   a. $\Delta S_{surr} = \dfrac{-\Delta H}{T} = \dfrac{-890 \text{ kJ}}{298 \text{ K}} = -3.0 \text{ kJ/K} = -3.0 \times 10^3 \text{ J/K}$

b. $\Delta S_{surr} = \dfrac{-\Delta H}{T} = \dfrac{43\ kJ}{298\ K} = 0.14\ kJ/K = 140\ J/K$

27. $\Delta G = \Delta H - T\Delta S$;   When ΔG is negative, then the process will be spontaneous.

a. $\Delta G = \Delta H - T\Delta S = 25 \times 10^3\ J - (300.\ K)(5.0\ J/K) = 24,000\ J$,   Not spontaneous

b. $\Delta G = 25,000\ J - (300.\ K)(100.\ J/K) = -5000\ J$,   Spontaneous

c.   Without calculating ΔG, we know this reaction will be spontaneous at all temperatures. ΔH is negative and ΔS is positive ($-T\Delta S < 0$). ΔG will always be less than zero with these sign combinations for ΔH and ΔS.

d. $\Delta G = -1.0 \times 10^4\ J - (200.\ K)(-40.\ J/K) = -2000\ J$,   Spontaneous

28. $\Delta G = \Delta H - T\Delta S$;   A process is spontaneous when $\Delta G < 0$. For the following, assume ΔH and ΔS are temperature independent.

a.   When ΔH and ΔS are both negative, ΔG will be negative below a certain temperature where the favorable ΔH term dominates. When $\Delta G = 0$, then $\Delta H = T\Delta S$. Solving for this temperature:

$$T = \dfrac{\Delta H}{\Delta S} = \dfrac{-25,000\ J}{-5.0\ J/K} = 5.0 \times 10^3\ K$$

At $T < 5.0 \times 10^3\ K$, this process will be spontaneous ($\Delta G < 0$).

b.   When ΔH and ΔS are both positive, ΔG will be negative above a certain temperature where the favorable ΔS term dominates.

$$T = \dfrac{\Delta H}{\Delta S} = \dfrac{25,000\ J}{5.0\ J/K} = 5.0 \times 10^3\ K$$

At $T > 5.0 \times 10^3\ K$, this process will be spontaneous ($\Delta G < 0$).

c.   When ΔH is positive and ΔS is negative, this process can never be spontaneous since ΔG can never be negative.

d.   When ΔH is negative and ΔS is positive, this process is spontaneous at all temperatures since ΔG will always be negative.

29.   At the boiling point, $\Delta G = 0$ so $\Delta H = T\Delta S$.

$$\Delta S = \dfrac{\Delta H}{T} = \dfrac{31.4\ kJ/mol}{(273.2 + 61.7)K} = 9.38 \times 10^{-2}\ kJ/K \bullet mol = 93.8\ J/K \bullet mol$$

30.   At the boiling point, $\Delta G = 0$ so $\Delta H = T\Delta S$. $T = \dfrac{\Delta H}{\Delta S} = \dfrac{58.51 \times 10^3\ J/mol}{92.92\ J/K \bullet mol} = 629.7\ K$

31.   a. $NH_3(s) \rightarrow NH_3(l)$; $\Delta G = \Delta H - T\Delta S = 5650$ J/mol - 200. K (28.9 J/K•mol)

$\Delta G = 5650$ J/mol - 5780 J/mol = -130 J/mol

Yes, NH_3 will melt since $\Delta G < 0$ at this temperature.

b.   At the melting point, $\Delta G = 0$ so $T = \dfrac{\Delta H}{\Delta S} = \dfrac{5650 \text{ J/mol}}{28.9 \text{ J/K•mol}} = 196$ K.

32.   Rhombic $\rightarrow$ Monoclinic; ΔH is (+) and ΔG is (-) above 95°C, thus ΔS must be positive.

At 95°C, $\Delta G = 0$; $\Delta S = \dfrac{\Delta H}{T} = \dfrac{(0.30 \times 10^3 \text{ J/mol})}{(95 + 273)\text{K}} = 0.82$ J/K•mol

Chemical Reactions: Entropy Changes and Free Energy

33.   a.   Decrease in disorder; $\Delta S°$(-)   b.   Increase in disorder; $\Delta S°$(+)

c.   Decrease in disorder ($\Delta n < 0$); $\Delta S°$(-)   d.   Increase in disorder ($\Delta n > 0$); $\Delta S°$(+)

For c and d, concentrate on the gaseous products and reactants. When there are more gaseous product molecules than gaseous reactant molecules ($\Delta n > 0$), then $\Delta S°$ will be positive (disorder increases). When Δn is negative then $\Delta S°$ is negative (disorder decreases).

34.   a.   Decrease in disorder ($\Delta n < 0$); $\Delta S°$(-)   b.   Decrease in disorder ($\Delta n < 0$); $\Delta S°$(-)

c.   Increase in disorder; $\Delta S°$(+)   d.   Increase in disorder; $\Delta S°$(+)

35.   a. $C_{12}H_{22}O_{11}$; Larger molecule, more parts, more disorder so larger S value.

b. H_2O (0°C); A substance at 0 K has S = 0. As temperature increases, S increases.

c. $H_2S(g)$; A gas has greater disorder than a liquid.

36.   a.   He (10 K); S = 0 at 0 K   b. N_2O; More complicated molecule

c. $H_2O(l)$:   The liquid state is more disordered than the solid state.

37.   a. $H_2(g)$ + 1/2 $O_2(g) \rightarrow H_2O(g)$;   Since there are more molecules of reactant gases as compared to product molecules of gas ($\Delta n < 0$), then $\Delta S°$ will be negative. $\Delta S° = \Sigma n_p S°_{products} - \Sigma n_r S°_{reactants}$

$\Delta S° = 1$ mol $H_2O(g)$(189 J/K•mol) - [1 mol $H_2(g)$(131 J/K•mol) + 1/2 mol $O_2(g)$(205 J/K•mol)]

$\Delta S° = 189$ J/K - 234 J/K = -45 J/K

b.   3 $O_2(g) \rightarrow 2 O_3(g)$;   Since Δn of gases is negative, then $\Delta S°$ will be negative.

$\Delta S° = 2$ mol(239 J/K•mol) - [3 mol(205 J/K•mol)] = -137 J/K

c. $N_2(g) + O_2(g) \rightarrow 2\,NO(g)$; Here Δn of gases = 2 - 2 = 0. We can't easily predict if $\Delta S°$ is positive or negative.

$\Delta S° = 2(211) - (192 + 205) = 25$ J/K

38. a. $H_2(g) + 1/2\,O_2(g) \rightarrow H_2O(l)$; Since Δn of gases is negative, then $\Delta S°$ will be negative.

$\Delta S° = 1$ mol $H_2O(l)(70.$ J/K•mol) - [1 mol $H_2(g)(131$ J/K•mol) + 1/2 mol $O_2(g)(205$ J/K•mol)]

$\Delta S° = 70.$ J/K - 234 J/K = -164 J/K

b. $N_2(g) + 3\,H_2(g) \rightarrow 2\,NH_3(g)$; Since Δn of gases is negative, then $\Delta S°$ will ne negative.

$\Delta S° = 2(193) - [1(192) + 3(131)] = -199$ J/K

c. $HCl(g) \rightarrow H^+(aq) + Cl^-(aq)$; The gaseous state dominates predictions of $\Delta S°$. Here the gaseous state is more disordered than the ions in solution so $\Delta S°$ will be negative.

$\Delta S° = 1$ mol $H^+(0) + 1$ mol $Cl^-(57$ J/K•mol) - 1 mol HCl(187 J/K•mol) = -130. J/K

39. $CS_2(g) + 3\,O_2(g) \rightarrow CO_2(g) + 2\,SO_2(g)$; $\Delta S° = S°_{CO_2} + 2\,S°_{SO_2} - [3\,S°_{O_2} + S°_{CS_2}]$

-143 J/K = 214 J/K + 2(248 J/K) - 3(205 J/K) - (1 mol)$S°_{CS_2}$, $S°_{CS_2} = 238$ J/K•mol

40. -144 J/K = (2 mol)$S°_{AlBr_3}$ - [2(28 J/K) + 3(152 J/K)], $S°_{AlBr_3} = 184$ J/K•mol

41. $P_4(s,\alpha) \rightarrow P_4(s,\beta)$

a. At T < -76.9°C, this reaction is spontaneous and the sign of ΔG is (-). At 76.9°C, $\Delta G = 0$ and above -76.9 °C, the sign of ΔG is (+). This is consistent with ΔH (-) and ΔS (-).

b. Since the sign of ΔS is negative, then the β form has the more ordered structure.

42. Enthalpy is not favorable, so ΔS must provide the driving force for the change. Thus, ΔS is positive. There is an increase in disorder, so the original enzyme has the more ordered structure.

43. a. A bond is broken which requires energy so ΔH is positive. Since there are more product molecules of gas than reactant molecules of gas ($\Delta n > 0$), then ΔS will be positive.

b. $\Delta G = \Delta H - T\Delta S$; For this reaction to be spontaneous, the favorable entropy term must dominate. The reaction will be spontaneous at higher temperatures where the ΔS term dominates.

44. Since there are more product gas molecules than reactant gas molecules ($\Delta n > 0$), then ΔS will be positive. From the signs of ΔH and ΔS, this reaction is spontaneous at all temperatures. It will cost money to heat the reaction mixture. Since there is no thermodynamic reason to do this, then the purpose of the elevated temperature must be to increase the rate of the reaction, i.e., kinetic reasons.

45.   a. $CH_4(g)$ + $2 O_2(g)$ → $CO_2(g)$ + $2 H_2O(g)$

ΔH_f°	-75 kJ/mol	0	-393.5	-242
ΔG_f°	-51 kJ/mol	0	-394	-229
S°	186 J/K•mol	205	214	189

Data from Appendix 4

$\Delta H^\circ = \Sigma n_p \Delta H_{f, products}^\circ - \Sigma n_r \Delta H_{f, reactants}^\circ$; $\Delta S^\circ = \Sigma n_p S_{products}^\circ - \Sigma n_r S_{reactants}^\circ$

$\Delta H^\circ = 2 \text{ mol}(-242 \text{ kJ/mol}) + 1 \text{ mol}(-393.5 \text{ kJ/mol}) - [1 \text{ mol}(-75 \text{ kJ/mol})] = -803 \text{ kJ}$

$\Delta S^\circ = 2 \text{ mol}(189 \text{ J/K•mol}) + 1 \text{ mol}(214 \text{ J/K•mol})$

$- [1 \text{ mol}(186 \text{ J/K•mol}) + 2 \text{ mol}(205 \text{ J/K•mol})] = -4 \text{ J/K}$

There are two ways to get ΔG°. We can use $\Delta G^\circ = \Delta H^\circ - T\Delta S^\circ$ (be careful of units):

$\Delta G^\circ = \Delta H^\circ - T\Delta S^\circ = -803 \times 10^3 \text{ J} - (298 \text{ K})(-4 \text{ J/K}) = -8.018 \times 10^5 \text{ J} = -802 \text{ kJ}$

or we can use ΔG_f° values where $\Delta G^\circ = \Sigma n_p \Delta G_{f, products}^\circ - \Sigma n_r \Delta G_{f, reactants}^\circ$:

$\Delta G^\circ = 2 \text{ mol}(-229 \text{ kJ/mol}) + 1 \text{ mol}(-394 \text{ kJ/mol}) - [1 \text{ mol}(-51 \text{ kJ/mol})]$

$\Delta G^\circ = -801 \text{ kJ}$ (Answers are the same within round off error)

b. $6 CO_2(g)$ + $6 H_2O(l)$ → $C_6H_{12}O_6(s)$ + $6 O_2(g)$

ΔH_f°	-393.5 kJ/mol	-286	-1275	0
S°	214 J/K•mol	70.	212	205

$\Delta H^\circ = -1275 - [6(-286) + 6(-393.5)] = 2802 \text{ kJ}$

$\Delta S^\circ = 6(205) + 212 - [6(214) + 6(70.)] = -262 \text{ J/K}$

$\Delta G^\circ = 2802 \text{ kJ} - (298 \text{ K})(-0.262 \text{ kJ/K}) = 2880. \text{ kJ}$

c.

$$P_4O_{10}(s) + 6 H_2O(l) \rightarrow 4 H_3PO_4(s)$$

ΔH_f° (kJ/mol)	-2984	-286	-1279
S° (J/K•mol)	229	70.	110.

$\Delta H^\circ = 4$ mol(-1279 kJ/mol) - [1 mol(-2984 kJ/mol) + 6 mol(-286 kJ/mol)] = -416 kJ

$\Delta S^\circ = 4(110.) - [229 + 6(70.)] = -209$ J/K

$\Delta G^\circ = \Delta H^\circ - T\Delta S^\circ = -416$ kJ - (298 K)(-0.209 kJ/K) = -354 kJ

d.

$$HCl(g) + NH_3(g) \rightarrow NH_4Cl(s)$$

ΔH_f° (kJ/mol)	-92	-46	-314
S° (J/K•mol)	187	193	96

$\Delta H^\circ = -314 - [-92 - 46] = -176$ kJ; $\Delta S^\circ = 96 - [187 + 193] = -284$ J/K

$\Delta G^\circ = \Delta H^\circ - T\Delta S^\circ = -176$ kJ - (298 K)(-0.284 kJ/K) = -91 kJ

46. $\Delta G^\circ = -58.03$ kJ - (298 K)(-0.1766 kJ/K) = -5.40 kJ

$$\Delta G^\circ = 0 = \Delta H^\circ - T\Delta S^\circ, \quad T = \frac{\Delta H^\circ}{\Delta S^\circ} = \frac{-58.03 \text{ kJ}}{-0.1766 \text{ kJ/K}} = 328.6 \text{ K}$$

ΔG° is negative below 328.6 K where the favorable ΔH° term dominates.

47. $$CH_4(g) + CO_2(g) \rightarrow CH_3CO_2H(l)$$

$\Delta H^\circ = -484 - [-75 + (-393.5)] = -16$ kJ; $\Delta S^\circ = 160 - [186 + 214] = -240.$ J/K

$\Delta G^\circ = \Delta H^\circ - T\Delta S^\circ = -16$ kJ - (298 K)(-0.240 kJ/K) = 56 kJ

This reaction is spontaneous only at temperatures below T = $\Delta H^\circ/\Delta S^\circ$ = 67 K (where the favorable ΔH° term will dominate). This is not practical. Substances will be in condensed phases and rates will be very slow at this extremely low temperature.

$CH_3OH(g) + CO(g) \rightarrow CH_3CO_2H(l)$

$\Delta H° = -484 - [-110.5 + (-201)] = -173 \text{ kJ}; \ \Delta S° = 160 - [198 + 240.] = -278 \text{ J/K}$

$\Delta G° = -173 \text{ kJ} - (298 \text{ K})(-0.278 \text{ kJ/K}) = -90. \text{ kJ}$

This reaction also has a favorable enthalpy and an unfavorable entropy term. This reaction is spontaneous at temperatures below $T = \Delta H°/\Delta S° = 622 \text{ K}$. The reaction of CH_3OH and CO will be preferred. It is spontaneous at high enough temperatures that the rates of reaction should be reasonable.

48. $C_2H_4(g) + H_2O(g) \rightarrow CH_3CH_2OH(l)$

$\Delta H° = -278 - (52 - 242) = -88 \text{ kJ}; \ \Delta S° = 161 - (219 + 189) = -247 \text{ J/K}$

When $\Delta G° = 0$, $\Delta H° = T\Delta S°$, $T = \dfrac{\Delta H°}{\Delta S°} = \dfrac{-88 \times 10^3 \text{ J}}{-247 \text{ J/K}} = 360 \text{ K}$

Since the signs of $\Delta H°$ and $\Delta S°$ are both negative, this reaction will be spontaneous at temperatures below 360 K (where the favorable $\Delta H°$ term will dominate).

$C_2H_6(g) + H_2O(g) \rightarrow CH_3CH_2OH(l) + H_2(g)$

$\Delta H° = -278 - (-84.7 - 242) = 49 \text{ kJ}; \ \Delta S° = 131 + 161 - (229.5 + 189) = -127 \text{ J/K}$

This reaction can never be spontaneous because of the signs of $\Delta H°$ and $\Delta S°$.

Thus the reaction $C_2H_4(g) + H_2O(g) \rightarrow C_2H_5OH(l)$ would be preferred.

49. $\begin{array}{ll} SO_3(g) \rightarrow SO_2(g) + 1/2\ O_2(g) & \Delta G° = -1/2\ (-142 \text{ kJ}) \\ S(s) + 3/2\ O_2(g) \rightarrow SO_3(g) & \Delta G° = -371 \text{ kJ} \end{array}$

$\begin{array}{ll} S(s) + O_2(g) \rightarrow SO_2(g) & \Delta G° = 71 \text{ kJ} - 371 \text{ kJ} = -300. \text{ kJ} \end{array}$

50. $\begin{array}{ll} 2\ CO_2(g) + H_2O(l) \rightarrow C_2H_2(g) + 5/2\ O_2(g) & \Delta G° = -(-1234 \text{ kJ}) \\ 2\ C(s) + 2\ O_2(g) \rightarrow 2\ CO_2(g) & \Delta G° = 2(-394 \text{ kJ}) \\ H_2(g) + 1/2\ O_2(g) \rightarrow H_2O(l) & \Delta G° = -237 \text{ kJ} \end{array}$

$\begin{array}{ll} 2\ C(s) + H_2(g) \rightarrow C_2H_2(g) & \Delta G° = 209 \text{ kJ} \end{array}$

51. $\Delta G° = \Sigma n_p \Delta G°_{f, \text{ products}} - \Sigma n_r \Delta G°_{f, \text{ reactants}}, \ -374 \text{ kJ} = -1105 \text{ kJ} - \Delta G°_{f, SF_4}, \ \Delta G°_{f, SF_4} = -731 \text{ kJ/mol}$

52. $7 \text{ kJ} = 3(-229 \text{ kJ}) - 1582 \text{ kJ} - (2 \text{ mol}) \Delta G°_{f, Al(OH)_3}, \ \Delta G°_{f, Al(OH)_3} = -1138 \text{ kJ/mol}$

53. $\Delta G° = \Sigma n_p \Delta G°_{f, \text{ products}} - \Sigma n_r \Delta G°_{f, \text{ reactants}}$

$\Delta G° = 1 \text{ mol}(-2698 \text{ kJ/mol}) + 6 \text{ mol}(-237 \text{ kJ/mol}) - [4 \text{ mol}(13 \text{ kJ/mol}) + 8 \text{ mol}(0)] = -4172 \text{ kJ}$

54.   a. $\Delta G° = 2 \text{ mol}(0) + 3 \text{ mol}(-229 \text{ kJ/mol}) - [1 \text{ mol}(-740. \text{ kJ/mol}) + 3 \text{ mol}(0)] = 53 \text{ kJ}$

b.   Since $\Delta G°$ is positive, then this reaction is not spontaneous at standard conditions and 298 K.

c. $\Delta G° = \Delta H° - T\Delta S°$, $\Delta S° = \dfrac{\Delta H° - \Delta G°}{T} = \dfrac{100.\ kJ - 53\ kJ}{298\ K} = 0.16\ kJ/K$

We need to solve for the temperature when $\Delta G° = 0$:

$$\Delta G° = 0 = \Delta H° - T\Delta S°,\ \Delta H° = T\Delta S°,\ T = \dfrac{\Delta H°}{\Delta S°} = \dfrac{100.\ kJ}{0.16\ kJ/K} = 630\ K$$

This reaction will be spontaneous ($\Delta G < 0$) at $T > 630$ K where the favorable entropy term will dominate.

Free Energy: Pressure Dependence and Equilibrium

55. $\Delta G = \Delta G° + RT \ln Q$; For this reaction: $\Delta G = \Delta G° + RT \ln \dfrac{P_{NO_2} \times P_{O_2}}{P_{NO} \times P_{O_3}}$

$\Delta G° = 1\ mol(52\ kJ/mol) + 1\ mol(0) - [1\ mol(87\ kJ/mol) + 1\ mol(163\ kJ/mol)] = -198\ kJ$

$\Delta G = -198\ kJ + \dfrac{8.3145\ J/K \bullet mol}{1000\ J/kJ}\ (298\ K)\ \ln \dfrac{(1.00 \times 10^{-7}\ atm)\ (1.00 \times 10^{-3}\ atm)}{(1.00 \times 10^{-6}\ atm)\ (2.00 \times 10^{-6}\ atm)}$

$\Delta G = -198\ kJ + 9.69\ kJ = -188\ kJ$

56. $\Delta G° = 3(0) + 2(-229) - [2(-34) + 1(-300.)] = -90.\ kJ$

$\Delta G = \Delta G° + RT \ln \dfrac{P_{H_2O}^2}{P_{H_2S}^2 \times P_{SO_2}} = -90.\ kJ + \dfrac{(8.3145)\ (298)}{1000}\ kJ\left[\ln \dfrac{(0.030)^2}{(1.0 \times 10^{-4})^2\ (0.010)}\right]$

$\Delta G = -90.\ kJ + 39.7\ kJ = -50.\ kJ$

57. $\Delta G = \Delta G° + RT \ln Q = \Delta G° + RT \ln\left(\dfrac{P_{SO_2}^2 \times P_{O_2}}{P_{SO_3}^2}\right)$

$\Delta G° = 2\ mol(-300.\ kJ/mol) - 2\ mol(-371\ kJ/mol) = 142\ kJ$

a.   These are standard conditions so $\Delta G = \Delta G°$ since $Q = 1$ and $\ln Q = 0$. Since $\Delta G°$ is positive, then the reverse reaction is spontaneous. The reaction shifts left to reach equilibrium.

b. $\Delta G = 142 \times 10^3\ J + 8.3145\ J/K \bullet mol\ (298\ K)\ \ln \dfrac{(1.00 \times 10^{-9})^2\ (1.00 \times 10^{-9})}{(1.00)^2}$

$\Delta G = 1.42 \times 10^5\ J - 1.54 \times 10^5\ J = -1.2 \times 10^4\ J$

Since ΔG is negative, then the forward reaction is spontaneous so the reaction shifts to the right to reach equilibrium.

58. $\Delta H^\circ = 2\ \Delta H^\circ_{f,\ NH_3} = 2(-46) = -92\ kJ;\quad \Delta G^\circ = 2\ \Delta G^\circ_{f,\ NH_3} = 2(-17) = -34\ kJ$

$\Delta S^\circ = 2(193\ J/K) - [192\ J/K + 3(131\ J/K)] = -199\ J/K;\quad \Delta G^\circ = -RT\ \ln K$

$K = \exp\ \dfrac{-\Delta G^\circ}{RT} = \exp\left(\dfrac{-(-34{,}000\ J)}{(8.3145\ J/K\bullet mol)\,(298\ K)}\right) = e^{13.72} = 9.1 \times 10^5$

Note: When determining exponents, we will round off after the calculation is complete. This helps eliminate excessive round off error.

a. $\Delta G = \Delta G^\circ + RT\ \ln\ \dfrac{P^2_{NH_3}}{P_{N_2} \times P^3_{H_2}} = -34\ kJ + \dfrac{(8.3145\ J/K\bullet mol)\,(298\ K)}{1000\ J/kJ}\ \ln\ \dfrac{(50.)^2}{(200.)\,(200.)^3}$

$\Delta G = -34\ kJ - 33\ kJ = -67\ kJ$

b. $\Delta G = -34\ kJ + \dfrac{(8.3145\ J/K\bullet mol)\,(298\ K)}{1000\ J/kJ}\ \ln\ \dfrac{(200.)^2}{(200.)\,(600.)^3}$

$\Delta G = -34\ kJ - 34.4\ kJ = -68\ kJ$

59. $NO(g) + O_3(g) \rightleftharpoons NO_2(g) + O_2(g);\quad \Delta G^\circ = \Sigma n_p \Delta G^\circ_{f,\ products} - \Sigma n_r \Delta G^\circ_{f,\ reactants}$

$\Delta G^\circ = 1\ mol(52\ kJ/mol) - [1\ mol(87\ kJ/mol) + 1\ mol(163\ kJ/mol)] = -198\ kJ$

$\Delta G^\circ = -RT\ \ln K,\ K = \exp\ \dfrac{-\Delta G^\circ}{RT} = \exp\left(\dfrac{-(-1.98 \times 10^5\ J)}{8.3145\ J/K\bullet mol\,(298\ K)}\right) = e^{79.912} = 5.07 \times 10^{34}$

Note: When determining exponents, we will round off after the calculation is complete. This helps eliminate excessive round off error.

60. $\Delta G^\circ = 2\ mol\,(-229\ kJ/mol) - [2\ mol\,(-34\ kJ/mol) + 1\ mol\,(-300.\ kJ/mol)] = -90.\ kJ$

$K = \exp\ \dfrac{-\Delta G^\circ}{RT} = \exp\left(\dfrac{-(-90. \times 10^3\ J)}{(8.3145\ J/K\bullet mol)\,(298\ K)}\right) = e^{36.32} = 5.9 \times 10^{15}$

Since there is a decrease in the number of moles of gaseous particles, ΔS° is negative. Since ΔG° is negative, then ΔH° must be negative. The reaction will be spontaneous at low temperatures (the favorable ΔH° term dominates at low temperatures).

61.   a. $\Delta H^\circ = 2\ mol(-92\ kJ/mol) - [1\ mol(0) + 1\ mol(0)] = -184\ kJ$

$\Delta S^\circ = 2\ mol(187\ J/K\bullet mol) - [1\ mol(131\ J/K\bullet mol) + 1\ mol(223\ J/K\bullet mol)] = 20.\ J/K$

$\Delta G^\circ = \Delta H^\circ - T\Delta S^\circ = -184 \times 10^3\ J - 298\ K\,(20.\ J/K) = -1.90 \times 10^5\ J = -190.\ kJ$

$\Delta G^\circ = -RT\ \ln K,\ \ln K = \dfrac{-\Delta G^\circ}{RT} = \dfrac{-(-1.90 \times 10^5\ J)}{8.3145\ J/K\bullet mol\,(298\ K)} = 76.683,\ K = e^{76.683} = 2.01 \times 10^{33}$

b. These are standard conditions so $\Delta G = \Delta G° = -190.$ kJ. Since ΔG is negative, then the forward reaction is spontaneous so the reaction shifts right to reach equilibrium.

62.　a.

	$\Delta H_f°$ (kJ/mol)	$S°$ (J/K•mol)
$NH_3(g)$	-46	193
$O_2(g)$	0	205
$NO(g)$	90.	211
$H_2O(g)$	-242	189
$NO_2(g)$	34	240.
$HNO_3(l)$	-174	156
$H_2O(l)$	-286	70.

$$4\ NH_3(g) + 5\ O_2(g) \rightarrow 4\ NO(g) + 6\ H_2O(g)$$

$$\Delta H° = 6(-242) + 4(90.) - [4(-46)] = -908\ \text{kJ}$$

$$\Delta S° = 4(211) + 6(189) - [4(193) + 5(205)] = 181\ \text{J/K}$$

$$\Delta G° = -908\ \text{kJ} - 298\ \text{K}\ (0.181\ \text{kJ/K}) = -962\ \text{kJ}$$

$$\Delta G° = -\ RT \ln K, \quad \ln K = \frac{-\Delta G°}{RT} = \left(\frac{-(-962 \times 10^3\ \text{J})}{8.3145\ \text{J/K•mol} \times 298\ \text{K}}\right) = 388$$

$\ln K = 2.303 \log K, \quad \log K = 168, \quad K = 10^{168}$ (an extremely large value)

$$2\ NO(g) + O_2(g) \rightarrow 2\ NO_2(g)$$

$$\Delta H° = 2(34) - [2(90.)] = -112\ \text{kJ}; \quad \Delta S° = 2(240.) - [2(211) + (205)] = -147\ \text{J/K}$$

$$\Delta G° = -112\ \text{kJ} - (298\ \text{K})(-0.147\ \text{kJ/K}) = -68\ \text{kJ}$$

$$K = \exp \frac{-\Delta G°}{RT} = \exp\left(\frac{-(-68,000\ \text{J})}{8.3145\ \text{J/K•mol}\ (298\ \text{K})}\right) = e^{27.44} = 8.3 \times 10^{11}$$

Note: When determining exponents, we will round off after the calculation is complete.

$$3\ NO_2(g) + H_2O(l) \rightarrow 2\ HNO_3(l) + NO(g)$$

$$\Delta H° = 2(-174) + (90.) - [3(34) + (-286)] = -74\ \text{kJ}$$

$$\Delta S° = 2(156) + (211) - [3(240.) + (70.)] = -267\ \text{J/K}$$

$$\Delta G° = -74\ \text{kJ} - (298\ \text{K})(-0.267\ \text{kJ/K}) = 6\ \text{kJ}$$

$$K = \exp \frac{-\Delta G°}{RT} = \exp\left(\frac{-6000\ \text{J}}{8.3145\ \text{J/K•mol}\ (298\ \text{K})}\right) = e^{-2.4} = 9 \times 10^{-2}$$

b. $\Delta G° = -RT \ln K$; $T = 825°C = (825 + 273)K = 1098$ K; We must determine $\Delta G°$ at 1098 K.

$\Delta G°_{1098} = \Delta H° - T\Delta S° = -908$ kJ $- (1098$ K$)(0.181$ kJ/K$) = -1107$ kJ

$K = \exp \dfrac{-\Delta G°}{RT} = \exp\left(\dfrac{-(-1.107 \times 10^6 \text{ J})}{8.3145 \text{ J/K•mol } (1098 \text{ K})} \right) = e^{121.258} = 4.589 \times 10^{52}$

c.   There is no thermodynamic reason for the elevated temperature since $\Delta H°$ is negative and $\Delta S°$ is positive. Thus, the purpose for the high temperature must be to increase the rate of the reaction.

63.   a. $\Delta G° = -RT \ln K = -\dfrac{8.3145 \text{ J}}{\text{K mol}}$ (298 K) $\ln (1.00 \times 10^{-14}) = 7.99 \times 10^4$ J $= 79.9$ kJ/mol

b. $\Delta G°_{313} = -RT \ln K = -\dfrac{8.3145 \text{ J}}{\text{K mol}}$ (313 K) $\ln (2.92 \times 10^{-14}) = 8.11 \times 10^4$ J $= 81.1$ kJ/mol

64.   a. $\Delta G° = 3(191.2) - 78.2 = 495.4$ kJ; $\Delta H° = 3(241.3) - 132.8 = 591.1$ kJ

$\Delta S° = \dfrac{\Delta H° - \Delta G°}{T} = \dfrac{591.1 \text{ kJ} - 495.4 \text{ kJ}}{298 \text{ K}} = 0.321$ kJ/K $= 321$ J/K

b. $\Delta G° = -RT \ln K$, $\ln K = \dfrac{-\Delta G°}{RT} = \dfrac{-495,400 \text{ J}}{8.3145 \text{ J/K•mol } (298 \text{ K})} = -199.942$

$K = e^{-199.942} = 1.47 \times 10^{-87}$

c.   Assuming $\Delta H°$ and $\Delta S°$ are temperature independent:

$\Delta G°_{3000} = 591.1$ kJ $- 3000.$ K $(0.321$ kJ/K$) = -372$ kJ

$\ln K = \dfrac{-(-372,000 \text{ J})}{8.3145 \text{ J/K•mol } (3000. \text{K})} = 14.914$, $K = e^{14.914} = 3.00 \times 10^6$

65.   The equation $\ln K = \dfrac{-\Delta H}{R}\left(\dfrac{1}{T}\right) + \dfrac{\Delta S°}{R}$ is in the form of a straight line equation $(y = mx + b)$.

Therefore, if we graph $\ln K$ vs $1/T$ we get a straight line with slope $= m = -\Delta H°/R$. For an endothermic process, the slope is negative since $\Delta H°$ is positive. A sketch of the plot would look like:

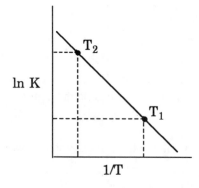

As we increase the temperature from T_1 to T_2 $(1/T_2 < 1/T_1)$, we can see that $\ln K$ and, hence, K increases. A larger value of K means some more reactants can be converted to products (reaction shifts right).

66.   A graph of ln K vs. 1/T will yield a straight line with slope equal to $-\Delta H°/R$ and y-intercept equal to $\Delta S°/R$ (see Exercise 16.65).

a.

Temp (°C)	T(K)	1000/T (K^{-1})	K_w	ln K_w
0	273	3.66	1.14×10^{-15}	-34.408
25	298	3.36	1.00×10^{-14}	-32.236
35	308	3.25	2.09×10^{-14}	-31.499
40.	313	3.19	2.92×10^{-14}	-31.165
50.	323	3.10	5.47×10^{-14}	-30.537

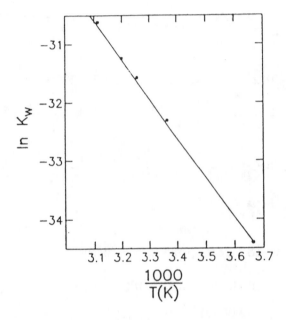

The straight line equation (from a calculator) is: $\ln K = -6.91 \times 10^3 \left(\dfrac{1}{T} \right) - 9.09$

Slope $= -6.91 \times 10^3$ K $= \dfrac{-\Delta H°}{R}$, $\Delta H° = -(-6.91 \times 10^3$ K $\times 8.3145$ J/K•mol$) = 5.75 \times 10^4$ J/mol

y-intercept $= -9.09 = \dfrac{\Delta S°}{R}$, $\Delta S° = -9.09 \times 8.3145$ J/K•mol $= -75.6$ J/K•mol

Additional Exercises

67.   It appears that the sum of the two processes has no net change. This is not so. By the second law of thermodynamics, ΔS_{univ} must have increased even though it looks as if we have gone through a cyclic process.

68.   As ΔS_{univ} continually increases, there will be less useful energy available for work. When no work can be done, the world ends.

69.   The introduction of mistakes is an effect of entropy. The purpose of redundant information is to provide a control to check the "correctness" of the transmitted information.

70.   At the boiling point, $\Delta G = 0$ so $\Delta H = T\Delta S$. $\Delta S = \dfrac{\Delta H}{T} = \dfrac{35.2 \times 10^3 \text{ J/mol}}{3680 \text{ K}} = 9.57 \text{ J/K·mol}$

71.   S (monoclinic) $\rightarrow$ S (rhombic); $\Delta H° = 0 - 0.30 = -0.30 \text{ kJ}$; $\Delta S° = 31.73 - 32.55 = -0.82 \text{ J/K}$

At the conversion temperature: $\Delta G° = 0$ so $\Delta H° = T\Delta S°$; $T = \dfrac{\Delta H°}{\Delta S°} = \dfrac{-3.0 \times 10^2 \text{ J}}{-0.82 \text{ J/K}} = 370 \text{ K}$

72.   a. $\Delta G° = -RT \ln K = -(8.3145 \text{ J/K·mol})(298 \text{ K}) \ln 0.090$, $\Delta G° = 6.0 \times 10^3 \text{ J/mol} = 6.0 \text{ kJ/mol}$

b.   H–O–H + Cl–O–Cl $\rightarrow$ 2 H–O–Cl

On each side of the reaction there are 2 H–O bonds and 2 O–Cl bonds. Both sides have the same number and type of bonds. Thus, $\Delta H \approx \Delta H° \approx 0$.

c. $\Delta G° = \Delta H° - T\Delta S°$, $\Delta S° = \dfrac{\Delta H° - \Delta G°}{T} = \dfrac{0 - 6.0 \times 10^3 \text{ J}}{298 \text{ K}} = -20. \text{ J/K}$

d.   For $H_2O(g)$, $\Delta H_f° = -242 \text{ kJ/mol}$ and $S° = 189 \text{ J/K·mol}$

$\Delta H° = 0 = 2 \, \Delta H_{f, \text{HOCl}}° - [1 \text{ mol}(-242 \text{ kJ/mol}) + 1 \text{ mol}(80.3 \text{ J/K/mol})]$, $\Delta H_{f, \text{HOCl}}° = -81 \text{ kJ/mol}$

$-20. \text{ J/K} = 2 \, S_{\text{HOCl}}° - [1 \text{ mol}(189 \text{ J/K·mol}) + 1 \text{ mol}(266.1 \text{ J/K·mol})]$, $S_{\text{HOCl}}° = 218 \text{ J/K·mol}$

e.   Assuming $\Delta H°$ and $\Delta S°$ are T independent: $\Delta G_{500}° = 0 - (500. \text{ K})(-20. \text{ J/K}) = 1.0 \times 10^4 \text{ J}$

$\Delta G° = -RT \ln K$, $K = \exp\left(\dfrac{-\Delta G°}{RT}\right) = \exp\left(\dfrac{-1.0 \times 10^4}{(8.3145)(500.)}\right) = e^{-2.41} = 0.090$

f. $\Delta G = \Delta G° + RT \ln \dfrac{P_{\text{HOCl}}^2}{P_{H_2O} \times P_{Cl_2O}}$; From part a, $\Delta G° = 6.0 \text{ kJ/mol}$.

We should express all P's in atm. However, we perform the pressure conversion the same number of times in the numerator and denominator, so the factors of 760 torr/atm will all cancel. Thus, we can use the pressures in units of torr.

$\Delta G = 6.0 \text{ kJ/mol} + \dfrac{(8.3145 \text{ J/K·mol})(298 \text{ K})}{1000 \text{ J/kJ}} \ln\left(\dfrac{(0.10)^2}{(18)(2.0)}\right) = 6.0 - 20. = -14 \text{ kJ/mol}$

73. $\Delta G_{CO}° = -RT \ln K_{CO}$ and $\Delta G_{O_2}° = -RT \ln K_{O_2}$

$\Delta G_{CO}° - \Delta G_{O_2}° = -RT \ln K_{CO} + RT \ln K_{O_2} = -RT \ln (K_{CO}/K_{O_2})$

$\Delta G_{CO}° - \Delta G_{O_2}° = -(8.3145 \text{ J/K·mol})(298 \text{ K}) \ln 210 = -13,000 \text{ J/mol} = -13 \text{ kJ/mol}$

74. $Ba(NO_3)_2(s) \rightleftharpoons Ba^{2+}(aq) + 2 \, NO_3^-(aq)$ $K = K_{sp}$; $\Delta G° = -561 + 2(-109) - (-797) = 18 \text{ kJ}$

$\Delta G° = -RT \ln K_{sp}$, $\ln K_{sp} = \dfrac{-\Delta G°}{RT} = \dfrac{-18,000 \text{ J}}{8.3145 \text{ J/K·mol} (298 \text{ K})} = -7.26$, $K_{sp} = e^{-7.26} = 7.0 \times 10^{-4}$

75. From our answer to Exercise 16.63, $\Delta G° = 79.9$ kJ/mol at 25°C.

$$\Delta G = \Delta G° + RT \ln ([H^+] [OH^-]) = 79.9 \text{ kJ/mol} + \frac{(8.3145 \text{ J/K·mol}) (298 \text{ K})}{1000 \text{ J/kJ}} \ln ([H^+] [OH^-])$$

a. $\Delta G = 79.9 + (8.3145 \times 298/1000) \ln [(1.00 \times 10^{-7}) \times (1.00 \times 10^{-7})] = 79.9 - 79.9 = 0$, At equilibrium.

b. $\Delta G = 0$, At equilibrium   c. $\Delta G = -34$ kJ, Shifts right

d. $\Delta G = 45.7$ kJ, Shifts left   e. $\Delta G = \Delta G° = 79.9$ kJ, Shifts left

Le Chatelier's principle gives the same results.

76. K^+ (blood) $\rightleftharpoons$ K^+ (muscle) $\Delta G° = 0$; $\Delta G = RT \ln \left(\dfrac{[K^+]_m}{[K^+]_b} \right)$; $\Delta G = w_{max}$

$$\Delta G = \frac{8.3145 \text{ J}}{\text{K mol}} (310. \text{ K}) \ln \left(\frac{0.15}{0.0050} \right), \ \Delta G = 8.8 \times 10^3 \text{ J/mol} = 8.8 \text{ kJ/mol}$$

At least 8.8 kJ of work must be applied.

Other ions will have to be transported in order to maintain electroneutrality. Either anions must be transported into the cells, or cations (Na^+) in the cell must be transported to the blood. The latter is what happens: [Na^+] in blood is greater than [Na^+] in cells as a result of this pumping.

77. a. $\Delta G° = - RT \ln K$, $K = \exp (-\Delta G°/RT) = \exp \left(\dfrac{-(-30,500 \text{ J})}{8.3145 \text{ J/K·mol} \times 298 \text{ K}} \right) = 2.22 \times 10^5$

b. $C_6H_{12}O_6(s) + 6 O_2(g) \rightarrow 6 CO_2(g) + 6 H_2O(l)$

$\Delta G° = 6 \text{ mol}(-394 \text{ kJ/mol}) + 6 \text{ mol}(-237 \text{ kJ/mol}) - 1 \text{ mol}(-911 \text{ kJ/mol}) = -2875 \text{ kJ}$

$$\frac{2875 \text{ kJ}}{\text{mol glucose}} \times \frac{1 \text{ mol ATP}}{30.5 \text{ kJ}} = 94.3 \text{ mol ATP}; \ 94.3 \text{ molecules ATP/molecule glucose}$$

This is an overstatement. The assumption that all of the free energy goes into this reaction is false. Actually only 38 moles of ATP are produced by metabolism of one mole of glucose.

78. a. $\Delta G° = -RT \ln K$

$$\ln K = \frac{-\Delta G°}{RT} = \frac{-14,000 \text{ J}}{(8.3145 \text{ J/K·mol}) (298 \text{ K})} = -5.65, \ K = e^{-5.65} = 3.5 \times 10^{-3}$$

b.
Glutamic acid + NH_3 → Glutamine + H_2O	$\Delta G° = 14$ kJ
ATP + H_2O → ADP + $H_2PO_4^-$	$\Delta G° = -30.5$ kJ

Glutamic acid + ATP + NH_3 → Glutamine + ADP + $H_2PO_4^-$ $\Delta G° = 14 - 30.5 = -17$ kJ

$$\ln K = \frac{-\Delta G^\circ}{RT} = \frac{-(-17,000 \text{ J})}{8.3145 \text{ J/K} \cdot \text{mol } (298 \text{ K})} = 6.86, \ K = e^{6.86} = 9.5 \times 10^2$$

79.   The light source for the first reaction is necessary for kinetic reasons. The first reaction is just too slow to occur unless a light source is available. The kinetics of a reaction are independent of the thermodynamics of a reaction. Even though the first reaction is more favorable thermodynamically (assuming standard conditions), it is unfavorable for kinetic reasons. The second reaction has a negative ΔG° value and is a fast reaction, so the second reaction occurs very quickly. When considering if a reaction will occur, thermodynamics and kinetics must both be considered.

Challenge Problems

80.   Arrangement I: $S = k \ln \Omega; \ \Omega = 1; \ S = k \ln 1 = 0$

       Arrangement II: $\Omega = 4; \ S = k \ln 4 = 1.38 \times 10^{-23} \text{ J/K} \ln 4, \ S = 1.91 \times 10^{-23} \text{ J/K}$

       Arrangement III: $\Omega = 6; \ S = k \ln 6 = 2.47 \times 10^{-23} \text{ J/K}$

81.   a.   From the plot, the activation energy of the reverse reaction is $E_a + (-\Delta G^\circ) = E_a - \Delta G^\circ$ (ΔG° is a negative number as drawn in the diagram).

$$k_f = A \exp\left(\frac{-E_a}{RT}\right) \text{ and } k_r = A \exp\left(\frac{-(E_a - \Delta G^\circ)}{RT}\right), \ \frac{k_f}{k_r} = \frac{A \exp\left(\dfrac{-E_a}{RT}\right)}{A \exp\left(\dfrac{-(E_a - \Delta G^\circ)}{RT}\right)}$$

       If the A factors are equal: $\dfrac{k_f}{k_r} = \exp\left(\dfrac{-E_a}{RT} + \dfrac{(E_a - \Delta G^\circ)}{RT}\right) = \exp\left(\dfrac{-\Delta G^\circ}{RT}\right)$

       From $\Delta G^\circ = -RT \ln K, \ K = \exp\left(\dfrac{-\Delta G^\circ}{RT}\right)$; Since K and $\dfrac{k_f}{k_r}$ are both equal to the same expression, then $K = \dfrac{k_f}{k_r}$.

       b.   The catalyst will change the activation energy of both the forward and reverse reaction (but not change ΔG°). Therefore, a catalyst must increase the rate of both the forward and reverse reactions.

82.   At equilibrium:

$$P_{H_2} = \frac{nRT}{V} = \frac{\left(\dfrac{1.10 \times 10^{13} \text{ molecules}}{6.022 \times 10^{23} \text{ molecules/mol}}\right)\left(\dfrac{0.08206 \text{ L atm}}{\text{mol K}}\right)(298 \text{ K})}{1.00 \text{ L}} = 4.47 \times 10^{-10} \text{ atm}$$

       The pressure of H_2 decreased from 1.00 atm to 4.47×10^{-10} atm. Essentially all of the H_2 and Br_2 has reacted. Therefore, $P_{HBr} = 2.00$ atm since there is a 2:1 mol ratio between HBr and H_2 in the balanced equation. Since we began with equal moles of H_2 and Br_2, then we will have equal moles of H_2 and Br_2 at equilibrium. Therefore, $P_{H_2} = P_{Br_2} = 4.47 \times 10^{-10}$ atm.

$$K = \frac{P_{HBr}^2}{P_{H_2} \times P_{Br_2}} = \frac{(2.00)^2}{(4.47 \times 10^{-10})^2} = 2.00 \times 10^{19}$$ Assumptions good.

$$\Delta G° = -RT \ln K = -(8.3145 \text{ J/K} \cdot \text{mol})(298 \text{ K}) \ln (2.00 \times 10^{19}) = -1.10 \times 10^5 \text{ J/mol}$$

$$\Delta S° = \frac{\Delta H° - \Delta G°}{T} = \frac{-103{,}800 \text{ J/mol} - (-1.10 \times 10^5 \text{ J/mol})}{298 \text{ K}} = 20 \text{ J/K} \cdot \text{mol}$$

83.   a. $\Delta G° = G_B° - G_A° = 11{,}718 - 8996 = 2722 \text{ J}$

$$K = \exp\left(\frac{-\Delta G°}{RT}\right) = \exp\left(\frac{-2722 \text{ J}}{(8.3145 \text{ J/K} \cdot \text{mol})(298 \text{ K})}\right) = 0.333$$

b.   Since Q = 1.00 > K, reaction shifts left. Let x = atm of B(g) which reacts to reach equilibrium.

$$A(g) \quad \rightleftharpoons \quad B(g)$$

	A(g)	B(g)
Initial	1.00 atm	1.00 atm
Equil.	1.00 + x	1.00 - x

$$\frac{1.00 - x}{1.00 + x} = 0.333, \ 1.00 - x = 0.333 + 0.333\,x, \ x = 0.50 \text{ atm}$$

$P_B = 1.00 - 0.50 = 0.50 \text{ atm}; \ P_A = 1.00 + 0.50 = 1.50 \text{ atm}$

c. $\Delta G = \Delta G° + RT \ln Q = \Delta G° + RT \ln (P_B/P_A)$

$\Delta G = 2722 \text{ J} + (8.3145)(298) \ln (0.50/1.50) = 2722 \text{ J} - 2722 \text{ J} = 0$ (carrying extra sig. figs.)

84.   From Exercise 16.65, $\ln K = \dfrac{-\Delta H°}{RT} + \dfrac{\Delta S°}{R}$, R = 8.3145 J/K•mol

For K at two temperatures T_2 and T_1, the equation is: $\ln \dfrac{K_2}{K_1} = \dfrac{\Delta H°}{R}\left(\dfrac{1}{T_1} - \dfrac{1}{T_2}\right)$

$\ln 10.0 = \dfrac{\Delta H°}{8.3145}\left(\dfrac{1}{300.0 \text{ K}} - \dfrac{1}{350.0 \text{ K}}\right)$, $2.30 = \dfrac{\Delta H°}{8.3145} (4.76 \times 10^{-4})$

$\Delta H° = 4.02 \times 10^4 \text{ J/mol} = 40.2 \text{ kJ/mol}$

85. $K_p = P_{CO_2}$; To insure Ag_2CO_3 from decomposing, P_{CO_2} should be greater than K_p.

From Exercise 16.65, $\ln K = \dfrac{-\Delta H°}{RT} + \dfrac{\Delta S°}{R}$. For two conditions of K and T, the equation is:

$$\ln \frac{K_2}{K_1} = \frac{\Delta H°}{R}\left(\frac{1}{T_1} - \frac{1}{T_2}\right)$$

Let $T_1 = 25\,°C = 298$ K, $K_1 = 6.23 \times 10^{-3}$ torr; $T_2 = 110.\,°C = 383$ K, $K_2 = ?$

$$\ln \frac{K_2}{6.23 \times 10^{-3} \text{ torr}} = \frac{79.14 \times 10^3 \text{ J/mol}}{8.3145 \text{ J/K}\bullet\text{mol}} \left(\frac{1}{298 \text{ K}} - \frac{1}{383 \text{ K}} \right)$$

$$\ln \frac{K_2}{6.23 \times 10^{-3}} = 7.1, \quad \frac{K_2}{6.23 \times 10^{-3}} = e^{7.1} = 1.2 \times 10^3, \quad K_2 = 7.5 \text{ torr}$$

To prevent decomposition of Ag_2CO_3, the partial pressure of CO_2 should be greater than 7.5 torr.

CHAPTER SEVENTEEN

ELECTROCHEMISTRY

Review of Oxidation - Reduction Reactions

13.   Oxidation: increase in oxidation number; loss of electrons

         Reduction: decrease in oxidation number; gain of electrons

14.   See Table 4.2 in Chapter 4 of the text for rules for assigning oxidation numbers.

   a.   H (+1), O (-2), N (+5)   b.   Cl (-1), Cu (+2)

   c.   O (0)   d.   H (+1), O (-1)

   e.   Mg (+2), O (-2), S (+6)   f.   Ag (0)

   g.   Pb (+2), O (-2), S (+6)   h.   O (-2), Pb (+4)

   i.   Na (+1), O (-2), C (+3)   j.   O (-2), C (+4)

   k. $(NH_4)_2Ce(SO_4)_3$ contains NH_4^+ ions and SO_4^{2-} ions. Thus, cerium exists as the Ce^{4+} ion.
        H (+1), N (-3), Ce (+4), S (+6), O (-2)

   l.   O (-2), Cr (+3)

15.   The species oxidized shows an increase in oxidation numbers and is called the reducing agent. The species reduced shows a decrease in oxidation numbers and is the oxidizing agent. The pertinent oxidation numbers are listed by the substance oxidized and the substance reduced.

	Redox?	Ox. Agent	Red. Agent	Substance Oxidized	Substance Reduced
a.	Yes	H_2O	CH_4	CH_4 (C, -4 → +2)	H_2O (H, +1 → 0)
b.	Yes	$AgNO_3$	Cu	Cu (0 → +2)	$AgNO_3$ (Ag, +1 → 0)
c.	Yes	HCl	Zn	Zn (0 → +2)	HCl (H, +1 → 0)

   d.   No; There is no change in any oxidation numbers.

442

16.   See Chapter 4.10 of the text for rules on balancing oxidation-reduction reactions.

a. $Cr \rightarrow Cr^{3+} + 3\ e^-$

$$NO_3^- \rightarrow NO$$
$$4\ H^+ + NO_3^- \rightarrow NO + 2\ H_2O$$
$$3\ e^- + 4\ H^+ + NO_3^- \rightarrow NO + 2\ H_2O$$

$$Cr \rightarrow Cr^{3+} + 3\ e^-$$
$$3\ e^- + 4\ H^+ + NO_3^- \rightarrow NO + 2\ H_2O$$

_____

$$4\ H^+(aq) + NO_3^-(aq) + Cr(s) \rightarrow Cr^{3+}(aq) + NO(g) + 2\ H_2O(l)$$

b. $(Al \rightarrow Al^{3+} + 3\ e^-) \times 5$

$$MnO_4^- \rightarrow Mn^{2+}$$
$$8\ H^+ + MnO_4^- \rightarrow Mn^{2+} + 4\ H_2O$$
$$(5\ e^- + 8\ H^+ + MnO_4^- \rightarrow Mn^{2+} + 4\ H_2O) \times 3$$

$$5\ Al \rightarrow 5\ Al^{3+} + 15\ e^-$$
$$15\ e^- + 24\ H^+ + 3\ MnO_4^- \rightarrow 3\ Mn^{2+} + 12\ H_2O$$

_____

$$24\ H^+(aq) + 3\ MnO_4^-(aq) + 5\ Al(s) \rightarrow 5\ Al^{3+}(aq) + 3\ Mn^{2+}(aq) + 12\ H_2O(l)$$

c. $(Ce^{4+} + e^- \rightarrow Ce^{3+}) \times 6$

$$CH_3OH \rightarrow CO_2$$
$$H_2O + CH_3OH \rightarrow CO_2 + 6\ H^+$$
$$H_2O + CH_3OH \rightarrow CO_2 + 6\ H^+ + 6\ e^-$$

$$6\ Ce^{4+} + 6\ e^- \rightarrow 6\ Ce^{3+}$$
$$H_2O + CH_3OH \rightarrow CO_2 + 6\ H^+ + 6\ e^-$$

_____

$$H_2O(l) + CH_3OH(aq) + 6\ Ce^{4+}(aq) \rightarrow 6\ Ce^{3+}(aq) + CO_2(g) + 6\ H^+(aq)$$

d. $$PO_3^{3-} \rightarrow PO_4^{3-}$$
$$(H_2O + PO_3^{3-} \rightarrow PO_4^{3-} + 2\ H^+ + 2\ e^-) \times 3$$

$$MnO_4^- \rightarrow MnO_2$$
$$(3\ e^- + 4\ H^+ + MnO_4^- \rightarrow MnO_2 + 2\ H_2O) \times 2$$

$$3\ H_2O + 3\ PO_3^{3-} \rightarrow 3\ PO_4^{3-} + 6\ H^+ + 6\ e^-$$
$$6\ e^- + 8\ H^+ + 2\ MnO_4^- \rightarrow 2\ MnO_2 + 4\ H_2O$$

_____

$$2\ H^+ + 2\ MnO_4^- + 3\ PO_3^{3-} \rightarrow 3\ PO_4^{3-} + 2\ MnO_2 + H_2O$$

Now convert to basic solution by adding 2 OH⁻ to <u>both</u> sides. $2\ H^+ + 2\ OH^- \rightarrow 2\ H_2O$ on the reactant side. After converting H^+ to OH^-, then simplify the overall equation by crossing off 1 H_2O on each side of the reaction. The overall balanced equation is:

$$H_2O(l) + 2\ MnO_4^-(aq) + 3\ PO_3^{3-}(aq) \rightarrow 3\ PO_4^{3-}(aq) + 2\ MnO_2(s) + 2\ OH^-(aq)$$

e. $$Mg \rightarrow Mg(OH)_2$$
$$2\ H_2O + Mg \rightarrow Mg(OH)_2 + 2\ H^+ + 2\ e^-$$

$$OCl^- \rightarrow Cl^-$$
$$2\ e^- + 2\ H^+ + OCl^- \rightarrow Cl^- + H_2O$$

$$2 H_2O + Mg \rightarrow Mg(OH)_2 + 2 H^+ + 2 e^-$$
$$2 e^- + 2 H^+ + OCl^- \rightarrow Cl^- + H_2O$$

$$OCl^-(aq) + H_2O(l) + Mg(s) \rightarrow Mg(OH)_2(s) + Cl^-(s)$$

The final overall reaction does not contain H^+ so we are done.

f.

$$H_2CO \rightarrow HCO_3^-$$
$$2 H_2O + H_2CO \rightarrow HCO_3^- + 5 H^+ + 4 e^-$$

$$Ag(NH_3)_2^+ \rightarrow Ag + 2 NH_3$$
$$(e^- + Ag(NH_3)_2^+ \rightarrow Ag + 2 NH_3) \times 4$$

$$2 H_2O + H_2CO \rightarrow HCO_3^- + 5 H^+ + 4 e^-$$
$$4 e^- + 4 Ag(NH_3)_2^+ \rightarrow 4 Ag + 8 NH_3$$

$$4 Ag(NH_3)_2^+ + 2 H_2O + H_2CO \rightarrow HCO_3^- + 5 H^+ + 4 Ag + 8 NH_3$$

Convert to basic solution by adding 5 OH^- to both sides ($5 H^+ + 5 OH^- \rightarrow 5 H_2O$). Then cross off 2 H_2O on both sides which gives the overall balanced equation:

$$5 OH^-(aq) + 4 Ag(NH_3)_2^+(aq) + H_2CO(aq) \rightarrow HCO_3^-(aq) + 3 H_2O(l) + 4 Ag(s) + 8 NH_3(aq)$$

Questions

17.   In a galvanic cell, a spontaneous redox reaction occurs which produces an electric current. In an electrolytic cell, electricity is used to force a redox reaction to occur that is not spontaneous.

18.   The salt bridge allows counter ions to flow into the two cell compartments to maintain electrical neutrality. Without a salt bridge, no sustained electron flow can occur.

19.   a.   Cathode: The electrode at which reduction occurs.

       b.   Anode: The electrode at which oxidation occurs.

       c.   Oxidation half-reaction: The half-reaction in which electrons are products. In a galvanic cell, the oxidation half-reaction always occurs at the anode.

       d.   Reduction half-reaction: The half-reaction in which electrons are reactants. In a galvanic cell, the reduction half-reaction always occurs at the cathode.

20.   a.   Purification by electrolysis is called electrorefining. See the text for a discussion of the electrorefining of copper. Electrorefining is possible because of the selectivity of the electrode reactions. The anode is made up of the impure metal. A potential is applied so just the metal of interest and all more easily oxidized metals are oxidized at the anode. The metal of interest is the only metal plated at the cathode due to the careful control of the potential applied. The metal ions that could plate out at the cathode in preference to the metal we are purifying will not be in solution, since these metals were not oxidized at the anode.

       b.   A more easily oxidized metal is placed in electrical contact with the metal we are trying to protect. It is oxidized in preference to the protected metal. The protected metal becomes the cathode electrode, thus, cathodic protection.

21.   As a battery discharges, E_{cell} decreases, eventually reaching zero. A charged battery is not at equilibrium. At equilibrium $E_{cell} = 0$ and $\Delta G = 0$. We get no work out of an equilibrium system. A battery is useful to us because it can do work as it approaches equilibrium.

22.   Standard reduction potentials are an intensive property, i.e., they do not depend on how many times the reaction occurs. As long as the concentrations of ions and gases are 1 M or 1 atm, then standard reduction potentials and standard oxidation potentials are a constant, and not dependent on the coefficients in the balanced equation.

23.   Both fuel cells and batteries are galvanic cells. However, fuel cells, unlike batteries, have the reactants continuously supplied and can produce a current indefinitely.

24.   Moisture must be present to act as a medium for ion flow between the anodic and cathodic regions. Salt provides ions necessary to complete the electrical circuit in the corrosion process. Together, salt and water make up the salt bridge in this spontaneous electrochemical process.

       Three methods discussed in the text to prevent corrosion are galvanizing, alloying and cathodic protection. Galvanizing coats the metal of interest (usually iron) with zinc which is an easily oxidized metal. Alloying mixes in metals which form durable, effective oxide coatings over the metal of interest. Cathodic protection connects, via a wire, a more easily oxidized metal to the metal we are trying to protect. The more active metal is preferentially oxidized, thus protecting our metal object from corrosion.

Galvanic Cells, Cell Potentials, Standard Reduction Potentials, and Free Energy

25.   A typical galvanic cell diagram is:

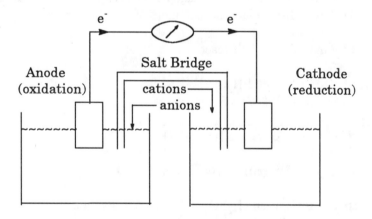

       The diagram for all cells will look like this. The contents of each half-cell compartment will be identified for each reaction, with all solute concentrations at 1.0 M and all gases at 1.0 atm. For Exercise 17.25 and 17.29, the flow of ions through the salt bridge was not asked for in the questions. If asked, however, cations always flow into the cathode compartment and anions always flow into the anode compartment. This is required to keep each compartment electrically neutral.

a. Table 17.1 of the text lists balanced reduction half-reactions for many substances. For this overall reaction, we need the Cl_2 to Cl^- reduction half-reaction and the Cr^{3+} to $Cr_2O_7^{2-}$ oxidation half-reaction. Manipulating these two half-reactions gives the overall balanced equation.

$$(Cl_2 + 2\ e^- \rightarrow 2\ Cl^-) \times 3$$
$$7\ H_2O + 2\ Cr^{3+} \rightarrow Cr_2O_7^{2-} + 14\ H^+ + 6\ e^-$$

$$7\ H_2O(l) + 2\ Cr^{3+}(aq) + 3\ Cl_2(g) \rightarrow Cr_2O_7^{2-}(aq) + 6\ Cl^-(aq) + 14\ H^+(aq)$$

The contents of each compartment is:

   Cathode: Pt electrode; Cl_2 bubbled into solution, Cl^- in solution

   Anode: Pt electrode; Cr^{3+}, H^+, and $Cr_2O_7^{2-}$ in solution

We need a nonreactive metal to use as the electrode in each case, since all of the reactants and products are in solution. Pt is a common choice. Another possibility is graphite.

b. $$Cu^{2+} + 2\ e^- \rightarrow Cu$$
$$Mg \rightarrow Mg^{2+} + 2e^-$$

$$Cu^{2+}(aq) + Mg(s) \rightarrow Cu(s) + Mg^{2+}(aq)$$

Cathode: Cu electrode; Cu^{2+} in solution

Anode: Mg electrode; Mg^{2+} in solution

26. Reference Exercise 17.25 for a typical galvanic cell diagram. The contents of each half-cell compartment is identified below with all solute concentrations at $1.0\ M$ and all gases at 1.0 atm.

a. Reference Table 17.1 for the balanced half-reactions.

$$5\ e^- + 6\ H^+ + IO_3^- \rightarrow 1/2\ I_2 + 3\ H_2O$$
$$(Fe^{2+} \rightarrow Fe^{3+} + e^-) \times 5$$

$$6\ H^+ + IO_3^- + 5\ Fe^{2+} \rightarrow 5\ Fe^{3+} + 1/2\ I_2 + 3\ H_2O$$

or $12\ H^+(aq) + 2\ IO_3^-(aq) + 10\ Fe^{2+}(aq) \rightarrow 10\ Fe^{3+}(aq) + I_2(s) + 6\ H_2O(l)$

   Cathode: Pt electrode; IO_3^-, I_2 and H_2SO_4 (H^+ source) in solution.

   Note: $I_2(s)$ would make a poor electrode since it sublimes.

   Anode: Pt electrode; Fe^{2+} and Fe^{3+} in solution

b. $$(Ag^+ + e^- \rightarrow Ag) \times 2$$
$$Zn \rightarrow Zn^{2+} + 2\ e^-$$

$$Zn(s) + 2\ Ag^+(aq) \rightarrow 2\ Ag(s) + Zn^{2+}(aq)$$

Cathode: Ag electrode; Ag^+ in solution

Anode: Zn electrode; Zn^{2+} in solution

27.   To determine E° for the overall cell reaction, we must add the standard reduction potential to the standard oxidation potential ($E°_{cell} = E°_{red} + E°_{ox}$). Reference Table 17.1 for values of standard reduction potentials. Remember that $E°_{ox} = -E°_{red}$ and that standard potentials are <u>not</u> multiplied by the integer used to obtain the overall balanced equation.

25a. $E°_{cell} = E°_{Cl_2 \rightarrow Cl^-} + E°_{Cr^{3+} \rightarrow Cr_2O_7^{2-}} = 1.36 \text{ V} + (-1.33 \text{ V}) = 0.03 \text{ V}$

25b. $E°_{cell} = E°_{Cu^{2+} \rightarrow Cu} + E°_{Mg \rightarrow Mg^{2+}} = 0.34 \text{ V} + 2.37 \text{ V} = 2.71 \text{ V}$

28.   26a. $E°_{cell} = E°_{IO_3^- \rightarrow I_2} + E°_{Fe^{2+} \rightarrow Fe^{3+}} = 1.20 \text{ V} + (-0.77 \text{ V}) = 0.43 \text{ V}$

26b. $E°_{cell} = E°_{Ag^+ \rightarrow Ag} + E°_{Zn \rightarrow Zn^{2+}} = 0.80 \text{ V} + 0.76 \text{ V} = 1.56 \text{ V}$

29.   Reference Exercise 17.25 for a typical galvanic cell design. The contents of each half-cell compartment is identified below with all solute concentrations at $1.0\,M$ and all gases at 1.0 atm. For each pair of half-reactions, the half-reaction with the largest standard reduction potential will be the cathode reaction and the half-reaction with the smallest reduction potential will be reversed to become the anode reaction. Only this combination gives a spontaneous overall reaction, i.e., a reaction with a positive overall standard cell potential.

a. $Cl_2 + 2\text{ e}^- \rightarrow 2\text{ Cl}^-$ E° = 1.36 V
                $2\text{ Br}^- \rightarrow Br_2 + 2\text{ e}^-$ -E° = -1.09 V
        _____
                $Cl_2(g) + 2\text{ Br}^-(aq) \rightarrow Br_2(aq) + 2\text{ Cl}^-(aq)$ $E°_{cell}$ = 0.27 V

The contents of each compartment is:

        Cathode: Pt electrode; $Cl_2(g)$ bubbled in, Cl^- in solution

        Anode: Pt electrode; Br_2 and Br^- in solution

b. $(2\text{ e}^- + 2\text{ H}^+ + IO_4^- \rightarrow IO_3^- + H_2O) \times 5$ E° = 1.60 V
                        $(4\text{ H}_2O + Mn^{2+} \rightarrow MnO_4^- + 8\text{ H}^+ + 5\text{ e}^-) \times 2$ -E° = -1.51 V
        _____
        $10\text{ H}^+ + 5\text{ IO}_4^- + 8\text{ H}_2O + 2\text{ Mn}^{2+} \rightarrow 5\text{ IO}_3^- + 5\text{ H}_2O + 2\text{ MnO}_4^- + 16\text{ H}^+$ $E°_{cell}$ = 0.09 V

This simplifies to:

        $3\text{ H}_2O(l) + 5\text{ IO}_4^-(aq) + 2\text{ Mn}^{2+}(aq) \rightarrow 5\text{ IO}_3^-(aq) + 2\text{ MnO}_4^-(aq) + 6\text{ H}^+(aq)$ $E°_{cell}$ = 0.09 V

Cathode: Pt electrode; IO_4^-, IO_3^-, and H_2SO_4 (as a source of H^+) in solution

Anode: Pt electrode; Mn^{2+}, MnO_4^- and H_2SO_4 in solution

30.   Reference Exercise 17.25 for a typical galvanic cell design. The contents of each half-cell compartment is identified below, with all solute concentrations at 1.0 M and all gases at 1.0 atm.

a. $(Ni^{2+} + 2 e^- \rightarrow Ni) \times 3$ $E° = -0.23$ V
            $(Al \rightarrow Al^{3+} + 3 e^-) \times 2$ $-E° = 1.66$ V
            _____
            $3 Ni^{2+}(aq) + 2 Al(s) \rightarrow 2 Al^{3+}(aq) + 3 Ni(s)$ $E°_{cell} = 1.43$ V

       Cathode: Ni electrode; Ni^{2+} in solution

       Anode: Al electrode; Al^{3+} in solution

b. $Co^{3+} + e^- \rightarrow Co^{2+}$ $E° = 1.82$ V
            $Fe^{2+} \rightarrow Fe^{3+} + e^-$ $-E° = -0.77$ V
            _____
            $Co^{3+}(aq) + Fe^{2+}(aq) \rightarrow Fe^{3+}(aq) + Co^{2+}(aq)$ $E°_{cell} = 1.05$ V

       Cathode: Pt electrode; Co^{3+} and Co^{2+} in solution

       Anode: Pt electrode; Fe^{2+} and Fe^{3+} in solution

31.   In standard line notation, the anode is listed first and the cathode is listed last. A double line separates the two compartments. By convention, the electrodes are on the ends with all solutes and gases towards the middle. A single line is used to indicate a phase change. We also included all concentrations.

25a. $Pt \mid Cr^{3+} (1.0\,M), Cr_2O_7^{2-} (1.0\,M), H^+ (1.0\,M) \mid\mid Cl_2 (1.0\ atm)\mid Cl^- (1.0\,M) \mid Pt$

25b. $Mg \mid Mg^{2+} (1.0\,M) \mid\mid Cu^{2+} (1.0\,M) \mid Cu$

29a. $Pt \mid Br^- (1.0\,M), Br_2 (1.0\,M) \mid\mid Cl_2 (1.0\ atm) \mid Cl^- (1.0\,M) \mid Pt$

29b. $Pt \mid Mn^{2+} (1.0\,M), MnO_4^- (1.0\,M), H^+ (1.0\,M) \mid\mid IO_4^- (1.0\,M), IO_3^- (1.0\,M), H^+ (1.0\,M) \mid Pt$

32.   26a. $Pt \mid Fe^{2+} (1.0\ M), Fe^{3+} (1.0\,M) \mid\mid IO_3^- (1.0\,M), I_2 (1.0\,M), H^+ (1.0\,M)\mid Pt$

       26b. $Zn \mid Zn^{2+} (1.0\,M) \mid\mid Ag^+ (1.0\,M) \mid Ag$

       30a. $Al \mid Al^{3+} (1.0\,M) \mid\mid Ni^{2+} (1.0\,M) \mid Ni$

       30b. $Pt \mid Fe^{2+} (1.0\,M), Fe^{3+} (1.0\,M) \mid\mid Co^{3+} (1.0\,M), Co^{2+} (1.0\,M) \mid Pt$

33.   a. $Pb^{2+} + 2 e^- \rightarrow Pb$ $E° = -0.13$ V
                   $Fe \rightarrow Fe^{2+} + 2 e^-$ $-E° = 0.44$ V
                   _____
                   $Pb^{2+}(aq) + Fe(s) \rightarrow Pb(s) + Fe^{2+}(aq)$ $E°_{cell} = 0.31$ V

b.
$$(MnO_4^- + 4 H^+ + 3 e^- \rightarrow MnO_2 + 2 H_2O) \times 2 \qquad E° = 1.68 \text{ V}$$
$$(Mn^{2+} + 2 H_2O \rightarrow MnO_2 + 4 H^+ + 2 e^-) \times 3 \qquad -E° = -1.21 \text{ V}$$

$$2 MnO_4^-(aq) + 3 Mn^{2+}(aq) + 2 H_2O(l) \rightarrow 5 MnO_2(s) + 4 H^+(aq) \qquad E°_{cell} = 0.47 \text{ V}$$

34.   a.
$$(H_2O_2 + 2 H^+ + 2 e^- \rightarrow 2 H_2O) \times 3 \qquad E° = 1.78 \text{ V}$$
$$2 Cr^{3+} + 7 H_2O \rightarrow Cr_2O_7^{2-} + 14 H^+ + 6 e^- \qquad -E° = -1.33 \text{ V}$$

$$3 H_2O_2(aq) + 2 Cr^{3+}(aq) + H_2O(l) \rightarrow Cr_2O_7^{2-}(aq) + 8 H^+(aq) \qquad E°_{cell} = 0.45 \text{ V}$$

b.
$$(2 H^+ + 2 e^- \rightarrow H_2) \times 3 \qquad E° = 0.00 \text{ V}$$
$$(Al \rightarrow Al^{3+} + 3 e^-) \times 2 \qquad -E° = 1.66 \text{ V}$$

$$6 H^+(aq) + 2 Al(s) \rightarrow 3 H_2(g) + 2 Al^{3+}(aq) \qquad E°_{cell} = 1.66 \text{ V}$$

35.   a. $2 Ag^+ + 2 e^- \rightarrow 2 Ag \qquad E° = 0.80 \text{ V}$
      $Cu \rightarrow Cu^{2+} + 2 e^- \qquad -E° = -0.34 \text{ V}$

$2 Ag^+ + Cu \rightarrow Cu^{2+} + 2 Ag \qquad E°_{cell} = 0.46 \text{ V}$ Spontaneous at standard conditions ($E°_{cell} > 0$).

b. $Zn^{2+} + 2 e^- \rightarrow Zn \qquad E° = -0.76 \text{ V}$
     $Ni \rightarrow Ni^{2+} + 2 e^- \qquad -E° = 0.23 \text{ V}$

$Zn^{2+} + Ni \rightarrow Zn + Ni^{2+} \qquad E°_{cell} = -0.53 \text{ V}$ Not spontaneous at standard conditions ($E°_{cell} < 0$).

36.   a.
$$(5 e^- + 8 H^+ + MnO_4^- \rightarrow Mn^{2+} + 4 H_2O) \times 2 \qquad E° = 1.51 \text{ V}$$
$$(2 I^- \rightarrow I_2 + 2 e^-) \times 5 \qquad -E° = -0.54 \text{ V}$$

$16 H^+ + 2 MnO_4^- + 10 I^- \rightarrow 5 I_2 + 2 Mn^{2+} + 8 H_2O \qquad E°_{cell} = 0.97 \text{ V}$ Spontaneous

b.
$$(5 e^- + 8 H^+ + MnO_4^- \rightarrow Mn^{2+} + 4 H_2O) \times 2 \qquad E° = 1.51 \text{ V}$$
$$(2 F^- \rightarrow F_2 + 2 e^-) \times 5 \qquad -E° = -2.87 \text{ V}$$

$16 H^+ + 2 MnO_4^- + 10 F^- \rightarrow 5 F_2 + 2 Mn^{2+} + 8 H_2O \qquad E°_{cell} = -1.36 \text{ V}$ Not spontaneous

37. $\Delta G° = -nFE°_{cell}$; Reference Exercises 17.25, 17.27 and 17.29 for balanced reactions and standard cell potentials. The balanced reactions are necessary to determine n, the moles of electrons transferred.

25a. $7 H_2O + 2 Cr^{3+} + 3 Cl_2 \rightarrow Cr_2O_7^{2-} + 6 Cl^- + 14 H^+ \qquad E°_{cell} = 0.03 \text{ V} = 0.03 \text{ J/C}$

$\Delta G° = -nFE°_{cell} = -(6 \text{ mol } e^-)(96,485 \text{ C/mol } e^-)(0.03 \text{ J/C}) = -1.7 \times 10^4 \text{ J} = -20 \text{ kJ}$

25b. $\Delta G° = -(2 \text{ mol } e^-)(96,485 \text{ C/mol } e^-)(2.71 \text{ J/C}) = -5.23 \times 10^5 \text{ J} = -523 \text{ kJ}$

29a. $\Delta G° = -(2 \text{ mol } e^-)(96,485 \text{ C/mol } e^-)(0.27 \text{ J/C}) = -5.21 \times 10^4 \text{ J} = -52 \text{ kJ}$

29b. $\Delta G° = -(10\ \text{mol}\ e^-)(96{,}485\ \text{C/mol}\ e^-)(0.09\ \text{J/C}) = -8.7 \times 10^4\ \text{J} = -90\ \text{kJ}$

38.   26a. $12\ H^+ + 2\ IO_3^- + 10\ Fe^{2+} \rightarrow 10\ Fe^{3+} + I_2 + 6\ H_2O;\quad E°_{cell} = 0.43\ V = 0.43\ \text{J/C}$

$\Delta G° = -nFE°_{cell} = -(10\ \text{mol}\ e^-)(96{,}485\ \text{C/mol}\ e^-)(0.43\ \text{J/C}) = -4.1 \times 10^5\ \text{J} = -410\ \text{kJ}$

26b. $\Delta G° = -(2\ \text{mol}\ e^-)(96{,}485\ \text{C/mol}\ e^-)(1.56\ \text{J/C}) = -3.01 \times 10^5\ \text{J} = -301\ \text{kJ}$

30a. $\Delta G° = -(6\ \text{mol}\ e^-)(96{,}485\ \text{C/mol}\ e^-)(1.43\ \text{J/C}) = -8.28 \times 10^5\ \text{J} = -828\ \text{kJ}$

30b. $\Delta G° = -(1\ \text{mol}\ e^-)(96{,}485\ \text{C/mol}\ e^-)(1.05\ \text{J/C}) = -1.01 \times 10^5\ \text{J} = -101\ \text{kJ}$

39.

$$Cl_2 + 2\ e^- \rightarrow 2\ Cl^- \qquad\qquad E° = 1.36\ V$$
$$(ClO_2^- \rightarrow ClO_2 + e^-) \times 2 \qquad -E° = -0.954\ V$$

$$\overline{2\ ClO_2^-(aq) + Cl_2(g) \rightarrow 2\ ClO_2(aq) + 2\ Cl^-(aq) \qquad E°_{cell} = 0.41\ V = 0.41\ \text{J/C}}$$

$\Delta G° = -nFE°_{cell} = -(2\ \text{mol}\ e^-)(96{,}485\ \text{C/mol}\ e^-)(0.41\ \text{J/C}) = -7.91 \times 10^4\ \text{J} = -79\ \text{kJ}$

40.   a.

$$(4\ H^+ + NO_3^- + 3\ e^- \rightarrow NO + 2\ H_2O) \times 2 \qquad\qquad E° = 0.96\ V$$
$$(Mn \rightarrow Mn^{2+} + 2\ e^-) \times 3 \qquad\qquad -E° = 1.18\ V$$

$$\overline{3\ Mn(s) + 8\ H^+(aq) + 2\ NO_3^-(aq) \rightarrow 2\ NO(g) + 4\ H_2O(l) + 3\ Mn^{2+}(aq) \qquad E°_{cell} = 2.14\ V}$$

$$5 \times (2\ e^- + 2\ H^+ + IO_4^- \rightarrow IO_3^- + H_2O) \qquad\qquad E° = 1.60\ V$$
$$2 \times (Mn^{2+} + 4\ H_2O \rightarrow MnO_4^- + 8\ H^+ + 5\ e^-) \qquad\qquad -E° = -1.51\ V$$

$$\overline{5\ IO_4^-(aq) + 2\ Mn^{2+}(aq) + 3\ H_2O(l) \rightarrow 5\ IO_3^-(aq) + 2\ MnO_4^-(aq) + 6\ H^+(aq) \qquad E°_{cell} = 0.09\ V}$$

b. Nitric acid oxidation (see above for $E°_{cell}$):

$\Delta G° = -nFE°_{cell} = -(6\ \text{mol}\ e^-)(96{,}485\ \text{C/mol}\ e^-)(2.14\ \text{J/C}) = -1.24 \times 10^6\ \text{J} = -1240\ \text{kJ}$

Periodate oxidation (see above for $E°_{cell}$):

$\Delta G° = -(10\ \text{mol}\ e^-)(96{,}485\ \text{C/mol}\ e^-)(0.09\ \text{J/C})(1\ \text{kJ}/1000\ \text{J}) = -90\ \text{kJ}$

41.   Since the cells are at standard conditions, then $w_{max} = \Delta G = \Delta G° = -nFE°_{cell}$. See Exercise 17.33 for the balanced overall equations and for $E°_{cell}$.

33a. $w_{max} = -(2\ \text{mol}\ e^-)(96{,}485\ \text{C/mol}\ e^-)(0.31\ \text{J/C}) = -6.0 \times 10^4\ \text{J} = -60.\ \text{kJ}$

33b. $w_{max} = -(6\ \text{mol}\ e^-)(96{,}485\ \text{C/mol}\ e^-)(0.47\ \text{J/C}) = -2.7 \times 10^5\ \text{J} = -270\ \text{kJ}$

42.   Since the cells are at standard conditions, then $w_{max} = \Delta G = \Delta G° = -nFE°_{cell}$. See Exercise 17.34 for the balanced overall equations and for $E°_{cell}$.

34a. $w_{max} = -(6\ \text{mol}\ e^-)(96{,}485\ \text{C/mol}\ e^-)(0.45\ \text{J/C}) = -2.6 \times 10^5\ \text{J} = -260\ \text{kJ}$

34b. w_{max} = -(6 mol e⁻)(96,485 C/mol e⁻)(1.66 J/C) = -9.61 × 10⁵ J = -961 kJ

43. $2 H_2O + 2 e^- \rightarrow H_2 + 2 OH^-$ $\Delta G° = \Sigma n_p \Delta G°_{f, products} - \Sigma n_r \Delta G°_{f, reactants}$ = 2(-157) -[2(-237)] = 160. kJ

$\Delta G° = -nFE°$, $E° = \dfrac{-\Delta G°}{nF} = \dfrac{-1.60 \times 10^5 J}{(2 \text{ mol e}^-)(96,485 \text{ C/mol e}^-)}$ = -0.829 J/C = -0.829 V

The two values agree (-0.83 V in Table 17.1).

44. $CH_3OH(l) + 3/2 O_2(g) \rightarrow CO_2(g) + 2 H_2O(l)$ $\Delta G° = 2(-237) + (-394) - [-166] = -702$ kJ

The balanced half-reactions are:

$H_2O + CH_3OH \rightarrow CO_2 + 6 H^+ + 6 e^-$ and $O_2 + 4 H^+ + 4 e^- \rightarrow 2 H_2O$

For 3/2 mol O_2, 6 moles of electrons will be transferred (n = 6).

$\Delta G° = -nFE°$, $E° = \dfrac{-\Delta G°}{nF} = \dfrac{-(-702,000 J)}{(6 \text{ mol e}^-)(96,485 \text{ C/mol e}^-)}$ = 1.21 J/C = 1.21 V

45. $\Delta G° = -nFE°$ = -(1 mol e⁻)(96,485 C/mol e⁻)(0.80 J/C)(1 kJ/1000 J) = -77 kJ

-77 kJ = $\Delta G°_{f, Ag}$ - [$\Delta G°_{f, Ag^+}$ + $\Delta G°_{f, e^-}$], -77 kJ = 0 - [$\Delta G°_{f, Ag^+}$ + 0], $\Delta G°_{f, Ag^+}$ = 77 kJ/mol

46. $Fe^{2+} + 2 e^- \rightarrow Fe$ $E° = -0.44 V = -0.44 J/C$

$\Delta G° = -nFE°$ = -(2 mol e⁻)(96,485 C/mol e⁻)(-0.44 J/C)(1 kJ/1000 J) = 85 kJ

85 kJ = 0 - [$\Delta G°_{f, Fe^{2+}}$ + 0], $\Delta G°_{f, Fe^{2+}}$ = -85 kJ

We can get $\Delta G°_{f, Fe^{3+}}$ two ways. Consider: $Fe^{3+} + e^- \rightarrow Fe^{2+}$ $E° = 0.77 V$

$\Delta G°$ = -(1 mol e)(96,485 C/mol e⁻)(0.77 J/C) = -74,300 J = -74 kJ

$Fe^{2+} \rightarrow Fe^{3+} + e^-$ $\Delta G° = 74$ kJ
$Fe \rightarrow Fe^{2+} + 2 e^-$ $\Delta G° = -85$ kJ
_____

$Fe \rightarrow Fe^{3+} + 3 e^-$ $\Delta G° = -11$ kJ, $\Delta G°_{f, Fe^{3+}}$ = -11 kJ/mol

or   consider: $Fe^{3+} + 3 e^- \rightarrow Fe$ $E° = -0.036 V$

$\Delta G°$ = -(3 mol e⁻)(96,485 C/mol e⁻)(-0.036 J/C) = 10,400 J ≈ 10. kJ

10. kJ = 0 - [$\Delta G°_{f, Fe^{3+}}$ + 0], $\Delta G°_{f, Fe^{3+}}$ = -10. kJ/mol; Round off error explains the 1 kJ
discrepancy.

47.   Good oxidizing agents are easily reduced. Oxidizing agents are on the left side of the reduction half-reactions listed in Table 17.1. We look for the largest, most positive standard reduction potentials to correspond to the best oxidizing agents. The ordering from worst to best oxidizing agents is:

$$Mg^{2+} < Fe^{2+} < Fe^{3+} < Cr_2O_7^{2-} < Cl_2 < MnO_4^-$$

$E°(V)$ -2.37   -0.44   0.77   1.33   1.36   1.68

48.   Good reducing agents are easily oxidized. The reducing agents are on the right side of the reduction half-reactions listed in Table 17.1. The best reducing agents have the most negative standard reduction potentials ($E°$) or the most positive standard oxidation potentials, $E°_{ox}$ $(= -E°)$.

$$F^- < Cr^{3+} < Fe^{2+} < H_2 < Zn < Li$$

$-E°(V)$ -2.87   -1.33   -0.77   0.00   0.76   3.05

49.   a. $2 H^+ + 2 e^- \rightarrow H_2$ $E° = 0.00$ V; $Cu \rightarrow Cu^{2+} + 2 e^-$ $-E° = -0.34$ V

        $E°_{cell} = -0.34$ V; No, H^+ cannot oxidize Cu to Cu^{2+} at standard conditions ($E°_{cell} < 0$).

        b. $2 H^+ + 2 e^- \rightarrow H_2$ $E° = 0.00$ V; $Mg \rightarrow Mg^{2+} + 2 e^-$ $-E° = 2.37$ V

        $E°_{cell} = 2.37$ V; Yes, H^+ can oxidize Mg to Mg^{2+} at standard conditions ($E°_{cell} > 0$).

        c. $Fe^{3+} + e^- \rightarrow Fe^{2+}$ $E° = 0.77$ V; $2 I^- \rightarrow I_2 + 2 e^-$ $-E° = -0.54$ V

        $E°_{cell} = 0.77 - 0.54 = 0.23$ V; Yes, Fe^{3+} can oxidize I^- to I_2.

        d. $Fe^{3+} + e^- \rightarrow Fe^{2+}$ $E° = 0.77$ V; $2 Br^- \rightarrow Br_2 + 2 e^-$ $-E° = -1.09$ V

        $E°_{cell} = 0.77 - 1.09 = -0.32$ V; No, Fe^{3+} cannot oxidize Br^- to Br_2.

50.   a. $H_2 \rightarrow 2 H^+ + 2 e^-$ $-E° = 0.00$ V; $Ag^+ + e^- \rightarrow Ag$ $E° = 0.80$ V

        $E°_{cell} = 0.80$ V; Yes, H_2 can reduce Ag^+ to Ag at standard conditions ($E°_{cell} > 0$).

        b. $H_2 \rightarrow 2 H^+ + 2 e^-$ $-E° = 0.00$ V; $Ni^{2+} + 2 e^- \rightarrow Ni$ $E° = -0.23$ V

        $E°_{cell} = -0.23$ V; No, H_2 cannot reduce Ni^{2+} to Ni at standard conditions ($E°_{cell} < 0$).

        c. $Fe^{2+} \rightarrow Fe^{3+} + e^-$ $-E° = -0.77$ V; $VO_2^+ + 2 H^+ + e^- \rightarrow VO^{2+} + H_2O$ $E° = 1.00$ V

        $E°_{cell} = 1.00 - 0.77 = 0.23$ V; Yes, Fe^{2+} can reduce VO_2^+ at standard conditions.

        d. $Fe^{2+} \rightarrow Fe^{3+} + e^-$ $-E° = -0.77$ V; $Cr^{3+} + e^- \rightarrow Cr^{2+}$ $E° = -0.50$ V

        $E°_{cell} = -0.50 - 0.77 = -1.27$ V; No, Fe^{2+} cannot reduce Cr^{3+} to Cr^{2+} at standard conditions.

51. $Br_2 + 2 e^- \rightarrow 2 Br^-$ $E° = 1.09$ V $La^{3+} + 3 e^- \rightarrow La$ $E° = -2.37$ V
       $2 H^+ + 2 e^- \rightarrow H_2$ $E° = 0.00$ V $Ca^{2+} + 2 e^- \rightarrow Ca$ $E° = -2.76$ V
       $Cd^{2+} + 2 e^- \rightarrow Cd$ $E° = -0.40$ V

   a.   Oxidizing agents are on the left side of the above reduction half-reactions. Br_2 is the best
         oxidizing agent (largest $E°$).

   b.   Reducing agents are on the right side of the reduction half-reactions. Ca is the best reducing
         agent (largest $-E°$).

   c. $MnO_4^- + 8 H^+ + 5 e^- \rightarrow Mn^{2+} + 4 H_2O$ $E° = 1.51$ V; Permanganate can oxidize Br^-, H_2, Cd
         and Ca at standard conditions. When MnO_4^- is coupled with these reagents, $E°_{cell}$ is positive.
         Note: La is not one of the choices given in the question or it would have been included.

   d. $Zn \rightarrow Zn^{2+} + 2 e^-$ $-E° = 0.76$ V;   Zinc can reduce Br_2 and H^+ since $E°_{cell} > 0$.

52. $Ce^{4+} + e^- \rightarrow Ce^{3+}$ $E° = 1.70$ V $Ni^{2+} + 2 e^- \rightarrow Ni$ $E° = -0.23$ V
       $Fe^{3+} + e^- \rightarrow Fe^{2+}$ $E° = 0.77$ V $Fe^{2+} + 2 e^- \rightarrow Fe$ $E° = -0.44$ V
       $Fe^{3+} + 3 e^- \rightarrow Fe$ $E° = -0.036$ V $Mg^{2+} + 2 e^- \rightarrow Mg$ $E° = -2.37$ V
       $Sn^{2+} + 2 e^- \rightarrow Sn$ $E° = -0.14$ V

   a. Ce^{4+} is the strongest oxidizing agent (largest $E°$).

   b.   Mg is the strongest reducing agent (largest $-E°$).

   c.   Yes, Ce^{4+} will oxidize Fe(s) to the soluble Fe^{3+} ion at standard conditions ($E°_{cell} > 0$).

   d. $2 H^+ + 2 e^- \rightarrow H_2$ $E° = 0.00$ V; Fe, Sn, and Mg can be oxidized by H^+ ($E°_{cell} > 0$).

   e. $H_2 \rightarrow 2 H^+ + 2 e^-$ $-E° = 0.00$ V; Ce^{4+} and Fe^{3+} can be reduced by H_2 ($E°_{cell} > 0$).

53.   a. $2 Br^- \rightarrow Br_2 + 2 e^-$ $-E° = -1.09$ V; $2 Cl^- \rightarrow Cl_2 + 2 e^-$ $-E° = -1.36$ V; $E° > 1.09$ V to oxidize
              Br^-; $E° < 1.36$ V to not oxidize Cl^-; $Cr_2O_7^{2-}$, O_2, MnO_2, and IO_3^- are all possible since when
              all of these oxidizing agents are coupled with Br^- give $E°_{cell} > 0$ and when coupled with Cl^- give
              $E°_{cell} < 0$ (assuming standard conditions).

       b. $Mn \rightarrow Mn^{2+} + 2 e^-$ $-E° = 1.18$; $Ni \rightarrow Ni^{2+} + 2 e^-$ $-E° = 0.23$ V; Any oxidizing agent with
              -0.23 V $> E° > -1.18$ V will work. $PbSO_4$, Cd^{2+}, Fe^{2+}, Cr^{3+}, Zn^{2+} and H_2O will be able to
              oxidize Mn but not oxidize Ni (assuming standard conditions).

54.   a. $Fe^{3+} + e^- \rightarrow Fe^{2+}$ $E° = 0.77$ V; $Fe^{2+} + 2 e^- \rightarrow Fe$ $E° = -0.44$ V. To reduce Fe^{3+} but not Fe^{2+},
              the reducing agent must have a standard oxidation potential ($E°_{ox} = -E°$) between -0.77 V and
              0.44 V (so $E°_{cell}$ is positive only for the Fe^{3+} reduction). The reducing agents are on the right
              side of the half-reactions in Table 17.1. The reagents at standard conditions which have $E°_{ox}$
              ($= -E°$) between -0.77 V and 0.44 V are H_2O_2, MnO_4^{2-}, I^-, Cu, OH^-, Hg (in 1 M Cl^-), Ag (in
              1 M Cl^-), H_2SO_3, H_2, Pb, Sn, Ni, and Cd.

b. $Ag^+ + e^- \rightarrow Ag$ $E° = 0.80$ V; $O_2 + 2 H^+ + 2 e^- \rightarrow H_2O_2$ $E° = 0.68$ V; From Table 17.1, only Fe^{2+} will reduce Ag^+ to Ag but will not reduce O_2 to H_2O_2 (assuming standard conditions). A potential of -0.77 V is required to oxidize Fe^{2+} to Fe^{3+}, which is large enough to oxidize Ag^+ but small enough not to oxidize O_2.

The Nernst Equation

55.
$$H_2O_2 + 2 H^+ + 2 e^- \rightarrow 2 H_2O \qquad\qquad E° = 1.78 \text{ V}$$
$$(Ag \rightarrow Ag^+ + e^-) \times 2 \qquad\qquad -E° = -0.80 \text{ V}$$

$$\overline{H_2O_2(aq) + 2 H^+(aq) + 2 Ag(s) \rightarrow 2 H_2O(l) + 2 Ag^+(aq) \qquad E°_{cell} = 0.98 \text{ V}}$$

a. A galvanic cell is based on spontaneous chemical reactions. At standard conditions, this reaction produces a voltage of 0.98 V. Any change in concentration that increases the tendency of the forward reaction to occur will increase the cell potential. Conversely, any change in concentration that decreases the tendency of the forward reaction to occur (increases the tendency of the reverse reaction to occur) will decrease the cell potential. Using Le Chatelier's principle, increasing the reactant concentrations of H_2O_2 and H^+ from 1.0 M to 2.0 M will drive the forward reaction further to right (will further increase the tendency of the forward reaction to occur). Therefore, E_{cell} will be greater than $E°_{cell}$.

b. Here, we decreased the reactant concentration of H^+ and increased the product concentration of Ag^+ from the standard conditions. This decreases the tendency of the forward reaction to occur which will decrease E_{cell} as compared to $E°_{cell}$ ($E_{cell} < E°_{cell}$).

56. The concentrations of Fe^{2+} in the two compartments are now 0.01 M and $1 \times 10^{-7} M$. The driving force for this reaction is to equalize the Fe^{2+} concentrations in the two compartments. This occurs if the compartment with $1 \times 10^{-7} M$ Fe^{2+} becomes the anode (Fe will be oxidized to Fe^{2+}) and the compartment with the 0.01 M Fe^{2+} becomes the cathode (Fe^{2+} will be reduced to Fe). Electron flow, as always for galvanic cells, goes from the anode to the cathode so electron flow will go from the right compartment ($[Fe^{2+}] = 1 \times 10^{-7} M$) to the left compartment ($[Fe^{2+}] = 0.01 M$).

57. At 25°C, $E_{cell} = E°_{cell} - \dfrac{0.0592}{n} \log Q$ where $Q = [Ag^+]^2 / [H_2O_2][H^+]^2$; See Exercise 17.55 for the overall balanced equation and for $E°_{cell}$.

a. $E_{cell} = 0.98 \text{ V} - \dfrac{0.0592}{2} \log \dfrac{(1.0)^2}{(2.0)(2.0)^2} = 0.98 \text{ V} - (-0.027 \text{ V}) = 1.01 \text{ V}$

b. $E_{cell} = 0.98 \text{ V} - \dfrac{0.0592}{2} \log \dfrac{(2.0)^2}{(1.0)(1.0 \times 10^{-7})^2} = 0.98 \text{ V} - 0.43 \text{ V} = 0.55 \text{ V}$

58. From Exercise 17.56, the 0.01 M Fe^{2+} compartment is the cathode and the $1 \times 10^{-7} M$ Fe^{2+} compartment is the anode.

$$Fe^{2+}(0.01 M) + 2 e^- \rightarrow Fe \qquad\qquad E° = -0.44 \text{ V}$$
$$Fe \rightarrow Fe^{2+}(1 \times 10^{-7} M) + 2 e^- \qquad\qquad -E° = 0.44 \text{ V}$$

$$\overline{Fe^{2+}(0.01 M) \rightarrow Fe^{2+}(1 \times 10^{-7} M) \qquad\qquad E°_{cell} = 0.00 \text{ V}}$$

As is always the case for a concentration cell, $E^\circ_{cell} = 0$. The driving force for the reaction is to equalize the ion concentrations in the two compartments.

$$E_{cell} = E^\circ_{cell} - \frac{0.0592}{n} \log Q = 0 - \frac{0.0592}{2} \log \frac{[Fe^{2+}]_{anode}}{[Fe^{2+}]_{cathode}}$$

$$E_{cell} = \frac{-0.0592}{2} \log \frac{1 \times 10^{-7}}{0.01} = 0.1 \text{ V}$$

59.   For concentration cells, the driving force for the reaction is the difference in ion concentrations between the anode and cathode. In order to equalize the ion concentrations, the anode always has the smaller ion concentration. The general set-up for this concentration cell is:

Cathode: $Ag^+(x\,M) + e^- \rightarrow Ag$ $E^\circ = 0.80$ V
Anode: $Ag \rightarrow Ag^+(y\,M) + e^-$ $-E^\circ = -0.80$ V
_____
                $Ag^+(\text{cathode}, x\,M) \rightarrow Ag^+(\text{anode}, y\,M)$ $E^\circ_{cell} = 0.00$ V

$$E_{cell} = E^\circ_{cell} - \frac{0.0592}{n} \log Q = \frac{-0.0592}{1} \log \frac{[Ag^+]_{anode}}{[Ag^+]_{cathode}}$$

For each concentration cell, we will calculate the cell potential using the above equation. Remember that the anode always has the smaller ion concentration.

a.   Since both compartments are at standard conditions ($[Ag^+] = 1.0\,M$) then $E_{cell} = E^\circ_{cell} = 0$ V. No voltage is produced since no reaction occurs. Concentration cells only produce a voltage when the ion concentrations are <u>not</u> equal.

b.   Cathode = $2.0\,M\,Ag^+$; Anode = $1.0\,M\,Ag^+$; Electron flow is always from the anode to the cathode so electrons flow to the right in the diagram.

$$E_{cell} = \frac{-0.0592}{n} \log \frac{[Ag^+]_{anode}}{[Ag^+]_{cathode}} = \frac{-0.0592}{1} \log \frac{1.0}{2.0} = 0.018 \text{ V}$$

c.   Cathode = $1.0\,M\,Ag^+$; Anode = $0.10\,M\,Ag^+$; Electrons flow to the left in the diagram.

$$E_{cell} = \frac{-0.0592}{n} \log \frac{[Ag^+]_{anode}}{[Ag^+]_{cathode}} = \frac{-0.0592}{1} \log \frac{0.10}{1.0} = 0.059 \text{ V}$$

d.   Cathode = $1.0\,M\,Ag^+$; Anode = $4.0 \times 10^{-5}\,M\,Ag^+$; Electrons flow to the left in the diagram.

$$E_{cell} = \frac{-0.0592}{1} \log \frac{4.0 \times 10^{-5}}{1.0} = 0.26 \text{ V}$$

e.   Since the ion concentrations are the same, then $\log([Ag^+]_{anode}/[Ag^+]_{cathode}) = \log(1.0) = 0$ and $E_{cell} = 0$. No electron flow occurs.

60.   As is the case in all concentration cells, $E^\circ_{cell} = 0$ and the smaller ion concentration always is in the anode compartment. The general Nernst equation for the $Zn \mid Zn^{2+} (x\,M) \mid\mid Zn^{2+} (y\,M) \mid Zn$ concentration cell is:

$$E_{cell} = E^\circ_{cell} - \frac{0.0592}{n} \log Q = \frac{-0.0592}{2} \log \frac{[Zn^{2+}]_{anode}}{[Zn^{2+}]_{cathode}}$$

a.   Since both compartments are at standard conditions ($[Zn^{2+}] = 1.0\,M$), then $E_{cell} = E^\circ_{cell} = 0$ V. No electron flow occurs.

b.   Cathode = $2.0\,M\ Zn^{2+}$; Anode = $1.0\,M\ Zn^{2+}$; Electron flow is always from the anode to the cathode so electrons flow to the right in the diagram.

$$E_{cell} = \frac{-0.0592}{2} \log \frac{[Zn^{2+}]_{anode}}{[Zn^{2+}]_{cathode}} = \frac{-0.0592}{2} \log \frac{1.0}{2.0} = 8.9 \times 10^{-3}\text{ V}$$

c.   Cathode = $1.0\,M\ Zn^{2+}$; Anode = $0.10\,M\ Zn^{2+}$; Electrons flow to the left in the diagram.

$$E_{cell} = \frac{-0.0592}{2} \log \frac{0.10}{1.0} = 0.030\text{ V}$$

d.   Cathode = $1.0\,M\ Zn^{2+}$; Anode = $4.0 \times 10^{-5}\,M\ Zn^{2+}$; Electrons flow to the left in the diagram.

$$E_{cell} = \frac{-0.0592}{2} \log \frac{4.0 \times 10^{-5}}{1.0} = 0.13\text{ V}$$

e.   Since both concentrations are equal then $\log (2.5/2.5) = \log 1.0 = 0$ and $E_{cell} = 0$. No electron flow occurs.

61.
$$5\,e^- + 8\,H^+ + MnO_4^- \rightarrow Mn^{2+} + 4\,H_2O \qquad\qquad E^\circ = 1.51\text{ V}$$
$$(Fe^{2+} \rightarrow Fe^{3+} + e^-) \times 5 \qquad\qquad -E^\circ = -0.77\text{ V}$$

$$\overline{8\,H^+(aq) + MnO_4^-(aq) + 5\,Fe^{2+}(aq) \rightarrow 5\,Fe^{3+}(aq) + Mn^{2+}(aq) + 4\,H_2O(l) \qquad E^\circ_{cell} = 0.74\text{ V}}$$

$$E_{cell} = E^\circ_{cell} - \frac{0.0592}{n} \log Q = 0.74\text{ V} - \frac{0.0592}{5} \log \frac{[Fe^{3+}]^5\,[Mn^{2+}]}{[Fe^{2+}]^5\,[MnO_4^-]\,[H^+]^8}; \quad \text{pH} = 4.0 \text{ so } H^+ = 1 \times 10^{-4}\,M$$

$$E_{cell} = 0.74 - \frac{0.0592}{5} \log \frac{(1 \times 10^{-6})^5\,(1 \times 10^{-6})}{(1 \times 10^{-3})^5\,(1 \times 10^{-2})\,(1 \times 10^{-4})^8}$$

$$E_{cell} = 0.74 - \frac{0.0592}{5} \log (1 \times 10^{13}) = 0.74\text{ V} - 0.15\text{ V} = 0.59\text{ V} = 0.6\text{ V} \text{ (1 sig. fig. due to concentrations)}$$

Yes, $E_{cell} > 0$ so reaction will occur as written.

62. $n = 2$ for this reaction (lead goes from $Pb \rightarrow Pb^{2+}$ in $PbSO_4$).

$$E = E° - \frac{0.0592}{2} \log \frac{1}{[H^+]^2[HSO_4^-]^2} = 2.04 \text{ V} - \frac{0.0592}{2} \log \frac{1}{(4.5)^2 (4.5)^2}$$

$E = 2.04 \text{ V} + 0.077 \text{ V} = 2.12 \text{ V}$

63. $Cu^{2+} + 2 \text{ e}^- \rightarrow Cu$ $E° = 0.34$ V
                        $Zn \rightarrow Zn^{2+} + 2 \text{ e}^-$ $-E° = 0.76$ V
_____
$Cu^{2+}(aq) + Zn(s) \rightarrow Zn^{2+}(aq) + Cu(s)$ $E°_{cell} = 1.10$ V

Since Zn^{2+} is a product in the reaction, then the Zn^{2+} concentration increases from $1.00 \, M$ to $1.20 \, M$. This means that the reactant concentration of Cu^{2+} must decrease from $1.00 \, M$ to $0.80 \, M$ (from the 1:1 mol ratio in the balanced reaction).

$$E_{cell} = E°_{cell} - \frac{0.0592}{n} \log Q = 1.10 \text{ V} - \frac{0.0592}{2} \log \frac{[Zn^{2+}]}{[Cu^{2+}]}$$

$$E_{cell} = 1.10 \text{ V} - \frac{0.0592}{2} \log \frac{1.20}{0.80} = 1.10 \text{ V} - 0.0052 \text{ V} = 1.09 \text{ V}$$

64. $(Ag^+ + \text{e}^- \rightarrow Ag) \times 2$ $E° = 0.80$ V
                        $Zn \rightarrow Zn^{2+} + 2 \text{ e}^-$ $-E° = 0.76$ V
_____
$2 \, Ag^+(aq) + Zn(s) \rightarrow Zn^{2+}(aq) + 2 \, Ag(s)$ $E°_{cell} = 1.56$ V

Zn^{2+} is a product in this spontaneous reaction at standard conditions and Ag^+ is a reactant. The Zn^{2+} concentration increases by $0.20 \, M$ while the Ag^+ concentration decreases by $2 \, (0.20) = 0.40 \, M$ (from the 2:1 mol ratio in the balanced equation).

$$E_{cell} = 1.56 \text{ V} - \frac{0.0592}{2} \log \frac{[Zn^{2+}]}{[Ag^+]^2} = 1.56 - \frac{0.0592}{2} \log \frac{1.20}{(0.60)^2}$$

$E_{cell} = 1.56 - 0.015 = 1.54 \text{ V}$

65. $Cu^{2+}(aq) + H_2(g) \rightarrow 2 \, H^+(aq) + Cu(s)$ $E°_{cell} = 0.34 \text{ V} - 0.00\text{V} = 0.34$ V and $n = 2$

Since $P_{H_2} = 1.0$ atm and $[H^+] = 1.0 \, M$: $E_{cell} = E°_{cell} - \frac{0.0592}{2} \log \frac{1}{[Cu^{2+}]}$

a. $E_{cell} = 0.34 \text{ V} - \frac{0.0592}{2} \log \frac{1}{2.5 \times 10^{-4}} = 0.34 \text{ V} - 0.11 = 0.23 \text{ V}$

b. $0.195 \text{ V} = 0.34 \text{ V} - \frac{0.0592}{2} \log \frac{1}{[Cu^{2+}]}$, $\log \frac{1}{[Cu^{2+}]} = 4.90$, $[Cu^{2+}] = 10^{-4.90} = 1.3 \times 10^{-5} \, M$

Note: When determining exponents, we will carry extra significant figures.

66. $2 Ag^+(aq) + Cu(s) \rightarrow Cu^{2+}(aq) + 2 Ag(s)$ $E°_{cell} = 0.80 - 0.34 = 0.46$ V and $n = 2$

Since $[Ag^+] = 1.0 M$, then $E_{cell} = 0.46$ V $- \dfrac{0.0592}{2} \log [Cu^{2+}]$

a. $E_{cell} = 0.46 - \dfrac{0.0592}{2} \log (1.3 \times 10^{-5}) = 0.46 + 0.14 = 0.60$ V

b. 0.58 V $= 0.46$ V $- \dfrac{0.0592}{2} \log [Cu^{2+}]$, $\log [Cu^{2+}] = -4.05$, $[Cu^{2+}] = 10^{-4.05} = 8.9 \times 10^{-5} M$

67. $Cu^{2+}(aq) + H_2(g) \rightarrow 2 H^+(aq) + Cu(s)$ $E°_{cell} = 0.34$ V $- 0.00$ V $= 0.34$ V and $n = 2$

Since $P_{H_2} = 1.0$ atm and $[H^+] = 1.0 M$: $E_{cell} = E°_{cell} - \dfrac{0.0592}{2} \log \dfrac{1}{[Cu^{2+}]}$

Use the K_{sp} expression to calculate the Cu^{2+} concentration in the cell.

$Cu(OH)_2(s) \rightleftharpoons Cu^{2+}(aq) + 2 OH^-(aq)$ $K_{sp} = 1.6 \times 10^{-19} = [Cu^{2+}][OH^-]^2$

From problem, $[OH^-] = 0.10 M$, so: $[Cu^{2+}] = \dfrac{1.6 \times 10^{-19}}{(0.10)^2} = 1.6 \times 10^{-17} M$

$E_{cell} = E°_{cell} - \dfrac{0.0592}{2} \log \dfrac{1}{[Cu^{2+}]} = 0.34$ V $- \dfrac{0.0592}{2} \log \dfrac{1}{1.6 \times 10^{-17}} = 0.34 - 0.50 = -0.16$ V

Since $E_{cell} < 0$, then the forward reaction is not spontaneous, but the reverse reaction is spontaneous. The Cu electrode becomes the anode and $E_{cell} = 0.16$ V for the reverse reaction. The cell reaction is: $2 H^+(aq) + Cu(s) \rightarrow Cu^{2+}(aq) + H_2(g)$.

68. $2 Ag^+(aq) + Cu(s) \rightarrow Cu^{2+}(aq) + 2 Ag(s)$ $E°_{cell} = 0.80 - 0.34 = 0.46$ V and $n = 2$

Since $[Ag^+] = 1.0 M$, then $E_{cell} = 0.46$ V $- \dfrac{0.0592}{2} \log [Cu^{2+}]$

Use the equilibrium reaction to calculate the Cu^{2+} concentration in the cell.

$Cu^{2+} + 4 NH_3 \rightleftharpoons Cu(NH_3)_4^{2+}$ $K = \dfrac{[Cu(NH_3)_4^{2+}]}{[Cu^{2+}][NH_3]^4} = 1.0 \times 10^{13}$

From the problem, $[NH_3] = 5.0 M$ and $[Cu(NH_3)_4^{2+}] = 0.010 M$:

$1.0 \times 10^{13} = \dfrac{0.010}{[Cu^{2+}](5.0)^4}$, $[Cu^{2+}] = 1.6 \times 10^{-18} M$

$E_{cell} = 0.46 - \dfrac{0.0592}{2} \log (1.6 \times 10^{-18}) = 0.46 - (-0.53) = 0.99$ V

69.   See Exercises 17.25, 17.27 and 17.29 for balanced reactions and standard cell potentials. Balanced reactions are necessary to determine n, the moles of electrons transferred.

25a. $7 H_2O + 2 Cr^{3+} + 3 Cl_2 \rightarrow Cr_2O_7^{2-} + 6 Cl^- + 14 H^+$ $E^\circ_{cell} = 0.03 V = 0.03 J/C$

$E_{cell} = E^\circ_{cell} - \dfrac{0.0592}{n} \log Q$: At equilibrium, $E_{cell} = 0$ and $Q = K$, so $E^\circ_{cell} = \dfrac{0.0592}{n} \log K$

$\log K = \dfrac{nE^\circ}{0.0592} = \dfrac{6(0.03)}{0.0592} = 3.04, \ K = 10^{3.04} = 1 \times 10^3$

Note: When determining exponents, we will round off after the calculation is complete in order to help eliminate excessive round off error.

25b. $\log K = \dfrac{2(2.71)}{0.0592} = 91.554, \ K = 3.58 \times 10^{91}$

29a. $\log K = \dfrac{2(0.27)}{0.0592} = 9.12, \ K = 1.3 \times 10^9$

29b. $\log K = \dfrac{10(0.09)}{0.0592} = 15.20, \ K = 2 \times 10^{15}$

70. $E^\circ_{cell} = \dfrac{0.0592}{n} \log K, \ \log K = \dfrac{nE^\circ}{0.0592}$

26a. $\log K = \dfrac{10(0.43)}{0.0592} = 72.64, \ K = 10^{72.64} = 4.4 \times 10^{72}$

26b. $\log K = \dfrac{2(1.56)}{0.0592} = 52.703, \ K = 5.05 \times 10^{52}$

30a. $\log K = \dfrac{6(1.43)}{0.0592} = 144.93 = 145, \ K \approx 10^{145}$

30b. $\log K = \dfrac{1(1.05)}{0.0592} = 17.736, \ K = 5.45 \times 10^{17}$

71.   a. $2 e^- + 2 H_2O \rightarrow 2 OH^- + H_2$ $E^\circ = -0.83 V$
                    $(Na \rightarrow Na^+ + e^-) \times 2$ $-E^\circ = 2.71 V$

$\overline{}$

$2 Na(s) + 2 H_2O(l) \rightarrow 2 NaOH(aq) + H_2(g)$ $E^\circ_{cell} = 1.88 V = 1.88 J/C$

$\Delta G^\circ = -nFE^\circ_{cell} = -(2 \ mol \ e^-)(96{,}485 \ C/mol \ e^-)(1.88 \ J/C) = -3.63 \times 10^5 \ J = -363 \ kJ$

From the Nernst equation: $E^\circ_{cell} = \dfrac{0.0592}{n} \log K, \ \log K = \dfrac{nE^\circ}{0.0592}$

$$\log K = \frac{2(1.88)}{0.0592} = 63.514, \ K = 10^{63.514} = 3.27 \times 10^{63}$$

Note: When determing exponents, we will round off after the calculation is complete.

b.

$$(2 \ H^+ + 2 \ e^- \rightarrow H_2 \ ) \times 3/2 \qquad\qquad E° = 0.00 \ V$$
$$Al \rightarrow Al^{3+} + 3 \ e^- \qquad\qquad -E° = 1.66 \ V$$

$$\overline{\ 3 \ H^+(aq) + Al(s) \rightarrow Al^{3+}(aq) + 3/2 \ H_2(g) \qquad E°_{cell} = 1.66 \ V\ }$$

$$\Delta G° = \ -(3 \ mol \ e^-)(96{,}485 \ C/mol \ e^-)(1.66 \ J/C) = -4.80 \times 10^5 \ J = -480. \ kJ$$

$$\log K = \frac{3(1.66)}{0.0592} = 84.122, \ K = 1.32 \times 10^{84}$$

c.

$$Br_2 + 2 \ e^- \rightarrow 2 \ Br^- \qquad\qquad E° = 1.09 \ V$$
$$2 \ I^- \rightarrow I_2 + 2 \ e^- \qquad\qquad -E° = -0.54 V$$

$$\overline{\ Br_2(aq) + 2 \ I^-(aq) \rightarrow 2 \ Br^-(aq) + I_2(s) \qquad E°_{cell} = 0.55 \ V\ }$$

$$\Delta G° = \ -(2 \ mol \ e^-)(96{,}485 \ C/mol \ e^-)(0.55 \ J/C) = -1.1 \times 10^5 \ J = -110 \ kJ$$

$$\log K = \frac{2(0.55)}{0.0592} = 18.58, \ K = 3.8 \times 10^{18}$$

72.   a. $Cu^+ + e^- \rightarrow Cu \qquad\qquad E° = 0.52 \ V$
$\qquad\qquad Cu^+ \rightarrow Cu^{2+} + e^- \qquad\qquad -E° = -0.16 \ V$

$$\overline{\ 2 \ Cu^+(aq) \rightarrow Cu^{2+}(aq) + Cu(s) \qquad E°_{cell} = 0.36 \ V; \ Spontaneous\ }$$

$$\Delta G° = \ -nFE°_{cell} = -(1 \ mol \ e^-)(96{,}485 \ C/mol \ e)(0.36 \ J/C) = -34{,}700 \ J = -35 \ kJ$$

$$E°_{cell} = \frac{0.0592}{n} \ \log K, \ \log K = \frac{nE°}{0.0592} = \frac{1(0.36)}{0.0592} = 6.08, \ K = 10^{6.08} = 1.2 \times 10^6$$

b. $Fe^{2+} + 2 \ e^- \rightarrow Fe \qquad\qquad E° = -0.44 \ V$
$\qquad\quad (Fe^{2+} \rightarrow Fe^{3+} + e^-) \times 2 \qquad\qquad -E° = -0.77 \ V$

$$\overline{\ 3 \ Fe^{2+}(aq) \rightarrow 2 \ Fe^{3+}(aq) + Fe(s) \qquad E°_{cell} = -1.21 \ V; \ Not \ spontaneous\ }$$

c. $HClO_2 + 2 \ H^+ + 2 \ e^- \rightarrow HClO + H_2O \qquad\qquad E° = 1.65 \ V$
$\qquad\quad HClO_2 + H_2O \rightarrow ClO_3^- + 3 \ H^+ + 2 \ e^- \qquad\qquad -E° = -1.21 \ V$

$$\overline{\ 2 \ HClO_2(aq) \rightarrow ClO_3^-(aq) + H^+(aq) + HClO(aq) \qquad E°_{cell} = 0.44 \ V; \ Spontaneous\ }$$

$$\Delta G° = \ -nFE°_{cell} = -(2 \ mol \ e^-)(96{,}485 \ C/mol \ e^-)(0.44 \ J/C) = -84{,}900 \ J = -85 \ kJ$$

$$\log K = \frac{nE°}{0.0592} = \frac{2(0.44)}{0.0592} = 14.86, \ K = 7.2 \times 10^{14}$$

73. a.
$$Fe^{2+} + 2\ e^- \rightarrow Fe \qquad\qquad E° = -0.44\ V$$
$$Zn \rightarrow Zn^{2+} + 2\ e^- \qquad\qquad -E° = 0.76\ V$$

$$Fe^{2+}(aq) + Zn(s) \rightarrow Zn^{2+}(aq) + Fe(s) \qquad E°_{cell} = 0.32\ V = 0.32\ J/C$$

b. $\Delta G° = -nFE°_{cell} = -(2\ mol\ e^-)(96,485\ C/mol\ e^-)(0.32\ J/C) = -6.2 \times 10^4\ J = -62\ kJ$

$$E°_{cell} = \frac{0.0592}{n}\ \log K, \quad \log K = \frac{nE°}{0.0592} = \frac{2(0.32)}{0.0592} = 10.81, \quad K = 10^{10.81} = 6.5 \times 10^{10}$$

c. $E_{cell} = E°_{cell} - \dfrac{0.0592}{n}\ \log Q = 0.32\ V - \dfrac{0.0592}{2}\ \log \dfrac{[Zn^{2+}]}{[Fe^{2+}]}$

$$E_{cell} = 0.32 - \frac{0.0592}{2}\ \log \frac{0.10}{1.0 \times 10^{-5}} = 0.32 - 0.12 = 0.20\ V$$

74. a.
$$Au^{3+} + 3\ e^- \rightarrow Au \qquad\qquad E° = 1.50\ V$$
$$(Tl \rightarrow Tl^+ + e^-) \times 3 \qquad\qquad -E° = 0.34\ V$$

$$Au^{3+}(aq) + 3\ Tl(s) \rightarrow Au(s) + 3\ Tl^+(aq) \qquad E°_{cell} = 1.84\ V$$

b. $\Delta G° = -nFE°_{cell} = -(3\ mol\ e^-)(96,485\ C/mol\ e^-)(1.84\ J/C) = -5.33 \times 10^5\ J = -533\ kJ$

$$\log K = \frac{nE°}{0.0592} = \frac{3(1.84)}{0.0592} = 93.243, \quad K = 10^{93.243} = 1.75 \times 10^{93}$$

c. $E_{cell} = 1.84\ V - \dfrac{0.0592}{3}\ \log \dfrac{[Tl^+]^3}{[Au^{3+}]} = 1.84 - \dfrac{0.0592}{3}\ \log \dfrac{(1.0 \times 10^{-4})^3}{1.0 \times 10^{-2}}$

$$E_{cell} = 1.84 - (-0.20) = 2.04\ V$$

75.
$$PbSO_4(s) + 2\ e^- \rightarrow Pb + SO_4^{2-} \qquad\qquad E° = -0.35\ V$$
$$Pb \rightarrow Pb^{2+} + 2\ e^- \qquad\qquad -E° = 0.13\ V$$

$$PbSO_4\ (s) \rightarrow Pb^{2+}(aq) + SO_4^{2-}(aq) \qquad\qquad E°_{cell} = -0.22\ V$$

For this overall reaction $K = K_{sp}$ so $E°_{cell} = \dfrac{0.0592}{n}\ \log K_{sp}.$

$$\log K_{sp} = \frac{nE°}{0.0592} = \frac{2(-0.22)}{0.0592} = -7.43, \quad K_{sp} = 10^{-7.43} = 3.7 \times 10^{-8}$$

76.
$$CdS + 2\ e^- \rightarrow Cd + S^{2-} \qquad E° = -1.21\ V$$
$$Cd \rightarrow Cd^{2+} + 2\ e^- \qquad -E° = 0.402\ V$$

$$CdS(s) \rightarrow Cd^{2+}(aq) + S^{2-}(aq) \qquad E°_{cell} = -0.81\ V \qquad K_{sp} = ?$$

$$\log K_{sp} = \frac{nE°}{0.0592} = \frac{2(-0.81)}{0.0592} = -27.36, \quad K_{sp} = 10^{-27.36} = 4.4 \times 10^{-28}$$

77. $e^- + AgI \rightarrow Ag + I^-$ $E°_{AgI} = ?$
       $Ag \rightarrow Ag^+ + e^-$ $-E° = -0.80$ V

       _____

       $AgI(s) \rightarrow Ag^+(aq) + I^-(aq)$ $E°_{cell} = E°_{AgI} - 0.80$ $K = K_{sp} = 1.5 \times 10^{-16}$

For this overall reaction:

$$E°_{cell} = \frac{0.0592}{n} \log K_{sp} = \frac{0.0592}{1} \log (1.5 \times 10^{-16}) = -0.94 \text{ V}$$

$E°_{cell} = -0.94$ V $= E°_{AgI} - 0.80$ V, $E°_{AgI} = -0.94 + 0.80 = -0.14$ V

78. $CuI + e^- \rightarrow Cu + I^-$ $E°_{CuI} = ?$
       $Cu \rightarrow Cu^+ + e^-$ $-E° = -0.52$ V

       _____

       $CuI(s) \rightarrow Cu^+(aq) + I^-(aq)$ $E°_{cell} = E°_{CuI} - 0.52$ V

For this overall reaction, $K = K_{sp} = 1.1 \times 10^{-12}$:

$$E°_{cell} = \frac{0.0592}{n} \log K_{sp} = \frac{0.0592}{1} \log (1.1 \times 10^{-12}) = -0.71 \text{ V}$$

$E°_{cell} = -0.71$ V $= E°_{CuI} - 0.52$, $E°_{CuI} = -0.19$ V

Electrolysis

79.   a. $Al^{3+} + 3 e^- \rightarrow Al$; 3 mol e^- are needed to produce 1 mol Al from Al^{3+}.

       $$1.0 \times 10^3 \text{ g Al} \times \frac{1 \text{ mol Al}}{26.98 \text{ g Al}} \times \frac{3 \text{ mol } e^-}{\text{mol Al}} \times \frac{96,485 \text{ C}}{\text{mol } e^-} \times \frac{1 \text{ s}}{100.0 \text{ C}} = 1.07 \times 10^5 \text{ s} = 30. \text{ hours}$$

       b. $$1.0 \text{ g Ni} \times \frac{1 \text{ mol Ni}}{58.69 \text{ g Ni}} \times \frac{2 \text{ mol } e^-}{\text{mol Ni}} \times \frac{96,485 \text{ C}}{\text{mol } e^-} \times \frac{1 \text{ s}}{100.0 \text{ C}} = 33 \text{ s}$$

       c. $$5.0 \text{ mol Ag} \times \frac{1 \text{ mol } e^-}{\text{mol Ag}} \times \frac{96,485 \text{ C}}{\text{mol } e^-} \times \frac{1 \text{ s}}{100.0 \text{ C}} = 4.8 \times 10^3 \text{ s} = 1.3 \text{ hours}$$

80. $MgCl_2(l) \rightarrow Mg^{2+} + 2 Cl^-$; 2 mol e^- are needed to produce 1 mol Mg from molten $MgCl_2$.

       $$1.00 \times 10^6 \text{ g Mg} \times \frac{1 \text{ mol Mg}}{24.31 \text{ g Mg}} \times \frac{2 \text{ mol } e^-}{1 \text{ mol Mg}} \times \frac{96,485 \text{ C}}{\text{mol } e^-} \times \frac{1 \text{ s}}{50.0 \text{ C}} = 1.59 \times 10^8 \text{ s} = 4.42 \times 10^4 \text{ hours}$$

81. $$15A = \frac{15 \text{ C}}{\text{s}} \times \frac{60 \text{ s}}{\text{min}} \times \frac{60 \text{ min}}{\text{h}} = 5.4 \times 10^4 \text{ C of charge passed in 1 hour}$$

       a. $$5.4 \times 10^4 \text{ C} \times \frac{1 \text{ mol } e^-}{96,485 \text{ C}} \times \frac{1 \text{ mol Co}}{2 \text{ mol } e^-} \times \frac{58.93 \text{ g Co}}{\text{mol Co}} = 16 \text{ g Co}$$

       b. $$5.4 \times 10^4 \text{ C} \times \frac{1 \text{ mol } e^-}{96,485 \text{ C}} \times \frac{1 \text{ mol Hf}}{4 \text{ mol } e^-} \times \frac{178.5 \text{ g Hf}}{\text{mol Hf}} = 25 \text{ g Hf}$$

c. $2 I^- \rightarrow I_2 + 2 e^-$; 5.4×10^4 C $\times \dfrac{1 \text{ mol } e^-}{96,485 \text{ C}} \times \dfrac{1 \text{ mol } I_2}{2 \text{ mol } e^-} \times \dfrac{253.8 \text{ g } I_2}{\text{mol } I_2} = 71$ g I_2

d. $CrO_3(l) \rightarrow Cr^{6+} + 3 O^{2-}$; 6 mol e^- are needed to produce 1 mol Cr from molten CrO_3.

5.4×10^4 C $\times \dfrac{1 \text{ mol } e^-}{96,485 \text{ C}} \times \dfrac{1 \text{ mol Cr}}{6 \text{ mol } e^-} \times \dfrac{52.00 \text{ g Cr}}{\text{mol Cr}} = 4.9$ g Cr

82. $Al_2O_3(l) \rightarrow 2 Al^{3+} + 3 O^{2-}$; 3 mol e^- are needed to produce 1 mol of Al from molten Al_2O_3.

11.5 h $\times \dfrac{60 \text{ min}}{h} \times \dfrac{60 \text{ s}}{\text{min}} \times \dfrac{5.00 \text{ C}}{s} \times \dfrac{1 \text{ mol } e^-}{96,485 \text{ C}} \times \dfrac{1 \text{ mol Al}}{3 \text{ mol } e^-} \times \dfrac{26.98 \text{ g Al}}{\text{mol Al}} = 19.3$ g Al

83.   600. s $\times \dfrac{5.00 \text{ C}}{s} \times \dfrac{1 \text{ mol } e^-}{96,485 \text{ C}} \times \dfrac{1 \text{ mol M}}{3 \text{ mol } e^-} = 1.04 \times 10^{-2}$ mol M where M = unknown metal

Atomic mass $= \dfrac{1.19 \text{ g M}}{1.04 \times 10^{-2} \text{ mol M}} = \dfrac{114 \text{ g}}{\text{mol}}$; The element is indium, In. Indium forms 3+ ions.

84.   74.6 s $\times \dfrac{2.50 \text{ C}}{s} \times \dfrac{1 \text{ mol } e^-}{96,485 \text{ C}} \times \dfrac{1 \text{ mol M}}{2 \text{ mol } e^-} = 9.66 \times 10^{-4}$ mol M

Atomic mass $= \dfrac{0.1086 \text{ g M}}{9.66 \times 10^{-4} \text{ mol M}} = \dfrac{112 \text{ g}}{\text{mol}}$

The metal is Cd. Cd has the correct atomic mass and does form Cd^{2+} ions.

85. F_2 is produced at the anode: $2 F^- \rightarrow F_2 + 2 e^-$

2.00 h $\times \dfrac{60 \text{ min}}{h} \times \dfrac{60 \text{ s}}{\text{min}} \times \dfrac{10.0 \text{ C}}{s} \times \dfrac{1 \text{ mol } e^-}{96,485 \text{ C}} = 0.746$ mol e^-

0.746 mol $e^- \times \dfrac{1 \text{ mol } F_2}{2 \text{ mol } e^-} = 0.373$ mol F_2; $PV = nRT$, $V = \dfrac{nRT}{P}$

$V = \dfrac{(0.373 \text{ mol}) (0.08206 \text{ L} \cdot \text{atm/K} \cdot \text{mol}) (298 \text{ K})}{1.00 \text{ atm}} = 9.12$ L F_2

K is produced at the cathode: $K^+ + e^- \rightarrow K$

0.746 mol $e^- \times \dfrac{1 \text{ mol K}}{\text{mol } e^-} \times \dfrac{39.10 \text{ g K}}{\text{mol K}} = 29.2$ g K

86.   The half-reactions for the electrolysis of water are:

$$(2 e^- + 2 H_2O \rightarrow H_2 + 2 OH^-) \times 2$$
$$2 H_2O \rightarrow 4 H^+ + O_2 + 4 e^-$$

_____

$$2 H_2O(l) \rightarrow 2 H_2(g) + O_2(g)$$

Note: $4 H^+ + 4 OH^- \rightarrow 4 H_2O$ and n = 4 for this reaction as it is written.

$$15.0 \text{ min} \times \frac{60 \text{ s}}{\text{min}} \times \frac{2.50 \text{ C}}{\text{s}} \times \frac{1 \text{ mol e}^-}{96,485 \text{ C}} \times \frac{2 \text{ mol H}_2}{4 \text{ mol e}^-} = 1.17 \times 10^{-2} \text{ mol H}_2$$

At STP, 1 mole of an ideal gas occupies a volume of 22.42 L (see Chapter 5 of the text).

$$1.17 \times 10^{-2} \text{ mol H}_2 \times \frac{22.42 \text{ L}}{\text{mol H}_2} = 0.262 \text{ L} = 262 \text{ mL H}_2$$

$$1.17 \times 10^{-2} \text{ mol H}_2 \times \frac{1 \text{ mol O}_2}{2 \text{ mol H}_2} \times \frac{22.42 \text{ L}}{\text{mol O}_2} = 0.131 \text{ L} = 131 \text{ mL O}_2$$

87. $$\frac{150. \times 10^3 \text{ g C}_6\text{H}_8\text{N}_2}{\text{h}} \times \frac{1 \text{ h}}{60 \text{ min}} \times \frac{1 \text{ min}}{60 \text{ s}} \times \frac{1 \text{ mol C}_6\text{H}_8\text{N}_2}{108.14 \text{ g C}_6\text{H}_8\text{N}_2} \times \frac{2 \text{ mol e}^-}{\text{mol C}_6\text{H}_8\text{N}_2} \times \frac{96,485 \text{ C}}{\text{mol e}^-}$$

$$= 7.44 \times 10^4 \text{ C/s or a current of } 7.44 \times 10^4 \text{ A}$$

88. $Al^{3+} + 3 \text{ e}^- \rightarrow Al$; 3 mol e$^-$ are needed to produce Al from Al^{3+}

$$2000 \text{ lb Al} \times \frac{453.6 \text{ g}}{\text{lb}} \times \frac{1 \text{ mol Al}}{26.98 \text{ g}} \times \frac{3 \text{ mol e}^-}{\text{mol Al}} \times \frac{96,485 \text{ C}}{\text{mol e}^-} = 1 \times 10^{10} \text{ C of electricity needed}$$

$$\frac{1 \times 10^{10} \text{ C}}{24 \text{ h}} \times \frac{1 \text{ h}}{60 \text{ min}} \times \frac{1 \text{ min}}{60 \text{ s}} = 1 \times 10^5 \text{ C/s} = 1 \times 10^5 \text{ A}$$

89. $$2.30 \text{ min} \times \frac{60 \text{ s}}{\text{min}} = 138 \text{ s}; 138 \text{ s} \times \frac{2.00 \text{ C}}{\text{s}} \times \frac{1 \text{ mol e}^-}{96,485 \text{ C}} \times \frac{1 \text{ mol Ag}}{\text{mol e}^-} = 2.86 \times 10^{-3} \text{ mol Ag}$$

$$[Ag^+] = 2.86 \times 10^{-3} \text{ mol Ag}^+/0.250 \text{ L} = 1.14 \times 10^{-2} \, M$$

90. $$0.50 \text{ L} \times 0.010 \text{ mol Pt}^{4+}/\text{L} = 5.0 \times 10^{-3} \text{ mol Pt}^{4+}$$

To plate out 99% of the Pt^{4+}, we will produce $0.99 \times 5.0 \times 10^{-3}$ mol Pt.

$$0.99 \times 5.0 \times 10^{-3} \text{ mol Pt} \times \frac{4 \text{ mol e}^-}{\text{mol Pt}} \times \frac{96,485 \text{ C}}{\text{mol e}^-} \times \frac{1 \text{ s}}{4.00 \text{ C}} = 480 \text{ s}$$

91. The metal ion with the most positive reduction potential will plate out first at the cathode. In this case, Pt will plate out first.

92. The reduction reactions and the cell potentials are:

$Fe^{2+} + 2 \text{ e}^- \rightarrow Fe$ $E = E° = -0.44$ V (standard conditions)

$Ag^+ + e^- \rightarrow Ag$ $E = E° - \dfrac{0.0592}{1} \log \dfrac{1}{[Ag^+]} = 0.80 - 0.0592 \log \dfrac{1}{0.010} = 0.68$ V

It is easier to plate out Ag from $1.0 \times 10^{-2} \, M \, Ag^+$ since this process has a more positive reduction potential than Fe^{2+} to Fe.

93.   Reduction occurs at the cathode and oxidation occurs at the anode. First determine all the species present, then look up pertinent reduction and/or oxidation potentials in Table 17.1 for all these species. The cathode reaction will be the reaction with the most positive reduction potential and the anode reaction will be the reaction with the most positive oxidation potential.

a.   Species present: K^+ and F^-; K^+ can be reduced to K and F^- can be oxidized to F_2 (from Table 17.1). The reactions are:

Cathode: $K^+ + e^- \rightarrow K$ $E° = -2.92$ V
Anode: $2 F^- \rightarrow F_2 + 2 e^-$ $-E° = -2.87$ V

b.   Species present: Cu^{2+} and Cl^-; Cu^{2+} can be reduced and Cl^- can be oxidized. The reactions are:

Cathode: $Cu^{2+} + 2 e^- \rightarrow Cu$ $E° = 0.34$ V
Anode: $2 Cl^- \rightarrow Cl_2 + 2 e^-$ $-E° = -1.36$ V

c.   Species present: Mg^{2+} and I^-; Mg^{2+} can be reduced and I^- can be oxidized. The reactions are:

Cathode: $Mg^{2+} + 2 e^- \rightarrow Mg$ $E° = -2.37$ V
Anode: $2 I^- \rightarrow I_2 + 2 e^-$ $-E° = -0.54$ V

94.   These are all in aqueous solutions so we must also consider the reduction and oxidation of H_2O in addition to the potential redox reactions of the ions present. For the cathode reaction, the species with the most positive reduction potential will be reduced and for the anode reaction, the species with the most positive oxidation potential will be oxidized.

a.   Species present: K^+, F^- and H_2O. Possible cathode reactions are:

$K^+ + e^- \rightarrow K$ $E° = -2.92$ V
$2 H_2O + 2 e^- \rightarrow H_2 + 2 OH^-$ $E° = -0.83$ V

Since it is easier to reduce H_2O than K^+ (assuming standard conditions), then H_2O will be reduced by the above cathode reaction.

Possible anode reactions are:

$2 F^- \rightarrow F_2 + 2 e^-$ $-E° = -2.87$ V
$2 H_2O \rightarrow 4 H^+ + O_2 + 4 e^-$ $-E° = -1.23$ V

Since H_2O is easier to oxidize than F^- (assuming standard conditions), then H_2O will be oxidized by the above anode reaction.

b.   Species present: Cu^{2+}, Cl^- and H_2O; Cu^{2+} and H_2O can be reduced. The reduction potentials are $E° = 0.34$ V for Cu^{2+} and $E° = -0.83$ V for H_2O (assuming standard conditions). Cu^{2+} will be reduced to Cu at the cathode ($Cu^{2+} + 2 e^- \rightarrow Cu$).

Cl^- and H_2O can be oxidized. The oxidation potentials are $-E° = -1.36$ V for Cl^- and $-E° = -1.23$ V for H_2O (assuming standard conditions). From the potentials, we would predict H_2O to be oxidized at the anode ($2 H_2O \rightarrow 4 H^+ + O_2 + 4 e^-$). Note: In real life, Cl^- is oxidized to Cl_2 when water is present due to a phenomenon called overvoltage (see Chapter 17.7 of the text). Since overvoltage is difficult to predict, we will generally ignore it.

c.   Species present: H_2O_2, H^+, Cl^- and H_2O; H_2O_2, H^+ and H_2O can be reduced. The possible cathode reactions are:

$$2 H_2O + 2 e^- \rightarrow H_2 + 2 OH^- \qquad E° = -0.83 \text{ V}$$
$$2 H^+ + 2 e^- \rightarrow H_2 \qquad E° = 0.00 \text{ V}$$
$$H_2O_2 + 2 H^+ + 2 e^- \rightarrow 2 H_2O \qquad E° = 1.78 \text{ V}$$

Reduction of H_2O_2 will occur at the cathode since $E°_{H_2O_2}$ is most positive.

H_2O_2, Cl^- and H_2O can all be oxidized. The possible anode reactions are:

$$2 Cl^- \rightarrow Cl_2 + 2 e^- \qquad -E° = -1.36 \text{ V}$$
$$2 H_2O \rightarrow O_2 + 4 H^+ + 4 e^- \qquad -E° = -1.23 \text{ V}$$
$$H_2O_2 \rightarrow O_2 + 2 H^+ + 2 e^- \qquad -E° = -0.68 \text{ V}$$

Oxidation of H_2O_2 will occur at the anode since $-E°_{H_2O_2}$ is more positive.

95.   Species present: Na^+, SO_4^{2-} and H_2O. From the potentials, H_2O is the more easily reduced than Na^+ and H_2O is more easily oxidized than SO_4^{2-}. The reactions are:

Cathode: $2 H_2O + 2 e^- \rightarrow H_2 + 2 OH^-$; Anode: $2 H_2O \rightarrow O_2 + 4 H^+ + 4 e^-$

96.   Species present: Mg^{2+}, I^- and H_2O: The only possible cathode reactions are:

$$2 H_2O + 2 e^- \rightarrow H_2 + 2 OH^- \qquad E° = -0.83 \text{ V}$$
$$Mg^{2+} + 2 e^- \rightarrow Mg \qquad E° = -2.37 \text{ V}$$

Reduction of H_2O will occur at the cathode since $E°_{H_2O}$ is more positive. The only possible anode reactions are:

$$2 I^- \rightarrow I_2 + 2 e^- \qquad -E° = -0.54 \text{ V}$$
$$2 H_2O \rightarrow O_2 + 4 H^+ + 4 e^- \qquad -E° = -1.23 \text{ V}$$

Oxidation of I^- will occur at the anode since $-E°_I$ is more positive.

97. $(Au(CN)_4^- + 3e^- \rightarrow Au + 4 CN^-) \times 4$
                              $(4 OH^- \rightarrow O_2 + 2 H_2O + 4 e^-) \times 3$

$$\overline{4 Au(CN)_4^-(aq) + 12 OH^-(aq) \rightarrow 4 Au(s) + 16 CN^-(aq) + 3 O_2(g) + 6 H_2O(l)}$$

$$1.00 \times 10^3 \text{ g Au} \times \frac{1 \text{ mol Au}}{197.0 \text{ g}} \times \frac{3 \text{ mol } O_2}{4 \text{ mol Au}} = 3.81 \text{ mol } O_2$$

$$V = \frac{nRT}{P} = \frac{3.81 \text{ mol} \left(\dfrac{0.08206 \text{ L atm}}{\text{mol K}} \right) (298 \text{ K})}{740. \text{ torr} \times \dfrac{1 \text{ atm}}{760 \text{ torr}}} = 95.7 \text{ L } O_2$$

98.   In the electrolysis of aqueous sodium chloride, H_2O is reduced in preference to Na^+ and Cl^- is oxidized in preference to H_2O. The anode reaction is $2\ Cl^- \rightarrow Cl_2 + 2\ e^-$ and the cathode reaction is $2\ H_2O + 2\ e^- \rightarrow H_2 + 2\ OH^-$. The overall reaction is $2\ H_2O(l) + 2\ Cl^-(aq) \rightarrow Cl_2(g) + H_2(g) + 2\ OH^-(aq)$.

From the 1:1 mol ratio between Cl_2 and H_2 in the overall balanced reaction, if 6.00 L of $H_2(g)$ are produced, then 6.00 L of $Cl_2(g)$ will also be produced since moles and volume of gas are directly proportional at constant T and P (see Chapter 5 of text).

Additional Exercises

99.   The half-reaction for the SCE is:

$$Hg_2Cl_2 + 2\ e^- \rightarrow 2\ Hg + 2\ Cl^- \qquad E_{SCE} = 0.242\ V$$

For a spontaneous reaction to occur, E_{cell} must be positive. Using the standard reduction potentials in Table 17.1 and the given SCE potential, deduce which combination will produce a positive overall cell potential.

a. $Cu^{2+} + 2\ e^- \rightarrow Cu \qquad E° = 0.34\ V$

$E_{cell} = 0.34 - 0.242 = 0.10\ V$; SCE is the anode.

b. $Fe^{3+} + e^- \rightarrow Fe^{2+} \qquad E° = 0.77\ V$

$E_{cell} = 0.77 - 0.242 = 0.53\ V$; SCE is the anode.

c. $AgCl + e^- \rightarrow Ag + Cl^- \qquad E° = 0.22\ V$

$E_{cell} = 0.242 - 0.22 = 0.02\ V$; SCE is the cathode.

d. $Al^{3+} + 3\ e^- \rightarrow Al \qquad E° = -1.66\ V$

$E_{cell} = 0.242 + 1.66 = 1.90\ V$; SCE is the cathode.

e. $Ni^{2+} + 2\ e^- \rightarrow Ni \qquad E° = -0.23\ V$

$E_{cell} = 0.242 + 0.23 = 0.47\ V$; SCE is the cathode.

100.   The potential oxidizing agents are NO_3^- and H^+. Hydrogen ion cannot oxidize Pt under either condition. Nitrate cannot oxidize Pt unless there is Cl^- in the solution. Aqua regia has both Cl^- and NO_3^-. The overall reaction is:

$$(NO_3^- + 4\ H^+ + 3\ e^- \rightarrow NO + 2\ H_2O) \times 2 \qquad\qquad E° =\ \ 0.96\ V$$
$$(4\ Cl^- + Pt \rightarrow PtCl_4^{2-} + 2\ e^-) \times 3 \qquad\qquad -E° = -0.755\ V$$

$$12\ Cl^-(aq) + 3\ Pt(s) + 2\ NO_3^-(aq) + 8\ H^+(aq) \rightarrow 3\ PtCl_4^{2-}(aq) + 2\ NO(g) + 4\ H_2O(l) \quad E°_{cell} = 0.21\ V$$

101. a. Possible reaction: $I_2(s) + 2\ Cl^-(aq) \rightarrow 2\ I^-(aq) + Cl_2(g)$ $E^\circ_{cell} = 0.54\ V - 1.36\ V = -0.82\ V$
          This reaction is not spontaneous at standard conditions. No reaction occurs.

      b. Possible reaction: $Cl_2(g) + 2\ I^-(aq) \rightarrow I_2(s) + 2\ Cl^-(aq)$ $E^\circ_{cell} = 0.82\ V$; This reaction is
          spontaneous at standard conditions. The reaction will occur.

      c. Possible reaction: $2\ Ag(s) + Cu^{2+}(aq) \rightarrow Cu(s) + 2\ Ag^+(aq)$ $E^\circ_{cell} = -0.46\ V$; No reaction
          occurs.

      d. Fe^{2+} can be oxidized or reduced, all of which have negative potentials. The other species present
          are H^+, SO_4^{2-}, H_2O, and O_2 from air. Only O_2 in the presence of H^+ has a large enough standard
          reduction potential to oxidize Fe^{2+} to Fe^{3+}. All other combinations, including the possible
          reduction of Fe^{2+}, give negative cell potentials. The reaction is: $4\ Fe^{2+}(aq) + 4\ H^+(aq) + O_2(g)$
          $\rightarrow 4\ Fe^{3+}(aq) + 2\ H_2O(l)$. $E^\circ_{cell} = 1.23 - 0.77 = 0.46\ V$; Spontaneous

102. 101b. $Cl_2(g) + 2\ I^-(aq) \rightarrow I_2(s) + 2\ Cl^-(aq)$ $E^\circ_{cell} = 0.82\ V = 0.82\ J/C$

      $\Delta G^\circ = -nFE^\circ_{cell} = -(2\ mol\ e^-)(96{,}485\ C/mol\ e^-)(0.82\ J/C) = -1.6 \times 10^5\ J = -160\ kJ$

      $E^\circ = \dfrac{0.0592}{n}\ \log K,\ \ \log K = \dfrac{nE^\circ}{0.0592} = \dfrac{2(0.82)}{0.0592} = 27.70,\ \ K = 10^{27.70} = 5.0 \times 10^{27}$

      101d. $4\ H^+(aq) + O_2(g) + 4\ Fe^{2+}(aq) \rightarrow 4\ Fe^{3+}(aq) + 2\ H_2O(l)$ $E^\circ_{cell} = 0.46\ V$

      $\Delta G^\circ = -nFE^\circ_{cell} = -(4\ mol\ e^-)(96{,}485\ C/mol\ e^-)(0.46\ J/C)(1\ kJ/1000\ J) = -180\ kJ$

      $\log K = \dfrac{4(0.46)}{0.0592} = 31.08,\ \ K = 1.2 \times 10^{31}$

103. $Al^{3+} + 3\ e^- \rightarrow Al$ $E^\circ = -1.66\ V$
              $Al + 6\ F^- \rightarrow AlF_6^{3-} + 3\ e^-$ $-E^\circ = 2.07\ V$
      _____
              $Al^{3+}(aq) + 6\ F^-(aq) \rightarrow AlF_6^{3-}(aq)$ $E^\circ_{cell} = 0.41\ V$ $K = ?$

      $\log K = \dfrac{nE^\circ}{0.0592} = \dfrac{3(0.41)}{0.0592} = 20.78,\ \ K = 10^{20.78} = 6.0 \times 10^{20}$

104. $Zn \rightarrow Zn^{2+} + 2\ e^-$ $-E^\circ = 0.76\ V$; $Fe \rightarrow Fe^{2+} + 2\ e^-$ $-E^\circ = 0.44\ V$

      It is easier to oxidize Zn than Fe, so the Zn will be oxidized protecting the iron of the *Monitor's* hull.

105. In a simplified view of the corrosion process, the half-reactions are:

              Cathode: $O_2 + 2\ H_2O + 4\ e^- \rightarrow 4\ OH^-$
              Anode: $(Fe \rightarrow Fe^{2+} + 2\ e^-) \times 2$
      _____
              $O_2(g) + 2\ H_2O(l) + 2\ Fe(s) \rightarrow 2\ Fe^{2+}(aq) + 4\ OH^-(aq)$

Since OH⁻ is a product in this reaction, then added H⁺ will react with OH⁻ to form H_2O. As OH⁻ is removed, this increases the tendency for the forward reaction to occur. Hence, corrosion is a greater problem under acidic conditions in our simplified view of this process.

106.   Aluminum has the ability to form a durable oxide coating over its surface. Once the HCl dissolves this oxide coating, Al is exposed to H⁺ and is easily oxidized to Al^{3+}, i.e., the Al foil disppears after the oxide coating is dissolved.

107.   Consider the strongest oxidizing agent combined with the strongest reducing agent from Table 17.1:

$$F_2 + 2 e^- \rightarrow 2 F^- \qquad\qquad E° = 2.87 \text{ V}$$
$$(Li \rightarrow Li^+ + e^-) \times 2 \qquad\qquad -E° = 3.05 \text{ V}$$

$$F_2(g) + 2 Li(s) \rightarrow 2 Li^+(aq) + 2 F^-(aq) \qquad E°_{cell} = 5.92 \text{ V}$$

The claim is impossible. The strongest oxidizing agent and reducing agent when combined only give $E°_{cell}$ of about 6 V.

108.   a. $O_2 + 2 H_2O + 4 e^- \rightarrow 4 OH^- \qquad\qquad E° = 0.40 \text{ V}$
              $(H_2 + 2 OH^- \rightarrow 2 H_2O + 2 e^-) \times 2 \qquad -E° = 0.83 \text{ V}$

$$2 H_2(g) + O_2(g) \rightarrow 2 H_2O(l) \qquad\qquad E°_{cell} = 1.23 \text{ V} = 1.23 \text{ J/C}$$

$\Delta G° = -nFE°_{cell} = -(4 \text{ mol e}^-)(96{,}485 \text{ C/mol e}^-)(1.23 \text{ J/C}) = -4.75 \times 10^5 \text{ J} = -475 \text{ kJ}$

   b. $\Delta H° = \Sigma n_p \Delta H°_{f, products} - \Sigma n_r \Delta H°_{f, reactants} = 2 \text{ mol}(-286 \text{ kJ/mol}) - [2(0) + 1(0)] = -572 \text{ kJ}$

$\Delta S° = 2 \text{ mol }(70. \text{ J/K}\bullet\text{mol}) - [2 \text{ mol }(131 \text{ J/K}\bullet\text{mol}) + 1 \text{ mol }(205 \text{ J/K}\bullet\text{mol})] = -327 \text{ J/K}$

   c.   At 90.°C (363 K): $\Delta G°_{90} = \Delta H° - T\Delta S° = -572 \text{ kJ} - (363 \text{ K})(-0.327 \text{ kJ/K}) = -453 \text{ kJ}$

$$\Delta G°_{90} = -nFE°_{90}, \quad E°_{90} = \frac{-\Delta G°_{90}}{nF} = \frac{-(-4.53 \times 10^5 \text{ J})}{(4 \text{ mol e}^-)(96{,}485 \text{ C/mol e}^-)} = 1.17 \text{ J/C} = 1.17 \text{ V}$$

At 0°C: $\Delta G°_0 = -572 \text{ kJ} - (273 \text{ K})(-0.327 \text{ kJ/K}) = -483 \text{ kJ}$

$$E°_0 = \frac{-\Delta G°_0}{nF} = \frac{-(-4.83 \times 10^5 \text{ J})}{(4 \text{ mol e}^-)(96{,}485 \text{ C/mol e}^-)} = 1.25 \text{ J/C} = 1.25 \text{ V}$$

109.   a. $O_2 + 2 H_2O + 4 e^- \rightarrow 4 OH^- \qquad\qquad E° = 0.40 \text{ V}$
              $(H_2 + 2 OH^- \rightarrow 2 H_2O + 2 e^-) \times 2 \qquad -E° = 0.83 \text{ V}$

$$2 H_2(g) + O_2(g) \rightarrow 2 H_2O(l) \qquad\qquad E°_{cell} = 1.23 \text{ V} = 1.23 \text{ J/C}$$

Since standard conditions are assumed, then $w_{max} = \Delta G°$ for 2 mol H_2O produced.

$\Delta G° = -nFE°_{cell} = -(4 \text{ mol e}^-)(96{,}485 \text{ C/mol e}^-)(1.23 \text{ J/C}) = -475{,}000 \text{ J} = -475 \text{ kJ}$

For 1.00×10^3 g H_2O produced, w_{max} is:

$$1.00 \times 10^3 \text{ g } H_2O \times \frac{1 \text{ mol } H_2O}{18.02 \text{ g } H_2O} \times \frac{-475 \text{ kJ}}{2 \text{ mol } H_2O} = -13,200 \text{ kJ} = w_{max}$$

The work done can be no larger than the free energy change. The best that could happen is that all of the free energy released goes into doing work, but this does not occur in any real process since there is always waste energy in any real process. Fuel cells are more efficient in converting chemical energy into electrical energy; they are also less massive. The major disadvantage is that they are expensive. In addition, $H_2(g)$ and $O_2(g)$ are an explositve mixture, if ignited; much more so than fossil fuels.

110. $C_2H_5OH(l) + 3 O_2(g) \rightarrow 2 CO_2(g) + 3 H_2O(l)$; In C_2H_5OH, C has a -2 oxidation state and in CO_2, C has a +4 oxidation state. Each mol of carbon loses 6 electrons. Since there are two mol of carbon in the balanced reaction, then n = 12 mol e^-.

$$w_{max} = -1.32 \times 10^6 \text{ J} = \Delta G = -nFE^\circ_{cell}, \; E_{cell} = \frac{-\Delta G}{nF}$$

$$E_{cell} = \frac{-(-1.32 \times 10^6 \text{ J})}{(12 \text{ mol } e^-)(96,485 \text{ C/mole})} = 1.14 \text{ J/C} = 1.14 \text{ V}$$

111.   If the metal, M, forms +1 ions, then the atomic mass of M would be:

$$\text{mol M} = 150. \text{ s} \times \frac{1.25 \text{ C}}{\text{s}} \times \frac{1 \text{ mol } e^-}{96,485 \text{ C}} \times \frac{1 \text{ mol M}}{1 \text{ mol } e^-} = 1.94 \times 10^{-3} \text{ mol M}$$

$$\text{Atomic mass of M} = \frac{0.109 \text{ g M}}{1.94 \times 10^{-3} \text{ mol M}} = 56.2 \text{ g/mol}$$

From the periodic table, the only metal with an atomic mass close to 56.2 g/mol is iron; but iron does not form stable +1 ions. If M forms +2 ions, then the atomic mass would be:

$$\text{mol M} = 150. \text{ s} \times \frac{1.25 \text{ C}}{\text{s}} \times \frac{1 \text{ mol } e^-}{96,485 \text{ C}} \times \frac{1 \text{ mol M}}{2 \text{ mol } e^-} = 9.72 \times 10^{-4} \text{ mol M}$$

$$\text{Atomic mass of M} = \frac{0.109 \text{ g M}}{9.72 \times 10^{-4} \text{ mol M}} = 112 \text{ g/mol}$$

Cadmium has an atomic mass of 112.4 g/mol and does form stable +2 ions. Cd^{2+} is a much more logical choice than Fe^+.

112. $15 \text{ kWh} = \frac{15000 \text{ J h}}{\text{s}} \times \frac{60 \text{ s}}{\text{min}} \times \frac{60 \text{ min}}{h} = 5.4 \times 10^7 \text{ J or } 5.4 \times 10^4 \text{ kJ}$ (Hall process)

To melt 1.0 kg Al requires: $1.0 \times 10^3 \text{ g Al} \times \frac{1 \text{ mol Al}}{26.98 \text{ g}} \times \frac{10.7 \text{ kJ}}{\text{mol Al}} = 4.0 \times 10^2 \text{ kJ}$

It is feasible to recycle Al by melting the metal because in theory, it takes less than 1% of the energy required to produce the same amount of Al by the Hall process.

Challenge Problems

113. $\Delta G^\circ = -nFE^\circ = \Delta H^\circ - T\Delta S^\circ$, $E^\circ = \dfrac{T\Delta S^\circ}{nF} - \dfrac{\Delta H^\circ}{nF}$

If we graph E° vs. T we should get a straight line (y = mx + b). The slope of the line is equal to $\Delta S^\circ/nF$ and the y-intercept is equal to $-\Delta H^\circ/nF$. From the equation above, E° will have a small temperature dependence when ΔS° is close to zero.

114.   a.   We can calculate ΔG° from $\Delta G^\circ = \Delta H^\circ - T\Delta S^\circ$ and then E° from $\Delta G^\circ = -nFE^\circ$; or we can use the equation derived in Exercise 17.113. For this reaction, n = 2 (see Exercise 17.62).

$$E^\circ_{-20} = \frac{T\Delta S^\circ - \Delta H^\circ}{nF} = \frac{(253\ K)\ (263.5\ J/K) + 315.9 \times 10^3\ J}{(2\ mol\ e^-)\ (96{,}485\ C/mol\ e^-)} = 1.98\ J/C = 1.98\ V$$

b. $E_{-20} = E^\circ_{-20} - \dfrac{RT}{nF} \ln Q = 1.98\ V - \dfrac{RT}{nF} \ln \dfrac{1}{[H^+]^2 [HSO_4^-]^2}$

$$E^\circ_{-20} = 1.98\ V - \frac{(8.3145\ J/K \cdot mol)\ (253\ K)}{(2\ mol\ e^-)\ (96{,}485\ C/mol\ e^-)} \ln \frac{1}{(4.5)^2 (4.5)^2} = 1.98\ V + 0.066\ V = 2.05\ V$$

c.   From Exercise 17.62, E = 2.12 V at 25°C. As the temperature decreases, the cell potential decreases. Also, oil becomes more viscous at lower temperatures, which adds to the difficulty of starting an engine on a cold day. The combination of these two factors result in batteries failing more often on cold days than on warm days.

115.   a. $\Delta G^\circ = \Sigma n_p \Delta G^\circ_{f,\ products} - \Sigma n_r \Delta G^\circ_{f,\ reactants} = 2(-480.) + 3(86) - [3(-40.)] = -582\ kJ$

From oxidation numbers, n = 6. $\Delta G^\circ = -nFE^\circ$, $E^\circ = \dfrac{-\Delta G^\circ}{nF} = \dfrac{-(-582{,}000\ J)}{6(96{,}485)\ C} = 1.01\ V$

$\log K = \dfrac{nE^\circ}{0.0592} = \dfrac{6(1.01)}{0.0592} = 102.365$, $K = 10^{102.365} = 2.32 \times 10^{102}$

b. $\qquad 3 \times (2\ e^- + Ag_2S \rightarrow 2\ Ag + S^{2-}) \qquad\qquad\qquad E^\circ_{Ag_2S} = ?$

$\qquad\qquad 2 \times (Al \rightarrow Al^{3+} + 3\ e^-) \qquad\qquad\qquad\qquad -E^\circ = 1.66\ V$

$\quad 3\ Ag_2S(s) + 2\ Al(s) \rightarrow 6\ Ag(s) + 3\ S^{2-}(aq) + 2\ Al^{3+}(aq) \qquad E^\circ_{cell} = 1.01\ V = E^\circ_{Ag_2S} + 1.66$

$E^\circ_{Ag_2S} = 1.01 - 1.66 = -0.65\ V$

116.   a. $Zn(s) + Cu^{2+}(aq) \rightarrow Zn^{2+}(aq) + Cu(s)$ $E^\circ_{cell} = 1.10\ V$; $E_{cell} = 1.10\ V - \dfrac{0.0592}{2} \log \dfrac{[Zn^{2+}]}{[Cu^{2+}]}$

$E_{cell} = 1.10\ V - \dfrac{0.0592}{2} \log \dfrac{0.10}{2.50} = 1.10\ V + 0.041\ V = 1.14\ V$

b. $10.0\ h \times \dfrac{60\ min}{h} \times \dfrac{60\ s}{min} \times \dfrac{10.0\ C}{s} \times \dfrac{1\ mol\ e^-}{96{,}485\ C} \times \dfrac{1\ mol\ Cu}{2\ mol\ e^-} = 1.87\ mol\ Cu\ produced$

The Cu^{2+} concentration decreases by 1.87 mol/L and the Zn^{2+} concentration will increase by 1.87 mol/L.

$[Cu^{2+}] = 2.50 - 1.87 = 0.63\ M;\ [Zn^{2+}] = 0.10 + 1.87 = 1.97\ M$

$E_{cell} = 1.10\ V - \dfrac{0.0592}{2} \log \dfrac{1.97}{0.63} = 1.10\ V - 0.015\ V = 1.09\ V$

c. $1.87\ \text{mol Zn consumed} \times \dfrac{65.38\ \text{g Zn}}{\text{mol Zn}} = 122\ \text{g Zn};\ \text{Mass of electrode} = 200. - 122 = 78\ \text{g Zn}$

$1.87\ \text{mol Cu formed} \times \dfrac{63.55\ \text{g Cu}}{\text{mol Cu}} = 119\ \text{g Cu};\ \text{Mass of electrode} = 200. + 119 = 319\ \text{g Cu}$

d. Three things could possibly cause this battery to go dead:

1. All of the Zn is consumed.
2. All of the Cu^{2+} is consumed.
3. Equilibrium is reached ($E_{cell} = 0$).

We began with 2.50 mol Cu^{2+} and 200. g Zn × 1 mol Zn/65.38 g Zn = 3.06 mol Zn. Cu^{2+} is the limiting reagent and will run out first. To react all the Cu^{2+} requires:

$2.50\ \text{mol Cu}^{2+} \times \dfrac{2\ \text{mol e}^-}{\text{mol Cu}^{2+}} \times \dfrac{96{,}485\ C}{\text{mol e}^-} \times \dfrac{1\ s}{10.0\ C} \times \dfrac{1\ h}{3600\ s} = 13.4\ h$

For equilibrium to be reached: $E = 0 = 1.10\ V - \dfrac{0.0592}{2} \log \dfrac{[Zn^{2+}]}{[Cu^{2+}]}$

$\dfrac{[Zn^{2+}]}{[Cu^{2+}]} = K = 10^{2(1.10)/0.0592} = 1.45 \times 10^{37}$

This is such a large equilibrium constant that virtually all of the Cu^{2+} must react to reach equilibrium. So, the battery will go dead in 13.4 hours.

117. $Pb^{2+} + 2\ e^- \rightarrow Pb$ $\qquad\qquad$ $E° = -0.13\ V$
$\qquad\quad$ $Zn \rightarrow Zn^{2+} + 2\ e^-$ $\qquad\qquad$ $-E° = 0.76\ V$

$Pb^{2+}(aq) + Zn(s) \rightarrow Zn^{2+}(aq) + Pb(s)$ $\qquad$ $E°_{cell} = 0.63\ V$

$Zn(OH)_2(s) \rightleftharpoons Zn^{2+}(aq) + 2\ OH^-(aq)\ \ K_{sp} = [Zn^{2+}][OH^-]^2;$ We must determine $[Zn^{2+}]$ in order to calculate K_{sp}. Using the Nernst equation for this cell:

$E_{cell} = E°_{cell} - \dfrac{0.0592}{n} \log Q,\ E_{cell} = 1.05\ V = 0.63\ V - \dfrac{0.0592}{2} \log \dfrac{[Zn^{2+}]}{[Pb^{2+}]}$

$1.05\ V = 0.63\ V - \dfrac{0.0592}{2} \log \dfrac{[Zn^{2+}]}{1.0},\ \log[Zn^{2+}] = -14.19,\ [Zn^{2+}] = 10^{-14.19} = 6.5 \times 10^{-15}\ M$

$K_{sp} = [Zn^{2+}][OH^-]^2 = (6.5 \times 10^{-15})(0.10)^2 = 6.5 \times 10^{-17}$

118.   a.

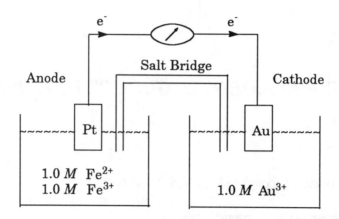

b. $Au^{3+}(aq) + 3\ Fe^{2+}(aq) \rightarrow 3\ Fe^{3+}(aq) + Au(s)$ $E^{\circ}_{cell} = 1.50 - 0.77 = 0.73$ V

$$E_{cell} = E^{\circ}_{cell} - \frac{0.0592}{n} \log Q = 0.73\ V - \frac{0.0592}{3} \log \frac{[Fe^{3+}]^3}{[Au^{3+}][Fe^{2+}]^3}$$

Since $[Fe^{3+}] = [Fe^{2+}] = 1.0\ M$: $0.31\ V = 0.73\ V - \dfrac{0.0592}{3} \log \dfrac{1}{[Au^{3+}]}$

$$\frac{3(-0.42)}{0.0592} = -\log \frac{1}{[Au^{3+}]}, \quad \log [Au^{3+}] = -21.28, \quad [Au^{3+}] = 10^{-21.28} = 5.2 \times 10^{-22}\ M$$

$Au^{3+} + 4\ Cl^- \rightleftharpoons AuCl_4^-$;   Since the equilibrium Au^{3+} concentration is so small, assume $[AuCl_4^-]$ $\approx [Au^{3+}]_o \approx 1.0\ M$, i.e., assume K is large so the reaction essentially goes to completion.

$$K = \frac{[AuCl_4^-]}{[Au^{3+}][Cl^-]^4} = \frac{1.0}{(5.2 \times 10^{-22})(0.10)^4} = 1.9 \times 10^{25};\ \text{Assumption good (K is large).}$$

119.   a. $E_{cell} = E_{ref} + 0.05916\ pH$, $0.480\ V = 0.250\ V + 0.05916\ pH$

$$pH = \frac{0.480 - 0.250}{0.05916} = 3.888;\ \text{Uncertainty} = \pm 1\ mV = \pm 0.001\ V$$

$$pH_{max} = \frac{0.481 - 0.250}{0.05916} = 3.905; \quad pH_{min} = \frac{0.479 - 0.250}{0.05916} = 3.871$$

So if the uncertainty in potential is ± 0.001 V, then the uncertainty in pH is ± 0.017 or about ± 0.02 pH units. For this measurement, $[H^+] = 10^{-3.888} = 1.29 \times 10^{-4}\ M$. For an error of $+ 1$ mV, $[H^+] = 10^{-3.905} = 1.24 \times 10^{-4}\ M$. For an error of -1 mV, $[H^+] = 10^{-3.871} = 1.35 \times 10^{-4}\ M$. So the uncertainty in $[H^+]$ is $\pm 0.06 \times 10^{-4}\ M = \pm 6 \times 10^{-6}\ M$.

b.   From part a, we will be within ± 0.02 pH units if we measure the potential to the nearest ± 0.001 V (1 mV).

CHAPTER EIGHTEEN

THE REPRESENTATIVE ELEMENTS: GROUPS 1A THROUGH 4A

Questions

1. The gravity of the earth is not strong enough to keep H_2 in the atmosphere.

2. Ammonia production; Hydrogenation of vegetable oils

3. Ionic, covalent, and metallic (or interstitial); The ionic and covalent hydrides are true compounds obeying the law of definite proportions and differ from each other in the type of bonding. The interstitial hydrides are more like solid solutions of hydrogen with a transition metal, and do not obey the law of definite proportions.

4. The small size of the Li^+ cation results in a much greater attraction to water. The attraction to water is not as great for the other alkali metal ions. Thus, lithium salts tend to absorb water.

5. Hydrogen forms many compounds in which the oxidation state is +1, as do the Group 1A elements. For example H_2SO_4 and HCl compared to Na_2SO_4 and NaCl. On the other hand, hydrogen forms diatomic H_2 molecules and is a nonmetal, while the Group 1A elements are metals. Hydrogen also forms compounds with a -1 oxidation state, which is not characteristic of Group 1A metals, e.g., NaH.

6. $4 KO_2(s) + 2 CO_2(g) \rightarrow 2 K_2CO_3(s) + 3 O_2(g)$; Potassium superoxide can react with exhaled CO_2 to produce O_2 which then can be breathed.

7. Group IA and IIA metals are all easily oxidized. They must be produced in the absence of materials (H_2O, O_2) that are capable of oxidizing them.

8. The acidity decreases. Solutions of Be^{2+} are acidic, while solutions of the other M^{2+} ions are neutral.

9. In graphite, planes of carbon atoms slide easily along each other. In addition, graphite is not volatile. The lubricant will not be lost when used in a high vacuum environment.

10. Quartz: Crystalline, long range order; The structure is an ordered arrangement of 12 membered rings, each containing 6 – Si and 6 – O atoms.

    Amorphous SiO_2:   No long range order; Irregular arrangement that contains many different ring sizes. See Chapter 10.5.

11.   The bonds in SnX_4 compounds have a large covalent character. SnX_4 acts as discrete molecules held together by weak dispersion forces. SnX_2 compounds are ionic and are held in the solid state by strong ionic forces. Since the intermolecular forces are weaker for SnX_4 compounds, they are more volatile.

12.   Group 3A elements have one fewer valence electron than Si or Ge. A p-type semiconductor would form.

13.   Size decreases from left to right and increases going down the periodic table. So going one element right and one element down would result in a similar size for the two elements diagonal to each other. The ionization energies will be similar for the diagonal elements since the periodic trends also oppose each other. Electron affinities are harder to predict, but atoms with similar size and ionization energy should also have similar electron affinities.

14.   The "inert pair effect" refers to the difficulty of removing the pair of s electrons from some of the elements in the fifth and sixth periods of the periodic table. As a result, multiple oxidation states are exhibited for the heavier elements of Groups 3A and 4A. Tl^+, Tl^{3+}, Pb^{2+}, and Pb^{4+} ions are all important to the chemistry of Tl and Pb.

Exercises

Group 1A Elements

15.   a. $\Delta H° = -110.5 - [-75 + (-242)] = 207$ kJ; $\Delta S° = 198 + 3(131) - [186 + 189] = 216$ J/K

        b. $\Delta G° = \Delta H° - T\Delta S°$; $\Delta G° = 0$ when $T = \dfrac{\Delta H°}{\Delta S°} = \dfrac{207 \times 10^3 \text{ J}}{216 \text{ J/K}} = 958$ K

             At $T > 958$ K and standard pressures, the favorable $\Delta S°$ term dominates and the reaction is spontaneous ($\Delta G° < 0$).

16.   For $3 \text{ Fe}(s) + 4 \text{ H}_2\text{O}(g) \rightarrow \text{Fe}_3\text{O}_4(s) + 4 \text{ H}_2(g)$

        a. $\Delta H° = -1117 - [4(-242)] = -149$ kJ; $\Delta S° = 146 + 4(131) - [3(27) + 4(189)] = -167$ J/K

        b. $\Delta G° = 0$ when $T = \dfrac{\Delta H°}{\Delta S°} = \dfrac{-149 \times 10^3 \text{ J}}{-167 \text{ J/K}} = 892$ K

             At $T < 892$ K and standard pressures, the favorable $\Delta H°$ term dominates and the reaction is spontaneous ($\Delta G° < 0$).

        For $\text{C}(s) + \text{H}_2\text{O}(g) \rightarrow \text{CO}(g) + \text{H}_2(g)$

        a. $\Delta H° = -110.5 - (-242) = 132$ kJ; $\Delta S° = 198 + 131 - [6 + 189] = 134$ J/K

        b. $T = \dfrac{\Delta H°}{\Delta S°} = \dfrac{132 \times 10^3 \text{ J}}{134 \text{ J/K}} = 985$ K

This reaction is spontaneous when the favorable $\Delta S°$ term dominates, which occurs at T > 985 K (assuming standard pressures).

17.   sodium oxide: Na_2O; sodium superoxide: NaO_2; sodium peroxide: Na_2O_2

18.   a. Rb_3N b. K_2O_2

19.   a. $Li_3N(s) + 3 \, HCl(aq) \rightarrow 3 \, LiCl(aq) + NH_3(aq)$

  b. $Rb_2O(s) + H_2O(l) \rightarrow 2 \, RbOH(aq)$

  c. $Cs_2O_2(s) + 2 \, H_2O(l) \rightarrow 2 \, CsOH(aq) + H_2O_2(aq)$

  d. $NaH(s) + H_2O(l) \rightarrow NaOH(aq) + H_2(g)$

20. $K(s) + O_2(g) \rightarrow KO_2(s)$; $16 \, K(s) + S_8(s) \rightarrow 8 \, K_2S(s)$

  $12 \, K(s) + P_4(s) \rightarrow 4 \, K_3P(s)$; $2 \, K(s) + H_2(g) \rightarrow 2 \, KH(s)$

  $2 \, K(s) + 2 \, H_2O(l) \rightarrow H_2(g) + 2 \, KOH(aq)$

21. $2 \, Li(s) + 2 \, C_2H_2(g) \rightarrow 2 \, LiC_2H(s) + H_2(g)$; This is an oxidation-reduction reaction.

22. $NaH(s) + H_2O(l) \rightarrow Na^+(aq) + OH^-(aq) + H_2(g)$; Acid-base: A proton is transferred from an acid, H_2O, to a base, H^-, forming the conjugate base of water, OH^-, and the conjugate acid of H^-, H_2. Oxidation-reduction: The oxidation state of H is -1 in NaH, +1 in H_2O, and zero in H_2. Thus, an electron is transferred from the hydride ion to a hydrogen in water when forming $H_2(g)$.

Group 2A Elements

23.   barium oxide: BaO; barium peroxide: BaO_2

24. $SrCO_3$; $MgSO_4$

25. $Mg_3N_2(s) + 6 \, H_2O(l) \rightarrow 2 \, NH_3(g) + 3 \, Mg^{2+}(aq) + 6 \, OH^-(aq)$

  $Mg_3P_2(s) + 6 \, H_2O(l) \rightarrow 2 \, PH_3(g) + 3 \, Mg^{2+}(aq) + 6 \, OH^-(aq)$

26. $2 \, Ca(s) + O_2(g) \rightarrow 2 \, CaO(s)$; $8 \, Ca(s) + S_8(s) \rightarrow 8 \, CaS(s)$

  $3 \, Ca(s) + N_2(g) \rightarrow Ca_3N_2(s)$; $6 \, Ca(s) + P_4(s) \rightarrow 2 \, Ca_3P_2(s)$

  $Ca(s) + H_2(g) \rightarrow CaH_2(s)$; $Ca(s) + 2 \, H_2O(l) \rightarrow H_2(g) + Ca(OH)_2(aq)$

27.

Geometry is trigonal planar about Be.

Be uses sp^2 hybrid orbitals.

N uses sp^3 hybrid orbitals.

$BeCl_2$ is a Lewis acid.

28. $BeCl_2(NH_3)_2$ would form in excess ammonia. $BeCl_2(NH_3)_2$ has $2 + 2(7) + 2(5) + 6(1) = 32$ valence electrons. A structure for this molecule can be drawn that obeys the octet rule. This is not the case for $BeCl_2NH_3$ (see Exercise 18.27).

29. $\dfrac{1 \text{ mg F}^-}{\text{L}} \times \dfrac{1 \text{ g}}{1000 \text{ mg}} \times \dfrac{1 \text{ mol F}^-}{19.00 \text{ g F}^-} = 5.3 \times 10^{-5} \, M \, F^- = 5 \times 10^{-5} \, M \, F^-$

$CaF_2(s) \rightleftharpoons Ca^{2+}(aq) + 2 \, F^-(aq)$ $K_{sp} = [Ca^{2+}][F^-]^2 = 4.0 \times 10^{-11}$; Precipitation will occur when $Q > K_{sp}$. Lets calculate $[Ca^{2+}]$ so that $Q = K_{sp}$.

$Q = 4.0 \times 10^{-11} = [Ca^{2+}]_o [F^-]_o^2 = [Ca^{2+}]_o (5 \times 10^{-5})^2$, $[Ca^{2+}] = 2 \times 10^{-2} \, M$

$CaF_2(s)$ will precipitate when $[Ca^{2+}]_o > 2 \times 10^{-2} \, M$. Therefore, hard water should have a calcium ion concentration of less than $2 \times 10^{-2} \, M$ to avoid precipitate formation.

30. $Mg(OH)_2(s) \rightleftharpoons Mg^{2+}(aq) + 2 \, OH^-(aq)$ $K_{sp} = 8.9 \times 10^{-12} = [Mg^{2+}][OH^-]^2$

Since buffer pH = 8.00, $[OH^-] = 1.0 \times 10^{-6} \, M$; $[Mg^{2+}](1.0 \times 10^{-6})^2 = 8.9 \times 10^{-12}$, $[Mg^{2+}] = 8.9 \, M$

Since there is a 1:1 mol ratio between Mg^{2+} and $Mg(OH)_2$, then the molar solubility of $Mg(OH)_2(s)$ is 8.9 mol/L.

31. $Ca^{2+} + 2 \, e^- \rightarrow Ca$; $1.00 \times 10^3 \text{ kg} \times \dfrac{1000 \text{ g}}{\text{kg}} \times \dfrac{1 \text{ mol Ca}}{40.08 \text{ g}} \times \dfrac{2 \text{ mol e}^-}{\text{mol Ca}} \times \dfrac{96,485 \text{ C}}{\text{mol e}^-} = 4.81 \times 10^9 \text{ C}$

$\dfrac{4.81 \times 10^9 \text{ C}}{8.00 \text{ h}} \times \dfrac{1 \text{ h}}{60 \text{ min}} \times \dfrac{1 \text{ min}}{60 \text{ s}} = \dfrac{1.67 \times 10^5 \text{ C}}{\text{s}} = 1.67 \times 10^5 \text{ A}$; $Ca^{2+} + 2 \, Cl^- \rightarrow Ca + Cl_2$

$1.00 \times 10^6 \text{ g Ca} \times \dfrac{1 \text{ mol Ca}}{40.08 \text{ g}} \times \dfrac{1 \text{ mol Cl}_2}{\text{mol Ca}} \times \dfrac{70.90 \text{ g Cl}_2}{\text{mol Cl}_2} = 1.77 \times 10^6 \text{ g of Cl}_2 = 1.77 \times 10^3 \text{ kg Cl}_2$

32.   Alkaline earth metals form +2 ions so 2 mol of e⁻ are transferred to form the metal, M.

$$\text{mol M} = 748 \text{ s} \times \frac{5.00 \text{ C}}{\text{s}} \times \frac{1 \text{ mol e}^-}{96,485 \text{ C}} \times \frac{1 \text{ mol M}}{2 \text{ mol e}^-} = 1.94 \times 10^{-2} \text{ mol M}$$

$$\text{atomic mass of M} = \frac{0.471 \text{ g M}}{1.94 \times 10^{-2} \text{ mol M}} = 24.3 \text{ g/mol}; \quad MgCl_2 \text{ was electrolyzed.}$$

Group 3A Elements

33.   a.   thallium(I) hydroxide   b.   indium(III) sulfide

        c.   Gallium(III) oxide, but is commonly called gallium oxide.

34.   a.   indium(III) phosphide   b.   thallium(III) fluoride

        c.   Gallium(III) arsenide, but is commonly called gallium arsenide.

35. $B_2H_6 + 3 O_2 \rightarrow 2 B(OH)_3$

36. $B_2O_3(s) + 3 Mg(s) \rightarrow 3 MgO(s) + 2 B(s)$

37. $Ga_2O_3(s) + 6 H^+(aq) \rightarrow 2 Ga^{3+}(aq) + 3 H_2O(l)$

        $Ga_2O_3(s) + 2 OH^-(aq) + 3 H_2O(l) \rightarrow 2 Ga(OH)_4^-(aq)$

        $In_2O_3(s) + 6 H^+(aq) \rightarrow 2 In^{3+}(aq) + 3 H_2O(l); \quad In_2O_3(s) + OH^-(aq) \rightarrow$ no reaction

38. $Al_2O_3(s) + 6 H^+(aq) \rightarrow 2 Al^{3+}(aq) + 3 H_2O(l); \quad Al_2O_3(s) + 3 H_2O(l) + 2 OH^-(aq) \rightarrow 2 Al(OH)_4^-(aq)$

39. $2 In(s) + 3 F_2(g) \rightarrow 2 InF_3(s); \quad 2 In(s) + 3 Cl_2(g) \rightarrow 2 InCl_3(s); \quad 4 In(s) + 3 O_2(g) \rightarrow 2 In_2O_3(s)$

        $2 In(s) + 6 HCl(aq) \rightarrow 3 H_2(g) + 2 InCl_3(aq)$ or $2 In(s) + 6 HCl(g) \rightarrow 3 H_2(g) + 2 InCl_3(s)$

40. $2 Al(s) + 2 NaOH(aq) + 6 H_2O(l) \rightarrow 2 Al(OH)_4^-(aq) + 2 Na^+(aq) + 3 H_2(g)$

Group 4A Elements

41.   a.   Linear about all carbons   b.   sp

42. CS_2 has $4 + 2(6) = 16$ valence electrons. C_3S_2 has $3(4) + 2(6) = 24$ valence electrons.

        :S═C═S: linear;   :S═C═C═C═S: linear

43.   a. $K_2SiF_6(s) + 4 K(l) \rightarrow 6 KF(s) + Si(s)$

b. K_2SiF_6 is an ionic compound, composed of K^+ cations and SiF_6^{2-} anions. The SiF_6^{2-} anion is held together by covalent bonds. The structure is:

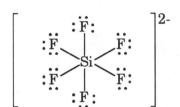

The anion is octahedral with Si utilizing d^2sp^3 hybrid orbitals to form the Si–F bonds.

44. $SiO_2(s) + 4\ HF(aq) \rightarrow SiF_4(g) + 2\ H_2O(l)$

45.   Lead is very toxic. As the temperature of the water increases, the solubility of lead will increase. Drinking hot tap water from pipes containing lead solder could result in higher lead concentrations in the body.

46. $Pb(OH)_2(s) \rightleftharpoons Pb^{2+} + 2\ OH^-$ $K_{sp} = 1.2 \times 10^{-15} = [Pb^{2+}][OH^-]^2$

Initial   s = solubility (mol/L)   0 $1.0 \times 10^{-7}\ M$
Equil.   s $1.0 \times 10^{-7} + 2s \approx 2s$

$K_{sp} = (s)(2s)^2 = 1.2 \times 10^{-15}$, $4s^3 = 1.2 \times 10^{-15}$, $s = 6.7 \times 10^{-6}$ mol/L; Assumption good.

$Pb(OH)_2(s)$ is more soluble in acidic solutions. Added H^+ reacts with OH^- to form H_2O. As OH^- is removed through this reaction, more $Pb(OH)_2(s)$ will dissolve.

47.   tin(II) fluoride

48.   lead(II) oxide; lead (IV) oxide

49.   The π electrons are free to move in graphite, thus giving it a greater conductivity (lower resistance). The electrons have the greatest mobility within sheets of carbon atoms, resulting in a lower resistance in the basal plane. Electrons in diamond are not mobile (high resistance). The structure of diamond is uniform in all directions; thus, there is no directional dependence of the resistivity.

50.   SiC would have a covalent network structure similar to diamond.

Additional Exercises

51. $K^+(out) \rightarrow K^+(in)$; $E = E° - \dfrac{0.0592}{1} \log\left(\dfrac{[K^+]_{in}}{[K^+]_{out}}\right)$; $E° = 0$ for this reaction

$E = -0.0592 \log\left(\dfrac{0.15}{5.0 \times 10^{-3}}\right) = -0.087$ V

$\Delta G =$ work $= -nFE = -(1\ mol\ e^-)(96{,}485\ C/mol\ e^-)(-0.087\ J/C) = 8400\ J = 8.4\ kJ =$ work

52.   In the gas phase, linear molecules would exist.

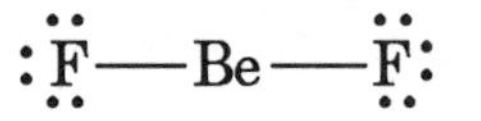

$$:\ddot{\ddot{F}}-Be-\ddot{\ddot{F}}:$$

In the solid state, BeF_2 would exist as a polymeric solid with a structure:

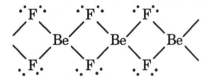

53.   Strontium and calcium are both alkaline earth metals so both have similar chemical properties. Since milk is a good source of calcium, strontium could replace some calcium in milk without much difficulty.

54.   The Be^{2+} ion is a Lewis acid and has a strong affinity for the lone pairs of electrons on oxygen in water. Thus, the compound is not dehydrated easily. The ion in solution is $Be(H_2O)_4^{2+}$. The acidic solution results from the reaction: $Be(H_2O)_4^{2+}(aq) \rightleftharpoons Be(H_2O)_3(OH)^+(aq) + H^+(aq)$

55.   Assuming $AlCl_3$ is covalent, then a single $AlCl_3$ has the Lewis structure:

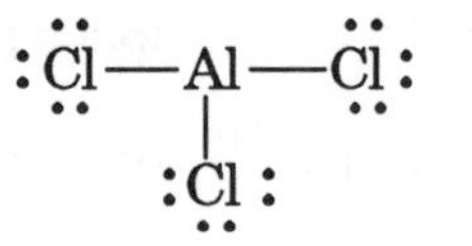

$$:\ddot{\ddot{Cl}}-Al-\ddot{\ddot{Cl}}:$$
$$\underset{\ddot{\ddot{Cl}}:}{|}$$

The Al has room for a pair of electrons. It can act as a Lewis acid. The only lone pairs are on Cl, so the dimer structure is:

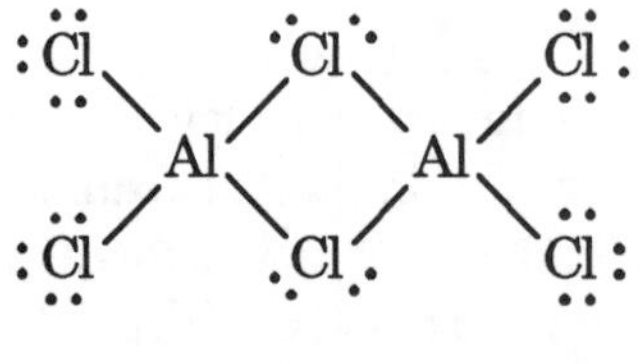

56.   Element 113: $[Rn]\ 7s^2 5f^{14} 6d^{10} 7p^1$; Element 113 would fall below Tl in the periodic table. Like Tl, we would expect element 113 to form +1 and +3 oxidation states in its compounds.

57. $LiAlH_4$: H, -1 (metal hydride); Li, +1; Al, $0 = +1 + x + 4(-1)$, $x = +3$ = oxidation state of Al

58. $Tl^{3+} + 2\ e^- \rightarrow Tl^+$ $E° = 1.25$ V
            $3\ I^- \rightarrow I_3^- + 2\ e^-$ $-E° = -0.55$ V
    _____
    $Tl^{3+} + 3\ I^- \rightarrow Tl^+ + I_3^-$ $E°_{cell} = 0.70$ V

In solution, Tl^{3+} can oxidize I^- to I_3^-. Thus, we expect TlI_3 to be thallium(I) triiodide.

59.   Ga(I): $[Ar]4s^2 3d^{10}$, no unpaired e^-; Ga(III): $[Ar]3d^{10}$, no unpaired e^-

Ga(II): $[Ar]4s^1 3d^{10}$, 1 unpaired e^-; Note: s electrons are lost before the d electrons.

If the compound contained Ga(II) it would be paramagnetic and if the compound contained Ga(I) and Ga(III) it would be diamagnetic. This can easily be determined by measuring the mass of a sample in the presence and in the absence of a magnetic field. Paramagnetic compounds will have an apparent greater mass in a magnetic field.

60.   a.   Out of 100.0 g of compound there are:

$$44.4 \text{ g Ca} \times \frac{1 \text{ mol}}{40.08 \text{ g}} = 1.11 \text{ mol Ca}; \quad 20.0 \text{ g Al} \times \frac{1 \text{ mol}}{26.98 \text{ g}} = 0.741 \text{ mol Al}$$

$$35.6 \text{ g O} \times \frac{1 \text{ mol}}{16.00 \text{ g}} = 2.23 \text{ mol O}$$

$$\frac{1.11}{0.741} = 1.50; \quad \frac{0.741}{0.741} = 1.00; \quad \frac{2.23}{0.741} = 3.01; \quad \text{Empirical formula is } Ca_3Al_2O_6.$$

b. $Ca_9Al_6O_{18}$

c.   There are covalent bonds between Al and O atoms in the $Al_6O_{18}{}^{18-}$ anion; sp^3 hybrid orbitals on aluminum overlap with sp^3 hybrid orbitals on oxygen to form the sigma bonds.

61.

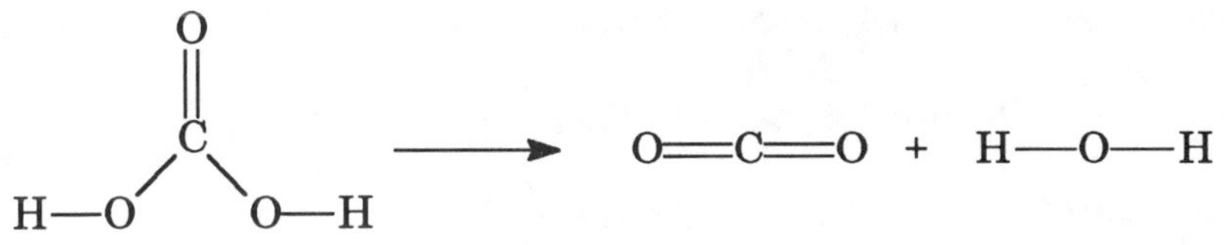

Bonds broken: 2 C – O (358 kJ/mol)   Bonds formed: 1 C $=$ O (799 kJ/mol)
                 1 O – H (467 kJ/mol)   1 O – H (467 kJ/mol)

$\Delta H = 2(358) + 467 - (799 + 467) = -83$ kJ; ΔH is favorable for the decomposition of H_2CO_3 to CO_2 and H_2O. ΔS is also favorable for the decomposition as there is an increase in disorder. Hence, H_2CO_3 will spontaneously decompose to CO_2 and H_2O.

62. Pb_3O_4: We assign -2 for the oxidation state of O. The sum of the oxidation states of Pb must be +8. We get this if two of the lead atoms are Pb(II) and one is Pb(IV). So 2/3 of the lead is Pb(II).

63.   From Table 18.14 in the text, Sn and Pb can reduce H^+ to H_2.

$Sn(s) + 2 H^+(aq) \rightarrow Sn^{2+}(aq) + H_2(g); \quad Pb(s) + 2 H^+(aq) \rightarrow Pb^{2+}(aq) + H_2(g)$

64.

$$CH_2(CH_2)_6CH_3$$

$$Sn$$

$$:\ddot{C}l \quad :\ddot{C}l: \quad CH_2(CH_2)_6CH_3$$

The compound is held together by covalent bonds. The structure is tetrahedral about the central Sn.

Challenge Problems

65. Na^+ can oxidize Na^- to Na. The purpose of the cryptand is to encapsulate the Na^+ ion so that it does not come in contact with the Na^- ion and oxidize it to sodium metal.

66. $SiCl_4(l) + 2 H_2O(l) \rightarrow SiO_2(s) + 4 H^+(aq) + 4 Cl^-(aq)$

$\Delta H° = -911 + 4(0) + 4(-167) - [-687 + 2(-286)] = -320.$ kJ

$\Delta S° = 42 + 4(0) + 4(57) - [240. + 2(70.)] = -110.$ J/K; $\Delta G° = \Delta H° - T\Delta S°$

$\Delta G° = 0$ when $T = \Delta H°/\Delta S° = -320. \times 10^3$ J/(-110. J/K) = 2910 K

Due to the favorable $\Delta H°$ term, this reaction is spontaneous at temperatures below 2910 K.

The corresponding reaction for CCl_4 is:

$CCl_4(l) + 2 H_2O(l) \rightarrow CO_2(g) + 4 H^+(aq) + 4 Cl^-(aq)$

$\Delta H° = -393.5 + 4(0) + 4(-167) - [-135 + 2(-286)] = -355$ kJ

$\Delta S° = 214 + 4(0) + 4(57) - [216 + 2(70.)] = 86$ J/K

Thermodynamics predicts that this reaction would be spontaneous at any temperature.

The answer must lie with kinetics. $SiCl_4$ reacts because an activated complex can form by a water molecule attaching to silicon in $SiCl_4$. The activated complex requires silicon to form a fifth bond. Silicon has low energy 3d orbitals available to expand the octet. Carbon will not break the octet rule, therefore, CCl_4 cannot form this activated complex. CCl_4 and H_2O require a different pathway to get to products. The different pathway has a higher activation energy and, in turn, the reaction is much slower. (See Exercise 18.67.)

67.   Carbon cannot form the fifth bond necessary for the transition state since carbon doesn't have low energy d orbitals available to expand the octet.

68.   White tin is stable at normal temperatures. Gray tin is stable at temperatures below 13.2°C. Thus for the phase change: Sn(gray) $\rightarrow$ Sn(white), ΔG is (-) at T > 13.2°C and ΔG is (+) at T < 13.2°C. This is only possible if ΔH is (+) and ΔS is (+). Thus, gray tin has the more ordered structure.

CHAPTER NINETEEN

THE REPRESENTATIVE ELEMENTS: GROUPS 5A THROUGH 8A

Questions

1. The reaction $N_2(g) + 3 H_2(g) \rightarrow 2 NH_3(g)$ is exothermic. Thus, the equilibrium constant K decreases as the temperature increases. Lower temperatures are favored for maximum yield of ammonia. However, at lower temperatures the rate is slow; without a catalyst the rate is too slow for the process to be feasible. The discovery of a catalyst increased the rate of reaction at a lower temperature favored by thermodynamics.

2. White phosphorus consists of discrete tetrahedral P_4 molecules. The bond angles in the P_4 tetrahedrons are only $60°$, which makes P_4 very reactive especially towards oxygen. Red and black phosphorus are covalent network solids. In red phosphorus the P_4 tetrahedra are bonded to each other in chains making them less reactive than white phosphorus. They need a source of energy to react with oxygen, such as when one strikes a match. Black phosphorus is crystalline with the P atoms tightly bonded to each other in the crystal and are fairly unreactive towards oxygen.

3. The pollution provides nitrogen and phosphorous nutrients so the algae can grow. The algae consume oxygen, causing fish to die.

4. In the upper atmosphere, O_3 acts as a filter for UV radiation: $O_3 \xrightarrow{h\nu} O_2 + O$

    O_3 is also a powerful oxidizing agent. It irritates the lungs and eyes, and at high concentration it is toxic. The smell of a "fresh spring day" is O_3 formed during lightning discharges. Toxic materials don't necessarily smell bad. For example, HCN smells like almonds.

5. Plastic sulfur consists of long S_n chains of sulfur atoms. As plastic sulfur becomes brittle, the long chains break down into S_8 rings.

6. In the presence of S^{2-}, sulfur forms polysulfide ions, S_n^{2-}, which are soluble like most species with charges, e.g., $S_8 + S^{2-} \rightleftharpoons S_9^{2-}$. Nitric acid oxidizes S^{2-} to S, which then causes sulfur to precipitate out of solution.

7. Fluorine is the most reactive of the halogens because it is the most electronegative atom and the bond in the F_2 molecule is very weak.

8.   One reason is that the H − F bond is stronger than the other hydrohalides, making it more difficult to form H^+ and F^-. The main reason HF is a weak acid is entropy. When $F^-(aq)$ forms from the dissociation of HF, there is a high degree of ordering that takes place as water molecules hydrate this small ion. Entropy is considerably more unfavorable for the formation of hydrated F^- than for the formation of the other hydrated halides. The result of the more unfavorable $\Delta S°$ term is a positive $\Delta G°$ value which leads to a K_a value less than one.

9.   Helium is unreactive and doesn't combine with any other elements. It is a very light gas and would easily escape the earth's gravitational pull as the planet was formed.

10. In Mendeleev's time none of the noble gases were known. Since an entire family was missing, no gaps seemed to appear in the periodic arrangement. Mendeleev had no evidence to predict the existence of such a family. The heavier members of the noble gases are not really inert. Xe and Kr have been shown to react and form compounds with other elements.

Exercises

Group 5A Elements

11. NO_4^{3-}

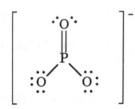

Both NO_4^{3-} and PO_4^{3-} have 32 valence electrons so both have similar Lewis structures. From the Lewis structure for NO_4^{3-}, the central N atom has a tetrahedral arrangement of electron pairs. N is small. There is probably not enough room for all 4 oxygen atoms around N. P is larger, thus, PO_4^{3-} is stable.

PO_3^-

PO_3^- and NO_3^- both have 24 valence electrons so both have similar Lewis structures. From the Lewis structure for PO_3^-, PO_3^- has a trigonal arrangement of electron pairs about the central P atom (two single bonds and one double bond). $P = O$ bonds are not particularly stable, while $N = O$ bonds are stable. Thus, NO_3^- is stable.

12.   a. PF_5;   N is too small and doesn't have low energy d-orbitals to expand its octet to form NF_5.

b. AsF_5;   I is too large to fit 5 atoms of I around As.

c. NF_3;   N is too small for three large bromine atoms to fit around it.

13.   a. $8 \ H^+(aq) + 2 \ NO_3^-(aq) + 3 \ Cu(s) \rightarrow 3 \ Cu^{2+}(aq) + 4 \ H_2O(l) + 2 \ NO(g)$

       b. $NH_4NO_3(s) \xrightarrow{Heat} N_2O(g) + 2 \ H_2O(g)$

       c. $NO(g) + NO_2(g) + 2 \ KOH(aq) \rightarrow 2 \ KNO_2(aq) + H_2O(l)$

14.   a. $P_4O_6(s) + 2 \ O_2(g) \rightarrow P_4O_{10}(s)$ b. $P_4O_{10}(s) + 6 \ H_2O(l) \rightarrow 4 \ H_3PO_4(aq)$

       c. $PCl_5(l) + 4 \ H_2O(l) \rightarrow H_3PO_4(aq) + 5 \ HCl(aq)$

15. NH_3: sp^3; N_2H_4: both Ns are sp^3; NH_2OH: sp^3; N_2: sp; N_2O: central N, sp

       NO: sp^2; N_2O_3: both Ns are sp^2; NO_2: sp^2; HNO_3: sp^2

16.   In the solid state: NO_2^+ and NO_3^-; In the gas phase: molecular N_2O_5

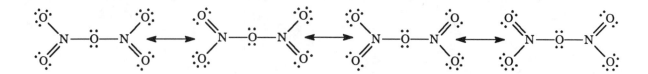

17. $27.37 \times 10^9 \ lb \ NH_3 \times \dfrac{1 \ kg}{2.2046 \ lb} \times \dfrac{1000 \ g}{1 \ kg} \times \dfrac{1 \ mol \ NH_3}{17.03 \ g \ NH_3} = 7.290 \times 10^{11} \ mol \ NH_3$

       $mol \ N_2 = 7.290 \times 10^{11} \ mol \ NH_3 \times \dfrac{1 \ mol \ N_2}{2 \ mol \ NH_3} = 3.645 \times 10^{11} \ mol \ N_2$; STP = 273.15 K and
       1.000 atm

       $V_{N_2} = \dfrac{nRT}{P} = \dfrac{3.645 \times 10^{11} \ mol \times \dfrac{0.08206 \ L \ atm}{K \cdot mol} \times 273.15 \ K}{1.000 \ atm} = 8.170 \times 10^{12} \ L \ N_2$

       Since T and P are constant, then moles and volume are directly proportional to each other. From the balanced reaction, there is a 3:1 mol ratio between H_2 and N_2, so the volume of H_2 required is 3 times the volume of N_2 required.

       $V_{H_2} = V_{N_2} \times \dfrac{3 \ mol \ H_2}{1 \ mol \ N_2} = 8.170 \times 10^{12} \ L \times 3 = 2.451 \times 10^{13} \ L \ H_2$

18. $P_{total} = 9.4 \times 10^4 \ Pa \times \dfrac{1 \ atm}{1.013 \times 10^5 \ Pa} \times \dfrac{760 \ torr}{atm} = 710 \ torr$

       $P_{total} = P_{N_2O} + P_{H_2O}$, $P_{N_2O} = 710 \ torr - 21 \ torr = 690 \ torr \times \dfrac{1 \ atm}{760 \ torr} = 0.91 \ atm$

       $2.6 \ g \ NH_4NO_3 \times \dfrac{1 \ mol \ NH_4NO_3}{80.05 \ g \ NH_4NO_3} \times \dfrac{1 \ mol \ N_2O}{mol \ NH_4NO_3} = 3.2 \times 10^{-2} \ mol \ N_2O$

$$V_{N_2O} = \frac{nRT}{P} = \frac{3.2 \times 10^{-2} \text{ mol N}_2\text{O} \times \dfrac{0.08206 \text{ L atm}}{\text{K} \bullet \text{mol}} \times 295 \text{ K}}{0.91 \text{ atm}} = 0.85 \text{ L}$$

19.

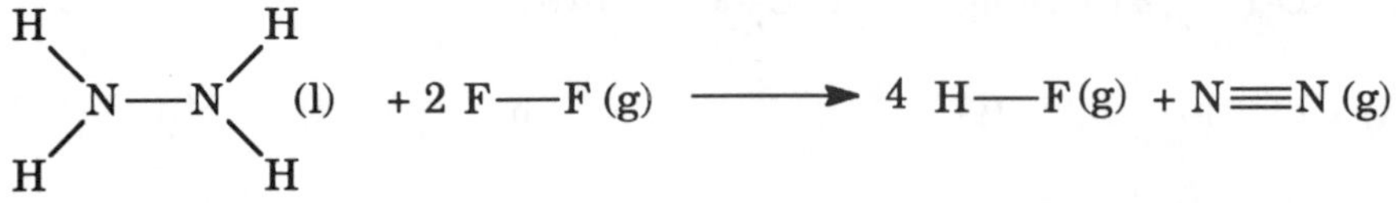

Bonds broken: Bonds formed:

    1 N – N (160. kJ/mol)   4 H – F (565 kJ/mol)
    4 N – H (391 kJ/mol)   1 N ≡ N (941 kJ/mol)
    2 F – F (154 kJ/mol)

$\Delta H = 160. + 4(391) + 2(154) - [4(565) + 941] = 2032 \text{ kJ} - 3201 \text{ kJ} = -1169 \text{ kJ}$

20.

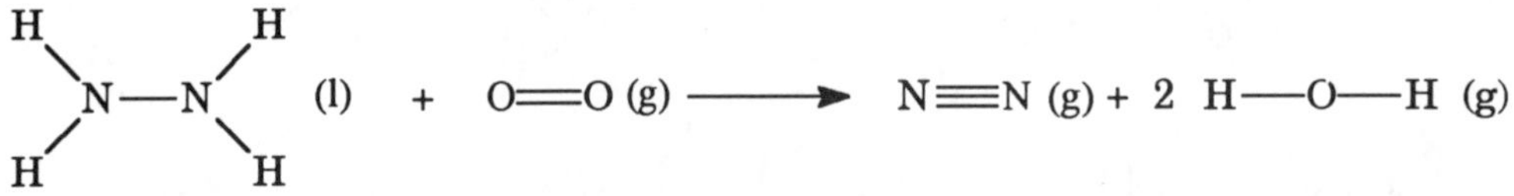

Bonds broken: Bonds formed:

    1 N – N (160. kJ/mol)   1 N ≡ N (941 kJ/mol)
    4 N – H (391 kJ/mol)   2 × 2 O – H (467 kJ/mol)
    1 O = O (495 kJ/mol)

$\Delta H = 160. + 4(391) + 495 - [941 + 4(467)] = 2219 \text{ kJ} - 2809 \text{ kJ} = -590. \text{ kJ}$

21. $1/2 \text{ N}_2(g) + 1/2 \text{ O}_2(g) \rightarrow \text{NO}(g)$ $\Delta G° = \Delta G°_{f, NO} = 87 \text{ kJ/mol}$; By definition, $\Delta G°_f$ for a compound equals the free energy change that would accompany the formation of 1 mol of that compound from its elements in their standard states. NO (and some other oxides of nitrogen) have weaker bonds as compared to the triple bond of N_2 and the double bond of O_2. Because of this, NO (and some other oxides of nitrogen) have higher (positive) free energies as compared to the relatively stable N_2 and O_2 molecules.

22. $\Delta H° = 2(90. \text{ kJ}) - [0 + 0] = 180. \text{ kJ}$; $\Delta S° = 2(211 \text{ J/K}) - [192 + 205] = 25 \text{ J/K}$

$\Delta G° = 2(87 \text{ kJ}) - [0] = 174 \text{ kJ}$

At the high temperatures in automobile engines, the reaction $N_2 + O_2 \rightarrow 2 \text{ NO}$ becomes spontaneous since the favorable $\Delta S°$ term will become dominate. In the atmosphere, even though $2 \text{ NO} \rightarrow N_2 + O_2$ is spontaneous at the cooler temperatures of the atmosphere, it doesn't occur because the rate is slow. Therefore, higher concentrations of NO are present in the atmosphere as compared to what is predicted by thermodynamics.

23.   M.O model:

NO^+: $(\sigma_{2s})^2(\sigma_{2s}*)^2(\pi_{2p})^4(\sigma_{2p})^2$, Bond order = (8 - 2)/2 = 3, 0 unpaired e^- (diamagnetic)

NO: $(\sigma_{2s})^2(\sigma_{2s}*)^2(\pi_{2p})^4(\sigma_{2p})^2(\pi_{2p}*)^1$, B.O. = 2.5, 1 unpaired e^- (paramagnetic)

NO^-: $(\sigma_{2s})^2(\sigma_{2s}*)^2(\pi_{2p})^4(\sigma_{2p})^2(\pi_{2p*})^2$, B.O. = 2, 2 unpaired e^- (paramagnetic)

Lewis structures: NO^+: $\left[\ :N\!\equiv\!\!\equiv\!O:\ \right]^+$

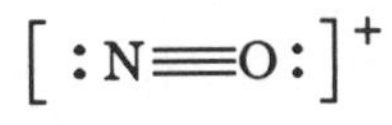

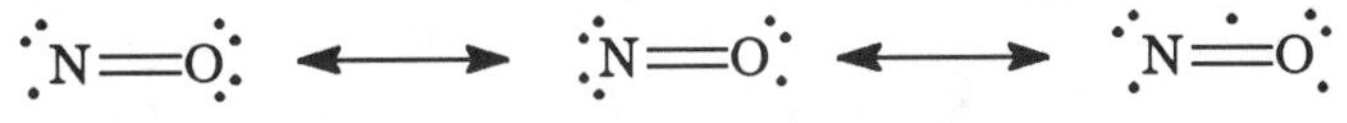

NO: $:\!\!N\!\!=\!\!O\!:$ $\longleftrightarrow$ $:\!\!N\!\!=\!\!O\!:$ $\longleftrightarrow$ $:\!\!N\!\!=\!\!O\!:$

NO^-: $\left[\ :\!\!N\!\!=\!\!O\!:\ \right]^-$

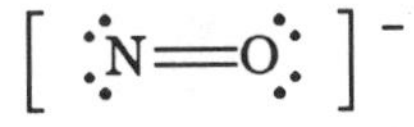

The two models only give the same results for NO^+ (a triple bond with no unpaired electrons). Lewis structure are not adequate for NO and NO^-. The M.O. model gives a better representation for all three species. For NO, Lewis structures are poor for odd electron species. For NO^-, both models predict a double bond but only the MO model correctly predicts that NO^- is paramagnetic.

24.   For $NCl_3 \rightarrow NCl_2 + Cl$, only the N – Cl bond is broken. For $O=N-Cl \rightarrow NO + Cl$, the NO bond gets stronger (bond order increases from 2.0 to 2.5) when the N – Cl bond is broken. This makes ΔH for the reaction smaller than just the energy necessary to break the N – Cl bond.

25.   a. $H_3PO_4 > H_3PO_3$; The strongest acid has the most oxygen atoms.

b. $H_3PO_4 > H_2PO_4^- > HPO_4^{2-}$; Acid strength decreases as protons are removed.

26.   TSP = Na_3PO_4; PO_4^{3-} is the conjugate base of the weak acid HPO_4^{2-} ($K_a = 4.8 \times 10^{-13}$). All conjugate bases of weak acids are effective bases ($K_b = K_w/K_a = 1.0 \times 10^{-14}/4.8 \times 10^{-13} = 2.1 \times 10^{-2}$). The weak base reaction of PO_4^{3-} with H_2O is: $PO_4^{3-} + H_2O \rightleftharpoons HPO_4^{2-} + OH^-$ $K_b = 2.1 \times 10^{-2}$.

27.   To complete the Lewis structures, just add lone pairs to the terminal atoms to complete the octets. See part d for the Lewis structures (32 valence electrons).

a.   Yes, both have 4 sets of electrons about the P. We would predict a tetrahedral structure for both.

b.   The hybridization is sp^3 for each P since both structures are tetrahedral.

c.   P has to use one of its d orbitals to form the π bond.

d.   The formal charges for the O and P atoms are next to the atoms in the following Lewis structures. In both structures all Cl atoms have a formal charge of zero.

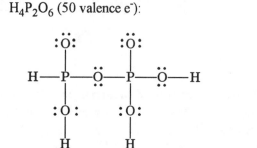

The structure with the P = O bond is favored on the basis of formal charge since it has a zero formal charge for all atoms in the structure.

28.   The acidic protons are attached to oxygen.

$H_4P_2O_6$ (50 valence e^-):

$H_4P_2O_5$ (44 valence e^-):

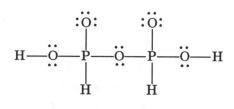

29.   Production of antimony:

$$2 Sb_2S_3(s) + 9 O_2(g) \rightarrow 2 Sb_2O_3(s) + 6 SO_2(g); \quad 2 Sb_2O_3(s) + 3 C(s) \rightarrow 4 Sb(s) + 3 CO_2(g)$$

Production of bismuth:

$$2 Bi_2S_3(s) + 9 O_2(g) \rightarrow 2 Bi_2O_3(s) + 6 SO_2(g); \quad 2 Bi_2O_3(s) + 3 C(s) \rightarrow 4 Bi(s) + 3 CO_2(g)$$

30. $4 As(s) + 3 O_2(g) \rightarrow As_4O_6(s); \quad 4 As(s) + 5 O_2(g) \rightarrow As_4O_{10}(s)$

$As_4O_6(s) + 6 H_2O(l) \rightarrow 4 H_3AsO_3(aq); \quad As_4O_{10}(s) + 6 H_2O(l) \rightarrow 4 H_3AsO_4(aq)$

Group 6A Elements

31. $O = O - O \rightarrow O = O + O$

Break $O - O$ bond: $\Delta H = \dfrac{146 \text{ kJ}}{\text{mol}} \times \dfrac{1 \text{ mol}}{6.022 \times 10^{23}} = 2.42 \times 10^{-22} \text{ kJ} = 2.42 \times 10^{-19} \text{ J}$

A photon of light must contain at least 2.42×10^{-19} J to break one $O - O$ bond.

$E_{photon} = \dfrac{hc}{\lambda}, \quad \lambda = \dfrac{hc}{E} = \dfrac{(6.626 \times 10^{-34} \text{ J s}) (2.998 \times 10^8 \text{ m/s})}{2.42 \times 10^{-19} \text{ J}} = 8.21 \times 10^{-7} \text{ m} = 821 \text{ nm}$

32.   From Figure 7.2 in the text, light from violet to green will work.

33.   a.   oxidation - reduction reaction

       b.   From Table 17.1 of the text, the reduction of NO_3^- to NO has a favorable (positive) reduction
            potential. Thus, NO(g) is the expected product.

       c.
$$(3e^- + 3 H^+ + HNO_3 \rightarrow NO + 2 H_2O) \times 2$$
$$(S^{2-} \rightarrow S + 2e^-) \times 3$$

$$6 H^+(aq) + 2 HNO_3(aq) + 3 S^{2-}(aq) \rightarrow 3 S(s) + 2 NO(g) + 4 H_2O(l)$$

34. $H_2SeO_4(aq) + 3 SO_2(g) \rightarrow Se(s) + 3 SO_3(g) + H_2O(l)$

35. SF_5^- has $6 + 5(7) + 1 = 42$ valence electrons.

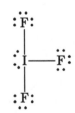

       square pyramid

36. $OTeF_5^-$ has $6 + 6 + 5(7) + 1 = 48$ valence electrons.

Group 7A Elements

37.   a. ClF_5, $7 + 5(7) = 42$ e$^-$ b. IF_3, $7 + 3(7) = 28$ e$^-$

       Square pyramid; d^2sp^3 T-shaped; dsp^3

c. $FBrO_2$, $7 + 7 + 2(6) = 26$ e⁻

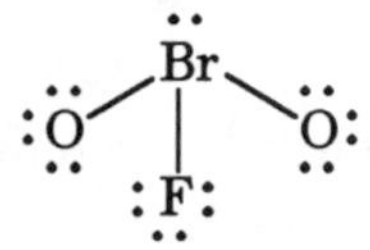

Trigonal pyramid; sp^3

38. O_2F_2 has $2(6) + 2(7) = 26$ valence e⁻; From the following Lewis structure, each oxygen atom has a tetrahedral arrangement of electron pairs. Therefore, bond angles ≈ 109.5 and each O is sp^3 hybridized.

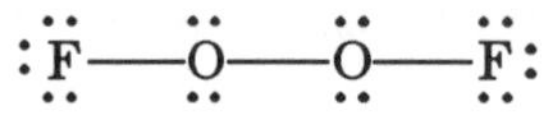

Formal Charge	0	0	0	0
Oxid. Number	-1	+1	+1	-1

Oxidation numbers are more useful. We are forced to assign +1 as the oxidation number for oxygen. Oxygen is very electronegative and +1 is not a stable oxidation state for this element.

39.   The balanced half-reactions and the overall balanced reaction for this oxidation-reduction reaction are:

$$2 \text{ e}^- + 2 \text{ H}^+ + XeF_2 \rightarrow Xe + 2 \text{ HF}$$
$$H_2O + BrO_3^- \rightarrow BrO_4^- + 2 \text{ H}^+ + 2 \text{ e}^-$$

$$H_2O(l) + BrO_3^-(aq) + XeF_2(aq) \rightarrow BrO_4^-(aq) + Xe(g) + 2 \text{ HF}(aq)$$

40.   a. $F_2 + H_2O \rightarrow HOF + HF$; $2 \text{ HOF} \rightarrow 2 \text{ HF} + O_2$; $HOF + H_2O \rightarrow HF + H_2O_2$ (dilute acid)

In dilute base, HOF exists as OF⁻ and HF exists as F⁻. The balanced reaction is:

$$(2\text{e}^- + H_2O + OF^- \rightarrow F^- + 2 \text{ OH}^-) \times 2$$
$$4 \text{ OH}^- \rightarrow O_2 + 2 \text{ H}_2O + 4\text{e}^-$$

$$2 \text{ OF}^- \rightarrow O_2 + 2 \text{ F}^-$$

b.   HOF: Assign +1 to H and -1 to F. The oxidation state of oxygen is then zero. Oxygen is very electronegative. A zero oxidation state is not very stable since oxygen is a very good oxidizing agent.

41. $ClO^- + H_2O + 2 \text{ e}^- \rightarrow 2 \text{ OH}^- + Cl^-$ $E° = 0.90$ V
              $2 \text{ NH}_3 + 2 \text{ OH}^- \rightarrow N_2H_4 + 2 \text{ H}_2O + 2 \text{ e}^-$ $-E° = 0.10$ V

$$ClO^-(aq) + 2 \text{ NH}_3(aq) \rightarrow Cl^-(aq) + N_2H_4(aq) + H_2O(l)$$ $E°_{cell} = 1.00$ V

Since $E°_{cell}$ is positive for this reaction, then at standard conditions ClO^- can spontaneously oxidize NH_3 to the somwhat toxic N_2H_4.

42.   A disproportionation reaction is an oxidation-reduction reaction in which one species will act as both an oxidizing agent and reducing agent. The species reacts with itself forming products with higher and lower oxidation states. For example, $2\ Cu^+ \rightarrow Cu + Cu^{2+}$ is a disproportionation reaction.

$HClO_2$ will disproportionate at standard conditions since $E°_{cell} > 0$:

$$
\begin{array}{ll}
HClO_2 + 2\ H^+ + 2\ e^- \rightarrow HClO + H_2O & E° = 1.65\ V \\
HClO_2 + H_2O \rightarrow ClO_3^- + 3\ H^+ + 2\ e^- & -E° = -1.21\ V \\
\hline
2\ HClO_2(aq) \rightarrow HClO(aq) + ClO_3^-(aq) + H^+(aq) & E°_{cell} = 0.44\ V
\end{array}
$$

Group 8A Elements

43.   Xe has one more valence electron than I. Thus, the isoelectric species will have I plus one extra electron substituted for Xe, giving a species with a net minus one charge.

   a. IO_4^- b. IO_3^- c. IF_2^- d. IF_4^- e. IF_6^- f. IOF_3^-

44.   a. XeO_3, $8 + 3(6) = 26\ e^-$ b. XeO_4, $8 + 4(6) = 32\ e^-$

       trigonal pyramid   tetrahedral

   c. $XeOF_4$, $8 + 6 + 4(7) = 42\ e^-$ d. $XeOF_2$, $8 + 6 + 2(7) = 28\ e^-$

       square pyramid   T-shaped

e. XeO_3F_2 has $8 + 3(6) + 2(7) = 40$ valence electrons.

trigonal bipyramid

45. XeF_2 can react with oxygen to produce explosive xenon oxides and oxyfluorides.

46. $10.0 \text{ m} \times 5.0 \text{ m} \times 3.0 \text{ m} = 1.5 \times 10^2 \text{ m}^3$; From Table 19.12, volume % Xe $= 9 \times 10^{-6}$.

$$1.5 \times 10^2 \text{ m}^3 \times \left(\frac{10 \text{ dm}}{\text{m}}\right)^3 \times \frac{1 \text{ L}}{\text{dm}^3} \times \frac{9 \times 10^{-6} \text{ L Xe}}{100 \text{ L air}} = 1 \times 10^{-2} \text{ L of Xe in the room}$$

$$PV = nRT, \ n = \frac{PV}{RT} = \frac{(1.0 \text{ atm}) (1 \times 10^{-2} \text{ L})}{(0.08206 \text{ L atm/mol·K}) (298 \text{ K})} = 4 \times 10^{-4} \text{ mol Xe}$$

$$4 \times 10^{-4} \text{ mol Xe} \times \frac{131.3 \text{ g}}{\text{mol}} = 5 \times 10^{-2} \text{ g Xe in the room}$$

$$4 \times 10^{-4} \text{ mol Xe} \times \frac{6.022 \times 10^{23} \text{ atoms}}{\text{mol}} = 2 \times 10^{20} \text{ atoms Xe in the room}$$

A 2 L breath contains: $2 \text{ L air} \times \dfrac{9 \times 10^{-6} \text{ L Xe}}{100 \text{ L air}} = 2 \times 10^{-7} \text{ L Xe}$

$$n = \frac{PV}{RT} = \frac{(1.0 \text{ atm}) (2 \times 10^{-7} \text{ L})}{(0.08206 \text{ L atm/mol·K}) (298 \text{ K})} = 8 \times 10^{-9} \text{ mol Xe}$$

$$8 \times 10^{-9} \text{ mol Xe} \times \frac{6.022 \times 10^{23} \text{ atoms}}{\text{mol}} = 5 \times 10^{15} \text{ atoms of Xe in a 2 L breath}$$

Additional Exercises

47.   As the halogen atoms get larger, it becomes more difficult to fit three halogen atoms around the small nitrogen atom, and the NX_3 molecule becomes less stable.

48. $\Delta H^\circ = 82 + 34 - [3(90.)] = -154 \text{ kJ}$; $\Delta S^\circ = 220. + 240. - [3(211)] = -173 \text{ J/K}$

$\Delta G^\circ = \Delta H^\circ - T\Delta S^\circ = -154 \text{ kJ} - 298 \text{ K}(-0.173 \text{ kJ/K}) = -102 \text{ kJ}$

$\Delta G^\circ = 0$ when $T = \dfrac{\Delta H^\circ}{\Delta S^\circ} = \dfrac{-154,000 \text{ J}}{-173 \text{ J/K}} = 890. \text{ K}$

At $T < 890. \text{ K}$ and standard pressures, the reaction is spontaneous since the favorable ΔH° term will dominates at these temperatures.

49.   OCN⁻ has 6 + 4 + 5 + 1 = 16 valence electrons.

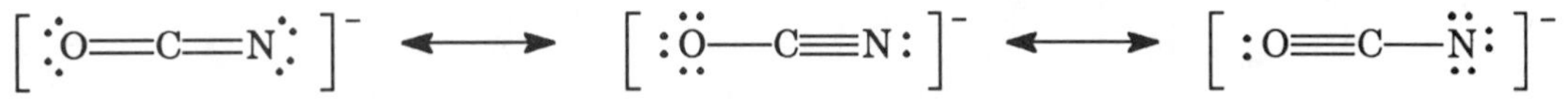

$$\left[\ddot{\underset{\cdot\cdot}{O}}=C=\ddot{N}\right]^{-} \longleftrightarrow \left[\ddot{\underset{\cdot\cdot}{O}}-C\equiv N:\right]^{-} \longleftrightarrow \left[:O\equiv C-\ddot{N}:\right]^{-}$$

Formal charge	0	0	-1		-1	0	0		+1	0	-2

Only the first two resonance structures should be important. The third places a positive formal charge on the most electronegative atom in the ion and a -2 formal charge on N.

CNO⁻:

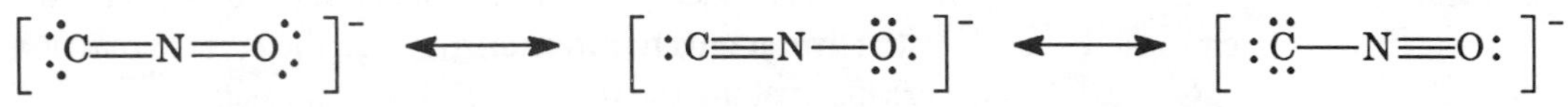

$$\left[:\ddot{C}=N=\ddot{\underset{\cdot\cdot}{O}}\right]^{-} \longleftrightarrow \left[:C\equiv N-\ddot{\underset{\cdot\cdot}{O}}:\right]^{-} \longleftrightarrow \left[:\ddot{\underset{\cdot\cdot}{C}}-N\equiv O:\right]^{-}$$

Formal charge	-2	+1	0		-1	+1	-1		-3	+1	+1

All of the resonance structures for fulminate (CNO⁻) involve greater formal charges than in cyanate (OCN⁻), making fulminate more reactive (less stable).

50.   AsCl₄⁺, 5 + 4(7) - 1 = 32 e⁻   AsCl₆⁻, 5 + 6(7) + 1 = 48 e⁻

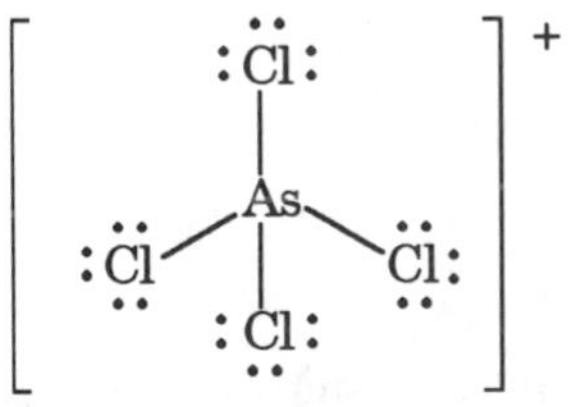

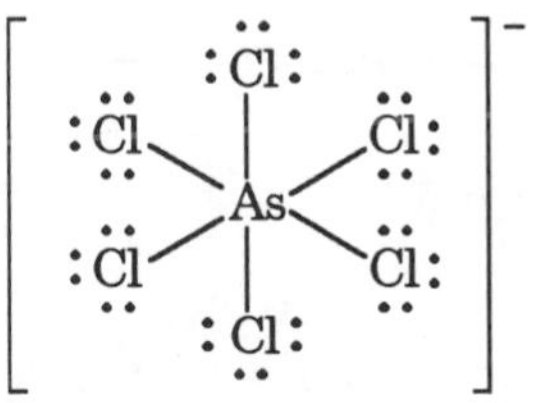

The reaction is a Lewis acid-base reaction. Chloride ion acts as a Lewis base when it is transferred from one AsCl₅ to another. Arsenic is the Lewis acid (electron pair acceptor).

51.   Hypochlorite can act as an oxidizing agent. For example, it is capable of oxidizing I⁻ to I₂. If a solution containing I⁻ turns brown when BiOCl is added, then BiOCl is bismuth(I) hypochlorite. The brown color indicates production of I₂. If the solution doesn't change color, then it is bismuthylchloride.

52. $2.42 \text{ eV} \times \dfrac{96.5 \text{ kJ/mol}}{\text{eV}} \times \dfrac{1 \text{ mol photons}}{6.022 \times 10^{23} \text{ photons}} \times \dfrac{1000 \text{ J}}{\text{kJ}} = \dfrac{3.88 \times 10^{-19} \text{ J}}{\text{photon}}$

$E = \dfrac{hc}{\lambda}, \ \lambda = \dfrac{hc}{E} = \dfrac{(6.626 \times 10^{-34} \text{ J s}) (2.998 \times 10^{8} \text{ m/s})}{3.88 \times 10^{-19} \text{ J}} = 5.12 \times 10^{-7} \text{ m} = 512 \text{ nm; Green light}$

53. TeF_5^- has $6 + 5(7) + 1 = 42$ valence electrons.

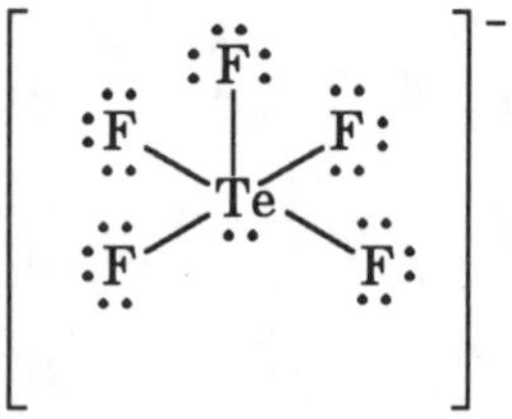

The lone pair of electrons around Te exerts a stronger repulsion than the bonding pairs, pushing the four square planar F's away from the lone pair and thus reducing the bond angles between the axial F atom and the square planar F atoms.

54.   a. $AgCl(s) \xrightarrow{h\nu} Ag(s) + Cl$; The reactive chlorine atom is trapped in the crystal. When light is removed, Cl reacts with silver atoms to reform AgCl, i.e., the reverse reaction occurs. In pure AgCl, the Cl atoms escape, making the reverse reaction impossible.

        b.   Over time chlorine is lost and the dark silver metal is permanent.

Challenge Problems

55.   For the reaction:

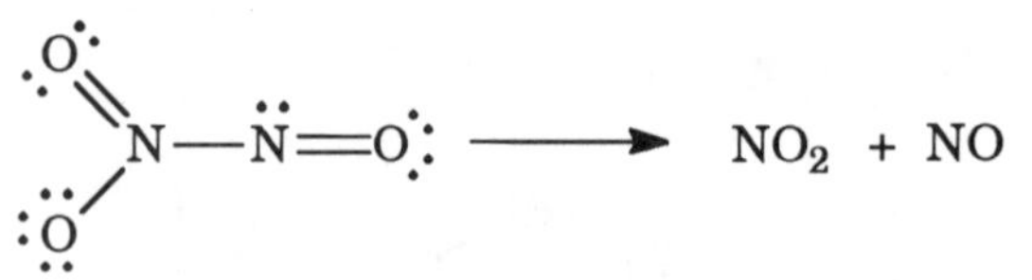

the activation energy must in some way involve the breaking of a nitrogen-nitrogen single bond. For the reaction:

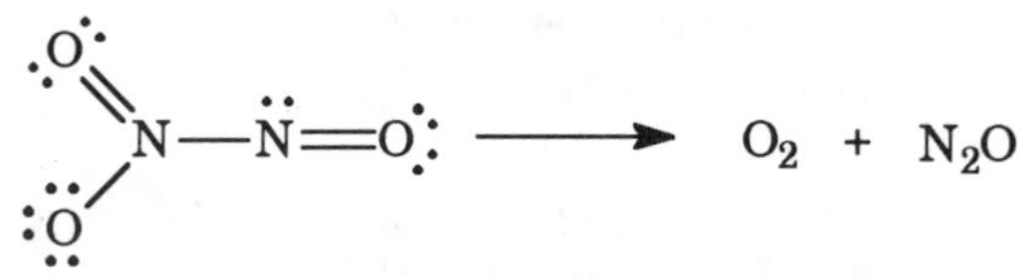

at some point nitrogen-oxygen bonds must be broken. N – N single bonds (160. kJ/mol) are weaker than N – O single bonds (201 kJ/mol). In addition, resonance structures indicate that there is more double bond character in the N – O bonds than in the N – N bond. Thus, NO_2 and NO are preferred by kinetics because of the lower activation energy.

56. $Mg^{2+} + P_3O_{10}^{5-} \rightleftharpoons MgP_3O_{10}^{3-}$ $K = 4.0 \times 10^8$; $[Mg^{2+}]_o = \dfrac{50. \times 10^{-3}\,g}{L} \times \dfrac{1\,mol}{24.31\,g} = 2.1 \times 10^{-3}\,M$

        $[P_3O_{10}^{5-}]_o = \dfrac{40.\,g\,Na_5P_3O_{10}}{L} \times \dfrac{1\,mol}{367.86\,g} = 0.11\,M$

Assume the reaction goes to completion since K is large. Then solve the back equilibrium problem to determine the small amount of Mg^{2+} present.

$$Mg^{2+} \quad + \quad P_3O_{10}^{5-} \quad \rightleftharpoons \quad MgP_3O_{10}^{3-}$$

Before	$2.1 \times 10^{-3}\,M$	$0.11\,M$	0
Change	-2.1×10^{-3}	$-2.1 \times 10^{-3} \quad \rightarrow$	$+2.1 \times 10^{-3}$	React completely
After	0	0.11	2.1×10^{-3}	New initial

x mol/L $MgP_3O_{10}^{3-}$ dissociates to reach equilibrium

Change	$+x$	$+x \quad \leftarrow$	$-x$
Equil.	x	$0.11 + x$	$2.1 \times 10^{-3} - x$

$$K = 4.0 \times 10^8 = \frac{[MgP_3O_{10}^{3-}]}{[Mg^{2+}][P_3O_{10}^{5-}]} = \frac{2.1 \times 10^{-3} - x}{x(0.11 + x)} \quad (\text{assume } x \ll 2.1 \times 10^{-3})$$

$$4.0 \times 10^8 \approx \frac{2.1 \times 10^{-3}}{x(0.11)}, \quad x = [Mg^{2+}] = 4.8 \times 10^{-11}\,M; \quad \text{Assumptions good.}$$

57. a. As we go down the family, K_a increases. This is consistent with the bond to hydrogen getting weaker.

   b. Po is below Te, so K_a should be larger. The K_a value for H_2Po should be on the order of 10^{-2}.

58. a.
$$2\,H^+ + 2\,e^- \rightarrow H_2 \qquad\qquad E° = \ 0.0\,V$$
$$ClO_3^- + H_2O \rightarrow ClO_4^- + 2\,H^+ + 2e^- \qquad -E° = -1.19\,V$$

$$ClO_3^-(aq) + H_2O(l) \rightarrow ClO_4^-(aq) + H_2(g) \qquad E°_{cell} = -1.19\,V$$

A minimum potential of 1.19 V must be applied assuming standard conditions.

   b. $3\,Al(s) + 3\,NH_4ClO_4(s) \rightarrow Al_2O_3(s) + AlCl_3(s) + 3\,NO(g) + 6\,H_2O(g)$

$$\Delta H° = -1676 + (-704) + 3(90.) + 6(-242) - [3(0) + 3(-295)] = -2677\,kJ$$

$$7 \times 10^8\,g\ NH_4ClO_4 \times \frac{1\,mol}{117.49\,g} \times \frac{-2677\,kJ}{3\,mol\ NH_4ClO_4} = -5 \times 10^9\,kJ \text{ of heat released}$$

CHAPTER TWENTY

TRANSITION METALS AND COORDINATION CHEMISTRY

Questions

5.   a.   Ligand: Species that donates a pair of electrons to form a covalent bond to a metal ion. Ligands act as Lewis bases (electron pair donors).

   b.   Chelate: Ligand that can form more than one bond to a metal ion.

   c.   Bidentate: Ligand that forms two bonds to a metal ion.

   d.   Complex ion: Metal ion plus ligands.

6.   A ligand must have at least one unshared pair of electrons.

7.   Both electrons in the bond originally came from one of the atoms in the bond.

8.   Since transition metals form bonds to species which donate lone pairs of electrons , then transition metals are Lewis acids (electron pair acceptors).

9.   a.   Isomers: Species with the same formulas but different properties. See text for examples of the following types of isomers.

   b.   Structural isomers: Isomers that have one or more bonds that are different.

   c.   Stereoisomers: Isomers that contain the same bonds but differ in how the atoms are arranged in space.

   d.   Coordination isomers: Structural isomers that differ in the atoms that make up the complex ion.

   e.   Linkage isomers: Structural isomers that differ in how one or more ligands are attached to the transition metal.

   f.   Geometric isomers: (Cis - trans isomerism); Stereoisomers that differ in the positions of atoms with respect to a rigid ring, bond, or each other.

g. Optical isomers: Stereoisomers that are nonsuperimposable mirror images of each other; that is, they are different in the same way that our left and right hands are different.

10. a. Ligand that will give complex ions with the maximum number of unpaired electrons.

b. Ligand that will give complex ions with the minimum number of unpaired electrons.

c. Complex with a minimum number of unpaired electrons (low-spin = strong-field).

d. Complex with a maximum number of unpaired electrons (high-spin = weak-field).

11. Cu^{2+}: $[Ar]3d^9$; Cu^+: $[Ar]3d^{10}$; Cu(II) is d^9 and Cu(I) is d^{10}. Color is a result of the electron transfer between split d orbitals. This cannot occur for the filled d orbitals in Cu(I). Cd^{2+}, like Cu^+, is also d^{10}. We would not expect $Cd(NH_3)_4Cl_2$ to be colored since the d orbitals are filled in this Cd^{2+} complex.

12. Sc^{3+} has no electrons in d orbitals. Ti^{3+} and V^{3+} have d electrons present. Color of transition metal complexes results from electron transfer between split d orbitals. If no d electrons are present, no electron transfer can occur and the compounds are not colored.

13. The d-orbital splitting in tetrahedral complexes is less than one-half the d-orbital splitting in octahedral complexes. There are no known ligands powerful enough to produce the strong-field case, hence all tetrahedral complexes are weak-field or high spin.

14. CN^- and CO form much stronger complexes with Fe(II) than O_2. Thus, O_2 cannot be transported by hemoglobin in the presence of CN^- or CO.

15. $Fe_2O_3(s) + 6\ H_2C_2O_4(aq) \rightarrow 2\ Fe(C_2O_4)_3{}^{3-}(aq) + 3\ H_2O(l) + 6\ H^+(aq)$; The oxalate anion forms a soluble complex ion with iron in rust (Fe_2O_3) which allows rust stains to be removed.

16. Most transition metals have unfilled d orbitals which creates a large number of valence electrons that can be removed. Stable ions of the representative metals are determined by how many s and p valence electrons can be removed. In general, representative metals lose all of the s and p valence electrons to form their stable ions. Transition metals generally lose the s electron to form +1 and +2 ions, but they can also lose some (or all) of the d electrons to form other oxidation states as well.

17. The lanthanide elements are located just before the 5d transition metals. The lanthanide contraction is the steady decrease in the atomic radii of the lanthanide elements when going from left to right across the periodic table. As a result of the lanthanide contraction, the sizes of the 4d and 5d elements are very similar (see Exercise 7.132). This leads to a greater similarity in the chemistry of the 4d and 5d elements in a given vertical group.

18. See text for examples.

a. Roasting: Converting sulfide minerals to oxides by heating in air below their melting points.

b. Smelting: Reducing metal ions to the free metal.

c. Flotation: Separation of mineral particles in an ore from the unwanted impurities. This process depends on the greater wetability of the mineral particles as compared to the unwanted impurities.

d. Leaching: The extraction of metals from ores using aqueous chemical solutions.

e. Gangue: The impurities (such as clay, sand or rock) in an ore.

19.   Advantages: cheap energy cost; less air pollution; Disadvantages: chemicals used in hydrometallurgy are expensive and sometimes toxic.

20.   In zone refining, a bar of impure metal travels through a heater. The impurities present are more soluble in the molten metal than in the solid metal. As the molten zone moves down a metal, the impurities are swept along with the liquid, leaving behind relatively pure metal.

Exercises

Transition Metals

21.   a. Ni: $[Ar]4s^23d^8$ b. Cd: $[Kr]5s^24d^{10}$

c. Zr: $[Kr]5s^24d^2$ d. Os: $[Xe]6s^24f^{14}5d^6$

22.   a. Sc: $[Ar]4s^23d^1$ b. Ru: $[Kr]5s^24d^6*$

c. Ir: $[Xe]6s^24f^{14}5d^7$ d. Mn: $[Ar]4s^23d^5$

*This is the expected electron configuration for Ru. The actual is $[Kr]5s^14d^7$.

23.   Transition metal ions lose the s electrons before the d electrons.

a. Ni^{2+}: $[Ar]3d^8$ b. Cd^{2+}: $[Kr]4d^{10}$

c. Zr^{3+}: $[Kr]4d^1$ Zr^{4+}: $[Kr]$ d. Os^{2+}: $[Xe]4f^{14}5d^6$; Os^{3+}: $[Xe]4f^{14}5d^5$

24.   a. Sc^{3+}: $[Ar]$ or $[Ne]3s^23p^6$ b. Ru^{2+}: $[Kr]4d^6$; Ru^{3+}: $[Kr]4d^5$

c. Ir^+: $[Xe]6s^14f^{14}5d^7$ or $[Xe]4f^{14}5d^8$ d. Mn^{2+}: $[Ar]3d^5$

Ir^{3+}: $[Xe]4f^{14}5d^6$

25.   Transition metal ions lose the s electrons before the d electrons. Also, Pt is an exception to the normal filling order of electrons (see Figure 7.27).

a. Co: $[Ar]4s^23d^7$ b. Pt: $[Xe]6s^14f^{14}5d^9$ c. Fe: $[Ar]4s^23d^6$

Co^{2+}: $[Ar]3d^7$ Pt^{2+}: $[Xe]4f^{14}5d^8$ Fe^{2+}: $[Ar]3d^6$

Co^{3+}: $[Ar]3d^6$ Pt^{4+}: $[Xe]4f^{14}5d^6$ Fe^{3+}: $[Ar]3d^5$

26.   Cr and Cu are exceptions to the normal filling order of electrons.

   a.   Cr: $[Ar]4s^1 3d^5$ b.   Cu: $[Ar]4s^1 3d^{10}$ c.   V: $[Ar]4s^2 3d^3$

        Cr^{2+}: $[Ar]3d^4$ Cu^+: $[Ar]3d^{10}$ V^{2+}: $[Ar]3d^3$

        Cr^{3+}: $[Ar]3d^3$ Cu^{2+}: $[Ar]3d^9$ V^{3+}: $[Ar]3d^2$

27.   a.   molybdenum(IV) sulfide; molybdenum(VI) oxide

        b. MoS_2, +4; MoO_3, +6 (Oxygen and sulfur are each in the -2 oxidation state.)

28.   pyrolusite, manganese(IV) oxide; rhodochrosite, manganese(II) carbonate

Coordination Compounds

29.   Ammonia solutions are basic. The precipitation reaction is $Cu^{2+}(aq) + 2\ OH^-(aq) \rightarrow Cu(OH)_2(s)$.
        As the concentration of NH_3 increases, the soluble complex ion $Cu(NH_3)_4{}^{2+}$ forms:
        $Cu(OH)_2(s) + 4\ NH_3(aq) \rightarrow Cu(NH_3)_4{}^{2+}(aq) + 2\ OH^-(aq)$.

30. $Hg^{2+}(aq) + 2\ I^-(aq) \rightarrow HgI_2(s)$ (orange precipitate)

        $HgI_2(s) + 2\ I^-(aq) \rightarrow HgI_4{}^{2-}(aq)$ (soluble complex ion)

31.   Only $[Cr(NH_3)_6]Cl_3$ will form a precipitate since only this compound will have Cl^- ions in solution.
        The Cl^- ions in the other compounds are ligands and are bound to the central Cr^{3+} ion. The Cl^- ions
        in $[Cr(NH_3)_6]Cl_3$ are counter ions needed to produce a neutral compound while the NH_3 molecules
        are the ligands bound to Cr^{3+}.

32. $BaCl_2$ gives no precipitate so $SO_4{}^{2-}$ must be in the coordination sphere. A precipitate with $AgNO_3$
        means that the Cl^- is not in the coordination sphere. Since there are only four ammonia molecules in
        the coordination sphere, then the $SO_4{}^{2-}$ must be acting as a bidentate ligand. The structure is:

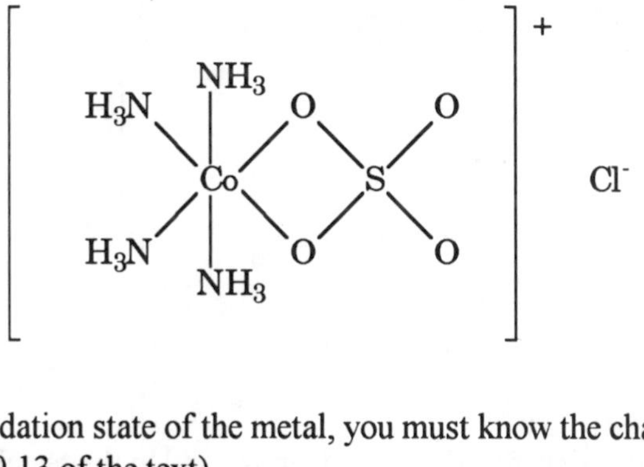

33.   To determine the oxidation state of the metal, you must know the charges of the various common
        ligands (see Table 20.13 of the text).

   a.   pentaamminechlororuthenium(III) ion   b.   hexacyanoferrate(II) ion

   c.   tris(ethylenediamine)manganese(II) ion   d.   pentaamminenitrocobalt(III) ion

34. a. tetrachloroferrate(III) ion   b. hexaamminenickel(II) ion

     c. tetrathiocyanatocobaltate(II) ion   d. hexaaquatitanium(III) ion

35. a. hexaamminecobalt(II) chloride   b. hexaaquacobalt(III) iodide

     c. potassium tetrachloroplatinate(II)   d. potassium hexachloroplatinate(II)

     e. pentaamminechlorocobalt(III) chloride   f. triamminetrinitrocobalt(III)

36. a. sodium tris(oxalato)nickelate(II)   b. potassium tetrachlorocobaltate(II)

     c. tetraamminecopper(II) sulfate

     d. chlorobis(ethylenediamine)thiocyanatocobalt(III) chloride

37. a. $[Co(C_5H_5N)_6]Cl_3$ b. $[Cr(NH_3)_5I]I_2$ c. $[Ni(NH_2CH_2CH_2NH_2)_3]Br_2$

     d. $K_2[Ni(CN)_4]$ e. $[Pt(NH_3)_4Cl_2][PtCl_4]$

38. a. $FeCl_4^-$ b. $[Ru(NH_3)_5H_2O]^{3+}$

     c. $[Pt(C_5H_5N)_5I]^{3+}$ d. $[Pt(NH_3)Cl_3]^-$

39. a.

cis   trans

Note: $C_2O_4^{2-}$ is a bidentate ligand. Bidentate ligands bond to the metal at two positions which are 90° apart from each other in octahedral complexes. Bidentate ligands do not bond to the metal at positions 180° apart.

b.

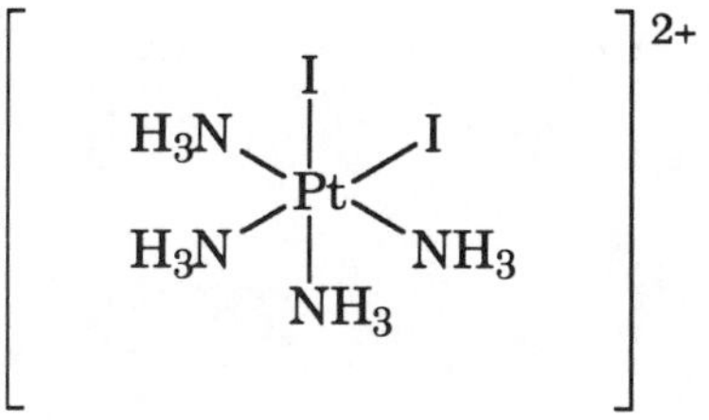

cis

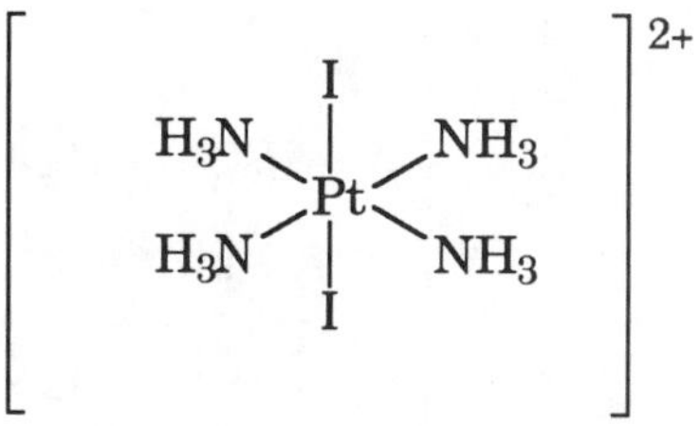

trans

c.

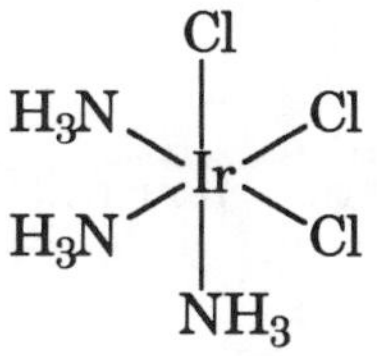

cis

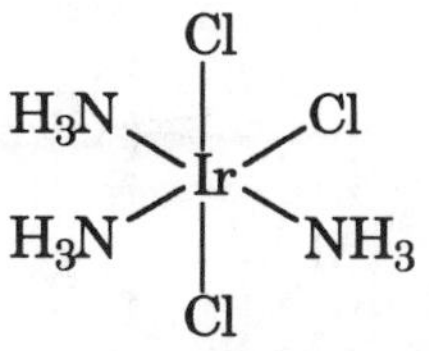

trans

d.

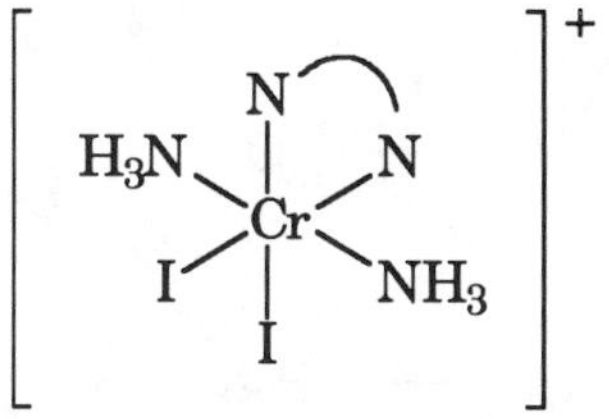

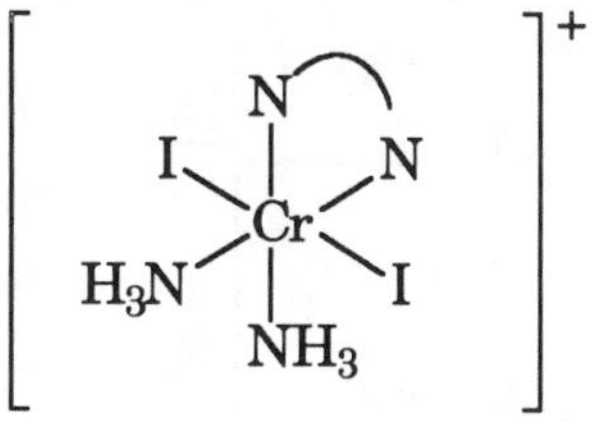

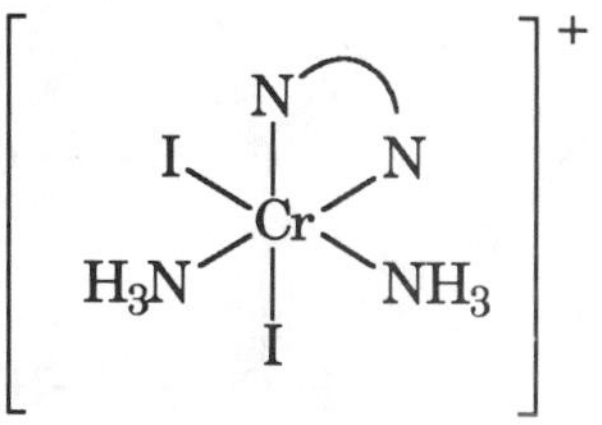

Note:    is an abbreviation for the bidentate ligand ethylenediamine ($H_2NCH_2CH_2NH_2$).

40.   a.   b.

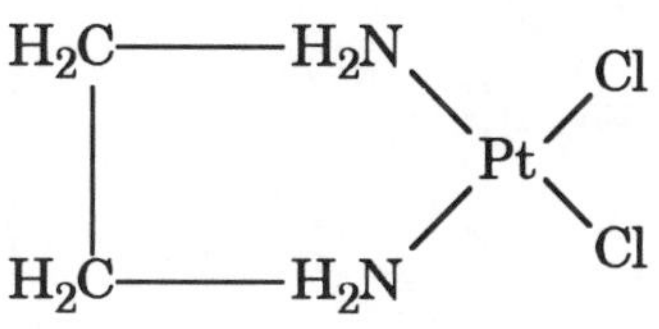

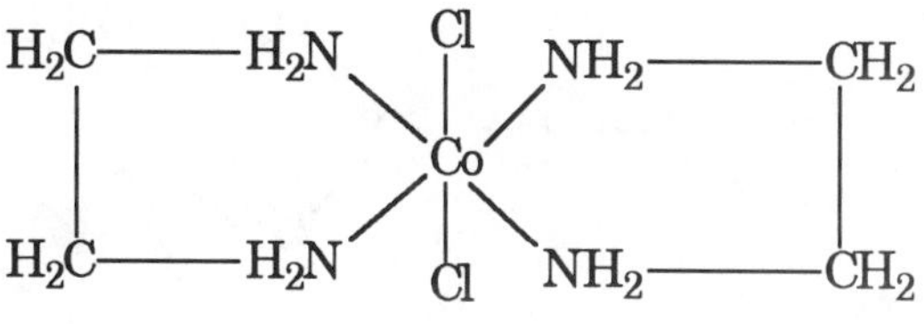

c. 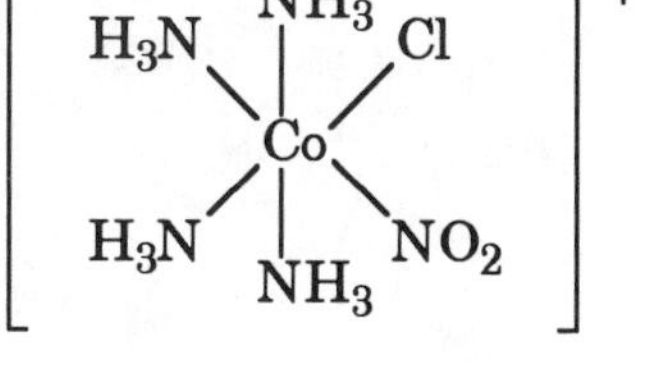   d.

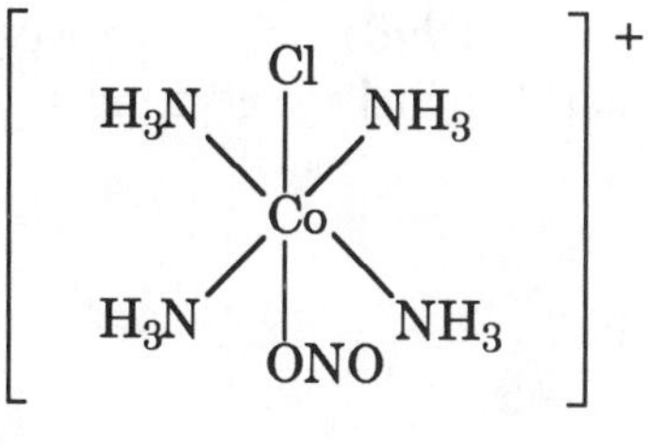

e.

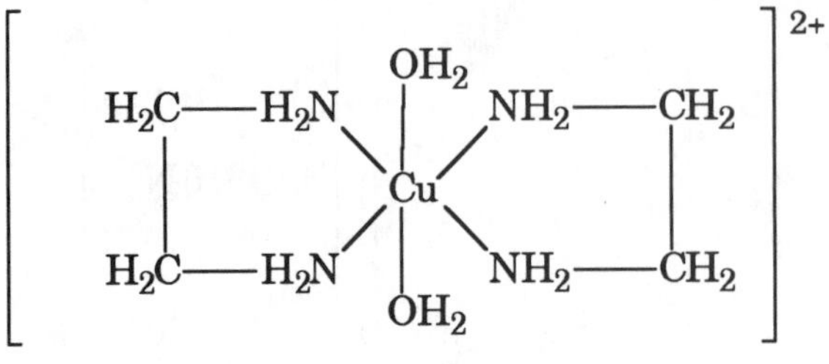

41.

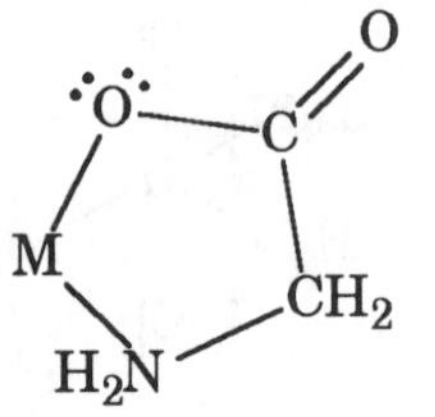

M = transition metal ion

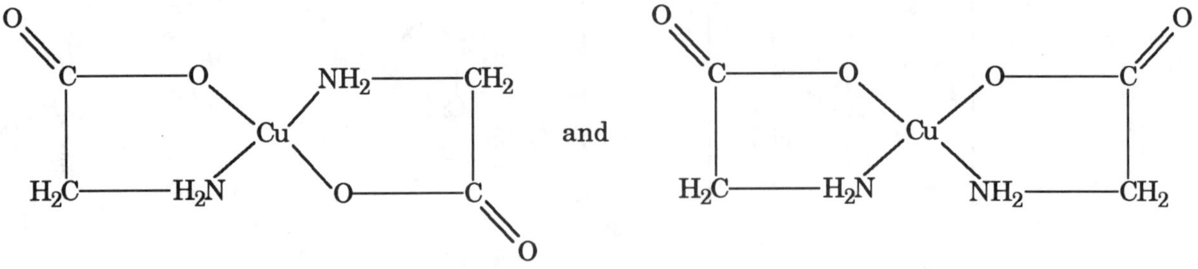

and

monodentate   bidentate

42.

43.   Linkage isomers differ in the way the ligand bonds to the metal. SCN⁻ can bond through the sulfur or through the nitrogen atom. NO_2^- can bond through the nitrogen or through the oxygen atom. OCN⁻ can bond through the oxygen or through the nitrogen atom. N_3^-, $NH_2CH_2CH_2NH_2$ and I⁻ are not capable of linkage isomerism.

44.

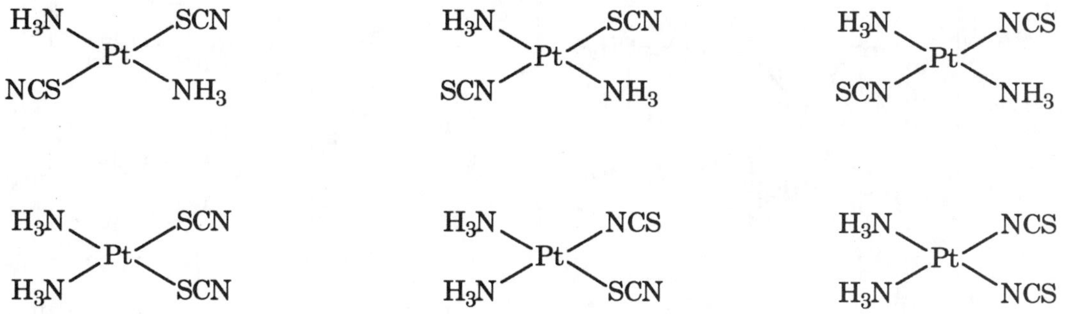

45. There are four geometrical isomers (labeled i-iv). Isomers iii and iv are optically active and the nonsuperimposable mirror images are shown.

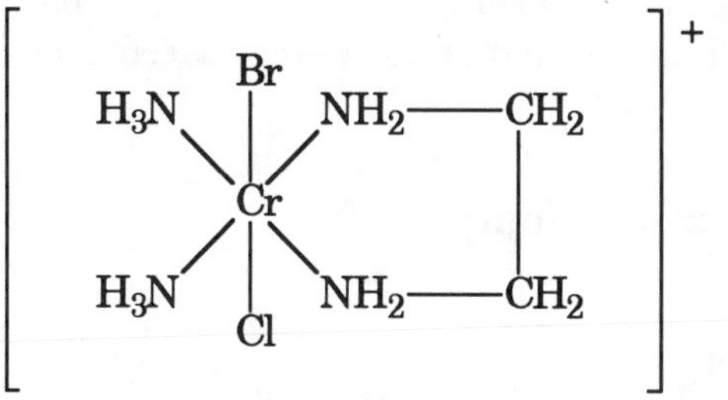

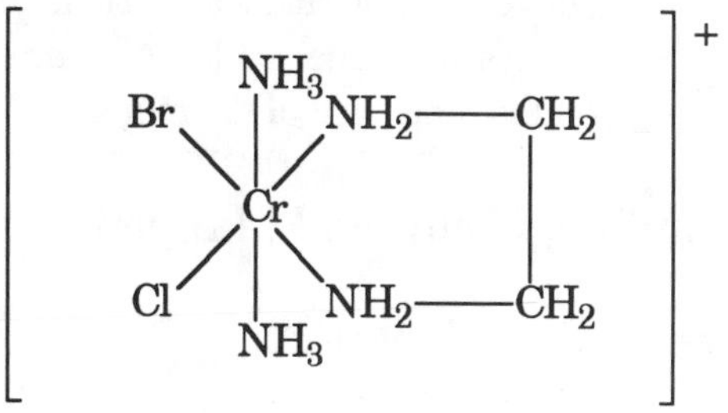

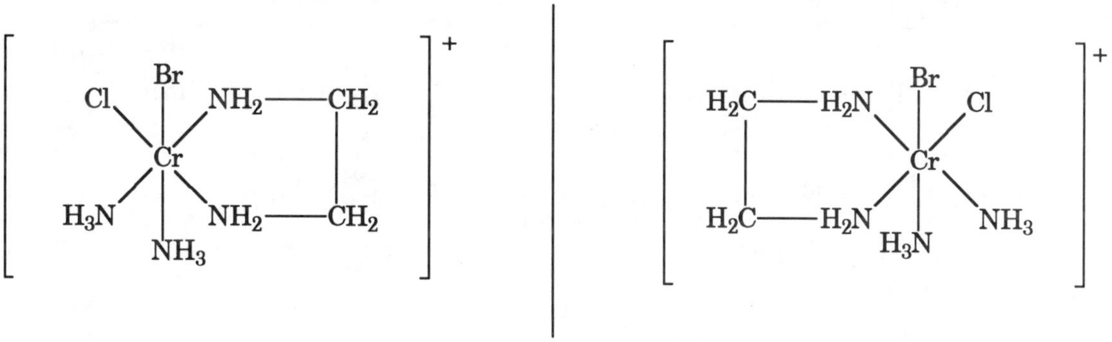

optically active   mirror   mirror image of iii
                                          (nonsuperimposable)

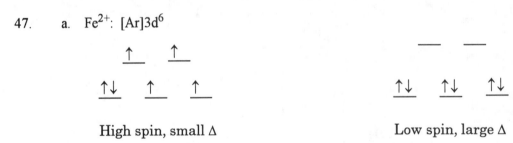

iv.

optically active   mirror   mirror image of iv
                                                              (nonsuperimposable)

46.   Similar to the molecules discussed in Figures 20.17 and 20.18 of the text, $Cr(acac)_3$ and cis-$Cr(acac)_2(H_2O)_2$ are optically active.   The mirror images of these two complexes are nonsuperimposable. There is a plane of symmetry in trans-$Cr(acac)_2(H_2O)_2$, so it is not optically active. A molecule with a plane of symmetry is never optically active as the mirror images are always superimposable. A plane of symmetry is a plane through a molecule where one side reflects on the other side of the molecule.

Bonding, Color, and Magnetism in Coordination Compounds

47.   a. Fe^{2+}: $[Ar]3d^6$

          High spin, small Δ Low spin, large Δ

       b. Fe^{3+}: $[Ar]3d^5$ c. Ni^{2+}: $[Ar]3d^8$

          High spin, small Δ

48.   a. Zn^{2+}: $[Ar]3d^{10}$

b. Co^{2+}: $[Ar]3d^7$

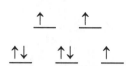

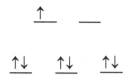

High spin, small Δ Low spin, large Δ

c. Ti^{3+}: $[Ar]3d^1$

49.   Since fluorine has a -1 charge as a ligand, then chromium has a +2 oxidation state in CrF_6^{4-}. The electron configuration of Cr^{2+} is: $[Ar]3d^4$. For four unpaired electrons, this must be a weak-field (high-spin) case where the splitting of the d-orbitals is small and the number of unpaired electrons is maximized. The crystal field diagram for this ion is:

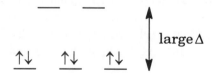

small Δ

50. NH_3 and H_2O are neutral ligands so the oxidation states of the metals are Co^{3+} and Fe^{2+}. Both have six d electrons ($[Ar]3d^6$). To explain the magnetic properties, we must have a strong-field for $Co(NH_3)_6^{3+}$ and a weak-field for $Fe(H_2O)_6^{2+}$.

Co^{3+}: $[Ar]3d^6$ Fe^{2+}: $[Ar]3d^6$

large Δ small Δ

Only this splitting of d-orbitals gives a diamagnetic $Co(NH_3)_6^{3+}$ (no unpaired electrons) and a paramagnetic $Fe(H_2O)_6^{2+}$ (unpaired electrons present).

51.   To determine the crystal field diagrams, you need to determine the oxidation state of the transition metal which can only be determined if you know the charges of the ligands (see Table 20.13). The electron configurations and the crystal field diagrams follow.

a. Ru^{2+}: $[Kr]4d^6$, no unpaired e^-

— —

↑↓   ↑↓   ↑↓

Low spin, large Δ

b. Ni^{2+}: $[Ar]3d^8$, 2 unpaired e^-

↑   ↑

↑↓   ↑↓   ↑↓

c. V^{3+}: $[Ar]3d^2$, 2 unpaired e^-

— —

↑   ↑   —

Note: Ni^{2+} must have 2 unpaired electrons, whether high-spin or low-spin, and V^{3+} must have 2 unpaired electrons, whether high-spin or low-spin.

52.   In both compounds, iron is in the +3 oxidation state with an electron configuration of $[Ar]3d^5$. Fe^{3+} complexes have one unpaired electron when a strong-field case and five unpaired electrons when a weak-field case. $Fe(CN)_6^{2-}$ is a strong-field case and $Fe(SCN)_6^{3-}$ is a weak-field case. Therefore, cyanide, CN^-, is a stronger field ligand than thiocyanate, SCN^-.

53.   Transition compounds exhibit the color complementary to that absorbed. From Table 20.16, $Ni(H_2O)_6Cl_2$ absorbs red light and $Ni(NH_3)Cl_2$ absorbs yellow-green light. $Ni(NH_3)_6Cl_2$ absorbs the shorter wavelength light which is the higher energy light ($E = hc/\lambda$). Therefore, Δ is larger for $Ni(NH_3)_6Cl_2$ which means that NH_3 is a stronger field ligand than H_2O. This is consistent with the spectrochemical series.

54.   All have cobalt in the +3 oxidation states, so any differences in light absorbtion must be due to the different ligands. The ion which absorbs light with the longest wavelength has the smallest crystal field splitting (Δ) since energy and wavelength of light are inversely proportional to each other ($E = hc/\lambda$). From the spectrochemical series in Chapter 20.6 of the text, I^- will produce the smallest Δ so CoI_6^{3-} will absorb light with the longest wavelength.

55. $CoCl_4^{2-}$; $CoCl_6^{4-}$ has an octahedral molecular structure and $CoCl_4^{2-}$ has a tetrahedral molecular structure (as are most Co^{2+} complexes with four ligands). Coordination complexes absorb light of energy equal to the energy difference between the split d-orbitals. Since the tetrahedral d-orbital splitting is less than one-half of the octahedral d-orbital splitting, then tetrahedral complexes will absorb lower energy light which corresponds to light with longer wavelengths ($E = hc/\lambda$).

56. Co^{2+}: $[Ar]3d^7$; The corresponding d-orbital splitting diagram for tetrahedral Co^{2+} complexes is:

↑   ↑   ↑

↑↓   ↑↓

All tetrahedral complexes are high spin since the d-orbital splitting is small. Ions with 2 or 7 d electrons should give the most stable tetrahedral complexes since they have the greatest number of electrons in the lower energy orbitals as compared to the number of electrons in the higher energy orbitals.

57.   Since the ligands are Cl⁻, then iron is in the +3 oxidation state. Fe^{3+}: $[Ar]3d^5$

Since all tetrahedral complexes are high spin, then there are 5 unpaired electrons in $FeCl_4^-$.

58.   In both complexes, nickel is in the +2 oxidation state: Ni^{2+}: $[Ar]3d^8$. The differences in unpaired electrons must be due to differences in molecular structure. $NiCl_4^{2-}$ is a tetrahedral complex and $Ni(CN)_4^{2-}$ is a square planar complex. The corresponding d-orbital splitting diagrams are:

$NiCl_4^{2-}$ $Ni(CN)_4^{2-}$

Metallurgy

59.   a. $\Delta H° = 2(-1117) + (-393.5) - [3(-826) + (-110.5)] = -39$ kJ

           $\Delta S° = 2(146) + 214 - [3(90.) + 198] = 38$ J/K

       b. $\Delta G° = \Delta H° - T\Delta S°$; T = 800. + 273 = 1073 K

           $\Delta G° = -39$ kJ $- 1073$ K$(0.038$ kJ/K$) = -39$ kJ $- 41$ kJ $= -80.$ kJ

60. $3 Fe + C \rightarrow Fe_3C$; $\Delta H° = 21 - [3(0) + 0] = 21$ kJ

       $\Delta S° = 108 - [3(27) + 6] = 21$ J/K

       $\Delta G° = \Delta H° - T\Delta S°$; When $\Delta H°$ and $\Delta S°$ are both positive, then the reaction is spontaneous at high temperatures where the favorable $\Delta S°$ term becomes dominant. Thus, to incorporate carbon into steel, high temperatures are needed for thermodynamic reasons but will also be beneficial for kinetic reasons (as temperature increases, the rate of the reaction will increase). The relative amount of Fe_3C (cementite) which remains in the steel depends on the cooling process. If the steel is cooled slowly, there is time for the equilibrium to shift back to the left; small crystals are of carbon form giving a relatively ductile steel. If cooling is rapid, there is not enough time for the equilibrium to shift back to the left; Fe_3C is still present in the steel and the steel is more brittle. Which cooling process occurs depends on the desired properties of the steel. The process of tempering fine-tunes the steel to the desired properties by repeated heating and cooling.

Additional Exercises

61.   a.   4 O atoms on faces × 1/2 O/face = 2 O atoms, 2 O atoms inside body, Total: 4 O atoms

8 Ti atoms on corners × 1/8 Ti/corner + 1 Ti atom/body center = 2 Ti atoms

Formula of the unit cell is Ti_2O_4. The empirical formula is TiO_2.

b. $$\overset{+4\ -2}{2\ TiO_2} + \overset{0}{3\ C} + \overset{0}{4\ Cl_2} \rightarrow \overset{+4\ -1}{2\ TiCl_4} + \overset{+4\ -2}{CO_2} + \overset{+2\ -2}{2\ CO}$$

Cl is reduced and C is oxidized. Cl_2 is the oxidizing agent and C is the reducing agent.

$$\overset{+4\ -1}{TiCl_4} + \overset{0}{O_2} \rightarrow \overset{+4\ -2}{TiO_2} + \overset{0}{2\ Cl_2}$$

O is reduced and Cl is oxidized. O_2 is the oxidizing agent and $TiCl_4$ is the reducing agent.

62.   a. $2\ CoAs_2(s) + 4\ O_2(g) \rightarrow 2\ CoO(s) + As_4O_6(s)$; As_4O_6, tetraarsenic hexoxide

b.   Using the half-reaction method to balance:

$$2\ e^- + 2\ H^+ + OCl^- \rightarrow Cl^- + H_2O$$
$$(3\ H_2O + Co^{2+} \rightarrow Co(OH)_3 + 3\ H^+ + e^-) \times 2$$

$$\overline{5\ H_2O + 2\ Co^{2+} + OCl^- \rightarrow Cl^- + 2\ Co(OH)_3 + 4\ H^+}$$ Add 4 OH^- to both sides then simplify:

$$4\ OH^-(aq) + H_2O(l) + 2\ Co^{2+}(aq) + OCl^-(aq) \rightarrow Cl^-(aq) + 2\ Co(OH)_3(s)$$

63. $(Au(CN)_2^- + e^- \rightarrow Au + 2\ CN^-) \times 2$ $E° = -0.60\ V$
          $Zn + 4\ CN^- \rightarrow Zn(CN)_4^{2-} + 2\ e^-$ $-E° = 1.26\ V$

$$\overline{2\ Au(CN)_2^-(aq) + Zn(s) \rightarrow 2\ Au(s) + Zn(CN)_4^{2-}(aq)}\qquad E°_{cell} = 0.66\ V$$

$\Delta G° = -nFE°_{cell} = -(2\ mol\ e^-)(96{,}485\ C/mol\ e^-)(0.66\ J/C) = -1.3 \times 10^5\ J = -130\ kJ$

$$E° = \frac{0.0592}{n}\ \log K,\ \ \log K = \frac{nE°}{0.0592} = \frac{2(0.66)}{0.0592} = 22.30,\ \ K = 10^{22.30} = 2.0 \times 10^{22}$$

Note: We carried extra significant figures to determine K.

64. $0.112 \text{ g Eu}_2\text{O}_3 \times \dfrac{304.0 \text{ g Eu}}{352.0 \text{ g Eu}_2\text{O}_3} = 0.0967 \text{ g Eu}; \quad \% \text{ Eu} = \dfrac{0.0967 \text{ g}}{0.286 \text{ g}} \times 100 = 33.8\% \text{ Eu}$

$\% \text{ O} = 100.00 - (33.8 + 40.1 + 4.71) = 21.4\% \text{ O}$

Out of 100.00 g of compound:

$33.8 \text{ g Eu} \times \dfrac{1 \text{ mol}}{152.0 \text{ g}} = 0.222 \text{ mol Eu}; \quad 40.1 \text{ g C} \times \dfrac{1 \text{ mol}}{12.01 \text{ g}} = 3.34 \text{ mol C}$

$4.71 \text{ g H} \times \dfrac{1 \text{ mol}}{1.008 \text{ g}} = 4.67 \text{ mol H}; \quad 21.4 \text{ g O} \times \dfrac{1 \text{ mol}}{16.00 \text{ g}} = 1.34 \text{ mol O}$

$\dfrac{3.34}{0.222} = 15.0, \quad \dfrac{4.67}{0.222} = 21.0, \quad \dfrac{1.34}{0.222} = 6.04$

The empirical and molecular formula is $\text{EuC}_{15}\text{H}_{21}\text{O}_6$. Since each acac⁻ is $\text{C}_5\text{H}_7\text{O}_2{}^-$, then an abbreviated molecular formula is Eu(acac)_3.

65.   a.   2; Forms bonds through the lone pairs on the two oxygen atoms.

        b.   3; Forms bonds through the lone pairs on the three nitrogen atoms.

        c.   4; Forms bonds through the two nitrogen atoms and the two oxygen atoms.

        d.   4; Forms bonds through the four nitrogen atoms.

66.   a.   In the following structures, we omitted the 4 NH_3 ligands coordinated to the outside cobalt atoms.

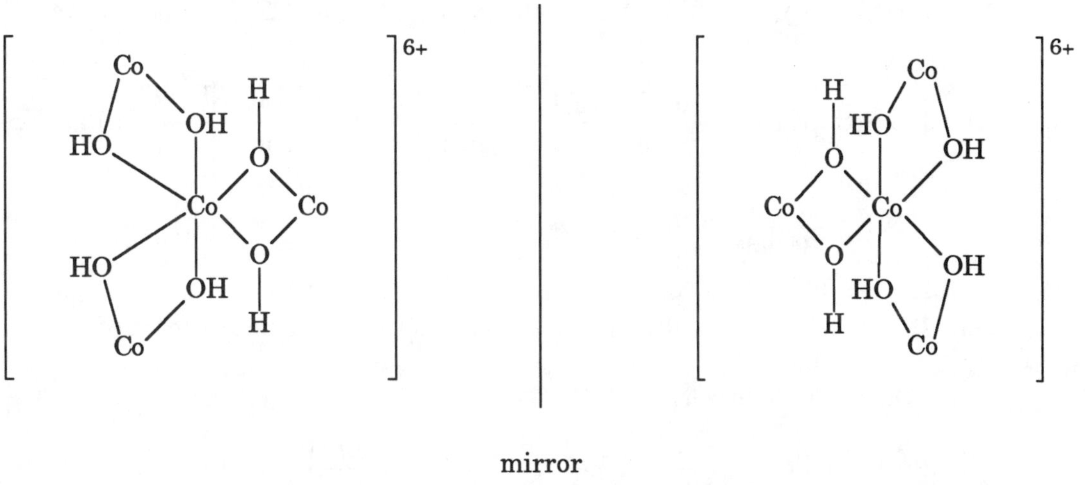

mirror

        b.   All are Co(III). The three "ligands" each contain 2 OH⁻ and 4 NH_3 groups. If each cobalt is in the +3 oxidation state, then each ligand has a +1 overall charge. The +3 charge from the three ligands along with the +3 charge of the central cobalt atom gives the overall complex a +6 charge. This is balanced by the -6 charge of the six Cl⁻ ions.

c. Co^{3+}: $[Ar]3d^6$; There are zero unpaired electrons if a low-spin (strong-field) case.

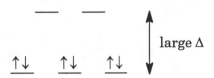

67. a. $Ru(phen)_3^{2+}$ exhibits optical isomerism [similar to $Co(en)_3^{3+}$ in Figure 20.17 of the text].

b. Ru^{2+}: $[Kr]4d^6$; Since there are no unpaired electrons, then Ru^{2+} is a strong-field (low-spin) case.

68. Octahedral Cr^{2+} complexes should be used. Cr^{2+}: $[Ar]3d^4$; High-spin (weak-field) Cr^{2+} complexes have 4 unpaired electrons and low-spin (strong-field) Cr^{2+} complexes have 2 unpaired electrons. Ni^{2+}: $[Ar]3d^8$; Octahedral Ni^{2+} complexes will always have 2 unpaired electrons, whether high or low-spin. Therefore, Ni^{2+} complexes cannot be used to distinguish weak from strong-field ligands by examining magnetic properties. Alternatively, the ligand field strengths can be measured using visible spectra. Either Cr^{2+} or Ni^{2+} complexes can be used for this method.

Challenge Problems

69. i. $0.0203 \text{ g } CrO_3 \times \dfrac{52.00 \text{ g Cr}}{100.0 \text{ g } CrO_3} = 0.0106 \text{ g Cr};$ $\% \text{ Cr} = \dfrac{0.0106}{0.105} \times 100 = 10.1\% \text{ Cr}$

ii. $32.93 \text{ mL HCl} \times \dfrac{0.100 \text{ mmol HCl}}{mL} \times \dfrac{1 \text{ mmol } NH_3}{mmol \text{ HCl}} \times \dfrac{17.03 \text{ mg } NH_3}{mmol} = 56.1 \text{ mg } NH_3$

$\% \text{ } NH_3 = \dfrac{56.1 \text{ mg}}{341 \text{ mg}} \times 100 = 16.5\% \text{ } NH_3$

iii. $73.53\% + 16.5\% + 10.1\% = 100.1\%$; The compound must be composed of only Cr, NH_3, and I.

Out of 100.00 of compound:

$10.1 \text{ g Cr} \times \dfrac{1 \text{ mol}}{52.00 \text{ g}} = 0.194$ $\qquad \dfrac{0.194}{0.194} = 1.00$

$16.5 \text{ g } NH_3 \times \dfrac{1 \text{ mol}}{17.03 \text{ g}} = 0.969$ $\qquad \dfrac{0.969}{0.194} = 4.99$

$73.53 \text{ g I} \times \dfrac{1 \text{ mol}}{126.9 \text{ g}} = 0.5794$ $\qquad \dfrac{0.5794}{0.194} = 2.99$

$Cr(NH_3)_5I_3$ is the empirical formula. Cr(III) forms octahedral complexes. So compound A is made of the octahedral $[Cr(NH_3)_5I]^{2+}$ complex ion and two I^- ions or $[Cr(NH_3)_5I]I_2$. Lets check this proposed formula using the freezing point data.

iv. $\Delta T_f = iK_f m$; For $[Cr(NH_3)_5I]I_2$, i = 3.0 (assuming complete dissociation).

$$m = \frac{0.601 \text{ g complex}}{1.000 \times 10^{-2} \text{ kg H}_2\text{O}} \times \frac{1 \text{ mol complex}}{517.9 \text{ g complex}} = 0.116 \text{ molal}$$

$\Delta T_f = 3.0 \times 1.86°C/\text{molal} \times 0.116 \text{ molal} = 0.65°C$

Since ΔT_f is close to the measured value, then this is consistent with the formula $[Cr(NH_3)_5I]I_2$.

70. $\underset{\text{II}}{(H_2O)_5Cr} - Cl - \underset{\text{III}}{Co(NH_3)_5} \rightarrow \underset{\text{III}}{(H_2O)_5Cr} - Cl - \underset{\text{II}}{Co(NH_3)_5} \rightarrow Cr(H_2O)_5Cl^{2+} + Co(II) \text{ complex}$

Yes; After the oxidation, the ligands on Cr(III) won't exchange. Since Cl^- is in the coordination sphere, then it must have formed a bond to Cr(II) before the electron transfer occurred (as proposed through the formation of the intermediate).

71.   No; In all three cases, six bonds are formed between Ni^{2+} and nitrogen, so ΔH values should be similar. $\Delta S°$ for formation of the complex ion is most negative for 6 NH_3 molecules reacting with a metal ion (7 independent species become 1). For penten reacting with a metal ion, 2 independent species become 1, so $\Delta S°$ is least negative of all three of the reactions. Thus, the chelate effect occurs because the more bonds a chelating agent can form to the metal, the more favorable $\Delta S°$ is for the formation of the complex ion and the larger the formation constant.

72.

The $d_{x^2-y^2}$ and d_{xy} orbitals are in the plane of the three ligands and should be destabilized the most. The amount of destabilization should be about equal when all the possible interactions are considered. The d_{z^2} orbital has some electron density in the xy plane (the doughnut) and should be destabilized a lesser amount as compared to the $d_{x^2-y^2}$ and d_{xy} orbitals. The d_{xz} and d_{yz} orbitals have no electron density in the plane and should be lowest in energy.

73.

The d_{z^2} orbital will be destabilized much more than in the trigonal planar case (see Exercise 20.72). The d_{z^2} orbital has electron density on the z-axis directed at the two axial ligands. The $d_{x^2-y^2}$ and d_{xy} orbitals are in the plane of the three trigonal planar ligands and should be destabilized a lesser amount as compared to the d_{z^2} orbital; only a portion of the electron density in the $d_{x^2-y^2}$ and d_{xy} orbitals is directed at the ligands. The d_{xz} and d_{yz} orbitals will be destabilized the least since the electron density is directed between the ligands.

74.

$$AgBr(s) \rightleftharpoons Ag^+ + Br^- \qquad\qquad K_{sp} = 5.0 \times 10^{-13}$$
$$Ag^+ + 2\,NH_3 \rightleftharpoons Ag(NH_3)_2^+ \qquad\qquad K_f = 1.7 \times 10^7$$

$$\overline{AgBr(s) + 2\,NH_3 \rightleftharpoons Ag(NH_3)_2^+ + Br^- \qquad K = 5.0 \times 10^{-13} \times 1.7 \times 10^7 = 8.5 \times 10^{-6}}$$

$$AgBr(s) \;+\; 2\,NH_3 \;\;\rightleftharpoons\;\; Ag(NH_3)_2^+ \;+\; Br^-$$

Initial		$3.0\,M$	0	0

If 0.10 mol of AgBr dissolves in 1.0 L, then the concentration changes are:

Change	-0.10	-0.20	$\rightarrow$	+0.10	+0.10
After		$2.8\,M$		$0.10\,M$	$0.10\,M$

If all the AgBr(s) dissolves, these would be the concentrations after dissolution. To check if all the AgBr(s) dissolves, use the reaction quotient.

$$Q = \frac{[Ag(NH_3)_2^+]_o[Br^-]_o}{[NH_3]_o^2} = \frac{(0.10)(0.10)}{(2.8)^2} = 1.3 \times 10^{-3}$$

Since $Q > K$ (8.5×10^{-6}), then the reaction will shift left producing AgBr(s). No, 0.10 mol of AgBr will not dissolve in 1.0 L of $3.0\,M$ NH$_3$.

CHAPTER TWENTY-ONE

THE NUCLEUS: A CHEMIST'S VIEW

Questions

1.   Fission: Splitting of a heavy nucleus into two (or more) lighter nuclei.

       Fusion: Combining two light nuclei to form a heavier nucleus.

       Fusion is more likely for elements lighter than Fe; fission is more likely for elements heavier than Fe.

2.   Characteristic frequencies of energies emitted in a nuclear reaction suggests that discrete energy levels exist in the nucleus. Extra stability of certain numbers of nucleons and the predominance of nuclei with even numbers of nucleons suggests that the nuclear structure might be described by using quantum numbers.

3.   The assumptions are that the ^{14}C level in the atmosphere is constant or that the ^{14}C level at the time the plant died can be calculated. A constant ^{14}C level is a poor assumption and accounting for variation is complicated. Another problem is that some of the material must be destroyed to determine the ^{14}C level.

4.   A nonradioactive substance can be put in equilibrium with a radioactive substance. The two materials can then be checked to see if all the radioactivity remains in the original material or if it has been scrambled by the equilibrium.

5.   No, coal fired power plants also pose risks. A partial list of risks are:

Coal	Nuclear
Air pollution	Radiation exposure to workers
Coal mine accidents	Disposal of wastes
Health risks to miners	Meltdown
(black lung disease)	Terrorists
	Public fear

6.   The maximum binding energy per nucleon occurs at about Fe. Smaller nuclei become more stable by fusing to form heavier nuclei. Larger nuclei form more stable nuclei by splitting to form lighter nuclei.

7.   For fusion reactions, a collision of sufficient energy must occur between two positively charged particles to initiate the reaction. This requires high temperatures. In fission, an electrically neutral neutron collides with the positively charged nucleus. This has a much lower activation energy.

8.   Moderator: Slows the neutrons to increase the efficiency of the fission reaction.

      Control rods: Absorbs neutrons to slow or halt the fission reaction.

Exercises

Radioactive Decay and Nuclear Transformations

9.   All nuclear reactions must be charge balanced and mass balanced. To charge balance, balance the sum of the atomic numbers on each side of the reaction and to mass balance, balance the sum of the mass numbers on each side of the reaction.

      a. $^{51}_{24}\text{Cr} + ^{\ 0}_{-1}\text{e} \rightarrow ^{51}_{23}\text{V}$

      b. $^{131}_{\ 53}\text{I} \rightarrow ^{\ 0}_{-1}\text{e} + ^{131}_{\ 54}\text{Xe}$

10.   a. $^{32}_{15}\text{P} \rightarrow ^{\ 0}_{-1}\text{e} + ^{32}_{16}\text{S}$

      b. $^{235}_{\ 92}\text{U} \rightarrow ^{4}_{2}\text{He} + ^{231}_{\ 90}\text{Th}$

11.   a. $^{3}_{1}\text{H} \rightarrow ^{3}_{2}\text{He} + ^{\ 0}_{-1}\text{e}$ c. $^{7}_{4}\text{Be} + ^{\ 0}_{-1}\text{e} \rightarrow ^{7}_{3}\text{Li}$

      b. $^{8}_{3}\text{Li} \rightarrow ^{8}_{4}\text{Be} + ^{\ 0}_{-1}\text{e}$ d. $^{8}_{5}\text{B} \rightarrow ^{8}_{4}\text{Be} + ^{\ 0}_{+1}\text{e}$

            $^{8}_{4}\text{Be} \rightarrow 2\,^{4}_{2}\text{He}$ e. $^{32}_{15}\text{P} \rightarrow ^{32}_{16}\text{S} + ^{\ 0}_{-1}\text{e}$

            ───────────────────

            $^{8}_{3}\text{Li} \rightarrow 2\,^{4}_{2}\text{He} + ^{\ 0}_{-1}\text{e}$

12.   a. $^{60}_{27}\text{Co} \rightarrow ^{60}_{28}\text{Ni} + ^{\ 0}_{-1}\text{e}$ b. $^{97}_{43}\text{Tc} + ^{\ 0}_{-1}\text{e} \rightarrow ^{97}_{42}\text{Mo}$

      c. $^{99}_{43}\text{Tc} \rightarrow ^{99}_{44}\text{Ru} + ^{\ 0}_{-1}\text{e}$ d. $^{239}_{\ 94}\text{Pu} \rightarrow ^{235}_{\ 92}\text{U} + ^{4}_{2}\text{He}$

13.   a. $^{207}_{\ 82}\text{Pb}$; Complete decay is 7α and 4β particles: $^{235}_{\ 92}\text{U} \rightarrow ^{207}_{\ 82}\text{Pb} + 7\,^{4}_{2}\text{He} + 4\,^{\ 0}_{-1}\text{e}$

b. $^{235}_{92}U \rightarrow ^{231}_{90}Th + ^4_2He \rightarrow ^{231}_{91}Pa + ^0_{-1}e \rightarrow ^{227}_{89}Ac + ^4_2He$

$^{215}_{84}Po + ^4_2He \leftarrow ^{219}_{86}Rn + ^4_2He \leftarrow ^{223}_{88}Ra + ^4_2He \leftarrow ^{227}_{90}Th + ^0_{-1}e$

$^4_2He + ^{211}_{82}Pb \rightarrow ^{211}_{83}Bi + ^0_{-1}e \rightarrow ^{207}_{81}Tl + ^4_2He \rightarrow ^{207}_{82}Pb + ^0_{-1}e$

The intermediate nucludes are:

$^{231}_{90}Th, ^{231}_{91}Pa, ^{227}_{89}Ac, ^{227}_{90}Th, ^{223}_{88}Ra, ^{219}_{86}Rn, ^{215}_{84}Po, ^{211}_{82}Pb, ^{211}_{83}Bi$ and $^{207}_{81}Tl$.

14.   a. $^{241}_{95}Am \rightarrow ^4_2He + ^{237}_{93}Np$

b. $^{241}_{95}Am \rightarrow 8\ ^4_2He + 4\ ^0_{-1}e + ^{209}_{83}Bi$; The final product is $^{209}_{83}Bi$.

c. $^{241}_{95}Am \rightarrow ^{237}_{93}Np + \alpha \rightarrow ^{233}_{91}Pa + \alpha \rightarrow ^{233}_{92}U + \beta \rightarrow ^{229}_{92}Th + \alpha \rightarrow ^{225}_{88}Ra + \alpha$

$^{213}_{84}Po + \beta \leftarrow ^{213}_{83}Bi + \alpha \leftarrow ^{217}_{85}At + \alpha \leftarrow ^{221}_{87}Fr + \alpha \leftarrow ^{225}_{89}Ac + \beta$

$^{209}_{82}Pb + \alpha \rightarrow ^{209}_{83}Bi + \beta$

The intermediate radionuclides are:

$^{237}_{93}Np, ^{233}_{91}Pa, ^{233}_{92}U, ^{229}_{90}Th, ^{225}_{88}Ra, ^{225}_{89}Ac, ^{221}_{87}Fr, ^{217}_{85}At, ^{213}_{83}Bi, ^{213}_{84}Po$, and $^{209}_{82}Pb$.

15.   Reference Table 21.2 of the text for potential radioactive decay processes. ^8B and ^9B contain too many protons or too few neutrons. Electron capture or positron production are both possible decay mechanisms that increase the neutron to proton ratio. Alpha particle production also increases the neutron to proton ratio, but it is not likely for these light nuclei. ^{12}B and ^{13}B contain too many neutrons or too few protons. Beta production lowers the neutron to proton ratio, so we expect ^{12}B and ^{13}B to be β-emitters.

16. $^{53}_{26}Fe$ has too many protons. It will undergo either positron production, electron capture and/or alpha particle production. $^{59}_{26}Fe$ has too many neutrons and will undergo beta particle production. (See Table 21.2 of the text.)

17.   a. $^1_1H + ^{14}_7N \rightarrow ^{11}_6C + ^4_2He$ b. $2\ ^3_2He \rightarrow ^4_2He + 2\ ^1_1H$

c. $^1_1H + ^1_1H \rightarrow ^2_1H + ^0_{+1}e$ (positron)   d. $^1_1H + ^{12}_6C \rightarrow ^{13}_7N$

18.   a. $_{95}^{240}\text{Am} + _{2}^{4}\text{He} \rightarrow _{97}^{243}\text{Bk} + _{0}^{1}\text{n}$ b. $_{92}^{238}\text{U} + _{6}^{12}\text{C} \rightarrow _{98}^{244}\text{Cf} + 6\,_{0}^{1}\text{n}$

       c. $_{98}^{249}\text{Cf} + _{8}^{18}\text{O} \rightarrow _{106}^{263}\text{Unh} + 4\,_{0}^{1}\text{n}$ d. $_{98}^{249}\text{Cf} + _{5}^{10}\text{B} \rightarrow _{103}^{257}\text{Lr} + 2\,_{0}^{1}\text{n}$

Kinetics of Radioactive Decay

19.   All radioactive decay follows first-order kinetics where $t_{1/2} = (\ln 2)/k$.

$$t_{1/2} = \frac{\ln 2}{k} = \frac{0.693}{1.0 \times 10^{-3}\,\text{h}^{-1}} = 690\,\text{h}$$

20.   For $t_{1/2} = 12{,}000$ yr:

$$k = \frac{\ln 2}{t_{1/2}} = \frac{0.693}{t_{1/2}} = \frac{0.693}{12{,}000\,\text{yr}} \times \frac{1\,\text{yr}}{365\,\text{d}} \times \frac{1\,\text{d}}{24\,\text{h}} \times \frac{1\,\text{h}}{3600\,\text{s}} = 1.8 \times 10^{-12}\,\text{s}^{-1}$$

$$\text{Rate} = kN = 1.8 \times 10^{-12}\,\text{s}^{-1} \times 6.02 \times 10^{23}\,\text{nuclei} = 1.1 \times 10^{12}\,\text{decays/s}$$

For $t_{1/2} = 12$ h:

$$k = \frac{0.693}{t_{1/2}} = \frac{0.693}{12\,\text{h}} \times \frac{1\,\text{h}}{3600\,\text{s}} = 1.6 \times 10^{-5}\,\text{s}^{-1}$$

$$\text{Rate} = kN = 1.6 \times 10^{-5}\,\text{s}^{-1} \times 6.02 \times 10^{23}\,\text{nuclei} = 9.6 \times 10^{18}\,\text{decays/s}$$

For $t_{1/2} = 12$ s:

$$\text{Rate} = kN = \frac{0.693}{12\,\text{s}} \times 6.02 \times 10^{23}\,\text{nuclei} = 3.5 \times 10^{22}\,\text{decays/s}$$

21.   Half-life for first-order kinetics is independent of concentration (or number of nuclides). To go from 1.00×10^{20} to 2.50×10^{19} nuclides represents 2 half-lives.

$$1.00 \times 10^{20} \underset{t_{1/2}}{\rightarrow} 5.00 \times 10^{19} \underset{t_{1/2}}{\rightarrow} 2.50 \times 10^{19}; \quad 2 \times t_{1/2} = 10.0\,\text{min},\ t_{1/2} = 5.00\,\text{min}$$

22.   12.5% of ^{60}Co remains. This decay represents 3 half-lives.

$$100.0\% \underset{t_{1/2}}{\rightarrow} 50.0\% \underset{t_{1/2}}{\rightarrow} 25.0\% \underset{t_{1/2}}{\rightarrow} 12.5\%; \quad \text{time} = 3 \times 5.26\,\text{yr} = 15.8\,\text{yr}$$

23.   Units for N and N_o are usually number of nuclei but can also be grams if the units are the same for both N and N_o. In this problem $m_o =$ the initial mass of $^{47}\text{Ca}^{2+}$ to be ordered.

$$k = \frac{\ln 2}{t_{1/2}};\ \ln\left(\frac{N}{N_o}\right) = -kt = \frac{-0.693\,t}{t_{1/2}},\ \ln\left(\frac{5.0\,\mu\text{g Ca}^{2+}}{m_o}\right) = \frac{-0.693\,(2.0\,\text{d})}{4.5\,\text{d}} = -0.31$$

$\dfrac{5.0}{m_o} = e^{-0.31} = 0.73, \; m_o = 6.8 \; \mu g$ of $^{47}Ca^{2+}$ needed initially

$6.8 \; \mu g \; ^{47}Ca^{2+} \times \dfrac{107.0 \; \mu g \; ^{47}CaCO_3}{47.0 \; \mu g \; ^{47}Ca^{2+}} = 15 \; \mu g \; ^{47}CaCO_3$ should be ordered at the minimum.

24.   a. $0.0100 \; Ci \times \dfrac{3.7 \times 10^{10} \text{ decays/s}}{Ci} = 3.7 \times 10^8 \text{ decays/s}; \; k = \dfrac{\ln 2}{t_{1/2}}$

Rate $= kN, \quad \dfrac{3.7 \times 10^8 \text{ decays}}{s} = \left(\dfrac{0.6931}{2.87 \, h} \times \dfrac{1 \, h}{3600 \, s} \right) \times N, \; N = 5.5 \times 10^{12}$ atoms of ^{38}S

5.5×10^{12} atoms $^{38}S \times \dfrac{1 \text{ mol } ^{38}S}{6.02 \times 10^{23} \text{ atoms}} \times \dfrac{1 \text{ mol } Na_2{}^{38}SO_4}{\text{mol } ^{38}S} = 9.1 \times 10^{-12} \text{ mol } Na_2{}^{38}SO_4$

$9.1 \times 10^{-12} \text{ mol } Na_2{}^{38}SO_4 \times \dfrac{148.0 \text{ g } Na_2{}^{38}SO_4}{\text{mol } Na_2{}^{38}SO_4} = 1.3 \times 10^{-9} \text{ g} = 1.3 \text{ ng } Na_2{}^{38}SO_4$

b.   99.99% decays, 0.01% left; $\ln \left(\dfrac{0.01}{100} \right) = -kt = \dfrac{-0.6931 \, t}{2.87 \, h}, \; t = 38.1$ hours ≈ 40 hours

25. $t = 52.0 \text{ yr}; \; k = \dfrac{\ln 2}{t_{1/2}}; \; \ln \left(\dfrac{N}{N_o} \right) = -kt = \dfrac{-0.6931 \times 52.0 \text{ yr}}{28.8 \text{ yr}} = -1.25, \; \left(\dfrac{N}{N_o} \right) = e^{-1.25} = 0.287$

28.7% of the ^{90}Sr remains as of July 16, 1997.

26. $\ln(N/N_o) = -kt; \; N = 0.020 \, N_o; \; t_{1/2} = (\ln 2)/k, \; k = 0.6931/12.3 \text{ yr} = 0.0563 \text{ yr}^{-1}$

$\ln(0.020) = -(0.0563 \text{ yr}^{-1})t, \; t = 69 \text{ yr}$

27. $t_{1/2} = 5730 \text{ yr}; \; k = (\ln 2)/t_{1/2}; \; \ln(N/N_o) = -kt$

$\ln \left(\dfrac{N}{N_o} \right) = \dfrac{-0.6931 \, t}{t_{1/2}} = \dfrac{-0.6931 \times 2200 \text{ yr}}{5730 \text{ yr}} = -0.27, \; \dfrac{N}{N_o} = e^{-0.27} = 0.76 = 76\%$ of ^{14}C remains

28. $t_{1/2} = 5730 \text{ yr}; \; k = (\ln 2)/t_{1/2}; \; \ln(N/N_o) = -kt; \; \ln \dfrac{15.1}{15.3} = \dfrac{-(\ln 2) \, t}{5730 \text{ yr}}, \; t = 109 \text{ yr}$

No; From ^{14}C dating, the painting was produced (at the earliest) during the late 1800s.

29.   Since 4.5×10^9 years is equal to the half-life, then one-half of the ^{238}U atoms will have been converted to ^{206}Pb. The numbers of atoms of ^{206}Pb and ^{238}U will be equal. Thus, the mass ratio is equal to the molar mass ratio:

$\dfrac{206}{238} = 0.866$

30.   a.   The decay of ^{40}K is not the sole source of ^{40}Ca.

b.   Decay of ^{40}K is the sole source of ^{40}Ar and that no ^{40}Ar is lost over the years.

c. $\dfrac{0.95 \text{ g } ^{40}\text{Ar}}{1.00 \text{ g } ^{40}\text{K}}$ = current mass ratio

0.95 g of ^{40}K decayed to ^{40}Ar. 0.95 g of ^{40}K is only 10.7% of the total ^{40}K that decayed, or:

0.107 (m) = 0.95 g, m = 8.9 g = total mass of ^{40}K that decayed

Mass of ^{40}K when the rock was formed was 1.00 g + 8.9 g = 9.9 g.

$\ln\left(\dfrac{1.00 \text{ g } ^{40}\text{K}}{9.9 \text{ g } ^{40}\text{K}}\right) = -kt = \dfrac{-(\ln 2)\, t}{t_{1/2}} = \dfrac{-0.6931\, t}{1.27 \times 10^9 \text{ yr}}$, t = 4.2 × 10^9 years old

d.   If some ^{40}Ar escaped then the measured ratio of ^{40}Ar/^{40}K is less than it should be. We would calculate the age of the rocks to be less than it actually is.

Energy Changes in Nuclear Reactions

31. $\Delta E = \Delta mc^2$, $\Delta m = \dfrac{\Delta E}{c^2} = \dfrac{3.9 \times 10^{23} \text{ kg m}^2/\text{s}^2}{(3.00 \times 10^8 \text{ m/s})^2} = 4.3 \times 10^6$ kg

The sun loses 4.3 × 10^6 kg of mass each second. Note: 1 J = 1 kg m^2/s^2

32. $\dfrac{1.8 \times 10^{14} \text{ kJ}}{\text{s}} \times \dfrac{1000 \text{ J}}{\text{kJ}} \times \dfrac{3600 \text{ s}}{\text{h}} \times \dfrac{24 \text{ h}}{\text{day}} = 1.6 \times 10^{22}$ J/day

$\Delta E = \Delta mc^2$, $\Delta m = \dfrac{\Delta E}{c^2} = \dfrac{1.6 \times 10^{22} \text{ J}}{(3.00 \times 10^8 \text{ m/s})^2} = 1.8 \times 10^5$ kg of solar material provides 1 day of solar energy to the earth.

$1.6 \times 10^{22} \text{ J} \times \dfrac{1 \text{ kJ}}{1000 \text{ J}} \times \dfrac{1 \text{ g}}{32 \text{ kJ}} \times \dfrac{1 \text{ kg}}{1000 \text{ g}} = 5.0 \times 10^{14}$ kg of coal is needed to provide the same amount of energy.

33.   We need to determine the mass defect, Δm, between the mass of the nucleus and the mass of the individual parts that make up the nucleus. Once Δm is known, we can then calculate ΔE (the binding energy) using $E = mc^2$. Note: 1 J = 1 kg m^2/s^2.

For $^{232}_{94}$Pu (94 e, 94 p, 138 n):

mass of ^{232}Pu nucleus = 3.85285 × 10^{-22} g - mass of 94 electrons

mass of ^{232}Pu nucleus = 3.85285 × 10^{-22} g - 94(9.10939 × 10^{-28}) g = 3.85199 × 10^{-22} g

Δm = 3.85199 × 10^{-22} g - (mass of 94 protons + mass of 138 neutrons)

Δm = 3.85199 × 10^{-22} g - [94(1.67262 × 10^{-24}) + 138(1.67493 × 10^{-24})] g = -3.168 × 10^{-24} g

For 1 mol of nuclei: $\Delta m = -3.168 \times 10^{-24}$ g/nuclei $\times 6.0221 \times 10^{23}$ nuclei/mol $= -1.908$ g/mol

$\Delta E = \Delta mc^2 = (-1.908 \times 10^{-3}$ kg/mol$)(2.9979 \times 10^8$ m/s$)^2 = -1.715 \times 10^{14} =$ J/mol

For $^{231}_{91}$Pa (91 e, 91 p, 140 n):

mass of ^{231}P nucleus $= 3.83616 \times 10^{-22}$ g - $91(9.10939 \times 10^{-28})$ g $= 3.83533 \times 10^{-22}$ g

$\Delta m = 3.83533 \times 10^{-22}$ g - $[91(1.67262 \times 10^{-24}) + 140(1.67493 \times 10^{-24})]$ g $= -3.166 \times 10^{-24}$ g

$$\Delta E = \Delta mc^2 = \frac{-3.166 \times 10^{-27} \text{ kg}}{\text{nuclei}} \times \frac{6.0221 \times 10^{23} \text{ nuclei}}{\text{mol}} \times \left(\frac{2.9979 \times 10^8 \text{ m}}{\text{s}} \right)^2$$

$$= -1.714 \times 10^{14} \text{ J/mol}$$

34.   Mass of proton = 1.00728 amu; mass of neutron = 1.00866 amu; mass of electron = 5.486×10^{-4} amu

Mass of nucleus = mass of atom - mass of electrons = 6.015126 - 3(0.0005486) = 6.013480 amu

$3\,^1_1$H + $3\,^1_0$n $\rightarrow\ ^6_3$Li; $\Delta m = 6.013480$ amu - $[3(1.00728) + 3(1.00866)]$ amu = -0.03434 amu

For 1 mol of ^{6}Li, the mass defect is -0.03434 g.

$\Delta E = \Delta mc^2 = -3.434 \times 10^{-2}$ g $\times \dfrac{1 \text{ kg}}{1000 \text{ g}} \times (2.9979 \times 10^8$ m/s$)^2 = -3.086 \times 10^{12}$ J/mol = binding energy

35.   1.0078 amu = mass of $^1_1$H atom; 1.0087 amu = mass of neutron; m_e = mass of electron

mass defect = Δm = mass of ^{24}Mg nucleus - [mass of 12 protons + mass of 12 neutrons]

$\Delta m = 23.9850$ amu - $12\,m_e$ - $[12(1.0078 - m_e) + 12(1.0087)]$; Mass of electrons cancel.

$\Delta m = 23.9850 - [12(1.0078) + 12(1.0087)] = -0.2130$ amu

$\Delta E = \Delta mc^2 = -0.2130$ amu $\times \dfrac{1.6606 \times 10^{-27} \text{ kg}}{\text{amu}} \times (2.9979 \times 10^8$ m/s$)^2 = -3.179 \times 10^{-11}$ J

$\dfrac{\text{BE}}{\text{nucleon}} = \dfrac{-3.179 \times 10^{-11} \text{ J}}{24 \text{ nucleons}} = -1.325 \times 10^{-12}$ J/nucleon

For ^{27}Mg (12 e, 12 p, 15 n):

$\Delta m = 26.9843 - 12\,m_e - [12(1.0078 - m_e) + 15(1.0087)] = -0.2398$ amu

$\Delta E = \Delta mc^2 = -0.2398$ amu $\times \dfrac{1.6606 \times 10^{-27} \text{ kg}}{\text{amu}} \times (2.9979 \times 10^8$ m/s$)^2 = -3.579 \times 10^{-11}$ J

$\dfrac{\text{BE}}{\text{nucleon}} = \dfrac{-3.579 \times 10^{-11} \text{ J}}{27 \text{ nucleons}} = -1.326 \times 10^{-12}$ J/nucleon

36.   For $_1^2 H$: mass defect = Δm = mass of $_1^2 H$ nucleus - mass of proton - mass of neutron; Lets determine the mass defect in a slightly different way than in Exercise 21.35. The mass of the $^2 H$ nucleus will equal the atomic mass of $^2 H$ minus the mass of the electron in a $^2 H$ atom. From the back of the text, the pertinent masses are: $m_e = 5.49 \times 10^{-4}$ amu, $m_p = 1.00728$ amu, $m_n = 1.00866$ amu.

$$\Delta m = 2.01410 \text{ amu} - 0.000549 \text{ amu} - [1.00728 \text{ amu} + 1.00866 \text{ amu}] = -2.39 \times 10^{-3} \text{ amu}$$

$$\Delta E = \Delta mc^2 = -2.39 \times 10^{-3} \text{ amu} \times \frac{1.6606 \times 10^{-27} \text{ kg}}{\text{amu}} \times (2.998 \times 10^8 \text{ m/s})^2 = -3.57 \times 10^{-13} \text{ J}$$

$$\frac{BE}{\text{nucleon}} = \frac{-3.57 \times 10^{-13} \text{ J}}{2 \text{ nucleons}} = -1.79 \times 10^{-13} \text{ J/nucleon}$$

For $_1^3 H$: $\Delta m = 3.01605 - 0.000549 - [1.00728 + 2(1.00866)] = -9.10 \times 10^{-3}$ amu

$$\Delta E = -9.10 \times 10^{-3} \text{ amu} \times \frac{1.6606 \times 10^{-27} \text{ kg}}{\text{amu}} \times (2.998 \times 10^8 \text{ m/s})^2 = -1.36 \times 10^{-12} \text{ J}$$

$$\frac{BE}{\text{nucleon}} = \frac{-1.36 \times 10^{-12} \text{ J}}{3 \text{ nucleons}} = -4.53 \times 10^{-13} \text{ J/nucleon}$$

37. $_1^1 H + _0^1 n \rightarrow 2 _1^1 H + _0^1 n + _{-1}^1 H$; mass $_{-1}^1 H$ = mass $_1^1 H$ = 1.00728 amu = mass of proton = m_p

$$\Delta m = 3 m_p + m_n - (m_p + m_n) = 2 m_p = 2(1.00728) = 2.01456 \text{ amu}$$

$$\Delta E = \Delta mc^2 = 2.01456 \text{ amu} \times \frac{1.66056 \times 10^{-27} \text{ kg}}{\text{amu}} \times (2.997925 \times 10^8 \text{ m/s})^2$$

$\Delta E = 3.00660 \times 10^{-10}$ J of energy is absorbed per nuclei or 1.81062×10^{14} J/mol nuclei.

The source of energy is the kinetic energy of the proton and the neutron in the particle accelerator.

38. $_1^2 H + _1^3 H \rightarrow _2^4 He + _0^1 n$; Mass of electrons cancel when determining Δm for this nuclear reaction.

$\Delta m = [4.00260 + 1.00866 - (2.01410 + 3.01605)]$ amu $= -1.889 \times 10^{-2}$ amu

For the production of one mol of $_2^4 He$: $\Delta m = -1.889 \times 10^{-2}$ g $= -1.889 \times 10^{-5}$ kg

$\Delta E = \Delta mc^2 = -1.889 \times 10^{-5}$ kg $\times (2.9979 \times 10^8$ m/s$)^2 = -1.698 \times 10^{12}$ J/mol

For 1 nuclei of $_2^4 He$:

$$\frac{-1.698 \times 10^{12}}{\text{mol}} \times \frac{1 \text{ mol}}{6.0221 \times 10^{23} \text{ nuclei}} = -2.820 \times 10^{-12} \text{ J/nuclei}$$

Detection, Uses, and Health Effects of Radiation

39.   The Geiger-Müller tube has a certain response time. After the gas in the tube ionizes to produce a "count," some time must elapse for the gas to return to an electrically neutral state. The response of the tube levels because at high activities, radioactive particles are entering the tube faster than the tube can respond to them.

40.   Not all of the emitted radiation enters the Geiger-Müller tube. The fraction of radiation entering the tube must be constant.

41.   All evolved oxygen in O_2 comes from water and not from carbon dioxide.

42.   Water is produced in this reaction by removing an OH group from one substance and H from the other substance. There are two ways to do this:

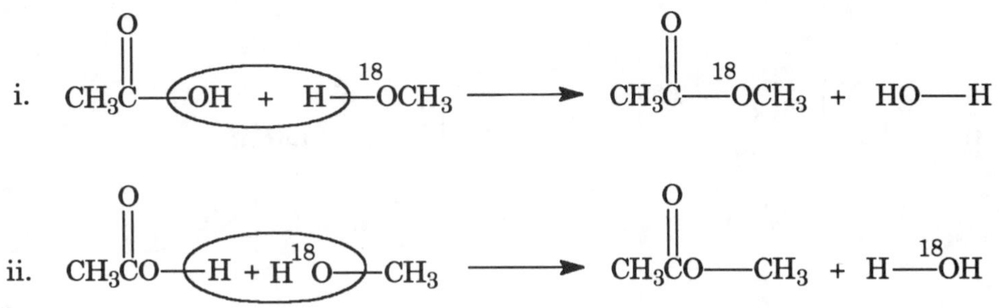

        Since the water produced is not radioactive, then methyl acetate forms by the first reaction where all the oxygen-18 ends up in methyl acetate.

43.   Release of Sr is probably more harmful. Xe is chemically unreactive. Strontium is in the same family as calcium and could be absorbed and concentrated in the body in a fashion similar to Ca. This puts the radioactive Sr in the bones: red blood cells are produced in bone marrow. Xe would not be readily incorporated in the body.

        The chemical properties determine where a radioactive material may be concentrated in the body or how easily it may be excreted. The length of time of exposure and what is exposed to radiation significantly affects the health hazard. (See exercise 21.44 for a specific example.)

44.   i) and ii) mean that Pu is not a significant threat outside the body. Our skin is sufficient to keep out the α particles. If Pu gets inside the body, it is easily oxidized to Pu^{4+} (iv), which is chemically similar to Fe^{3+} (iii). Thus, Pu^{4+} will concentrate in tissues where Fe^{3+} is found. One of these is the bone marrow where red blood cells are produced. Once inside the body, α particles cause considerable damage.

Additional Exercises

45.   The most abundant isotope is generally the most stable isotope. The periodic table predicts that the most stable isotopes for exercises a - d are ^{39}K, ^{56}Fe, ^{23}Na and ^{204}Tl. (Reference Table 21.2 of the text for potential decay processes.)

a.   Unstable; ^{45}K has too many neutrons and will undergo beta particle production.

b.   Stable

c.   Unstable; ^{20}Na has too few neutrons and will most likely undergo electron capture or positron production. Alpha particle production makes too severe of a change to be a likely decay process for the relatively light ^{20}Na nuclei. Alpha particle production usually occurs for heavy nuclei.

d.   Unstable; ^{194}Tl has too few neutrons and will undergo electron capture, positron production and/or alpha particle production.

46. $\ln(N/N_o) = -kt;\ \ k = (\ln 2)/t_{1/2}\ ;\ \ N = 0.001 \times N_o$

$$\ln\left(\frac{0.001 \times N_o}{N_o}\right) = \frac{-(\ln 2)\,t}{24{,}100\ \text{yr}},\ \ \ln 0.001 = -2.88 \times 10^{-5}\,t,\ \ t = 2 \times 10^5\ \text{yr} = 200{,}000\ \text{yr}\ (!)$$

47. $k = \dfrac{\ln 2}{t_{1/2}};\ \ \ln\left(\dfrac{N}{N_o}\right) = -kt = \dfrac{-0.6931\,t}{t_{1/2}},\ \ \ln\left(\dfrac{N}{15.3}\right) = \dfrac{-0.693\,(15{,}000\ \text{yr})}{5730\ \text{yr}} = -1.8$

$\dfrac{N}{15.3} = e^{-1.8} = 0.17,\ \ N = 15.3 \times 0.17 = 2.6$ counts per minute per g of C

If we had 10. mg C, we would see:

$$10.\ \text{mg} \times \frac{1\ \text{g}}{1000\ \text{mg}} \times \frac{2.6\ \text{counts}}{\text{min g}} = \frac{0.026\ \text{counts}}{\text{min}}$$

It would take roughly 40 min to see a single disintegration. This is too long to wait and the background radiation would probably be much greater than the ^{14}C activity. Thus, ^{14}C dating is not practical for very small samples.

48. $\Delta m = -2(5.486 \times 10^{-4}\ \text{amu}) = -1.097 \times 10^{-3}\ \text{amu}$

$\Delta E = \Delta mc^2 = -1.097 \times 10^{-3}\ \text{amu} \times \dfrac{1.6606 \times 10^{-27}\ \text{kg}}{\text{amu}} \times (2.9979 \times 10^8\ \text{m/s})^2 = -1.637 \times 10^{-13}\ \text{J}$

$E_{photon} = 1/2(1.637 \times 10^{-13}\ \text{J}) = 8.185 \times 10^{-14}\ \text{J} = hc/\lambda$

$\lambda = \dfrac{hc}{E} = \dfrac{6.6261 \times 10^{-34}\ \text{J s} \times 2.9979 \times 10^8\ \text{m/s}}{8.185 \times 10^{-14}\ \text{J}} = 2.427 \times 10^{-12}\ \text{m} = 2.427 \times 10^{-3}\ \text{nm}$

49. $20{,}000\ \text{ton TNT} \times \dfrac{4 \times 10^9\ \text{J}}{\text{ton TNT}} \times \dfrac{1\ \text{mol}\ ^{235}\text{U}}{2 \times 10^{13}\ \text{J}} \times \dfrac{235\ \text{g}\ ^{235}\text{U}}{\text{mol}\ ^{235}\text{U}} = 940\ \text{g}\ ^{235}\text{U} \approx 900\ \text{g}\ ^{235}\text{U}$

This assumes that all of the ^{235}U undergoes fission.

Challenge Problems

50.   The radiation may cause nuclei in the metal to undergo nuclear reaction. This changes the identity of the element and the new atom may not fit very well in the crystal lattice. One result is the crystal becomes brittle. One needs to look for materials that do not undergo nuclear reactions when subjected to radiation (particularly neutrons) or ones that produce products of a similar atomic size as the original atoms.

51.   Assuming that the radionuclide is long lived enough such that no significant decay occurs during the time of the experiment, the total counts of radioactivity injected are:

$$0.10 \text{ mL} \times \frac{5.0 \times 10^3 \text{ cpm}}{\text{mL}} = 5.0 \times 10^2 \text{ cpm}$$

Assuming that the total activity is uniformly distributed only in the rats blood, the blood volume is:

$$V \times \frac{48 \text{ cpm}}{\text{mL}} = 5.0 \times 10^2 \text{ cpm}, \quad V = 10.4 \text{ mL} = 10. \text{ mL}$$

52.   a.   From Table 17.1: $2 \text{ H}_2\text{O} + 2 \text{ e}^- \rightarrow \text{H}_2 + 2 \text{ OH}^-$ $E° = -0.83$ V

$$E°_{cell} = E°_{\text{H}_2\text{O}} - E°_{\text{Zr}} = -0.83 \text{ V} + 2.36 \text{ V} = 1.53 \text{ V}$$

Yes, the reduction of H_2O to H_2 by Zr is spontaneous at standard conditions since $E°_{cell} > 0$.

b. $(2 \text{ H}_2\text{O} + 2 \text{ e}^- \rightarrow \text{H}_2 + 2 \text{ OH}^-) \times 2$
      $\text{Zr} + 4 \text{ OH}^- \rightarrow \text{ZrO}_2 \cdot \text{H}_2\text{O} + \text{H}_2\text{O} + 4 \text{ e}^-$
      _____
      $3 \text{ H}_2\text{O(l)} + \text{Zr(s)} \rightarrow 2 \text{ H}_2\text{(g)} + \text{ZrO}_2 \cdot \text{H}_2\text{O(s)}$

c. $\Delta G° = -nFE° = -(4 \text{ mol e}^-)(96{,}485 \text{ C/mol e}^-)(1.53 \text{ J/C}) = -5.90 \times 10^5 \text{J} = -590. \text{ kJ}$

$$E = E° - \frac{0.0592}{n} \log Q; \quad \text{At equilibrium, } E = 0 \text{ and } Q = K.$$

$$E° = \frac{0.0592}{n} \log K, \quad \log K = \frac{4(1.53)}{0.0592} = 103, \quad K \approx 10^{103}$$

d. $1.00 \times 10^3 \text{ kg Zr} \times \dfrac{1000 \text{ g}}{\text{kg}} \times \dfrac{1 \text{ mol Zr}}{91.22 \text{ g Zr}} \times \dfrac{2 \text{ mol H}_2}{\text{mol Zr}} = 2.19 \times 10^4 \text{ mol H}_2$

$2.19 \times 10^4 \text{ mol H}_2 \times \dfrac{2.016 \text{ g H}_2}{\text{mol H}_2} = 4.42 \times 10^4 \text{ g H}_2$

$$V = \frac{nRT}{P} = \frac{(2.19 \times 10^4 \text{ mol})(0.08206 \text{ L atm/mol} \cdot \text{K})(1273 \text{ K})}{1 \text{ atm}} = 2 \times 10^6 \text{ L H}_2$$

e. Probably yes; Less radioactivity overall was released by venting the H_2 than what would have been released if the H_2 exploded inside the reactor (as happened at Chernobyl). Neither alternative is pleasant, but venting the radioactive hydrogen is the less unpleasant of the two alternatives.

53. a. ^{12}C; It takes part in the first step of the reaction but is regenerated in the last step. ^{12}C is not consumed so it is not a reactant.

b. ^{13}N, ^{13}C, ^{14}N, ^{15}O, and ^{15}N are the intermediates.

c. $4\,^1_1H \rightarrow \,^4_2He + 2\,^0_{+1}e$; $\Delta m = 4.00260$ amu $- 2\,m_e + 2\,m_e - [4(1.00782$ amu $- m_e)]$

$\Delta m = 4.00260 - 4(1.00782) + 4(0.000549) = -0.02648$ amu for 4 protons reacting

For 4 mol of protons, $\Delta m = -0.02648$ g and ΔE for the reaction is:

$$\Delta E = \Delta mc^2 = -2.648 \times 10^{-5} \text{ kg} \times (2.9979 \times 10^8 \text{ m/s})^2 = -2.380 \times 10^{12} \text{ J}$$

For 1 mol of protons reacting: $\dfrac{-2.380 \times 10^{12} \text{ J}}{4 \text{ mol } ^1H} = -5.950 \times 10^{11}$ J/mol ^1H

CHAPTER TWENTY-TWO

ORGANIC CHEMISTRY

Questions

1. There is only one consecutive chain of C-atoms in the molecule. They are not all in a true straight line since the bond angles at each carbon are the tetrahedral angles of 109.5°.

2. Structural isomers: Same formula but different bonding, either in the kinds of bonds present or the way in which the bonds connect atoms to each other.

    Geometrical isomers: Same formula and same bonds but differ in the arrangement of atoms in space about a rigid bond or ring.

3. Substitution: An atom or group is replaced by another atom or group.

    e.g., H in benzene is replaced by Cl. $C_6H_6 + Cl_2 \xrightarrow{\text{catalyst}} C_6H_5Cl + HCl$

    Addition: Atoms or groups are added to a molecule.

    e.g., Cl_2 adds to ethene. $CH_2 = CH_2 + Cl_2 \rightarrow CH_2Cl - CH_2Cl$

4. a. Addition polymer: Polymer that forms by adding monomer units together (usually by reacting double bonds). Teflon, polyvinyl chloride and polyethylene are examples of addition polymers.

    b. Condensation polymer: Polymer that forms when two monomers combine by eliminating a small molecule (usually H_2O or HCl). Nylon and dacron are examples of condensation polymers.

    c. Copolymer: Polymer formed from more than one type of monomer. Nylon and dacron are also copolymers.

5. A thermoplastic polymer can be remelted; a thermoset polymer cannot be softened once it is formed.

6. The physical properties depend on the strengths of the intermolecular forces between adjacent polymer chains. These forces are affected by chain length and extent of branching.

    longer chains = stronger intermolecular forces; branched chains = weaker intermolecular forces

7.   Plasticizer compounds make a polymer more flexible by inserting themselves between adjacent polymer chains which weakens the intermolecular forces. Crosslinking makes a polymer more rigid by bonding adjacent polymer chains together.

8.   The regular arrangement of the methyl groups in the isotactic chains allows adjacent polymer chains to pack together very closely. This leads to stronger intermolecular forces between chains as compared to atactic polypropylene where the packing of polymer chains is not as efficient.

9.   a.   Replacement of hydrogens with halogens results in fewer H and OH radicals in the flame. In addition, as halogen atoms are released in a flame, they react readily with H radicals by forming hydrogen halides. Thus, halogens act as free radical scavengers.

        b.   With aromatic groups present, it is more difficult to "chip off" pieces of the polymer in the pyrolysis zone due to the increased strength of the bonds in aromatic rings. In addition, polymers with aromatic rings tend to char naturally, which inhibits combustion.

        c.   It is more difficult to "chip off" fragments of the polymer in the pyrolysis zone. The crosslinked polymer chars instead of burns.

10.   In nylon, hydrogen bonding interactions occur due to the presence of N–H bonds in the polymer. For a given polymer chain length, there are more N–H groups in Nylon-46 as compared to Nylon-6. Hence, Nylon-46 forms a stronger polymer as compared to Nylon-6 due to the increased hydrogen bonding interactions.

11.   Polyvinyl chloride contains some polar C–Cl bonds as compared to only nonpolar C–H bonds in polyethylene. The stronger interparticle forces would be found in polyvinyl chloride since there are dipole-dipole forces present in PVC that are not present in polyethylene.

12.   a.   Cracking is a process whereby large molecules are broken down to smaller ones by breaking carbon-carbon bonds.

        b.   Alkylation is a process whereby small molecules are combined to form larger molecules.

        c.   Isomerization involves converting straight chain alkanes into branched chain alkanes.

        d.   Catalytic reforming involves converting alkanes and cycloalkanes into aromatic compounds.

        Note: Cracking and alkylation are used to increase the quantity of gasoline and isomerization and catalytic reforming are used to increase the octane rating of gasoline.

Exercises

Hydrocarbons

13.   i.   ii. CH_3
                                                                                                    |
        CH_3–CH_2–CH_2–CH_2–CH_2–CH_3 CH_3——CH——CH_2——CH_2——CH_3

iii.

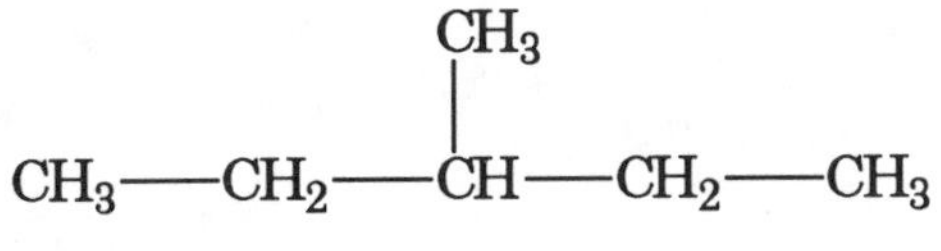

$$CH_3 - CH_2 - \overset{\overset{\displaystyle CH_3}{|}}{CH} - CH_2 - CH_3$$

iv.

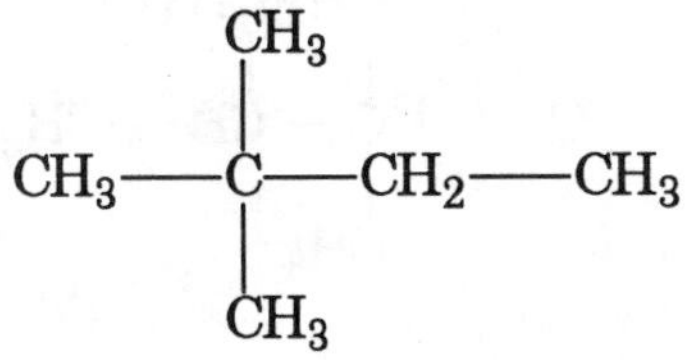

$$CH_3 - \overset{\overset{\displaystyle CH_3}{|}}{\underset{\underset{\displaystyle CH_3}{|}}{C}} - CH_2 - CH_3$$

v.

$$CH_3 - \overset{\overset{\displaystyle CH_3}{|}}{CH} - \overset{\overset{\displaystyle CH_3}{|}}{CH} - CH_3$$

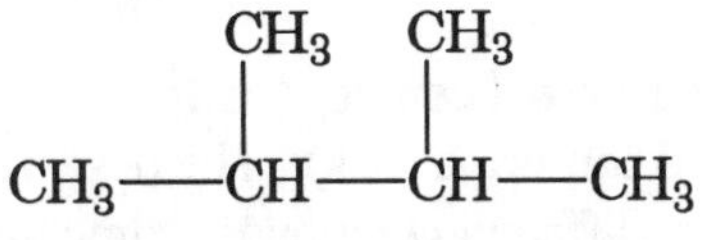

All other possibilities are identical to one of these five compounds.

14.   i.

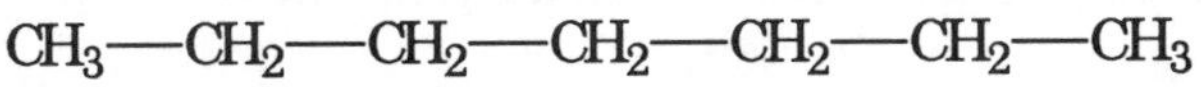

$$CH_3 - CH_2 - CH_2 - CH_2 - CH_2 - CH_2 - CH_3$$

ii.

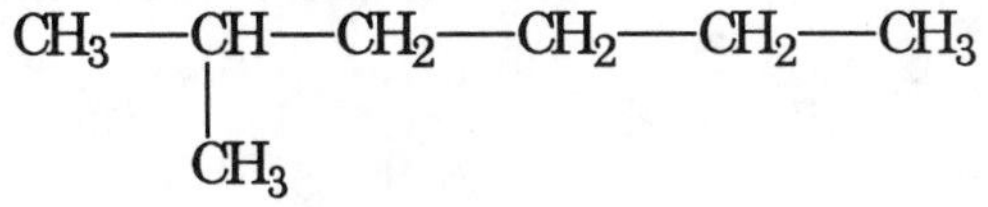

$$CH_3 - \overset{}{\underset{\underset{\displaystyle CH_3}{|}}{CH}} - CH_2 - CH_2 - CH_2 - CH_3$$

iii.

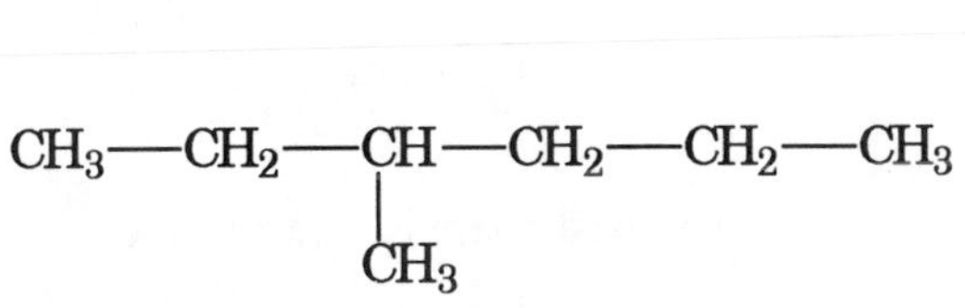

$$CH_3 - CH_2 - \overset{}{\underset{\underset{\displaystyle CH_3}{|}}{CH}} - CH_2 - CH_2 - CH_3$$

iv.

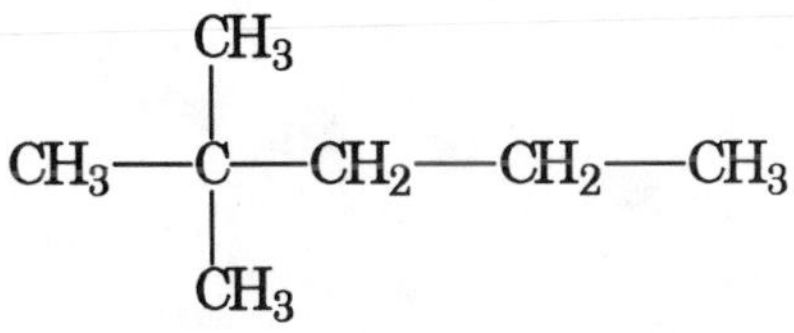

$$CH_3 - \overset{\overset{\displaystyle CH_3}{|}}{\underset{\underset{\displaystyle CH_3}{|}}{C}} - CH_2 - CH_2 - CH_3$$

v.

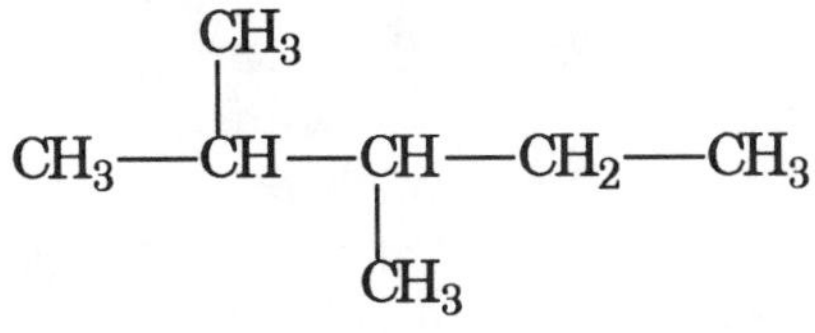

$$CH_3 - \overset{\overset{\displaystyle CH_3}{|}}{CH} - \overset{}{\underset{\underset{\displaystyle CH_3}{|}}{CH}} - CH_2 - CH_3$$

vi.

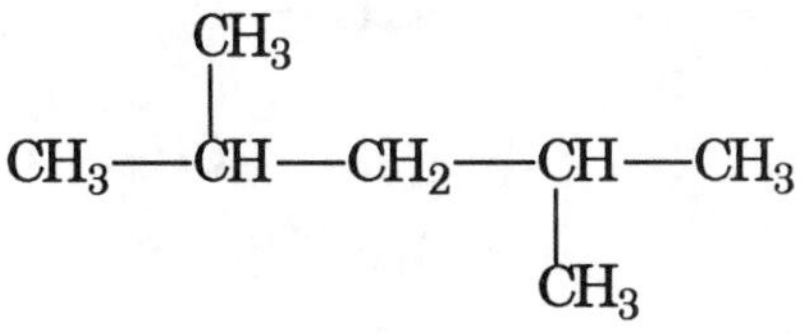

$$CH_3 - \overset{\overset{\displaystyle CH_3}{|}}{CH} - CH_2 - \overset{}{\underset{\underset{\displaystyle CH_3}{|}}{CH}} - CH_3$$

vii.

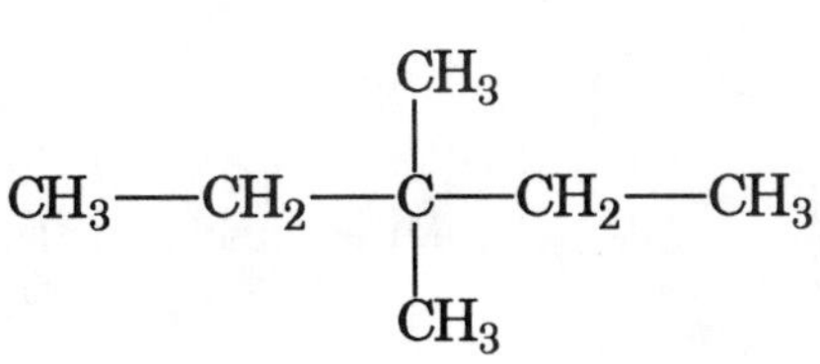

$$CH_3 - CH_2 - \overset{\overset{\displaystyle CH_3}{|}}{\underset{\underset{\displaystyle CH_3}{|}}{C}} - CH_2 - CH_3$$

viii.

$$CH_3 - CH_2 - \overset{}{\underset{\underset{\underset{\displaystyle CH_3}{|}}{\underset{\displaystyle CH_2}{|}}}{CH}} - CH_2 - CH_3$$

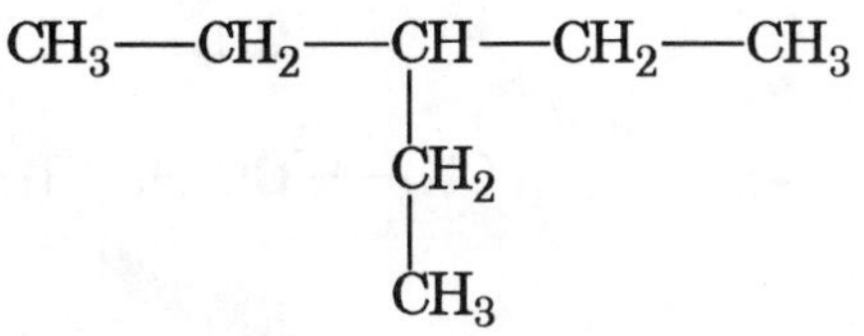

ix.

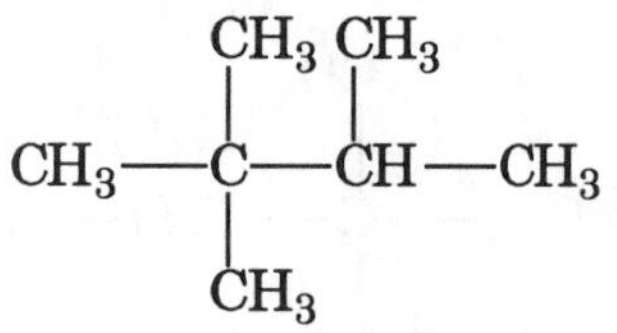

$$CH_3-\underset{\underset{\displaystyle CH_3}{|}}{\overset{\overset{\displaystyle CH_3}{|}}{C}}-\underset{}{\overset{\overset{\displaystyle CH_3}{|}}{CH}}-CH_3$$

15.   See Exercise 22.13 for the structures. The names of structure i - v respectively, are: hexane (or n-hexane), 2-methylpentane, 3-methylpentane, 2,2-dimethylbutane and 2,3-dimethylbutane.

Note: A difficult task in Exercise 22.13 is recognizing different compounds from compounds that differ by rotations about one or more C–C bonds (called conformations). The best way to distinguish different compounds is to name them. Different name = different compound; same name = same compound so it is not an isomer.

16.   See Exercise 22.15 for the structures. The names of structures i - ix respectively, are: heptane (or n-heptane), 2-methylhexane, 3-methylhexane, 2,2-dimethylpentane, 2,3-dimethylpentane, 2,4-dimethylpentane, 3,3-dimethylpentane, 3-ethylpentane and 2,2,3-trimethylbutane.

17.   a.   b.

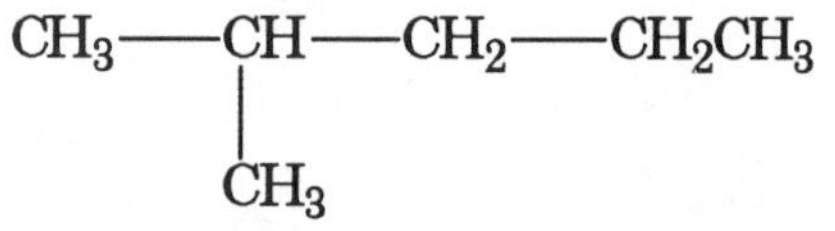

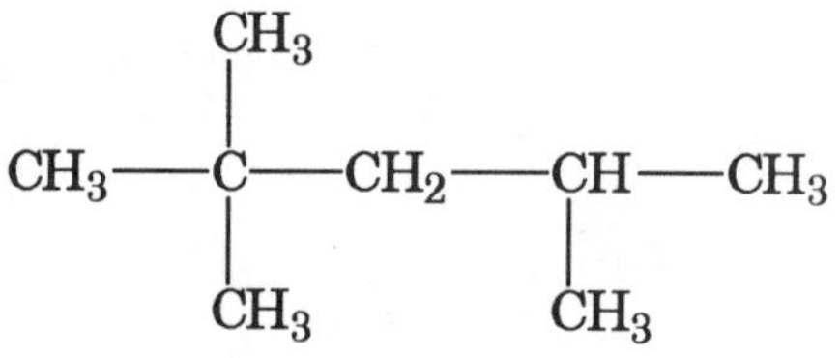

$$CH_3-\underset{\underset{\displaystyle CH_3}{|}}{CH}-CH_2-CH_2CH_3$$

$$CH_3-\underset{\underset{\displaystyle CH_3}{|}}{\overset{\overset{\displaystyle CH_3}{|}}{C}}-CH_2-\underset{\underset{\displaystyle CH_3}{|}}{CH}-CH_3$$

c. 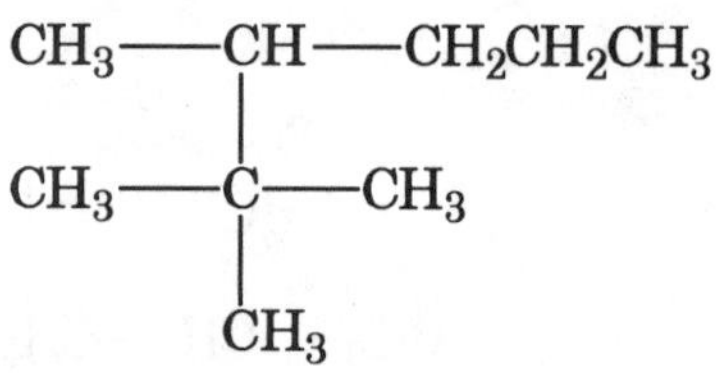   d.   The longest chain is 6 carbons long.

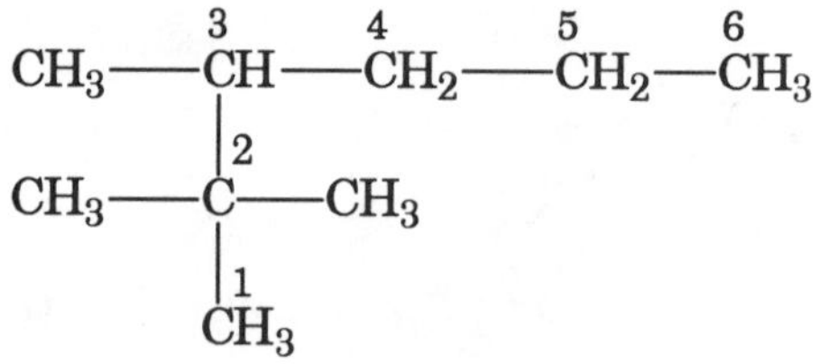

$$CH_3-\underset{}{CH}-CH_2CH_2CH_3$$
$$CH_3-\overset{}{C}-CH_3$$
$$CH_3$$

$$CH_3-\overset{3}{CH}-\overset{4}{CH_2}-\overset{5}{CH_2}-\overset{6}{CH_3}$$
$$CH_3-\overset{2}{C}-CH_3$$
$$\overset{1}{CH_3}$$

2,2,3-trimethylhexane

18.

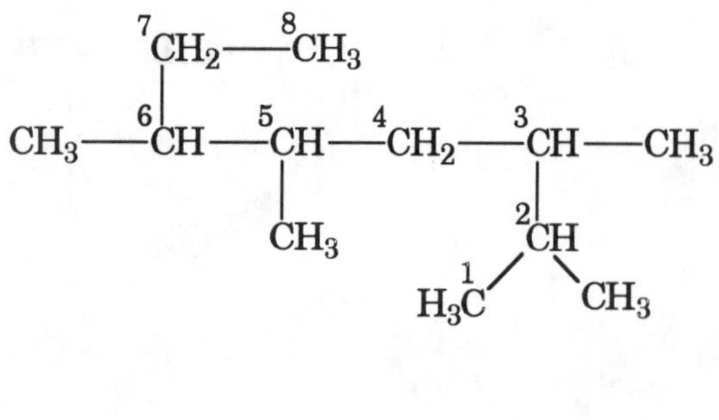

$$\overset{7}{CH_2}-\overset{8}{CH_3}$$
$$CH_3-\overset{6}{CH}-\overset{5}{CH}-\overset{4}{CH_2}-\overset{3}{CH}-CH_3$$
$$CH_3 \quad \overset{2}{CH}$$
$$H_3\overset{1}{C} \quad CH_3$$

2,3,4,6-tetramethyloctane

19. a. 2,3,3-trimethylhexane   b.   8-ethyl-2,5,5-trimethyldecane   c.   3-methylhexane

Note: For alkanes always identify the longest carbon chain for the base name first, then number the carbons to give the lowest overall numbers for the substituent groups.

20. a. methylcyclopropane   b.   tert-butylcyclohexane

c. 1,2,4-trimethylcyclopentane

21. a. 1-butene   b.   2-methyl-2-butene   c.   2,5-dimethyl-3-heptene

Note: The multiple bond is assigned the lowest number possible.

22. a. 2-methyl-1-butene   b.   2-propyl-1-hexene
                                                       (Want longest chain that contains all multiple
                                                       bonds for the base name.)

c. 3-methyl-1-butyne

23. a. $CH_3–CH_2–CH=CH–CH_2–CH_3$ b. $CH_3–CH=CH–CH=CH–CH_2CH_3$

c.

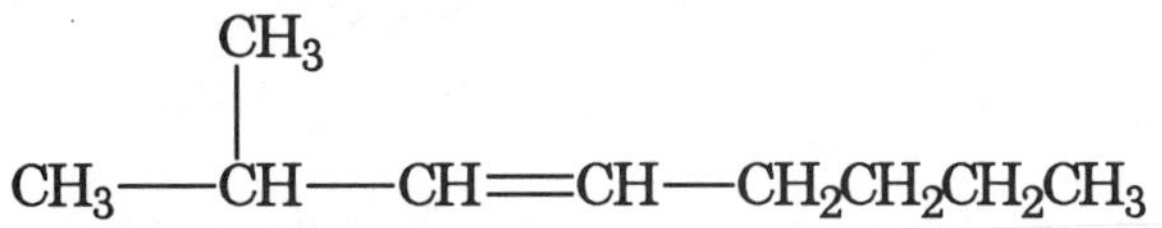

$$CH_3—\underset{\underset{\displaystyle CH_3}{|}}{CH}—CH=CH—CH_2CH_2CH_2CH_3$$

24.

a. $HC\equiv C—CH_2—\underset{\underset{\displaystyle CH_3}{|}}{CH}—CH_3$ b. $H_2C=\underset{\underset{\displaystyle CH_3}{|}}{\overset{\overset{\displaystyle CH_3}{|}}{C}}—\overset{\overset{\displaystyle CH_3}{|}}{C}—CH_2CH_2CH_3$

c. $CH_3CH_2—\underset{\underset{\displaystyle CH_2CH_3}{|}}{CH}—CH=CH—CH_2CH_2CH_2CH_2CH_3$

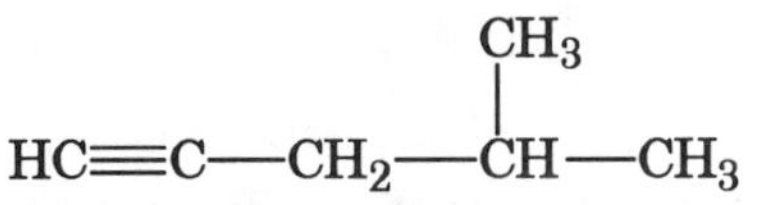

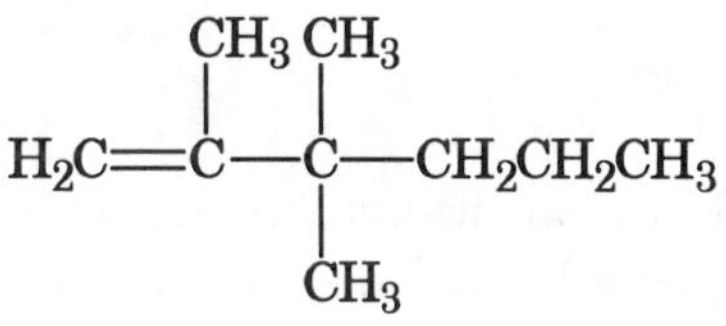

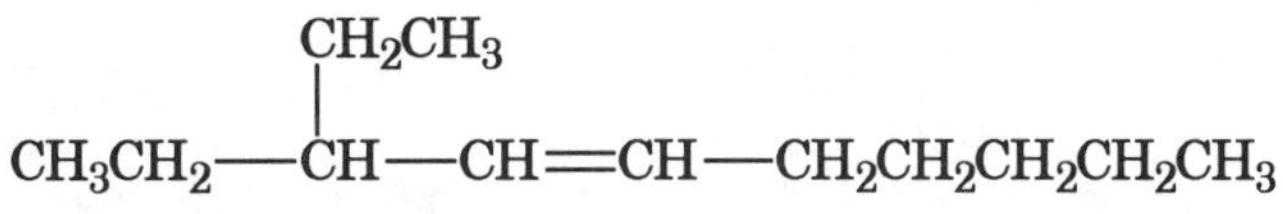

25. a. 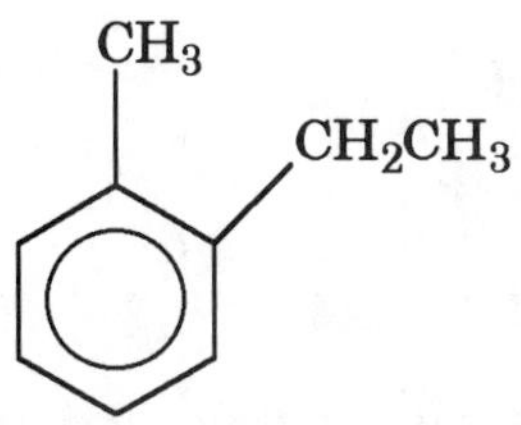   b.

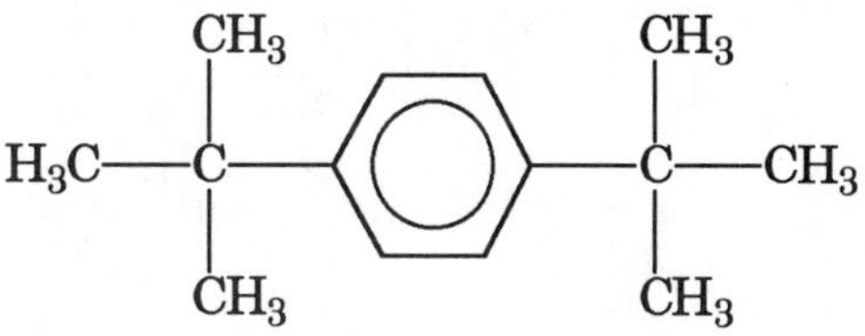

c.

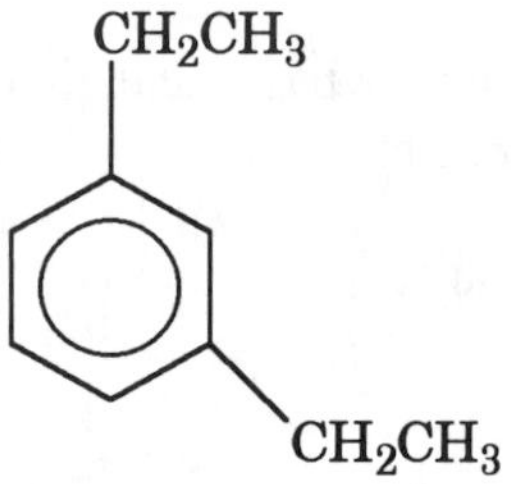

CH_2CH_3

CH_2CH_3

d.

CH_2—CH=CH—CH_3

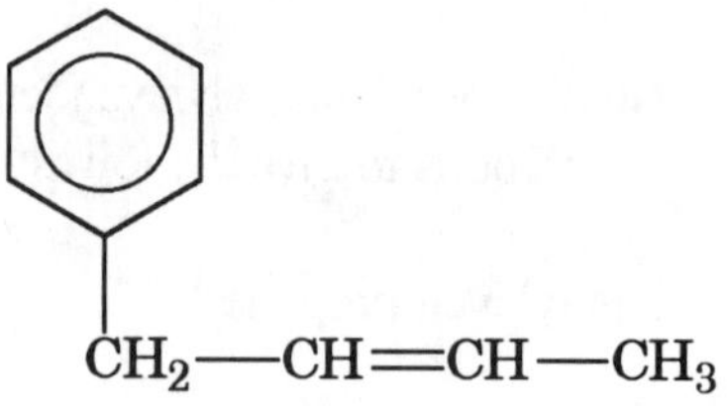

26.   isopropylbenzene or 2-phenylpropane

27.   a.   1,3-dichlorobutane   b.   1,1,1-trichlorobutane

        c.   2,3-dichloro-2,4-dimethylhexane   d.   1,2-difluoroethane

28.   a.   3,4-dimethylcyclopentene   b.   chloroethene

        c.   1,2-dimethylcyclopentene   d.   1,1-dichlorocyclohexane

        e.   chlorobenzene   f.   chlorocyclohexane

        g.   3-chlorocyclohexene

        Note: If the location of double bond is not given in the name, then it is assumed to be located between C_1 and C_2.

Isomerism

29.   To exhibit cis-trans isomerism, each carbon in the double bond must have two structurally different groups bonded to it. In Exercise 22.21, this only occurs for compound c. The cis and trans isomers for 21c are:

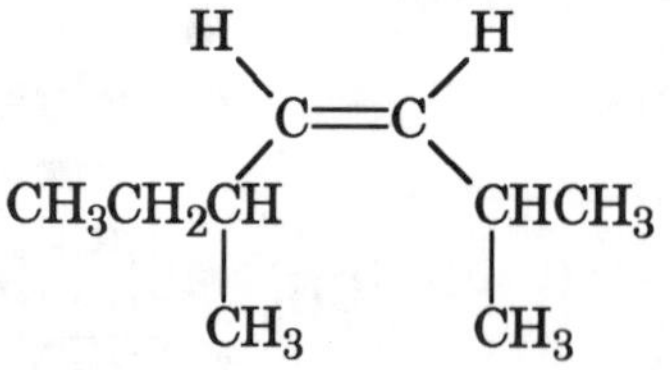

cis 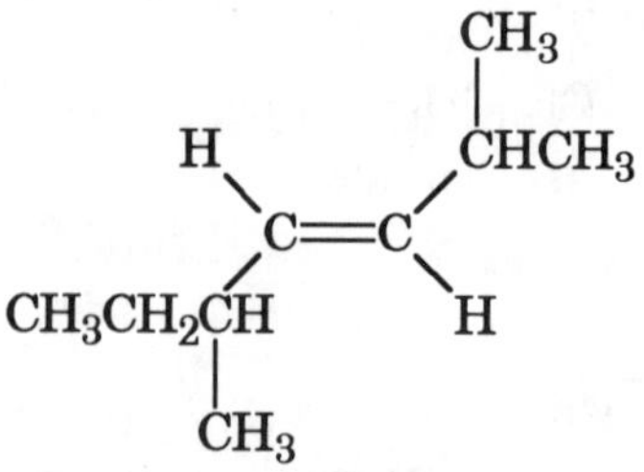   trans

        Similarly, all the compounds in Exercise 23.23 can also exhibit cis-trans isomerism.

In the other compounds in Exercise 23.21, each carbon in the double bond does <u>not</u> contain two different groups. In 21a, the first carbon in the double bond contains two H atoms and in 21b, the first carbon in the double bond contains 2 CH_3 groups. To illustrate that these compounds do not exhibit cis-trans isomerism, lets look at the potential cis-trans isomers for the compound in Exercise 23.21a.

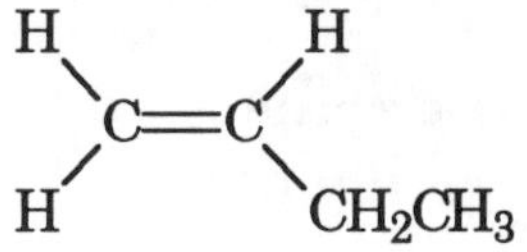

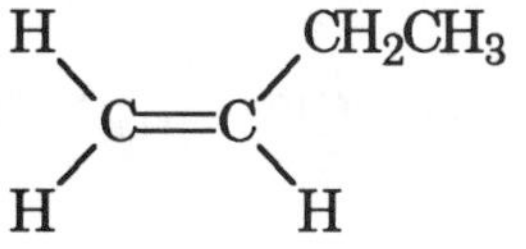

These are the same compounds; they only differ by a simple rotation of the molecule. Therefore, they are <u>not</u> isomers of each other.

30.   In Exercise 22.22, none of the compounds can exhibit cis-trans isomerism and in Exercise 22.24, only 3-ethyl-4-decene can exhibit cis-trans isomerism.

31. C_5H_{10} has the general formula for alkenes, C_nH_{2n}. To distinguish the different isomers from each other, we will name them. Each isomer must have a different name.

$$CH_2{=}CHCH_2CH_2CH_3 \qquad\qquad CH_3CH{=}CHCH_2CH_3$$

1-pentene   2-pentene

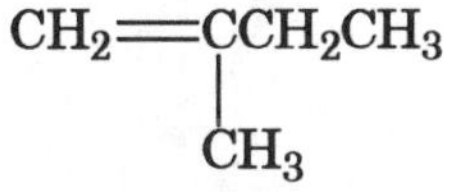

2-methyl-1-butene 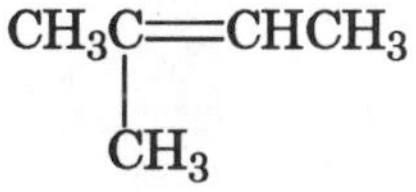   2-methyl-2-butene

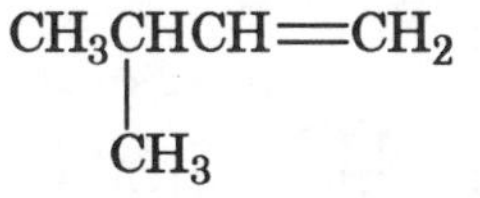

3-methyl-1-butene

32.   Only 2-pentene exhibits cis-trans isomerism. The isomers are:

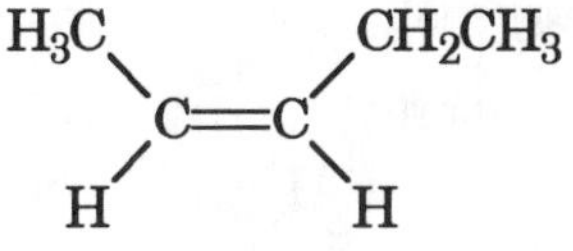

cis 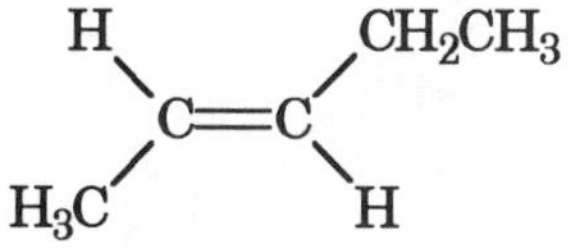   trans

The other isomers of C_5H_{10} do not contain carbons in the double bond that each contain two different groups attached.

33.   To help distinguish the different isomers, we will name them.

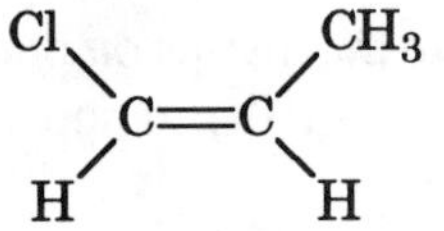

cis-1-chloro-1-propene

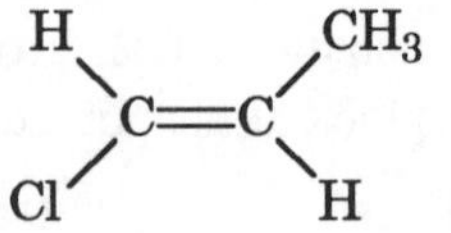

trans-1-chloro-1-propene

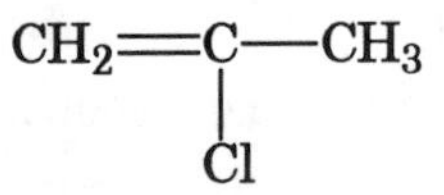

$$CH_2{=}C{-}CH_3$$
$$\quad\quad\ \ |$$
$$\quad\quad\ \ Cl$$

2-chloro-1-propene

$$CH_2{=}CH{-}CH_2$$
$$\quad\quad\quad\quad\ |$$
$$\quad\quad\quad\quad\ Cl$$

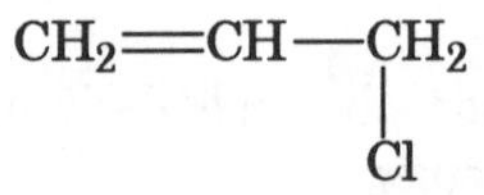

3-chloro-1-propene

34. $Cl_2C{=}CH{-}CH_3$ $CH_2{=}CCl{-}CH_2Cl$ $CH_2{=}CH{-}CHCl_2$

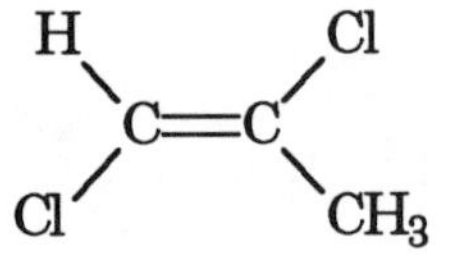

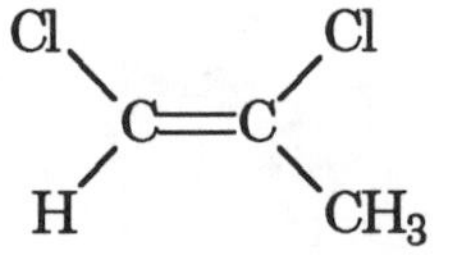

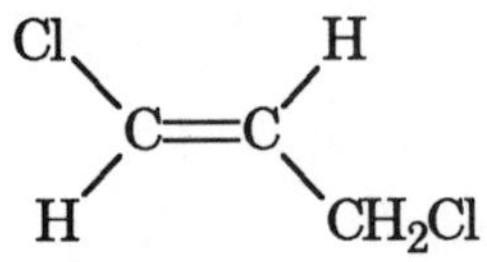

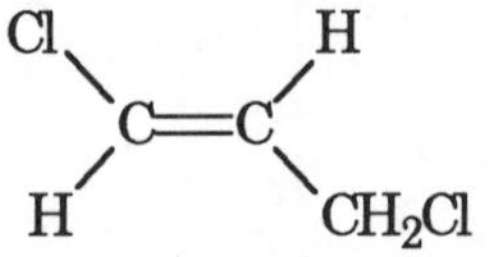

35.

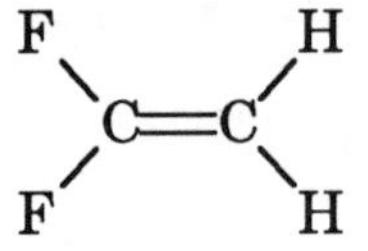

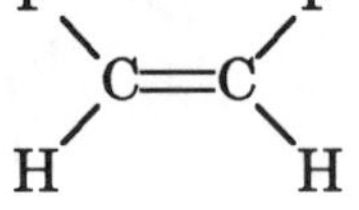

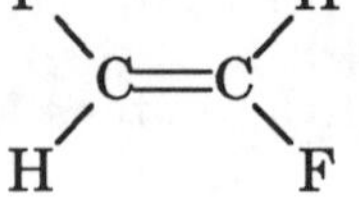

36. $HCBrCl{-}CH{=}CH_2$

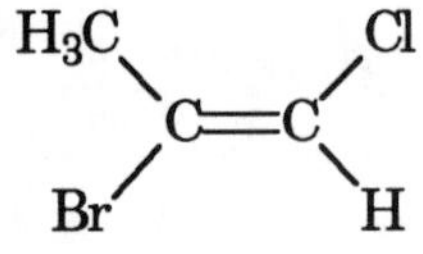

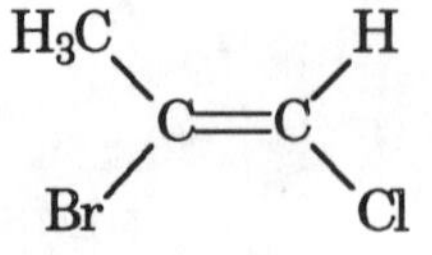

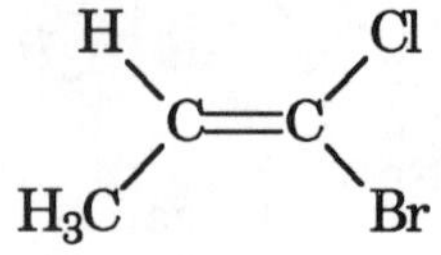

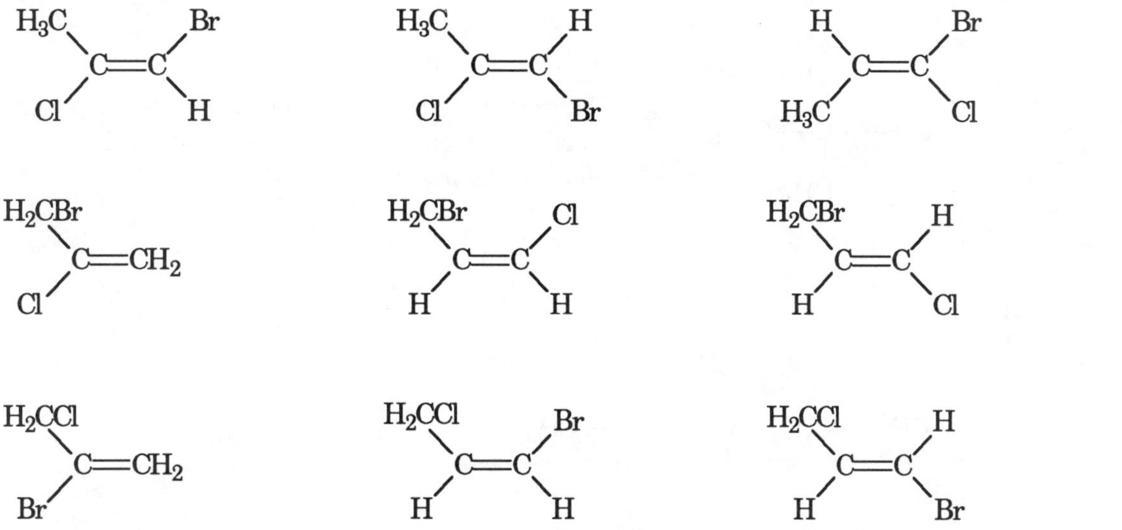

Note: 1-bromo-1-chlorocyclopropane, cis-1-bromo-2-chlorocyclopropane and trans-1-bromo-2-chlorocyclopropane are the ring structures that are isomers of bromochloropropene. We did not include the ring structures in the answer since their base name is not bromochloropropene.

37.

a. 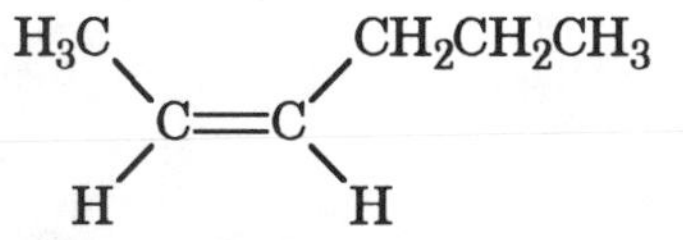   b. 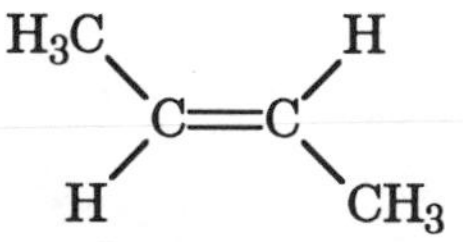   c.

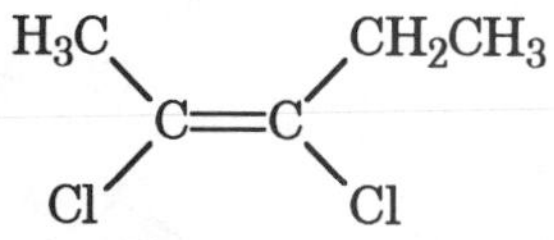

38.   a. cis-2-butene   b. cis-3-methyl-2-heptene   c. cis-1,2-dichlorpropene

Note: In general, cis-trans designations refers to the relative positions of the largest groups. In compound b, the largest group off the first carbon in the double bond is CH_3 and the largest group off the second carbon in the double bond is $CH_2CH_2CH_2CH_3$. Since their relative placement is on the same side of the double bond, then this is the cis isomer.

39.   a. $CH_3^*–CH_2^*–CH_2–CH_3$; There are two "types" of hydrogens in n-butane (see asterisks). Thus they are two monochloro isomers of n-butane (1-chlorobutane and 2-chlorobutane).

b.

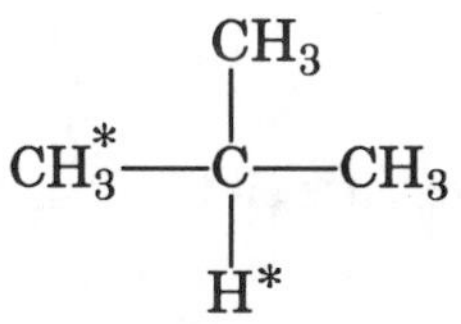

There are two types of hydrogens in 2-methylpropane (see *). Thus, there are two monochloro isomers of 2-methylpropane (1-chloro-2-methylpropane and 2-chloro-2-methylpropane).

c.

$$CH_3 \underset{\underset{CH_3}{|}}{\overset{\overset{CH_3}{|}}{C}} CH_3$$

There is only one type of hydrogen in this compound so there is only one monochloro isomer (1-chloro-2,2-dimethylpropane).

d.

$$CH_3^* \underset{\underset{H^*}{|}}{\overset{\overset{CH_3^*}{|}}{CH_2^* \overset{}{C}}} CH_2 \underset{}{CH_2} CH_3$$

There are four types of hydrogens in this compound (see *) so they are four monochloro isomers possible.

40.   a.

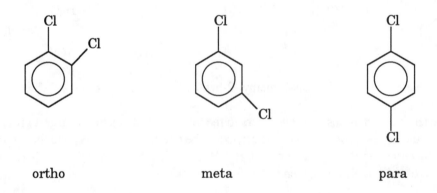

ortho   meta   para

b.   There are three trichlorobenzenes (1,2,3-trichlorobenzene, 1,2,4-trichlorobenzene and 1,3,5-trichlorobenzene).

c.   The meta isomer will be very difficult to synthesize.

d.   1,3,5-trichlorobenzene will be the most difficult to synthesize since all Cl groups are meta to each other in this compound.

Functional Groups

41.   Reference Table 22.5 for the common functional groups.

a.   ketone   b.   aldehyde   c.   carboxylic acid   d.   amine

42.   a.   b.

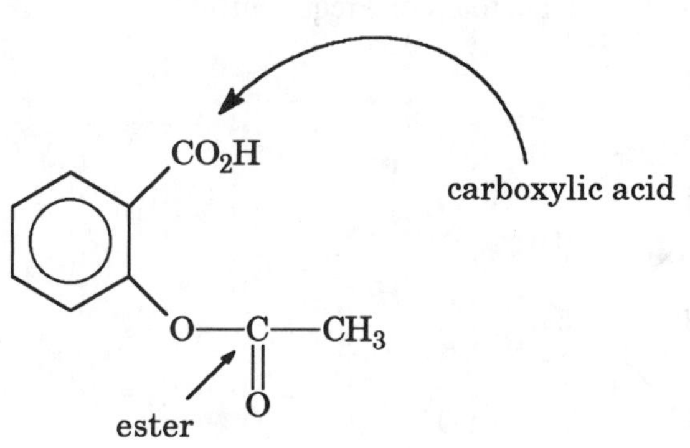

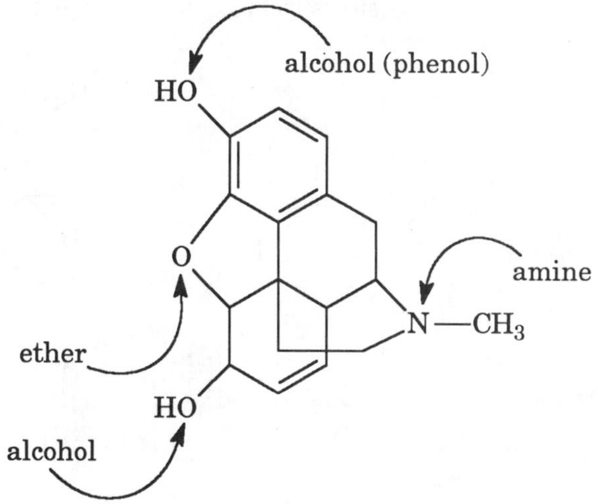

c.

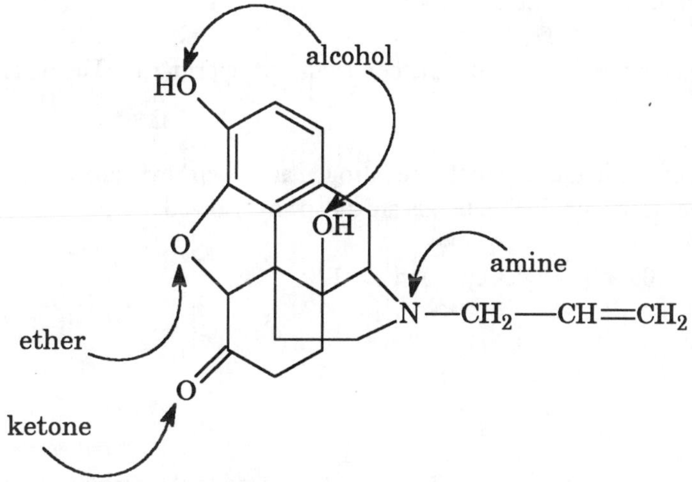

43.   a.

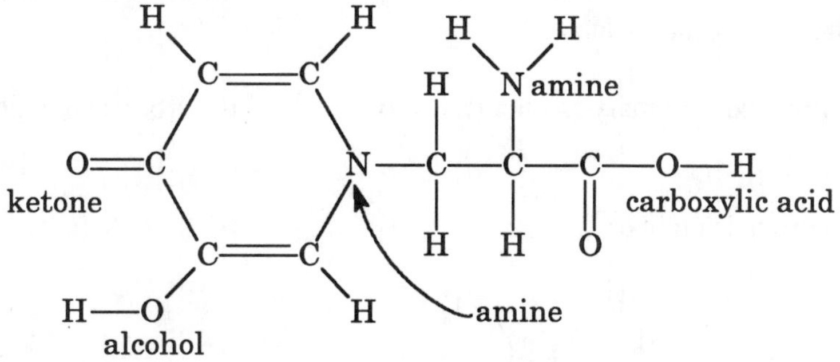

b.   5 carbons in ring and the carbon in –CO$_2$H: sp^2; the other two carbons: sp^3

c.   24 sigma bonds; 4 pi bonds

44.   Hydrogen atoms are usually omitted from ring structures. In organic compounds, the carbon atoms
       generally form four bonds. With this in mind, the following structure has the missing hydrogen
       atoms included in order to give each carbon atom the four bond requirement.

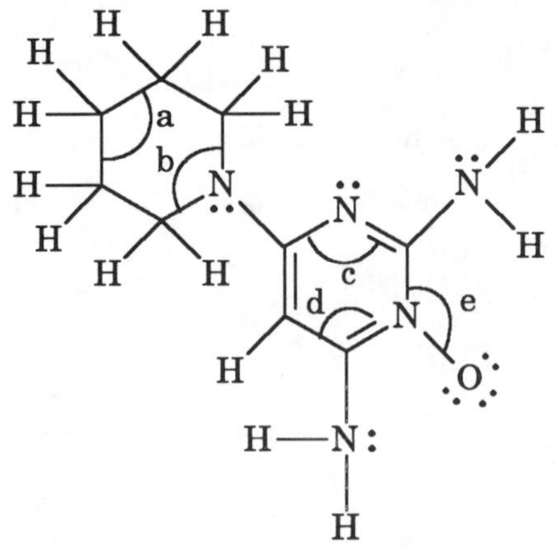

a.   Minoxidil would be more soluble in acidic solution. The nitrogens with lone pairs can be
      protonated, forming a water soluble cation.

b.   The two nitrogens in the ring with double bonds are sp^2 hybridized. The other three N's are sp^3
      hybridized.

c.   The five carbon atoms in the ring with one nitrogen are all sp^3 hybridized. The four carbon
      atoms in the other ring with double bonds are all sp^2 hybridized.

d.   Angles a and b ≈ 109.5°; Angles c, d, and e ≈ 120°

e.   31 sigma bonds

f.   3 pi bonds

45.   a.   2-methyl-2-propanol; Since there are 3 R groups attached to the carbon containing the OH
             group, then this is a tertiary alcohol.

       b.   cyclobutanol; Since the carbon containing the OH group is bonded to two other carbons (2 R
             groups), then this is a secondary alcohol.

       c.   2,2-dimethyl-1-propanol; primary alcohol (1 R group bonded to carbon containing the OH
             group)

46.   The C with the −OH group is number 1.

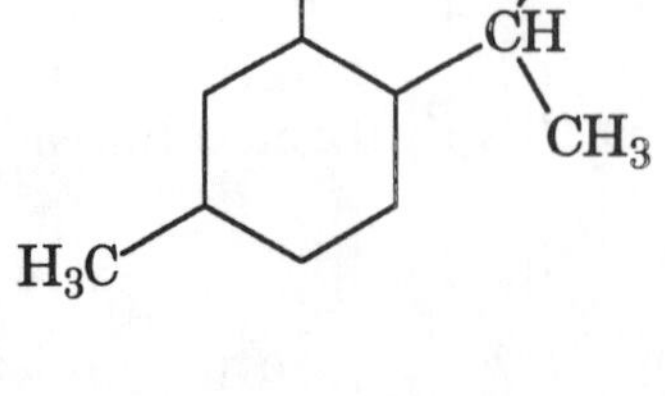

47.

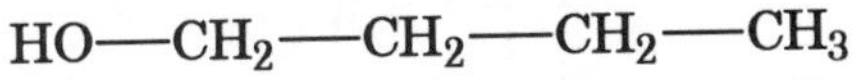

HO—CH$_2$—CH$_2$—CH$_2$—CH$_3$

1-butanol

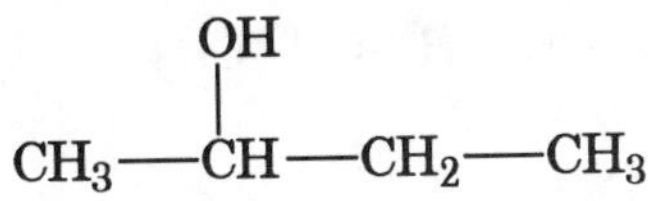

CH$_3$—CH—CH$_2$—CH$_3$ (with OH above CH)

2-butanol

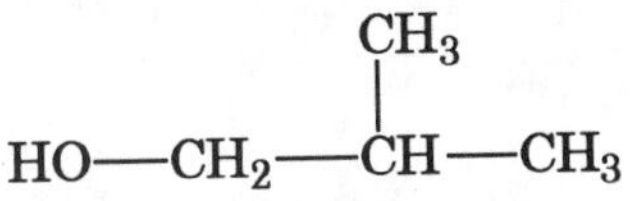

HO—CH$_2$—CH—CH$_3$ (with CH$_3$ above CH)

2-methyl-1-propanol

CH$_3$—C—CH$_3$ (with CH$_3$ above C and OH below C)

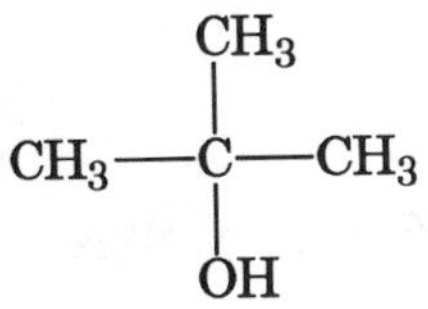

2-methyl-2-propanol

There are three possible ethers with the formula C$_4$H$_{10}$O. They are:

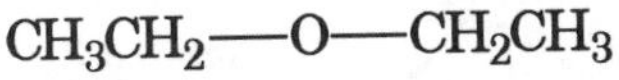

CH$_3$CH$_2$—O—CH$_2$CH$_3$

diethyl ether

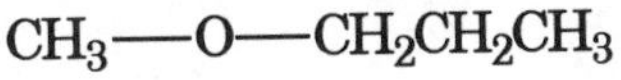

CH$_3$—O—CH$_2$CH$_2$CH$_3$

methylpropyl ether

CH$_3$—O—CH (with CH$_3$ above and CH$_3$ below)

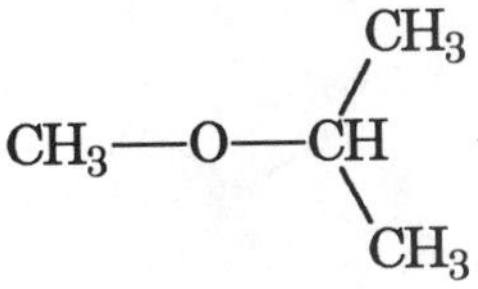

isopropylmethyl ether

48.   Two possible aldehydes:

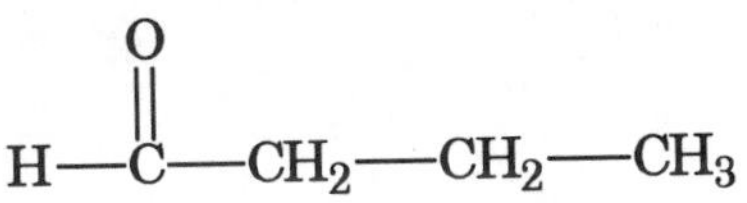

H—C—CH$_2$—CH$_2$—CH$_3$ (with O double bonded to C)

butanal

H—C—CH—CH$_3$ (with O double bonded to C, and CH$_3$ below CH)

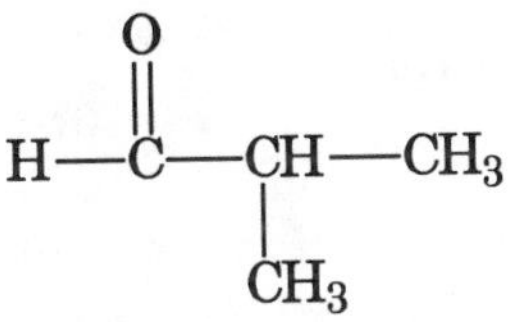

2-methylpropanal

One possible ketone: CH$_3$—C—CH$_2$—CH$_3$ (with O double bonded to C)   2-butanone

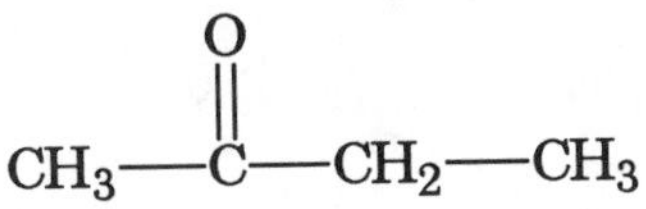

49.   a.   1-bromo-2-propanone

b.   2-methylbutanal

50.   a.   propanoic acid

b.   3-methylbutanoic acid

Reactions of Organic Compounds

51.   a. $CH_2 = CH_2 + Br_2 \rightarrow CH_2Br - CH_2Br$

b. $C_6H_6 + Br_2 \xrightarrow{FeBr_3} C_6H_5Br + HBr$

c. $CH_2{=}CH{-}CH_2{-}CH{=}CH_2 + 2\,H_2 \xrightarrow{Pt} CH_3{-}CH_2{-}CH_2{-}CH_2{-}CH_3$

52. a.

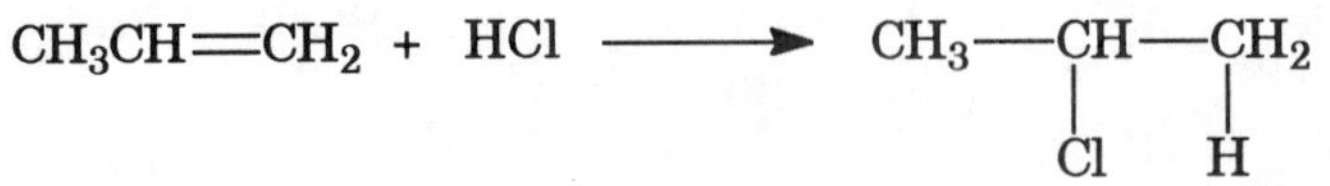

$$CH_3CH{=}CH_2 + HCl \longrightarrow CH_3{-}\underset{\underset{Cl}{|}}{CH}{-}\underset{\underset{H}{|}}{CH_2}$$

b.

$$CH_3CH{=}CH_2 + H_2O \longrightarrow CH_3{-}\underset{\underset{OH}{|}}{CH}{-}\underset{\underset{H}{|}}{CH_2}$$

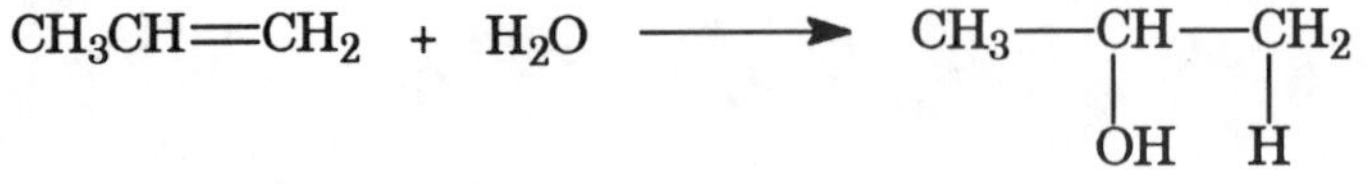

c.

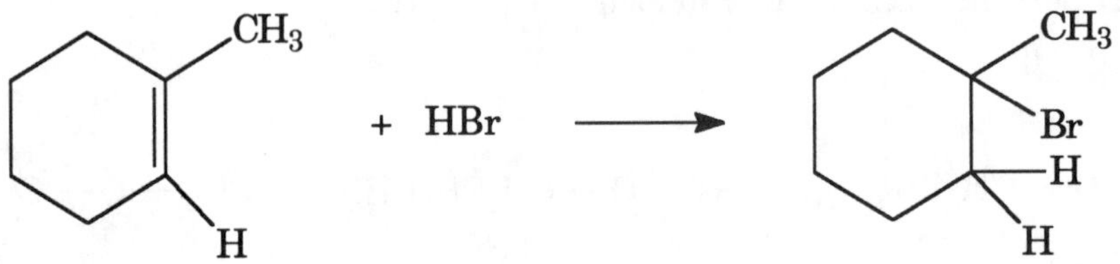

d.

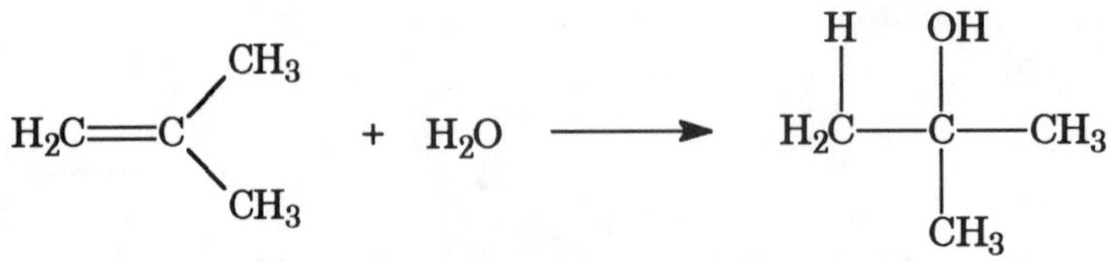

e.

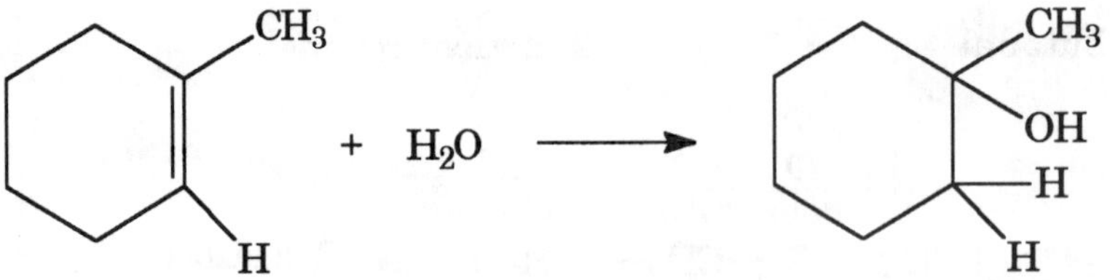

f.

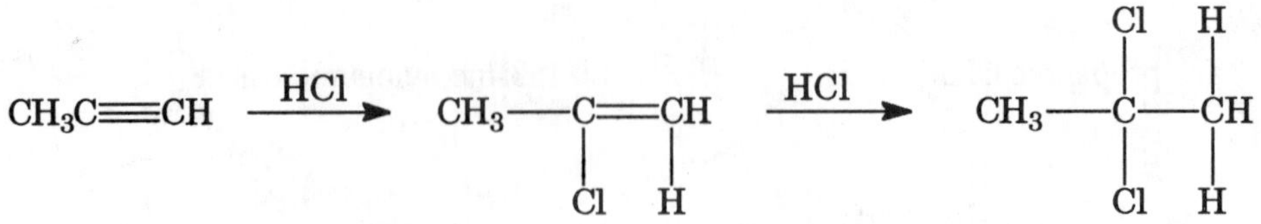

53.

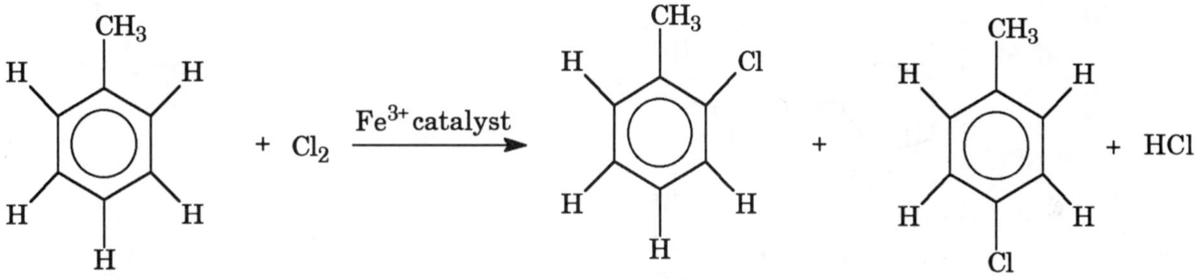

ortho   para

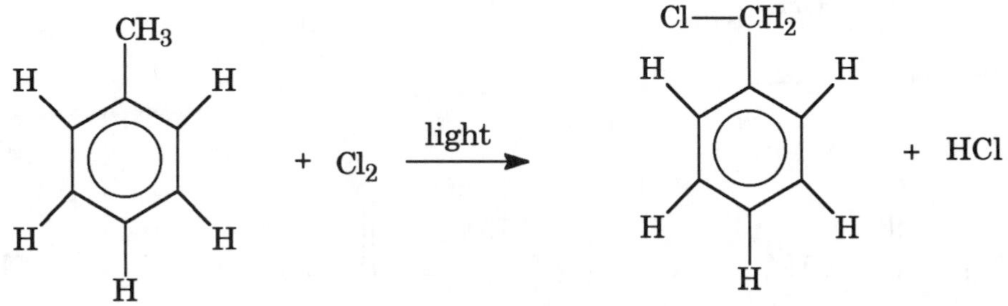

To substitute for the benzene ring hydrogens, an iron(III) catalyst must be present. Without this special iron catalyst, the benzene ring hydrogens are unreactive. To substitute for an alkane hydrogen, light must be present. For toluene, the light catalyzed reaction substitutes a chlorine for a hydrogen in the methyl group attached to the benzene ring.

54.   When $CH_2{=}CH_2$ reacts with HCl, there is only one possible product, chloroethane. When Cl_2 is reacted with CH_3CH_3 (in the presence of light), there are six possible products because any number of the six hydrogens in ethane can be substituted for by Cl. The light-catalyzed substitution reaction is very difficult to control, hence, it is not a very efficient method to produce monochlorinated alkanes.

55.   Primary alcohols (a and e) are oxidized to aldehydes which can be oxidized further to carboxylic acids. Secondary alcohols (b and c) are oxidized to ketones and tertiary alcohols (d) do not undergo this type of oxidation reaction. For the primary alcohols (a and e), we listed both the aldehyde and the carboxylic acid as possible products.

a.   b.

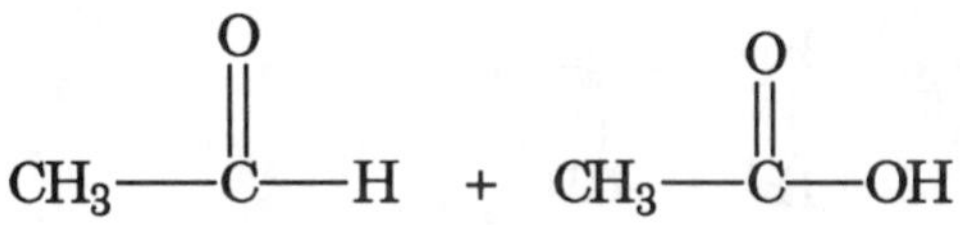

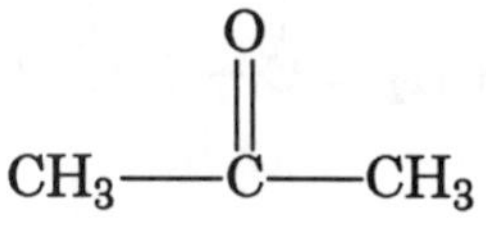

c.

d.

No reaction occurs

e.

56.

a.

b.

c.

57.   a. $CH_3CH = CH_2 + Br_2 \rightarrow CH_3CHBrCH_2Br$ (Addition reaction of Br_2 with propene)

b. $CH_3C \equiv CH + H_2 \xrightarrow{\text{catalyst}} CH_3CH = CH_2 + Br_2 \rightarrow CH_3CHBrCH_2Br$

Hydrogenation of propyne followed by the addition reaction of Br_2 could yield the desired product.

58.

a.

Oxidation of 2-propanol yields acetone (2-propanone).

b.

Addition of H_2O to 2-methylpropene would yield tert-butyl alcohol (2-methyl-2-propanol) as the major product.

c. $CH_3CH_2CH_2OH \xrightarrow{KMnO_4} CH_3CH_2\overset{\overset{\displaystyle O}{\|}}{C}\!-\!OH$

Oxidation of 1-propanol would eventually yield propanoic acid.
Propanal is produced first in this reaction which is then
oxidized to propanoic acid.

59.   When an alcohol is reacted with a carboxylic acid, an ester is produced.

a. $CH_3\overset{\overset{\displaystyle O}{\|}}{C}\!-\!OH + HO\!-\!CH_3 \longrightarrow CH_3\overset{\overset{\displaystyle O}{\|}}{C}\!-\!O\!-\!CH_3 + H_2O$

b. $H\!-\!\overset{\overset{\displaystyle O}{\|}}{C}\!-\!OH + HO\!-\!CH_2CH_2CH_3 \longrightarrow H\!-\!\overset{\overset{\displaystyle O}{\|}}{C}\!-\!O\!-\!CH_2CH_2CH_3 + H_2O$

60.   Reaction of a carboxylic acid with an alcohol can produce these esters.

a.

$CH_3\overset{\overset{\displaystyle O}{\|}}{C}\!-\!OH + HOCH_2CH_2CH_2CH_3 \longrightarrow CH_3\overset{\overset{\displaystyle O}{\|}}{C}\!-\!O\!-\!CH_2CH_2CH_2CH_3 + H_2O$
ethanoic acid   butanol   butyl acetate or butyl ethanoate

b.

$CH_3CH_2CH_2\overset{\overset{\displaystyle O}{\|}}{C}\!-\!OH + HO\!-\!CH_2CH_3 \longrightarrow CH_3CH_2CH_2\overset{\overset{\displaystyle O}{\|}}{C}\!-\!OCH_2CH_3 + H_2O$
butanoic acid   ethanol   ethyl butyrate or ethyl butanoate

Polymers

61.   The backbone of the polymer contains only carbon atoms which indicates that Kel-F is an addition
polymer. The smallest repeating unit of the polymer and the monomer used to produce this polymer
are:

Note: Condensation polymers generally have O or N atoms in the backbone of the polymer.

62.   a. repeating unit: $-(CHF-CH_2)_n-$ monomer: $CHF=CH_2$

     b.

        repeating unit: $-(OCH_2CH_2\overset{\displaystyle O}{\overset{\|}{C}})_n-$ monomer: $HO-CH_2CH_2-CO_2H$

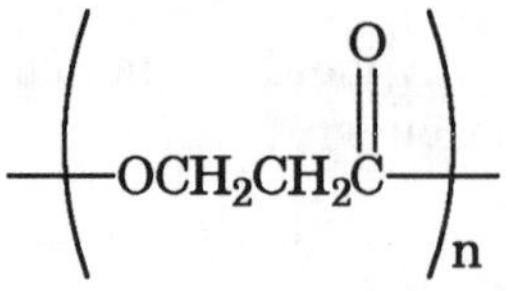

     c. repeating unit:

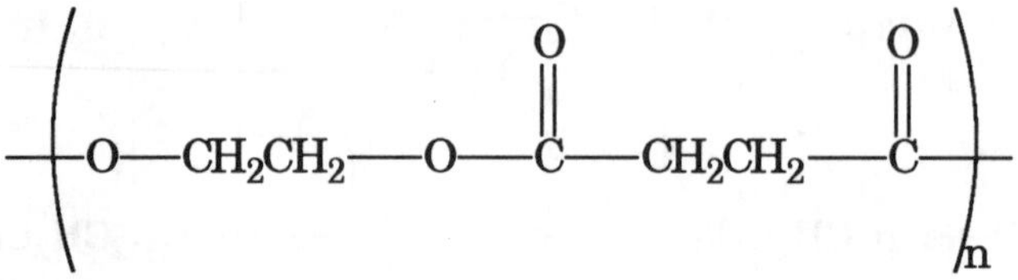

$$-\left(O-CH_2CH_2-O-\overset{\displaystyle O}{\overset{\|}{C}}-CH_2CH_2-\overset{\displaystyle O}{\overset{\|}{C}}\right)_n-$$

copolymer of:
$HOCH_2CH_2OH$ and
$HO_2CCH_2CH_2CO_2H$

     d. monomer: e. monomer:

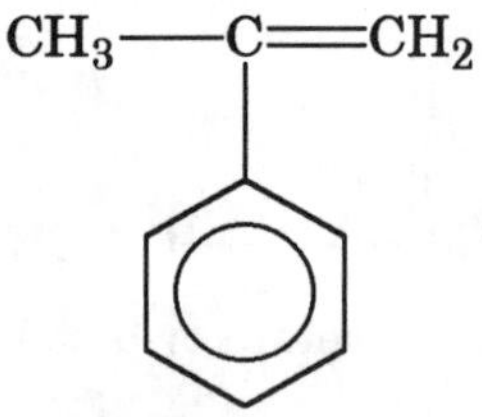

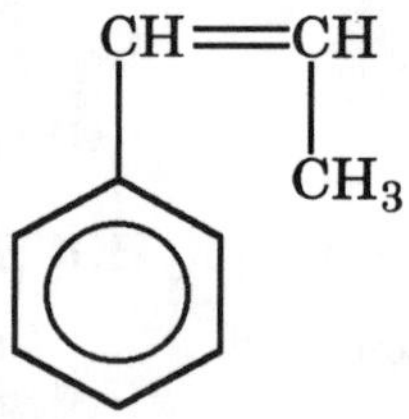

$CH_3-C=CH_2$ (with phenyl attached to C) $CH=CH$ with CH_3 and phenyl

     f. monomer: $CClF=CF_2$

     g. copolymer of:

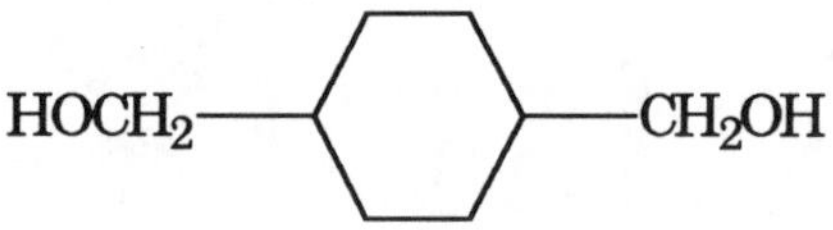

$HOCH_2-$(cyclohexane)$-CH_2OH$ and HO_2C-(benzene)$-CO_2H$

Addition polymers: a, d, e and f; Condensation polymers: b, c and g; Copolymer: c and g

63.

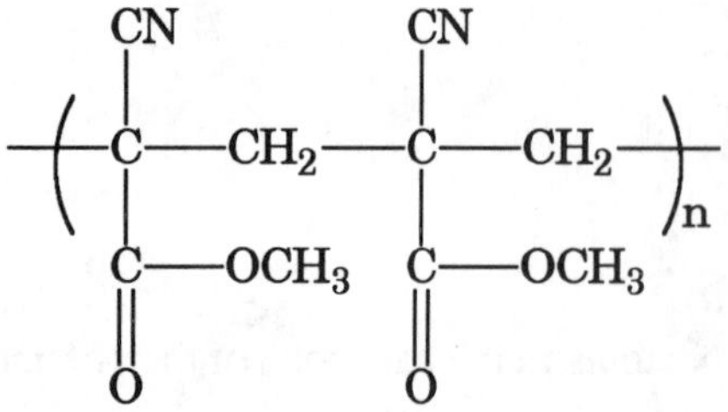

$$-\left(\overset{\displaystyle CN}{\underset{\underset{\displaystyle O}{\overset{\|}{C}-OCH_3}}{C}}-CH_2-\overset{\displaystyle CN}{\underset{\underset{\displaystyle O}{\overset{\|}{C}-OCH_3}}{C}}-CH_2\right)_n-$$

Super glue is an addition polymer formed
by reaction of the C=C bond in methyl
cyanoacrylate.

64.   a.   2-methyl-1,3-butadiene

b.

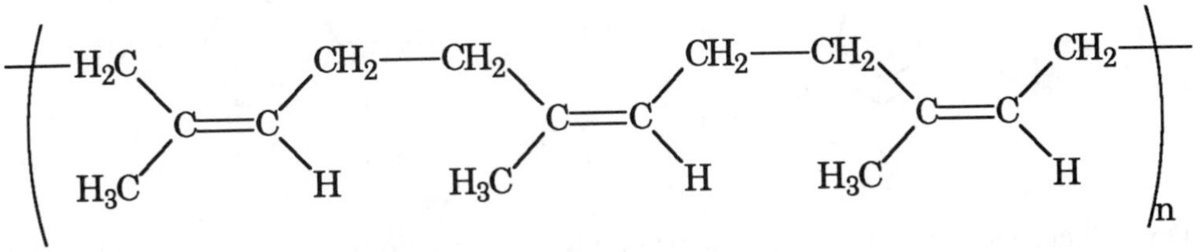

cis-polyisoprene (natural rubber)

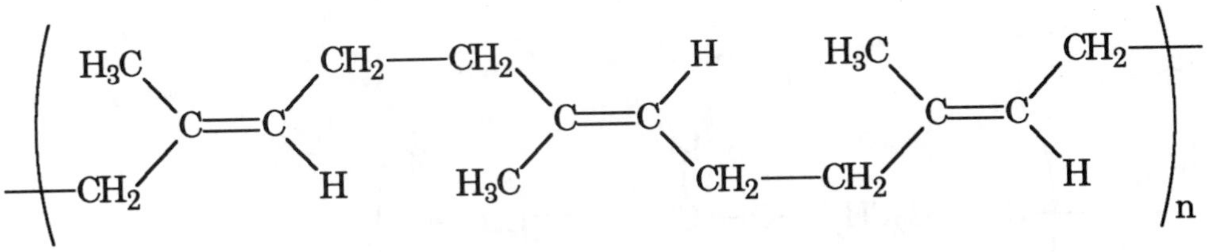

trans-polyisoprene (gutta percha)

65. H_2O is eliminated when Kevlar forms. Two repeating units of Kevlar are:

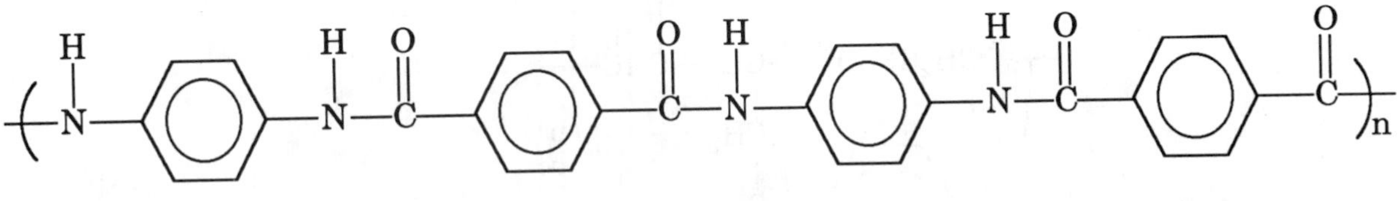

66.   This condensation polymer forms by elimination of water. The ester functional group repeats hence the term polyester.

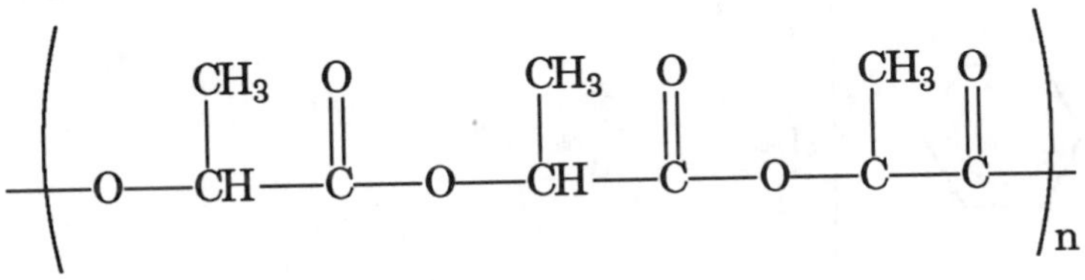

67.   This is a condensation polymer where two molecules of H_2O form when the monomers link together.

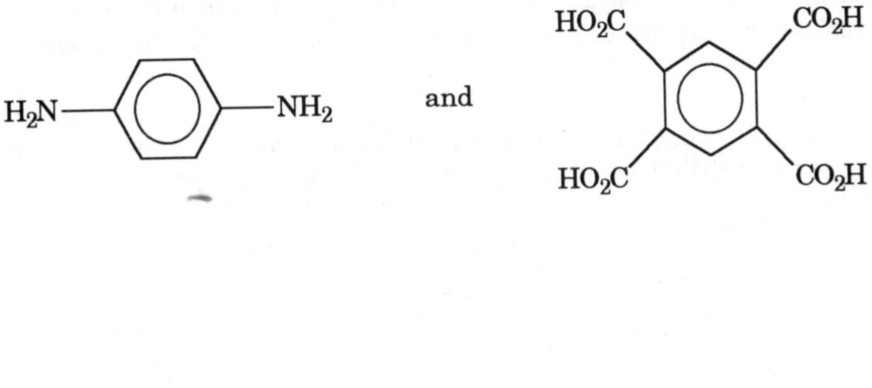

68.

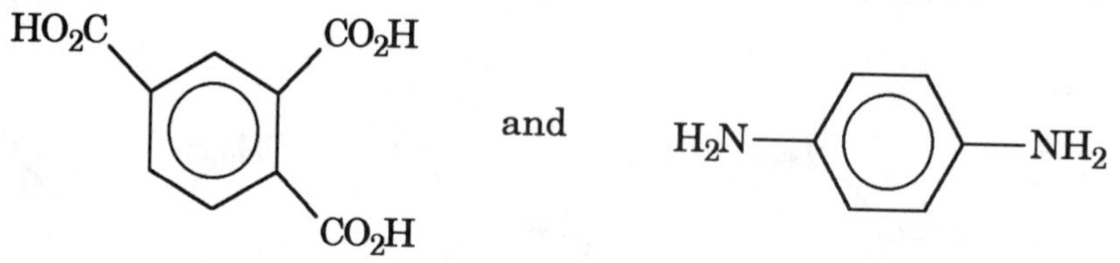

and

69.   Divinylbenzene has two reactive double bonds which are both used when divinylbenzene inserts itself into two adjacent polymer chains. The chains cannot move past each other because the crosslinks bond adjacent polymer chains together making the polymer more rigid.

70.   a.

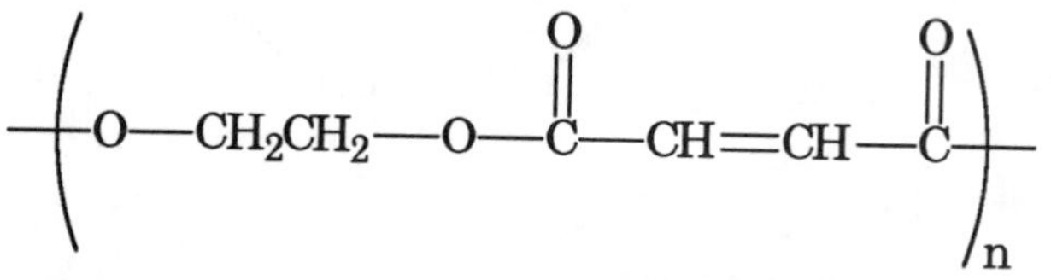

b.

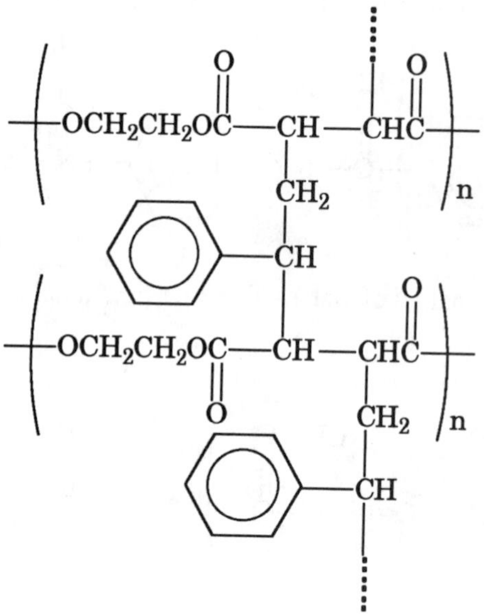

71.   a.   The polymer formed using 1,2-diaminoethane will exhibit relatively strong hydrogen bonding interactions between adjacent polymer chains. Since hydrogen-bonding is not present in the ethylene glycol polymer (a polyester polymer forms), then the 1,2-diaminoethane polymer will be stronger.

      b.   The presence of rigid groups (benzene rings or multiple bonds) makes the polymer stiffer. Hence, the monomer with the benzene ring will produce the more rigid polymer.

c.   Polyacetylene will have a double bond in the carbon backbone of the polymer.

$$n \ HC\equiv CH \longrightarrow \left(\!\!-CH\!=\!CH\!-\!\right)_n$$

The presence of the double bond in polyacetylene will make polyacetylene a more rigid polymer than polyethylene. Polyethylene doesn't have $C=C$ bonds in the backbone of the polymer (the double bonds in the monomers react to form the polymer).

72.   At low temperatures, the polymer is coiled into balls. The forces between poly(lauryl methacrylate) and oil molecules will be minimal and the effect on viscosity will be minimal. At higher temperatures, the chains of the polymer will unwind and become tangled with the oil molecules, increasing the viscosity of the oil. Thus, the presence of the polymer counteracts the temperature effect and the viscosity of the oil remains relatively constant.

Additional Exercises

73. CH_2Cl-CH_2Cl, 1-2-dichloroethane; There is free rotation about the C–C single bond which doesn't lead to different compounds. $CHCl=CHCl$, 1,2-dichloroethene; There is no rotation about the $C=C$ double bond. This creates the cis and trans isomers which are different compounds.

74.   a.   Two monochloro products are formed: $CH_2ClCH_2CH_3$ and $CH_3CHClCH_3$

      b.   Four dichloro products are formed:

      $CHCl_2CH_2CH_3$, $CH_3CCl_2CH_3$, $CH_2ClCHClCH_3$ and $CH_2ClCH_2CH_2Cl$

75.   The cis isomer has the CH_3 groups on the same

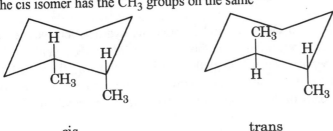

cis   trans

76.   a.   Acetone: Aldehyde that is an isomer of acetone:

b.   2-propanol:

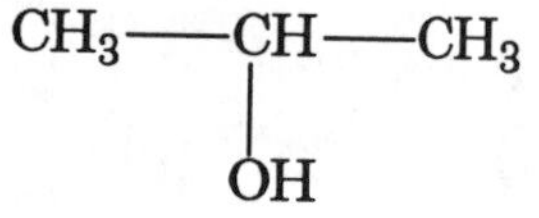

$$CH_3 \text{—} CH \text{—} CH_3$$
$$\phantom{CH_3 \text{—} }OH$$

ether: $CH_3\text{–}O\text{–}CH_2\text{–}CH_3$ ethylmethyl ether

c.   cis-2-butene:

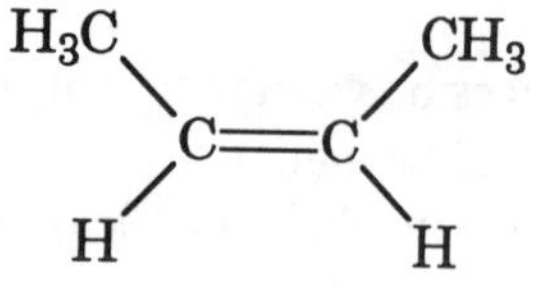

trans-2-butene:

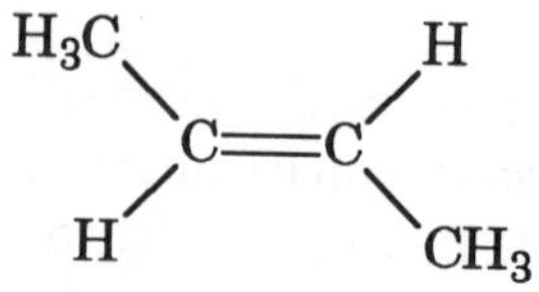

d.   trimethylamine:

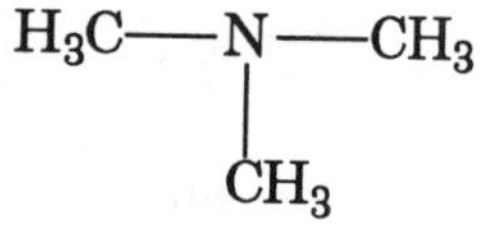

$$H_3C \text{—} N \text{—} CH_3$$
$$\phantom{H_3C \text{—} }CH_3$$

primary amine:

$CH_3CH_2CH_2NH_2$
propylamine or 1-aminopropane

e.   Ethylmethylamine (secondary amine):

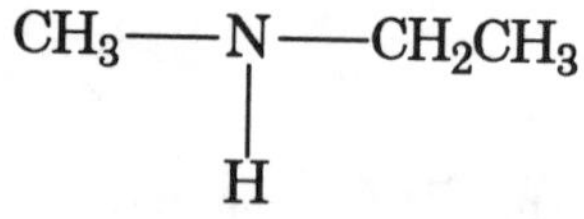

$$CH_3 \text{—} N \text{—} CH_2CH_3$$
$$\phantom{CH_3 \text{—} }H$$

f.   1-propanol (primary alcohol): $CH_3CH_2CH_2OH$

77.   Alcohols consist of two parts, the polar OH group and the nonpolar hydrocarbon chain attached to the OH group. As the length of the nonpolar hydrocarbon chain increases, the solubility of the alcohol decreases in water, a very polar solvent. In methyl alcohol (methanol), the polar OH group can override the effect of the nonpolar CH_3 group and methyl alcohol is soluble in water. In stearyl alcohol, the molecule consists mostly of the long nonpolar hydrocarbon chain so it is insoluble in water.

78. $CH_3CH_2CH_2CH_2CH_2CH_2CH_2COOH + OH^- \rightarrow CH_3\text{–}(CH_2)_6\text{–}COO^-$

Octanoic acid is more soluble in 1 M NaOH. Added OH^- will remove the acidic proton from octanoic acid, creating a charged species. As is the case with any substance with an overall charge, solubility in water increases.

79.   In the presence of H^+, the amine group is protonated creating a positive charge on morphine $(R_3N + H^+ \rightarrow R_3NH^+)$. By treating morphine with HCl, an ionic compound results which is more soluble in water and in the blood stream than the neutral covalent form of morphine.

80.

polychloropene:

$$\left(\!-CH_2-\overset{\overset{\displaystyle Cl}{|}}{C}=CH-CH_2-\!\right)_n$$

polyisoprene:

$$\left(\!-CH_2-\overset{\overset{\displaystyle CH_3}{|}}{C}=CH-CH_2-\!\right)_n$$

polynitrile:

$$\left(\!-CH_2-\underset{\underset{\displaystyle C\equiv N}{|}}{CH}-CH_2-CH=CH-CH_2-\!\right)_n$$

polybutadiene:

$$\left(\!-CH_2-CH=CH-CH_2-\!\right)_n$$

81.   a.

$$H_2N-\bigcirc-NH_2 \quad and \quad HO_2C-\bigcirc-CO_2H$$

b.   Repeating unit:

$$\left(\!-\overset{\overset{\displaystyle H}{|}}{N}-\bigcirc-\overset{\overset{\displaystyle H}{|}}{N}-\overset{\overset{\displaystyle O}{||}}{C}-\bigcirc-\overset{\overset{\displaystyle O}{||}}{C}-\!\right)_n$$

The two polymers differ in the substitution pattern on the benzene rings. The Kevlar chain is straighter and there is more efficient hydrogen bonding between Kevlar chains than between Nomex chains.

82.   Polyacrylonitrile:

$$\left(\!-CH_2-\underset{\underset{\displaystyle N\equiv C}{|}}{CH}-\!\right)_n$$

The CN triple bond is very strong and will not easily break in the combustion process. A likely combustion product is the toxic gas hydrogen cyanide, HCN(g).

83.   As the car ages the plasticizers escape from the seat covers: i) The waxy coating is the escaped plasticizer; ii) The new car smell is the smell of the plasticizers (di-octylphthalate); iii) Loss of plasticizer causes and vinyl to become brittle.

Challenge Problems

84.   For the reaction: $3 \, CH_2 = CH_2(g) + 3 \, H-H(g) \rightarrow 3 \, CH_3 - CH_3(g)$

Bonds broken: Bonds formed:

3 C = C (614 kJ/mol)   3 C – C (347 kJ/mol)
3 H – H (432 kJ/mol)   6 C – H (413 kJ/mol)

$\Delta H = 3(614) + 3(432)$ - $[3(347) + 6(413)] = -381$ kJ

From enthalpies of formation: $\Delta H° = 3 \, \Delta H°_{f, \, C_2H_6} - 3 \, \Delta H°_{f, \, C_2H_4} = 3(-84.7) - 3(52) = -410.$ kJ

The two values agree fairly well.

For $C_6H_6(g) + 3 \, H_2(g) \rightarrow C_6H_{12}(g)$ and starting with one of the Lewis structures for benzene, we would get the same ΔH from bond energies as the first reaction since the same number and type of bonds are broken and formed as in the previous reaction. $\Delta H = -381$ kJ.

From enthalpies of formation: $\Delta H° = -90.3$ kJ - (82.9 kJ) = -173.2 kJ

There is about a 208 kJ discrepancy. Benzene is more stable by about 208 kJ/mol (lower in energy) than we expect from bond energies. This extra stability is evidence for resonance stabilization.

85.

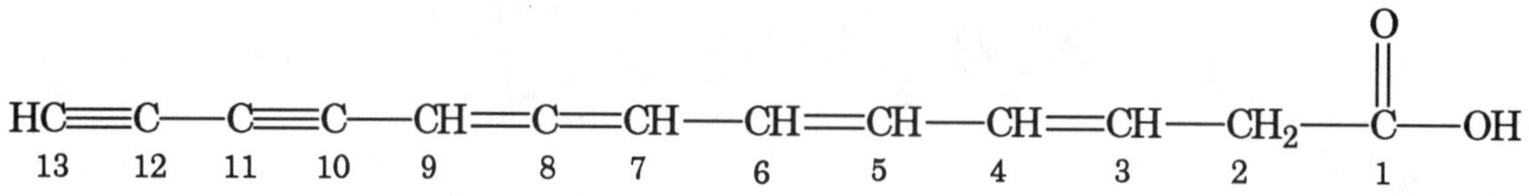

86.

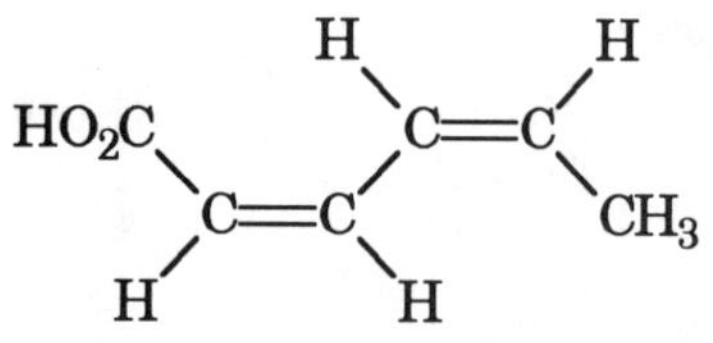

cis-2-cis-4-hexadienoic acid 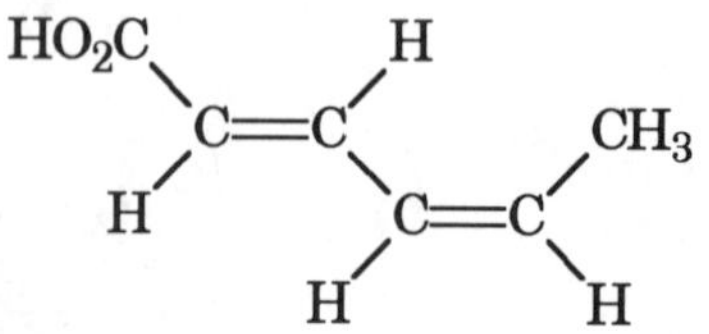   trans-2-cis-4-hexadienoic acid cis-2-trans-4-hexadienoic acid 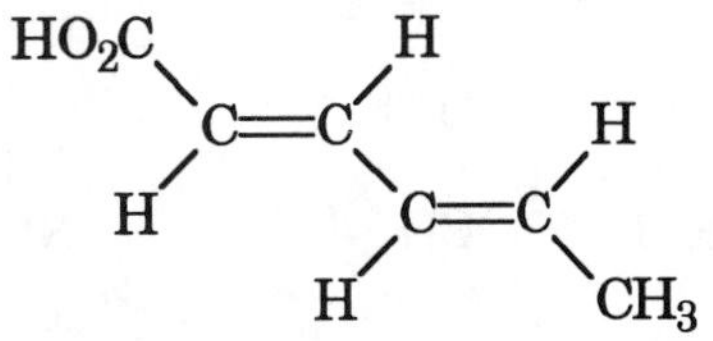   trans-2-trans-4-hexadienoic acid

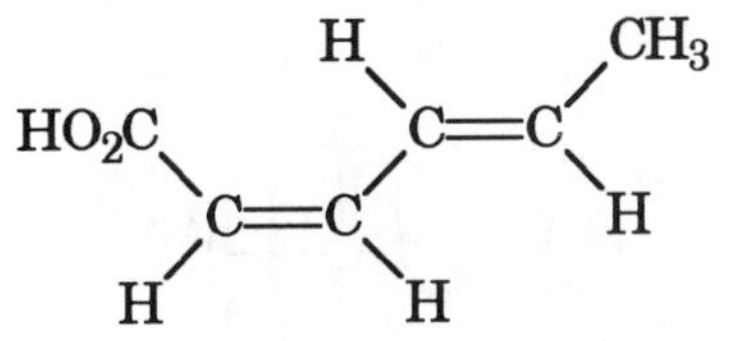

87.   Out of 100.00 g:

$$71.89 \text{ g C} \times \frac{1 \text{ mol C}}{12.01 \text{ g C}} = 5.986 \text{ mol} \approx 6 \text{ mol C}$$

$$12.13 \text{ g H} \times \frac{1 \text{ mol H}}{1.008 \text{ g H}} = 12.03 \text{ mol} \approx 12 \text{ mol H} \qquad \text{The empirical formula is } C_6H_{12}O.$$

$$15.98 \text{ g O} \times \frac{1 \text{ mol O}}{16.00 \text{ g O}} = 0.9988 \text{ mol} \approx 1 \text{ mol O}$$

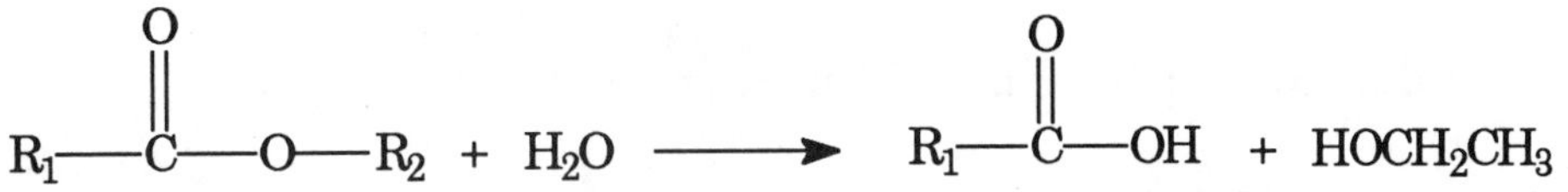

R_2 must be CH_3CH_2 since CH_3CH_2OH is one of the products. The molar mass of $-CO_2H$ is ≈ 45 g/mol, so the mass of R_1 is 172 - 45 = 127. If R_1 is $CH_3-(CH_2)_n-$, then 15 + n(14) = 127 and n = 112/14 = 8. Ethyl caprate is:

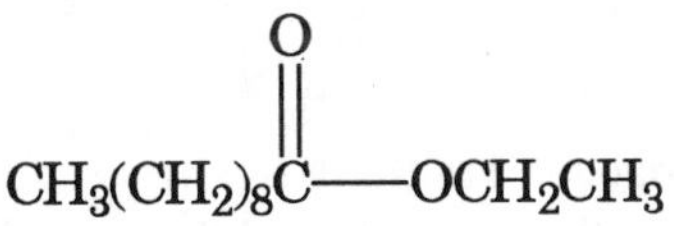

The molecular formula of $C_{12}H_{24}O_2$ agrees with the empirical formula.

$$CH_3(CH_2)_8\overset{\displaystyle O}{\overset{\|}{C}}\!\!-\!\!OCH_2CH_3$$

88.   a.

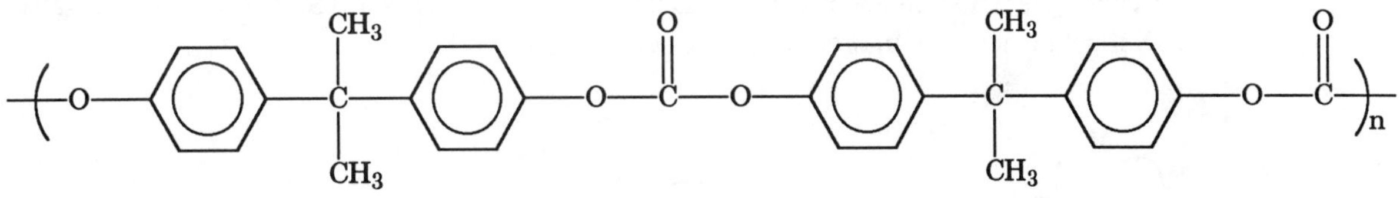

       b.   Condensation; HCl is eliminated when the polymer bonds form.

89.

$$\left(\text{OCH}_2\text{CH}_2\text{O}\overset{\displaystyle O}{\text{C}}\text{N}\!\!-\!\!\underset{H}{}\quad\overset{\displaystyle O}{\text{N}\text{C}}\text{OCH}_2\text{CH}_2\text{O}\overset{\displaystyle O}{\text{C}}\text{N}\!\!-\!\!\underset{H}{}\quad\overset{\displaystyle O}{\text{N}\text{C}}\right)_n$$

90.   a.   For (i), we need 1 mol of furfural to produce one mole of diamine.

furfural: $C_5H_4O_2$; $1 \text{ mol } C_5H_4O_2 \times \dfrac{96.08 \text{ g}}{\text{mol } C_5H_4O_2} \times \dfrac{1 \text{ lb}}{453.6 \text{ g}} \times \dfrac{\$0.79}{\text{lb}} = \$0.17 = 17¢$

Furfural cost is 17¢ per mol of diamine produced.

For (ii), we need 1 mol C_4H_6 to produce 1 mol of diamine.

$1 \text{ mol } C_4H_6 \times \dfrac{54.09 \text{ g}}{\text{mol } C_4H_6} \times \dfrac{1 \text{ lb}}{453.6 \text{ g}} \times \dfrac{\$0.12}{\text{lb}} = \$0.014 = 1.4¢$

Butadiene cost is 1.4¢ per mol of diamine produced.

The butadiene process is economically advantageous.

   b.   Oil prices would have to increase by a factor of 17/1.4 = 12 for the costs to equalize.

91.   a.   The temperature of the rubber band increases when it is stretched.

   b.   Exothermic since heat is released.

   c.   As the chains are stretched, they line up more closely together resulting in stronger London dispersion forces between the chains. Heat is released as the strength of the intermolecular forces increases.

   d.   Stretching is not spontaneous so ΔG is positive. $\Delta G = \Delta H - T\Delta S$; Since ΔH is negative then ΔS must be negative in order to give a positive ΔG.

   e.

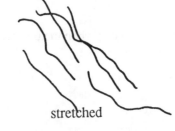

unstretched   stretched

The structure of the stretched polymer is more ordered (lower S).

CHAPTER TWENTY THREE

BIOCHEMISTRY

Questions

1. Primary: The amino acid sequence in the protein. Covalent bonds (peptide linkages) are the forces that link the various amino acids together in the primary structure.

    Secondary: Includes structural features known as α-helix or pleated sheet. Both are maintained mostly through hydrogen bonding interactions.

    Tertiary: The overall shape of a protein, long and narrow or globular. Maintained by hydrophobic and hydrophilic interactions, such as salt linkages, hydrogen bonds, disulfide linkages and dispersion forces.

2. The secondary and tertiary structures are changed by denaturation.

3. Both denaturation and inhibition reduce the catalytic activity of an enzyme. Denaturation changes the three-dimensional structure of an enzyme. Inhibition involves the attachment of an incorrect molecule at the active site, preventing the substrate from interacting with the enzyme.

4. All amino acids can act as both a weak acid and a weak base; this is the requirement for a buffer. The weak acid is the carboxylic end of the amino acid and the weak base is the amine end of the amino acid.

5. Hydrogen bonding occurs between the -OH groups of starch and water molecules.

6. Structural isomers: Same formula, but a different attachment of atoms in the molecule (different bonds).

    Geometrical isomers: (Cis-trans isomerism); Same formula and same bonds, but different spatial arrangement of some atoms with respect to a rigid ring, bond or each other.

    Optical isomers: Same formula and same bonds, but the compounds are nonsuperimposable mirror images of each other.

7. They all contain nitrogen atoms with lone pairs of electrons.

8. DNA: Deoxyribose sugar; double stranded; A, T, G and C are the major bases.

    RNA: Ribose sugar; single stranded; A, G, C and U are the bases.

9.   A deletion may change the entire code for a protein, thus giving an entirely different sequence of amino acids. A substitution will change only one single amino acid in a protein.

10.   Organic solvents are generally nonpolar solvents. Lipids are nonpolar and will be soluble in organic (nonpolar) solvents. Carbohydrates contain several -OH groups capable of hydrogen bonding and are soluble in polar solvents (water).

11.   A polyunsaturated fat contains more than one carbon-carbon double bond.

12.   Triglycerides (fats in the human body) are hydrolyzed to glycerol and fatty acids (soap) in the presence of base. This is how soap is produced. The products of the reaction (soap) feel slippery.

Exercises

Proteins and Amino Acids

13.   They are both hydrophilic amino acids because both contain highly polar R groups.

14.   Crystalline amino acids exist as zwitterions, $^{+}H_3NCRHCOO^{-}$ held together by ionic forces. The ionic interparticle forces are strong. Before the temperature gets high enough to melt the solid, the amino acid decomposes.

15.   a.   Aspartic acid and phenylalanine make up aspartame.

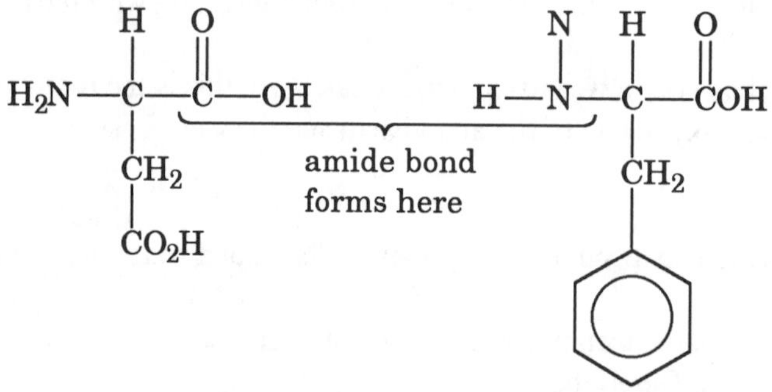

       b.   Aspartame contains the methyl ester of phenylalanine. This ester can hydrolyze to form methanol:

$$R\text{–}CO_2CH_3 + H_2O \rightleftharpoons RCO_2H + HOCH_3$$

16.

ser - ala   ala - ser

17.   a.   Six tripeptides are possible. They are (from NH_2 to CO_2H end):

gly-phe-ala, gly-ala-phe, phe-gly-ala, phe-ala-gly, ala-phe-gly, ala-gly-phe

b.   Three tripeptides are possible. They are (from NH_2 to CO_2H end):

ala-ala-gly, ala-gly-ala, gly-ala-ala

18.   a.   Six tetrapeptides are possible. From NH_2 to CO_2H end:

phe-phe-gly-gly, gly-gly-phe-phe, gly-phe-phe-gly,

phe-gly-gly-phe, phe-gly-phe-gly, gly-phe-gly-phe

b.   Twelve tetrapeptides are possible. From NH_2 to CO_2H end:

phe-phe-gly-ala, phe-phe-ala-gly, phe-gly-phe-ala,

phe-gly-ala-phe, phe-ala-phe-gly, phe-ala-gly-phe,

gly-phe-phe-ala, gly-phe-ala-phe, gly-ala-phe-phe

ala-phe-phe-gly, ala-phe-gly-phe, ala-gly-phe-phe

19.   There are 5 possibilities for the first amino acid, 4 possibilities for the second amino acid, 3 possibilities for the third amino acid, 2 possibilities for the fourth amino acid and 1 possibility for the last amino acid. The number of possible sequences is:

$5 \times 4 \times 3 \times 2 \times 1 = 5! = 120$ different pentapeptides

20.   For a dipeptide there are $5 \times 5 = 25$ different cases. For a tripeptide there are $5 \times 5 \times 5 = 125$ different cases. So for 25 amino acids in the polypeptide, there are $5^{25} = 2.98 \times 10^{17}$ different polypeptides. This assumes an unlimited number of each of the 5 different amino acids.

21.   a.   Ionic: Need NH_2 on side chain of one amino acid with CO_2H on side chain of the other amino acid. The possibilities are:.

NH_2 on side chain = His, Lys or Arg; CO_2H on side chain = Asp or Glu

b.   Hydrogen bonding: Need N–H or O–H bond in side chain. The hydrogen bonding interaction occurs between the X– H bond and a carbonyl group from any amino acid.

X–H $\cdots\cdots$ O $=$ C (carbonyl group)

Ser Asn   Any amino acid
Glu Thr
Tyr Asp
His Gln
Arg Lys

c.   Covalent: Cys – Cys (forms a disulfide linkage)

d.   London dispersion: All amino acids with nonpolar R groups. They are:

Gly, Ala, Pro, Phe, Ile, Trp, Met, Leu and Val

e.   Dipole-dipole: Need side chain with OH group. Tyr, Thr and Ser all could form this specific dipole-dipole force with each other since all contain an OH group in the side chain.

22.   Phenylalanine - isoleucine: London disperson forces; Aspartic acid - lysine: ionic
(See Exercise 23.21.)

23.   Glutamic acid: R = $-CH_2CH_2CO_2H$; Valine: R = $-CH(CH_3)_2$; A polar side chain is replaced by a nonpolar side chain. This could affect the tertiary structure of hemoglobin and the ability of hemoglobin to bind oxygen.

24.   Glutamic acid: R = $-CH_2CH_2COOH$; Glutamine: R = $-CH_2CH_2CONH_2$; The R groups only differ by OH vs NH_2. Both of these groups are capable of forming hydrogen bonding interactions so the change in intermolecular forces is minimal. Thus, this change is not critical as the secondary and tertiary structure of hemoglobin should not be greatly affected.

Carbohydrates

25.   Of the pentoses, D-ribulose has only two chiral carbon atoms while the other pentoses have three chiral carbon atoms (see the following structure where the asterisks denotes the various chiral carbon atoms). Of the hexoses, D-fructose has only three chiral carbon atoms versus four chiral carbon atoms in the other hexoses. Note: A chiral carbon atom has four different substituent groups attached.

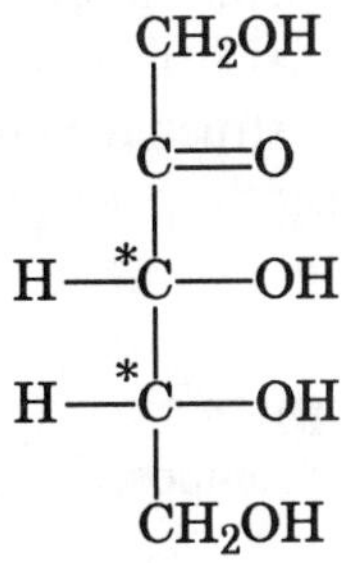
CH2OH
C=O
H—*C—OH
H—*C—OH
CH2OH

D-Ribulose

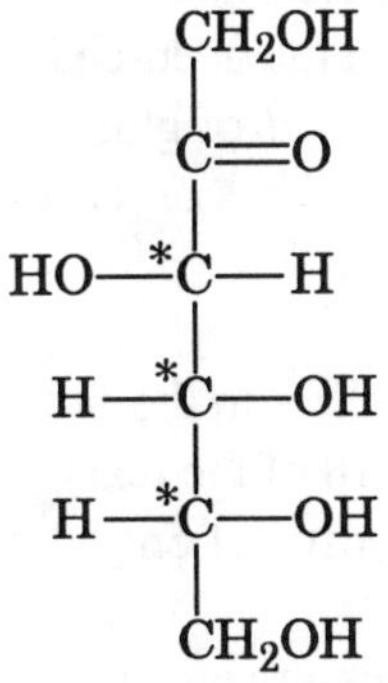
CH2OH
C=O
HO—*C—H
H—*C—OH
H—*C—OH
CH2OH

D-Fructose

26. The chiral carbon atoms are marked with asterisks.

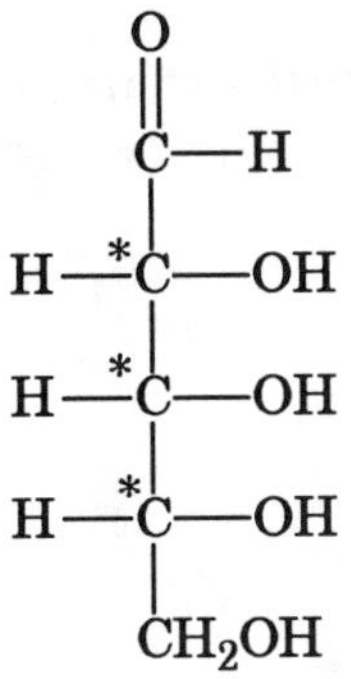
O
C—H
H—*C—OH
H—*C—OH
H—*C—OH
CH2OH

D-Ribose

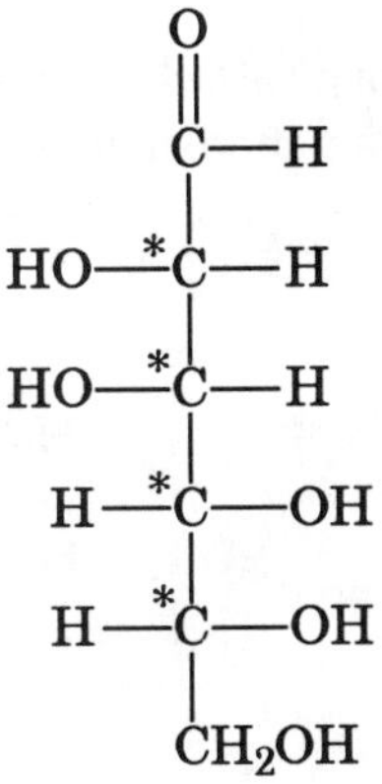
O
C—H
HO—*C—H
HO—*C—H
H—*C—OH
H—*C—OH
CH2OH

D-Mannose

27. See Figures 23.17 and 23.18 of the text for examples of the cyclization process.

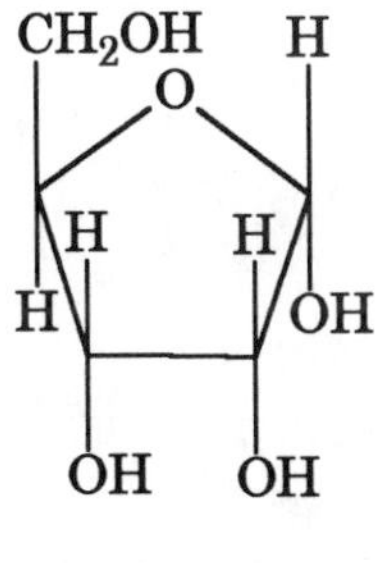
CH2OH
H
O
H
H
H
OH
OH
OH

D-Ribose

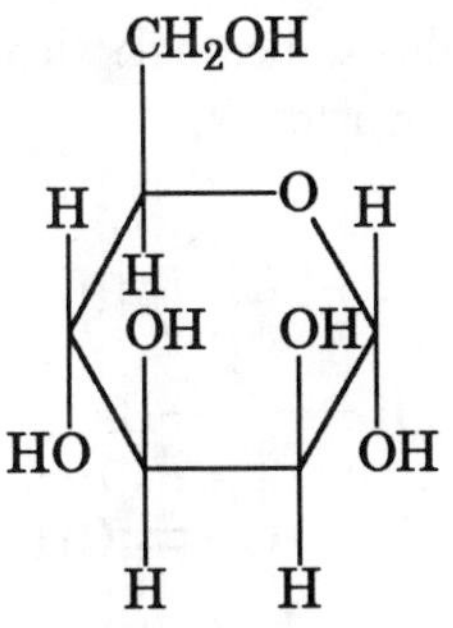
CH2OH
H
O
H
H
OH
OH
HO
OH
H
H

D-Mannose

28.   This is an example of Le Chatelier's principle at work. For the equilibrium reactions between the various forms of glucose, reference Figure 23.18 of the text. The chemical tests involve reaction of the aldehyde group found only in the open-chain structure. As the aldehyde group is reacted, the equilibrium between the cyclic forms of glucose and the open-chain structure will shift to produce more of the open-chain structure. This process continues until either the glucose or the chemicals used in the tests run out.

29.   The α and β forms of glucose differ in the orientation of a hydroxy group on one specific carbon in the cyclic forms (see Figure 23.18 of the text). Starch is a polymer composed of only α-D-glucose and cellulose is a polymer composed of only β-D-glucose.

30.   Humans do not possess the necessary enzymes to break the β-glycosidic linkages found in cellulose. Cows, however, do possess the necessary enzymes to break down cellulose into the β-D-glucose monomers and therefore, can derive nutrition from cellulose.

Optical Isomerism and Chiral Carbon Atoms

31.   A chiral carbon has four different groups attached to it. A compound with a chiral carbon is optically active. Isoleucine and threonine contain more than the one chiral carbon atom (see asterisks).

isoleucine   threonine

32.   There is no chiral carbon atom in glycine since it contains no carbon atoms with four different groups bonded to it.

33.   Only one of the isomers is optically active. The chiral carbon in this optically active isomer is marked with an asterisk.

34.

The fourth group bonded to each of the three chiral carbon atoms (see asterisks) is a hydrogen atom.

Nucleic Acids

35. The complimentary base pairs in DNA are cytosine (C) and guanine (G), and thymine (T) and adenine (A). The complimentary sequence is: T-A-C-G-C-C-G-T-A

36. For each letter, there are 4 choices; A, T, G, or C. Hence, the total number of codons is $4 \times 4 \times 4 = 64$.

37. Uracil will H-bond to adenine.

38. The tautomer could hydrogen bond to guanine, forming a G–T base pair instead of A–T.

39.   Base pair:

      RNA   DNA

      A . . . . T

      G C

      C . . . . G

      U . . . . A

a. Glu: CTT, CTC   Val: CAA, CAG, CAT, CAC

    Met: TAC   Trp: ACC

    Phe: AAA, AAG   Asp: CTA, CTG

b. DNA sequence for Met - Met - Phe - Asp - Trp:

      TAC - TAC - AAA - CTA - ACC
              or   or
             AAG CTG

c. Due to phe and asp, there is a possibility of four different DNA sequences.

d.

    C—T—T—A—C—C—A—A—A
      Glu - Trp - Phe

e. C - T - C - A - C - C - A - A - A

    C - T - T - A - C - C - A - A - G

    C - T - C - A - C - C - A - A - G

40.   In sickle cell anemia, glutamic acid is replaced by valine. DNA codons: Glu: CTT, CTC; Val: CAA, CAG, CAT, CAC; Replacing a T with an A in the code for Glu will code for Val.

      CTT → CAT or CTC → CAC
      Glu   Val   Glu   Val

Lipids and Steroids

41.

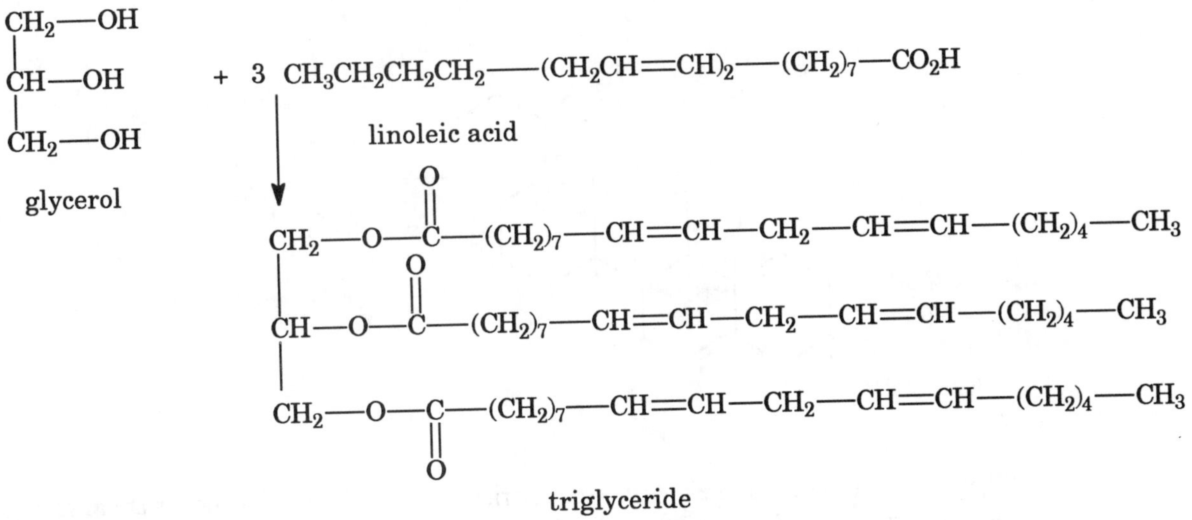

$$CH_2—OH$$
$$CH—OH \quad + \quad 3 \ CH_3CH_2CH_2CH_2—(CH_2CH=CH)_2—(CH_2)_7—CO_2H$$
$$CH_2—OH$$

glycerol

linoleic acid

$$CH_2—O—\overset{O}{\overset{\|}{C}}—(CH_2)_7—CH=CH—CH_2—CH=CH—(CH_2)_4—CH_3$$
$$CH—O—\overset{O}{\overset{\|}{C}}—(CH_2)_7—CH=CH—CH_2—CH=CH—(CH_2)_4—CH_3$$
$$CH_2—O—\underset{O}{\overset{\|}{C}}—(CH_2)_7—CH=CH—CH_2—CH=CH—(CH_2)_4—CH_3$$

triglyceride

42.

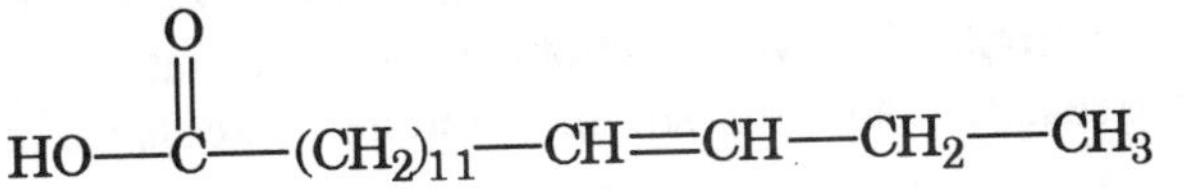

$$HO—\overset{O}{\overset{\|}{C}}—(CH_2)_{11}—CH=CH—CH_2—CH_3 \qquad \text{16 carbon omega-3 fatty acid}$$

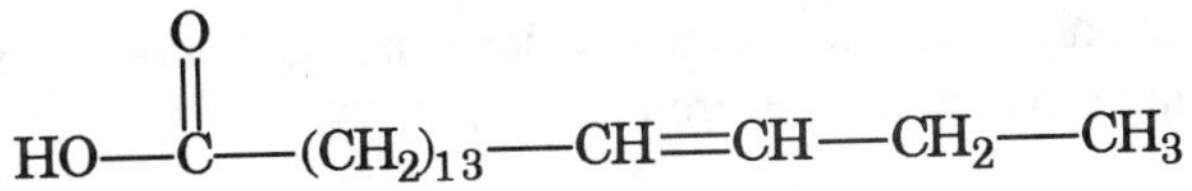

$$HO—\overset{O}{\overset{\|}{C}}—(CH_2)_{13}—CH=CH—CH_2—CH_3 \qquad \text{18 carbon omega-3 fatty acid}$$

43. Hydrogenation converts unsaturated fats (double bonds present) to saturated fats (no double bonds present). Unsaturated fats occur as oily liquids and saturated fats occur as solids. Therefore, hydrogenation will solidify fats. The triglyceride in Exercise 23.41 contains 6 carbon-carbon double bonds, hence, 6 mol of H_2 are required to completely hydrogenate 1 mol of this triglyceride.

44. The prefix mono means one. Hence, monounsaturated fats contain fatty acids which only have one carbon-carbon double bond. From Table 23.4 of the text, only oleic acid would be able to form monounsaturated fats.

45. A detergent molecule has a polar (or ionic) head group and a long nonpolar hydrocarbon tail.

46. Hard water contains Ca^{2+} and Mg^{2+} ions. These ions form precipitates with the fatty acids in soap. The formation of these precipitates (known as soap scum) greatly reduces the soap's cleaning efficiency. In the following reaction, we used Ca^{2+} as the hard water ion and stearic acid as the fatty acid. Mg^{2+} with other fatty acids react in a similar fashion.

$$Ca^{2+}(aq) + 2 \ CH_3–(CH_2)_{16}–COO^-(aq) \rightarrow Ca[CH_3–(CH_2)_{16}–COO^-]_2(s)$$

47.   No; Chloresterol compounds are naturally occuring compounds in all humans and are essential for human life. Some important compounds produced from cholesterol are bile acids, steroid hormones and vitamin D.

48.

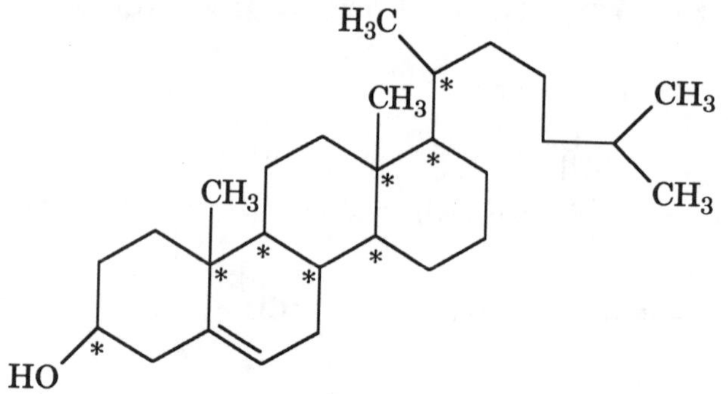

Eight chiral carbon atoms are present (see asterisks). Note: Many C–H bonds are missing from this shorthand structure. Since carbon atoms form four bonds in organic structures, then the number of missing C–H bonds at each carbon atom can easily be determined.

49.   See Figure 23.33b of the text for the structure. Since vitamin D_3 is composed almost entirely of carbon and hydrogen, then it is essentially nonpolar and will be soluble in nonpolar fat. Fat soluble vitamins will accumulate in the body more readily than water soluble vitamins. Excess water soluble vitamins will pass through the body.

50.   See Figures 23.33d and e of the text. The only difference in the two structures is that an –OH group in testosterone is replaced by the following group in progesterone:

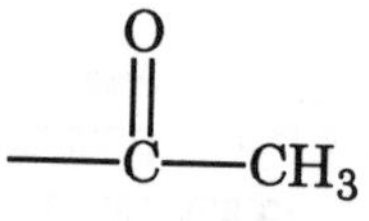

Additional Exercises

51.   Glutamic acid: Monosodium glutamate:

One of the two acidic protons in the carboxylic acid groups is lost to form MSG. Which proton is lost is impossible for you to predict.

In MSG, the acidic proton from the carboxylic acid in the R group is lost, allowing formation of the ionic compound.